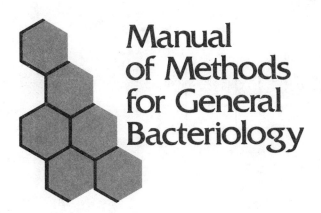

Manual
of Methods
for General
Bacteriology

Manual of Methods for General Bacteriology

PHILIPP GERHARDT, *Editor-in-Chief*
Department of Microbiology and Public Health
Michigan State University
East Lansing, MI 48824

R. G. E. MURRAY, *Editor, I. Morphology*
Department of Microbiology and Immunology
University of Western Ontario
London, Ont., Canada N6A 5C1

RALPH N. COSTILOW, *Editor, II. Growth*
Department of Microbiology and Public Health
Michigan State University
East Lansing, MI 48824

EUGENE W. NESTER, *Editor, III. Genetics*
Department of Microbiology
University of Washington
Seattle, WA 98195

WILLIS A. WOOD, *Editor, IV. Metabolism*
Department of Biochemistry
Michigan State University
East Lansing, MI 48824

NOEL R. KRIEG, *Editor, V. Systematics*
Biology Department
Virginia Polytechnic Institute
Blacksburg, VA 24061

G. BRIGGS PHILLIPS, *Editor, VI. Laboratory Safety*
Health Industry Manufacturers Association
1030 15th Street, N.W.
Washington, DC 20005

AMERICAN SOCIETY FOR MICROBIOLOGY
Washington, DC 20006   1981

Copyright © 1981
American Society for Microbiology
1913 I St., N.W.
Washington, DC 20006

Library of Congress Cataloging in Publication Data

Main entry under title:

Manual of methods for general bacteriology.

   Includes bibliographies and index.
   1. Bacteriology—Technique. 2. Bacteriology—Laboratory manuals.
I. Gerhardt, Philipp, 1921- II. American Society for Microbiology.
QR65.M26   589.9′0028   80-22275

ISBN 0-914826-29-8 (*cloth binding*)
ISBN 0-914826-30-1 (*flexible binding*)

CONTENTS

AUTHORS

Brian Austin. 21. Numerical Taxonomy
Department of Microbiology, University of Maryland, College Park, Maryland 20742; 301/454-2848

W. Emmett Barkley. 24. Containment and Disinfection
Division of Safety, National Institutes of Health, Bethesda, Maryland 20205; 301/496-1357

Barbara J. Brown. 13. Gene Mutation
Department of Molecular and Population Genetics, University of Georgia, Athens, Georgia 30602; 404/542-5011

Bruce C. Carlton. 13. Gene Mutation
Department of Molecular and Population Genetics, University of Georgia, Athens, Georgia 30602; 404/542-5011

Roger M. Cole. 4. Electron Microscopy
Laboratory of Streptococcal Diseases, National Institute of Allergy and Infectious Diseases, Bethesda, Maryland 20205; 301/496-4278

Rita R. Colwell. 21. Numerical Taxonomy
Department of Microbiology, University of Maryland, College Park, Maryland 20742; 301/454-2848

Ralph N. Costilow. II. Introduction to Growth, 6. Biophysical Factors in Growth
Department of Microbiology and Public Health, Michigan State University, East Lansing, Michigan 48824; 517/355-6516

Jorge H. Crosa. 15. Plasmids
Department of Microbiology, University of Washington, Seattle, Washington 98195; 206/543-8772

Roy Curtiss III. 14. Gene Transfer
Department of Microbiology, University of Alabama in Birmingham, Birmingham, Alabama 35294; 205/934-2430

Walter J. Dobrogosz. 18. Enzymatic Activity
Department of Microbiology, North Carolina State University, Raleigh, North Carolina 27607; 919/737-2393

R. N. Doetsch. 3. Determinative Methods of Light Microscopy
Department of Microbiology, University of Maryland, College Park, Maryland 20742; 301/454-2848

Stephen W. Drew. 10. Liquid Culture
Department of Chemical Engineering, Virginia Polytechnic Institute, Blacksburg, Virginia 24061. Present address: Merck & Co., Inc., Rahway, New Jersey 07065; 201/574-6912

Stanley Falkow. 15. Plasmids
Department of Microbiology, University of Washington, Seattle, Washington 98195; 207/543-1444

Philipp Gerhardt. Methodology for General Bacteriology, 9. Solid Culture, 25. Diluents and Biomass Measurement
Department of Microbiology and Public Health, Michigan State University, East Lansing, Michigan 48824; 517/353-1780

Robert L. Gherna. 12. Preservation
American Type Culture Collection, Rockville, Maryland 20852; 301/881-2600

Beverly M. Guirard. 7. Biochemical Factors in Growth
Department of Microbiology, University of Texas, Austin, Texas 78712; 512/471-5543

R. S. Hanson. 17. Chemical Composition
Department of Bacteriology, University of Wisconsin, Madison, Wisconsin 53706. Present address: Gray Freshwater Biological Institute, University of Minnesota, St. Paul, Minnesota 55108; 612/372-3719

John L. Johnson. 22. Genetic Characterization
Anaerobe Laboratory, Virginia Polytechnic Institute, Blacksburg, Virginia 24061; 703/961-7127

Arthur L. Koch. 11. Growth Measurements
Department of Microbiology, Indiana University, Bloomington, Indiana 47401; 812/337-5036

Michael S. Korczynski. 23. Sterilization
Hospital Products Division, Abbott Laboratories, North Chicago, Illinois 60064; 312/937-3560

Noel R. Krieg. 8. Enrichment and Isolation, 9. Solid Culture, V. Introduction to Systematics, 20. General Characterization
Biology Department, Virginia Polytechnic Institute, Blacksburg, Virginia 24061; 703/961-5912

R. E. Marquis. 19. Permeability and Transport
Microbiology Department, University of Rochester, Rochester, New York 14642; 716/275-3411

R. G. E. Murray. I. Introduction to Morphology, 1. Light Microscopy, 2. Specimen Preparation for Light Microscopy
Department of Microbiology and Immunology, University of Western Ontario, London, Ontario, Canada N6A 5C1; 519/679-3751

Eugene W. Nester. III. Introduction to Genetics
Department of Microbiology, University of Washington, Seattle, Washington 98195; 206/543-5483

G. Briggs Phillips. VI. Introduction to Laboratory Safety
Health Industry Manufacturers Association, Washington, DC 20005; 202/452-8240

J. A. Phillips. 17. Chemical Composition
Department of Bacteriology, University of Wisconsin, Madison, Wisconsin 53706; 608/262-3381

Terry J. Popkin. 4. Electron Microscopy
Laboratory of Streptococcal Diseases, National Institute of Allergy and Infectious Diseases, Bethesda, Maryland 20205; 301/496-4278

C. F. Robinow. 1. Light Microscopy, 2. Specimen Preparation for Light Microscopy
Department of Microbiology and Immunology, University of Western Ontario, London, Ontario, Canada N6A 5C1; 519/679-3751

Carl A. Schnaitman. 5. Cell Fractionation
Department of Microbiology, University of Virginia, Charlottesville, Virginia 22903; 804/924-2476

Robert M. Smibert. 20. General Characterization
Anaerobe Laboratory, Virginia Polytechnic Institute, Blacksburg, Virginia 24061; 703/961-7125

Esmond E. Snell. 7. Biochemical Factors in Growth
Department of Microbiology, University of Texas at Austin, Austin, Texas 78712; 512/471-5543

Willis A. Wood. IV. Introduction to Metabolism, 16. Physical Methods
Department of Biochemistry, Michigan State University, East Lansing, Michigan 48824; 517/355-1605

REVIEWERS

Marion Alexander
John E. Algie
Arthur I. Aronson
Manfred E. Bayer
Robert E. Benoit
Terrance J. Beveridge
John A. Breznak
Thomas D. Brock
Marvin P. Bryant
Richard O. Burns
Ercole Canale-Parola
James Champoux
Don B. Clewell
Thomas R. Corner
Pierre-Marc Daggett
Eugene A. Delwiche
Richard Donovick
Howard C. Douglas
Frederick A. Eiserling
Derek C. Ellwood
Robert K. Finn
Cecil W. Forsberg
Rolf G. Freter
David T. Gibson
Neal B. Groman
Lynne Haroun
Robert J. Heckly

Peter Hirsch
Shung-Chang Jong
Elliot Juni
June J. Lascelles
Edward R. Leadbetter
Robert A. MacLeod
Robert P. Mortlock
Nels H. Nelson
Leo N. Ornston
Crellin Pauling
Michael J. Pelczar, Jr.
Robert M. Pengra
Irving J. Pflug
Allen T. Phillips
Sydney C. Rittenberg
William R. Romig
Frank P. Simione
Ellen M. Simon
James T. Staley
Diane E. Taylor
James M. Tiedje
Donald J. Tipper
H. Edwin Umbarger
Paul J. Van Demark
William G. Walter
Helen R. Whiteley
J. Gregory Zeikus

PREFACE

The *Manual of Methods for General Bacteriology* was planned to meet the need for a compact, moderately priced handbook of reliable, basic methods for practicing general bacteriology in the laboratory.

The scope of the *Manual* covers all the taxa of bacteria, and the approach is generalized for pure and applied bacteriology. Parasitic and pathogenic bacteria are treated without emphasis in context with the natural prevalence of saprophytes. No attempt is made to include specialized methods for applied fields, or to encompass all of microbiology. Viruses are included only with bacteriophage as vectors in gene transfer; algae, with cyanobacteria (blue-green algae); and yeast, molds, and protozoa, with passing mention.

The *Manual* was written for serious students who are beyond the elementary level and who need to acquire skill in laboratory procedures for research and applications of bacteriology. Typical users will be senior undergraduate or starting graduate science majors in an advanced course. Teachers will find the book useful for designing laboratory exercises, and research workers will use it for conducting experiments. Laboratory workers in industry and federal agencies will find much practical information here.

By nature and style, this is a "how to" manual. Principles are introduced sufficient to understand why and how the method is conducted. Procedures are described step-by-step with enough detail so that the user can carry out the method without further reference to another source. Common problems, pitfalls, and precautions are pointed out. Examples, illustrations, tabulations, and schematics are used wherever possible. Selections are made of the methods used most frequently and reliably; for less important methods, references are given to guide the user to original literature sources for complete descriptions. As much as possible, a method is generalized so that it might be applied for use with an unknown, newly isolated bacterium. However, a method often is exemplified with a specific bacterium familiar to the author or widely used.

The methods described herein are believed to be reliable recommendations offered by competent scientists, but in no way are they to be regarded as "best" or "standard." Sponsorship and publishing by the American Society for Microbiology is not to be regarded as an endorsement, nor are the methods in any way "official" or "approved." The authors frequently suggest commercial sources for equipment and materials, but these suggestions are not meant either to endorse or to exclude a particular product, and any omissions are unintentional.

The book is organized into six sections corresponding to main subjects of general bacteriology and then into chapters corresponding to relevant areas of methodology. Each chapter is independent and self-contained, with a table of contents to help the reader see the organization in outline and find a specific method. A decimal numbering system is used to facilitate identification, cross-referencing, and indexing. Headings are topical and brief so as to enable easy identification of a method. Literature citations are listed at the end of each chapter: first, general references with full citations and annotations; next, specific articles with abbreviated citations. Lists of commercial sources are included in some chapters.

The external review process was extensive and rigorous, and the editors served not simply to collect manuscripts but literally to edit them. When the review and rewrite process was well along, manuscripts for the entire book were refereed in three distinguished university departments of general microbiology. Finally, the editor-in-chief scrutinized the entire book for content and conformity. This review and editing process was enormously helpful, if often painful. Further criticism is welcomed: through use we can learn what needs to be added, deleted, or modified in a future edition.

The history of the *Manual* goes back to 1923, when H. J. Conn first authored and published methods manuals for the Society as short leaflets. Conn later served primarily in an editorial capacity and as chairman of a Committee on Bacteriological Technique which produced a looseleaf compilation entitled *Manual of Methods for Pure Culture Study of Bacteria*. The Committee established a precedent for voluntary authorship, and sales of the manual became a major portion of Society income. In 1957, led by M. J. Pelczar, Jr., the Committee rewrote, retitled, and arranged to publish commercially the *Manual of Microbiological Methods* as a book.

The present venture was undertaken in response to a continuing demand for the 1957 book and the need for a thorough updating of its content. Page allocations were made to authors at the outset to produce a single volume of minimum size and verbiage. Low cost was sought so the book would be affordable by students. This is possible because the Society now is its own publisher and reaches a large and growing membership (over 30,000 in 1980); thus, it can publish books with large press runs at low unit costs. Furthermore, the Society is a tax-exempt, nonprofit organization; and the editors and authors serve without personal pay or royalty.

Sincere and appreciative acknowledgment is made to everyone who has contributed; the authors and editors for splendid cooperation and excellent effort in writing the book; the many reviewers, most of whose names are included in the foregoing list, for diligent and constructive criticism; Erwin F. Lessel and his Lederle Laboratories staff, who volunteered to rework the artwork and graphics for uniform style and quality; the universities and other institutions which allowed the considerable time, staff support, and often direct costs for preparing and editing manuscripts; Cynthia Lounsbery, who created the artwork for the cover; and the staff of the ASM Publications Office, particularly Estella B. Bradley, who copy edited and indexed the book, Cheryl Cross, Chief of the Books Section, and Gisella Pollock, Acting Managing Editor. All of us were privileged to share in this contribution to the advancement of bacteriology.

Philipp Gerhardt

Methodology for General Bacteriology

PHILIPP GERHARDT

The science of general bacteriology, like that of any generalized field, deals with the entire group rather than with particular types of bacteria, and its methodology embodies the procedures and techniques which are basic and widely applicable in laboratory investigation of their nature and uses. In this manual, all of the bacteria described in *Bergey's Manual of Determinative Bacteriology* are encompassed and no particular ones are emphasized. The methods in this manual should be useful in any environment where bacteria are found and should be applicable for any practical purpose in which bacteria are involved.

Bacteriology became a science only after unique methods were developed, and they are responsible for its continuing influence on and expansion into subsequently developed fields such as virology, immunology, and molecular biology. Koch's introduction of pure culture technique and Pasteur's use of immunological response and chemical analysis remain as influential now as then.

The methodology for general bacteriology is defined in this manual with a sectioning of content which reflects that of several exemplary textbooks in the field. Thus, the content includes sections on morphology, growth, genetics, metabolism, and systematics, plus laboratory safety. These sections then are divided into chapters which correspond to main topics of methods for general bacteriology. Often the divisions are arbitrary, so some overlap of methods occurs; this is managed with cross-referencing in the text and indexing at the end, facilitated by a decimal number identification system for each method. General references and specific articles are cited for the methods in each chapter.

The main general resource for the reader seeking a method not included here is the multivolume and multiauthor work *Methods in Microbiology* (5). The Meynells' *Theory and Practise in Experimental Bacteriology* (4) provides an excellent resource for many procedures, and the Society of American Bacteriologists' predecessor *Manual of Microbiological Methods* (8) continues to be useful. Methods for medical microbiology and immunology, many of which are applicable in general bacteriology, are obtainable in a number of references, including the companion manuals *Manual of Clinical Microbiol-*ogy (3) and *Manual of Clinical Immunology* (6). The *CRC Handbook of Microbiology* (2) is a generally helpful reference, as is the *Dictionary of Microbiology* (7). Information on further commercial sources is obtainable from the annual *Yellow Pages of Instrumentation, Equipment and Supplies* (1) and the annual buyers' guide issues of journals such as *American Laboratory*, *Chemical Engineering*, and *Science*.

LITERATURE CITED

1. **Industrial Research/Development.** 1979. Yellow pages of instrumentation, equipment and supplies. Industrial Research/Development Publishers, Barrington, Ill.
2. **Laskin, A. I., and H. Lechevalier.** 1978–1980. CRC handbook of microbiology, 2nd ed. CRC Press, Inc., Cleveland.
   Vol. 1. Bacteria. *Introductory descriptions occupy the first 120 pages; the next 340 pages provide useful descriptions of the main groups of bacteria; the last 220 pages contain usual handbook information, including a list of culture collections.*
   Vol. 2. Fungi, algae, protozoa and viruses.
   Vol. 3. Amino acids and proteins.
   Vol. 4. Carbohydrates, lipids and minerals.
3. **Lennette, E. H., A. Balows, W. H. Hausler, Jr., and J. P. Truant (ed.).** 1980. Manual of clinical microbiology, 3rd ed. American Society for Microbiology, Washington, D.C.
4. **Meynell, G. G., and E. Meynell.** 1970. Theory and practise in experimental bacteriology, 2nd ed. University Press, Cambridge, England.
   An excellent, compact (347 pages) resource book containing the rationale and procedure for many methods in general bacteriology. There are sections on growth; culture media; oxygen, carbon dioxide, and anaerobiosis; sterilization; light microscopy; quantitation; and genetic technique.
5. **Norris, J. R., and D. W. Ribbons (ed.).** 1969–1979. Methods in microbiology, vol. 1–13. Academic Press, Inc., New York.
   Volumes 1, 2, 3A, 5A, 5B, 6A, 6B, 7A, 7B, 8, and 9 include methods for general bacteriology. Volume 4 covers mycology, and volumes 10 to 13 describe typing methods for the epidemiology of pathogenic bacteria. A cumulative table of contents for the preceding volumes is contained in each volume. The style of presentation varies with each author but usually is that of a comprehensive review with extensive references to original papers.

6. **Rose, N. R., and H. Friedman (ed.).** 1980. Manual of clinical immunology, 2nd ed. American Society for Microbiology, Washington, D.C.

7. **Singleton, P., and D. Sainsbury.** 1978. Dictionary of microbiology. John Wilson & Sons, Inc., New York.

   A useful 481-page encyclopedia of terms, tests, techniques, taxa, concepts, and other topics. Entries range from short definitions to concise reviews and cover the broad field of pure and applied microbiology, including biochemistry and genetics.

8. **Society of American Bacteriologists, Committee on Bacteriological Technique.** 1957. Manual of microbiological methods. McGraw-Hill Book Co., Inc., New York.

   This 315-page book continues to be useful particularly because of its chapters on staining methods, preparation of media, and routine tests for the identification of bacteria.

Section I

MORPHOLOGY

Introduction to Morphology

R. G. E. MURRAY

Microscopes and the need for effective techniques of microscopy provide the fundamental definition of microbiology as a branch of biological science. Bacteriology and the description of bacteria have taken full advantage of the development of sophisticated as well as simple methods of microscopy because these ubiquitous procaryotic organisms are, indeed, small cells. And now that the components of cells can be resolved to the level of macromolecules and functional components, and processes can be studied in cell fractions, the techniques of high-resolution analysis have multiplied the usefulness and applications of electron microscopy. But all levels of microscopy play a part in the study of bacteria and of cell-free components, in nature as in culture.

The approaches that a student will find useful and that have the best chance of success are described in this section. The intent is to provide, to a reasonable degree, the fruits of the experience of others to complement the basic understanding that the reader brings to the examination of bacteria. The main concern is to make the microscope (both the light and the electron microscope) more useful, effective, and rewarding. The need for illustration is minimal since the basic characteristics and design of the instruments are profusely depicted in most texts and in the references given with each chapter. Therefore, we describe as clearly as possible the procedures that we have found to work and draw attention to the pitfalls. We try to point out that much good work can be done with simple equipment and that this becomes easier with practice, for which there is no substitute.

The light microscope is available to and needed by all microbiologists, and because effective utilization is so desirable but so seldom attained, some detailed advice is given in Chapter 1—everyone is expected to know all about it! Uses of the light microscope to get structural information and to identify a bacterium are described in Chapters 2 and 3, respectively. The electron microscope is a much more specialized instrument and is usually operated, and guarded, by specialized professionals who instruct, assist, and advise; therefore, more attention is given to preparative aspects of the methodology in Chapter 4. Both approaches are needed to follow the progress of cell disintegration and cell fractionation, as described in Chapter 5—the electron microscope is as essential to monitoring the cleanliness of subcellular fractions as it is to establishing the relation of structure and function.

Chapter 1

Light Microscopy

R. G. E. MURRAY AND C. F. ROBINOW

Microscopy is made easy, interesting, and accurate if at least four different kinds of instruments are readily available. It is best if they are permanently set up for work and maintained in good order. These are (i) a rugged, well-equipped but uncomplicated instrument for "encounters of the first kind," (ii) a phase-contrast instrument, (iii) an optimally equipped and adjusted microscope for the resolution of detail and the best of photomicrography, and (iv) an instrument equipped for fluorescence microscopy. Some combinations are possible, and a more limited selection may suit individual needs.

A vast variety of microscopes are available. Those from the first-rank manufacturers are all of comparable quality, and a more than adequate range of accessories is available. The differences involve, mostly, the support structures, and choice is likely to be based on convenience factors and personal preferences. The really solid instruments are expensive. With care, good work can be done with first-class optics fitted to a general-purpose or even a student microscope stand. However, it is hard to equal the modern instruments of the best manufacturers. Other choices may involve technical requirements and it is *always* advisable to make direct comparisons of the possible instruments and the alternative accessories on your own material under your conditions of work. Therefore, it is not necessary to be specific about makers. What matters is an awareness that a good microscope (whatever its age), a discerning choice of components, and a knowledge of how to use them enhance the pleasure derived from the work, save time, and improve accuracy. The use of poorly adjusted and poorly illuminated micro-

scopes is so widespread that few biologists are any longer aware of what microscopes can do and are astounded when given a demonstration of the best resolution obtainable with even a student microscope when it is used in accord with the principles outlined in part 1.1 or in parts 1.3 and 1.3.1.

1.1. SIMPLE MICROSCOPY

The microscope assigned to general use in the classroom and in the research or routine laboratory can be of variable vintage, quality, and condition. The applications to bacteriology are simple, and the questions answered with its help are as follows: Are there objects of bacterial size and staining properties? Are they consistent in size and shape? Are they gram positive or gram negative? Is there more than one kind of organism or cell present? ... and so on to more sophisticated questions. These primary observations can usually be made without attaining the highest resolution, so a simplified procedure suffices as long as the user remembers that, if higher resolution is demanded, almost every step can be improved.

The microscope (Fig. 1) should be equipped with:

1. Oculars (eyepieces) of at least 10× magnification. However, serious observations on bacterial structure require more enlarging (12× or 15×) compensating oculars. These fit in the microscope tube, which may be fixed in length or adjustable (to 180 mm or whatever tube length the manufacturer recommends).

2. Low-power, dry objectives (10× and 40×). These are essential for looking at growing cul-

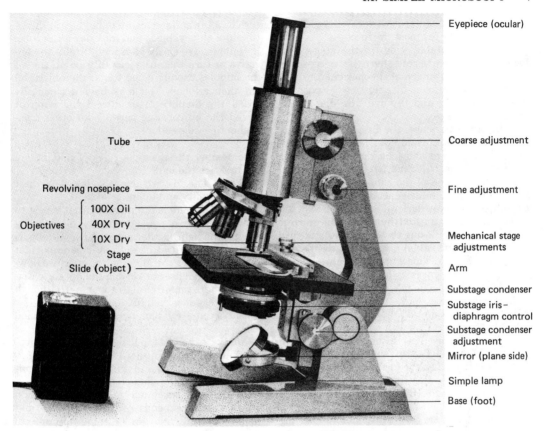

Eyepiece (ocular)

Tube

Coarse adjustment

Revolving nosepiece

Fine adjustment

Objectives { 100X Oil
40X Dry
10X Dry

Mechanical stage
adjustments

Stage

Slide (object)

Arm

Substage condenser

Substage iris–
diaphragm control

Substage condenser
adjustment

Mirror (plane side)

Simple lamp

Base (foot)

FIG. 1. *Rugged everyday microscope and its parts.*

tures and for identifying rewarding areas in preparations.

3. An achromatic oil-immersion objective (90
to 100×). This is essential to bacteriologists.

4. A stage with a mechanical slide-holding
device or with simple spring clips. If a mechanical stage is provided, let it be easily removable
so that petri plates or staining dishes may be
examined with low powers.

5. A substage condenser of the "improved
Abbe type" (one that has three lenses, is aplanatic, and has a numerical aperture of 1.3 to 1.4
is preferable). This type is essential for oil-immersion objectives because it provides the quality and quantity of light they require. The condenser should have an iris diaphragm (aperture
diaphragm) incorporated below the lower lens.

6. A light source, either built-in and of limited
adjustability or (preferably) external and movable. In the latter case an adjustable mirror
(plane on one face and concave on the other) is
fixed under the condenser, and the light is directed up the optical axis of the microscope by
tilting. There are many lamps from which to

choose, and a frosted bulb behind a "bull's-eye"
lens is adequate for much work. However, a
lamp capable of projecting a sharp image of the
light source and provided with a filter carrier
and a field stop iris diaphragm allows much
improved images for the study of stained preparations, especially when Koehler illumination
(1.3.1) is approximated. Coiled-filament projection lamps (giving an inhomogeneous image)
must be focused on a frosted-glass diffusing
screen in the filter carrier of the lamp; the images
are then interspersed by tilting the reflector in
the lamp housing and defocused a little to give
a more evenly and brightly lit surface as the
source of light.

The steps in operating the microscope are as
follows:

1. With the eyepiece and 10× objective in
place, the condenser fully up and almost level
with the stage, and the aperture diaphragm (condenser) fully open, tilt the mirror to provide
illumination up the optical axis. Adjust the objective (coarse control wheel) to focus the light

source, and then center it with the mirror. Use the *plane* mirror for most microscopy.

2. Place a suitable slide on the stage, and focus the objective to get a sharp (not necessarily well illuminated) image of the specimen, which must then be improved by adjusting the position of the condenser and the aperture diaphragm for the best image.

3. In our experience, the best way to determine the optimum condenser position is to *remove* the eyepiece and look down the tube at the back of the objective lens. While looking, adjust the position of the condenser to give the fullest and most homogeneous illumination of the objective. Any lack of centration can be recognized and adjusted by tilting the mirror. Reduce glare by closing the condenser aperture diaphragm (if available) *just* enough to impinge on the bright disk. This whole procedure applies to *all* objective lenses as long as they are first focused on the specimen.

4. Replace the eyepiece to examine the specimen, and make any final adjustments for the best illumination.

5. Turn the nosepiece to any other objective desired. Many microscope makers provide parfocal objectives (i.e., the focal position for one is nearly identical to that for all the others). Clearance is sufficient for the dry lenses. If the lenses are mixed as to makes or otherwise mismatched, it is wise to raise the tube a little before focusing the oil-immersion objectives, which have close focal distances. The clearance will be even smaller when the specimen is mounted under a cover slip.

6. Apply a drop of oil to the slide (center of specimen) before placing it on the stage, if you are going direct to oil immersion. If you have scanned for a rewarding area with a lower power, it is convenient to apply the oil drop with the nosepiece halfway between the objectives and before swinging the oil-immersion objective into position over the drop.

7. Using and focusing the oil-immersion objective takes a little practice and care (see 1.5 and 1.6), even if it has a spring-loaded front end to protect both lens and slide from the ham-fisted operator. While looking from aside the slide, lower the objective onto the oil drop and down just a little more. Using the *coarse focusing wheel* and watching in the eyepiece, look for the specimen while lowering *gently*. When the specimen is found, the *fine focus wheel* can be used. Adjust illumination as above, and subjectively determine the optimum.

8. If the image is hazy, moves, or is only dim and unsharp, there is something wrong. The objective may need to have the front lens cleaned (lens paper, please). Oil-immersion ob-

jectives often suffer because of bubbles in the oil (take out the eyepiece and look down the tube; the bubbles are easily identified). Raise the objective or turn the nosepiece to wipe off the oil (the bubbles usually come up on the objective) and then refocus. You may have accidentally closed the diaphragm or moved the lamp or moved the mirror. See part 1.7 for other troubleshooting suggestions.

Remember these simple rules of microscopy and practice them:

Rule 1. The focused image requires the best possible quality and quantity of light.

Rule 2. If there is too much light, it is best to reduce it with a neutral density filter or by moving the light farther away. With oil-immersion (high-aperture) objectives, it is important to avoid reducing the illuminating aperture (lowering the condenser or closing the aperture diaphragm more than 10% of the opening), although this stratagem is helpful for low-aperture dry objectives and for looking at living, unstained cells.

Rule 3. If there is too little light for the objective you wish to use, either there is something in the way or you need a more appropriate illumination system (lamp and condenser).

Rule 4. The specimen must be part of a (nearly) homogeneous optical system and mounted in oil or water or some appropriate medium. No details can be discerned in dried, stained films of bacteria viewed with dry objectives.

Rule 5. Keep lenses clean and free from dust, fingerprints, noseprints, and the grime from oily eyelashes.

Single, living bacteria, yeasts, or fungal hyphae are clearly seen even through the back of a petri dish by using a 10× objective, and the nature and condition of what is being examined is established without further manipulation. Two points about work with petri dishes are worth making here: (i) the best illumination for observations made through the back of a dish is provided by the *concave* mirror, and (ii) the distortion of images due to scratches in the glass or unevenness of shape of the bottom of the dish can be cancelled out over areas of interest by a drop of immersion oil topped with a cover slip to give a homogeneous light path and an unflawed surface.

1.2. PHASE-CONTRAST MICROSCOPY

An instrument that is permanently set up for phase-contrast microscopy is preferable to one that sees alternating service with ordinary optics on the same stand. Phase microscopy is demanding of light, and much more has to be provided than is adequate for the examination of stained specimens with ordinary optics. Use a green filter to reduce unavoidable chromatic aberration.

Centration of the phase plate in the condenser, in respect to the analyzer plate in the back focal plane of the objective, is essential. The centration is achieved by imaging both with a focusing eyepiece ("telescope") and by centering the visible rings deriving from the phase plates using the two adjusting knobs on the sides of the condenser.

It is one of the advantages of phase-contrast over the now less available dark-field microscopy that the former utilizes the full aperture of the objective lens. It pays, therefore, to make use of the best of phase-contrast objectives. This in turn implies the use of a substage condenser of suitably high aperture and, as before, the use of 15× or even 20× compensating eyepieces which fully utilize the fine resolution that the objectives are capable of giving. Medium-power, phase-contrast, oil-immersion 40× or 60× objectives are obtainable and provide crisp, rewarding images for work with fungal mycelia and other large cells. For the very best results, the condenser and slide must be connected by oil when using 90× or 100× objectives with numerical apertures higher than 1.0. However, adequate phase-contrast microscopy can be attained without the oiled condenser. This is used for counting bacteria, protozoa, and spores, for checks on motility, or for determining shape and form. If intracellular detail of living organisms is wanted, attention will have to be paid to the refractility of the medium in which the cells are examined, and this is most easily provided by dissolving gelatin (15 to 30%) in the medium. This equalizes or brings closer together the refractive index of contents and medium, since phase retardation is proportional to the refractive index times the light path. The "halos" are also reduced.

1.3. HIGH-RESOLUTION MICROSCOPY

The "research microscope" comes into its own for the resolution of fine detail at high magnification. Its use is not called for in determining the outcome of a Gram test, but it will help to settle questions about flagella and matters requiring the perception of fine detail. For such purposes it is essential to pay close attention to the quality of the optical components. Achieving high resolution as well as freedom from chromatic and spherical aberration requires attention to basic principles.

Resolution (R) is the smallest distance between points of detail in material which will still appear as a distinct gap in the images, visual or photographic. $R = \lambda/2$ n.a. where λ is the wavelength of the light used and n.a. stands for numerical aperture, a measure of the light-gathering power of the objective. The lowest values of R require the use of light of short wavelengths, but, unfortunately, the most effective wavelengths (namely, ultraviolet of around 365 nm) are not perceived by the eye and do not pass through glass. Quartz and fluorite optics (permeable in various degrees to ultraviolet) and indirect methods of observation are required to take advantage of the short wavelength of ultraviolet light, as is the case for fluorescence microscopy. The best results with visible light are obtained in the green/yellow region of the spectrum because it is the wavelength (close to the mercury green spectral line) for which microscope objectives are designed to transmit with a minimum of aberrations. So much for the λ part of the expression for resolution.

The **numerical aperture** of an oil-immersion objective is usually given as a figure of 1.30 or 1.32, and the best may be 1.4. It represents the sine of one-half of the angle described by the cone of light admitted by the objective lens, multiplied by the refractive index of the medium through which light passes on its way to the lens. The maximum value of the sine function does not exceed 1.0, but the refractive index of the medium through which light passes to the objective can be raised to that of optical glass (namely, 1.5150) by the use of immersion oil. It is this fact that allows us to raise the numerical aperture to values above 1.0 and, provided certain other conditions are met, to utilize to the fullest the resolving power of the costly objectives with numerical aperture values of 1.30 or 1.32 engraved on their glittering housings.

High-aperture **achromatic** or **apochromatic objectives** are spherically correct for one or for three colors in the center of the field of view (where most of the work is done at high magnification). The two kinds work almost equally well, price notwithstanding, especially when narrow band-pass (interference) filters are used.

The objectives best suited to the task in hand should be complemented by oculars of commensurate quality. What is needed are **compensating eyepieces**, so called because "they are designed to correct the lateral chromatic error of magnification inherent in objective lenses of high corrections and high apertures" (3). Experience amply proves the soundness of the advice of Shillaber (9) that "the compensating ocular should be of high power, preferably 15× or 20×. The ability of apochromatic objectives to take a high-power eyepiece should be fully utilized;

otherwise, the fine detail that they are capable of bringing out in a photomicrograph is likely to be lost." For those who wear glasses with major corrections, it is convenient and helpful to buy **high-eyepoint oculars** which are made so that glasses may be worn while observing.

Optimum performance of a highly resolving objective depends on **optimum illumination**. The field of view needs to be evenly filled with light of a brightness sufficient for photography. The light from coiled filaments is unevenly distributed over their surface and, therefore, over their image unless diffused by ground glass. The best structureless, uniformly bright sources of light are provided by small mercury or zirconium arcs and by the less expensive conventional projection lamps with tungsten bead or ribbon filaments that can be imaged to fill the condenser. The **Koehler system of illumination** (1.3.1), which is widely used, requires that the microscope lamp be equipped with two additional items: a lens system capable of projecting a sharp image of the light source into the entrance plane of the substage condenser and, in front of this lens, an iris diaphragm, the so-called field stop (see below for Koehler drill). **Neutral density filters** (aluminized optical flats) allow the brightness to be adjusted for comfortable vision, and a color filter restricts the wavelength used. Two kinds of color filters are available: expensive "**interference filters**" of narrow band pass consist of two half-silvered glass plates facing each other across a precisely measured gap, and **Wratten filters** (Kodak) consist of colored gelatin mounted between two 5-cm-square flats of optically polished glass. Wratten filters deteriorate with time because air creeps in from the sides, but they come in a wide range of colors and densities and are most valuable accessories. A red filter (Wratten no. 29) or an orange filter (Wratten no. 22) proves helpful in making dense structures transparent or enhancing structures stained with blue dyes. A green filter (Wratten no. 58B) is the most generally useful because it is the kind of light to which our retina is maximally sensitive and because many commonly used histological stains are red and the contrast is greatly enhanced when seen through a green screen.

Glass-covered mirrors (i.e., mirrored on the back of the glass) do well in everyday work and are durable. However, they are not the best for optimum illumination because the front and back of the glass that covers the reflecting metal produce separate and overlapping reflections of the light source or of the opening of the field-stop when the Koehler procedure is followed and add undesirable scattered light to the final image. **First-surface mirrors** lack a protective covering and give but a single reflection.

A most important station on the path of the light from lamp to object to eye is the **substage condenser**. Objective apertures of 1.30 to 1.32 will be fully utilized only if the numerical aperture of the condenser is equally high or higher. Such condensers are usually also achromatic or aplanatic, thus matching the optical properties of the objectives described above. The role of the refractive index in the expression for numerical aperture needs to be remembered at this point: the resolving power of objectives of high numerical aperture is utilized only when the medium through which light passes after it has entered the condenser and has passed through the object and the lens system of the objective is homogeneous, glasslike all the way. In other words, effective resolution requires that immersion oil be interposed between the condenser and the object slide, not just between the slide and the objective lens.

With eyepiece, objectives, and condenser of matching optical quality, and with a first-surface mirror receiving filtered green light from a bright and homogeneous source, the research microscope is ready for work and needs only a rewarding specimen preparation to show its worth.

1.3.1. Koehler Illumination

The steps allowing achievement of Koehler (nearly optimum) illumination for an oil-immersion objective are as follows:

1. Focus the image of the lamp filament, using a white card placed on the mirror under the substage condenser, with the lamp so placed that the image (preferably, a ribbon filament) completely covers the opening of the condenser observed in the mirror when the card is taken away.

2. Insert appropriate filters in the light path for visual work. Fully open all diaphragms. If a coil-filament lamp is used, place a ground-glass diffusing screen close to the lens of the lamp (see 1.1).

3. Oil the condenser (use one drop of high-viscosity oil), and bring it up to meet a similar drop on the underside of the object slide (see 1.5).

4. Approximate a reasonable centration and illumination in the back lens of a low-power objective, which is in focus on the preparation, and adjust the mirror and condenser for central homogeneous and maximum illumination while looking down the tube (eyepiece removed). Air bubbles in the oil, or other general problems, will be obvious and should be rectified (see 1.1).

5. Replace the eyepiece, and scan the preparation for a rewarding or useful area.

6. Apply oil to the slide (1.5), change to the oil-immersion objective, and focus on the specimen.

7. Adjust the field (lamp) diaphragm to a minimum opening, and focus its margin with the substage condenser in the plane of the specimen (i.e., objective and condenser are now *both* in focus). Then readjust the field diaphragm so that its image just impinges on the field of view.

Note: If the margins of the field diaphragm give an uneven color fringe when using white light, the condenser needs centration. To center, turn the two adjustment knobs to give movement towards the *red* side, centering the aperture with the mirror as you move, and stop when the color fringes are symmetrical.

8. Removing the ocular (which should be *at least* 12 or 15×), look down the tube and restrict the aperture (condenser) diaphragm to abolish glare or reflections. This restriction should not be more than one-tenth the diameter of the back lens; excess restriction lowers resolution.

9. Replace the eyepiece and insert filters appropriate to observation or photography.

10. The same principles apply to dry lenses, but oil immersion of the condenser is then unnecessary as long as the condenser is of adequate quality.

1.4. FLUORESCENCE MICROSCOPY

Fluorescence microscopy employs all the principles of optics described above as requirements for the research microscope. The differences in practice and design relate to generating and transmitting wavelengths of light suitable to the excitation of fluorochromes and natural fluorescence as far as the object and the detection of the wavelengths exhibited by the fluorescence of that object. Because the excitation process usually requires short wavelengths in the near-ultraviolet range, the lamp (a high-pressure mercury-vapor arc lamp), the condenser lenses, and any lens between lamp and object must be made of material (usually fluorite) appropriate to passing that range of wavelengths. A first-surface mirror is essential (to avoid a glass layer). The immersion oil for the objective and condenser must be nonfluorescent (e.g., a special synthetic formulation or sandalwood oil).

Most important to the microscopist is the arrangement of filters in the light path, which vary in mechanical arrangement from maker to maker but which all have the following principles:

1. Between objective and ocular there must be a **barrier or ultraviolet-stopping filter**, which protects the eye from harmful levels of ultraviolet transmission.

2. Between lamp and object there must be exciter filters and a stop filter. **Exciter filters** are interference filters with a narrow band pass (e.g., 365 nm) appropriate to the fluorochrome in use. They are opaque to most visible light so that the background of a specimen is black (or very deep blue) to the eye. A "stop" filter withholds ultraviolet light but transmits visible light for finding and preliminary focusing at an ordinary level of illumination.

The illumination systems, however, take several forms: (i) transmitted light through a substage condenser, (ii) dark-field illumination using a specialized substage condenser, and (iii) incident (epi-) illumination through the objective. The second and third systems have the advantage of an effective dark-field system because the exciting beam cannot provide excessive stray light to the eye and interfere with detection of the weaker fluorescent light from the specimen. In the third, the objective acts as the condenser and, therefore, there is no centration problem. The weak fluorescence suffers minimum attenuation due to specimen thickness. The exciter beam is reflected down to the objective from a side port in the microscope tube by a beam-splitting mirror that reflects the exciting wavelength but transmits visible light, thus acting as a barrier filter.

The cells or materials to be examined react with fluorochromes (e.g., xanthenes, acridines, quinolines, etc.) as stains and with substrates that release fluorochromes (e.g., fluorescein) to allow antigen localization or recognition. The manner of using the microscope and its filters and precautions in its use are best learned from an experienced operator and the manual provided by the microscope maker. The results are best recorded on film, either color or black and white.

The following advice derives from the small light response (less than 1:100) in fluorescent emission, the small proportion of the specimen (in most cases) that fluoresces, and the tendency for the response of most fluorochromes to fade with time. In brief, the light levels are low.

1. Work in a dark or well-screened room.

2. Do not use any unnecessary lenses or filters (no ground glass), and keep the light paths as short as the system allows.

3. If there is an auxiliary lens, check the position giving the most light. Open the condenser (aperture) diaphragm completely.

4. Use (oculars) eyepieces of low magnification (8×) because the intensity of fluorescence perceived decreases exponentially with the total magnification.

5. Use high-speed films for photomicrography.

6. Use incident (epi-) illumination, if available, rather than transmitted light.

1.5. IMMERSION OILS

Immersion oils are designed for use with oil-immersion lenses and provide the correct refractive environment for the front lens of the objective. They also generate a homogeneous system, approximating the refractive index of glass, and they "clarify" the dried bacteriological slide preparations by permeating and embedding the stained cells.

The commercially formulated oils recommended by microscope manufacturers have been tested to be sure that they do not have acidic or solvent effects on lenses or their mountings; substitutes are not recommended and may be a false economy. Choices of properties can be made for special purposes, such as "very high viscosity" oil to fill wide gaps or to stay in place on horizontal or inverted microscopes, and "low fluorescence" oil for fluorescence microscopy. For general purposes we prefer immersion oil of a fairly high viscosity (we use type B, high-viscosity oil of 1,250 centipoises from Cargille Laboratories Inc., Cedar Grove, NJ 07009), which stays more or less where it is placed and creeps less than low-viscosity oils. The only problem with this level of viscosity is that bubbles are more easily entrained, but this is avoidable.

Oil-immersion objectives should always be cleaned after use or, at least, at the end of a day's work. The modern nonoxidizing oils are less of a problem than the traditional cedarwood oil (which hardens and leaves annoying traces), but they still alter slowly and may also creep where not wanted. *CAUTION*: They are safe to use, but before 1972 oils were often formulated with polychlorinated biphenyl (PCB) compounds, which are now believed to be carcinogenic and toxic, so look out for old bottles!

Never use more oil than necessary (usually one drop will do). A good "oil bottle" will help to reduce mess. Use either a squeeze bottle or a glass bottle with a broad base containing an applicator (glass rod or wire loop) attached to the cap. Among the best is a double-chambered bottle in which the stopper is formed into a vessel for oil and the base is made to hold xylol for cleaning up.

Oiling the condenser to the slide can be the messiest procedure of all for the unhandy. It demands a high-viscosity oil and some simple precautions both when mounting and when demounting a preparation, as follows:

1. Place a drop of oil on the top lens of the condenser and then lower it below stage level. Avoid bubbles in placing the oil drop.

2. Place a drop of oil on the underside of the slide below the area you wish to examine.

3. Set the slide down on the microscope stage with the drop central in the hole, and hold the slide there while adjusting the mechanical stage or clips to keep it in place.

4. Raise the condenser so that drop meets drop to the point that the top lens is fully covered.

5. With the appropriate objective in focus, now adjust the condenser to give Koehler illumination (1.3.1).

6. When demounting a preparation, lower the condenser before picking up the slide. Lateral movement or wide scanning may lead to oil being deposited and creeping under the stage. This needs to be remembered and recognized when cleaning the top lens (with lens paper) after use.

7. Only condensers with integral or screw-on top elements are suitable for immersion. Movable "swing-out" top elements on some "universal" condensers (i.e., those that can be slid aside with a lever for low-power work) must be oiled with great care or not at all.

1.6. CARE OF MICROSCOPES

Microscopes are remarkably tough and durable. With reasonable care and some protection from the elements (particularly the hostile acid-laden air of some laboratories), they will last a lifetime or more. Follow these points of care:

1. Protect the microscope from dust and grit. A cover should be put over the microscope when it is not in use, and the best is a rigid, transparent and all-enclosing glass or plastic "bell." Do not allow dust to accumulate anywhere; it drifts into lenses and mechanisms.

2. The moving parts, especially the rackwork and gears, need cleaning and new grease at very long intervals (the grease for model train gears from a hobby shop works well). Do *not* use thin oil on gears or bearing surfaces; the tube or condenser may sink by its own weight.

3. Clean the stage regularly, and mop up any spills.

4. Keep the lenses clean and, especially, clean up after a session of oil-immersion work. *Never* use the finger instead of an appropriate lens tissue (see below).

5. Do not attempt to repair an objective lens or the complex mechanism of fine adjustments. These are best left to professionals. Good microscopes respond to a professional cleaning and adjustment every 20 years or so! Microscope dealers usually have a repair facility or can recommend one.

6. Keep the tube closed at all times with an

eyepiece, and keep all objective mounts filled or plugged, to minimize dust in the tube.

The "old soldiers" among microscopes can be cleaned up, repaired if necessary, and put back to good use. The old brass bodies (how elegant they are!) can be rubbed down with a lightly oiled cloth and then a dry cloth. Modern finishes just need to be kept clean.

Optical surfaces need special care and some general rules can be given:

1. Keep a good supply of lens tissue ("lens paper"), a soft brush (artist's watercolor brush, 0.6 cm), and a nebulizer bulb (with a short narrow rubber tube attached) near to your microscopy area. Keep them scrupulously clean and in a dust-free box.

2. Nonadherent dust can be easily removed by using the brush, puffing with the rubber bulb, or wiping with lens tissue after breathing on the optical surface.

3. Finger marks or other adherent grease and dirt must be polished off with lens tissue doubled over a finger and just moistened with xylol or benzene. Do not flood lenses with solvents, and *never* use alcohol, ether, or acetone for fear of penetrating the cement between lenses. If the lens has a raised mount, the edges can be cleaned with lens tissue wrapped around an applicator stick.

4. If xylol does not remove dirt, try moistening the tissue with distilled water.

5. Oil and grease can also be efficiently removed with a freshly broken piece of polystyrene foam (common packing material) by pressing it against the objective front lens and rotating it (H. Pabst quoted by James [7]). The foam has lipophilic properties. Xylol dissolves the foam, so do not use both.

6. The usual way of cleaning oil off objective and condenser lenses is to wipe most of it away with lens tissue and finish with tissue just moistened with xylol.

7. Never take an objective apart to clean the elements because the inter-lens distances are too critical. Dust can be puffed from the back lens or lifted with the super-clean brush.

8. Dust on oculars is a constant problem and produces apparent dark spots which move round as you rotate the lens. The eyepieces are quite simple (top and bottom lenses with a fixed diaphragm between), and both elements unscrew. If cleaning the external top and bottom surfaces does not remove the spot, then there may be particles on the inner surfaces. Be careful of "bloomed" lens surfaces (antireflective coating which looks iridescent blue, and use a brush or tissue with care.

9. *Keep fingers off optical surfaces!*

10. Clean lenses make for a dramatic improvement in image quality.

1.7. TROUBLESHOOTING

The list of technical problems, causes, and remedies shown in Table 1 is taken (with permission) from the excellent book by James (7). If it does not provide for your problem, you must consult books, an experienced microscopist, or an instrument technican to identify the problem, whether mechanical, optical, or preparational. If the problem is optical in nature, then work along the light path in systematic order from the lamp and then through the condenser, specimen, objective, and ocular. Adjust each and determine the effect of adjustment on the problem; this makes Table 1 useful to you.

1.8. COMFORTABLE MICROSCOPY

Long hours at a microscope are needlessly tiring if the microscopist cannot sit comfortably and upright while looking into the instrument. Inclined eyepiece tubes are no help if the table height and chair height are inappropriate and cause stooping or straining. Chairs should be adjustable (including stools in teaching laboratories). For the very tall, a block of wood can be cut and used on a table of fixed height to raise the microscope to just the right position. The whole object of the adjustments is to allow the microscopist to look down the tube with minimal deflection of the neck or back. A comfortable position minimizes strain and the misery that can result.

1.9. PHOTOMICROGRAPHY

All types of microscopy have the possibility of making exact records on film, whatever the wavelength of the radiation used. The technical requirements are important, in all cases, so that the most faithful record may be obtained. Every detail can be seen for control purposes unless extraordinary wavelengths are being used. There are excellent books available that amplify and explain the requirements (3, 5, 7, 9).

Three types of apparatus are in most general use:

1. A roll film camera back (usually for 35-mm film) with integral enclosure, including an ocular. Above it, a beam splitter giving access to the image through a side arm is a common arrangement for the purpose of focusing. The whole device rests on the microscope tube.

2. Integral cameras have been developed by a number of microscope makers and are complete with built-in light and timing controls. These are usually adequate to their purpose and have their own rule books.

TABLE 1. *Common defects in light microscopy and their causes and remedies (7)*

Defect	Possible causes	Remedies
Coarse adjustment is too stiff	Mechanism has been faultily adjusted	With many stands, easy to adjust (often by moving the two control knobs in opposite directions)
	Dirt in rackwork	Clean and put on new grease
Tube or stage sinks spontaneously under its own weight (image drifts out of focus)	Incorrect adjustment of rackwork and/or lubrication with too thin oil	As above
	Faulty adjustment of focus control	As with first entry
Micrometer movement is blocked to one side	Fine adjustment at the end of its travel	Bring a 10× objective into position with the revolving nosepiece; set the fine focus control at the middle of its range and then refocus with the coarse adjustment
Drift of focus with the slightest movement with the fine adjustment (especially with oil-immersion objectives)	Objective insufficiently screwed into the revolving nosepiece	Evident
	Surface of the cover slip stuck to the objective by the layer of oil	Use less viscous immersion oil; clip specimen firmly
Veiled, spotty image	Dirt or grease on the *eyepiece* (spots move when the eyepiece is rotated in the tube) or *objective*; contaminations on *cover slip* (spots move when specimen is shifted) or on any surface of the illumination apparatus	Cleaning where necessary
Sharply focused spots or specks in the image which change and disappear in moving the condenser up and down	Dirt near light source, or a diffusing screen in front of it with critical illumination, or at the cover plate of a built-in illumination or a filter near to it with Koehler illumination	As above; when the contaminated surface cannot be reached, change the focusing of the condenser slightly
Hazy image, which cannot be brought sharply into focus	Wrong immersion (oil instead of air, air instead of oil, air bubble in oil), transparent contamination on objective front lens	Use correct immersion; clean where necessary
	Cover glass too thick, too thick layer of mounting medium	Use of objective with correction collar, or (better) immersion objective
	Irregularly distributed remnants of immersion oil on the cover glass, when using high-power dry objective	Clean with dry cloth or paper tissue; beware of xylene, as this may weaken or dissolve the mounting medium
	Slide upside down on the stage (only with high-power objectives)	Turn slide; make sure that label is not stuck to the wrong side of the slide
Object field partially illuminated	Filter holder partially in light path Objective not clicked into position Condenser (or swing-out lens) not in optical axis	Evident

TABLE 1.—*Continued*

Defect	Possible causes	Remedies
Object field unevenly illuminated	Mirror not correctly in position Condenser not centered (with critical illumination)	
	Irregularity in light source and/or diffusing screen (with critical illumination)	Move condenser slightly up and down; use ground glass in front of light source
Drift of a cloud across the field; after this, image out of focus (oil immersion)	Air bubble in the immersion oil; oil in image space with a dry objective	Wipe off the oil from the specimen and set up anew; clean slide and objective carefully
Sharply delineated bright spots in the image	Transversal reflections in the interior of the microscope (often sickle or ring shaped)	Try another eyepiece, use correct Koehler illumination
	Longitudinal reflections in the tube, causing more round light spots	Use lenses with antireflection coatings, change combination objective-eyepiece
Unsharp bright spots in the image	Contaminations at a lens surface, upper or under side of the object, or air in immersion oil of condenser (differentiate as explained before)	When the localization of the contamination cannot be traced, the effect can be reduced often by opening the condenser diaphragm somewhat more

3. An old-fashioned light-tight bellows is sometimes used. It is located on a stand that can be placed next to a microscope with a light-excluding sleeve at the bottom of the bellows and with its mate to fit on top of the microscope tube. A plate carrier is fitted at the top and can be made for either cut film or instant (Polaroid) film.

In all cases the success of photomicrography and the resulting photographic print depends on:

1. A first-class preparation appropriately mounted and exactly focused.

2. Good optics, properly aligned and illuminated (Koehler principle works the best) to give the best possible image, and a suitable choice of color filters.

3. A camera-microscope association that is free from vibration.

4. Proper film and paper, and processing appropriate to giving optimum grain size with adequate contrast and gray scale.

5. An appropriate final magnification.

Both item 1 and item 2 have been discussed in detail. Item 3 does present some problems because plugging photographic devices into the microscope tubes tends to generate and transmit vibrations. Partly, this is because pressing the shutter release first removes the beam splitter and then activates the shutter; these actions generate persisting vibrations. Shutters for timing of exposures are best put in the light path between lamp and microscope. Vibration is no problem with the bellows cameras as long as the light collar around the eyepiece allows the camera and the microscope to be independent and not to touch each other. Vibrations still need to be kept to a minimum possible level, and a sturdy, heavy table that is not attached to a wall helps.

Film should be a panchromatic film of ASA 60 to 100 for general purposes (to keep the exposures at a manageable level), but for phase photomicrography a faster film (ASA 300 to 400) is a great help because the light levels are much lower. An exposure meter allows repetition of values established by test.

If your department or laboratory is discarding an old bellows-type photomicrographic camera, latch onto it because it is the best apparatus for the highest quality work, even if it isn't the most convenient.

When taking and printing micrographs, you have to think about the final magnification and the detail you want to show. Use the film processing, according to manufacturer's instructions, that enhances contrast. The film grain in the developed negative dictates that you should not enlarge the film more than 2 to 2.5 times in printing. Good microscopy of most ordinary bacteria allows a total magnification of 1,500×, which is needed for fine detail. This provides one reason that the larger images taken on cut film in an old-fashioned camera, for which enlargements of 1.25 to 1.5 times are usually sufficient, are better than the much smaller images on 35-mm film. If the micrographs are not as sharp and as good as they appeared in the properly adjusted microscope, try again. Focusing is critical, whatever the camera used. The trick with bellows cameras is to use a plain glass insert and to focus on this with a focusing telescope adjusted to focus on a mark on the inside of the glass. You can still use the traditional ground glass and black hood if you wish.

Prints should be enlarged to a format that allows easy visibility of the important detail, with the object occupying most of the frame.

1.10. SOURCES OF INFORMATION

Books on light microscopy and photomicrography are readily available and range from encyclopedias to paperbacks. However, books can only give basic principles. Satisfying microscopy can be learned only through direct instruction and practical experience, not a small part of which is the art of preparing worthwhile specimens. No single book is directed toward microscopy for bacteriologists, but the principles are not different. The list presented below (1–10) is representative of the types of material available.

1.11. LITERATURE CITED

1. **Anonymous.** Photography through the microscope. Eastman Kodak Co., Rochester, N.Y.
   This is an example of the booklets produced by manufacturers involved in aspects of microscopy that are obtainable from such firms and their agents in updated versions. They are generally excellent.
2. **Barer, R.** 1968. Lecture notes on the use of the microscope. Blackwell Scientific Publications, Oxford.
   The advice of a master microscopist.
3. **Bradbury, S.** 1976. The optical microscope in biology. Edward Arnold, London.
   This is a short paperback book that deals clearly with matters of resolution and modern forms of microscopy.
4. **Cargille, J. J.** 1975. Immersion oil and the microscope. Technical Reprint 10-1051. R. P. Cargille Laboratories, Cedar Grove, N.J.
   Another booklet produced by a manufacturer. See comment above (1).
5. **Culling, C. F. A.** 1974. Modern microscopy—elementary theory and practice. Butterworths & Co., London.
   Another short paperback book. See comment above (3).
6. **Engle, C. E. (ed.).** 1968. Photography for the scientist. Academic Press, Inc., New York.
   General aspects of scientific photography, including photomicrography.
7. **James, J.** 1976. Light microscopic techniques in biology and medicine. Martinus Nijhoff Medical Division, Amsterdam.
   A fine modern book on theory and practice, with emphasis on the latter. It provides good advice on special and advanced techniques and on their application, including phase-contrast, interference, dark-field, polarization, and fluorescence microscopy.
8. **Möllring, F. K.** Microscopy from the very beginning. Carl Zeiss, Oberkochen, Federal Republic of Germany.
   Another booklet produced by a manufacturer. See comment above (1).
9. **Shillaber, C. P.** 1944. Photomicrography in theory and practice. John Wiley & Sons, Inc., New York.
   Nothing is likely to replace this classic text, which deals exhaustively but readably with the properties of objective lenses, oculars, and condensers. It sets out the practice of good illumination, weighs the advantages of different mounting media, and deals at the same time with theoretical and practical bench microscopy.
10. **Slayter, E. M.** 1970. Optical methods in biology. John Wiley & Sons, Inc., New York.
    A source book for the theoretical bases of most forms of microscopy and leading into analytical processes including diffraction, spectroscopy, and related optical techniques. It is concerned with principles and not practice.

Chapter 2

Specimen Preparation for Light Microscopy

R. G. E. MURRAY AND C. F. ROBINOW

The advent of electron microscopic methods has diminished in part the drive to gain the maximum of cytological information from light microscopic methods. Yet these methods are very useful in the general study of bacteria and in bridging the gaps in the level of resolution and the detection of components between the living cell one one hand and the images gained from the electron microscope on the other. The preparation of specimens for light microscopy is not difficult, and the results are very rewarding aesthetically and scientifically.

Useful items that are mentioned or implied in a number of places in this chapter are as follows:

1. Cover slip forceps. Forceps with flat, non-serrated gripping surfaces bent at an angle of 30° are convenient for picking up cover slips. Curved, pointed, and nonserrated jeweler's forceps are also useful for this purpose.

2. A knife for cutting agar blocks. All sorts of tools will do this, but real knives or scalpels are ruined by sterilizing in a flame. Therefore, obtain some thick Nichrome wire and hammer a narrow leaf shape at the tip on an anvil. This wire, cut fairly short, is held in a loop holder and is sterilized in a flame.

3. Columbia staining jars. These make for easy handling of cover slip preparations during storage, rinsing, and staining, and are made to take 22-mm-square cover slips (A. H. Thomas Co., Philadelphia, Pa.).

2.1. LIVING CULTURES

Hanging-block preparations suspended from a cover slip in a hole in a thick slide made of clear plastic are very useful and informative (see 3.2.3). The bacteria are still surrounded by the medium of growth, and they lie flat against the cover slip. Because they are usually spread from a colony to give separation of the cells, the natural arrangement is not preserved, and because of the thick block and optical problems, resolution is not of a high order. Despite these shortcomings, the technique is readily adapted to following the effects of toxic chemicals and antibiotics on growth and form. For longer observation under better conditions for microscopy, slide cultures are superior to hanging blocks and can be made thin enough to be studied under effective optical conditions. It is best to use phase microscopy, although much can be learned by using ordinary optics.

2.1.1. Agar Slide

Dip clean, sterile slides twice (or more times for thicker layers) into molten agar of appropriate constitution in a petri dish. Wipe the agar from the underside with dampened tissue. Inoculate a selected area of the upper surface under a dissecting (stereoscopic) microscope, using a finely drawn out glass fiber on the end of a Pasteur pipette or glass rod. Wet the tip of the fiber with peptone-water, lightly touch it to a colony or culture of the organism, immediately move it to a drop of peptone-water on the fresh agar surface, and spread. Trim away the agar around the select area, place a cover slip on top, and seal the edges with wax. If the square of nutrient agar is smaller than the cover slip, there is a supply of air and minimal drying. Appropriate modifications and circumstances can be contrived for anaerobes. These cultures can give adequate information on shape, size, and growth habit.

2.1.2. Gelatin-Agar Slide

Changes in internal detail are made visible in phase microscopy by using a solid nutrient agar medium of high refractility (1), which can be provided by adding 14 to 18% gelatin (9). The higher concentrations are suitable for gram-positive bacteria, while gram-negative bacteria may need lower concentrations. Polyvinylpyrrolidone (up to 30%) can be used to replace the gelatin (11). This sort of technique allowed Mason and Powelson (3) to follow and photograph the patterns of nuclear division in growing bacteria.

If necessary, slide cultures can be fixed for staining procedures and even prepared for electron microscopy, but such flat embeddings (2, 7) need considerable skill in handling and sectioning. The cover slip, with undisturbed culture and medium attached, must be removed from the slide for appropriate fixation and embedding in plastic. This can be facilitated by fashioning a "handle" of a strip of Mylar tape or a self-stick label to allow the cover slip to be lifted up with the medium.

2.2. FIXED PREPARATIONS

2.2.1. Bouin Fixation

The most lifelike preparations are obtained by fixing cells in situ, and these preparations are the most useful for the description of shape and growth habits. Grow the bacteria on a solid medium in a petri dish that has been inoculated to provide, after a *few* cell divisions, either a lawn of cells in a single layer or else separate microcolonies in which the cells have not piled upon each other. Check the appropriate state by low-power (10× objective) microscopy during growth. If preparations must be made from fluid media, transfer an appropriate amount of growth to an area on a well-dried agar medium and allow the fluid to sink into the agar plate. In each case, cut blocks of agar from a region showing appropriate distribution of cells. For the best results, cut a block about 1 cm square and invert it, *organism side down*, onto a 22-mm-square cover slip without sliding or pushing it on the glass. The cells are then in contact with the glass and retain their arrangement. Pick up the cover slip with the block attached, and slip it gently into a petri dish containing enough Bouin fixative to cover the block completely.

Bouin fixative.
    Saturated aqueous picric acid, 75 ml
    Formalin, 25 ml
    Glacial acetic acid, 5 ml
Mix and store on a stoppered bottle. The solubility of picric acid is 1.4 g/100 ml at 20°C, and the saturated solution should be kept in stock for preparing the fixative when needed.

Several blocks can be fixed in a dish. Leave the block undisturbed for enough time for the fixative to penetrate completely (generally, 45 min is minimum and 4 h is maximum) and to fix the cells to the cover slip. The block may come off by itself. Separation may be assisted by holding the cover slip down with forceps and by flicking the block upwards with the point of a knife, avoiding a sliding movement. Then rinse the cover slip in water and transfer it to 70% ethanol, where it may be kept until staining is undertaken.

The Bouin-fixed preparation is best stained with dilute solutions of basic dyes. Appropriate stains are 0.01% solutions of crystal violet, Thionine, or methylene blue, applied for 1 to 5 min as determined by trials. Other stains for cytoplasmic inclusions may be used (see 3.3.11). The basic dyes stain the ribosome-rich cytoplasm with advantage on these better-preserved cells, and the nucleoplasms are unstained so that they are revealed, negatively, in conformations that are remarkably similar to those seen in phase-contrast microscopy of living cells. Unfortunately, Bouin fixation is not appropriate for staining the nucleoplasms with Giemsa stain even after ribonuclease treatment or acid hydrolysis.

The cell walls are unstained (most convincingly shown in filamentous *Bacillus* sp.), but they can be displayed effectively by treating the Bouin-fixed cells with 5% tannic acid (wt/vol in water) for 20 min, washing them, and then staining with 0.01% crystal violet for a sufficient time to give intense staining of the wall. Pretreatment with a saturated solution of mercuric chloride for 5 min also allows staining of the surface with basic dyes such as crystal violet, Thionine, and Victoria blues (10). However, no method is entirely reliable, and difficulties are the greatest with gram-negative bacteria.

Mount the preparations in water or diluted stain (which is helpful if the intensity of staining is low for photomicrography), blot gently to remove excess water from the top and edges of the cover slip, and seal to the slide with wax, Vaspar, or nail varnish. The preparations can be dehydrated and made permanent by using a resin mountant, but the result is usually disappointing. Permanence is best achieved by photomicrography, for which these preparations are ideal.

Some cells do not adhere strongly to the cover slip after fixation through the agar or in the impression film technique (see below). This problem can be reduced by first coating the cover slip surface intended for the cells with a thin film of serum or egg albumin and letting it dry. This film may provide faint background staining in the final preparation, but the result is very satisfactory. A similar coating of polyly-sine, which provides a positively charged inter-face, should be as effective or more so (4).

2.2.2. Vapor Fixation

Appropriate distributions of cells on the sur-face of agar blocks, whether grown there or transferred from other media, are fixed conven-iently for light microscopy purposes by vapors of osmium tetroxide or of aldehydes such as formaldehyde or glutaraldehyde. This will not be effective in destroying all enzymes and, there-fore, is not necessarily sufficient fixation for cy-tochemical purposes. The procedure for OsO_4 fixation will be described because it is the most generally useful. The critical steps are as follows:

1. Prepare a 1 or 2% solution of osmium te-troxide (wt/vol) in distilled water. The chemical comes in weighed, crystalline form in a sealed ampoule, which is broken and dropped into the appropriate volume of water. The chemical is slow to go into solution, so one must make it up 1 or 2 days ahead of need. *CAUTION:* OsO_4 is irritating, and the vapor is harmful to eyes and mucous membranes; therefore, use a fume hood. Store the chemical in a cool, dark place and preferably in brown bottles to prevent formation of a black, inactive product.

2. Place agar blocks, cells upward and sup-ported on a slide or a portion thereof, in a closed jar containing a few milliliters of the OsO_4 solu-tion. Use a bed of glass beads to support the slide above the fixative.

3. Allow the vapor to be absorbed for 1.5 to 2 min.

4. Remove the slide and place a cover slip on the block. Hold the edge of the cover slip with a cover slip forceps, and lift the coverslip quickly and vertically to separate it from the block. Alternatively, invert the block onto a cover slip and then flick it off with a knife point so that some of the cells are taken up on the cover slip.

5. Immediately put the cover slip into 70% ethanol in a Columbia jar, in which it can be stored until ready for staining.

Staining procedures can be those suggested for Bouin-fixed preparations. Again, the cyto-plasm stains directly with basic dyes, such as Thionine, and the nucleoplasms do not stain. The latter, however, show as narrower clefts in the stained cell than is the case in the living cell in phase-contrast microscopy. The OsO_4-fixed preparation is suitable for the staining of nucleo-plasms with Giemsa stain after either digestion with ribonuclease (100 μg/ml in 1 mM $MgSO_4$ for 30 min) or hydrolysis with acid. The latter procedure is as follows:

1. Put a cover slip bearing fixed cells in a 1 N solution of HCl at 60°C for 6 to 8 min.

2. Wash the cover slip in water.

3. Stain it with Giemsa solution (10 drops of stain per 10 ml of 0.067 M phosphate buffer, pH 6.6) for 10 min or more for most gram-negative species.

4. Rinse with buffer and examine to see if adequately stained. (A water-immersion objec-tive is helpful but not essential for this step.)

5. Mount in water or diluted stain. The latter is useful for gram-negative bacteria because the cytoplasm can be very pale.

The nucleoplasms stain blue-purple, and the cytoplasm is somewhat pink. Step 1 of this pro-cedure is, of course, the basis for the Feulgen test, and the deoxyribonucleic acid-containing nucleoplasms, after hydrolysis, stain red with Schiff reagent (8).

Impression preparations, fixed in this way, can be very helpful in the study of changes in cells treated in manifold ways, whether in agar or fluid cultures, e.g., cell changes following phage infection (5). One should note, however, that nucleoplasms respond to the cationic environ-ment at the time of fixation, and this, if exces-sive, leads to chromatin condensation (6).

2.3. LITERATURE CITED

1. **Barer, R., R. F. A. Ross, and S. Tkczk.** 1953. Refractometry of living cells. Nature (London) **171**:720–724.
   Basic statements on refractive index of media for phase microscopy.
2. **Girbardt, M.** 1965. Eine Zielschnittmethode für Pilzzellen. Mikroscopie **20**:254, 264.
   Useful advice on flat embedding of slide cultures.
3. **Mason, D. J., and D. M. Powelson.** 1956. Nu-clear division as observed in live bacteria by a new technique. J. Bacteriol. **71**:474–479.
   A prime example of bacterial nuclear division sequences followed by phase photomicrogra-phy.
4. **Mazia, D., G. Schatten and W. Sale.** 1975. Adhesion of cells to surfaces coated with poly-lysine. J. Cell Biol. **66**:198–200.
5. **Murray, R. G. E., and J. F. Whitfield.** 1953. Cytological effects of infection with T5 and some related phages. J. Bacteriol. **65**:715–726.
   The use of light microscope cytology in phage infections.
6. **Murray, R. G. E., and J. F. Whitfield.** 1956. The effects of the ionic environment on the chromatin structures of bacteria. Can. J. Micro-biol. **2**:245–260.
   Cautionary observations on the effect of cations on nuclear structure.
7. **Patton, A. M., and R. Marchant.** 1978. An ultra-structural study of septal development in hy-phae of *Polyporus biennis*. Arch. Microbiol. **118**:271–277.
   Another example of flat embedding techniques.
8. **Piekarski, G.** 1937. Zytologische Untersuchungen

an Bakterien mit Hilfe der Feulgenschen Nuclealreaktion. Arch. Mikrobiol. **8:**428–439.
The application of the Feulgen reaction to delineation of bacterial nuclei.

9. **Robinow, C. F.** 1975. The preparation of yeasts for light microscopy. Methods Cell Biol. **11:**1–22.
Practical advice on using gelatin-agar slide cultures.

10. **Robinow, C. F., and R. G. E. Murray.** 1953. The differentiation of cell wall, cytoplasmic membrane and cytoplasm of gram-positive bacteria by selective staining. Exp. Cell Res. **4:**390–407.
Demonstration of cell walls by staining methods.

11. **Schaechter, M., J. P. Williamson, J. R. Hood, and A. L. Koch.** 1962. Growth, cell and nuclear divisions in some bacteria. J. Gen. Microbiol. **29:**421–434.
An alternative method for increasing the refractive index of media for phase microscopy.

Chapter 3

Determinative Methods of Light Microscopy

R. N. DOETSCH

To characterize a bacterium, specific morphological details usually are determined by means of light microscopy. Some of the methods employed are time honored and trace their genesis to the early days of bacteriological science.

Much useful information may be gleaned from these classical techniques. However, it is important to understand methodological limitations relating to details of form and structure because *any* laboratory manipulation may introduce some alteration, albeit small in most cases. In any event, data from microscopic observations of bacteria are necessary, but usually not sufficient, factors for identifying them (see Chapter 20). Nevertheless, errors in identification often are traceable to mistakes in judging the shape, Gram reaction, and motility of a new isolate.

3.1. SAMPLING

It is difficult to observe living bacteria *directly* in natural habitats, and information so obtained may be meager. There are several reasons for this:

1. Not all environments contain large populations of bacteria per unit mass. Organisms in marine and lake waters, for example, usually must be concentrated by filtration through cellulose or polycarbonate disks, or by centrifugation, in order to obtain sufficient numbers for study.

2. Certain environments may support too many bacteria per unit mass. Sewage sludge and feces contain an astonishing array of bacteria and other microorganisms intermixed with particulate materials. In these cases, the material must be diluted with one of the many available dipolar ionic buffers. Certain soils and marine muds contain only a few bacteria mixed with much opaque colloidal matter, some of which may be hard to distinguish from bacterial forms.

3. Most bacteria constituting the natural flora of an environment do not exhibit particularly

distinctive morphological features. Exceptions to this are, for example, star-shaped *Prosthecomicrobium*, sheathed *Sphaerotilus*, or the trichomes of *Caryophanon*.

Since bacteria in aqueous suspension have optical properties similar to those of water and are difficult to observe by use of ordinary transmission light microscopy, specimens should be examined by positive or negative phase-contrast or dark-field microscopy. If the latter are not available, the condenser must be lowered and the illumination must be reduced to increase the contrast obtainable by ordinary microscopy.

The handling of specimens, loops, slides with droplets on them, and cover slips, and many of the procedures in making and manipulating preparations for microscopy, offer manifold opportunities for the contamination of fingers, benches, clothing, and equipment. Appropriate precautions and attention to technical details and hazards are essential, and are the more important when working with pathogens. Remember that even fixed preparations of organisms containing endospores may retain a level of viability and, therefore, hazard.

Tools for handling bacterial specimens include a large loop (3-mm diameter), a small loop (1-mm diameter), and a straight inoculating needle. Fashion all from platinum or nichrome (the latter is preferable for the needle) and of such a length that, when mounted in a handle of glass or aluminum, they are easily controlled for purposes of streaking plates, sampling colonies, and inoculating tubes of media.

3.2. PREPARATION

Several methods are appropriate to the examination of living bacterial cells. Cultures grown in liquid media can be used with the proviso that dilution with sterile medium is generally required for reducing the population for microscopic study to useful numbers. Cultures grown on solid media can be suspended with a sterile loop or needle in a drop of the sterile liquid medium or a diluent, *not tap water*, to a faint turbidity (2.5.1). Remember when examining bacteria that the age of the culture and the physicochemical characteristics of the medium may cause nonheritable (phenotypic) modifications in both the extraprotoplasmic components (flagella and capsules) and those lying within the protoplast (inclusions), all of which are important as determinative characteristics. Furthermore, some laboratory concoctions may favor development of mutants having heritably altered morphological features from the wild

type. These changes may be stable or unstable, but it is obvious that cultural conditions must be clearly defined when describing bacterial form and structure since a range of morphological variations are possible with varying culture age and medium composition.

Special methods for the examination of growing cultures of bacteria were described in Chapter 2. Three commonly used methods for the examination simply of living cells, as needed for usual determinative purposes, are described as follows.

3.2.1. Wet Mounts of Living Cells

In preparing "wet mounts" of specimens, place 1 drop (ca. 0.05 to 0.1 ml) of sample on a *clean* and "degreased" (by prior heating for 20 min at 400°C) microscope slide (0.8 to 1.0 mm thick) and cover with a glass cover slip. The latter should be 22-mm square and from 0.13 to 0.16 mm thick (no. 1 or 1½ is satisfactory). To prevent convection currents, "drifting," and drying, seal the edges of the cover slip to the slide by applying Vaspar (a mixture of equal parts of petrolatum and paraffin wax), petrolatum, birthday-candle wax (using the wick as a brush after melting the wax—not lighting it—in a pilot flame), or clear nail polish.

The motility of *strictly* aerobic bacteria can be observed only for a brief time in these preparations because the bacteria cease moving once the oxygen is depleted. The inclusion of small air bubbles prolongs activity.

Phase microscopy is recommended for examining wet mounts, and the thinnest possible film should be used for best results. Placing a piece of blotting paper over the cover slip before sealing assists in drawing off sufficient fluid to give a satisfactory thin film for viewing and dries the exposed surfaces.

Heat from the light source may interfere with optimum movement, and it may be advisable to interpose a heat filter if observations are to be made for a long period of time. A green filter is easy on the observer's eyes, as it utilizes the best corrections of the lenses, and also is useful since white light may affect motility. Heating of specimens also can be prevented by use of an intercepting optical trough filled with a 5-cm layer of water or Mohr's solution (50 g of ferric ammonium sulfate, 1.3 ml of 33% [vol/vol] sulfuric acid, and 250 ml of distilled water).

3.2.2. Hanging-Drop Mounts of Living Cells

Make a hanging-drop mount by placing a drop of specimen at the center of a clean cover slip, which is then inverted over the well of a depression slide in such a manner that the drop does not run down the cover slip. The drop of sample must be sufficiently small so that it does not contact the bottom of the well. Apply a small drop of water at the edge of the cover slip to provide an adequate seal and maintain the cover slip in place.

Hanging-drop preparations have certain disadvantages in that the concavity of the well, curvature of the drop, and increased thickness of the depression slide introduce optical aberrations. The varying depth of focus required for viewing the sample from edge to center makes continuous focusing on a single organism, especially if motile, difficult, and it is best to focus on the edge of the drop. Some of these problems are avoided by using slides with flat-bottomed wells. The wells of these slides may be filled with sample and covered with a cover slip as described above.

3.2.3. Hanging-Block Mounts of Living Cells

A useful device is a slide made of clear plastic with a hole (ca. 1 cm in diameter) drilled through its center. Place a cover slip with a hanging drop over the hole, and hold it in place by a touch of immersion oil or water to the edge of the cover slip. Thin growth on agar or parts of colonies spread over an adjacent sterile area may also be attached as a "hanging block." The chamber may be closed if desired by attaching a cover slip to the underside.

3.2.4. Negative Staining

Negative staining provides the simplest and often the quickest means of gaining information on refractile inclusions such as sulfur and poly-β-hydroxybutyrate granules and spores (12). Thoroughly mix a loopful of 7% (wt/vol) aqueous nigrosin (India ink and Congo red are alternatives) with bacteria on a cover slip; spread the mixture into a thin film and air dry it. Then place the cover slip *face down* on a glass microscope slide and hold it in place with one or two spots of wax melted off a birthday candle.

These "air-mounted" preparations reveal bacteria unstained and standing out brightly against a blue-black background. They can even be used to follow the progress and extent of cell disruption and disintegration (see Chapter 5). Negatively stained preparations should not be used for making cell length and width measurements (see below) because they may include the capsule or slime layer outside the cell wall. Furthermore, the negatively charged particles of colloidal nigrosin do not react with the bacterial surface because, at physiological pH, they also are

negatively charged. When films of these colloids dry, the similar charge on both bacterium and colloidal particle causes the edge of the dark film to dry at a small but significant distance from the actual cell boundary. This causes the visible profile of the bacterium to be somewhat larger than the organism itself, even if no capsule is present.

Nigrosin solutions can acquire contaminating organisms; a little formaldehyde helps to prevent this and does not harm the solution.

3.2.5. Simple Staining

Morphological studies of bacteria are generally done on fixed and stained preparations. When a single dye is used, the process is referred to as "simple staining." Spread the specimen to be examined on a glass microscope slide over an area of about 1 cm^2. One loopful of a culture grown in a liquid medium, or one drop of sterile distilled water in which is suspended a *small* amount of culture from a solid medium taken up with an inoculating needle or loop, is satisfactory. A just-visible turbidity is enough. (Broth cultures are not always satisfactory because of the protein precipitates formed during fixation procedures; therefore, Formalin-fixed, centrifuged, and washed preparations may be more useful in many cases.) When the preparation is dry, "heat fix" by passing the slide, organism up, several times over the flame of a Bunsen burner to kill surviving organisms and to assure attachment of the film to the slide. Drying and heating induce shrinkage and distortion of bacteria, and may destroy cell associations; keep these effects in mind when making morphological interpretations.

It is more accurate, albeit more time-consuming, to make chemically fixed preparations. Fixation with Formalin (5%, vol/vol) for a few minutes is simple and often adequate, particularly for Gram staining (3.3.5). See Chapter 2 for details of fixation with osmium tetroxide (2.2.2) and Bouin fixative (2.2.1). For the latter, use bacteria grown for several hours into microcolonies or spread to give a single layer of cells on a solid medium in a petri dish. Cut out a block of agar-bearing bacteria, fix it in situ, and transfer it to a cover slip, which is then stained appropriately with simple basic dyes or treated to gain cytochemical information. Apply these techniques to young cultures, mere hours after inoculation on agar, forming a lawn of close-set microcolonies. Flooding the agar with an appropriate suspension and tilting the plate for aspiration of the excess with a Pasteur pipette works the best. Then dry and incubate the plate. *Wet films* of bacteria placed on cover slips can be fixed prior to drying by exposing them to 2% (wt/vol) aqueous OsO_4 fumes for 2 to 3 min. But it is much better to spread a loopful of cell suspension over the surface of a block of sterile agar and then fix the cells in an appropriate fashion after the fluid has soaked into the agar, preparatory to staining.

With the above techniques, the cover slip carries the bacteria, and the glass microscope slide serves as a base. Cytological advantage is gained if the chemically fixed and lightly stained organisms are *mounted in water and not allowed to dry at any time in the processing.* OsO_4-fixed bacteria do not stain well with simple stains unless treated with a second fixative such as Schaudinn fixative (6.6 ml of saturated aqueous $HgCl_2$, 33.3 ml of absolute ethanol, and glacial acetic acid added to make 2% [vol/vol] before use; 6.9 g of $HgCl_2$ in 100 ml of water at 20°C makes a saturated solution).

Intensely stained cells, for purposes of overall morphology, are obtained by immersing the heat-fixed or chemically fixed preparations briefly in solutions of basic dyes such as crystal violet (10 s) and methylene blue (30 s), and then gently rinsing them with tap or distilled water and blotting dry. Prepare these stains as follows.

Crystal violet (Hucker formula).

> Crystal violet, 2 g
> Ethanol, 95% (vol/vol), 20 ml
> Ammonium oxalate, 1% (wt/vol) aqueous,
>   80 ml

Dissolve the crystal violet in the ethanol. Then add the ammonium oxalate solution and allow to stand for 48 h before use.

Methylene blue (Loeffler formula).

> Methylene blue chloride, 1.6 g
> Ethanol, 95% (vol/vol), 100 ml
> Potassium hydroxide, 0.01% (wt/vol) aqueous, 100
>   ml

Prepare a saturated alcoholic methylene blue solution by adding the dye to the alcohol. Thirty milliliters of this is then added to the potassium hydroxide.

If the cells stain lightly or not at all with one of the above, it may be necessary to stain them with carbolfuchsin, and they may even be acid fast (3.3.6). A few seconds of exposure to this stain is usually enough for most bacteria; it is a very intense dye and overdoes the staining for most purposes. Mycobacteria, however, usually require heat or a surfactant and time to allow the stain to penetrate the waxy cell wall.

3.2.6. Permanent Mounts

Stained films of bacteria on cover slips may be mounted by placing them in water on a glass slide and sealing the cover slip edges with petrolatum, clear nail polish, or candle wax. These preparations are temporary, remaining useful for only a few days, but provide excellent material for photomicrography. For extended preservation, stained and dried films of bacteria on cover slips should be cleared in xylene and then placed on a drop of Canada balsam on a slide; excess balsam should be avoided but can be wiped away with a tissue moistened with xylene. Neutral mounting media consisting of a plastic (polystyrene) dissolved in solvents (toluene, xylene) and containing a plasticizer (tricresyl phosphate) are marketed under such names as Flotexx, DPX, and Permount. Flo-texx may be painted onto a stained film on a glass microscope slide without need for a cover slip. Permount is neutral and does not become acid or discolor with age, nor does it tend to trap bubbles under the cover slip.

3.3. CHARACTERIZATION

The methods described below have been found satisfactory for revealing characteristics of determinative value for a large number of different bacteria. These techniques, however, ought not be considered insusceptible to improvements or modifications; indeed, the literature abounds with statements of minor alterations found necessary to obtain the best results in specific situations. Cytologically rewarding preparations require great care in fixation, staining, handling, and microscopy; the precautions are outlined here and in Chapters 1 and 2. The following are sound routine methods, but the procedures, especially those for cytoplasmic inclusions, are the most effective if applied to chemically fixed preparations, with drying avoided at any stage.

3.3.1. Size

Measurements of bacterial lengths and widths cannot be made with the degree of precision one might desire because of certain unavoidable technical difficulties. Boundaries of living bacteria do not appear sharp when examined by phase-contrast or bright-field microscopy. The boundaries of the organism in phase microscopy are obscured by halos, which interfere also with the visibility of internal anatomical details. The problem can be reduced by mounting the organisms in a medium of greater refractive index, which is simply done by using solutions of gelatin from 15 to 30% (wt/vol) (9). The precise amount giving best results must be determined by experiment for each kind of organism and detail.

A less precise method involves drying, ahead of time, a thin film of 15 to 20% (wt/vol) gelatin on slides and using these to make wet mounts, which swell the gelatin and embed the cells in it, immobilizing them and allowing better phase relations between organism and surround.

Organisms in fixed and stained preparations all suffer some degree of distortion, depending upon the technique used; hence, one obtains an *approximation* of true dimensions. Dried, stained films, even if prepared from Formalin-fixed material, do not reveal the cell wall. Tannic acid-crystal violet staining (see 2.2.1) does reveal the cell wall and makes bacteria appear appreciably wider than they look when live in a wet medium or stained after drying on glass. This staining procedure does not work well with gram-negative bacteria.

Microscopic measurements are made using either a calibrated ocular micrometer or a filar micrometer eyepiece. The **ocular micrometer**, which consists of a circular glass disk upon which a series of arbitrarily marked regular gradations have been placed, is set on the field diaphragm of the ocular after the top element has been unscrewed. A **stage micrometer** is a glass slide specially marked in lines exactly 10 μm apart, and when brought into focus, the ocular micrometer lines superimpose on the calibrated stage micrometer lines. If, for example, 10 lines of the ocular micrometer occupy the space between two adjacent lines of the stage micrometer, then each space on the former would equal 1.0 μm. After calibration, the stage micrometer is no longer required, but new calibrations must be made for each objective used.

The **filar micrometer eyepiece** has a measuring line controlled by a micrometer screw knob. Each *complete* revolution of the knob moves the hairline 1.0 mm and gives direct readings to 10 μm. Moving the hairline over a known distance of the stage micrometer, and counting the number of divisions on the micrometer knob drum, gives a figure showing micrometers per drum division.

3.3.2. Shape

For determinative and descriptive purposes (Chapter 20), individual bacterial shapes are designated as straight, curved, spiral, coccobacillary, branching, pleomorphic, square ended, round ended, tapered, and fusiform. Perhaps star shaped, stalked, and lobular should be added to these classical terms. Arrangements of individual bacteria are described as single, pairs, short chains (<five bacteria), long chains (>five bacteria), packets, tetrads, octets, clumps, and filaments.

When assigning the descriptive terms used in all textbooks, it is assumed that natural groupings are not disturbed. For example, excessive shaking may break long chains or cause clumping of single cells. The breakup of long chains or other fragile associations can be minimized by adding formaldehyde to the culture to give 1 to 2% (vol/vol); the culture is allowed to react for 15 min, centrifuged, and resuspended in water for making films or negative stains. This is highly recommended for streptococci. The influence of culture age and medium composition also must be taken into account.

3.3.3. Division Mode

The mode of division employed by a given bacterium is ordinarily not seen at first glance, and continuous observation of the living organism in a suitable medium is required (5, 10). Simple staining procedures may not adequately reveal whether division is binary, ternary, budding, or by fragmentation; it may be necessary to employ cell wall staining and even electron microscopy for determining this property.

3.3.4. Motility

Translational movement of bacteria by flagellar propulsion may be observed in wet mounts of specimens using, in most cases, the low-power or high-dry objectives. Bacteria vary in their translational velocity, and slow organisms must be differentiated from those showing only Brownian motion. Some flagellated nonmotile mutants ($fla^+ mot^-$) have been described, and in order to ascertain the presence of flagella in doubtful cases, as well as to determine flagellar distribution (polar, peritrichous, lateral), staining procedures (3.3.10) are required.

When a standard condenser and an oil-immersion objective are used, the light must be reduced by the aperture diaphragm and the condenser must be lowered to improve contrast. Low-power views of motility are the most effective (and the most dramatic) in dark-field microscopy. The specialized condensers needed for this technique may not be available to all, but a quite adequate dark field is possible by the stratagem of using a phase microscope oil-immersion condenser, with the oil-immersion phase plate in place, for illuminating the specimen observed with *standard* (*not* phase) 10× or 45× objectives. The condenser is not immersed, but its position for best dark field is discovered by trial. This trick makes it possible to see even spirochetes under low power!

Some procaryotic protists exhibit a peculiar type of translational movement known as "**gliding**" when in contact with a solid surface. This is a comparatively slow and stately, intermittent progression parallel to the longitudinal axis of the organism. Gliding is characterized by frequent directional changes and the absence of external locomotory organelles (8). Gliding motility can be suspected if microscopic (high-power, dry objective) examination of the margin of colonies shows cells singly or in groups isolated from their margin. Gliding is often not obvious on the primary medium of isolation; it is usually facilitated by solid media containing very small amounts of the nutrients appropriate to the group of bacteria involved.

Gliding should be differentiated from **swarming**, the latter being an expression of flagellar motility under special conditions. Swarming is spreading produced on relatively dry agar surfaces in which the organisms move in the form of large rafts or microcolonies (6). A continuously shifting pattern of organisms from short, tonguelike extensions and isolated groups, to interlacing bands interspersed with empty areas, may be observed. "Tracks" or trails, often considered indicative of gliding motility, are apparent in these areas and are well preserved by Bouin fixation in situ (2.2.1). Young, active cultures of filamentous gliders form wavy, curly, or spiral patterns. Translocation by gliding movement occurs at velocities around 10 to 15 μm/s.

3.3.5. Gram Staining

Gram staining is the most important differential technique applied to bacteria. In theory, it should be possible to divide bacteria into two groups, gram positive and gram negative; in practice, there are instances when a given bacterium is gram variable. Numerous modifications of Gram staining have been published since the method was first developed by Christian Gram in 1884.

The film of bacteria on the slide can be made directly from a liquid or solid culture as in simple staining (3.2.5), but is best made from a Formalin-fixed, washed sample. The few extra minutes spent suspending the bacteria in 5% (vol/vol) Formalin and concentrating and washing them by centrifugation pay off in good preservation of size and shapes, solid staining, and the absence of messy precipitates from liquid culture media. In either case, air dry and heat fix the film as for simple staining. It is important to standardize the Gram-staining procedure (4). Two techniques that give equivalent results are those of Hucker and of Burke.

Hucker method for Gram staining.

Solution A for staining reagent:
　Crystal violet (certified 90% dry content), 2.0 g
　Ethanol, 95% (vol/vol), 20 ml

Solution B for staining reagent:
　Ammonium oxalate, 0.8 g
　Distilled water, 80 ml
Mix A and B to obtain the crystal violet staining reagent. Store for 24 h and filter through paper before using.
Mordant:
　Iodine, 1.0 g
　Potassium iodide, 2.0 g
　Distilled water, 300 ml
Grind the iodine and potassium iodide in a mortar, and add water slowly with continuous grinding until the iodine is dissolved. Store in amber bottles.
Decolorizing solvent:
　Ethanol, 95% (vol/vol)
Counterstain:
　Safranin O (2.5% [wt/vol] in 95% [vol/vol] ethanol), 10 ml
　Distilled water, 100 ml

Procedure:
1. Immerse the air-dried, heat-fixed film for 1 min with the crystal violet staining reagent.
2. Wash the film in a gentle and indirect stream of tap water for 2 s.
3. Immerse the film in the iodine mordant for 1 min.
4. Wash the film in a gentle and indirect stream of tap water for 2 s, and then blot the film dry with absorbent paper.
5. Immerse the film in 95% (vol/vol) ethanol for 30 s with agitation, and then blot the film dry with absorbent paper.
6. Immerse the film for 10 s with the counterstain.
7. Wash the film in a gentle and indirect stream of tap water until no color appears in the effluent, and then blot the film dry with absorbent paper.

Burke method for Gram staining.

Solution A:
　Crystal violet (certified 90% dye content), 1.0 g
　Distilled water, 100 ml
Solution B:
　Sodium bicarbonate, 1.0 g
　Distilled water, 100 ml
Iodine mordant:
　Iodine, 1.0 g
　Potassium iodide, 2.0 g
　Distilled water, 100 ml
Prepare this as for Hucker's modification.
Decolorizing solvent (Caution—highly flammable):
　Ethyl ether, 1 volume
　Acetone, 3 volumes
Counterstain:
　Safranin O (85% dry content), 2.0 g
　Distilled water, 100 ml

Procedure:
1. Immerse the air-dried, heat-fixed film with

solution A, add 2 or 3 drops of solution B, and let stand for 2 min.

2. Rinse off the film with the iodine mordant and then cover the film with fresh iodine mordant for 2 min.

3. Wash the film with a gentle and indirect stream of tap water for 2 s; blot *around* the film with absorbent paper, but do not allow the film to dry.

4. Add the decolorizing solvent dropwise to the film, with the slide slanted, until no color appears in the drippings (less than 10 s), and allow the film to air dry.

5. Immerse the film for 5 to 10 s with the counterstain.

6. Wash the film in a gentle and indirect stream of tap water until no color appears in the effluent, and then blot the film dry with absorbent paper.

With either technique, gram-positive bacteria appear blue or violet and gram-negative bacteria appear red.

Gram-negative bacteria may seem to be gram positive if the film is too thick and the decolorization is not completed. Gram-positive organisms, on the other hand, may seem gram negative if the film is over-decolorized; this happens particularly if the culture is in the maximum stationary phase of growth. Some *Bacillus* species are gram-positive for only a few divisions after spore germination. Furthermore, gram-positive organisms will seem gram negative if the integrity of their cell walls is breached mechanically, by death, by autolysis or enzyme action (e.g., muramidase), or by drying on glass and wetting afterwards. It is well to prepare light films (faint turbidity) of young, actively growing cultures for best results since older cultures tend to give variable reactions. A wise precautionary measure is to use known gram-positive and gram-negative organisms as controls.

3.3.6 Acid-Fast Staining

A second important tinctorial property of bacteria is that of not being readily decolorized with acid-alcohol after staining with hot solutions of carbolfuchsin. Described as being "acid fast," this small group of bacteria include the actinomycetes, mycobacteria, some related bacteria, and dormant endospores.

Ziehl-Neelsen method for acid-fast staining.

Carbolfuchsin stain:
  Basic fuchsin, 0.3 g
  Ethanol, 95% (vol/vol), 10 ml
  Phenol, heat-melted crystals, 5 ml
  Distilled water, 95 ml

Dissolve the basic fuchsin in the ethanol; then add the phenol dissolved in the water. Mix, and let stand for several days. Filter before use.
Decolorizing solvent:
  Ethanol, 95% (vol/vol), 97 ml
  Hydrochloric acid (concentrated), 3 ml
Counterstain:
  Methylene blue chloride, 0.3 g
  Distilled water, 100 ml

Procedure:

1. Place the slide containing an air-dried and heat-fixed film on a slide carrier over a trough. Immerse the film with the carbolfuchsin stain. Carefully heat the underside of the slide with a Bunsen burner or on a hot plate until steam rises (without boiling!). Keep the preparation in this condition for 5 min, with intermittent heating as needed (overheating causes spattering of the stain and cracking of the slide). Wash the film with a gentle and indirect stream of tap water until no color appears in the effluent.

2. Immerse the film with the decolorizing solvent. Immediately wash off with tap water, as above. Repeat the decolorizing and washing until the film appears faintly pink.

3. Immerse the film with the counterstain for 20 to 30 s, wash with tap water as above, blot the film dry with absorbent paper, and examine.

Acid-fast bacteria appear red and non-acid-fast bacteria (and other organisms) appear blue.

It is also possible to determine acid fastness of an organism by use of fluorescence techniques. One advantage of this procedure is that slides containing specimens of suspected mycobacteria, for example, may be screened using a 60× rather than a 100× objective; hence, an *entire* slide may be screened in a short time.

Truant method for acid-fast staining (13).

Fluorescent staining reagent:
  Auramine O, CI 41000, 1.50 g
  Rhodamine B, CI 749, 0.75 g
  Glycerol, 75 ml
  Phenol (heat-melted crystals), 10 ml
  Distilled water, 50 ml
Mix the two dyes well with 25 ml of the water and the phenol. Add the remaining water and glycerol, and again mix. Filter the resulting fluorescent staining reagent through glass wool, and store at 4°C or room temperature.
Decolorizing solvent:
  Ethanol, 70% (vol/vol), 99.5 ml
  Hydrochloric acid (concentrated), 0.5 ml
Counterstain:
  Potassium permanganate, 0.5 g
  Distilled water, 99.5 g

Procedure:

1. Immerse a lightly heat-fixed film with the

fluorescent staining reagent for 15 min at 20 to 37°C.

2. Wash the film with a gentle and indirect stream of distilled water until no color appears in the effluent.

3. Immerse the film with the decolorizing solvent for 2 to 3 min; then wash with distilled water as above.

4. Immerse the film with the counterstain for 2 to 4 min.

5. Wash the film with distilled water as above, and blot dry with absorbent paper.

Examine with a fluorescence microscope equipped with a BG-12 exciter filter and an OG-1 barrier filter. Bacteria that are acid fast appear as fluorescent yellow-orange cells in a dark field.

3.3.7 Endospores

Endospores of living bacteria, when viewed unstained, appear edged in black and shine brightly in a plane slightly above true focus. However, do not assume that any highly refractile body within a bacterium is an endospore, particularly if information concerning heat resistance is lacking (see 20.1.52). A direct test is provided by a visible phenomenon (12) occurring in many endospores after immersion in an acid oxidizer (e.g., 0.1% $KMnO_4$ in 0.3 N HNO_3 or 0.3 N HCl). The spore cortex ruptures, and a portion of the spore body, now readily stainable with basic dyes, is herniated through the aperture. The dramatic suddenness of the event (up to 20 min in the solution) makes it deserve the term "popping test." The test can be conveniently performed by mounting a dried film of spores on a cover slip in the reagent for 10 to 20 min; the consequent "popping" is visible without staining, although this does make it more easily visible.

There are several satisfactory methods for staining endospores in bacteria. The simplest is a negative stain, which yields lifelike and life-preserving preparations. In this method, a loopful of 7% (wt/vol) aqueous nigrosin is mixed on a cover slip with a loopful of culture, spread into a thin film, and air dried. The cover slip is inverted and placed film side down on a microscope slide and maintained in place with several spots of candle wax. Endospores appear as highly refractile spherical or ellipsoidal bodies both within and outside the bacterial soma (12).

Endospores strongly resist application of simple dyes, but once stained are quite resistant to decolorization. A useful positive staining method for bacterial endospores is that of Dorner.

Dorner method for staining endospores.

1. Air dry or heat fix the organism on a glass slide, and cover with a square of blotting paper or toweling cut to fit the slide.

2. Saturate the blotting paper with carbolfuchsin (3.3.6) and steam for 5 to 10 min as described, keeping the paper moist by addition of more dye as required. Alternatively, the slides may be steamed by placing them on a rack over a container of boiling water.

3. Remove the blotting paper and decolorize the film with acid-alcohol for 1 min; rinse with tap water, and blot dry.

4. Dry a thin, even film of saturated aqueous nigrosin solution (10 g of nigrosin, 100 ml of distilled water; immerse for 30 min in boiling water, cool, and add 0.5 ml of Formalin to preserve; filter before use) on the slide and examine.

Vegetative cells are colorless, the endospores are red, and the background is black.

A variation of Dorner's technique is to mix in a test tube an aqueous suspension of bacteria with an equal volume of carbolfuchsin. The tubes are immersed in a boiling-water bath for 10 min. Then a loopful of 7% (wt/vol) aqueous nigrosin is mixed on a glass slide with one loopful of the boiled carbolfuchsin-organism suspension and air dried in a thin film. The results are as indicated above, but the procedure is less messy.

Schaeffer-Fulton method for staining endospores.

The Schaeffer-Fulton technique employs 0.5% (wt/vol) aqueous malachite green instead of carbolfuchsin. The specimen is air dried on a glass slide, heat fixed, covered with filter paper, blotting paper, or paper towel, saturated with the dye, and placed over boiling water for 5 min. The slide is washed in tap water, and the film is counterstained with safranin as in Gram staining for 30 s, again followed by washing and blotting dry. Endospores are bright green and vegetative cells are brownish red.

3.3.8 Cysts

Bacterial cysts (found only in *Azotobacter* sp.) stain weakly with simple stains and generally appear as spherical bodies surrounded by thick, but poorly staining, walls. The following method (14) is useful for demonstrating *Azotobacter* cysts, including forms developed prior to the appearance of a mature cyst.

Staining cysts.

Reagent:
   Glacial acetic acid, 8.5 ml
   Sodium sulfate (anhydrous), 3.25 g
   Neutral red, 200 mg
   Light green S. F. yellowish, 200 mg
   Ethanol, 95% (wt/vol), 50 ml
   Distilled water, 100 ml
Add the chemicals and dyes to the water with

continuous stirring for 15 min. Filter through a membrane filter (pore diameter, 0.5 μm).

Procedure:
Immerse the specimen in the reagent and examine the wet preparation. Vegetative azotobacters are yellowish green; early encystment stages appear with a darker green cytoplasm somewhat receded from the outer cell wall, from which it is divided by a brownish-red layer. In mature cysts the central body appears dark green and it is separated by the unstained intine from the outer, brownish-red exine.

3.3.9. Capsules and Slime Layers

Capsules and slime layers are produced, by bacteria capable of forming them, under specific cultural conditions. They are best demonstrated in wet preparations because the highly hydrated colloidal polymers constituting them are easily distorted and shrunk by drying and fixation.

Duguid's method is the best and simplest of the three that follow.

Duguid method for staining capsules.

1. Place a large loopful of India ink on a clean slide and mix in a loopful of culture; then place a glass cover slip on this in such a way that only part of the mixture is covered.
2. Press firmly down on the cover glass, using several thicknesses of blotting paper, until a thin, brownish-colored fluid is seen. Examine with high-dry and oil-immersion lens systems.

Capsules appear as clear zones around the refractile organism and the brownish-black background.

Hiss method for staining capsules.

1. Mix a loopful of suspension with a drop of normal horse serum or skim milk on a glass slide.
2. Air dry and *gently* heat fix the film.
3. Cover the film with crystal violet stain (crystal violet, 0.1 g; distilled water, 100 ml), and heat until steam rises.
4. Wash off the crystal violet with a 20% (wt/vol) aqueous solution of copper sulfate ($CuSO_4 \cdot 5H_2O$). Blot dry, and examine.

Capsules appear faint blue and the organisms are dark purple.

Anthony method for staining capsules.

1. Stain an air-dried film for 2 min with a 1% (wt/vol) aqueous solution of crystal violet.
2. Wash off the dye with a 20% (wt/vol) aqueous solution of copper sulfate ($CuSO_4 \cdot 5H_2O$), drain, and blot dry.

Capsules appear light blue, and organisms are dark purple.

3.3.10. Flagella

Although Koch devised a technique for staining flagella over a century ago, no easy or constantly reliable method yet is available. Since the width of bacterial flagella lies below the limits of resolution for transmission light microscopy (flagella are around 10 to 30 nm in diameter), it is necessary to "tar and feather" them, so to speak, in order to make them visible. In general, nonmotile bacteria do not possess flagella, but there are cases of organisms having "paralyzed" flagella (*fla⁺ mot⁻*).

Gray method for staining flagella.

Solution A:
  Tannic acid (20% [wt/vol] aqueous), 2 ml
  Potassium alum [$KAl(SO_4)_2 \cdot 12H_2O$], saturated aqueous, 5 ml
  Mercuric chloride, saturated aqueous, 2 ml
  Basic fuchsin (3% [wt/vol] in 95% [wt/vol] ethanol), 0.4 ml
The dye is added to the other ingredients, preferably immediately before use, and the solution is filtered.
Solution B:
  Ziehl's carbolfuchsin (3.3.6)

Procedure:
1. For ordinary organisms, it is effective to grow them on an agar slant of appropriate nutrient properties and to add, 0.5 h before sampling, a volume of peptone-water into which the organisms swim. In any case, a fluid medium is most appropriate. The specimen then is centrifuged to free it from medium constituents, washed, and recentrifuged. The pellet is *gently* (to prevent loss of flagella) resuspended with 10% (vol/vol) aqueous Formalin to produce a light, faint turbidity.
2. Allow a loopful of suspension to run down a tilted (45° angle) glass slide and air dry.
3. Cover the film by filtering solution A over it and allowing to remain for 6 min. The exact time will have to be determined experimentally.
4. Wash solution A off the slide with distilled water.
5. Place a piece of blotting paper over the film, and flood the slide with solution B for 3 min.
6. Remove the blotting paper, and wash the slide with distilled water, gently blot dry, and examine.

Flagella and bacterial bodies stain red.

An important part in the success of this technique depends upon the use of *absolutely clean* and grease-free slides. Either heat a slide by holding it in a Bunsen flame until it shows yellow around the edges and allow it to cool, or degrease by heating loosely packed slides at 400°C for 20 min in a muffle furnace. Remove the slides after cooling, and use immediately or store in dust-free containers. A slurry of Bon Ami cleanser, allowed to dry on slides in a thin film and removed with a Kleenex tissue, also leaves slides very clean. Solutions or solvents such as chromic acid or 50% (vol/vol) alcohol are not recommended for cleaning slides to be used for staining flagella.

Leifson method for staining flagella.

Solution 1:
  Sodium chloride, 1.5 g
  Distilled water, 100 ml
Solution 2:
  Tannic acid, 3.0 g
  Distilled water, 100 ml
Solution 3:
  Pararosaniline acetate, 0.9 g
  Pararosaniline hydrochloride, 0.3 g
  Ethanol, 95% (vol/vol), 100 ml
Mix equal volumes of solutions 1 and 2; then add 2 volumes of this to one part of solution 3. Keeps under refrigeration for 1 to 2 months.

Procedure:
1. Prepare an air-dried specimen on a slide as in Gray's method, steps 1 and 2.
2. Using a wax glass-marking pencil, draw a rectangle around the film.
3. Place 1 ml of the dye solution on the slide without allowing any to flow over the wax lines. Leave for 7 to 15 min; the "best" time must be determined experimentally.
4. As soon as a golden film develops on the dye surface, and a precipitate appears throughout the film as determined by illumination under the slide, remove the stain by "floating off" the film with flowing tap water. Air dry and examine.
Bacterial bodies and flagella stain red.

3.3.11. Cytoplasmic Inclusions

Many bacteria grown under certain cultural conditions produce, as a result of metabolic reactions, deposits within the cytoplasm which are termed "inclusions." Among these are deposits of fat, poly-β-hydroxybutyrate, polyphosphate, starchlike polysaccharides, sulfur, and various crystals. Some of these are polymerized waste products and others are food reserves. Several staining procedures are used to reveal these inclusions, which are also refractile.

Staining poly-β-hydroxybutyrate.

A convenient method for demonstrating these inclusions in bacteria is as follows:
1. Prepare a heat-fixed film of the specimen on a slide, and immerse in a solution of 0.3%

(wt/vol) Sudan black B made up in ethylene glycol. Filter and stain for 5 to 15 min (determine time experimentally).
2. Drain, and air dry the slide on blotting paper.
3. Immerse and withdraw the slide several times in xylene, and blot dry.
4. Counterstain for 5 to 10 s with 0.5% (wt/vol) aqueous safranin.
5. Rinse the slide with tap water, blot dry, and examine.
Poly-β-hydroxybutyrate inclusions appear as blue-black droplets, and cytoplasmic parts of the organism appear pink.

Staining polyphosphate.

Inclusions consisting of unbranched structures with the elementary composition $M_{n+2}P_nO_{3n+1}$ may be detected as follows.
1. Prepare a heat-fixed film of the specimen on a glass slide.
2. Stain for 10 to 30 s in a solution of Loeffler's methylene blue or 1% (wt/vol) toluidine blue.
3. Rinse the slide with tap water, blot dry, and examine.
Polyphosphate granules appear as deep-blue to violet spheres, and the remaining cytoplasm appears light blue with methylene blue. Toluidine blue stains metachromatically, and the granules are red in a blue cytoplasm.

Periodate-Schiff method for staining polysaccharide.

"Glycogen-like" materials may be revealed by using the following technique.
1. Prepare a heat-fixed film of suspension on a glass slide.
2. Cover the slide with a periodate solution (4% [wt/vol] aqueous periodic acid, 20 ml; 0.2 M aqueous sodium acetate, 10 ml; and ethanol, 95% [vol/vol], 70 ml; protect the solution from light!) for 5 min.
3. Wash the slide in 70% (vol/vol) ethanol.
4. Cover the slide with the following reducing solution for 5 min: ethanol, 300 ml; 2 N hydrochloric acid, 5 ml; potassium iodide, 10 g; sodium thiosulfate pentahydrate, 5 ml; distilled water, 200 ml. Add the ethanol and then the hydrochloric acid to the solution of potassium iodide and sodium thiosulfate in distilled water. Stir, allow the sulfur precipitate to settle, and then decant the supernatant.
5. Wash the slide with 70% (wt/vol) ethanol.
6. Stain with the following solution (Schiff's reagent) for 15 to 45 min (determine experimentally). Dissolve 2 g of basic fuchsin in 400 ml of

boiling distilled water, cool to 50°C, and filter through paper. Add 10 ml of 2 N hydrochloric acid and 4 g of potassium metabisulfite to the filtrate. Stopper tightly and allow to stand for 12 h in a cool, dark place. Add about 10 ml of 2 N hydrochloric acid until the reagent when dried on a glass slide does not show a pink tint. Store in the dark.

7. Wash the slide several times in a solution consisting of 2 g of potassium metabisulfite and 5 ml of concentrated hydrochloric acid in 500 ml of distilled water.

8. Wash the slide with tap water, and counterstain with a 0.002% (wt/vol) aqueous solution of malachite green for 2 to 5 s.

9. Wash the slide in tap water, blot dry, and examine.

Polysaccharides stain red, and other cytoplasmic components are green.

Alcian blue method for staining polysaccharide.

Alcian blue also may be used to reveal polysaccharides. A 1% (wt/vol) solution of Alcian blue in 95% (vol/vol) ethanol is used, with carbolfuchsin as counterstain. The procedure is as follows.

1. Heat fix a film of specimen on a glass slide.
2. Stain for 1 min with a 1:9 dilution (in water) of Alcian blue prepared as above.
3. Wash the film in tap water and dry.
4. Counterstain (care must be taken not to overstain!) briefly with carbolfuchsin (3.3.6), and wash immediately with tap water. Air dry the film and examine.

Polysaccharides appear blue, and other cytoplasmic constituents are red.

3.4. SPECIAL TECHNIQUES

A number of techniques not routinely employed on first look at specimens are included in this section.

3.4.1. Spirochetes

Spirochetes do not stain well with "ordinary" stains, but free-living spirochetes show up well in nigrosin films. Furthermore, a number are of dimensions (width) quite near the limits of light microscope resolution. Ordinarily, these bacteria are best observed in life by direct dark-field or phase microscopy. It is useful to know that, in films containing spirochetes and stained with Giemsa, the organisms fluoresce bright golden-yellow when observed with dark-field illumination.

The following positive stain generally gives good results.

Fontana method for staining spirochetes.

Solution A (fixative):
  Acetic acid, 1 ml
  Formalin, 2 ml
  Distilled water, 100 ml
Solution B (mordant):
  Phenol, 1 g
  Tannic acid, 5 g
  Distilled water, 100 ml
Solution C:
  Add a solution of 10% (wt/vol) aqueous ammonium hydroxide (drops) to a 0.5% (wt/vol) aqueous solution of silver nitrate until the precipitate which initially forms just redissolves.
Solution D:
  Ethanol, absolute

Procedure:
1. Using slides cleaned in the manner described for making suitable flagellar stains, air dry a film of specimen on a slide.
2. Fix the film in solution A for 1 to 2 min.
3. Rinse the film in solution D for 3 min.
4. Cover the film with solution B, and heat until steam rises for 30 s.
5. Wash the film with distilled water and dry.
6. Cover the film with solution C, and heat until steam rises and the film appears brown. Wash the film, dry, and examine.

Spirochetes appear brownish black.

3.4.2. *Mycoplasma*

Mycoplasmas are probably the smallest bacteria observable with light microscopic techniques and are hard to recognize individually because of extreme pleomorphism. The characteristic "fried-egg" colonies are visible by direct microscopy of agar plates, and this can be assisted by Dienes' stratagem of examining after placing a drop of methylene blue on the growth and covering this with a cover slip. Indeed, the use of phase microscopy with objects close to the limits of resolution makes optical measurements difficult because of halo artifacts and the inherent characteristics of phase-contrast image formation, in which circular objects appear smaller than they really are. Morphological and motility studies of mycoplasmas involve the use of wet mounts, and either dark-field illumination or phase-contrast methods are recommended. If stained preparations of mycoplasmas are required, some modification of the Giemsa stain (see below) is preferred, and it must be applied to previously fixed organisms, preferably by fixing the organisms through an agar block onto a cover slip with Bouin solution (see 2.2.1).

3.4.3. *Rickettsia*

Rickettsiae are obligately intracellular parasites (except for *Rickettsia quintana*) that are generally rod shaped, occur in pairs, and show marked pleomorphism. Rickettsiae may be seen in films of infected tissue by direct microscopic examination after staining by the Giemsa or Gimenez procedures. Most rickettsiae will appear as purple pleomorphic coccobacillary organisms varying in length from 0.25 to 2.0 μm. Pairs and short chains are most frequently observed. *Coxiella burnetii* is the smallest (0.25 μm by 1.0 μm), and it appears as a bipolarly staining rod in the cytoplasm of infected cells. Rickettsiae of typhus and spotted fever are generally larger (0.3 to 0.6 μm by 1.2 μm). Spotted fever rickettsiae are found in the nuclei of infected cells, as well as in the cytoplasm, and often appear to be surrounded by a "halo," sometimes mistaken for a capsule.

Infected tissues may be examined directly by fluorescence microscopy using specific antisera labeled with fluorescein isothiocyanate. With the Gram stain, rickettsiae appear gram negative.

Giemsa method for staining rickettsiae.

Stock solution:
  Giemsa powder, 0.5 g
  Glycerol, 33 ml
  Absolute methanol (acetone-free), 33 ml
Dissolve the Giemsa powder in the glycerol at 55 to 60°C for 1.5 to 2 h. Add the methanol, mix thoroughly, and allow to stand. Store the sediment-free reagent at room temperature.

For use, dilute 1 part of the stock solution with 40 to 50 parts of neutral distilled water or buffered water (pH 6.8).

Procedure:

1. Air dry the specimen on a glass slide, fix in absolute methanol for 5 min, and again air dry.
2. Cover the slide with freshly prepared Giemsa stain for 1 h.
3. Rinse the slide in 95% (vol/vol) ethanol to remove any excess dye, air dry it, and examine for the presence of basophilic intracytoplasmic organisms.

Gimenez method for staining rickettsiae.

Stock solution:
  Basic fuchsin, 10 g
  Ethanol, 95% (vol/vol), 100 ml
  Phenol, 4% (wt/vol) aqueous, 250 ml
  Distilled water, 650 ml
Dissolve the dye in ethanol, and then add the other ingredients. Allow the stain to stand at 37°C for 48 h before use.

To use, dilute the stock solution 1:2.5 with phosphate buffer (pH 7.45) prepared by mixing 3.5 ml of 0.2 M NaH_2PO_4, 15.5 ml of 0.2 M Na_2HPO_4,

and 19 ml of distilled water. The solution is then filtered. It will keep for 3 to 4 days but should be filtered before each use. A 0.8% (wt/vol) aqueous solution of malachite green oxalate also is required.

Procedure:

1. Heat fix the specimen, and cover it with the staining solution for 1 to 2 min.
2. Wash the film with tap water, and stain for 5 to 10 s with malachite green.
3. Wash the film again with tap water, and make a second application of malachite green.
4. Rinse the slide thoroughly in tap water, blot it dry, and examine.

Rickettsiae appear as reddish-staining bacteria against a green background.

3.4.4. Bacteria in water

Direct microscopic examination of marine water samples generally reveals few bacteria. Filtration techniques are used to concentrate organisms for estimation of total number per unit volume, as well as for direct visualization of different morphological types. One filtration procedure (7) is as follows.

Polycarbonate filters 25 mm in diameter (pore diameter, 0.2 μm) are prestained for 5 min in an 0.2% (wt/vol) solution of Irgalen black (color index acid black no. 107, Union Carbide Corp., New York, N.Y.) in 2% (vol/vol) acetic acid and then are rinsed in cell-free distilled water and placed wet on a cell-free glass filter apparatus (Millipore Corp., Bedford, Mass.). Prestained filters, dried and stored after rinsing, are wetted with cell-free distilled water before use in the appropriate filter holder. A seawater sample, fixed by bringing the concentration of neutralized (with $BaCO_3$) glutaraldehyde to a concentration of 0.1% (vol/vol), is added to the filter funnel. Then acridine orange (40% dye content) at a concentration of 0.1% (wt/vol) in 0.02 M tris(hydroxymethyl)aminomethane, pH 7.2 (at 20°C), is added to the sample to make a final concentration of 0.01% (wt/vol). After staining for 3 min, the seawater sample-acridine orange is drawn through the filter membrane by suction (125 mmHg). The membrane is then removed from the filter apparatus and placed over a drop of immersion oil (type I, F, or A; Cargille Laboratories, Inc., Cedar Grove, N.J.) on a glass microscope slide. Another drop of oil is placed on top of the membrane, followed by a glass cover slip. The steps must be performed rapidly to prevent the filter membrane from becoming dry. The preparation is examined with a standard microscope equipped with an epifluorescent illumination system including a 100-W halogen lamp, a BG-12 excitation filter, an LP-510 barrier filter, and an FT-510 beam splitter. Bacteria on the filter will fluoresce green, and individual morphological types may be observed.

Another and time-honored technique, popularized by A. T. Henrici, involves suspending slides or cover slips in the water being sampled for hours or days and retrieving them for staining and microscopy. This selects for organisms with the property of attaching to surfaces, and many organisms from both fresh- and saltwater do this.

3.4.5. Bacteria in soil

Nonselective techniques are available for direct visualization of soil bacteria (3, 11). Special glass capillaries of rectangular cross section are available as Microslides in various widths (Carlson Scientific Inc., Matteson, IL 60443). These have been developed with one wall quite thin (0.17 mm). The "peloscope," consisting of a set of four or five bundles of five rectangular capillary tubes 1 to 2 cm in length, is filled with sterile distilled water, attached to a holder (with a scale marked in millimeters) by means of a rubber band, and placed vertically into the soil sample. After suitable periods, the peloscope is removed and examined microscopically to observe the "microbial landscapes" that grow into the capillaries. It is recommended that the peloscope be carefully washed with water, and the set of cells is then removed from the holder. The outside of each capillary is wiped clean with lens paper and then is observed, thin wall up. Peloscopic capillaries may be made into permanent preparations by use of special techniques (11).

Another simple means of obtaining a selection (but by no means complete) of soil organisms for microscopic examination consists of placing the soil sample in the bottom of a watch glass or small dish, adding water from *below* by use of a Pasteur pipette until a water surface is formed above the sample, and floating cover slips or electron microscope grids on the surface to sample the organisms in the surface film for examination.

3.5 LITERATURE CITED

3.5.1. General References

1. **Norris, J. R., and D. W. Ribbons (ed.).** 1971. Methods in microbiology, vol. 5A. Academic Press, Inc., New York.
   The material presented in the first part of this volume is particularly useful as a source of information supplementary to that presented here. The relevant chapters are as follows:
   I. Microscopy and micrometry, by L. B. Quesnel, p. 1–103
   II. Staining bacteria, by J. R. Norris and H. Swain, p. 105–144.
   III. Techniques involving optical brightening agents, by A. M. Paton and S. M. Jones, p. 135–144.
   IV. Motility, by T. Iino and M. Enomoto, p. 145–163.
2. **Paik, G., and M. T. Suggs.** 1974. Reagents, stains, and miscellaneous test procedures, p. 930–950. *In* E. H. Lennette, E. H. Spaulding, and J. P. Truant (ed.), Manual of clinical microbiology, 2nd ed. American Society for Microbiology, Washington, D.C.
   This provides some additional procedures and appropriate references for medical microbiology.

3.5.2. Specific References

3. **Aristovskaya, T. V.** 1973. The use of capillary techniques in ecological studies in microorganisms, p. 47–52. *In* T. Rosswall (ed.), Modern methods in the study of microbial ecology. Swedish National Science Research Council, Stockholm.
4. **Bartholomew, J. W.** 1962. Variables influencing results, and the precise definition of steps in gram staining as a means of standardizing the results obtained. Stain Technol. **37:**139–155.
5. **Casida, L.** 1972. Interval scanning photomicrography of microbial cell populations. Appl. Microbiol. **23:**190–193.
6. **Henrichsen, J.** 1972. Bacterial surface translocation: a survey and a classification. Bacteriol. Rev. **36:**478–503.
7. **Hobbie, J. E., R. J. Daley, and S. Jasper.** 1977. Use of Nuclepore filters for counting bacteria by fluorescence microscopy. Appl. Environ. Microbiol. **33:**1225–1228.
8. **MacRae, T. H., and H. D. McCurdy.** 1976. The isolation and characterization of gliding motility mutants of *Myxococcus xanthus.* Can. J. Microbiol. **22:**1282–1292.
9. **Mason, D. J., and D. M. Powelson.** 1956. Nuclear division as observed in live bacteria by a new technique. J. Bacteriol. **71:**474–479.
10. **Noller, E. C., and N. N. Durham.** 1968. Sealed aerobic slide culture for photomicrography. Appl. Microbiol. **16:**439–440.
11. **Perfil'ev, B. V., and D. R. Gabe.** 1969. Capillary methods of investigating micro-organisms. Oliver and Boyd, Edinburgh.
12. **Robinow, C. F.** 1960. Morphology of bacterial spores, their development and germination, p. 207–248. *In* I. C. Gunsalus and R. Y. Stanier (ed.), The bacteria, vol. 1. Academic Press, Inc., New York.
13. **Truant, J. P., W. A. Brett, and W. Thomas.** 1962. Fluorescence microscopy of tubercle bacilli stained with auramine and rhodamine. Henry Ford Hosp. Med. Bull. **10:**287–296.
14. **Vela, G. R., and O. Wyss.** 1964. Improved stain for visualization of *Azotobacter* encystment. J. Bacteriol. **87:**476–477.

Chapter 4

Electron Microscopy

ROGER M. COLE AND TERRY J. POPKIN

An individual without experience who acquires an electron microscope (EM) has to gain the experience of others by working with them and, preferably, the identical instrument, by absorbing working manuals and reference books, and, perhaps, by taking a course. The methods we supply are for those who have access to an EM unit and need to make bacteriological studies. They also might benefit from a course, but they can learn the necessary procedures at the console with the regular operator and from guided reading and general texts (1–14). We will not second-guess that instructor—each unit has its rights.

Unlike light microscopes, which everyone in bacteriology is expected to use to the best of her or his ability, the EM is a specialized instrument looked after by specialists and maintained, because of the expense of parts and labor, by service contracts with the manufacturer. Most instruments have multiple users to justify the capital expense and the complex operating problems. A unit often provides access to facilities for section cutting, shadow-casting, freeze-cleaving and freeze-etching, and perhaps other kinds of microscopy, including scanning electron microscopy. Those in charge of units serving other users establish rules and standards of operation, which protect the integrity of the equipment and the work that comes out of it, and are strict in enforcing them. Experience says that casual users of the EM fit the description given by a museum guard of its young patrons: "Lady, they will do *anything*." To this one must add, "and not tell you what it was." This means that the user must be considerate of the instrument (some parts are delicate and it *does* matter what knobs you twist), the user must be considerate of the next user (a "golden rule"), and the user must be scrupulously honest about anything he does (or does not do), to assist troubleshooting or maintenance.

These are some things to remember:

1. See that you are checked out on every instrument and instrumental procedure that you need, not only to know how to do those things but also so that you and the responsible operator know your limitations at each step in your training.

2. If it doesn't work, behaves strangely, or requires unusual force, *stop* in a safe position, think, and seek help.

3. Do not be tempted to undertake repairs: own up and confess.

4. Keep records, even if that is not the habit of the unit, and see that your operating times are recorded and every malfunction is reported.

5. Make your own notes of the steps in procedures, have them with you when you work, and use a *checklist* which should include, e.g.:

34

Positions of switches and settings for start-up (also the positions at turn-off)
Start-up procedures
Readiness requirements
Alignment procedures for everyday operation
  Centering of objective aperture
  Condenser adjustment
  Astigmatism compensation
Setting for wide-field scanning
Operating procedures and camera operation
Warning signs and responses to them
Standby and safety actions
Turn-off procedure and settings

Good preparations, procedures, patience, and operational skills with the EM will be rewarded by good micrographs and will by appreciated by both the supervisor and the microscope.

The value of electron microscopy lies in its capacity to resolve objects that cannot be resolved by optical microscopy with visible or ultraviolet light. The short wavelength of electrons, which can be shortened in direct relationship to the accelerating voltage applied, permits the resolution or discrimination of objects as small as or separated by 2 Ångstroms (0.2 nanometer [nm] or 0.0002 micrometer [μm]), or even less, whereas the limit of light optical resolution lies close to 0.2 μm, depending on the wavelength of light used. In practice, the best modern transmission electron microscopes probably achieve a resolution somewhere between 0.5 and 1.0 nm under the very best conditions of operation (i.e., regarding vacuum, lens alignments, acceleration voltage, cleanliness, physical stability, and specimen preparation). But biological specimens are not infinitely thin; the practical limit of real resolution in sections is about 2.5 nm, and in negatively stained preparations it is about 1.0 nm. It takes skill and a well-tuned instrument to do better. Thick specimens and support films increase electron scattering, and resolution deteriorates markedly.

It is useful to recall some of the limitations of electron microscopy. Because specimens must be examined in a high vacuum in order to permit electron flow, they obviously will be dried and dead. The nature of electron penetration into and passage through biological specimens is such that the degree of contrast seen or photographed is poor and, contrary to resolution, decreases with an increase in accelerating voltage. Consequently, the specimen must be as thin as possible and selectively stained or contrasted by salts of heavy metals that are electron dense.

Because of the increased resolution, useful magnification of the EM can also be greatly increased over that obtainable by optical microscopy. Instruments in current use achieve enormous direct magnifications on the viewing screen or photographic film, but most specimens, unless there is need to resolve macromolecules, are well served by magnifications on the screen or film of 5,000 to 15,000× (subsequent enlargements of 5 to 10 times are more than adequate), which routinely produce more stable images. The quality of the results depends not only on conditions of operation already mentioned, but also on effective preparations, operator skill, and photographic techniques in the darkroom.

The value of the increased resolution of electron microscopy in general bacteriology is the ability to elucidate details of bacterial structure provided that the different preparative methods and the choice of EM techniques are suitable to the structure involved. For example, the use of salts of heavy metals to surround the bacterium and to penetrate into surface irregularities supplies contrast by differential impedance of electrons, thus producing the effect known as **negative staining**, analogous to the use of nigrosine in optical microscopy. However, unless there are breaks in the cell wall or membrane to allow penetration of the contrasting stain, no intracytoplasmic features can be seen by this method. Cellular components may be studied by purposeful rupture and fractionation before negative staining.

Internal detail in situ is assessed by sectioning a plastic embedding of the bacteria, which are first chemically fixed and then stained with salts of heavy metals to supply needed differential contrast. In all instances, some electrons pass through the specimen and others are scattered by components of structure so that an image can be recorded on the fluorescent viewing screen or film. Because this involves transmission of electrons (or at least some of them), this type of electron microscope is known as the **transmission electron microscope** (TEM). Microscopes of this type possess the widest variety of possible applications in microbiology (53–63).

A more recent technical development is the **scanning electron microscope** (SEM). Some modern instruments are constructed so as to combine the capabilities for both transmission and scanning, with features allowing a rapid changeover from one to the other. In the SEM, as inferred, the electron beam rapidly scans the surface of the specimen to induce radiations that form a display image through mediation of a cathode-ray tube, in a process similar to formation of a television picture. The specimen must be first chemically fixed, critically dried to avoid distortion, and vacuum coated with gold or another heavy metal. The resolution achieved by SEMs has been markedly improved since their introduction, but is still less (at best about 3.0

nm in commercially available models) than that routinely available by TEMs. The SEM can picture surfaces only, but does so directly in a three-dimensional mode that avoids the tedious preparation of replicas of intact, fragmented, or frozen and cleaved bacteria used in transmission electron microscopy for similar purposes. In bacteriology (64–75), it is most useful for discerning surface appendages and other surface structure, and for defining shape and topographic relationships as in colonies or on surfaces of infected tissues.

This chapter describes a limited number of methods that, from our own experience, we consider useful to the worker who has need of basic anatomical information about bacteria. Emphasis is on the use of negatively stained preparations and thin-sectioned materials for examination by transmission electron microscopy. No attempt is made to describe the operations and maintenance of the EM (15–19) or of other instruments. Materials mentioned and their sources are given (4.6 and 4.7). It is assumed that functional instruments and experienced operators will be available for assistance and instruction, and the aim here is to supply the student with helpful information on preparative techniques and interpretation. Although methods and instrumentations other than those emphasized will be mentioned, details will be found only by reference to selected texts, other manuals, and articles in journals. References for general reading are supplied (1–14; more specialized information sources are listed subsequently).

4.1. TRANSMISSION ELECTRON MICROSCOPY

4.1.1. Negative Staining (20–33)

Grids (20–23).

Standard grids, of a diameter (2.3 or 3.0 mm) appropriate to the microscope employed, are commercially available. Although furnished in several metals and mesh sizes, the most commonly used are made of copper, and the size most generally useful is 200 mesh (mesh size indicates lines per inch). The grids have both a dull (matt) and a shiny side. Before applying support films, the grids must be clean and, therefore, may be treated with acetone and/or other solvents. However, supply houses can furnish grids that are already satisfactorily cleaned.

Support films (20–23).

Although naked grids may be used to support sections cut from epoxy resins (see below), liquid samples for negative staining must be supported on thin films covering the holes in the grid. Film materials commonly used are nitrocellulose or collodion (Parlodion), polyvinyl Formol (Formvar), or carbon. Films of the first two substances, which are plastics, are frequently stabilized by a thin additional layer of carbon. Formvar is the most satisfactory plastic-film material, and we employ it as follows:

1. Have available the materials listed in part 4.6.1.

2. Prepare a working solution of 0.20% Formvar in ethylene dichloride, using the 0.5% stock solution and solvent. Mix thoroughly while cleaning slides.

3. Wash the microscope slides with soap (or detergent) and water, and rinse thoroughly in distilled or deionized water. *Blot* dry with lint-free wipes; *never wipe* dry. (This minimizes static charge buildup.) Cleaned slides may be stored indefinitely by wrapping them in folds of lint-free wipes.

4. Fill one Coplin jar with the working solution of 0.2% Formvar. Add 4 or 5 drops of ethylene dichloride to the other jar, and cover it. The latter creates an atmosphere of the solvent to facilitate subsequent even drying of the Formvar-coated slides.

5. Place two or three slides in the Coplin jar containing the Formvar solution. Allow to stand for 3 to 5 min. Withdraw one slide at a time and immediately place in the "drying" jar. When dry, withdraw and place on a lint-free wipe. With an alcohol-cleaned razor edge, score the Formvar layer on the slide around the entire perimeter of both sides.

6. Fill the tape-covered glass dish with deionized water to form a convex meniscus at the top. Keep a clean glass rod in contact on all sides with the top edge of the dish, and slide it across the top from edge to edge to skim off the meniscus and remove all surface dust. (The bowl may be placed in a tray to catch the overflow of this operation.)

7. *Carefully*, while holding the coated slide vertically over the tray, drop 10% hydrofluoric acid from a Pasteur pipette on the score marks on the Formvar-coated slide. (Discard the pipette after each use.) *Immediately*, but *slowly*, immerse the slide vertically into the water. The Formvar films on each side will slowly release from the glass surface and float off onto the meniscus of the water surface. Inspect the films visually, preferably under a fluorescent light source, to judge uniformity and thickness. They should appear gray to silver-gray, representing thicknesses of ±60 nm according to interference color charts. Theoretically ideal thicknesses of 10 to 20 nm are probably

rarely achieved. If uniformity or thickness of films is unsatisfactory, repeat the above steps with appropriate modifications. If films are not uniform, the coated slides need to be dried longer and/or immersed in the water more slowly. If the film is too thick, dilute the Formvar solution with ethylene dichloride. If too thin, add a little of the 0.5% stock solution of Formvar and repeat the process. (The use of hydrofluoric acid etches the Formvar from the glass and replaces, in a more reproducible fashion, the usual procedure of breathing on Formvar coatings to free them from the glass.)

8. Once satisfactory films are obtained on the water surface, carefully place the grids (dull side down) onto the film. Avoid using film areas with obvious irregularities, in order to maintain a degree of consistency.

9. To retrieve the coated grids, cut a 9.0-cm-diameter circle of filter paper (Whatman no. 1) in half, hold it over the Formvar strip of grids, and lower it gently onto the surface until it has uniformly absorbed water. Push down under the meniscus while inverting the paper, and then come back up through the water surface. The grids will now lie between the filter paper and the Formvar film.

10. Place the semicircle of filter paper with grids in a petri dish on another piece of filter paper, and cover the dish partially with its lid while drying.

Note: Some filter papers are too "hairy," and the fibers damage the film; avoid these.

After drying, lift the grids by the edge with finely pointed grid forceps, and the plastic film, if suitable, will break cleanly around the perimeter of the grid while being retained over the grid bars and holes. Films that lift away from the filter paper with the grid, or do not break cleanly and evenly, are too thick; some areas of the film may show this property while others are satisfactory. Films that have large holes (which are usually avoided by use of hydrofluoric acid instead of the exhaling technique during the stripping procedure) may fail to hold liquid well and will be obvious on examination in the microscope. Holes also result from too high an ambient humidity during the drying process. Films that are too thin will tear readily in the electron beam.

Because of ionization and consequent changes in electron-static charge, plastic films expand and contract to cause specimen drift as the electron beam is intensified and decreased. Any resulting movement will degrade the photographic image in proportion to the magnification employed. Often, adjustments of beam intensity

and a brief wait will establish a level of stability. We find that Formvar films prepared as described are quite satisfactory for routine purposes with magnifications up to 50,000×—provided that there is careful operator attention to conditions of operation and to duration of photographic exposure. For fine detail and high magnifications, however, stability of plastic films is improved by deposition of a thin (2.0 to 3.0 nm) layer of carbon, and this procedure gives the best results. Thin films of pure carbon avoid drift and improve resolution because of a more even and finer substructure. They are prepared by dissolving away most of the plastic of a carbon-coated plastic film or by stripping carbon films from glass or freshly cleaved mica surfaces onto water, in a fashion similar to that used for plastic films. Grids with carbon films may also be purchased from commercial suppliers. Although of great advantage for fine work with subcellular particles and viruses, they are delicate and brittle, and require very careful handling. Their preparation involves the deposition of very fine particles of carbon by evaporation from carbon electrodes in a vacuum (10^{-5} to 10^{-6} mmHg [1.33 to 0.13 nPa]). The operation is carried out in a high-vacuum evaporator, a device of which several different commercial models are available. Modern devices have simple or automatic valving and operating systems, as well as nearly automatic grid-coating cycles, and are essential components of every EM laboratory. For descriptions of operation and uses (which include the shadowing of specimens or their replicas with heavy metals), the reader is referred to specialized sources of information (44, 45). It is assumed that an instrument and an operator/instructor will be available.

Staining.

More properly, negative "staining" is a process of embedding the bacterium or other particle in a thin film of a dilute solution of a heavy metal salt, which then is dried to a glass. There are many variations among the stains used and methods of applying them (see below). The standard procedure in our laboratory is a two-step method of first applying the specimen and then staining with 2% aqueous ammonium molybdate, as follows:

1. Place the coated grid, film side up, by one edge on the edge of a triply folded strip of sticky tape (sticky side out, throughout) pressed onto a glass microscope slide.

2. With a Pasteur pipette, place on the grid a drop of the bacterial or other suspension to be examined. The drop should be of a size to just round up on the entire grid surface, being held

by the meniscus. (Nonaqueous suspensions, or those containing highly surface-active substances such as detergents, will not be properly retained on the grid.)

3. Allow the drop to remain for 1 min or longer to permit adherence of the bacteria or other particles to the film surface. Trial and error will determine the optimum time, which is also affected by the nature and charge (hydrophilic properties) of the film and the size and nature of the particles. To improve both adherence and distribution of particles, the hydrophilicity of the support film can be improved by use of coated grids that have been stored in a refrigerator or recently exposed to ultraviolet irradiation, to "glow discharge" in a high-vacuum evaporator, or to discharge from a Tesla coil, in order to minimize electrostatic charges. We find that the use of a Zerostat gun, just prior to placing the specimen drop on the grid, is also effective.

4. With the point of a small triangle cut from no. 1 Whatman filter paper, draw up vertically most of the specimen drop. Then at once use the *torn* edge of the base of the paper triangle, applied to one edge of the grid, to rapidly withdraw the rest of the fluid.

5. Without permitting the grid surface to dry, immediately apply a drop of the 2% ammonium molybdate.

6. After 30 to 60 s, withdraw the drop of stain in the fashion described above. (Staining times may require some experimentation, especially with different sorts of specimens.)

7. Allow to dry and examine in the EM.

There are several alternative methods for negative staining of suspensions. One involves mixing equal volumes (or other ratios) of stain and suspension, placing a single drop on the grid, and then withdrawing excess fluid with filter paper, either immediately or after an appropriate interval of time (e.g., 30 s). It is usually best to do this immediately after mixing, since interactions of stain and particles can, with time, produce confusing results. Sometimes it is advantageous to pick up a thin film of the mixture in a platinum-iridium loop just bigger than the grid (e.g., 3.2 mm) and to break this flat film on a grid resting on blotting paper, to supply a more uniform particle distribution. The grid is allowed to dry and examined. If more than a thin film is applied, the excess must be withdrawn with filter paper.

Another negative staining method is to float the grid successively on a drop of the suspension, a drop of a fixative or other agent if desired, several drops of wash water (see below, Washing), and a drop of stain, followed as usual by filter paper withdrawal (step 6 above). The drops are conveniently handled on a sheet of dental wax that has been rinsed in distilled or deionized water to remove dust and soluble salts.

Occasionally, stain-suspension mixtures are sprayed onto grids, but this requires special apparatus and is most useful for quantitation of viruses or other small particles (26).

A particularly useful technique for bacteria is to place the coated grid surface in contact with a bacterial lawn or colony, followed by staining immediately on withdrawal. This permits examination of the natural state of the bacteria and their topographic relationships, undisturbed by pipetting or other shearing procedures that may remove flagella, pili, or other delicate structures. The procedure is also useful for a quick determination of the morphological type of phage producing a given plaque on a bacterial indicator lawn.

Washing.

A useful accessory procedure after application of a suspension, but before negative staining (in the two-step method), is to wash the grid to eliminate excessive debris, salt, and protein that is precipitable by some stains and obscures the detail obtainable by others. Washing can be done by touching the coated grid surface rapidly and repeatedly in succession to the surfaces of several fresh drops of distilled water or other fluid. Another method is to place the grid, film side down, on the wet surface of a thin bent strip of no. 1 Whatman filter paper as it withdraws the washing fluid by capillarity over the edge and down from a completely filled small beaker or petri dish into an empty surrounding dish. In either example, the grid is stained immediately after removing any excess fluid with the torn edge of filter paper.

In addition to water, various solutions may be tried as washes, but one must keep in mind the properties of the particle to be visualized as well as the necessity to avoid remaining salts or other substances that may interfere with staining. The fluid that is most generally useful, because it appears unreactive with bacteria, mycoplasmas, membranes, viruses, and most intracellular components, and because any excess is volatile in the vacuum of the electron microscopy, is 1% aqueous ammonium acetate. In fact, for initial examination of most microorganisms from liquid cultures, we centrifuge the culture and suspend the pellet in a small volume of the ammonium acetate for placement on the grid. (For centrifuging bacteria for any purpose, including steps of fixation and washing in preparation for embedment [see part 4.1.2], bench-top machines such as the Brinkmann [Eppendorf] microfuge are the most rapid and convenient. In the latter, volumes in the conical plastic tubes must be 1.5 ml or less.)

Stains.

Many different heavy metal salts have been used as negative stains. In our experience, ammonium molybdate is the most useful because it spreads well, does not react with proteins or most other components in a suspension, and can be readily adjusted in tonicity as required (32). Solutions of sodium or potassium phosphotungstate are perhaps the most widely used and are generally satisfactory at a neutral pH. Salts of uranium are widely used because their smaller molecules allow good penetration into surface irregularities and therefore improve the definition of fine detail. Uranyl acetate, which is generally available in the EM laboratory because it is also used for staining of sections, is useful in concentrations of 1% or less—but its pH is very low (about 2.5), and even if partially neutralized just before use, it begins to precipitate above pH 4.5. It may stain some structures positively, and both positive and negative staining may be seen on different areas of the same grid. In addition, it acts as a fixative for proteins and nucleic acids, contracts and rounds membrane-bounded cells (mycoplasmas, for example), disrupts some protein structures at pH 4.5, and should be used only with well-washed and clean preparations. Uranyl oxalate is particularly useful because it can be used at a neutral pH. Uranyl formate, sodium silicotungstate, and various other salts are also sometimes used. Salts of uranium are slightly radioactive and toxic, and care in both use and disposal is advised.

Some workers recommend the addition to some stains (i.e., phosphotungstate—not uranium salts) of very small amounts of wetting agents, such as serum albumin, starch, or sucrose, to improve spreading (27). Not all investigators find this useful or necessary, especially if high magnifications are to be used and uncluttered backgrounds are desired.

Fixation.

To avoid structural alterations and distortion due to exposure to solutions varying in pH and osmolality or to the stains themselves, or by drying in the stain, it is sometimes useful to chemically fix microorganisms before negative staining. This is usually done by use of glutaraldehyde in low concentrations, or sometimes by exposure of the suspension on the grid to glutaraldehyde vapors in a closed vessel. Fixed bacteria require washing before staining—either directly or by the grid washing procedures mentioned above. Even so, the result is often a degree of positive staining and retention of stain around the fixed microorganism to the extent that much less concentrated solutions of stain than usual may be required.

Pitfalls, problems, precautions, and hints.

Many suggestions have already been noted in the appropriate places in the foregoing sections. It is assumed that the availability of an EM implies the presence of one or more experienced operators familiar with operation and maintenance of vacuum evaporators, ultramicrotomes, and the microscope. For examination of negatively stained specimens, it is of particular importance to balance contrast and resolution by use of an appropriate accelerating voltage (usually 60 to 80 kV). Pointed filaments of the electron gun are useful, but not essential. Contamination of microscope column, apertures, and specimen by evaporated stain or other substances is minimized by attention to proper use of the liquid nitrogen-cooled anti-contamination device and employment of self-cleaning gold objective apertures. Specimen drift is avoided by patience, experimentation with beam intensity, and/or use of carbon-coated plastic or pure carbon support films. Beam damage to, or alteration of, fine structure—as well as "puddling" of stain about or on the specimen—can be minimized by learning to achieve focus as quickly as possible in the lowest possible beam intensity and making the shortest possible photographic exposure that is consistent with achievement of desired contrast according to the system of photographic film development in use. A short exposure at a relatively low magnification usually produces the best results: by trial and error, each microscopist will determine the magnifications most easily focused and producing the most consistent results with the microscope employed and the type of object under study.

Solutions of stains, washes (ammonium acetate), or even deionized or distilled water—depending on condition and time of storage or piping supply—may become contaminated with bacteria. Avoid the problem, at least when suspected, by EM examination of the fluids and by keeping working stains (we routinely maintain 2% solutions of ammonium molybdate, potassium phosphotungstate, and uranyl acetate) in a rack of small vials that are filled as needed, with sterile precautions, from stock bottles. Contaminated solutions may be filtered, but are best made up anew to avoid confusion by subparticles that may pass most filters. Solutions of uranyl acetate require protection from light by covering of the container with foil and should be filtered (0.2 μm) if a precipitate develops. A convenient method for storage and delivery is the use of a plastic syringe with attached Swinney filter.

The distribution of stain and particles on the support film, though minimized by the methods already mentioned, may be unsatisfactory; it is therefore useful and timesaving to prepare two or more grids of the same material. A grid should be searched thoroughly at low magnification. A satisfactory grid prepared by the drop withdrawal method will show a gradient of particle and stain distribution from one side to the other, within which a satisfactory balance of particle distribution and staining can be found. If the particles are in low concentration in the original suspension or if drop withdrawal techniques are unsatisfactory, particles and stain may be found only on one small area of the grid.

Ammonium molybdate produces less contrast to the eye viewing the focusing screen than does phosphotungstate or uranyl acetate, and its use therefore may require some practice. However, the photographic negatives (with appropriate development) will have more contrast than anticipated while presenting a greater range of printable tones. Even molybdate tends to heavily surround most bacteria with an almost impenetrable mass of stain, and short staining times and less concentrated solutions (0.1 to 0.5%) are often required for good results with these large objects.

Expectations.

What sort of information on the nature of structure of bacteria can be obtained by negative staining techniques? The method is the simplest and fastest that can be used (short of direct examination, which is almost useless because of lack of contrast), and it should, therefore, be tried first. The size and shape of the microorganism can be quickly ascertained, the presence of flagella (perhaps already suggested by motility studies) can be affirmed, and their distribution and fine structure can be determined. Other surface appendages such as pili and fimbriae, or the presence of a slime layer or capsular materials, can be defined. Even the differences between walled and wall-less membrane-bounded procaryotes (mycoplasmas or bacterial L-forms) become apparent.

Determination of the nature of the cell envelope is aided by adjunct procedures to rupture and fractionate the microorganisms so that periodic fine structure of wall layers can be clearly seen. Such procedures also permit separation and visualization of components such as membranes, mesosomes, other intracytoplasmic membrane systems, gas vacuoles, metabolic granules, ribosomes, flagella, pili, etc. (Shear forces engendered by syringing or excessive pipetting can result in loss of flagella and pili from the cell surface.) The stain will penetrate the periplasm between wall and membrane and, by thus extending into any membranous invaginations, delineate the mesosomes that permit a quick determination (usually accurate) of whether or not a bacterium is gram positive.

Negative staining also is a convenient means of following the course of cell disruption or of effects of treatments with agents such as enzymes, antibiotics, surface-active agents, etc.

Directly, or preferably after induction, negative staining can be used for rapid assessment of lysogeny by detection of bacteriophages or their unassembled components, and it is, of course, indispensable in studying virus ultrastructure, stages of development, and many virus-host interactions.

Examples of negative staining results, as well as of other examination by transmission electron microscopy, may be found in references 53–63.

4.1.2. Sectioning (34–43)

Procaryotic cells to be examined by section must be chemically fixed, washed, dehydrated for embedment in a material that can be firmly solidified for the cutting of very thin sections, and then stained with solutions of heavy metal salts for examination by electron microscopy. Although many different combinations can be used, we find it generally adequate to use sequential fixation in glutaraldehyde and osmium tetroxide, or in osmium tetroxide alone, followed by embedment in Spurr's medium (43). There are choices of embedding plastics, but Spurr's medium works reasonably well for routine purposes.

Fixation with glutaraldehyde-osmium.

1. To the cell suspension or liquid culture add an equal volume of 5% glutaraldehyde in 0.2 M sodium cacodylate buffer, pH 7.4 (final concentration, 2.5% glutaraldehyde). Alternatively, centrifuge the cells and suspend the pellet in 2.5% glutaraldehyde in cacodylate. (Cacodylate is toxic; see 4.6 and 4.7 for sources and formulations for these and other materials.) A wash in fresh medium or an appropriate buffer may precede this step. For some procaryotes, especially some mycoplasmas, it may be preferable to avoid cacodylate and make up the glutaraldehyde—perhaps in a different concentration—in fresh culture medium to balance osmolality and/ or pH.

2. Allow to fix for 2 to 4 h at room tempera-

ture. Actually, the time is flexible: any time beyond 20 min is usually adequate, and we have often successfully processed bacteria shipped in glutaraldehyde during several days across the country.

3. Wash the pellet twice in Kellenberger Veronal-acetate buffer, centrifuging between washes.

4. Suspend the washed pellet in 1% osmium tetroxide for 6 to 16 h at room temperature. (The glutaraldehyde-fixed pellet may be enrobed in agar prior to this step: see steps 6–8.) (*CAUTION*: Volatile and toxic fixatives and other substances should be used in a chemical hood or at least with the protection of safety goggles and respirator masks to avoid exposure of corneal and respiratory tract epithelia. If the odor of osmium tetroxide is detectable, the eyes may be subject to dangerous vapors. For insurance, a commercially available eyewash station [Nalge Co.] should be available, properly labeled, and strategically located.)

5. Centrifuge and wash the now black pellet once by suspension in Veronal-acetate buffer. Recentrifuge.

6. Make up 2% Noble agar in the Veronal-acetate buffer and cool to 45°C. Add 2 or 3 drops to the pellet and mix quickly with a warm Pasteur pipette.

7. With the pipette, quickly dispense the suspension onto an alcohol-cleaned glass slide and allow to solidify. The bacteria should be in good concentration and evenly distributed. With a clean razor blade, dice into 1-mm cubes.

8. Suspend the cubes in 0.5% aqueous uranyl acetate for 16 h at 4°C. (The uranyl acetate solution should be filtered [0.2 μm, Millipore Corp.] to prevent crystallization at this temperature.)

Fixation with osmium.

1. Centrifuge the cell suspension or culture.

2. Wash the pellet by suspension in fresh culture medium or Veronal-acetate buffer, centrifuge, and suspend in tryptone medium or tryptic soy broth. (The use of tryptone or tryptic soy broth is optional, but usually improves definition of ribosomes and nucleoid [58, 59, 62]).

3. Centrifuge and suspend the pellet in 0.1% osmium tetroxide in Veronal-acetate buffer. Fix for 2 to 4 h at room temperature.

4. Proceed as in steps 3 through 8 above.

Dehydration and embedment with Spurr's medium.

1. Place the uranyl acetate-treated agar cubes in two washes of 70% ethanol for 15 min each.

(Use 2- to 3-ml volumes in 4.5-ml [1-dram] screw-capped glass vials.)

2. Transfer to 80% and 95% ethanol, successively, for 15 min each.

3. Transfer into 100% ethanol for 30 min.

4. Replace with 1.0 ml of fresh 100% ethanol. Add 1.0 ml of complete Spurr's medium, and mix thoroughly but gently. Allow to stand, with occasional mixing, for 30 min or until cubes sink to the bottom of the vial. (The medium is most easily dispensed by use of disposable plastic syringes.)

5. Add 1.0 ml of Spurr's medium, and repeat the process.

6. Replace the mixture with 2.0 ml of Spurr's medium, and allow to stand for 1 to 2 h until cubes settle to bottom.

7. Insert typed labels into size 00 Beem capsules (in an appropriate holder), and add to each an equal volume of complete Spurr's medium. (We use 0.55 ml, so that the final block fits precisely into the LKB microtome chucks.)

8. With a sharpened wooden applicator stick, place one infiltrated cube (from step 6) in each filled Beem capsule. The cube should sink to the bottom of the capsule within 1 min.

9. With caps *off* (to allow volatilization of any residual solvent), place the capsules in a 70°C oven for polymerization for 16 h. Immediately on removal, the blocks in the capsules may feel slightly spongy, but they will harden to optimal cutting quality after 2 to 4 h at room temperature.

Other fixatives and embedding media.

Many different fixatives, or combinations and sequences thereof, have been devised for special purposes. These include permanganates and various aldehydes; principles of their use, including effects of pH, buffers, tonicity, concentration, temperature, and duration of fixation, are discussed in numerous books (see 34–39). Similarly, numerous embedding media are employed (34–39), alone or in mixtures, and the selection of that to be used will depend in part on which is available. The most common is probably Epon 812 (although there are recent hints that it may soon be unavailable from the original producer and be replaced by somewhat similar formulations from other manufacturers). Others in common use include Durcupan, Araldite, Maraglas, Vestopal, various methacrylates for special purposes, and numerous epoxy resins. The reader is referred to the abundant literature and to local experts for advice on the use and properties of these. We have chosen here to detail use of only one representative procedure that we find most satisfactory.

Block trimming.

Before sectioning, hardened blocks in Beem or other capsules must be removed and trimmed to proper size and shape. Embedments may be made in plain gelatin capsules or in flat molds, and various methods of trimming have been described (42). Newer models of ultramicrotomes often incorporate trimming devices, and separate specimen trimmers are also available. The preparatory procedure is as follows:

1. With pliers, press firmly over the entire capsule to loosen the enclosed block. Now press from the pyramidal end, and proceed upward to force out the cured block. (Alternatively, blocks can be released by cutting the capsule lengthwise on two sides with a razor blade.) Mount the block in a holder, pyramidal end up, and place the holder in a microtome chuck.

2. Under a binocular microscope, use an alcohol-cleaned razor blade to carefully trim the top of the pyramid (which will be the front or cutting face when mounted in the ultramicrotome) until the cell mass has been reached. Cut down and away from this surface on all four sides at an angle of about 30° (the angle of the pyramid formed by the block in the capsule is about 60°) to leave a small (face about 1.0 mm square) truncated pyramid that encompasses the specimen to be sectioned. With care and practice, this may be adequate trimming—especially if the face is cut in the shape of a trapezoid. However, grooves inevitably left by the razor blade are rather large and run at right angles to the face: as a result, water may run onto the face during sectioning and interfere with good results. We prefer to further fine trim the face and pyramidal sides in the microtome, so that any grooves on the sides run parallel to the edges of the block face. For fine trimming with microtome, etc., see reference 42.

Sectioning.

Because procedures differ in each laboratory, as well as with the type and mode of operation of the ultramicrotome used, we will not attempt to describe the sectioning process here. Descriptions of methods, operation of microtomes, and pitfalls are thoroughly described in many texts and manuals of electron microscopy as well as in pamphlets and operating instructions available from the instrument makers (see 39, 40, 42). In addition, some firms (e.g., LKB Instruments) present seminars or short courses on ultramicrotomy on request. We assume that the student of bacterial anatomy will have available the services of an ultramicrotomist who can instruct or, at the least, can furnish (from the student's block) sections already mounted on grids and ready for staining.

Knives.

Both glass and diamond knives are used for cutting ultrathin sections. Glass knives, which are obviously less expensive, are prepared by manual methods or (more precisely and in quantity) by use of knife-making apparatus such as the LKB 7800 Knife Maker. The making of glass knives is described in texts (34–40, 42) and in manuals accompanying the instruments. Such knives are the only ones used for trimming and block facing. For sectioning, the choice of knife depends on financial limitations, operator choice and experience, and the polymer used for embedment. In some instances, glass knives are preferable and produce superior results. A diamond knife produces more consistent results as the operator becomes familiar with its edge and other qualities and when a given standard embedment is consistently used. Edge angle requirements may differ with each type of embedment, and the appropriate included angle must be specified when purchasing a diamond knife. An angle of 42° to 45° is satisfactory for most conventional embedments of bacteria (Epon, Maraglas, Araldite), as well as the somewhat softer formulation of Spurr's medium that we use most often. With such embedment, careful handling, and caution in operation, we can use a single diamond knife of good quality for 5 years or more. Naturally, knife life will also depend on the volume of cutting done.

Staining with uranyl acetate.

Because of en bloc staining prior to embedment in the procedure described, uranyl acetate staining of sections is not usually needed. If desired for extra contrast, or if en bloc staining is not used, proceed as follows:

1. Fill an appropriate number of *caps* of size 00 Beem capsules to a positive meniscus with 1% uranyl acetate, made up in either water or 70% ethanol. (For aqueous solutions, drops may be placed on a sheet of clean dental wax.)

2. Float each grid-bearing section, section surface down, on the surface of the stain, and withdraw a little fluid with a Pasteur pipette to center the grid in the cap.

3. Cover with a petri dish lid, and allow to stain for 30 min.

4. Remove the grids with grid tweezers, immerse three times in distilled water to wash, and dry on filter paper.

Staining with lead citrate.

Some workers, for some purposes, use uranium staining only, but the degree of contrast may be relatively feeble, and overstaining can produce an objectionable granularity. Lead citrate (see 4.6.4) is commonly used to supply contrast and for the staining of some components that are not well demarcated by uranium alone. It is conveniently stored in and dispensed from a plastic syringe, thus excluding air. We use it as follows.

1. Prepare a staining chamber by coating the bottom of a glass petri dish with melted dental wax and allow to solidify. A plastic petri dish may be used without the need for wax coating, since the only purpose is to allow "beading" and separation of drops of stain.

2. Fill three small glass weighing bottles with twice-boiled distilled or deionized water, for use as washes.

3. In the staining chamber, place a 35 by 100 mm plastic tissue culture dish on a slightly larger disk of filter paper. Fill the dish with NaOH pellets, and saturate the paper with 0.02 N NaOH.

4. Dispense a number of drops of the lead citrate solution on the wax or plastic surface. Cover immediately with the lid to avoid formation of insoluble lead carbonate by reaction with CO_2 in the air. The NaOH atmosphere in the dish also prevents this by absorbing CO_2 to form sodium carbonate.

5. When ready, raise the lid and invert a section-bearing grid (section surface down) on each rounded drop. Quickly cover the dish and allow to stain for 3 to 5 min. (Experience will determine the optimum time.)

6. Remove each grid with tweezers and, while holding, immerse rapidly and repeatedly (five times each) in the water in each of the wash bottles. Touch the grid edge to filter paper to drain and place on filter paper to dry.

Note: For unknown reasons, brief uranyl acetate treatment followed by lead citrate leads to much more intense lead uptake.

4.1.3. Direct Microscopy

Bacterial preparations may be examined directly in the TEM without the benefits of contrast enhancement by negative stains. The microorganisms may be grown on, or can be placed on, Formvar or collodion support films and examined at any stage of growth. However, information obtained by such means is limited and of no advantage unless one is interested in the inherent electron opacity of the specimen. The minimal increase in time involved in negative staining will yield much more structural information.

Other methods of direct examination include high-resolution dark-field electron microscopy and electron-optical phase microscopy. Because of special instrument requirements and limitation of their applications to particular types of specimens, these techniques are of little impor-

tance to investigators beginning the study of bacterial structure. Similarly, high-voltage electron microscopy, which with anticipated refinements may permit examination of thicker specimens without excessive beam damage, is a specialized procedure. (The interested reader is referred to Literature Cited, especially Hayat [7], vol. 3, chapters 3 and 4.)

4.1.4. Metal Shadowing (44, 45)

Because of the growing popularity and availability of SEMs, the use of direct metal shadowing has declined. However, like scanning electron microscopy, it is a method of eliciting the surface structure and dimensions of bacteria that can be useful because of the greater resolving power of the TEM, and it can be readily employed if an SEM is not available.

Shadow-casting involves the deposition of electron-dense materials at specific angles onto the specimen while in a vacuum evaporator. Those areas shielded from deposition of the metal by the angle are left uncoated and, being less electron dense, permit passage of electrons that thus delineate such areas as black shadows on the photographic negative after it is processed. Since this appears as a positive image, it is customary (though not always done) to make contact reversal negatives and to make the final prints from these to reproduce the positive image. (The techniques for shadowing are well described in chapter 5 of reference 11 and chapters 4 and 5, vol. 2, of reference 7.) Although direct shadowing with evaporated metals may be applied to whole cells or their components, bacteriophages, or other suspensions, the technique is also required for examination of replicas made of intact preparations or of freeze-fractured and freeze-etched materials (see below). In addition to delineating surface irregularities or periodic structures, the technique can be used to measure the vertical height of the specimen by measuring shadow length and calculating geometrically from the known shadow angle. This process is aided by inclusion on the grid of a known standard such as latex spheres of a given size, for purposes of calibration.

Shadowing is conducted as follows.

1. The specimen to be examined should be free from extraneous debris. It may or may not be fixed first.

2. Place the material on a Formvar-coated grid and allow to dry. Attach the grid to double-sticky tape on a glass microscope slide, together with a small piece of fine filter paper to serve as a visual indicator of the amount of metal deposited during the evaporation. (Fold the paper partially, to provide coated and noncoated sides for comparison.)

3. Place the slide on the specimen support head of the vacuum evaporator. (We use a rotating support head, Fullam no. 1205, in a Varion VE 10 evaporator containing a liquid nitrogen cooling trap to improve vacuum efficiency. Instruments, and methods of varying the shadow angle between specimen grid and the filament, will be different in each laboratory.)

4. Wrap platinum-palladium wire around the tungsten V-filament held between the electrodes. The V is formed from standard 30-mil tungsten wire (Fullam no. 1623), and we wrap on it 1 mm of 80% Pt–20% Pd wire (Fullam no. 1221). Before doing this, it is essential to first heat the tungsten filament to white heat in vacuo; then reopen the evaporator and affix the Pt-Pd wire. (This procedure degasses the filament, to improve vacuum during subsequent metal evaporation and to relieve strains produced in the wire during its bending, thus prolonging filament life.)

5. With the Pt-Pd wire wound around the apex of the tungsten V-filament, place this metal source approximately 12 cm above the grids at the desired angle. Establish a good vacuum (10^{-5} mmHg) in the evaporator, turn up the controls to heat the metal quickly to the melting point, and hold it there until all is evaporated or until the shadow on the visual filter paper indicator appears correct. This judgment requires some practice and experience.

6. Sometimes, it is useful to use shadowing with tungsten oxide to produce a minimal background and finer granularity than that obtained with Pt-Pd. To do this, heat the tungsten V-filament *in air* until it is coated with yellow tungsten oxide (WO_2). Scrape off these crystals with a scalpel, except for those at the apex, place the filament between the electrodes in the evaporator, and proceed with shadowing as described.

4.1.5. Replicating (44, 45)

Replicas are casts which, because of their fragile nature and lack of electron opacity, are unsuitable for direct examination and must be shadowed with metal. They are thin films of electron-transparent materials (usually collodion, Formvar, or polyvinyl alcohol) that are cast in contact with, and thus delineate directly, the surfaces of the structures being examined. Pseudoreplicas are those that, in the process of stripping by floating the film off the casting surface, may pick up and include some of the particles under examination.

Replicas are essential to examination of freeze-fractured and freeze-etched surfaces and are prepared (usually with platinum-carbon shadowing) while the fractured specimens are maintained at −100°C. Although useful in examining cell membrane interiors and for viewing specimens that need not be fixed or otherwise chemically altered, freeze-fracture techniques require specialized equipment and experience, and a description of methods is beyond the scope of this manual (46–48).

4.1.6. Nucleic Acids

Since the description by Kleinschmidt and Zahn (74), the spreading of nucleic acid molecules on protein or other monolayers for examination after uranium staining and rotary metal shadowing has come into widespread usage and modification. It is an essential method in many aspects of molecular and genetic microbiology and is well described in appropriate texts (71–73, 75) and in many specialized publications. For determination of plasmid DNA homology, see 15.3.5.

4.2. SCANNING ELECTRON MICROSCOPY

The features and advantages of the SEM have been noted above in the introductory section. Its value to the bacteriologist lies in its ability to define three-dimensional forms of microorganisms, to discriminate surface features (within its limits of resolution), and to demonstrate topographic relationships of bacteria growing in various associations. Preparation of specimens involves fixation on a surface (e.g., a cover slip) that can be affixed to the specimen holder, sometimes washing or other treatment, drying by the critical point or other methods, and usually in vacuo coating with gold or another heavy metal prior to examination. Because of the time and expertise needed for these procedures and the unavailability of a SEM in many laboratories, scanning is not likely to be used for a "first look" by the student of bacterial structure when negative staining can more easily and rapidly supply much of the same information. For those with appropriate facilities and interest in applying scanning electron microscopy to particular problems, a number of references are supplied (49–52, 64–70).

4.3. PHOTOGRAPHY

The working and reference material of the electron microscopist consists of electron micrographs, and the proper taking and processing of photographs is therefore essential. Most microscopes are designed for "plates" or roll film, or both. Roll film (35 mm) is cheap, effective, and easily (inexpensively) stored. Glass plates are easily broken, are heavy, and need a lot of expensive storage facilities. Glass plates are now largely replaced by emulsions on cut films of Estar base, which provide dimensionally stable and effective substitutes that are readily filed.

The major problem with all of these films is that the emulsion and the backing absorb and retain water vapor. Therefore, the prudent operator keeps a series of aluminum desiccators with solid lightproof lids. These latter are made of sheet aluminum tapped to take a vacuum needle valve; film cut to appropriate lengths, or separated stacks of cut film, are placed on a support grid for drying. A tray in the bottom contains phosphorus pentoxide, which has to be stirred at each opening in the darkroom and replaced as needed. A modest vacuum and the chemical drier are effective in drying the film during storage and thus prevent long pump-down times in the EM due to outgassing. Some microscopes have a built-in desiccator, but this is seldom sufficient for the storage advised above.

In our experience, 35-mm Fine Grain Positive film (Kodak) is most satisfactory for routine purposes when the microscope is equipped to handle this type of roll film. For plate cameras, we use Kodak Electron Image film no. 4489—an Estar-based sheet film of excellent dimensional stability and tear resistance that replaces glass plates and a somewhat similar film (no. 4463) that is no longer available. The beginning worker should refer to Kodak Data Release p-198, revised February 1980, for characteristics of this film and its development; indeed, the basic message is that it is important to first review the recommendations and data supplied by the manufacturer of any film (or paper) considered for use. Experience will then indicate an appropriate routine exposure, development, and the degree of contrast desired in the negative.

Skill in focusing to achieve sharp negatives is obtained by experience, but it is helpful to recall that, at high magnifications, the exact focus is not necessarily the best for printing the negative. It is therefore a good habit to take three to five "through-focus" pictures of each field desired, trying to have focus near the middle step. Each step ranges from 0.05 μm (at 30,000× on the film) to 0.4 μm (at 4,000× on the film).

Printing of electron micrographs requires more than ordinary attention to detail. A high-quality enlarger is essential, and many choices are available; remember that point source condenser enlargers, while most critical, will also show every scratch in the nonemulsion side of the film! Automated photosensor exposure devices (e.g., the Lektra PTM-4A Densitimer) are helpful for reproducibility, once calibrated for each paper used. The choice of papers should be reasonably wide because of some unavoidable variation among negatives and emphasis desired. This is met by use of a multigrade paper system with appropriate filters or a collection of bromide papers of grades of "hardness" 2 to 6. There is a trend toward supplying many emulsions as resin-coated (RC) papers which speed processing and drying and are of particular advantage in avoiding the need for heated drum dryers. Large units may employ automated processors, stabilization papers, contrast-controlled enlargers, and the like.

Because microscopes, exposure arrangements, films, and processing differ among EM laboratories, we make no attempt here to present detailed procedures. The student should consult with experienced investigators within his/her own unit and may wish also to peruse the articles on this subject that are furnished in the references (76–82).

4.4. INTERPRETATION

The viewer must be able to relate what is seen in the EM or EM photos to what is already known (or not known) about bacterial anatomy. In this broad sense of "interpretation," then, it is useful to have some idea of the field, and it is assumed that the student will have been exposed, through courses or some reading, to the general features of microbial ultrastructure. Unfortunately, pictorial atlases, in which one can rapidly and readily compare different genera or general types of microorganisms, are rare or hard to find. Useful references, if available, include those by Avakyan, Katz, and Pavlova (53) (text in Russian, but bacterial names in English), articles in Fuller and Lovelock (54, 56, 57, 60, 63), Greenhalgh and Evans on fungi (55), Lickfeld on transmission electron microscopy of bacteria (61), and Kormendy (66) and Yoshii et al. (70) on microbiological scanning electron microscopy. In addition, or otherwise, the student may rely on the advice of a preceptor in searching the now abundant literature.

In a more precise sense, "interpretation" may be construed as getting the most information out of the electron micrograph. Usually, this refers to details of fine periodic structure such as in flagella or accessory protein layers of the cell envelope. For such advanced work, acquisition of a satisfactory micrograph is only the first step, and methods of image enhancement and reconstruction—including computer processing—can be utilized. Such techniques require sophisticated optical and other equipment not available in every EM unit, and descriptions of procedures are beyond the limited aims of this manual. The interested investigator will seek the advice of local experts and read the pertinent literature (76–82).

4.5. SPECIFIC IDENTIFICATION

The identities and positions of bacterial antigens, either within or on the surfaces of cells, are often matters of research importance. Sometimes, it is also essential to know the cellular site of enzymes and their activities or to follow the synthesis or incorporation of cellular components within the cell. Realization of these aims requires methods of labeling that enable microscopic visualization of added substances, of their specific binding to other components, or of their chemical interactions. Several different techniques are employed: most are lengthy, some are technically complicated, and a few are difficult of satisfactory achievement. Because of these problems, we make no attempt to present details of these advanced techniques and limit our remarks to principles and references.

4.5.1. Antigens

The most widely used methods for detection of bacterial antigens in situ employ antiserum or antibody fractions labeled with fluorescent dyes such as fluorescein or rhodamine (see 1.4). Binding which is exclusively to surface antigens of intact bacteria is recognized by optical dark-field microscopy with appropriately filtered ultraviolet illumination, to give visible and characteristic emission of the fluorochrome. The method has diagnostic and taxonomic value, and many labeled antisera are commercially available. A number of treatises on theory and practice have been published (83–85).

The principle of the fluorescent-antibody method has been adapted to electron microscopy by coupling antibodies or haptens to substances that are electron dense (e.g., ferritin, or iron-containing protein) or that can be made electron dense by a reaction subsequent to the antigen-antibody binding (e.g., horseradish peroxidase coupled to antibody and then allowed to react after binding with diaminobenzidine). Many combinations using antisera with different specificities, antibody fragments, haptens, visual markers such as small viruses, and the like are possible. These methods, which are applicable to whole bacteria and may be viewed directly in the EM, permit a more precise discrimination of the distribution of *surface* antigens than is possible by optical microscopy. The advantage is increased by applying the labeled antibody prior to fixation, embedment, and sectioning, so that the label is viewed in thin sections. In some instances, substances with known specificities, other than antibodies, can be similarly labeled and employed: plant lectins such as concanavalin A, which bind to specific sugar residues, are most commonly used (88).

However, the detection of *intracellular* antigens (89) within intact bacteria is a more difficult problem than locating surface antigens because antibodies fail to pass the intact cell envelope. Consequently, methods for solving this sort of problem rely on first cutting thin sections of bacteria and then floating the sections on solutions of labeled antibody before washing, sometimes staining, and viewing in the EM. A major difficulty is the embedding medium to be used: it must permit access and binding of *aqueous* solutions of antibody, and a number of substances (methacrylates, cross-linked albumin, variously etched plastics or epoxies) have been tried. None is satisfactory in all respects, and there is a need for continued investigation. For a basic view of techniques and problems in immunocytochemistry or immuno-electron microscopy, the interested student may wish to begin with the references supplied (87–91).

4.5.2. Enzymes

The precise localization of enzymatic activities as they occur within a cell is the longstanding subject of histochemistry and now of cytochemistry at the level of electron microscopy. Methods differ with the type of enzyme and, therefore, with the possibilities for devising an artificial or labeled substrate that is normally transported and allowed to react to give a product visible in either the optical microscope or the EM. The subject area is a large one which is of itself productive of a good deal of research that cannot even be summarized in this limited volume: the reader is referred to available texts (86) and to the several extant journals dealing exclusively with histochemistry and cytochemistry.

4.5.3. Autoradiography

Another large subject which, like antibody labeling and histochemistry, has progressed from the optical to the electron microscopic level is that of following the intracellular fate, distribution, or partition of substances labeled with radioactive atoms. The technique is applicable to a wide variety of problems but, especially in bacteria (because of their small size), requires careful attention to conditions and improved emulsions that will minimize grain size. Principles and methods are fully described in several books (92–95).

4.6. MATERIALS

Commercial sources of the materials listed below are indicated by the letters in parentheses, which refer to the letters in the next part (4.7).

4.6.1. For Preparation of Plastic Support Films

Formvar, powder (i, j, n, r)
Formvar solutions (i, j, r)
Ethylene dichloride (h, i, j, n, o, r)
Clean glass microscope slides
Lint-free photowipes (i, j, n, p)
Large metal (or plastic) pan or tray
Large glass dish with straight sides. The best dish is one measuring 200 by 80 mm (culture dish, Vitro) (n, s). The top edge must be even and smooth. The bottom and sides must be permanently covered with black electrical tape (to permit seeing the floating films against a black background).
Black electrical tape
Glass rod longer than the diameter of the dish
Coplin glass staining jars (a, h, n, s)
Pasteur pipettes and rubber bulbs
Hydrofluoric acid, 10% (in Teflon container) (h)
Razor blades (alcohol washed)
Filter paper, Whatman no. 1, 9.0-cm diameter (a, h)
Petri dishes, standard, 9.0-cm diameter
Scissors
Grid tweezers, no. 5 Dumont or equivalent (i, j, n, o, q, r). (A loop or two of Nichrome wire placed around the base can be pushed forward to clamp the tips on the grid, thus preventing accidental dropping as well as the cost of expensive tweezers designed for this purpose.)
Copper grids, precleaned (f, i, j, k, n, o, q, r)
Light source, fluorescent

4.6.2. For Preparing Carbon Films

High-vacuum evaporators and accessories (b, i, j)

4.6.3. For Negative Staining

Coated grids, prepared as described or from commercial sources (i, j, r)
Sticky tape (Time Tape) (a, h, n)
Glass microscope slides
Zerostat gun (d, n)
Pasteur pipettes
Ammonium acetate (h, o)
Ammonium molybdate (h, o, r)
Potassium phosphotungstate (f, i, j, n, o, r)
Uranyl acetate (f, h, i, j, o, q, r)

4.6.4. For Fixing, Embedding, Sectioning, and Staining of Sections

Glutaraldehyde (f, h, i, j, n, o, r)
Sodium cacodylate buffer:
  May be purchased as buffer or as a kit with premeasured ingredients (r). Otherwise, make up sodium cacodylate (f, h, i, j, n, o, r) to 0.2 M in water and adjust the pH to 7.2 with HCl. (*Caution:* Cacodylate

is toxic; avoid contact and inhalation of arsenical vapor.)

Veronal-acetate buffer:
  May be purchased as the buffer (j, o) or made up from sodium Veronal (o) according to the following formula for *stock solution*:
    Sodium Veronal (barbitone sodium), 2.94 g
    Sodium acetate, hydrated, 1.94 g
    Sodium chloride, 3.40 g
    Distilled water, 100.00 ml
  From this, prepare the *working buffer* daily:
    Stock solution, 25.0 ml
    Distilled water, 96.5 ml
    HCl, 1 N, 3.5 ml
    $CaCl_2$, 1 N (110.98 g/liter), 1.25 ml
  Dispense the 25 ml of stock solution into a 125-ml bottle. In a 100-ml graduated cylinder, add 90 ml of water and 3.5 ml of 1 N HCl, and bring to 100 ml with water. Add to the stock solution in the bottle, add 1.25 ml of 1 N $CaCl_2$, and stir. (*Note:* If cells to be washed in this buffer were previously exposed to phosphates as in culture media or a phosphate buffer, it is necessary to have available working buffer made *without* $CaCl_2$ for use as initial washes to avoid precipitates of calcium phosphate. Complete buffer is then used as a final wash before proceeding.)

Safety goggles
Osmium tetroxide (f, h, i, j, n, o, r)
Noble agar (c)
Uranyl acetate (f, h, i, j, n, o, r)
Tryptone (c)
Tryptic soy broth (c)
Glass vials, screw capped, 1 dram (n, s)
Respirator mask, Gasfoe (l)
Disposable plastic containers, 4.5 oz (ca. 130 ml) (a, g, h)
Spurr's medium:
  Supplied as a kit (i, j, o) or as ingredients (r) to be used as follows to make the stock mixture.
  Stock mixture:
    ERL-4206 (vinylcyclohexene dioxide), 30 g or 27.0 ml
    DER-736 (diglycidyl ether of propylene glycol epoxy resin), 24 g or 22.8 ml
    NSA (nonenylsuccinic anhydride), 78 g or 76.5 ml
  In a chemical hood or wearing an appropriate mask, add the three components in the above order to a disposable container (Falcon no. 4013), mixing as added on a magnetic stirrer. Mix for 5 min at ambient temperature. This stock mixture, which lacks accelerator, can be stored tightly covered at $-4°C$ for 3 to 5 months. If so stored, allow ample time to equilibrate to room temperature on removal and before opening, to avoid water of condensation that may destroy the mixture.
  Complete mixture:
    To make the complete mixture, put 22 ml of stock mixture in a disposable container, add 0.24 ml of the accelerator S-1 (DMAE, or dimethylaminoethanol), and stir on a magnetic stirrer for several

minutes at room temperature. (For each bacterial pellet or other sample to be embedded—*not* each agar cube resulting from the sample—allow at least 6 ml of complete mixture. This volume will take care of mixtures in steps 4 to 6 of the dehydration and embedment procedure as well as provide for filling of three Beem capsules. Obviously, if more than three blocks from each specimen are desired, more mixture will be needed. Also, if more than three pellets or specimens are to be handled at a time, the initial volume of the complete mixture will need to be greater than 22 ml; make up in the same proportions.) The useful pot life of complete mixture is only 2 days.

Beem capsules (catalog no. 23200, size 00) (i, j, n, o, r)
Drying oven
Centrifuge, Brinkmann 3200 (Eppendorf)
Pliers
Razor blades, single edge
Ultramicrotome
Plate glass
Glass knife maker
Diamond knife (e)
Dental wax
Glass weighing bottles
Plastic tissue culture dish, 35 by 100 mm
NaOH pellets
Petri dishes, glass or plastic
Lead citrate:
  Available ready-made from commercial sources (i, j, n, o, r). It can also be prepared by the method of Reynolds (J. Cell Biol. **17**:208–212, 1963) as follows. To a 50-ml volumetric flask, add:
    Lead nitrate, 1.33 g
    Sodium citrate, 1.76 g
    Distilled water, 30.00 g
  Shake for 1 min, and allow to stand at ambient temperature for 30 min. Add 8 ml of freshly prepared 1 N NaOH, and dilute to 50 ml with water.

4.7. COMMERCIAL SOURCES

a. Curtin-Matheson Scientific, Inc:, 10727 Tucker St., Beltsville, MD 20705
b. Denton Vacuum Inc., Cherry Hill Industrial Center, Cherry Hill, NJ 08003
c. Difco Laboratories, P.O. Box 1958A, Detroit, MI 48232
d. Discwasher, Inc., 1407 N. Providence Rd. Columbia, MO 65201 (or local photo stores)
e. Dupont Instruments, Biomedical Division, Newtown, CT 06470
f. Electron Microscopy Sciences, P.O. Box 251, Fort Washington, PA 19034
g. Falcon Labware Division, 1950 Williams Dr., Oxnard, CA 93030
h. Fisher Scientific Co., 1 Reagent Lane, P.O. Box 375, Fairlawn, NJ 07410
i. Ernest F. Fullam, Inc., P.O. Box 444, Schenectady, NY 12301

j. Ladd Research Industries, Inc., P.O. Box 901, Burlington, VT 05401
k. LKB Instruments, Inc., 12221 Parklawn Dr., Rockville, MD 20852
l. Mine Safety Appliances Co., Pittsburgh, PA 15208
m. Nalge Co., Division of Sybion Corp., Rochester, NY 14602
n. Ted Pella Co., P.O. Box 510, Tustin, CA 92680
o. Polysciences, Inc., Paul Valley Industrial Park, Warrington, PA 18976
p. Sorg Paper Co., Middletown, Ohio 45042
q. SPI Supplies, P.O. Box 342, West Chester, PA 19380
r. Tousimis Research Corp., P.O. Box 2189, Rockville, MD 20852
s. Wheaton Scientific (Vitro), 1000 North 10th St., Millville, NJ 08332

4.8. LITERATURE CITED

4.8.1. General References

1. **Benedetti, E. L., and P. Favard (ed.).** 1973. Freeze-etching: techniques and applications. Société Française de Microscopie Electronique, Paris.
  A good single source of information on this technique.
2. **Dawes, C. J.** 1971. Biological techniques in electron microscopy. Barnes and Noble, Inc., New York.
  A very useful small paperback with methods, formulas, appendix, and general references.
3. **Fuller, R., and D. W. Lovelock (ed.).** 1976. Microbial ultrastructure. The use of the electron microscope. Academic Press, Inc., New York.
  Good examples of results of different techniques applied to microorganisms are included.
4. **Glauert, A. M. (ed.).** 1972. Practical methods in electron microscopy, vol. 1–6. American Elsevier Publishing Co., Inc., New York.
  Under this title, six volumes have been published through 1977. Each is divided into parts (e.g., Ultramicrotomy, Staining Methods for Sectioned Materials, Autoradiography and Immunocytochemistry, etc.), and some of these are available separately as paperbacks that are handy in the laboratory.
5. **Goldstein, J. I., and H. Yakowitz (ed.).** 1975. Practical scanning electron microscopy. Plenum Press, New York.
  Despite its title, this volume is subtitled Electron and Ion Microprobe Analysis and is largely devoted to physics and theory and application to non-biological specimens.
6. **Harris, R. J. C. (ed.).** 1962. The interpretation of ultrastructure. Academic Press, Inc., New York.
  Contains some useful early methods.
7. **Hayat, M. A.** 1970. Principles and techniques of electron microscopy. Biological applications, vol. 1–9. Van Nostrand Reinhold Co., New York.
  Through 1978, nine volumes of this same title, dealing principally with transmission electron microscopy, have been published. In the latter eight, edited by Hayat, the contents are contributed chapters by other authors. Some are useful to microbiologists and are cited separately.

8. **Hayat, M. A.** 1972. Basic electron microscopy techniques. Van Nostrand Reinhold Co., New York.
   A good presentation of beginning techniques.
9. **Hayat, M. A.** 1975. Positive staining for electron microscopy. Van Nostrand Reinhold Co., New York.
10. **Hayat, M. A. (ed.).** 1974. Principles and techniques of scanning electron microscopy. Biological applications, vol. 1–6. Van Nostrand Reinhold Co., New York.
    Six volumes of this title have been published through 1978. As in the previous series on transmission electron microscopy, the contributed chapters often deal with specific and restricted areas, few of which are of immediate applicability by the microbiologist—but they are of value for those interested in extending their electron microscope horizons. Sections pertinent to microbiology are individually cited in other sections of these references.
11. **Kay, D. (ed.).** 1965. Techniques for electron microscopy, 2nd ed. F. A. Davis Co., Philadelphia.
    A thoroughly useful book in which methods are carefully described.
12. **Koehler, J. K. (ed.).** 1973. Advanced techniques in biological electron microscopy. Springer-Verlag, New York.
    Discusses a few techniques in detail: of value to the beginner in acquiring perspective.
13. **Meek, G. A.** 1976. Practical electron microscopy for biologist, 2nd ed. John Wiley & Sons, Inc., New York.
    Covers most methods quite completely.
14. **Wells, O. C.** 1974. Scanning electron microscopy. McGraw-Hill Book Co., New York.
    Mostly theory. Only Chapter 12 ("Histological and Cytological Methods for the SEM in Biology and Medicine," by A. Boyde) touches on pertinent techniques, but does not specifically deal with microorganisms.

4.8.2. Specific References by Subject

Electron microscope theory and practice.

15. **Agar, A. W.** 1965. *In* D. Kay (ed.), Techniques for electron microscopy, 2nd ed., p. 1–42. F. A. Davis Co., Philadelphia.
16. **Agar, A. W., R. H. Alderson, and D. Chescoe.** 1974. *In* A. M. Glauert (ed.), Practical methods in electron microscopy, vol. 2, p. 1–190. American Elsevier Publishing Co., New York.
17. **Dawes, C. J.** 1971. Biological techniques in electron microscopy, p. 1–15. Barnes and Noble, Inc., New York.
18. **Siegel, B. M.** 1964. *In* B. M. Siegel (ed.), Modern developments in electron microscopy, p. 1–79. Academic Press, Inc., New York.
19. **Wischnitzer, S.** 1973. *In* M. A. Hayat (ed.), Principles and techniques of electron microscopy. Biological applications, vol. 3, p. 3–50. Van Nostrand Reinhold Co., New York.

Grids and support films.

20. **Baumeister, W., and M. Hahn.** 1978. *In* M. A. Hayat (ed.), Principles and techniques of electron microscopy. Biological applications, vol. 8, p. 1–112. Van Nostrand Reinhold Co., New York.
21. **Bradley, D. E.** 1965. *In* D. Kay (ed.), Techniques for electron microscopy, 2nd ed., p. 58–74. F. A. Davis Co., Philadelphia.
22. **Dawes, C. J.** 1971. Biological techniques in electron microscopy, p. 107–114. Barnes and Noble, Inc., New York.
23. **Hayat, M. A.** 1970. Principles and techniques of electron microscopy. Biological applications, vol. 1, p. 323–333. Van Nostrand Reinhold Co., New York.

Negative staining.

24. **Anderson, T. F.** 1962. *In* R. J. C. Harris (ed.), The interpretation of ultrastructure, p. 251–262. Academic Press, Inc., New York.
25. **Bradley, D. E.** 1965. *In* D. Kay (ed.), Techniques for electron microscopy, 2nd ed., p. 75–83, 87–89. F. A. Davis Co., Philadelphia.
26. **Brenner, S., and R. W. Horne.** 1959. Biochim. Biophys. Acta **34**:103–110.
27. **Dawes, C. J.** 1971. Biological techniques in electron microscopy, p. 146–148. Barnes and Noble, Inc., New York.
28. **Haschemeyer, R. H., and R. J. Myers.** 1972. *In* M. A. Hayat (ed.), Principles and techniques of electron microscopy. Biological applications, vol. 2, p. 101–147. Van Nostrand Reinhold Co., New York.
29. **Horne, R. W.** 1965. *In* D. Kay (ed.), Techniques for electron microscopy, 2nd ed., p. 328–355. F. A. Davis Co., Philadelphia.
30. **Huxley, H. E., and G. Zubay.** 1960. J. Mol. Biol. **2**:10–18.
31. **Kay, D.** 1976. *In* J. R. Norris and D. W. Ribbons (ed.), Methods in microbiology, vol. 9, p. 177–214. Academic Press, Inc., New York.
32. **Muscatello, U., and R. W. Horne.** 1968. J. Ultrastruct. Res. **25**:73–83.
33. **Valentine, R. C., and R. W. Horne.** 1962. *In* R. J. C. Harris (ed.), The interpretation of ultrastructure, p. 263–277. Academic Press, Inc., New York.

Sectioned preparations (fixation, dehydration, embedment, trimming, sectioning, staining).

34. **Dawes, C. J.** 1971. Biological techniques in electron microscopy, p. 17–105. Barnes and Noble, Inc., New York.
35. **Glauert, A. M.** 1965. *In* D. Kay (ed.), Techniques for electron microscopy, 2nd ed., p. 166–212. F. A. Davis Co., Philadelphia.
36. **Glauert, A. M.** 1965. *In* D. Kay (ed.), Techniques for electron microscopy, 2nd ed., p. 254–282. F. A. Davis Co., Philadelphia.
37. **Glauert, A. M.** 1975. *In* A. M. Glauert (ed.), Practical methods in electron microscopy, vol. 1, p. 1–216. American Elsevier Publishing Co., New York.
38. **Glauert, A. M., and R. Phillips.** 1965. *In* D. Kay (ed.), Techniques for electron microscopy, 2nd ed., p. 213–253. F. A. Davis Co., Philadelphia.

39. **Hayat, M. A.** 1970. Principles and techniques of electron microscopy. Biological applications, vol. 1, p. 3-51. Van Nostrand Reinhold Co., New York.

40. **Knobler, R. L., J. G. Stempak, and M. Laurencin.** 1978. *In* M. A. Hayat (ed.), Principles and techniques of electron microscopy. Biological applications, vol. 8, p. 113-155. Van Nostrand Reinhold Co., New York.

41. **Luft, J. H.** 1973. *In* J. K. Koehler (ed.), Advanced techniques in biological electron microscopy, p. 1-34. Springer-Verlag, New York.

42. **Reid, N.** 1975. *In* A. M. Glauert (ed.), Practical methods in electron microscopy, vol. 3, p. 217-353. American Elsevier Publishing Co., New York.

43. **Spurr, A. R.** 1969. J. Ultrastruct. Res. **26:**31-43.

Replicas, metal shadowing, etc.

44. **Bradley, D. E.** 1965. *In* D. Kay (ed.), Techniques for electron microscopy, 2nd ed., p. 96-152. F. A. Davis Co., Philadelphia.

45. **Henderson, W. J., and K. Griffiths.** 1972. *In* M. A. Hayat (ed.), Principles and techniques of electron microscopy. Biological applications, vol. 2, p. 151-217. Van Nostrand Reinhold Co., New York.

Freeze-etch, freeze-fracture.

46. **Benedetti, E. L., and P. Favard (ed.).** 1973. Freeze-etching: techniques and applications. Société Française de Microscopie Electronique, Paris.

47. **Bullivant, S.** 1973. *In* J. K. Koehler (ed.), Advanced techniques in biological electron microscopy, p. 67-112. Springer-Verlag, New York.

48. **Koehler, J. K.** 1972. *In* M. A. Hayat (ed.), Principles and techniques of electron microscopy. Biological applications, vol. 2, p. 53-98. Van Nostrand Reinhold Co., New York.

Scanning electron microscopy.

49. **Black, J. T.** 1974. *In* M. A. Hayat (ed.), Principles and techniques of scanning electron microscopy. Biological applications, vol. 1, p. 1-43. Van Nostrand Reinhold Co., New York.

50. **Cohen, A. L.** 1974. *In* M. A. Hayat (ed.), Principles and techniques of scanning electron microscopy. Biological applications, vol. 1, p. 44-112. Van Nostrand Reinhold Co., New York.

51. **Hayat, M. A., and B. R. Zirkin.** 1973. *In* M. A. Hayat (ed.), Principles and techniques of electron microscopy. Biological applications, vol. 3, p. 311-319. Van Nostrand Reinhold Co., New York.

52. **Hayes, T. L.** 1973. *In* J. K. Koehler (ed.), Advanced techniques in biological electron microscopy, p. 153-214. Springer-Verlag, New York.

Transmission electron microscopy in microbiology.

53. **Avakyan, A. A., P. N. Katz, and I. B. Pavlova.** 1972. Atlas of bacteria pathogenic for man and animals (in Russian). Meditsyna, Moscow.

54. **Glauert, A. M., M. J. Thornley, K. J. I. Thorne, and U. B. Sleytr.** 1976. *In* R. Fuller and D. W. Lovelock (ed.), Microbial ultrastructure. The use of the electron microscope, p. 31-47. Academic Press, Inc., New York.

55. **Greenhalgh, G. N., and L. V. Evans.** 1971. *In* C. Booth (ed.), Methods in microbiology, vol. 4, p. 518-565. Academic Press, Inc., New York.

56. **Highton, P. I.** 1976. *In* R. Fuller and D. W. Lovelock (ed.), Microbial ultrastructure. The use of the electron microscope, p. 161-173. Academic Press, Inc., New York.

57. **Hodgkiss, W., J. A. Short, and P. D. Walker.** 1976. *In* R. Fuller and D. W. Lovelock (ed.), Microbial ultrastructure. The use of the electron microscope, p. 51-71. Academic Press, Inc., New York.

58. **Kellenberger, E.** 1962. *In* R. J. C. Harris (ed.), The interpretation of ultrastructure, p. 233-246. Academic Press,] , New York.

59. **Kellenberger, E., Ryter, and J. Sechaud.** 1958. J. Biophys. chem. Cytol. **4:**671-678.

60. **Lawn, A. M.** 1976. *In* R. Fuller and D. W. Lovelock (ed.), Microbial ultrastructure. The use of the electron microscope, p. 73-86. Academic Press, Inc., New York.

61. **Lickfeld, K. G.** 1976. *In* J. R. Norris and D. W. Ribbons (ed.), Methods in microbiology, vol. 9, p. 127-176. Academic Press, Inc., New York.

62. **Ryter, A., and E. Kellenberger.** 1958. Z. Naturforsch. Teil B **13:**597-605.

63. **Walker, P. D., J. A. Short, and G. Roper.** 1976. *In* R. Fuller and D. W. Lovelock (ed.), Microbial ultrastructure. The use of the electron microscope, p. 118-146. Academic Press, Inc., New York.

Scanning electron microscopy in microbiology.

64. **Bulla, L. A., Jr., G. St. Julian, C. W. Hesseltine, and F. L. Baker.** 1973. *In* J. R. Norris and D. W. Ribbons (ed.), Methods in microbiology, vol. 8, p. 2-33. Academic Press, Inc., New York.

65. **Kondo, I.** 1978. *In* M. A. Hayat (ed.), Principles and techniques of scanning electron microscopy. Biological applications, vol. 6, p. 309-316. Van Nostrand Reinhold Co., New York.

66. **Kormendy, A. C.** 1975. *In* M. A. Hayat (ed.), Principles and techniques of scanning electron microscopy. Biological applications, vol. 3, p. 82-108. Van Nostrand Reinhold Co., New York.

67. **Nanninga, N.** 1973. *In* E. L. Benedetti and P. Favard (ed.), Freeze etching: techniques and applications, p. 151-179. Société Française de Microscopie Electronique, Paris.

68. **Nickerson, A. W., L. A. Bulla, Jr., and C. P. Kurtzman.** 1974. *In* M. A. Hayat (ed.), Principles and techniques of scanning electron microscopy. Biological applications, vol. 1, p. 159-180. Van Nostrand Reinhold Co., New York.

69. **Passmore, S. M., and B. Bole.** 1976. *In* R. Fuller and D. W. Lovelock (ed.), Microbial ultrastructure. The use of the electron microscope, p. 19-29. Academic Press, Inc., New York.

70. **Yoshii, Z., J. Tokumaga, and J. Tawara.** 1975.

Atlas of scanning electron microscopy in microbiology. Igaku Shoin, Ltd., Tokyo (The Williams & Wilkins Co., Baltimore).

Electron microscopy of nucleic acids.

71. **Bradley, D. E.** 1965. *In* D. Kay (ed.), Techniques for electron microscopy, 2nd ed., F. A. Davis Co., Philadelphia.
72. **Davis, R. W., M. Simon, and N. Davidson.** 1971. Methods Enzymol. **21:**413–428.
73. **Kay, D.** 1976. *In* J. R. Norris and D. W. Ribbons (ed.), Methods in microbiology, vol. 9, p. 177–214. Academic Press, Inc., New York.
74. **Kleinschmidt, A. K., and R. K. Zahn.** 1959. Z. Naturforsch. Teil B **14:**770–779.
75. **Thomas, J. O.** 1978. *In* M. A. Hayat (ed.), Principles and techniques of electron microscopy. Biological applications, vol. 9, p. 64–83. Van Nostrand Reinhold Co., New York.

Electron microscope photography, interpretation, and image enhancement.

76. **Agar, A. W., R. H. Alderson, and D. Chescoe.** 1974. *In* A. M. Glauert (ed.), Practical methods in electron microscopy, vol. 2, p. 191–276. American Elsevier Publishing Co., New York.
77. **Farnell, G. C., and R. B. Flint.** 1975. *In* M. A. Hayat (ed.), Principles and techniques of electron microscopy. Biological applications, vol. 5, p. 19–61. Van Nostrand Reinhold Co., New York.
78. **Frank, J.** 1973. *In* J. K. Koehler (ed.), Advanced techniques in biological electron microscopy, p. 215–274. Springer-Verlag, New York.
79. **Hawkes, P. W.** 1978. *In* M. A. Hayat (ed.), Principles and techniques of electron microscopy. Biological applications, vol. 8, p. 262–306. Van Nostrand Reinhold Co., New York.
80. **Horne, R. W., and R. Markham.** 1972. *In* A. M. Glauert (ed.), Practical methods in electron microscopy, vol. 1, p. 327–435. American Elsevier Publishing Co., New York.
81. **Markham, R., S. Frey, and G. J. Hills.** 1963. Virology **20:**88–102.
82. **Peters, K.-R.** 1977. *In* M. A. Hayat (ed.), Principles and techniques of electron microscopy. Biological applications, vol. 7, p. 118–143. Van Nostrand Reinhold Co., New York.

Fluorescent antibody.

83. **Goldman, M.** 1968. Fluorescent antibody methods. Academic Press Inc., New York.

84. **Kawamura, A., Jr.** 1977. Fluorescent antibody techniques and their applications. University Park Press, Baltimore.
85. **Walker, P. D., I. Batty, and R. O. Thomson.** 1971. *In* J. R. Norris and D. W. Ribbons (ed.), Methods in microbiology, vol. 5A, p. 219–254. Academic Press, Inc., New York.

Cytochemistry.

86. **Glauert, A. M.** 1965. *In* D. Kay (ed.), Techniques for electron microscopy, 2nd ed., p. 274–282. F. A. Davis Co., Philadelphia.

Immunocytochemistry.

87. **Glauert, A. M.** 1965. *In* D. Kay (ed.), Techniques for electron microscopy, 2nd ed., p. 292–303. F. A. Davis Co., Philadelphia.
88. **Hayat, M. A.** 1975. Positive staining for electron microscopy, p. 194–209. Van Nostrand Reinhold Co., New York.
89. **Kraehenbuhl, J. P., and J. D. Jamieson.** 1974. Int. Rev. Exp. Pathol. **13:**1–53.
90. **Walker, P. D., I. Batty, and R. O. Thomas.** 1971. *In* J. R. Norris and D. W. Ribbons (ed.), Methods in microbiology, vol. 5A, p. 219–254. Academic Press, Inc., New York.
91. **Williams, M. A.** 1977. *In* A. M. Glauert (ed.), Practical methods in electron microscopy, vol. 6, p. 1–217. Elsevier/North Holland, Inc., New York.

Autoradiography.

92. **Glauert, A. M.** 1965. *In* D. Kay (ed.), Techniques for electron microscopy, 2nd ed., p. 282–292. F. A. Davis Co., Philadelphia.
93. **Salpeter, M. M., and L. Bachmann.** 1972. *In* M. A. Hayat (ed.), Principles and techniques of electron microscopy. Biological applications, vol. 2, p. 221–279. Van Nostrand Reinhold Co., New York.
94. **Salpeter, M. M., and F. A. McHenry.** 1972. *In* J. K. Koehler (ed.), Advanced techniques in biological electron microscopy, p. 113–152. Springer-Verlag, New York.
95. **Williams, M. A.** 1977. *In* A. M. Glauert (ed.), Practical methods in electron microscopy, vol. 6, p. 1–217. Elsevier/North Holland, Inc., New York.

Chapter 5

Cell Fractionation

CARL A. SCHNAITMAN

Subcellular fractionation schemes must be "custom tailored," not only to fit the peculiarities of individual organisms, but also to fit the kind of organelles which are sought or the type of information on subcellular localization which is required. A discussion of each of the many elaborate schemes which are used with bacteria and other microorganisms is beyond the scope of this manual. Instead, I have attempted to describe some of the common techniques of cell breakage and fractionation which are applied to bacteria, and I have included a few examples in which these techniques are applied to the fractionation of *Escherichia coli*.

5.1. BREAKAGE TECHNIQUES

The choice of an appropriate method of cell breakage is the most important decision to be made in designing a cell fractionation scheme. The following questions must be kept in mind when choosing a breakage method:

1. Is the breakage method appropriate to the type of organism? There is no single breakage method which works well with all types of bacteria. For example, the French pressure cell, which is the method of choice for breakage of most gram-negative bacteria, is not very effective in breaking gram-positive cocci. The Braun MSK shaker, which can cause extensive damage of gram-negative bacteria, is quite effective in breaking gram-positive cocci.

2. Will the breakage method damage subcellular organelles? An example that illustrates this potential problem is the technique of ultrasonic disintegration. Although this method is effective in breaking many types of bacterial cells, it often causes extensive damage to subcellular organelles.

3. Is the necessary equipment available? A number of mechanical devices which are effective in breaking bacterial cells are expensive, not available on a routine basis, or not suitable for processing the amount of material required.

4. Which procedure is the simplest? Procedures which require prolonged incubation or which involve many complicated steps result in poor recovery or loss of enzymatic activity.

With any breakage procedure, some method should be used to monitor the extent of cell breakage. Viable counts are often used for this purpose, but this is probably not the best method since cells can lose viability without release of intracellular contents. Light microscopy of wet mounts is a good method, but only if done quantitatively by counting cells which remain intact in some kind of counting chamber. This is necessary because, with many procedures, the subcellular fragments are too small to be resolved with a light microscope. Phase-contrast microscopy or negative staining with nigrosin or India ink can be used to distinguish between intact cells and empty "ghosts" or large cell envelope fragments. When procedures that involve conversion of cells to spheroplasts or protoplasts in osmotically stabilized medium are employed, it is essential to determine the percentage of conversion in a counting chamber rather than simply to examine wet mounts. This is because lysed spheroplasts or protoplasts are often difficult to see by light microscopy, and qualitative comparison of the protoplasts or spheroplasts that have remained intact with the number of normal cells can be misleading.

Another effective monitoring procedure is to remove samples during the course of cell breakage or at the end of the breakage procedure and centrifuge these samples for a time and at a speed which are sufficient to pellet intact cells. The supernatants are then examined for protein, ultraviolet-absorbing material (i.e., nucleic acids), or specific cytoplasmic enzymes to determine the extent to which these have been released from the cells.

The following parts describe some of the specific methods which can be used for cell breakage. Also see parts 18.1.3. and 22.1.

5.1.1. Pressure Shearing

Pressure shearing is probably the most widely used and useful method of cell breakage. A sample of bacterial suspension (5 to 40 ml of a suspension, up to 30% cells by volume) is placed in a steel cylinder fitted with a piston and a small relief valve connected to an outlet tube. This entire assembly is placed in a 10-ton hydraulic press. When the cell piston is forced down, the cells are broken by the high shear forces generated as the suspension passes through the small orifice of the relief valve. This method has several advantages. Cooling is no problem, since the press cylinder and piston can be precooled and heat is generated only when the cells pass through the relief valve. The effluent temperature will rise by 5 to 10°C, but the effluent can be rapidly cooled if it is collected in a metal tube or beaker in an ice bath. Breakage is virtually instantaneous, and the broken suspension is not subjected to additional shear forces which could damage subcellular particles. If a satisfactory hydraulic press is used, very reproducible conditions can be obtained. The extent of breakage is not influenced by the density of the cell suspension, the growth phase at which the cells were harvested, or the breakage medium in which the cells are suspended.

The French pressure cell assembly consists of a cylinder, piston, relief valve, and a loading stand (American Instrument Co. [Aminco], Rockville, Md.). Older models of this cell were fitted with a steel-needle relief valve which was subject to wear and damage during operation. A replacement valve fitted with a nylon ball is available (Aminco) and should be used instead; this allows the piston to travel to the bottom of the cylinder without damage and provides more uniform breakage. A motor-driven hydraulic press which can be preset to deliver a constant force over a range of piston speeds is also required; suitable hydraulic presses can be obtained (Aminco; or Enerpac, Inc., Butler, Wis.).

The cell suspension is loaded into the pressure cell and placed in the hydraulic press. The relief valve is closed, and the hydraulic press piston is lowered onto the cell piston and allowed to reach maximum force. The relief valve is opened slowly to the point where the piston moves downward slowly.

For breakage of *E. coli* cells, cultures are harvested, washed once, and suspended at 0.1 culture volume in 0.01 M HEPES (N-2-hydroxyethylpiperazine-N'-2-ethanesulfonic acid), pH 7.4, at 0°C. A small amount of pancreatic ribonuclease and deoxyribonuclease (approximately 0.1 mg/ml) is added, and the suspension is broken with the French pressure cell operated at a cell pressure of 20,000 lb/in^2. After breakage, $MgCl_2$ is added to give a final concentration of 1 mM, and unbroken cells are removed by centrifugation at 5,000 × g for 5 min. The purpose of the Mg^{2+} is to allow deoxyribonuclease to act, and it is added after breakage to prevent stabilization of ribosomes. Ribonuclease acts in the absence of Mg^{2+}. Other buffers such as phosphate may be used, but tris(hydroxymethyl)aminomethane (Tris) buffer or chelating agents should be avoided to prevent outer membrane damage.

The French pressure cell is effective for breakage of gram-negative bacteria and some gram-positive bacteria (particularly the gram-positive bacilli). Endospores and gram-positive cocci are not broken by this device, and ballistic disintegration or lysozyme digestion of the cell wall is more effective for breaking these resistant cells.

The Ribi cell fractionator is similar in principle to the French pressure cell, except that the Ribi device provides higher pressure (which is more effective with endospores and some of the more resistant cells), a cooling system, and a containment suitable for pathogens. The Ribi fractionator is no longer obtainable commercially.

5.1.2. Ultrasonic Disintegration

A variety of ultrasonic probe devices can be used to break cells. The rapid vibration of the probe tip causes "cavitation," which means the formation of microscopic gas bubbles moving at high velocity in the vicinity of the tip. The high shear forces generated by these rapidly moving bubbles result in cell breakage. This breakage method is tempting since the probe devices are inexpensive and effective. However, there are inherent disadvantages which limit the usefulness of ultrasonic disintegration in cell fractionation studies.

The major disadvantage is that breakage is not instantaneous, and a cell suspension must be treated for 30 s to several minutes in order to break a reasonable proportion of the cells. During this time, subcellular particles released from broken cells are subject to the same high shear forces as unbroken cells. Membrane vesicles are degraded to small lipoprotein fragments which can no longer be sedimented in the ultracentrifuge, and extensive redistribution of membrane proteins occurs between various membranes (as, for example, between the inner and outer membranes of gram-negative bacteria).

In addition, it is difficult to control the temperature of the sample during breakage. Foaming can cause protein denaturation, and the cavitation phenomenon promotes oxidation of oxygen-sensitive enzymes and unsaturated lipids. Problems of foaming and temperature control can be minimized by subjecting the sample to several short bursts rather than one continuous treatment, with cooling periods between the bursts. Oxidation problems can be minimized by covering the sample with argon gas during treatment. It is very difficult to obtain reproducible breakage, since the effectiveness of the probe depends upon sample viscosity and upon the size, shape, and composition of the sample vessel (for example, a glass beaker is far more effective than a plastic one, which will absorb some of the ultrasonic energy). *CAUTION:* Operators should wear ear protection, unless the probe is enclosed in a sound-absorbing cabinet.

Although ultrasonic treatment is not a good method for primary cell breakage, it is useful for lysis of spheroplasts (as in the procedure of Osborn and Munson [8] for separation of inner and outer membranes of gram-negative bacteria), as a method for dispersing clumps of subcellular organelles, and for suspending centrifuge pellets. In these procedures, very short bursts of ultrasonic energy are required, and subcellular organelles are not damaged.

5.1.3. Ballistic Disintegration

The term ballistic disintegration may be applied to a variety of methods in which bacteria are broken by the shear forces developed when a suspension of cells together with small glass or plastic beads is shaken or agitated violently. A variety of commercial devices have been used for this procedure, and they are described in considerable detail in reviews by Salton (10) and by Hughes et al. (4).

Two devices which were widely used in early studies were the Mickle shaker and a shaker fitted to the shaft of an International centrifuge (14). Neither of these devices had any provision for cooling the sample during shaking, and it was necessary to interrupt the shaking frequently to cool the sample container. These devices are no longer available, and have been replaced by the Braun MSK tissue disintegrator (B. Braun Melsungen Apparatebau, Melsungen, Federal Republic of Germany), which is available from many laboratory supply houses in the United States. The Braun disintegrator has a 65-ml sample container which is shaken horizontally at 2,000 to 4,000 oscillations per min. Cooling is provided by a stream of liquid CO_2 delivered to the sample container. Disintegration of most samples can be accomplished in 3 to 5 min at a temperature of less than 4°C.

The Braun disintegrator is the method of choice for isolation of cell walls and particulate enzymes from gram-positive bacteria. The following procedure described by Work (14) can be used with most organisms.

A cell suspension (30 ml, containing 20 to 50 mg [dry weight]/ml) in a suitable buffer is mixed with 20 ml of Ballotini no. 12 beads in the sample container (an air space of at least 20% of the container volume is essential for breakage), and CO_2 is passed through the apparatus for 0.5 min to cool the container. The container is shaken at 3,000 strokes per min for 3.5 to 5.0 min, depending on the species. The beads may be removed by low-speed centrifugation or by passage through coarse-grade sintered glass. Some workers prefer to use plastic beads as described by Ross (9) instead of glass beads, to minimize enzyme denaturation. When the procedure described above is employed for the isolation of cell walls, the sample should be treated promptly to inactivate autolytic enzymes (14).

Bacterial endospores may be broken with the Braun disintegrator by shaking dry spores with dry glass beads (5 g of dry spores plus 15 g of beads) as described above and then suspending the sample in buffer, as described by Steinberg (12).

5.1.4. Solid Shearing

Cells may be broken by grinding cell paste or lyophilized cells with abrasives such as alumina, or by the solid shear developed when a frozen suspension or paste is forced through a small hole or slit. Hughes et al. (4) have provided a complete description of these methods. A commercially available device for solid shear of frozen cell paste is the X-Press cell disintegrator (LKB-Produkter AB, Bromma, Sweden). This device is effective in breaking both gram-positive and gram-negative cells. However, the apparatus is limited in capacity and somewhat expensive and difficult to use, and has no particular advantage over the French press or the Braun disintegrator.

5.1.5. Muramidase Digestion

Egg white lysozyme may be used to prepare protoplasts of the gram-positive bacteria that are sensitive to this enzyme. Lysozyme digestion is normally carried out in hypotonic sucrose dilution (0.3 to 0.5 M) at neutral or slightly alkaline pH. Treatment for 30 min at lysozyme concentrations of 0.1 to 1.0 mg/ml should result in complete protoplast formation. Phase microscopy is used to monitor protoplast formation. Low levels of Mg^{2+} or Ca^{2+} (in the range of 0.5 to 5 mM) may be required to stabilize the protoplasts. Lysozyme digestion can be carried out either in the cold or at room temperature.

Muramidases which have a broader substrate specificity than that of egg white lysozyme are available commercially, and these may be effective in digesting the cell wall of the gram-positive organisms that are resistant to egg white lysozyme. These include lysostaphin, a muramidase from *Staphylococcus staphylolyticus* (Sigma Chemical Co., St. Louis, Mo.), and a fungal muramidase from *Chalaropsis* sp. (Miles Laboratories, Inc., Elkhart, Ind.). Caution should be used with these enzymes, since they are not as highly purified as egg white lysozyme and are often contaminated with protease, lipase, or nuclease activity.

Unlike gram-positive bacteria, virtually all gram-negative bacteria have a peptidoglycan structure which is sensitive to lysozyme. However, the outer membrane of gram-negative bacteria is impermeable to lysozyme, and the integrity of the outer membrane must be destroyed before lysozyme will act. This can be done by freezing and thawing, by pretreatment with ethylenediaminetetraacetic acid (EDTA), or in some cases by treatment with membrane-active antibiotics such as polymyxin B. Digestion of the peptidoglycan of gram-negative bacteria does not result in removal of the outer membrane; hence, the osmotically sensitive structures which are formed are termed spheroplasts instead of protoplasts.

Spheroplasts of enteric bacteria can be prepared by the procedure of Osborn and Munson (8), as follows. Cultures are centrifuged and suspended in one-tenth of the culture volume in cold 0.75 M sucrose in 10 mM Tris-acetate buffer, pH 7.8. Lysozyme is added to give a final concentration of 0.1 mg/ml, and the suspension is incubated on ice for 2 min. Conversion to spheroplasts is accomplished by slowly diluting the suspension over a period of 8 to 10 min by the addition (via a peristaltic pump) of 2 volumes of cold 1.5 mM EDTA. The suspension is gently agitated during dilution, and conversion to spheroplasts is monitored by phase microscopy.

Protoplast and spheroplast formation is markedly affected by a number of factors, including the growth phase at which organisms are harvested, the kind of washing procedures used after harvesting, and the type of culture medium. Osborn and Munson (8) note, in reference to the procedure above, that cells grown in rich medium form better spheroplasts and that exposure to cold buffer or Tris prior to exposure to the sucrose solution is deleterious. As a general rule, cultures harvested during the mid-log phase are most suitable for spheroplasts or protoplasts.

5.1.6. Osmotic Lysis

Rapid dilution (10- to 20-fold) with dilute buffer or distilled water effectively lyses spheroplasts or protoplasts. To a certain extent, the size of the vesicles produced is determined by the speed and extent of the dilution. A procedure which is often used is to pellet the protoplasts or spheroplasts by centrifugation and then to suspend the pellet in the lysing buffer. Such pellets can be suspended rapidly and uniformly with a syringe (particularly the spring-loaded variety supplied with an Oxford syringe pipettor) fitted with a large-diameter needle or canula. Several rapid strokes with the syringe will provide even and rapid suspension.

In intact cells, the deoxyribonucleic acid (DNA) is folded into a structure called a nucleoid. This structure is stabilized by ribonucleic acid, and even the slightest trace of ribonuclease will allow the nucleoid to unfold during lysis, resulting in a very viscous mass of released DNA. This viscosity problem can be overcome in several ways. The DNA can be sheared by brief ultrasonic treatment, or by treating the suspension of lysed cells in a laboratory blender. Deoxyribonuclease can be included in the lysing solution, but in order for this enzyme to work, some Mg^{2+} (at least 0.1 mM) must be provided. When spheroplasts formed in the presence of EDTA are being lysed, excess Mg^{2+} is required.

5.1.7. Freezing and Thawing

Freezing and thawing may be used to render

gram-negative cells sensitive to lysozyme or detergents. This procedure can be applied to the large-scale isolation of membranes or subcellular organelles.

The following procedure is effective for the large-scale isolation of membranes from *E. coli*. A thick suspension of washed cells (about 30% cells by volume) in 0.02 M Tris buffer, pH 7.8, containing 5 mM EDTA, 0.25 M sucrose, and 0.5 mg of lysozyme per ml is placed in a flask and shell frozen by swirling in a dry ice-acetone bath. The flask is then thawed in warm water until the ice is just melted, and the contents are poured into 20 volumes of cold 0.02 M Tris buffer, pH 7.8, containing 0.5 mM $MgCl_2$ and 0.1 mg of deoxyribonuclease per ml. This material is immediately treated in a laboratory blender for about 20 s to shear the DNA and disperse the cells. Unbroken cells are removed by centrifugation at 5,000 × g for 5 min. If the pellet is substantial, it can be suspended again in the Tris-EDTA-sucrose solution described above, and the freezing-thawing-dilution cycle is repeated. The envelope material, consisting of inner and outer membranes, is recovered by centrifugation at 27,000 × g for 20 min.

5.2. FRACTIONATION TECHNIQUES

5.2.1. Differential Centrifugation

The techniques of differential centrifugation, rate-zonal centrifugation, and equilibrium density gradient centrifugation are widely used in cell fractionation. Detailed discussion of both the theoretical and practical aspects of centrifugation is provided in a very comprehensive review by Sykes (13).

In differential centrifugation, samples are centrifuged for a given time at a given speed, after which the supernatant is decanted. This technique is useful for separation of particles with very different sedimentation velocities. For example, centrifugation for 5 to 10 min at 3,000 to 5,000 × g will pellet intact bacterial cells while leaving most cell fragments in the supernatant. Cell wall fragments and large membrane structures can be pelleted by centrifugation at 20,000 to 50,000 × g for 20 min, whereas centrifugation at 200,000 × g for 1 h is required to pellet small membrane vesicles or ribosomes.

5.2.2. Rate-Zonal Centrifugation

Zonal centrifugation is an effective means of separating subcellular structures which have a similar buoyant density but which differ in shape or particle mass. Examples include the separation of ribosomal subunits, the separation of different classes of polysomes, and the separation of various forms of DNA molecules. Centrifugation is carried out either in swinging-bucket rotors or specially designed zonal rotors, and a shallow gradient (usually sucrose) is employed to prevent convection in the tubes or the chamber of the zonal rotor during centrifugation. The sample is applied as a zone or narrow band at the top of the gradient. For subcellular particles, a sucrose gradient of 15 to 40% (wt/vol) sucrose is commonly used, and centrifugation at 100,000 × g for 1 to 4 h is sufficient for separation of most subcellular particles.

5.2.3. Equilibrium Density Gradient Centrifugation

Equilibrium density gradient centrifugation separates particles on the basis of buoyant density instead of sedimentation velocity. This technique is widely used to separate various membrane fractions, since membrane fragments derived from the same subcellular membrane may differ greatly in size (and, hence, in sedimentation velocity) but should have the same buoyant density. The sample is centrifuged in a solute density gradient (sucrose gradients are commonly used for membranes and organelles with a density of less than 1.3 g/ml; tartrate and cesium chloride are used for denser structures such as viruses) until an equilibrium state is reached at which each particle has migrated to a point in the gradient where the particle has the same density as the surrounding solution. Since sucrose solutions are relatively viscous, preformed gradients are generally employed. Cesium chloride solutions have a low viscosity so that preformed gradients of this solute are difficult to prepare, and the technique of isopycnic density gradient centrifugation is used. In this latter technique, the sample is mixed with enough cesium chloride to provide a density equal to the average density of the subcellular particles. This homogeneous suspension is placed in the centrifuge, and the cesium chloride gradient is formed during centrifugation as a result of the sedimentation of the cesium chloride in the centrifugal field.

Since the sedimentation velocity of a particle becomes progressively smaller as the particle approaches the region in the gradient where it has the same density as the solution, very long centrifugation times are required in order to approach equilibrium. This is particularly true for small membrane vesicles such as chromatophores or cytoplasmic membrane fragments from cells broken with a French press, since the sedimentation velocity of these vesicles is low even in the absence of the gradient. If centrifugal forces on the order of 100,000 to 200,000 $\times g$ are used, at least 24 h will be required for a reasonable separation, and periods as long as 72 h are needed for critical separation.

5.2.4. Sucrose Gradients

Sucrose concentrations in gradients are reported in the literature in several ways: density, molar concentration of sucrose, percent sucrose by weight (%, wt/wt), and percent sucrose per unit volume (%, wt/vol, or grams/100 ml). Standard chemistry handbooks contain tables listing the concentrative properties of aqueous sucrose solutions which relate to these various units. The most useful unit is percent sucrose by weight. When preparing sucrose solutions, one may assume that water or dilute buffers have a density of 1 g/ml. Therefore, to prepare a 54% (wt/wt) sucrose solution, for example, one would dissolve 54 g of sucrose in 46 ml of water. This will give 100 g of solution, and since the density of a 54% (wt/wt) sucrose solution is 1.2451, the final volume will be 80.3 ml. Preparing solutions in this manner avoids the necessity of having to make viscous solutions up to fixed volumes.

A variety of commercial gradient-forming devices are available, but these are unnecessary since satisfactory gradients can be prepared by layering a series of sucrose solutions in a centrifuge tube and allowing the gradient to "age" overnight in order for diffusion to produce a linear gradient. It is not necessary to "age" gradients which will be used for centrifuge runs of 24 h or longer, since diffusion during centrifugation will result in a linear gradient. We customarily prepare sucrose gradients as a series of five to seven layers. When wettable tubes such as cellulose nitrate or polycarbonate are used, each layer can be added by allowing it to run down the side of the tube from a pipette held at an angle against the side of the tube. When nonwettable tubes such as polyallomer or polypropylene are used, one must keep the pipette tip in contact with the meniscus while adding the solution in order to avoid mixing.

The simplest method of fractionating sucrose gradients is to pump out the gradients with a peristaltic pump. The centrifuge tube is clamped in a ring stand (we use a block of clear plastic drilled to fit the tube which is clamped to the stand), and a stainless steel capillary tube is clamped to the ring stand above the centrifuge tube. The capillary tube is attached to the pump and is lowered into the centrifuge tube until it almost touches the bottom of the tube. The gradient is then pumped out into a fraction collector or into graduated tubes.

5.2.5. Detergents

Detergents have three uses in cell fractionation. When nonmembranous organelles such as ribosomes, nucleoids, etc., are sought, detergents provide a gentle means of lysing the cells once the integrity of the peptidoglycan (gram-positive bacteria) or outer membrane (gram-negative bacteria) has been damaged. Detergents are used to selectively solubilize the cytoplasmic membrane of gram-negative bacteria while leaving the outer membrane intact. Detergents can also be used to remove membrane contamination from ribosomes, polysomes, or gram-positive cell walls.

Detergents are amphipathic molecules, meaning that the molecules have both hydrophilic and hydrophobic regions, and they are sparingly soluble in water. At very low concentrations, detergents will form true solutions in water. As the concentration is increased, additional molecules of detergent will aggregate to form micelles in which the hydrophilic regions are exposed to water and the hydrophobic regions are shielded from water on the inside of the micelle. The concentration at which micelles begin to form as the amount of detergent added to water is increased is designated as the critical micelle concentration, or CMC. The CMC and the size and shape of the detergent micelle are characteristic of each detergent. An excellent review by Helenius and Simons (3) summarizes the properties of many of the detergents used in cell fractionation.

Detergents may be grouped into three classes, which differ in micelle properties, protein binding, and response to other solutes. These classes are ionic detergents, nonionic detergents, and bile salts. Each presents unique problems and advantages in cell fractionation.

Ionic detergents.

The most commonly used ionic detergents are sodium dodecyl sulfate (sodium lauryl sulfate, SDS), sodium N-lauryl sarcosinate (Sarkosyl), alkyl benzene sulfonates (common household detergents), and quaternary amine salts such as cetyl trimethylammonium bromide (CETAB). Ionic detergents tend to form small micelles (molecular weights of about 10,000) and exhibit a rather high CMC (the CMC for SDS is about 0.2% at room temperature in dilute buffers). The CMC and the micelle solubility of ionic detergents are strongly influenced by the ionic strength of the solution and the nature of the counterions present. For example, a 10% solution of SDS is stable down to about 17°C, whereas a similar solution of Tris dodecyl sulfate is stable at 0°C. Potassium dodecyl sulfate is soluble only at elevated temperatures, and K^+ must be excluded from all buffers when this detergent is used.

Detergents such as SDS which have a hydrophilic group that is strongly ionized are not affected by pH and are not precipitated by 5% trichloroacetic acid. Ionic detergents bind strongly to proteins, and in the case of SDS this usually results in unfolding and irreversible denaturation of proteins. Although ionic detergents can be removed by dialysis, this is not practical because of extensive protein binding.

Nonionic detergents.

The nonionic detergents include Triton X-100, Nonidet P-40 (NP-40), Tween 80, and octyl glucoside. In general, these detergents have a high micelle molecular weight (50,000 or greater) and a low CMC (0.1% or less), which limit the usefulness of these detergents in gel filtration or gel electrophoresis. The solution properties of these detergents are not strongly affected by pH or ionic strength, although they may be precipitated by 5% trichloroacetic acid. These detergents bind only to hydrophobic proteins and generally do not cause denaturation or loss of biological activity.

Bile salts.

Bile salts are salts of sterol derivatives, such as sodium cholate, deoxycholate, or taurocholate. Because of the poor packing ability of the bulky sterol nucleus, these detergents form small micelles (often just a few molecules), and unlike other detergents, the micelle molecular weight is a function of the detergent concentration. As these detergents are salts of very insoluble weak acids with pK_a's in the range of 6.5 to 7.5, they must be used at alkaline pH. To avoid solubility problems, stock solutions are often prepared by dissolving the free acid in excess NaOH. Ionic composition, pH, and total detergent concentration must all be maintained constant when these detergents are used.

Problems in the use of detergents.

Since most detergents are employed at concentrations well above the CMC, and since detergents act by forming mixed micelles with lipids or by binding to proteins, the ratio of detergent to protein or lipid is much more important than the actual detergent concentration. As a general rule of thumb, one must use at least 2 to 4 mg of detergent per mg of sample protein to ensure a satisfactory excess of detergent. For example, if 2% Triton X-100 is used for membrane solubilization, the maximum protein content of the sample should be less than 5 to 10 mg/ml.

Triton X-100, one of the most useful nonionic detergents, presents problems in protein assays because it has an aromatic residue which prevents measurement of absorbance at 280 nm and because it forms a cloudy precipitate in chemical protein assays. We usually overcome this problem by labeling cultures with a small amount of [^{3}H]leucine. If this is done in minimal or defined media, care must be taken to add sufficient carrier leucine so that the amount of isotope incorporated per milligram of protein is constant throughout growth. With *E. coli*, 20 to 40 μg of unlabeled leucine per ml is sufficient carrier to provide uniform labeling. When labeling is not practical, proteins can be assayed in the presence of Triton X-100 by the modification of the Lowry procedure described by Schaechterle and Pollack (11), in which an excess of SDS is added to the sample. This forms stable mixed micelles with the Triton which do not interfere with the assay.

Triton X-100 and other similar nonionic detergents are soluble in ethanol-water mixtures, and proteins dissolved in these detergents may be freed from detergent by ethanol precipitation. The sample is placed on ice, and 2 volumes of ice-cold absolute ethanol are added with stirring. The sample is then allowed to stand overnight in a freezer, and the protein precipitate is collected by centrifugation. Efficient precipitation requires a protein concentration of at least 0.2 mg/ml. Dilute samples may be concentrated with an Amicon ultrafiltration apparatus with a PM-30 filter, although this is limited by the fact that this will also concentrate detergent micelles.

SDS can be removed from samples by acetone precipitation. Six volumes of anhydrous acetone are stirred into the sample at room temperature, and the precipitate is recovered by centrifugation. The precipitate is then washed several times with acetone-water (6:1). Since the precipitate is often waxy and difficult to work with, we commonly disperse it in water with a Potter homogenizer and then lyophilize it.

5.2.6. Polyacrylamide Gel Electrophoresis

Polyacrylamide gel electrophoresis, in the presence of SDS, is the simplest and most effective technique for examination of the polypeptide profile of subcellular fractions. To a large extent this technique has replaced enzymatic and chemical analysis as a technique for establishing the purity and homogeneity of subcellular fractions.

A variety of commercially available or homemade devices can be employed for polyacrylamide gel electrophoresis. We favor those devices which allow thin slabs (generally 1.0 mm or less in thickness) to be run, for reasons of both resolution and convenience. Thin slabs provide superior resolution because the heat generated during electrophoresis is easily dissipated and because the gels may be rapidly fixed after electrophoresis to minimize diffusion of protein bands. Slabs allow side-by-side comparison of multiple samples and can easily be preserved by drying on filter-paper sheets. Radioactive samples can be visualized by radioautography or photofluorography, and gels dried on filter paper can easily be sliced with scissors or a paper cutter and, after rehydration of the dried slices, counted in a scintillation counter. When very long gels are not required, electrophoresis, fixing and staining, and drying can all be accomplished within a single day.

There are also a variety of buffer systems which have been employed for SDS-gel electrophoresis. The systems which offer the best resolution are "stacking" buffer systems, in which the upper buffer contains a rapidly moving ionic species which moves through the gel as a zone or front. Proteins are compressed by this moving front into a very thin band at the time they enter the running gel, whereupon they are retarded by the sieving properties of the gel. The system described below is a modification of the stacking buffer system introduced by Laemmli (5). Also see part 16.5.1.

The separation, or running, gel consists of 11.5% acrylamide, 0.2% bis-acrylamide, and 0.1% SDS in 0.375 M Tris adjusted to pH 8.8 with HCl. Polymerization is initiated by deaerating the solution and adding 0.08% tetramethylethylenediamine and 0.015% persulfate. The running gel is overlaid with water until polymerization occurs. This is poured off, and a stacking, or sample, gel consisting of 4.5% acrylamide, 0.12% bis-acrylamide, and 0.1% SDS in 0.125 M Tris adjusted to pH 6.8 with HCl and polymerized with 0.125% tetramethylethylenediamine and 0.05% ammonium persulfate is cast. A Teflon comb is inserted into the stacking gel before polymerization to form the sample wells. After polymerization, the sample wells are rinsed several times with the upper electrophoresis buffer containing a small amount of bromophenol blue. This removes unreacted persulfate and stains the sides of the sample wells slightly so that they can be seen during sample loading. The upper electrophoresis buffer consists of 0.182 M glycine, 0.0255 M Tris (final pH, 8.3), and 0.1% SDS. The lower electrophoresis buffer is the same as the upper buffer except that SDS is omitted.

Samples are dissolved in stacking gel buffer to which 12.5% glycerol, 1.25% SDS, and 1.25% 2-mercaptoethanol have been added. The sample is dissolved in this buffer and heated for 5 min in a boiling-water bath prior to electrophoresis in order to completely dissociate and denature all proteins. Bromophenol blue or phenol red can also be added to the sample to serve as a tracking dye. The final samples should contain 1 to 5 mg of protein per ml, and with small wells (3.0 by 0.75 mm) 5 to 25 μg of protein is sufficient. Since the conductivity of the gel changes as the stacking buffer moves through it, constant-voltage or constant-power (wattage) power supplies should be used instead of constant current. An initial voltage of 50 V is used until the sample has entered the separation gel, at which point it is increased to 145 V for the remainder of the run. For best resolution, the gel should be maintained at 25°C.

The most common problem associated with polyacrylamide electrophoresis is distorted bands resulting from improper polymerization. To avoid improper polymerization, gel solutions should be thoroughly deaerated prior to casting the gels, and the upper surface of the separation gel should be rinsed several times shortly after polymerization has occurred to remove any remaining unpolymerized gel. Electrophoresis reagents vary in purity, and it is necessary to increase or decrease the amount of ammonium persulfate in order to achieve a constant polymerization time. A polymerization time of about 30 min is optimum for the separation gel. Longer times result in a soft gel or incomplete polymerization, and shorter times can result in uneven polymerization and wavy bands. Ammonium persulfate is very hygroscopic and somewhat unstable. Instead of weighing out small amounts for each gel, it is preferable to prepare a 50-mg/ml stock solution which is frozen in 1-ml amounts. Such a stock solution is stable for several months in a freezer.

Smeared bands can also result from lipids or glycolipids (a common example is lipopolysaccharide) present in the sample. These lipids bind a great deal of SDS and can actually deplete the SDS available to bind to proteins migrating through the gel. This problem can be overcome by increasing the amount of SDS in the upper electrophoresis buffer to 1% or more.

Gels are stained by the method of Fairbanks et al. (2). Gels are soaked for 1 h each, with gentle agitation, in the following: (i) 525 ml of 95% isopropanol, 200 ml of glacial acetic acid, 1.0 g of Coomassie blue, and 1,275 ml of water; (ii) 210 ml of 95% isopropanol, 200 ml of glacial acetic acid, 0.1 g of Coomassie blue, and 1,590 ml of water; (iii) 200 ml of glacial acetic acid, 0.05 g of Coomassie blue, and 1,800 ml of water; and (iv) 10% glacial acetic acid. When the gel is fully destained, it should be soaked in 10% acetic acid containing 1% glycerol prior to drying.

5.3. FRACTIONATION EXAMPLE: *ESCHERICHIA COLI*

5.3.1. Complete Fractionation

The scheme illustrated in Fig. 1 can be used to determine the subcellular localization of an enzyme or protein in *E. coli* or other enteric bacteria. This scheme takes advantage of the fact that, in the presence of low levels of Mg^{2+} (contributed in this case by the cells themselves and the breakage buffer), the outer membrane is insoluble in Triton X-100 while the cytoplasmic membrane is completely solubilized.

When cells are broken by shear forces, some of the cytoplasmic membrane becomes fragmented extensively (probably into short, "open" fragments of membrane) and cannot be sedimented in the ultracentrifuge. This is the reason for the incubation of the first cytoplasmic supernatant fraction (Supernatant I, Fig. 1) at 21°C.

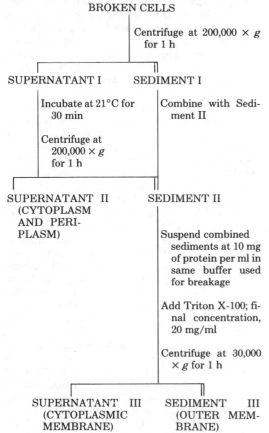

FIG. 1. *Scheme for fractionating Escherichia coli cells to determine subcellular localization of an enzyme or protein. The cells are broken with a French press in HEPES buffer, and intact cells are removed by sedimentation at 5,000 × g for 5 min.*

Incubation at a temperature above the phase transition temperature of the membrane lipid allows these tiny fragments to aggregate or fuse into larger membrane fragments which can then be sedimented (6). In some cases, this non-sedimentable membrane fraction can represent as much as 20 to 30% of the cytoplasmic membrane. Its formation is not unique to the French press—similar fragments are generated by ultrasonic treatment, or even by rapid lysis of protoplasts or spheroplasts.

In this type of fractionation scheme, there is no way to distinguish whether soluble proteins are derived from the cytoplasm or periplasm. This distinction requires a separate experiment in which cells are subjected to osmotic shock by the procedure of Neu and Hepple (7) to release periplasmic proteins.

If the final cytoplasmic membrane fraction (Supernatant III) is to be examined by SDS-gel electrophoresis, the protein should be freed from Triton X-100 by ethanol precipitation as previously described. Unless large excesses of SDS are added to the sample and upper electrophoresis buffers, the Triton X-100 will form mixed micelles with the SDS and interfere with the resolution of the proteins on gels.

5.3.2. Inner and Outer Membranes

The cells are broken with a French press, and the cell extract is freed from whole cells by centrifugation at 5,000 × g for 5 min as described previously. The crude envelope fraction is sedimented by centrifugation at 200,000 × g for 1 h, and the sediment is suspended in 0.01 M HEPES buffer, pH 7.4, at approximately 20 mg of protein per ml.

Sucrose gradients are prepared in the same buffer. For a Beckman SW 27 rotor (37-ml volume), we use five 7-ml layers, having the following amounts of sucrose (wt/wt): 35%, 40%, 45%, 50%, and 55%. Two milliliters of the envelope suspension is layered over the gradient and centrifuged at maximum speed (131,000 × g) for 36 to 48 h. The small amount of sediment at the bottom of the tube is discarded. The outer membrane appears as either a single opalescent band or a pair of similar bands at about 52% (wt/wt) sucrose. The cytoplasmic membrane is a translucent, yellowish band at about 36% (wt/wt) sucrose. Gradients are pumped from the tubes, located by determining the absorbance of the fractions at 280 nm, and peak fractions of each membrane are pooled. These pooled fractions are diluted 1:3 with HEPES buffer as above, containing 1 mM $MgCl_2$, and are centrifuged at 200,000 × g for 2 h to sediment the membrane particles. A variety of biochemical markers can be used to assay cross-contamination. Cytochromes or succinic dehydrogenase are convenient markers for the cytoplasmic membrane. The "major" outer membrane proteins (1) visualized on SDS-polyacrylamide gels are useful markers for the outer membrane, as are specific lipopolysaccharide sugars or phospholipases (8). We have obtained the best results with cultures grown on minimal media with generation times of 1 h or more. Membranes from cells grown on very rich media appear to be more difficult to separate, perhaps as a result of an increased number of zones of adhesion between the outer and cytoplasmic membranes in rapid growing cells.

5.4. LITERATURE CITED

1. Bassford, P. J., Jr., D. L. Diedrich, C. A. Schnaitman, and P. Reeves. 1977. Outer membrane proteins of *Escherichia coli*. VI. Protein alteration in bacteriophage-resistant mutants. J. Bacteriol. **131**:608–622.
2. Fairbanks, G., T. L. Steck, and D. F. Wallach. 1971. Electrophoretic analysis of the major polypeptides of the human erythrocyte membrane. Biochemistry **10**:2606–2617.
3. Helenius, A., and K. Simons. 1975. Solubilization of membranes by detergents. Biochim. Biophys. Acta **415**:29–79.
4. Hughes, D. E., J. W. T. Wimpenny, and D. Lloyd. 1971. The disintegration of microorganisms, p. 1–54. *In* J. R. Norris and D. W. Ribbons (ed.), Methods in microbiology, vol. 5B. Academic Press, Inc., New York.
5. Laemmli, U. 1970. Cleavage of structural proteins during assembly of bacteriophage T4. Nature (London) **222**:293–298.
6. MacGregor, C. H., and C. Schnaitman. 1973. Reconstitution of nitrate reductase activity and formation of membrane particles from cytoplasmic extracts of chlorate-resistant mutants of *Escherichia coli*. J. Bacteriol. **114**:1164–1176.
7. Neu, H. C., and L. A. Heppel. 1965. The release of enzymes from *Escherichia coli* by osmotic shock and during the formation of spheroplasts. J. Biol. Chem. **240**:3685–3692.
8. Osborn, M. J., and R. Munson. 1974. Separation of the inner (cytoplasmic) and outer membranes of gram negative bacteria. Methods Enzymol. **31A**:642–653.
9. Ross, J. W. 1963. Continuous-flow mechanical cell disintegrator. Appl. Microbiol. **11**:33–35.
10. Salton, M. R. 1974. Isolation of cell walls from gram positive bacteria. Methods Enzymol. **31A**:653–667.
11. Schaechterle, G. E., and R. L. Pollack. 1973. A simplified method for quantified assay of small amounts of protein in biological material. Anal. Biochem. **51**:654–655.
12. Steinberg, W. 1974. Properties and developmental roles of the lysyl- and tryptophanyl-transfer ribonucleic acid synthetases of *Bacillus subtilis*: common genetic origin of the corresponding spore and vegetative enzymes. J. Bacteriol. **118**:70–82.
13. Sykes, J. 1971. Centrifugal techniques for the isolation and characterization of sub-cellular components from bacteria, p. 55–207. *In* J. R. Norris and D. W. Ribbons (ed.), Methods in microbiology, vol. 5B. Academic Press, Inc., New York.
14. Work, E. 1971. Cell walls, p. 361–418. *In* J. R. Norris and D. W. Ribbons (ed.), Methods in microbiology, vol. 5A. Academic Press, Inc., New York.

Section II

GROWTH

Introduction to Growth

RALPH N. COSTILOW

Starting with the elegant work of Robert Koch in the mid-19th century, methods for growing bacteria have been a primary concern of bacteriologists. Literally thousands of procedures and media have been developed for growing various species and strains for various purposes. However, all of these involve certain unifying concepts.

Thus, for any bacterium to be propagated for any purpose, it is necessary to provide the appropriate biophysical environment and biochemical nutrients. In Chapter 6 there are described methods to control biophysical factors such as pH, temperature, oxygen, anaerobiosis, and others. Chapter 7 includes nutrient requirements, compositions of defined and undefined media, and quantitative growth-response assays. The crafty manipulation of biophysical, biochemical, and biological factors is of primary importance in the enrichment from nature of specific types of bacteria and thence their isolation as pure cultures (Chapter 8).

The manner in which bacteria are cultivated varies widely, depending not only on the requirements of the organism but also upon the use to be made of the culture. Solidified media are used widely for isolating pure cultures of bacteria, estimating viable populations, and a variety of other purposes. Gelling agents and special techniques for solid and semisolid media are described in Chapter 9. In Chapter 10 there are described various liquid culture systems (batch, continuous, dialysis, and other special culture systems), harvesting and cleaning procedures, and growth yield calculations.

Growth is usually defined as the orderly increase in all biochemical components of an individual. With bacteria, the most obvious result of growth is the increase in the numbers of cells, i.e., multiplication. In Chapter 11 there are described various methods for measuring the numbers of cells and for employing mathematical tools in statistical and growth calculations.

Modern techniques for preservation of pure cultures are described in Chapter 12. The *Journal of Irreproducible Results* could be filled with data resulting from the failure of investigators adequately to preserve the specific strains of bacteria used by them. Some journals now insist that any culture used in a published report be deposited in a recognized culture collection. Unfortunately, this does not always eliminate the problem: the culture may have undergone mutation and selection during frequent transfers.

Among the many satisfactory methods for growing bacteria, the authors here have selected those which they have found most valuable. Where possible, highly condensed summaries are included, notably, the tables of medium composition in Chapter 7. In some areas, the methods commonly used are so extensive and variable that the authors chose to present them in terms of general principles along with selected examples and references. Common sources of error are also presented.

Chapter 6

Biophysical Factors in Growth

RALPH N. COSTILOW

Some biophysical factors affecting bacterial growth are controlled primarily by the constituents of the culture medium (pH, water activity, and osmotic pressure) and others, by the external environment (temperature, oxygen, and pressure supply). Another major factor (oxidation-reduction potential) is controlled by both the environment and the medium. All of these factors influence the growth rate, biomass produced, metabolic pattern, and chemical composition of bacteria. The control of pH, temperature, and oxygen supply is critical with every bacterial culture, whereas control of oxidation-reduction potential is of major importance in culturing obligately anaerobic bacteria.

6.1. pH

pH is a measure of the hydrogen ion (H^+) activity. In dilute solutions, the activity coefficient is very close to 1, and the H^+ activity is essentially equal to the concentration. In such instance, pH $\approx -\log [H^+]$. The pH scale ranges from 0 ($[H^+] = 1$ M) to 14 ($[H^+] = 10^{-14}$ M). Although some bacteria grow at pH 1.0 and others at pH 11, most species grow over a relatively narrow range of pH, and most grow best near neutrality (pH 7.0).

6.1.1. Measurement

Use a pH meter with a glass electrode for accurate pH determinations. See part 16.2.1 for a description of the hydrogen-ion electrode, procedures used for calibration of pH meters, and proper care of electrodes. Also see Munro (9) for details of the measurement and control of pH. Where accurate pH measurements are desired, observe the following precautions:

1. Adjust the temperature of the buffer used for standardization of the pH meter to the same temperature as the sample. The pH of buffers changes with temperature; e.g., the pH of standard phosphate buffer is 6.98 at 0°C, 6.88 at 20°C, and 6.84 at 37°C. See *Handbook of Chemistry and Physics* (16) for pH values of standard buffers at different temperatures.

2. If the sample is to be stirred with a magnetic stirrer during measurement, stir the calibrating buffer in the same manner.

3. Determine the pH of media after sterilization. Autoclaving media frequently results in a significant change in the pH. Even filtration may have a significant effect.

4. Check the pH of media that have been stored, just prior to use.

5. Some pH electrodes do not give accurate

pH readings with tris(hydroxymethyl)amino-methane (Tris) buffers. Be sure that the electrodes used with Tris buffers are recommended for such use by the manufacturer.

In instances where precision is not required, such as in preparation of routine media, the pH may be checked by use of pH indicator dye solutions or paper. By proper selection of either, the pH can be estimated within ±0.2 pH unit. Indicator solutions and papers, and their corresponding color standards for various pH ranges, are available from most scientific supply houses. Common pH indicators and their useful pH ranges are listed in Table 1.

A number of pH indicators are frequently incorporated into culture media to demonstrate pH changes during growth of bacteria. Select the appropriate indicator for the pH range of interest, and make sure that it does not inhibit growth of the organism. All of these pH indicators are poorly soluble in water and are used in media at very low concentrations (less than 0.01%).

6.1.2. Buffers

Many bacteria utilize and/or produce significant amounts of acidic or basic ions in solution. Unless acid production and consumption are balanced, large changes in pH may occur during growth. The pH of cultures is controlled to some extent by incorporating buffers (i.e., substances that resist change in pH) into the medium. Complete pH control is achieved by the automatic addition of acid or base.

pH buffers usually are mixtures of weak acids and their conjugate bases. In complex media, the acidic and basic groups of organic molecules (such as proteins, peptides, and amino acids) serve as buffers. Because of the variety of compounds present in such media, there may be some buffering action over a wide range of pH. However, the buffering capacity at any given pH

varies widely with the types of organic molecules present and their concentration.

In cultivating many bacteria, it is necessary to include a buffer system in the medium. Consider the following in selecting a buffer to use:

1. The pH desired.
2. Possible inhibitory or toxic effects of the buffer.
3. Possible utilization of the buffer by the bacterial culture.
4. Possible binding of di- and trivalent metal ions by the buffer.

Weak acids and bases buffer most effectively at the pH where they are 50% dissociated. This pH is equal to the pK_a (the negative logarithm of the dissociation constant) of the acid or base. The effective range of a buffer is ±1 pH unit from the pK_a. A number of compounds that have been used as buffers in growth media are listed in Table 2, along with their useful pK_a values at 25°C. Note that salts of citric, phthalic, and succinic acids buffer over a wide range of pH because these acids have more than one dissociable hydrogen ion. Buffers containing the compounds in Table 2 are prepared by one of the following general procedures:

1. Prepare a solution of the acid or base at twice the desired concentration, and titrate with either NaOH or HCl to the desired pH, using a pH meter. Dilute to final volume. For example, to prepare 1 liter of 0.1 M Tris-hydrochloride buffer at pH 7.6, make 500 ml of a 0.2 M solution of Tris, titrate with HCl to pH 7.6, and then dilute to 1 liter. Dilution will have only a small effect on the final pH. If this effect is important, make a final correction before adding the last few milliliters of water during dilution.

2. Prepare equimolar concentrations of the acidic and basic components of the system. Mix the two solutions in appropriate proportions to yield the desired pH. Use a magnetic stirrer and pH meter, and titrate one solution with the other. If the desired pH is near the pK, the ratio

TABLE 1. *pH indicators and their useful ranges of pH*

Indicator[a]	pH range	Color	
		Acid	Alkali
Thymol blue	8.0–9.6	Yellow	Blue
Phenol red	6.8–8.4	Yellow	Red
Bromothymol blue	6.0–7.6	Yellow	Blue
Bromocresol purple	5.2–6.8	Yellow	Purple
Methyl red	4.4–6.0	Yellow	Red
Bromocresol green	3.8–5.4	Yellow	Blue
Bromophenol blue	3.0–4.6	Yellow	Blue

[a] Prepare 0.04% solutions by solubilizing 0.1 g in the smallest possible volume (10 to 30 ml) of 0.01 N NaOH, and dilute to 250 ml.

TABLE 2. *pK values of compounds used in buffers*

Compound	Useful pK_a at 25°C
Citric acid	3.13, 4.76, 6.39
Phthalic acid	2.89, 5.51
Barbituric acid	4.01
Oxalic acid	4.19
Succinic acid	4.16, 5.61
Acetic acid	4.75
Monobasic phosphate	7.21
Tris(hydroxymethyl)aminomethane	8.08
Boric acid	9.24
Glycine	9.87

of acidic solution to basic solution is about 1:1; if 1 pH unit above the pK, the ratio is about 1:10; and if 1 pH unit below the pK, the ratio is about 10:1. For example, to prepare 100 ml of a 0.1 M sodium phosphate buffer at pH 7.0, mix solutions of 0.1 M NaH_2PO_4 (monobasic) and 0.1 M Na_2HPO_4 (dibasic) in approximately equal proportions, and then adjust to pH 7.0 by adding more of the appropriate solution as determined with a pH meter. To prepare about 100 ml of a phosphate buffer at pH 6.0, titrate 95 ml of the monobasic solution with the dibasic solution.

The most commonly used buffers in bacterial growth media are phosphate, Tris-hydrochloride, citrate, and acetate. These four systems will buffer over almost the entire range of pH permitting bacterial growth. Solutions used for their preparation and the appropriate volumes required to buffer at various pH values are given in Table 3. As noted above, the actual pH of any buffer is temperature dependent. Tables for pre-

TABLE 3. *Formulas for buffers frequently used in bacteriological media*[a]

pH	Value of x to be used in buffer formulas			
	Acetate buffer[b]	Citrate-phosphate buffer[c]	Phosphate buffer[d]	Tris-hydrochloride buffer[e]
4.0	41.0	30.7	—	—
4.2	36.8	29.4	—	—
4.4	30.5	27.8	—	—
4.6	25.5	26.7	—	—
4.8	20.0	25.2	—	—
5.0	14.8	24.3	—	—
5.2	10.5	23.3	—	—
5.4	8.8	22.2	—	—
5.6	4.8	21.0	—	—
5.8	—	19.7	46.0	—
6.0	—	17.9	43.85	—
6.2	—	16.9	40.75	—
6.4	—	15.4	36.75	—
6.6	—	13.6	31.25	—
6.8	—	9.1	25.5	—
7.0	—	6.5	19.5	—
7.2	—	—	14.0	44.2
7.4	—	—	9.5	41.4
7.6	—	—	6.5	38.4
7.8	—	—	4.25	32.5
8.0	—	—	2.65	26.8
8.2	—	—	—	21.9
8.4	—	—	—	16.5
8.6	—	—	—	12.2
8.8	—	—	—	8.1
9.0	—	—	—	5.0

[a] Data adapted from reference 3.
[b] Formula: x ml of 0.2 M acetic acid + (50 − x) ml of 0.2 M sodium acetate, diluted to 100 ml.
[c] Formula: x ml of 0.1 M citric acid + (50 − x) ml of 0.2 M Na_2HPO_4, diluted to 100 ml.
[d] Formula: x ml of 0.2 M $NaHPO_4$ + (50 − x) ml of 0.2 M Na_2HPO_4, diluted to 100 ml.
[e] Formula: 50 ml of 0.2 M tris(hydroxymethyl)-aminomethane + x ml of 0.2 M HCl, diluted to 100 ml.

paring other buffers and buffers at different temperatures are available elsewhere (2, 3, 9, 16).

When cultures are incubated in atmospheres enriched in CO_2, the carbonic acid bicarbonate buffer is very important. The concentration of carbonic acid (H_2CO_3) in the medium is directly dependent on the partial pressure of CO_2 in the atmosphere. The (Henderson-Hasselbach) equation for this system is

$$pH = pK' + \log \frac{HCO_3^-}{CO_2}$$

The pK' value is about 6.4 at 20°C. Therefore, the molar ratio of CO_2 to bicarbonate (HCO_3^-) is about 1:1 at this pH, and CO_2-bicarbonate buffers are effective over a pH range of about 5.4 to 8.0. If the pH of the medium is below 5.0, essentially no bicarbonate is present, and an increase in CO_2 concentration in the atmosphere will result in a decrease in pH due to formation of carbonic acid. In an atmosphere of 50 to 100% CO_2, the medium must be very alkaline to maintain pH levels near neutrality. The high salt concentration which results may be toxic to some bacteria.

After appropriate substitution, the Henderson-Hasselbach equation can be solved to determine the amount of bicarbonate to add to a medium for buffering at a given pH when the atmosphere contains various percentages of CO_2. For this calculation, use the equation

$$\log(HCO_3^-) = pH - pK' + \log P(\alpha)(\%CO_2)(5.87 \times 10^{-7})$$

where P is the atmospheric pressure in millimeters of Hg, α is the solubility of CO_2 in milliliters per milliliter of water, and $\%CO_2$ is the percent CO_2 in the atmosphere. If atmospheric pressure is between 720 and 760 mmHg (ca. 96.0 and 101.3 kPa), the variation has little effect. Therefore, in most instances the following simplified equation can be used:

$$\log(HCO_3^-) = pH - pK' + \log \alpha(\%CO_2)(4.35 \times 10^{-4})$$

Values for α and pK' at various temperatures are given in Table 4. See Umbreit, Burris, and

TABLE 4. *Solubility of CO_2 in water and pK' values for carbonic acid at various temperatures*[a]

Temp (°C)	CO_2 solubility[b]	pK'
20	0.878	6.392
25	0.759	6.365
30	0.665	6.348
35	0.592	6.328
40	0.530	6.312

[a] Data from Umbreit et al. (35).
[b] Milliliters of CO_2 per milliliter of water at 1 atm.

Stauffer (35) for a detailed discussion of carbonate buffers.

The pH of media used for growing bacteria that produce large amounts of acid (e.g., lactic acid bacteria) cannot be controlled by soluble buffers. However, the acid can be neutralized as fast as it is formed by the addition of finely ground $CaCO_3$ (chalk) to the medium. In fact, a good way of visualizing acid production by colonies growing on agar plates is to add 0.3% finely ground chalk to the agar. Make sure that the chalk is well suspended in the agar when the plate is poured. Colonies which produce acid will have clear zones around them. The acid can also be withdrawn by use of a dialysis culture system (10.3.4).

6.1.3. Continuous Control

In instances where the pH of a culture must be controlled within a narrow range, an automatic pH control system should be employed. Such a system should have a controller with a pH meter, steam-sterilizable pH electrodes, controls for setting the desired pH limit, and pumps for addition of acid and base. Most systems are linked to a recorder to provide a continuous record. Automatic systems are available from a number of sources (e.g., New Brunswick Scientific Co., New Brunswick, N.J.; The VirTis Co., Inc., Gardiner, N.Y.; and Cole-Palmer Instrument Co., Chicago, Ill.). Steam-sterilizable pH electrodes, both dual and combination, are available from a number of sources (e.g., Beckman Instruments, Inc., Fullerton, Calif.; Ingold Electrodes, Lexington, Mass.; and Leeds and Northrup, North Wales, Pa.).

Check pH control systems at frequent intervals for accuracy, using a separate pH meter. Problems which may be encountered are (i) improper or inadequate grounding of the system, (ii) aging of the pH glass electrode due to steam sterilization, and (iii) contamination of the reference-electrode diaphragm by continuous exposure to constituents of the media or metabolic products of the bacteria. The most common contaminants of diaphragms are proteins and silver sulfide. The latter is formed by reaction of the silver chloride in the reference electrode with sulfur-containing substances in the medium. Remove the proteins by soaking the electrode in a protease (e.g., pronase) solution, and remove the silver sulfide deposits with an acidic solution of thiourea.

6.2. WATER ACTIVITY AND OSMOTIC PRESSURE

Water must be available to cells for metabolism and growth. However, the mere presence of water in a medium does not ensure its availability, which is determined by the water activity (a_w) of the medium. The a_w represents the fraction of the total water molecules that are available and is equal to the ratio of the vapor pressure of the solution to that of pure water (p/p_0), which is equal to the fractional relative humidity (RH/100) of the atmosphere above the medium at equilibrium (10–12; 14). Temperature variation within the range in which bacteria grow has little effect on the a_w of the medium.

The a_w is related to the true osmolality of the medium. This relationship is

$$a_w = 55.51/(vm\phi + 55.51)$$

where v is the number of ions generated per molecule of solute, m is the molality of solute, and ϕ is the molal osmotic coefficient. One kilogram of water contains 55.51 mol.

The osmotic pressure (π) of a solution, expressed in atmospheres, is dependent on the concentration of solute as follows:

$$\pi = \frac{vn\phi RT}{V'}$$

where n is the number of moles of solute present, R is the gas constant (0.0821 liter atm mol^{-1} deg^{-1}), T is the absolute temperature (°C + 273°C), and V' is the volume of solvent. The relationship between π and a_w is

$$\pi = -\frac{RT}{\bar{V}} \ln a_w$$

where $\bar{V}$ is the volume of 1 mol of water. For a more detailed presentation of these relationships, see references 10, 12, and 14.

The minimum a_w at which bacteria grow varies widely, but the optimum values for most species are greater than 0.99. Some halophilic bacteria require sodium ions and grow best in the presence of high concentrations of sodium chloride ($a_w \approx 0.80$). Variations in a_w may affect growth rates, cell composition, and metabolic activities of bacteria. Many solutes have very specific effects at concentrations below that required to limit the availability of water. Therefore, determinations of limiting a_w values should be conducted in media adjusted with more than one solute. See references 12, 14, and 32 for reviews of the effect of a_w on bacteria.

A number of methods have been developed for measurement of water activity. Special instruments available for this purpose are the Brady Array (Thunder Scientific Co., Albuquerque, N.M.), the Relative Humidity Indicator, model 400D (General Eastern Corp., Watertown, Mass.), the Hygrometer (Hygrodynamics, Silver Spring, Md.), and the Sina-Scope (Sina Ltd., Zurich, Switzerland; marketed in the

United States by Beckman Instruments, Inc., Cedar Grove, N.J.). Advantages and disadvantages of these instruments were reviewed recently (11, 14).

Results of a collaborative study (27) of methods for determining a_w indicated that there is considerable variability among the values obtained by different procedures and that the significance of values reported beyond the second decimal place is doubtful. A statistical analysis (34) of measurements made with the electronic Sina-Scope instrument (over the range of 0.755 to 0.967 a_w) indicated a variation of less than ±0.01.

The most commonly used instruments for measuring osmotic pressure are based on the reduction of vapor pressure (e.g., Wescan Instruments, Santa Clara, Calif.) or freezing point depression (e.g., Advanced Instruments, Needham Heights, Mass.). Other instruments measure osmotic pressures developed across a semipermeable membrane (e.g., Schleicher & Schuell Co., Keene, N.H.). While some of these are calibrated for direct readout, several standard solutions should be run simultaneously with unknowns. Instruments are available to measure different ranges of osmotic pressure between 0 and 4 osmolal (the molality of an ideal solution having the same a_w and osmotic pressure as the solution in question).

The a_w and osmotic pressure of bacteriological media are best controlled by adding non-nutrient solutes such as sodium chloride, potassium chloride, sodium sulfate, or mixtures of such salts. Scott (32) detailed methods for calculating the a_w of nutrient media and for calculating the required concentrations of salt(s) to attain a given a_w value.

Considerations of osmotic pressure relative to bacterial permeability and transport are presented elsewhere in this manual (Chapter 19).

6.3. TEMPERATURE

The temperature of incubation dramatically affects the growth rate of bacteria since it affects the rates of all cellular reactions. In addition, the temperature may affect the metabolic pattern, the nutritional requirements, and the composition of bacterial cells.

The growth rates of most bacteria respond to temperature in a manner similar to the response shown in Fig. 1, top. The useful range of temperature near the optimum is usually quite narrow, and the maximum temperature for growth is only a few degrees (3 to 5°C) above the optimum. In contrast, the minimum temperature for growth may be 20 to 40°C below the optimum. It is frequently necessary to incubate a bacterial

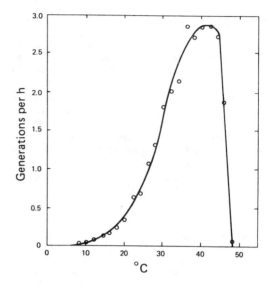

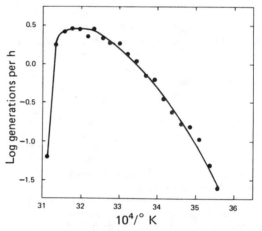

FIG. 1. *Effect of temperature on the growth rate of Escherichia coli, as indicated by a linear plot (top) and an Arrhenius plot (bottom). Data from Ingraham (26).*

culture for a prolonged period to observe the growth rate at minimum temperatures; for example, the *Escherichia coli* culture studied by Ingraham (26) required over 41 h for one generation at 8°C, as compared to 21 min at the optimum temperature range.

Over a limited range of temperature below the optimum, the changes in growth rates of a bacterium are comparable to the responses of chemical reaction rates to temperature. Thus, if the logarithm of the growth rate is plotted against the reciprocal of the absolute temperature (an Arrhenius plot) (Fig. 1, bottom), a linear slope is observed in a limited range (10, 26, 31). For most bacteria, the temperature coefficient (Q_{10}) for the growth rate in this limited range is about 2; i.e., the growth rate doubles with a 10°C increase in temperature. The slope of an Arrhenius plot of growth rates becomes essentially zero at a range near the optimum and vertical at ranges near the maximum and minimum temperatures for growth.

Generally, bacteria not only grow slowly but also die rapidly at temperatures a little above the maximum for growth. Consequently, incubation of a bacterial culture at temperatures above its optimum requires precise temperature control. To be safe, therefore, incubate a culture below its optimum to an extent determined by the variability of the incubator.

The temperature within ordinary convection-type air incubators will vary several degrees Celsius, or even more if opened frequently. Water-jacketed incubators maintain a more constant temperature, but variations in temperature of 1°C are not uncommon. For precise temperature control, use stirred water baths; in these, temperature variations are less than a few hundredths of a degree Celsius. For accurate measurements of the effect of temperature on the growth rate of bacteria, incubation in a water bath is essential.

Practical procedures and precautions relative to temperature control are as follows:

1. Measure temperatures at various positions in incubators, using thermometers inserted into flasks or tubes of water.

2. Air convection incubators may vary widely in temperature and humidity. Use circulation fans, and check to see that they are operating efficiently. Position cultures in the incubators so that "dead" air spaces are not created. Do not open such incubators more frequently than absolutely necessary.

3. Surface cooling due to evaporation of media must be controlled in critical studies of temperature effects. When possible, use tightly stoppered culture containers. Where free air exchange is necessary, as for aerobic bacteria, use incubators with provision to obtain high relative humidities. This is exceedingly important at high incubation temperatures.

4. Use 50% polyethylene glycol antifreeze or light silicone oil in baths used for high-temperature incubators.

To assess the temperature response of a bacterium, measure growth rates at a number of carefully controlled temperatures. Plot these data as illustrated in Fig. 1. If the general shape of the resulting curve is in great variance with the normal pattern, recheck some of the more critical determinations.

A number of temperature-gradient incubators have been described and were reviewed by Patching and Rose (31). These provide for incubation temperatures which exceed the entire temperature range of growth for a bacterium. Different designs allow for incubation of organisms with various oxygen requirements. If large numbers of determinations are to be made, the construction and use of a gradient incubator may be desirable. A commercial source for a temperature gradient incubator is not known, but one can be constructed easily in a machine shop.

6.4. PRESSURE

High hydrostatic pressure adversely affects the growth and survival of bacteria. However, little or no effect on most species is observed at pressures from 1 to 100 atm, and no pure cultures of obligatory barophilic bacteria have been described (8). Therefore, studies of cultures under high pressures are limited, and the methodology will not be described here. Marquis (8) described experimental procedures for applying high pressures to cultures. Taylor and Jannasch (33) solved a major problem in this area by developing an apparatus which allows sampling and additions to cultures under high hydrostatic pressure without exposure of the cultures to pressure changes.

6.5. OXYGEN

Aerobic and facultative bacteria utilize dissolved oxygen, and their metabolism and growth rates are dependent on the concentration of oxygen in solution. Dissolved oxygen is a nutrient as much as is glucose or ammonia. However, unlike most nutrients, oxygen is relatively insoluble (<10 mg/liter). Therefore, oxygen may quickly become limiting in liquid bacterial cultures unless special precautions are taken to ensure that it is supplied and dissolved continuously during growth.

Principal factors to consider in supplying dissolved oxygen to liquid cultures (aeration) are as follows:

1. Availability of oxygen to the gas-liquid interface. The opening of the culture vessel must be sufficiently large, and stoppered or covered with porous material so as to allow maximum exchange of the atmospheres inside and outside the vessel.

2. Maintenance of large gaseous surface to liquid volume ratios. In static cultures, this can only be accomplished by use of very shallow liquid volumes. Either rotary or reciprocal shaking of flasks greatly increases the surface area of a given volume of medium, and this effect can be further enhanced, when rotary shaking is

used, by creation of turbulence in the flask either by making indentations ("dimples") in the side of the flask or by placing a stainless-steel coil-spring in the broth. The oxygen-solution rate obtained in a shaken flask decreases rapidly as the volume of culture increases.

With large volumes of cultures (>0.5 to 1 liter), it is usually necessary to force air through (sparge) the liquid. The efficiency of oxygen dissolution in sparged cultures depends primarily on the number and smallness of the air bubbles, and the time of contact between the bubbles and the liquid. Use of gas spargers with small pore sizes will result in small bubbles, but may induce foaming. The time of contact can be maximized by using culture vessels with high height to diameter ratios, by stirring the cultures, and by using baffles in the flask.

See Chapter 10 for more details of aeration systems for cultivating bacteria. Pirt (10) and Wang et al. (15) have reviewed the principles of aeration in detail.

6.5.1. Dissolved Oxygen

Dissolved oxygen is most conveniently measured polarographically with an oxygen electrode. The principles involved and operating procedures are described elsewhere in this manual (16.2.3). Procedures for using such an instrument for determining the rates of oxygen solution in cultures are described in detail by Pirt (10), Wang et al. (15), and Hitchman (4).

6.5.2. Oxygen Absorption Rate

Oxygen absorption rates (OAR) are expressed in millimoles per liter per minute and are readily determined chemically by measuring the rate of oxidation of sulfite to sulfate by dissolved oxygen in the presence of a copper catalyst. This method will provide information on the efficiency of an aeration system. However, the actual OAR in broth cultures may be quite different from that observed in sodium sulfite solutions. Procedures for measuring OAR in media in the presence or absence of biomass are described elsewhere (1, 10, 15). The following is a simple method (23) for measuring OAR by sulfite oxidation:

1. Instead of a growth medium use the appropriate volume of a sodium sulfite solution containing 0.001 M copper sulfate at the same temperature and under the same conditions of aeration to be used for cultures. The appropriate concentration of sodium sulfite to use depends on the aeration conditions under study (see below).

2. At various intervals after initiating aeration, pipette duplicate 5-ml samples of the sulfite solution into 1 by 10 inch (2.5 by 25 cm) test tubes each containing a small pellet of dry ice. The rising CO_2 stirs the sample during titration and blankets the sample to prevent further oxidation. If partial freezing occurs, warm the sample before completing the titration.

3. Add a few drops of a starch solution, and titrate to a permanent blue endpoint with a freshly standardized solution of iodine. Standardize against a standard sodium thiosulfate solution. The normality of the iodine solution should be about one-fifth that of the sulfite solution.

4. Calculate the OAR as follows:

$$OAR = \frac{\text{ml of titration difference} \times \text{N of iodine}}{4}$$
$$\times \frac{1,000 \text{ ml}}{5 \text{ ml}} \times \frac{1}{\text{min}}$$

Note that the concentrations of sulfite and iodine most appropriate for any given efficiency of aeration may be calculated by substitution in this equation. For example, if the OAR is about 3.0, the use of 0.15 N iodine solution would result in a titration difference of about 4 ml for a 10-min interval. The initial sodium sulfite concentration should be about five times the iodine concentration, or 0.75 N.

6.6. ANAEROBIOSIS

There is no sharp line of demarcation between aerobic and anaerobic environments. There are bacteria which grow either with or without oxygen (**facultatives**) and those which prefer low oxygen tensions for growth (**microaerophiles**). *Spirillum volutans* is an obligate microaerophile in that it requires the presence of oxygen at low concentrations (~3 to 9% O_2) to initiate growth. See 9.3.1. for methods of cultivating microaerophilic bacteria.

While obligate **anaerobes** are usually defined as bacteria which are unable to grow in the presence of dissolved oxygen, in practice bacteria which are unable to grow on or near the surface of solid or semisolid media in air at atmospheric pressure are considered to be anaerobic. Tolerant anaerobic bacteria are able to grow well on the surface of agar plates with low but significant levels of oxygen in the atmosphere (24), whereas more stringent anaerobic bacteria die almost immediately on exposure to such an environment. Removing oxygen by boiling or replacing oxygen with another gas will not permit the growth of the latter type of bacteria.

The **oxidation-reduction (redox) potential (Eh)** provides the most useful scale for measuring the degree of anaerobiosis (6). Simply stated, the Eh is a measure of the tendency of a solution to give or take up electrons (i.e., to be oxidized or reduced). Measurements of Eh are expressed in units of electrical potential difference (i.e., volts): the more positive the number of volts, the higher the relative concentration of oxidant to reductant in solution, and vice versa. The capacity of a medium to resist changes in Eh is called **poising**, analogous to the buffering of pH.

In normal laboratory media, oxygen is primarily responsible for raising the Eh, and positive Eh values resulting from dissolved oxygen inhibit all anaerobic bacteria. However, positive Eh values created by the presence of other chemicals in a medium may not affect the growth of even stringent anaerobes (30, 36). Therefore, no specific tolerances for Eh can be set for various anaerobic bacteria. However, in standard media, most anaerobic bacteria are inhibited at Eh values higher than −100 mV. Some will not initiate growth at potentials higher than −330 mV; the theoretical concentration of oxygen at this Eh is 1.48×10^{-56} molecules per liter (6). Even at a potential of −140 mV, the calculated oxygen level is only 1.5×10^{11} molecules per liter (10), or less than 1 molecule of oxygen per cell in many bacterial cultures. This illustrates the importance of creating reducing conditions beyond those achieved by merely removing oxygen from media for the cultivation of the more stringent anaerobic bacteria.

6.6.1. Eh Measurement

The Eh of a solution is most accurately measured electrometrically. Detailed procedures and precautions for such determinations are described in a review by Jacob (7). The measurements are not believed to be sufficiently meaningful to the general bacteriologist to be described here.

Redox dyes are used much more widely to estimate the Eh of media and of cultures, especially in anaerobic bacteriology. The useful dyes are reversibly oxidized and reduced, and are colored in the oxidized state and colorless in the reduced state. Each dye becomes reduced at a different Eh, and the Eh at which it is 50% oxidized or reduced at pH 7.0 is the **standard redox potential** (E_0').

Some useful dyes and their standard redox potentials are listed in Table 5. The total range of Eh covered (completely oxidized to completely reduced) by a redox dye having a two-electron charge is about 120 mV at constant pH. For example, methylene blue ($E_0' = 11$ mV at pH 7.0) would be almost completely oxidized (full color) at an Eh of 71 mV and almost com-

TABLE 5. *Standard redox potentials of dyes at pH 7.0 and 30°C[a]*

Dye	E_0' (mV)
Methylene blue	11
Toluidine blue	−11
Indigotetrasulfonate	−46
Resorufin[b]	−51
Indigotrisulfonate	−81
Indigodisulfonate	−125
Indigomonosulfonate	−160
1,5-Anthraquinone sulfate	−200
Phenosafranine	−252
Benzyl viologens	−359

[a] Data from Jacob (7), where a more complete list is to be found.
[b] Product of first reduction step of reasurin.

pletely reduced (colorless) at an Eh of −49 mV (7).

The redox potential of dyes varies with the pH. For example, the Eh at which methylene blue is 50% reduced is 101, 11, and −50 mV at pH 5.0, 7.0, and 9.0, respectively (8). The exact change in the E_0' with change in pH is variable among dyes. However, the E_0' for most dyes will increase by 30 to 60 mV per pH unit of decrease in pH, and vice versa.

A number of redox dyes are toxic to certain bacteria even at very low concentrations. Unless a dye is known to be nontoxic, use it only in control tubes or flasks of medium handled in exactly the same manner as the medium used for cultures. In such instances, use the dyes at the lowest possible concentration because the dye may alter the Eh of certain media in which the redox is weakly poised (7).

Many anaerobic bacteria are able to rapidly reduce the Eh of a medium, and their ability to grow in a given medium will depend on the physiological state and size of the inoculum. A lag period after inoculation may be observed while the inoculated cells reduce the potential to a level at which they can grow. This can best be observed by incorporation of a nontoxic Eh indicator into the medium. Resazurin is widely used for this purpose at a concentration of 0.1 to 0.2 mg/100 ml. This indicator first undergoes an irreversible reduction step to resorufin, which is pink at pH values near neutrality. The E_0' of resorufin is −51 mV at pH 7.0, so it becomes colorless at an Eh of about −110 mV (7).

When very strict anaerobic conditions are required, phenosafranine ($E_0' = -252$ mV) may be incorporated into the medium if it does not prove to be inhibitory. Titanium III citrate is both a reducing agent ($E_0' = -480$ mV) and an Eh indicator (37); it becomes colorless when completely oxidized (its use will be described along with other reducing agents below).

6.6.2. Reducing Agents

Reducing agents are added to most anaerobic media to depress and poise the redox potential at optimum levels. Such agents must be nontoxic at the concentration used and result in final potentials low enough for the particular organism under study. The reducing agents most widely used in anaerobic cultures are listed in Table 6. All of these compounds have an E_0' low enough to completely reduce resazurin, but only those with an E_0' less than -300 mV should be used in media for the most stringent anaerobes.

For maximum effectiveness, prepare oxygen-free stock solutions, sterilize, and store the reducing compounds under O_2-free gas (see procedures for preparation of reduced media, below). Add appropriate concentrations of the reducing compound to the medium just prior to use.

Bacteria which are facultative to oxygen may be used to reduce the Eh of media for anaerobic bacteria. Growth of a bacterium such as *E. coli* in the medium prior to inoculation will scavenge all residual oxygen and reduce the Eh to low levels if the culture is completely protected from oxygen. The *E. coli* may be heat killed prior to inoculation with the anaerobe. It is necessary to be sure that the facultative bacterium used does not interfere with the growth of the anaerobe either by competing for nutrients or by producing toxic products.

6.6.3. Techniques for Nonstringent Anaerobes

Procedures for preparing media and for cultivating and transferring nonstringent anaerobic bacteria (including most clostridia) are not difficult (13). All operations can be conducted in the normal atmosphere by taking a few precautions to prevent excessive exposure of media and bacterial cells to oxygen. Following are some useful procedures for such cultures:

1. Prepare media in containers which provide for a small surface to volume ratio for the medium.

2. Screw-cap bottles or vials completely filled with freshly boiled media are excellent for the prolonged incubation of non-gas-producing anaerobes (e.g., photosynthetic bacteria).

3. Make liquid media semisolid by adding 0.05 to 0.1% agar. This reduces convection currents.

4. Pour a thick layer of Vaspar (50% petrolatum, 50% paraffin) on the surface of inoculated media in tubes.

TABLE 6. *Reducing agents for anaerobic media*

Compounds	E_0' (mV)	Concn in media
Sodium thioglycolate[a]	< -100	0.05%
Cysteine·HCl[a]	-210 (22)	0.025%
Dithiothreitol[a]	-330 (22)	0.05%
H$_2$ + palladium chloride (18)	-420	
Titanium III·citrate[b] (37)	-480	0.5–2 mM
Na$_2$S·9H$_2$O[a] (or H$_2$S)	-571 (6)	0.025%
Cysteine·HCl + Na$_2$S· 9H$_2$O[a]		0.025% + 0.025%

[a] Stock solutions may be autoclaved and stored under O_2-free gas.

[b] Add 5 ml of 15% titanium trichloride solution to 50 ml of 0.2 M sodium citrate, and neutralize with a saturated sodium bicarbonate solution. Filter sterilize, and add 30 ml/liter of sterile oxygen-free medium in the appropriate concentration. The titanium III citrate complex is blue-violet when reduced and colorless when oxidized.

5. Autoclave media promptly after combining the ingredients with a reducing agent.

6. Do not store media for prolonged periods at any temperature, and never store in a refrigerator. Oxygen increases in solubility as the temperature decreases.

7. Include resazurin (0.1 mg/100 ml) in the medium. If the top one-third of the medium is pink when ready for use, boil and cool the container before inoculation.

8. Use large inocula when possible.

9. Dilute cultures in tubes of growth medium or in freshly autoclaved diluent that contains a reducing agent.

10. Some aerotolerant anaerobes (e.g., *Clostridium perfringens*) may be diluted and spread on the surface of a fresh agar medium containing a reducing agent in a petri dish, provided that the petri dish is incubated in a jar or chamber free from oxygen. However, cells of some common anaerobes will not tolerate even this brief exposure to air.

11. Petri plates with freshly prepared media inoculated with cells of aerotolerant anaerobes or with spores of anaerobes must be incubated in an anaerobic environment. One of the following systems may be used.

Vacuum desiccator: Use only for reasonably aerotolerant anaerobes. Evacuate three times to a vacuum (about 50 cm of Hg), refilling with the gas desired. One may use nitrogen, argon, hydrogen, or mixtures of these, together with 5 to 10% carbon dioxide (some anaerobes require carbon dioxide).

Anaerobic jars with catalyst: The jar used most commonly is the Torbal, model A-3 (The Torsion Balance Co., Clayton, N.J.). Evacuate as above and fill three times with either H$_2$ or a gas mixture that contains a significant percentage of H$_2$. Check the catalyst at intervals to ensure that it is active by exposing the cold catalyst to a stream of H_2; an active catalyst will become hot within 1 min. Reactivate catalysts by heating at 160°C for 1 h.

Bags or jars with H_2 + CO_2 generators: The system most commonly used is the GasPak (BBL Microbiology Systems, Cockeysville, Md.). No evacuation is required. After activation of the gas generation unit with water, close the bag or jar quickly and observe for condensation of water within 30 min. Methylene blue indicator strips may be sealed in transparent bags or jars and observed for decoloration. Check and reactivate the catalyst as described above.

6.6.4. Hungate Technique for Stringent Anaerobes

Many anaerobes found in the intestinal tract, rumen, and other natural anaerobic environments require very low redox potentials to initiate growth. Extraordinary precautions must be taken to protect media and cells from even brief exposure to oxygen. The fundamentals of two general procedures for doing this are described in this and the following part.

A roll tube technique was described by Hungate (25) for the isolation and maintenance of pure cultures of stringently anaerobic bacteria, and many modifications of it have been developed. Its major advantages are that it requires little special apparatus and allows the use of unique atmospheres for cultivating specific groups such as the methanogens (17). Clear, well-illustrated descriptions of various modifications were published by Hungate (6), Macy, Snellen, and Hungate (28), Wolfe (17), and Miller and Wolin (29). The *Anaerobe Laboratory Manual* (5) published by the Virginia Polytechnic Institute and State University details a modification for handling large numbers of anaerobic cultures rapidly. Essentially all the supplies and equipment required for the "V.P.I. Anaerobic Culture System" are commercially available (Bellco Glass, Inc., Vineland, N.J.).

Even the most detailed descriptions of some modifications of the Hungate technique are frequently difficult to master without demonstration. It is best to visit a laboratory where the technique is in use. Therefore, only the basic steps and simple procedures used in the technique will be described here. They are as follows.

Removal of oxygen from gases.

The gases that are used to replace air are treated to remove all oxygen. This may be accomplished in one of the following ways:

1. Pass the gas through a column or oven containing heated (350°C) copper (6). When the copper turns black, it is completely oxidized. Regenerate the copper to the reduced form by purging it with gas containing at least 3% H_2. (*CAUTION:* Do not use pure H_2 unless all O_2 is swept from the column first, or an explosion may result.)

2. Use about 3% H_2 in the gas, and pass the gas through a Deoxo purifier (Matheson Scientific, Inc., Rutherford, N.J.) inserted in the line.

3. Bubble the gas through a solution of titanium III citrate prepared as described in Table 6. Use a gas washing bottle. The solution will be colorless when completely oxidized, and must be replaced.

Preparation of prereduced media.

Combine the heat-stable ingredients of the medium (omit the reducing agent) in a round-bottom flask and add a glass boiling chip. Boil the solution gently while passing a stream of oxygen-free gas over the surface. Continue the gassing while the medium cools and is dispensed or stoppered. Choose carbon dioxide, argon, or either of these gases plus 3% hydrogen depending on the medium and anaerobic species to be cultivated. Use a probe such as illustrated in Fig. 2 for the gassing. If the medium is to be used soon after autoclaving, the reducing agent may be added at this time provided it is heat stable; otherwise, it should be added just before inoculation. Dispense the medium into tubes or bottles equipped with black rubber stoppers (butyl rubber is best) while exercising the following precautions:

1. Maintain a constant flow of oxygen-free gas over the surface of the medium during trans-

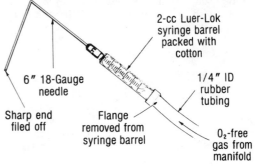

FIG. 2. *Gassing probe (cannula) for Hungate technique. At least two are needed. After assembly, autoclave the unit to sterilize the cotton. Thereafter, sterilize the needle in a flame. Adjust the flow rate of CO_2 or other oxygen-free gas (not more than 3% H_2) by directing it at a Bunsen burner flame; the flow should make a visible dent in the flame. Tubes or flasks must be gassed constantly when open.*

fer, and flush each tube or bottle to receive medium with the gas before and during the transfer.

2. Fill the pipette used for transferring medium with the gas prior to drawing medium into it. Use a rubber mouth tube attached to the pipette or a pipette bulb for transfer. Do not draw gas or liquid into the pipette faster than gas is entering the vessel.

3. After transfer, stopper the tube or bottle *without* allowing the entrance of air. (This is a critical step and requires practice.) Place the stopper alongside the needle in the mouth of the tube and continue gassing for a few seconds. Withdraw the needle rapidly while pushing in the stopper. Seat the stopper, using a twisting motion.

Alternatively, the flask of oxygen-free medium may be stoppered with a butyl rubber stopper as described above; the stopper is clamped or wired in place, and the medium is autoclaved in bulk (21). It must then be dispensed as described above, but using aseptic technique. This procedure increases the risks of contamination and of permitting traces of oxygen to enter the sterile medium.

For sterilization, the stoppers of tubes or bottles must be held in position. If normal-shaped rubber stoppers are used, they must be clamped in place. A special press for holding an entire rack of stoppered culture tubes is available (Bellco Glass, Inc.). Clamps for single tubes may be constructed from coat hangers (17). Hungate-type anaerobic tubes equipped with a flanged butyl rubber stopper and a screw cap with a 9-mm opening (Bellco Glass, Inc., no. 2047) to hold the stopper in place do not require a clamp. Tubes or bottles with serum bottle necks equipped with a butyl rubber septum stopper may be used (29), and the stopper may be held in place with an aluminum seal (Wheaton Scientific, Millville, N.J.).

Inoculation and transfer.

When either Hungate tubes (6) or tubes or bottles having serum bottle necks (29) are used, inoculation and transfer of cultures may be accomplished with disposable syringes and hypodermic needles. Before use, the syringes are washed with sterile oxygen-free gas by drawing in and expelling gas that flows through a sterile tube or bottle. Tubes having regular-shaped rubber stoppers are inoculated by quickly inserting a sterile gassing cannula in beside the stopper as it is removed, adding inoculum, and restoppering as described above. The specimen or culture from which the inoculum is taken should also be under a stream of oxygen-free gas and should be adjacent to the tube of media to be inoculated.

Fill serological or Pasteur pipettes with sterile gas prior to transfer. Transfer can also be made with stainless-steel or platinum loops. The V.P.I. Anaerobic Culture System (Bellco Glass, Inc.) is very useful for rapid inoculation of different media with one culture (5).

Roll tubes.

Instead of petri dishes, roll tubes are used for the isolation of single colonies and for the estimation of viable populations of stringent anaerobes. Prepare agar medium as above and dispense in anaerobic tubes. Cool the melted agar to 45°C while the stoppers are still clamped in place. If the tubes of medium have been prepared in advance, be sure that stoppers are clamped before boiling. Transfer samples of specimens or dilutions made in prereduced medium or diluent to tubes as described above, mix with media but avoid frothing, and then roll the tubes under a cold-water tap until the agar solidifies. Try to coat the walls of the tube uniformly. A mechanical spinner that simplifies the procedure is described in the *Anaerobe Laboratory Manual* (5) and is available commercially (Bellco Glass, Inc.).

Roll tubes may be streaked with a specimen by starting at the bottom of a prepared roll tube under a gassing cannula and rotating the tube as the streaking loop is drawn straight up (see 8.4.1). A device for rotating a tube for streaking is also available (Bellco Glass, Inc.).

6.6.5. Anaerobic Chambers for Stringent Anaerobes

Aranka et al. (19) and Aranka and Freter (18) described a flexible plastic anaerobic glove-box chamber which is as efficient as the roll tube method in isolating anaerobic bacteria. This chamber, commonly known as the "Freter anaerobic chamber," has been tested and used in many laboratories. Even the very sensitive methanogenic bacteria can be safely handled in such a way (20). A number of glove-box chambers are now available from different commercial sources (e.g., Coy Laboratory Products, Inc., Ann Arbor, Mich.).

The primary advantages of chambers are that: they permit use of standard bacteriological techniques including spread plates, replica plating, and antibiotic sensitivity testing; they allow preparation of media in a conventional manner; and they require no special training to operate. However, a sizable initial investment is required, and a significant amount of laboratory space is occupied. Also, there is some inconvenience in working with gloves, and it is necessary to anticipate media needs well in advance of use.

The combined use of the Hungate techniques and a properly functioning anaerobic chamber makes it possible to do almost any type of common experiment with the most stringent anaerobic bacteria.

Following are some precautions and considerations in the use of an anaerobic chamber:

1. Chambers are available which can be heated to serve also as an incubator. At 30 or 37°C chamber temperature, one's hands become hot, and perspiration makes it difficult to remove the gloves. Even with a chamber at room temperature, it is helpful to use nylon gloves under the rubber gloves when working for long periods. In many cases, it is preferable to have a small incubator inside the chamber. Alternatively, inoculated petri dishes or other cultures may be placed in jars, which can be sealed and removed from the chamber for incubation.

2. Gas mixtures used in the chamber must contain 10% hydrogen. The most commonly used mixtures are 5% CO_2 and 10% H_2 in nitrogen, or 10% H_2 in nitrogen. If the bacteria under study require CO_2, necessary precautions should be taken to ensure that the CO_2 does not become limiting. Mixtures may be purchased (Matheson Scientific), but they are expensive. A gas-mixing apparatus is recommended (Matheson Scientific).

3. Open flames cannot be used in the chamber. Install an electric incinerator (e.g., Bacticinerator, Scientific Products).

4. Installation of a forced-air filter such as those used in germfree hoods (Standard Safety Equipment Co., Palatine, Ill.) will help prevent contamination.

5. Change the palladium catalyst at frequent intervals. Once a week is usually adequate, but twice weekly may be necessary when there is heavy usage. Unless poisoned by H_2S or high moisture levels, the catalyst can be regenerated indefinitely by heating at 160°C for 2 h. Cool and return the catalyst to the chamber promptly after rejuvenation. Check the catalyst for activity occasionally by directing a stream of H_2 over the cold catalyst, which should heat up quickly. (CAUTION: Never expose hot catalyst to hydrogen because a violent reaction may occur.)

6. H_2S irreversibly poisons the catalyst. Whenever possible, grow H_2S-producing cultures in closed containers and open them outside the glove box. To remove H_2S from the chamber atmosphere, fill a tray with activated charcoal (8 to 12 mesh) on a layer of cheesecloth, and place under the palladium catalyst tray to trap any H_2S. This treatment reportedly will reduce the H_2S level in a chamber from 100 μl/liter to 1 μl/liter in 30 min (20).

7. Place a relative humidity indicator and a large tray of silica gel in the chamber. When the humidity exceeds 50%, the silica gel should be changed. Recycle the gel by heating at 160°C for 1 h.

8. When possible, keep agar plates in plastic bags and keep media in screw-cap vials or bottles to minimize evaporation.

9. Keep the anaerobic chamber under positive pressure. There is some gas exchange through common plastics, and even more through polycarbonate plastics or rubber. If the chamber is used frequently, the concentration of H_2 will remain high enough in the chamber as a result of opening and flushing the entry lock. It is necessary to add gas mixture about twice a month to maintain good positive pressure. Loss of pressure is readily observed in flexible plastic chambers.

10. Watch for leaks. Whenever the positive pressure is lost unusually fast or when the humidity stays high, it is likely that there is a leak. This is also indicated when an Eh indicator such as resazurin or phenosafranine in reduced media fails to remain colorless. The most common place for a leak to develop is in a rubber glove. This can be checked by placing a large beaker of water inside the chamber and immersing each glove in it. If gloves are replaced as directed by the manufacturer, significant gas exchange will not occur. Leaks in the plastic may occur around seams and are difficult or impossible to find visually. An electronic leak detector (Coy Laboratory Products, Inc.) is very useful in locating leaks.

11. A medium prepared in a conventional manner outside the chamber should be placed in the chamber as soon as cooled. Hot medium will boil out of tubes or flasks due to the vacuum in the entry lock unless the medium is in tightly sealed containers.

12. Before use, hold media which have been prepared in a conventional manner in the chamber for 1 or 2 days, or until the Eh indicator is reduced. Also, equilibrate glassware and plasticware for a day before use.

13. Prereduced media prepared by the Hungate technique may be used promptly when placed in the chamber. For some organisms such as the methanogens, it may be necessary to use prereduced media.

14. Protect the neck of culture tubes and flasks after sterilization. They cannot be flamed in the chamber.

6.7. LITERATURE CITED

6.7.1. General References

1. **Brown, D. E.** 1970. Aeration in the submerged culture of micro-organisms, p. 127–174. In J. R. Norris and D. W. Ribbons (ed.), Methods in microbiology, vol. 2. Academic Press, Inc., New York.

2. **Datta, S. P., and A. K. Grzybowski.** 1961. pH and acid-base equilibria, p. 19–58. *In* C. Long, E. J. King, and W. M. Speery (ed.), Biochemists' handbook. D. Van Nostrand Co., Inc., New York.
   Excellent treatment of pH and buffers including ionic strength and buffer value. Also, extensive tables of buffers, dissociation constants of weak acids and bases, and of pH indicators.
3. **Gomori, G.** 1955. Preparation of buffers for use in enzyme studies. Methods Enzymol. **1**:138–146.
   Contains tables of various buffer systems.
4. **Hitchman, M. L.** 1978. Measurement of dissolved oxygen. (Chemical analysis, vol. 49) John Wiley & Sons, Inc., New York.
5. **Holdeman, L. V., E. P. Cato, and W. E. C. Moore.** 1977. Anaerobe laboratory manual, 4th ed. Virginia Polytechnic Institute and State University, Blacksburg.
   Excellent descriptions and illustrations of anaerobic techniques.
6. **Hungate, R. E.** 1969. A roll tube method for cultivation of strict anaerobes, p. 117–132. *In* J. R. Norris and D. W. Ribbons (ed.), Methods in microbiology, vol. 3B. Academic Press, Inc., New York.
   Good description of Hungate technique.
7. **Jacob, H.-E.** 1970. Redox potential, p. 92–123. *In* J. R. Norris and D. W. Ribbons (ed.), Methods in microbiology, vol. 2. Academic Press, Inc., New York.
   Describes electrometric measurement of redox potential and the effects of microbes on the potential.
8. **Marquis, R. E.** 1976. High-pressure microbial physiology. Adv. Microb. Physiol. **14**:159–241.
   A good review of how high pressures affect microorganisms and their activities.
9. **Munro, A. L. S.** 1970. Measurement and control of pH values, p. 39–89. *In* J. R. Norris and D. W. Ribbons (ed.), Methods in microbiology, vol. 2. Academic Press, Inc., New York.
   Contains several useful tables for preparing buffers at different ionic strengths.
10. **Pirt, S. J.** 1975. Principles of microbe and cell cultivation. John Wiley & Sons, Inc., New York.
    Excellent chapters on oxygen supply and demand, and on effects of temperature, hydrogen ion concentration, and water activity.
11. **Prior, B. A.** 1979. Measurement of water activity in foods: a review. J. Food Protect. **42**:668–674.
    Reviews and evaluates methods for measuring water activity.
12. **Scott, W. J.** 1957. Water relations of food spoilage microorganisms. Adv. Food Res. **3**:84–123.
    Principles of water activity, relationship to osmolality, methods of control of a_w, and effect on bacteria.
13. **Sutter, V. L., D. M. Citron, and S. M. Finegold.** 1980. Wadsworth anaerobic bacteriology manual, 3rd ed. C. V. Mosby Co., St. Louis.
    An excellent, updated, how-to manual written mainly for clinical laboratories. A useful appendix includes a list of pitfalls in anaerobic work.
14. **Troller, J. A., and J. H. B. Christian.** 1978.

Water activity and food. Academic Press, Inc., New York.
    Excellent authoritative, well-written new monograph with pertinent chapters on basic concepts, methods, enzyme reactions, microbial growth, and microbial survival.
15. **Wang, D. I. C., C. L. Cooney, A. L. Demain, P. Dunnill, A. E. Humphrey, and M. D. Lilly.** 1979. Fermentation and enzyme technology, p. 157–193. John Wiley & Sons, Inc., New York.
    Describes principles and procedures for measurement of oxygen transfer.
16. **Weast, R. C. (ed.).** 1976. Handbook of chemistry and physics, 57th ed., p. D-133–D-137, D-147–D-152. CRC Press, Cleveland, Ohio.
    Tables of buffers, pH of standard buffers at different temperatures, acid base indicators, and dissociation constants of acids and bases.
17. **Wolfe, R. S.** 1971. Microbial formation of methane. Adv. Microb. Physiol. **6**:107–146.
    Illustrated description of the Hungate technique.

6.7.2. Specific Articles

18. **Aranka, A., and R. Freter.** 1972. Am. J. Clin. Nutr. **25**:1329–1334.
19. **Aranka, A., S. A. Syed, E. B. Kenney, and R. Freter.** 1969. Appl. Microbiol. **17**:568–576.
20. **Balch, W. E., and R. S. Wolfe.** 1976. Appl. Environ. Microbiol. **32**:781–791.
21. **Bryant, M. P., and L. A. Brukey.** 1953. J. Dairy Sci. **36**:205–217.
22. **Cleland, W. W.** 1964. Biochemistry **3**:480–482.
23. **Corman, J., H. M. Tsuchiya, H. J. Koepsell, R. G. Benedict, S. E. Kelley, V. H. Feger, R. G. Dworschak, and R. W. Jackson.** 1957. Appl. Microbiol. **5**:313–318.
24. **Gordon, J., R. A. Holman, and J. W. McLeod.** 1953. J. Pathol. Bacteriol. **66**:527–537.
25. **Hungate, R. E.** 1950. Bacteriol. Rev. **14**:1–49.
26. **Ingraham, J. L.** 1958. J. Bacteriol. **76**:75–80.
27. **Labuza, T. P., K. Acott, S. R. Tatini, R. Y. Lee, J. Flink, and W. McCall.** 1976. J. Food Sci. **41**:910–917.
28. **Macy, J. M., J. E. Snellen, and R. E. Hungate.** 1972. Am. J. Clin. Nutr. **25**:1318–1323.
29. **Miller, T. L., and M. J. Wolin.** 1974. Appl. Microbiol. **27**:985–987.
30. **Onderdonk, A. B., J. Johnston, J. W. Mayhew, and S. L. Gorbach.** 1976. Appl. Environ. Microbiol. **31**:168–172.
31. **Patching, J. W., and A. H. Rose.** 1970. *In* J. R. Norris and D. W. Ribbons (ed.), Methods in microbiology, vol. 2, p. 23–38. Academic Press, Inc., New York.
32. **Scott, W. J.** 1953. Aust. J. Biol. Sci. **6**:549–564.
33. **Taylor, C. D., and H. W. Jannasch.** 1976. Appl. Environ. Microbiol. **32**:355–359.
34. **Troller, J. A.** 1977. J. Food Sci. **42**:86–90.
35. **Umbreit, W. W., R. H. Burris, and J. F. Stauffer.** 1964. Manometric techniques, 4th ed. Burgess Publishing Co., Minneapolis.
36. **Walden, W. C., and D. J. Hentges.** 1975. Appl. Microbiol. **30**:781–785.
37. **Zehnder, A. J. B., and K. Wuhrmann.** 1976. Science **194**:1165–1166.

Chapter 7

Biochemical Factors in Growth

BEVERLY M. GUIRARD AND ESMOND E. SNELL

Like other living organisms, all bacteria require for growth an exogenous energy source and accessory compounds that supply carbon, nitrogen, sulfur, and inorganic ions in forms that can be used for synthesis of their cell materials. Individual species, however, vary tremendously in the array of compounds necessary to serve these functions. The **autotrophic** bacteria have simple nutritional requirements: water, carbon dioxide, and appropriate inorganic salts suffice; their energy is derived from light or by the oxidation of one or more inorganic elements or compounds. Examples are the bacteria in the genus *Nitrobacter*, which assimilate CO_2 and obtain their energy from the oxidation of nitrites to nitrates. The **heterotrophic** bacteria obtain their energy from the oxidation or dissimilation of reduced carbon compounds. Glucose is most commonly supplied in bacteriological media for this purpose, but essentially all naturally occurring compounds of carbon and many synthetic ones not found in nature can be utilized as an energy source by one or another heterotrophic bacterium.

Heterotrophic bacteria utilize organic compounds for two distinct purposes: first, as an energy source in which the organic compound is oxidized or dissimilated with the release of energy and a variety of waste products (CO_2, organic acids, etc.) and, second, as the source of organic materials from which cell components are derived directly or synthesized by energy-consuming reactions. A given compound (e.g., glucose) may sometimes serve both roles. *Escherichia coli,* for example, grows on a simple medium containing only glucose and inorganic salts. Other bacteria, for example, the lactic acid bacteria, have more complex requirements. In addition to inorganic salts and an energy source (usually glucose), they must be supplied with a variety of organic compounds (vitamins, amino acids, etc.) which they are unable to synthesize; such compounds are termed **growth factors.** Organisms that require the addition of such a compound to their growth medium are **auxotrophic** for that compound. Bacteria with all gradations of complexity in their requirements for the recognized nutrients are known. Other bacteria have yet unidentified growth requirements which must be supplied with complex natural materials such as yeast extract, peptone, rumen fluid, and blood serum. In addition to this array of **saprophytic** bacteria that can be grown in vitro, the obligately **parasitic** bacteria grow only in the presence of living animal, plant, or other bacterial cells which provide yet unidentified nutrients or substitute for metabolic deficiencies.

The requirement of a heterotrophic bacterium for a particular growth factor may vary quantitatively depending on the composition of the medium. For example, the requirement of *Streptococcus faecalis* for vitamin B_6 may vary from as much as 10 μg/liter in the absence of one or more amino acids to zero in the presence of a complete supplement of amino acids. By study of the effect on growth of interactions of medium components, important information concerning the function of growth factors has accrued.

Preliminary to listing crude and defined media for various representative bacteria, we discuss briefly below the nature of the requirement for certain compounds utilized for synthesis of cell materials and give some practical information concerning their use in the preparation of growth media (for references and a detailed discussion, see 70, 191). Additional information on bacterial nutrition may be found in Koser's text on vitamin requirements of yeast and bacteria (114), in *Methods in Microbiology* (122), and in the *CRC Handbook Series in Nutrition and Food* (172). The American Type Culture Collection *Catalogue of Strains* (4) gives undefined media for propagating the strains offered in their catalog.

7.1. NUTRIENT REQUIREMENTS

7.1.1. Carbon, Nitrogen, and Sulfur

The requirement of bacteria for carbon compounds has been discussed in the preceding paragraphs. Most bacteria so far studied utilize ammonia for the synthesis of the nitrogenous components of the cell. Many, however, also require nitrogenous growth factors, which must be supplied in the medium. Many form ammonia by reduction of nitrate or nitrite or by degradation of organic nitrogenous compounds. A few (e.g., *Azotobacter*, *Rhizobium*, purple sulfur bacteria) are able to reduce elemental nitrogen to ammonia. Most bacteria can synthesize essential sulfur compounds from sulfate ion; for those that cannot, sulfur is supplied in the medium as H_2S, cystine, or methionine.

7.1.2. Inorganic Ions

The inorganic ion requirements have been extensively studied. These studies have been complicated by: (i) the ubiquitous occurrence of metal ions, (ii) the small amounts of many mineral elements required to permit growth, (iii) the difficulty of removing these ions from media in order to demonstrate a requirement, (iv) the sparing effects of one metal ion on the requirement for another, (v) the antagonistic effects between certain related metal ions, and (vi) the occurrence of metal ions in many biological products as chelates of varying stability, thus affecting their availability to bacteria. For a detailed treatment of this topic, the extensive precautions necessary for its study, and the techniques involved, the reviews of Hutner et al.

(91), Knight (112), and Snell (205) should be consulted.

For practical purposes, the inorganic ion requirements of most bacteria are satisfied by the addition of K^+, Mg^{2+}, Mn^{2+}, Fe^{2+} or Fe^{3+}, PO_4^{3-}, and SO_4^{2-} to defined media. Trace amounts of several other ions (e.g., Zn^{2+}, Cu^{2+}, Co^{2+}, Ca^{2+}, Cl^-, molybdate, etc.) frequently are also required, but are usually supplied in sufficient amounts either as contaminants of other medium components or by the use of tap water in preparing media. Complex natural media (i.e., those containing peptones, yeast extract, meat infusions, and similar substances) contain an abundance of metal ions.

In preparing concentrated stock solutions of inorganic salts for dilution into synthetic media, it is frequently useful to store them at low pH or to add minimal amounts of nontoxic, chelating agents (e.g., ethylenediaminetetraacetic acid or nitrilotriacetic acid) to prevent precipitation. The presence of chelating agents in the final medium is an important factor in determining the availability of metal ions, and the concentrations of these chelators must be carefully balanced with those of trace metals if abundant and sustained growth is to be attained (91). In addition to chelating agents already mentioned, many medium components (amino acids, citrate, malate, etc.) also function in this capacity.

The nutritionally essential metal ions serve bacteria mainly as activators or cofactors of a variety of enzymes (145). Inorganic ions (principally Na^+ and K^+) also function in transport of materials across cell membranes (155) and in the regulation of protein synthesis (131). They are also components of protein complexes which play important roles in bacterial metabolism, e.g., iron in the heme prosthetic group of cytochrome, an electron-transporting protein.

7.1.3. Vitamins

Vitamins play catalytic roles within the cell, usually as components of coenzymes or as prosthetic groups of enzymes. Those that are required by one species or another and some of their interrelationships with other components of the growth medium are discussed briefly below.

p-Aminobenzoic acid.

p-Aminobenzoic acid serves as a precursor for the biosynthesis of folic acid and thereby influences also the metabolism of thymine, methionine, serine, the purine bases, and vitamin B_{12}. The presence of these compounds in the medium may decrease or eliminate the requirement of a given bacterium for *p*-aminobenzoic acid in the medium.

Folic acid group.

Several different forms of folic acid occur naturally and differ in their availability for individual bacteria. Pteroyl-L-glutamic acid and its N^{10}-formyl derivative are fully active for Lactobacillus casei and S. faecalis; pteroyltriglutamic acid, while fully active for L. casei, is only slightly active for S. faecalis; pteroylheptaglutamic acid does not support growth of either bacterium. N^5-Formyltetrahydropteroylglutamic acid (leucovorin) is active for L. casei, S. faecalis, and Pediococcus cerevisiae. Folic acid functions as a cofactor in synthesis of thymine, purine bases, serine, methionine, and pantothenic acid; the requirement for it is frequently eliminated if these compounds are added to the medium. Folic acid is only slightly soluble in water, but the ammonium salt is very soluble. The vitamin is stable to autoclaving.

Biotin.

Biotin participates in several biosynthetic reactions that require CO_2 fixation, including synthesis of oxalacetate and of fatty acids. Biocytin (N^ϵ-biotinyllysine) also occurs naturally; it is as active as biotin for L. casei but not for Lactobacillus arabinosus. Biotin is slightly soluble, but its salts are freely soluble in water. Biotin is stable to autoclaving and to acids, but it is readily oxidized to the sulfoxide and sulfone. These oxidation products also promote growth of certain bacteria, but only if they can convert them to biotin. Oxidation is a problem only in extremely dilute (<1 μg of biotin per ml) solution and only if other reducing agents are absent.

Nicotinic acid and its derivatives.

Most nicotinic acid auxotrophs utilize nicotinic acid and its amide interchangeably. The coenzyme forms of the vitamin, nicotinamide adenine dinucleotide (NAD) and nicotinamide adenine dinucleotide phosphate (NADP), function in oxidoreduction reactions in cellular metabolism. A few organisms (e.g., Haemophilus influenzae) cannot synthesize NAD from nicotinic acid; such organisms require the preformed coenzyme. The latter is destroyed by autoclaving, acids, and alkalies; when required, it is sterilized by filtration through bacterial filters and added to the separately autoclaved medium. Nicotinic acid and its amide are stable to autoclaving at pH 7.0, the pH of most growth media.

Pantothenic acid and related compounds.

Pantothenic acid is a component of coenzyme A (CoA) and the acyl carrier protein (ACP), which serve as acyl group carriers in metabolism. A few bacteria utilize the intact coenzyme as a growth factor, but many more require pantothenic acid or pantetheine for growth. The interrelationship of these compounds is shown in the following diagram:

$$\text{Pantoic acid} + \beta\text{-alanine} \rightarrow$$
$$\text{pantothenate} \rightarrow 4'\text{-phosphopantothenate} \rightarrow$$
$$4'\text{-phosphopantothenylcysteine} \rightarrow$$
$$4'\text{-phosphopantetheine} \rightarrow$$
$$\uparrow$$
$$\text{pantetheine}$$
$$\text{dephosphoCoA} \rightarrow \text{CoA} \xrightarrow{} \text{ACP}$$
$$\text{apoprotein}$$

The requirement of individual species for one or another of these related compounds derives from differences in their synthetic abilities and also from differences in their transport capabilities.

Free pantothenic acid and many of its salts are very hygroscopic; it is marketed as the only slightly hygroscopic calcium salt which is freely soluble in water and is stable to autoclaving. Like its more complex derivatives, pantothenic acid is readily hydrolyzed, and thus its growth-promoting activity is destroyed (for all bacteria requiring the intact vitamin) by acid or alkaline hydrolysis.

Riboflavin and its derivatives.

Although relatively few bacterial species require preformed riboflavin, they all apparently contain the vitamin, and some synthesize it in large enough quantities to serve as useful commercial sources.

Riboflavin 5'-phosphate (flavin mononucleotide, FMN) and flavin adenine dinucleotide (FAD) are the major coenzyme forms of riboflavin; they function in oxidoreduction reactions. All three compounds are equally active in supporting growth of a riboflavin auxotroph, L. casei.

Riboflavin is slightly soluble in cold water, but is made easily soluble by warming at the concentration recommended for preparing the stock vitamin solution (see 7.3.1.). It is destroyed by visible light, especially at neutral pH or above; aqueous solutions should be protected from light. A dilute solution (~100 μg/ml) in 0.02 M acetic acid may be used as a stock solution. Riboflavin is stable to autoclaving.

Thiamine.

A large number of bacteria are auxotrophic for thiamine, its precursors, or its coenzyme form, thiamine pyrophosphate (TPP). TPP functions in the decarboxylation of α-keto acids and in the transketolase reaction.

Since thiamine is cleaved into its component moieties when aqueous solutions at pH 5.0 or above are autoclaved, the solutions should be filter sterilized or autoclaved separately at a pH < 5.0 if the intact vitamin is required.

Vitamin B₆.

Pyridoxine, pyridoxal, and pyridoxamine and their phosphorylated derivatives compose the vitamin B₆ group. There is much species variation in the growth response of vitamin B₆ auxotrophs to these various forms. Pyridoxal is most widely used as a growth factor, but sufficient amounts of it are formed during autoclaving if sufficient pyridoxine is included in the medium. A few bacteria require pyridoxamine 5′-phosphate (PMP) for growth; for most auxotrophs, however, the phosphorylated forms are inactive, apparently because they lack the necessary transport systems. Pyridoxal 5′-phosphate and PMP participate in a large number of reactions involving the synthesis and degradation of the naturally occurring α-amino acids.

Although all three forms of the vitamin are stable to heat, pyridoxal and pyridoxamine react with many naturally occurring compounds; if it is important that they remain unchanged, they should be sterilized separately from medium components. All forms of the vitamin are labile to light, especially at alkaline pH.

Vitamin B₁₂ (cobalamine).

Vitamin B₁₂ appears to be synthesized in nature exclusively by microorganisms, which serve together with their by-products as the commercial source of the vitamin. The coenzyme form of vitamin B₁₂ (cobalamine coupled to adenine nucleoside) is as active as the vitamin itself in promoting growth. The coenzyme functions in a number of isomerization reactions in metabolism and also (directly or indirectly) in the biosynthesis of deoxyribonucleosides, methionine, and perhaps other compounds.

Vitamin B₁₂ coenzyme is unstable to light and should be filter sterilized. In contrast, vitamin B₁₂ itself is relatively stable to both light and autoclaving.

Lipoic acid (thioctic acid).

Lipoic acid is an essential growth factor for several lactic acid bacteria and for *Butyribacterium rettgeri*. An essential role in pyruvate and α-ketoglutarate oxidation is related to the fact that many lipoate auxotrophs do not require the growth factor in media that contain acetate. Lipoic acid, dihydrolipoic acid, and a variety of mixed disulfides all have about equal growth-promoting activities for *S. faecalis*. The compound is stable to acid and to autoclaving but is labile to oxidizing agents. Lability to oxidation

does not usually present a problem because of the presence of more reactive reducing agents (e.g., glucose) during preparation of growth media.

Iron porphyrins.

Iron porphyrins, ferrichrome, and related Fe-binding compounds are found in all aerobic bacteria examined and are components of an electron transport system leading to oxygen. A few bacteria (e.g., *H. influenzae*) apparently cannot synthesize the porphyrins and require a supply in the medium.

A group of iron chelators collectively known as **siderochromes,** of which ferrichrome and the terregens factor are examples, are required by several species of bacteria. They are replaceable in the medium by much larger quantities of hemin and are thought to function catalytically as carriers of iron.

Vitamin K.

Some *Mycobacterium* and *Fusiformis* species require vitamin K or related compounds for growth. The function is not yet fully known. This fat-soluble vitamin is destroyed by light, alkalies, and reducing agents. It is best added to media as menadione (2-methyl-1,4-napthoquinone) in ethanol solution, which is stable to autoclaving but labile to light and alkali.

Choline.

Choline has been shown to be required only by some pneumococci, apparently as a precursor of certain cellular lipids.

7.1.4. Amino Acids, Peptides, and Proteins

Bacteria very frequently require one, several, or many of the naturally occurring α-amino acids. Although this requirement stems from loss of the ability to synthesize such amino acids, this loss may be conditional rather than absolute. For example, an amino acid required in the absence of vitamin B₆ may be synthesized in its presence, and an amino acid required when an excess of a second amino acid is present may be dispensable in its absence.

Even in nonconditional auxotrophs, the addition of an excess of one amino acid may result in an increased requirement for another amino acid. This effect derives from the competition of the two amino acids for a single transport system. Such antagonistic interrelationships can be bypassed by the addition of an appropriate peptide of the limiting amino acid, because the peptide is absorbed via an independent transport system. This phenomenon has led to occasional reports of the requirement for specific peptide growth factors. However, there is as yet no known instance of the obligatory requirement for a preformed peptide for bacterial growth, although peptides are usually readily utilized (and frequently desirable) sources of essential amino acids.

The reported requirement of certain bacteria for specific proteins requires further study. The proteins may be acting solely as a source of amino acids, as detoxicants, or in other ways.

Except for glutamine, which must be filter sterilized for addition to culture media, amino acids and peptides are stable to autoclaving. All except tryptophan, glutamine, and asparagine are stable during acid hydrolysis of a protein. Cysteine is frequently destroyed during acid hydrolysis. All naturally occurring α-amino acids except cystine are sufficiently soluble in warm water for media preparation. Cystine must be dissolved by warming with a small amount of hydrochloric acid. Tyrosine, although difficultly soluble, becomes more soluble in the presence of additional amino acids.

7.1.5. Purines and Pyrimidines

Many bacteria are incapable of synthesizing purines and pyrimidines and require an external source. The interrelationships of these compounds with folic acid and vitamin B_{12} have been mentioned. Guanine and xanthine are dissolved by warming with a minimal amount of hydrochloric acid. The other compounds are soluble in hot water.

7.1.6. Fatty Acids and Related Compounds

Essential fatty acids, particularly the unsaturated fatty acids, are sometimes toxic, even when supplied in the low concentrations at which they promote growth. They are frequently added to media together with detoxifying materials such as serum albumin, proteins, Tweens, lecithin, etc. The requirement for these compounds shows group specificity rather than absolute specificity, i.e., any of several active fatty acids are frequently incorporated unchanged into essential lipid material.

Mevalonic acid, required by some species of *Lactobacillus*, functions as a precursor of isoprenoid compounds (carotenes, sterols, ubiquinone, etc.). Unlike most bacteria, whose cells contain little or no sterols, mycoplasmas require sterols (e.g., cholesterol) for growth.

Fatty acids and other lipids are generally stable to autoclaving, but their addition to an aqueous medium may present problems. They should be dissolved in minimal amounts of ethanol or emulsified in a solution of Tween.

7.1.7. Miscellaneous Compounds

Several additional compounds serve as growth factors, either essential or stimulatory, for one or more species of bacteria. Among these are the polyamines (putrescine, spermine, spermidine), carnitine, substances produced when glucose is autoclaved with phosphate or in the presence of amino acids, and several unidentified factors present in yeast extract, serum, and other natural materials.

7.1.8. Determination of Nutrient Requirements

The use of bacteria as test organisms in microbiological assays and many other biological studies requires a knowledge of their nutritional requirements and the effects of other medium components on these requirements.

To prepare a growth medium for a newly isolated bacterium, an **undefined medium** is first made up with crude components; the recipe typically includes an enzymatic hydrolysate of protein to provide the nitrogen source as amino acids, a sugar (usually glucose) to provide a carbon and energy source, various salts to satisfy inorganic ion requirements, and yeast extract to ensure that needs for vitamins are met. As the nutritional requirements of the bacterium become better understood, a **semidefined medium** can be described, in which known compounds are used for the obvious requirements and a small amount of yeast extract is added to take care of the obscure requirements. Eventually, a completely **synthetic** or **defined medium** can be described, in which all the nutritional requirements are fulfilled in a known biochemical and biophysical environment. The selection of a given medium will depend upon the purpose of the growth experiment. Complex defined media are expensive, and it is frequently practical to revert to an undefined medium for many uses of bacterial cultures.

A general stepwise procedure for determining the nutritional requirements of a fastidious heterotrophic bacterium is as follows:

1. Grow the bacterium on a medium (either defined or undefined) known to support its growth.

2. If one begins with an undefined medium, vary singly the concentrations of all components (from zero to greater than the concentration at which they are present in the preliminary medium) to determine their optimum concentrations and whether they are required or stimulatory.

3. Using optimum concentrations, replace the main complex component of the medium

that provides nitrogen (e.g., casein) initially with a complete mixture of amino acids at levels similar to those used in an assay medium (see Table 14).

4. If this mixture permits growth, vary singly the concentration of each amino acid (from zero to greater than the concentration present in the test medium) to determine the optimum concentration of each in the presence of all the others. If any are inhibitory, omit.

5. Replace the complex growth factor component (e.g., yeast extract) with a complete assortment of known vitamins, and then vary the level of each constituent of the mixture singly, as was done with the amino acids. If yeast extract or peptone served in the preliminary medium as the source of both amino acids and vitamins, try to replace it with a complete mixture of amino acids plus a complete assortment of known vitamins, and then vary the level of each component of the mixture singly.

6. Failure to grow on the resulting defined medium may reflect the existence of an as yet unidentified growth factor, the presence of an imbalance of known growth factors in the medium, or the requirement for known growth factors (e.g., peptides, vitamin derivatives, fatty acids, nucleotides, inorganic ions, etc.) which were supplied by the crude supplements, but not by the mixtures of the synthetic compounds tested. Detailed discussions of the nutritional requirements of bacteria (e.g., reference 70) should be consulted in that event.

7. If one has begun with a defined medium and growth after step 2 is not abundant, replace the nitrogen source with 10% hydrolyzed casein. If the replacement improves growth, proceed as in step 4. If growth still is not adequate, add 1% yeast extract. If improvement in growth results, proceed as in step 5.

7.2. COMPOSITIONS OF MEDIA

It is impossible to list all of the media used for cultivation of bacteria. The media tabulated here have been reported to support the growth of pure cultures of the indicated genera and are representative of what can be used with related ones.

7.2.1. Key to Media for Representative Bacteria

Table 1 lists 192 species of bacteria representative of each of the 19 groups that are listed in the 8th edition of *Bergey's Manual of Determinative Bacteriology*. Also given are the optimal growth temperature and a number-letter key to the composition of one or more media reported in the literature to support the growth of each

species. The superscript letters (in lowercase) refer to footnotes in which details are provided about cultivation of that bacterium. The number-letter keys (in uppercase) refer to the media in Tables 2 through 13, where the compositions of the media are compiled. The key letter A refers to an undefined medium in Tables 2 through 7, B refers to a semidefined medium in Tables 2 through 7 (a few with the B code are in Tables 12 and 13), and C refers to a synthetic medium in Tables 8 through 13.

The tables include only those bacteria which have been cultured in vitro in pure culture. Most of the media given are not minimal media; i.e., they frequently contain compounds that are not specifically required or are present in amounts far in excess of minimal requirements. The fact that a given bacterium has been cultured in a cited medium also should not be taken to mean that the medium is an optimal one for that bacterium, since only rarely have detailed nutritional studies been conducted.

7.2.2. Undefined and Semidefined Media

Tables 2 through 7 list the compositions of undefined media (code letter A) and semidefined media (code letter B) for the bacteria listed in Table 1. Prepared media that are commercially available are marked with an asterisk. For solid media, add 1.5 to 2.0% agar to the broth media. Check the footnotes (superscript lowercase letters) for additional components and critical methodology.

7.2.3. Defined Media

Tables 8 through 13 list the compositions of defined media (code letter C) and a few semidefined media (code letter B) for the individual bacteria listed in Table 1. Check the footnotes (superscript lowercase letters) for additional components and critical methodology.

7.3. QUANTITATIVE GROWTH-RESPONSE ASSAYS

Bacteria have been widely used in quantitative assays for a wide variety of substances that promote or inhibit growth: vitamins, amino acids, trace elements, pyrimidine bases, polyamines, deoxyribonucleosides, deoxyribonucleotides, drugs, and antibiotics and other chemotherapeutic agents (70, 105a). Although instrumental methods (e.g., amino acid analysis with an amino acid analyzer) have partially replaced some of these bacteriological methods, the latter are still useful in many applications. In some instances (e.g., the assay for vitamin B_6), bacteriological assays are differentially sensitive to different forms of a vitamin and are not readily

TABLE 1. *Key to media and cultural requirements for representative groups and species of bacteria*

Group, genus, and species[a]	Optimal growth temp (°C)	Number-letter key to media[b]	References
I. Phototrophic Bacteria:			
Amoebobacter roseus[c]	25–30	1C (Table 8)	158, 224
Chlorobium limicola[c, d]	25–30	1B, 2C (Tables 2, 8)	123, 124
Chlorobium thiosulfatophilum[c, d]	25–30	1B, 2C (Tables 2, 8)	123, 124
Chromatium okenii[c, d]	25–30	3C (Table 8)	59
Rhodomicrobium vannielii[c]	30	2A, 4C (Tables 2, 8)	17, 52, 154
Rhodopseudomonas sphaeroides[c, h]	25–30	2A, 4C (Tables 2, 8)	17, 35, 154
Rhodospirillum rubrum[c, f, g]	25	2A, 4C (Tables 2, 8)	17, 154
II. Gliding Bacteria:			
Archangium gephyra[i]	18–32	3A (Table 2)	80, 141
Beggiatoa alba[i]	20	4A, 5C (Tables 2, 8)	115, 189
Chondromyces crocatus[i, j]	28–30	3A (Table 2)	140
Cystobacter fuscus[i]	30	3A (Table 2)	142
Flexibacter flexilis[i]	20–25	5A, 5B (Table 2)	130
Flexithrix dorotheae[i]	35	5A, 6C (Tables 2, 8)	129
Myxococcus xanthus[i]	30	6B, 7C (Tables 2, 8)	53, 80
Nannocystis exedens[i]	30	6B (Table 2)	174
Saprospira grandis[i]	30–37	5A, 5B (Table 2)	128, 130
Sporocytophaga myxococcoides[i]	30	8C (Table 8)	194, 212
Stigmatella aurantiaca	30	7B (Table 2)	175
III. Sheathed Bacteria:			
Leptothrix ochracea[i]	20–25	8A, 9C (Tables 2, 8)	150, 229
Sphaerotilus natans[i]	30	9A, 10C (Tables 2, 8)	50, 219, 229
Streptothrix hyalina[i, k]	25–27	8A (Table 2)	151
IV. Budding and/or Appendaged Bacteria:			
Ancalomicrobium adetum[i]	30	10A, 11C (Tables 2, 8)	211
Asticcacaulis excentricus	25–30	11A, 12C (Tables 2, 8)	162
Caulobacter henricii[i, l]	30	11A, 12C (Tables 2, 8)	162
Caulobacter halobacteroides[i, m]	25	11A, 12C (Tables 2, 8)	162
Caulobacter vibrioides[i, n]	30	11A, 12C (Tables 2, 8)	162
Gallionella ferruginea[o, p]	25	13C (Table 8)	74, 150
Hyphomicrobium vulgare[i, q]	25–30	13C (Table 8)	86
Hyphomonas polymorpha[i]	37	12A (Table 2)	86, 183
Pedomicrobium ferrugineum[o]		12C (Table 8)	5
Prosthecomicrobium pneumaticum[i, r]	30	10A, 11C (Table 2, 8)	211
V. Spirochetes:			
Leptospira canicola[i]	29	13B, 14C (Tables 3, 8)	97, 127, 192
Leptospira pomona[i]	29	13B, 14C (Tables 3, 8)	97, 127, 192
Leptospira grippotyphosa[i]	29	13B, 14C (Tables 3, 8)	97, 127, 192
Reiter treponeme[c, s]	37	14A, 15C (Tables 3, 9)	217, 218
Spirochaeta stenostrepta[c, t]	30–37	15A (Table 3)	85
VI. Spiral and Curved Bacteria:			
Campylobacter fetus[o]	37	16A (Table 3)	127, 190
Spirillum volutans[o, u]	30	17A (Table 3)	29, 117
VII. Gram-Negative Aerobic Rods and Cocci:			
Acetobacter aceti	30	18A (Table 3)	46, 93
Agrobacterium rhizogenes[v]	25–28	18A, 16C (Tables 3, 9)	46, 215
Azomonas agilis[w]	20–30	17C (Table 9)	223
Azotobacter chroococcum[w]	20–30	17C (Table 9)	12, 58
Beijerinckia indica[w, x]	30	18C (Table 9)	12, 13
Bordetella pertussis[y]	35–37	19A (Table 3)	127, 185
Brucella melitensis[z]	34–37	16A, 19C (Tables 3, 9)	65, 127, 178, 186
Derxia gummosa	25	20C (Table 9)	96
Francisella tularensis[aa]	37	20A (Table 3)	127
Gluconobacter oxydans	25–30	18A (Table 3)	45
Halobacterium salinarium	37	21A (Table 3)	103
Methylomonas methanica[bb]	30	22B (Table 3)	54

TABLE 1—*continued*

Group, genus, and species[a]	Optimal growth temp (°C)	Number-letter key to media[b]	References
Pseudomonas aeruginosa	37	23A, 21C, (Tables 3, 9)	213
Pseudomonas fluorescens	25–30	23A, 21C (Tables 3, 9)	213
Pseudomonas lemoignei[cc]	30	23A, 21C, (Tables 3, 9)	213
Pseudomonas maltophilia[dd]	35	23A, 21C, (Tables 3, 9)	213
Pseudomonas putida	25–30	23A, 21C (Tables 3, 9)	213
Pseudomonas testosteroni[ee]	30	23A, 21C (Tables 3, 9)	213
Rhizobium leguminosarum	20	24A, 22C (Tables 3, 9)	153
Xanthomonas campestris[ff]	28	18A, 16C (Tables 3, 9)	214
VIII. Gram-Negative Facultatively Anaerobic Rods:			
Actinobacillus lignieresii	37	25A (Table 4)	127, 159
Aeromonas hydrophila	30	26A, 23C (Tables 4, 9)	119, 188
Cardiobacterium hominis[gg]	30–37	12A, 24C (Tables 2, 9)	127, 144, 198, 199
Chromobacterium violaceum[hh]	30–35	12A, 25C (Tables 2, 9)	48, 127, 203
Citrobacter freundii	30–37	26C (Table 9)	242
Edwardsiella tarda	30–37	12A (Table 2)	109
Erwinia amylovora[ii]	27–30	26A (Table 4)	234
Escherichia coli	37	26A, 27C (Tables 4, 9)	133, 230
Flavobacterium aquatile	20–25	27A (Table 4)	147, 239
Haemophilus influenzae	30	27A (Table 4)	243, 247
Haemophilus parainfluenzae	37	27A, 28C (Tables 4, 10)	84, 247
Klebsiella pneumoniae	35–37	28A, 27C (Tables 4, 9)	104
Lucibacterium harveyi	25–30	29A (Table 4)	82
Pasteurella multocida	37	12A, 29C (Tables 2, 10)	100
Proteus vulgaris	34–37	28A, 30C (Tables 4, 10)	98
Salmonella typhimurium	37	30A, 27C (Tables 4, 9)	187, 228
Serratia marcescens[jj]	30–37	30A, 27C (Tables 4, 9)	44
Shigella dysenteriae	37	31A (Table 4)	156, 244
Streptobacillus moniliformis[kk]	37	12A (Table 2)	79
Vibrio cholerae[ll]	18–37	32A, 31C (Tables 4, 10)	9, 26
Yersinia pestis	30–37	33A, 32C (Tables 4, 10)	56, 177
Zymomonas mobilis[mm]	30	34A, 33C (Tables 4, 10)	14
IX. Gram-Negative Anaerobic Bacteria:			
Bacteroides fragilis	37	35A (Table 4)	15
Bacteroides ruminicola[nn]	30	36A, 34C (Tables 4, 10)	161
Butyrivibrio fibrisolvens[nn]	30–37	36A, 34C (Tables 4, 10)	20, 22
Desulfovibrio desulfuricans	25–30	37A, 35C (Tables 4, 10)	78, 163
Fusobacterium nucleatum	37	38A (Table 5)	6, 7
Lachnospira multiparus	30–45	38A, 34C (Tables 5, 10)	20, 23
Leptotrichia buccalis	25–30	39A (Table 5)	19, 92
Selenomonas sputigena	35–40	12A (Table 2)	
Succinimonas amylolytica	30–37	40A, 34C (Tables 5, 10)	20, 24
Succinivibrio dextrinosolvens	30–37	40A (Table 5)	23
X. Gram-Negative Cocci and Coccobacilli:			
Acinetobacter calcoaceticus[i, oo]	30–32	24A, 21C (Tables 3, 9)	11
Branhamella catarrhalis[i, pp]	37	30A, 21C (Tables 4, 9)	176
Lampropedia hyalina[i]	30	30A, 36C (Tables 4, 10)	167
Moraxella lacunata[i]	30	41A (Table 5)	10
Neisseria gonorrhoeae[i, qq]	32–37	27A, 37C (Tables 4, 10)	30, 49, 67, 68, 106, 120, 121
Neisseria meningitidis[i]	37	42A, 38C (Tables 5, 10)	31, 32
XI. Gram-Negative Anaerobic Cocci:			
Acidaminococcus fermentans	30–37	43B, 39C (Tables 5, 10)	181
Megasphaera elsdenii	35–38	44A (Table 5)	72, 182
Veillonella parvula	37	45A, 40C (Tables 5, 10)	179, 180

TABLE 1—*continued*

Group, genus, and species[a]	Optimal growth temp (°C)	Number-letter key to media[b]	References
XII. Gram-Negative Chemolithotropic Bacteria:			
Nitrobacter winogradskyi[i]	25–30	41C (Table 11)	246
Nitrococcus mobilis[i]	25–30	42C (Table 11)	238
Nitrosococcus nitrosus[i, e]	20–25	41C (Table 11)	245
Nitrosolobus multiformis[i, rr]	25–30	43C (Table 11)	237
Nitrosomonas europaea[i, rr]	25–30	43C (Table 11)	236
Nitrosospira briensis[i, rr]	25–30	43C (Table 11)	235
Nitrospina gracilis[i, ss]	25–30	42C (Table 11)	238
Sulfolobus acidocaldarius[i, tt]	70	44C (Table 11)	193
Thiobacillus thiooxidans[i]	28–30	44C (Table 11)	231
Thiobacillus thioparus[i]	28	46C (Table 11)	75
XIII. Methane-Producing Bacteria:			
Methanobacterium ruminantium[c]	39	46A (Table 5)	202
Methanococcus vannielii[c, vv]	37–42	47C (Table 11)	99, 210
Methanosarcina methanica[c, uu]	30–37	47C (Table 11)	209
XIV. Gram-Positive Cocci:			
Aerococcus viridans[o, ww]	30	30A, 48C (Tables 4, 11)	146
Leuconostoc mesenteroides[xx]	30	47A, 49C (Tables 5, 11)	62
Micrococcus luteus[i]	35	30A, 50C (Tables 4, 11)	1, 157
Pediococcus cerevisiae[o]	30	48A, 51C (Tables 5, 11)	94, 134
Ruminococcus flavefaciens[c]	35–45	49A, 52C (Tables 5, 11)	2, 3
Sarcina ventriculi[c]	30–37	50A, 53C (Tables 5, 11)	28
Staphylococcus aureus[yy]	30–37	51B, 54C (Tables 6, 12)	61, 136
Streptococcus equinus[xx, zz]	37	48A, 55C (Tables 5, 12)	216
Streptococcus faecalis[xx]	37	48A, 56C (Tables 5, 12)	168
Streptococcus pyogenes[xx]	37	52A, 57C (Tables 6, 12)	143
XV. Endospore-Forming Rods and Cocci:			
Bacillus cereus[xx]	30–37	30A, 58C (Tables 4, 12)	111
Bacillus subtilis[i]	37	30A, 59C (Tables 4, 12)	111
Clostridium barkeri[c]	30	12A, 53A (Tables 2, 6)	208
Clostridium botulinum[c]	30–40	30A, 60C (Tables 4, 12)	108, 135
Clostridium tetani[c]	37	42A, 61C (Tables 5, 12)	105
Desulfotomaculum nigrificans[c]	55	54A (Table 6)	27, 164
Sporolactobacillus inulinus[o]	35	55A, 62C (Tables 6, 12)	110, 110a
Sporosarcina ureae[i]	30	56A (Table 6)	227
XVI. Gram-Positive Asporogenous Rod-Shaped Bacteria:			
Caryophanon latum[i]	26	63C (Table 12)	—[aaa]
Erysipelothrix rhusiopathiae[i]	35	57A (Table 6)	90
Lactobacillus acidophilus[xx, bbb]	37	58A, 64B (Tables 6, 12)	40, 51, 81
Lactobacillus bulgaricus[xx, bbb]	37	58A, 64B (Tables 6, 12)	40
Lactobacillus casei[xx]	37	58A, 65C (Tables 6, 12)	107
Lactobacillus delbrueckii[xx, bbb, ccc]	37	58A, 64B (Tables 6, 12)	40, 168
Lactobacillus fermentum[xx, ddd]	37	58A, 64B (Tables 6, 12)	40, 168
Lactobacillus heterohiochii[xx, bbb, eee]	37	58A, 64B (Tables 6, 12)	195, 221, 222
Lactobacillus leichmannii[xx, bbb]	37	58A, 64B (Tables 6, 12)	40
Lactobacillus plantarum[xx, ddd]	37	58A, 64B (Tables 6, 12)	40
Lactobacillus 30a[xx]	40	59A, 66C (Tables 6, 12)	33, 71
Listeria monocytogenes[i]	37	12A, 67C (Tables 2, 13)	57, 241
XVII. Actinomycetes and Related Bacteria:			
Actinobifida dichotomica[i]	50–58	30A (Table 4)	41, 116
Actinomyces bovis[c]	37	60A (Table 6)	64, 89, 160
Actinoplanes philippinensis[i]	24–26	61A (Table 6)	37, 220
Amorphosporangium auranticolor[i]	25–27	61A (Table 6)	42
Ampullariella regularis[i]	18–35	61A (Table 6)	42, 184
Arachnia propionica[xx]	37	60A (Table 6)	89, 160
Arthrobacter terregens[i]	20	62A, 68C (Tables 6, 13)	25, 173

TABLE 1—*continued*

Group, genus, and species[a]	Optimal growth temp (°C)	Number-letter key to media[b]	References
Bacterionema matruchotii[xx, fff]	37	63A, 56C (Tables 0, 12)	102
Bifidobacterium bifidum[c, ggg]	36–38	30A (Table 4)	197
Cellulomonas flavigena[i]	30	64A (Table 7)	73
Corynebacterium bovis[i, hhh]	37	30A, 69C (Tables 4, 13)	196
Corynebacterium diphtheriae[i]	37	12A, 70C (Tables 2, 13)	149
Dactylosporangium aurantiacum[i]	37	30A, 71C (Tables 4, 13)	226
Dermatophilus congolensis[i, xx]	37	65A (Table 7)	66
Eubacterium saburreum[c, iii]	37	12A (Table 2)	87
Geodermatophilus obscurus[i]	24–28	66A (Table 7)	132
Kitasatoa purpurea[i]	27	61A, 72C (Tables 6, 13)	138, 166
Kurthia zopfii[i]	25	32A (Table 4)	60
Microbispora rosea[i]		67A (Table 7)	42
Microellobosporia cinerea[i]	26	69A, 73C (Tables 7, 13)	125
Micromonospora chalcea[i]	24–26	68A (Table 7)	95
Micropolyspora brevicatena[i]	28–37	69A, 73C (Tables 7, 13)	125
Mycobacterium phlei[i]	28–45	70A, 74C (Tables 7, 13)	165
Mycobacterium tuberculosis H37[i, jjj]	37	75B (Table 13)	118, 137
Nocardia asteroides[i]	37	30A, 76C (Tables 4, 13)	34
Pilimelia terevasa[i]	30	61A (Table 6)	101
Planobispora longispora[i]	28–37	30A, 72C (Tables 4, 13)	225
Propionibacterium acnes[c, kkk]	35	12A, 71A (Tables 2, 7)	147a, 148
Propionibacterium freudenreichii[c]	30–37	30A, 77C (Tables 4, 13)	47
Pseudonocardia thermophila[i]	40–50	30A, 72A (Tables 4, 7)	83
Rothia dentocariosa[i]	35–37	52A (Table 6)	63
Spirillospora albida[i]	25	61A (Table 6)	39
Sporichthya polymorpha[xx]	28–37	70A (Table 7)	126
Streptomyces albolongus[i]	25–35	70A, 72C (Tables 7, 13)	228
Streptomyces aureofaciens[i, lll]		61A, 30A (Tables 4, 6)	232
Streptomyces erythraeus[i, lll]	25	61A (Table 6)	232
Streptomyces griseus[i]	25	30A (Table 4)	232
Streptomyces niveus[i, lll]		61A (Table 6)	232
Streptomyces noursei[i]		61A (Table 6)	232
Streptomyces venezuelae[i]	25–28	30A, 72C (Tables 4, 13)	55, 232, 234
Streptosporangium roseum[i]	18–35	61A (Table 6)	38
Streptoverticillium baldaccii[i]	27	30A (Table 4)	8
Thermoactinomyces vulgaris	57	30A (Table 4)	233
Thermomonospora curvata	50	67A (Table 7)	83
XVIII. Rickettsiae:[mmm]			
Bartonella bacilliformis[i, nnn]	28	12A (Table 2)	240
Rickettsia quintana[i, ooo]	37	12A (Table 2)	152
XIX. Mycoplasmas:[ppp]			
Acholeplasma laidlawii[xx, ggg]	37	33A, 78B (Tables 4, 13)	170, 171
Mycoplasma mycoides[xx, rrr]	33–37	73A, 79B (Tables 7, 13)	77, 200
Spiroplasma citri[xx]	32	74A (Table 7)	36
Thermoplasma acidophilum[i]	59	75A (Table 7)	43, 201

[a] Bacteria are grouped in accordance with the 8th edition of *Bergey's Manual of Determinative Bacteriology*.

[b] The compositions of media referred to by number-letter key are given in Tables 2–13, as indicated. Undefined or crude media are designated A, semidefined media are designated B, and defined or synthetic media are designated C.

[c] Anaerobic. Anaerobic conditions are achieved by one of the methods described in Chapter 6.

[d] Vitamin B_{12} required by some strains (2 µg per liter of medium).

[e] For this species, modify medium 41C, Table 11, to contain an ammonium salt in place of nitrite; the energy-yielding reaction is the oxidation of NH_3 to NO_3^-. Atmospheric CO_2 is the carbon source.

[f] If carbon sources other than malate are used (e.g., ethanol or butyrate) CO_2 may also be necessary.

[g] Medium 4C will also support growth of *R. molischianum* and *R. tenue* (J. Gibson, R. Tabita, and W. Whitman, personal communication). Neither of these bacterial strains requires biotin.

[h] Requires thiamine and nicotinic acid also (1 mg of each vitamin per liter of medium). This medium will also support growth of *R. capsulata* (whose vitamin requirement is satisfied by thiamine only) and *R. palustris*, which requires *p*-aminobenzoic acid (0.1 mg/liter of medium) in addition to biotin (W. Whitman, J. Gibson, and R. Tabita, personal communication).

[i] Aerobic.

[j] Growth is stimulated by and initially requires an extract from bacterial cells.

TABLE 1—*continued*

[k] *S. hyalina* (151) can be grown on a synthetic medium with glucose, lactose, or sucrose as carbon source, with nitrate or ammonium compounds as nitrogen source, and containing vitamins B_1 and B_{12}. Better growth is obtained with organic nitrogen sources (glutamate or peptone).

[l] Requires vitamin B_{12} (40 µg/liter) instead of biotin.

[m] *C. halobacteròides* requires the addition of 2% NaCl to all growth media.

[n] The vitamin requirement of *C. vibrioides* is satisfied by riboflavin (100 µg/liter).

[o] Microaerophilic.

[p] Carbon dioxide serves as carbon source. Aerate liquid medium with CO_2 for 10 to 15 s before pouring onto a sterilized water-agar slant containing freshly precipitated FeS. Prepare FeS by mixing equimolar quantities of ferrous ammonium sulfate and sodium sulfide in boiling distilled water. Wash several times by decantation with sterile distilled water. Instead of incorporating it in agar, FeS may be introduced by means of a pipette and allowed to settle at the bottom of the culture tube.

[q] CO_2 required; stimulated by humus or soil extracts.

[r] Vitamin requirements: in addition to biotin, thiamine (5 mg/liter) and vitamin B_{12} (0.1 mg/liter) are required.

[s] Nonpathogenic form of virulent *Treponema pallidum*; the latter has not been successfully cultured in vitro.

[t] Greater cell yields are obtained by buffering the medium with 0.033 M N-2-hydroxyethylpiperazine-N'-2-ethanesulfonic acid (HEPES), piperazine-N,N'-bis(2-ethanesulfonic acid) (PIPES), or N-tris(hydroxymethyl)methyl-2-aminoethanesulfonic acid (TES) or by intermittent additions of $NaHCO_3$. Attempts to grow the organism on a completely synthetic medium have been unsuccessful. The yeast extract concentration can be reduced to 0.05%, but not eliminated. The vitamin requirement (biotin, riboflavin, and vitamin B_{12}) is probably more extensive than reported, since the assay medium contains 0.5% yeast extract, a rich source of vitamins. (HEPES, etc., are obtainable from Sigma or Calbiochem.)

[u] A defined medium has also been described which supports growth of this bacterium under microaerobic conditions (18).

[v] *Agrobacterium rubi* requires, in addition, pantothenate (100 µg/liter) and nicotinic acid (100 µg/liter). *A. tumefaciens* grows well in the unsupplemented glucose-NH_4Cl-salts medium.

[w] In nitrogen-deficient media, these microorganisms fix atmospheric nitrogen; a trace of molybdate is necessary for nitrogen fixation (replaceable by vanadate for *Azotobacter*). Ammonia, nitrate, or amino acids can be utilized as the nitrogen source by *A. chroococcum*.

[x] Extremely pH tolerant (range, pH 4 to 10).

[y] Boil peeled potatoes (125 g) in a glycerol-water mixture until the potatoes are soft; strain them through gauze, allow to stand until the supernatant is clear, and decant; add salt and make up to 1 liter. Add 15 g of agar and heat to dissolve. Sterilize in 250-ml amounts; aseptically add 50 ml of defibrinated sheep blood to each 250 ml of agar base and mix. For liquid media, omit the agar.

[z] This medium also supports moderate to poor growth of strains of *B. suis* and *B. abortus*. Brucella broth (127), however, is suitable for culturing these species.

[aa] Aseptically add 25 ml of packed, human blood cells or 50 ml of defibrinated rabbit blood to the autoclaved and cooled medium.

[bb] Atmosphere of 40% methane and 60% air; methanol is the only other known carbon source.

[cc] Only 6 of 146 compounds tested can serve as sources of carbon. They are: β-hydroxybutyrate, acetate, butyrate, valerate, pyruvate, and succinate.

[dd] Restricted to a few carbohydrates for a carbon source. Methionine is a required amino acid.

[ee] Carbohydrates (including glucose) are not utilized as carbon sources; however, a large number of organic compounds (including saturated dicarboxylic acids, glycolate, and testosterone) can serve.

[ff] Some strains require or are stimulated by methionine (0.02%) and/or glutamic acid (0.1%) and a few, by nicotinic acid (100 µg/liter). No biotin requirement has been demonstrated; in all other respects the nutritional requirements resemble those of *Agrobacterium rhizogenes* (46).

[gg] Growth is dependent upon an atmosphere of high humidity (198, 199). Adenine, adenosine, riboflavin, and flavin adenine mononucleotide are markedly inhibitory.

[hh] Of the amino acids, only methionine is essential; however, other amino acids stimulate growth. The methionine requirement can be satisfied by either the D or the L isomer or, partially, by vitamin B_{12}. Growth on the completely synthetic medium is not as good as in the presence of yeast extract.

[ii] Add sucrose (5%) to the nutrient agar.

[jj] Grows on Koser's citrate medium, which contains the same components as the defined medium cited and is available commercially (Difco).

[kk] Add 10% serum to nutrient broth for a liquid medium. Growth on solid media is favored by moisture.

[ll] Aerobic, but some strains will grow anaerobically also.

[mm] Anaerobicity is obtained by flushing with argon. Growth and growth rates on the synthetic medium are about half those on complex medium. The addition of a vitamin supplement and other factors does not improve growth on the synthetic medium.

[nn] Stock cultures are maintained on 40% rumen fluid-glucose-cellobiose-starch agar.

[oo] The defined medium differs from that cited in the use of 0.02 M phosphate buffer and carbon sources other than D-glucose. Acetate, butyrate, alcohols, amino acids, pentose sugars, hydrocarbons, and aromatic compounds can serve as carbon sources.

[pp] The medium differs from that cited by the addition of amino acids (see footnote w in Table 10). Lactate and succinate are the principal carbon and energy sources utilized by *B. catarrhalis*.

[qq] Much heterogeneity has existed in reports on nutritional requirements of *N. gonorrhoeae*. Besides the high incidence of naturally occurring variants, this bacterium readily undergoes mutation, and frequent transfer in the laboratory tends to select for the less fastidious strains, resulting in populations which differ genetically from the primary isolates (C. E. Lankford, personal communication).

[rr] Oxidizes ammonia to nitrite and fixes CO_2; cannot be grown in seawater.

[ss] Markedly inhibited by most organic compounds at 10^{-3} M.

[tt] *S. acidocaldarius* is a facultative autotroph; it grows more rapidly on mineral base plus 0.1% yeast extract than on sulfur.

[uu] The studies were made on a slightly contaminated culture; further purification was not attempted because of the very slow rate of development of the *Methanosarcina*. The authors apparently felt that the results of their experiments were not affected by this contamination.

TABLE 1—*continued*

vv To 1 liter of basal medium add the following: 15 g of sodium formate and (after autoclaving) a sterile solution containing 0.5 g of potassium pyruvate. The medium has been revised by Jones and Stadtman (99) and supplemented with selenite (1 μM) and tungstate (100 μM).

ww Plus 0.5% glucose.

xx Facultative anaerobe.

yy Strains vary from facultative anaerobes to aerobes.

zz Formerly known as *Leuconostoc mesenteroides* P-60, an erroneous classification (139).

aaa R. A. Kele and E. McCoy, Bacteriol. Proc., p. 66, 1970.

bbb For good growth, this species requires the addition of the following components to 1 liter of medium 58A: 300 mg of ascorbic acid and 5 ml of Tween 80.

ccc At least one *L. delbrueckii* strain (namely, *L. delbrueckii* 730) requires pyridoxamine phosphate (0.002 mg/liter) and thymidine (4 mg/liter) in addition.

ddd Add 20 μg of calcium pantothenate in place of pantethine.

eee In the absence of acetate, this microorganism requires a growth factor called hiochic acid or mevalonic acid. (available as the lactone from Sigma Chemical Co.). It is supplied in the medium at a concentration of 1 μg/liter.

fff Hemin (0.2 mg/liter) stimulates growth under anaerobic conditions. CO_2 is also stimulatory.

ggg CO_2 required for growth.

hhh All *C. bovis* strains require long-chain unsaturated fatty acids (palmitoleic, ricinoleic, or oleic acid); 0.1% Tween 80 effectively supplies this requirement.

iii Blood agar is required for continued subculture. The following is recommended as a liquid medium: nutrient broth enriched with 0.3% yeast extract, 0.1% cysteine·HCl, 1% glucose, and 5% human ascitic fluid. For another strain of *Eubacterium*, the addition of substrate amounts of arginine to a peptone-based medium markedly increases growth (207).

jjj The genus includes parasites, saprophytes, and intermediate forms. Saprophytic strains grow in simple media; others require more complex media or supplements (e.g., mycobactin), and still others have not been cultivated outside living cells. Growth is slow on all media.

kkk Boil all liquid media for 10 min and cool rapidly just prior to inoculation. Incubate all cultures under an atmosphere of 80% N_2, 10% H_2, and 10% CO_2 in anaerobic jars.

lll These bacteria grow in Czapek's medium without supplementation with peptone.

mmm These bacteria, except the few noted, are obligate parasites and do not survive outside animal cells. They can be grown in chicken embryo yolk sacs and in cell cultures of vertebrate tissues.

nnn Can be cultivated in vitro on nonliving media. Serum is essential and hemoglobin is favorable to growth. Glutathione and ascorbic acid are sometimes stimulatory. Most often grown on semisolid medium (1% agar) but will also grow on solid media such as blood agar. *B. bacilliformis* can also grow in tissue culture in primary explants of rat and guinea pig lung, bone marrow, or spleen. In embryonated eggs, it grows in the chorioallantoic and yolk fluids. Optimal pH of the culture medium is 7.8.

ooo To blood agar base add hemin, 40 μg/liter (or hemoglobin, 1 mg/liter). Incubate in an atmosphere of 95% air–5% CO_2. A detoxifying substance such as bovine albumin (2 g/liter), starch (2 g/liter), or charcoal (0.2 g/liter) is also necessary.

ppp See a recent review (169) for more information on the nutrition of this group of bacteria.

qqq Heart infusion broth is enriched with 20% sterile, heated (56°C for 30 min) horse serum plus 10% (vol/vol) fresh yeast extract solution (25 g of yeast cells in 100 ml of solution).

rrr All strains require sterols and long-chain fatty acids for growth.

replaced. Until a growth factor is isolated and characterized, its effect on the growth of an appropriate assay organism often is the only means of following its purification from a natural material.

Quantitative growth-response assays in principle are based on the arithmetic (linear) response in the extent (not the rate) of growth as the amount of a limiting nutrient is increased within a limited range.

The reproducibility of microbiological assays is usually given as ±10%; however, under carefully controlled conditions it may exceed ±3%. In general, an assay for an essential nutrient can be performed with a higher degree of accuracy than for a stimulatory one.

For vitamin and amino acid assays, species of bacteria from the genera *Lactobacillus, Streptococcus,* and *Pediococcus* have been particularly useful since they have complex requirements; i.e., they require the addition of a large number of nutrients to the growth medium. One bacterium can be used for the assay of several vitamins or amino acids. For instance, an assay

method employing *Streptococcus equinus* ATCC 8042 (formerly known as *Leuconostoc mesenteroides* P-60) permits the determination of all amino acids except alanine (216); the latter amino acid can be determined by use of the same medium with the addition of folinic acid (see footnote *b*, Table 12) with *P. cerevisiae* 8081 as the test bacterium.

Details of a bacteriological assay method will be illustrated with the method of Steele et al. (216) for the determination of arginine using *S. equinus.*

7.3.1. Amino Acids (Arginine)

Arginine is one of 17 amino acids which can be determined using *S. equinus.* Procedures for other amino acids, vitamins, etc., are similar. References to specific methods for other growth factors are given in the review by Guirard and Snell (70). Bacterial mutants auxotrophic for one or more compounds have also been very useful in studying the pathways of synthesis and degradation of vitamins and amino acids.

TABLE 2. *Compositions of undefined (A) and semidefined (B) media 1 through 12 for the bacteria listed in Table 1*

Component	Amt (in units given in column 1)/liter for medium:												
	1B[a]	2A	3A	4A[b]	5A[c]	5B[d]	6B	7B	8A	9A	10A[e]	11A[b]	12A
Glucose[f] (g)					1.0	1.0		5.4	1.0		1.0		
Glycerol (g)										10			
(NH₄)₂SO₄ (g)	1.0											0.5	
KNO₃ (g)					0.5	0.1							
Na₂HPO₄ (g)											5.24	1.74	
KH₂PO₄ (g)		1.13						1.4	0.04		2.77	1.06	
K₂HPO₄ (g)	1.0	0.48	0.5			0.25			0.04				
MgSO₄·7H₂O	0.5	1.0	0.5			0.1	1.0	3.0	0.075			0.1	
NaCl (g)	0–30												
CaCl₂·2H₂O (g)		0.2				0.1	1.0		0.05				
FeSO₄·7H₂O (mg)	0.5												
Trace elements					Yes[g]	Yes[g]							
Vitamins						Yes[h]	Yes[i]						
Beef extract (g)			1.0										
Yeast extract (g)		3.0	1.0		5.0					1.0		0.05	
Acid-hydrolyzed casein (e.g., Casamino Acids) (g)						1.0							
Enzyme-hydrolyzed casein (e.g., Casitone) (g)		20					0.5	10		5.0			
Peptone (g)		3.0	1.0						0.5		1.0	0.05	
Tryptone (g)					5.0								
Special media													Yes[j]
pH	7.3	6.8	7.0	7.0	7.0	7.5	7.2	7.0–7.2	7.0[k]	7.0	7.0	7.0	7.0–7.4

[a] NaHCO₃ (2 g/liter) serves as a source of CO₂ and Na₂S·9H₂O (1.0 g/liter) is used as a reducing agent and source of S. Use tap water in place of distilled water to supply trace elements.

[b] Use tap water in place of distilled water.

[c] Use filtered seawater for marine strains; it can be replaced by a commercial seawater substitute or by a mixture of the following salts, per liter: NaCl, 25 g; MgSO₄·7H₂O, 5 g; CaCl₂·2H₂O, 1.0 g; KCl, 1.0 g. Add sodium glycerophosphate (0.1 g/liter) and tris(hydroxymethyl)aminomethane (Tris) buffer (Sigma) (1.0 g/liter).

[d] Medium used for freshwater strains. Supplement the medium with sodium glycerophosphate (0.1 g/liter) and Tris buffer (Sigma) (1.0 g/liter).

[e] Twenty milliliters of Hutner's vitamin-free mineral base. Hutner's mineral base as modified by Cohen-Bazire et al. (35) has the following composition: nitriloacetic acid (dissolved and neutralized with KOH), 10 g; MgSO₄·7H₂O, 14.45 g; CaCl₂·2H₂O, 3.33 g; (NH₄)₆Mo₇O₂₄·24H₂O, 9.25 mg; FeSO₄·7H₂O, 99 mg; metals 44 (see below), 50 ml; distilled water to 1 liter; pH adjusted to 6.6 to 6.8. Metals 44 contains, in milligrams per 100 ml: ethylenediaminetetraacetic acid, 250; ZnSO₄·7H₂O, 1,095; FeSO₄·7H₂O, 500; MnSO₄·H₂O, 154; CuSO₄·5H₂O, 39.2; Co(NO₃)₂·6H₂O, 24.8; Na₂B₄O₇·10H₂O, 17.7; add a few drops of 6 N H₂SO₄ to retard precipitation.

[f] For most media it is recommended that glucose be sterilized (either by autoclaving or by filtering through appropriate membranes) separately from other components of the medium and then combined aseptically.

[g] Add 1 ml of the following trace element solution per liter of medium: soluble salts of B, Fe, Mn (each as element, 0.5 mg/liter); soluble salts of Co, Cu, Mo, Zn (each as element, 0.01 mg/liter).

[h] Thiamine hydrochloride (1 mg/liter) and vitamin B₁₂ (1.0 μg/liter of final medium).

[i] Vitamin B₁₂ (500 μg/liter).

[j] Suspend 40 g of blood agar base (available commercially) in 1 liter of distilled water, heat to boiling, autoclave, cool, and (while still liquid) add aseptically 5% sterile defibrinated blood. Blood agar base has the following composition: infusion from beef heart, 500 g; tryptose (Difco), 10 g; sodium chloride, 5 g; agar (Difco), 15 g. Add 10 g of succinic acid per liter.

[k] For *Leptothrix ochracea*, maintain pH at ~7.0 during incubation by addition of sterile 0.5 N NaOH.

Organism.

S. equinus ATCC 8042 is obtainable from the American Type Culture Collection, 12301 Parklawn Drive, Rockville, MD 20852. Maintain a stock culture of the bacterium in stabs on a medium composed of 1.5% agar, 1% glucose, and 1% yeast extract. Transfer the culture to fresh medium once a month. Incubate the stabs at 37°C for 24 h. Make at least two transplants, one to be reserved for use only when the next monthly transfer is due and the other for use in preparing inocula as needed for assay.

Complete basal medium.

The composition of the basal medium is given in Table 14. Because of the complexity of the medium, there are numerous ways in which it can be compounded, and various workers may prefer to alter that recommended here. It is convenient to mix various groups of ingredients (e.g., amino acids, mineral salts, purine and pyrimidine bases, vitamins) that can be stored together either in solid form or as stock solutions and then to combine all of the ingredients in appropriate amounts to prepare the desired

TABLE 3. *Compositions of undefined (A) and semidefined (B) media 13 through 24 for the bacteria listed in Table 1*[a]

Component	\multicolumn{12}{Amt (in units given in column 1)/liter for medium:}
	13B*	14A[b]	15A[c]	16A*[d,e]	17A[f]	18A[g]	19A*[p]	20A*[h]	21A[i]	22B	23A[j]	24A
Glucose[k] (g)		5.0	5.0	1.0		100		25			1–3.0	10
Glycerol (g)	0.1						20					
$(NH_4)_2SO_4$ (g)	0.25			1.0							1.0	
KNO_3 (g)									0.1	2.5		0.6
Na_2HPO_4 (g)	1.0									0.21	5.24	0.45
KH_2PO_4 (g)	0.3								0.05	0.09	2.77	
K_2HPO_4 (g)									0.05			
$MgSO_4 \cdot 7H_2O$ (g)				1.0					20	0.2		0.1
$MnSO_4 \cdot H_2O$ (g)									0.005			
NaCl (g)	1.0	2.5		5.0	0.002	5.6		5.0	120			
$CaCl_2 \cdot 2H_2O$ (g)												0.2
$FeSO_4 \cdot 7H_2O$ (mg)					2.0				5.0	1.0		10
Trace elements									Yes[l]	Yes[m]		
Vitamins	Yes[n]									Yes[o]		
Yeast extract (g)		5.0	2.0	2.0		2.0			5.0		5.0	3.0
Acid-hydrolyzed casein (e.g., Casamino Acids) (g)									15	1.0		
Enzyme-hydrolyzed casein (e.g., Casitone) (g)		15										
Peptone (g)			2.0		10							
pH	7.4	7.4	7.6	7.0		5.4–6.3	7.0	6.8	6.8–7.0	7.0	6.8	7.5

[a] Prepared media that are available commercially are marked with an asterisk.

[b] The medium also includes the following components: cysteine hydrochloride, 0.75 g; sodium thioglycolate, 0.5 g. Add 1.0 ml of filter-sterilized rabbit serum to 9 ml of sterile basal medium.

[c] Add 0.5 g of sodium thioglycolate per liter of medium. Greater yields are obtained by buffering the medium with 0.033 M N-2-hydroxyethylpiperazine-N'-ethanesulfonic acid (HEPES), N-tris(hydroxymethyl)methyl-2-aminoethanesulfonic acid (TES), or piperazine-N,N'-bis(2-ethanesulfonic acid) (PIPES) or by intermittent additions of $NaHCO_3$. (Buffers are obtainable from Calbiochem or Sigma.)

[d] Additional components: $NaHSO_3$, 0.1 g; peptic digest of animal tissue USP*, 10 g; pancreatic digest of casein USP*, 10 g. Medium is obtainable commercially as brucella broth.

[e] For *Campylobacter*, add 0.16% agar to the brucella broth* to attain partial anaerobicity.

[f] Add succinic acid (10 g/liter).

[g] Add $CaCO_3$ (30 g/liter).

[h] Combine the listed ingredients with the following: cysteine hydrochloride, 1.0 g; pancreatic digest of heart muscle*, 3.0 g; papaic digest of soya meal*, 10 g; and thiamine hydrochloride, 0.05 mg. Make up to 1 liter.

[i] Add 2 g of sodium succinate and 0.5 g of sodium citrate per liter.

[j] Twenty milliliters of Hutner's vitamin-free mineral base. Hutner's mineral base as modified by Cohen-Bazire et al. (35) has the following composition: nitriloacetic acid (dissolved and neutralized with KOH), 10 g; $MgSO_4 \cdot 7H_2O$, 14.45 g; $CaCl_2 \cdot 2H_2O$, 3.33 g; $(NH_4)_6Mo_7O_{24} \cdot 24H_2O$, 9.25 mg; $FeSO_4 \cdot 7H_2O$, 99 mg; metals 44 (see below), 50 ml; distilled water to 1 liter; pH adjusted to 6.6 to 6.8. Metals 44 contains, in milligrams per 100 ml: ethylenediaminetetraacetic acid, 250; $ZnSO_4 \cdot 7H_2O$, 1,095; $FeSO_4 \cdot 7H_2O$, 500; $MnSO_4 \cdot H_2O$, 154; $CuSO_4 \cdot 5H_2O$, 39.2; $Co(NO_3)_2 \cdot 6H_2O$, 24.8; $Na_2B_4O_7 \cdot 10H_2O$, 17.7; add a few drops of 6 N H_2SO_4 to retard precipitation.

[k] For most media it is recommended that glucose be sterilized (either by autoclaving or filtering through appropriate membranes) separately from other components of the medium and then combined aseptically.

[l] Add (mg/liter): $CoSO_4 \cdot 7H_2O$, 0.05; $Na_2MoO_4 \cdot 2H_2O$, 0.2; $ZnSO_4$, 0.2; $CuSO_4 \cdot 5H_2O$, 0.01.

[m] Add (μg/liter): Cu (as $CuSO_4 \cdot 5H_2O$), 50; B (as H_3BO_3), 10; Mn (as $MnSO_4 \cdot 4H_2O$), 10; Zn (as $ZnSO_4 \cdot 7H_2O$), 70; and Mo (as MoO_3), 10.

[n] Thiamine hydrochloride, 5 mg. To nine parts of the sterile basal medium add one part of filter-sterilized albumin-fatty acid supplement* which has the following composition: bovine albumin fraction V, 20 g; $CaCl_2 \cdot 2H_2O$, 0.01 g; $MgCl_2 \cdot 6H_2O$, 0.01 g; $ZnSO_4 \cdot 7H_2O$, 8.0 mg; $CuSO_4 \cdot 5H_2O$, 0.6 mg; $FeSO_4 \cdot 7H_2O$, 0.1 g; vitamin B_{12}, 0.4 mg; Tween 80, 2.5 ml; distilled water to 200 ml.

[o] Add pantothenic acid, calcium salt (0.2 g/liter); an unidentified substance extracted from agar is also necessary for growth (3 ml/liter, 90 mg of solids; it is not necessarily a vitamin). No growth occurs on complex media.

[p] Add an extract of potatoes prepared by boiling peeled, sliced potatoes in a mixture of glycerol and water until they are soft. Strain, allow to settle, and decant.

quantity of medium at twice its final concentration ("double-strength" medium). For assay purposes, the particular compound to be determined is omitted from such mixtures. The mixtures are prepared as follows:

Amino acids. Weigh out the amino acids in amounts designated in Table 14, minus the amino acid to be assayed. Combine, grind in a mortar, and store them in a well-stoppered bottle in a dark place. Mixtures for assaying all of the amino acids desired can be prepared and stored in this way.

In general, only the L isomers of amino acids are active. The DL isomers are sometimes used for convenience; if a medium calls for x mg of an L isomer and only the DL isomer is available, use $2x$ mg of the latter.

Mineral salts. Dissolve 25 g of KH_2PO_4 and 25 g of K_2HPO_4 in distilled water, and make up to 250 ml. This solution is known as "salts A." Combine the remaining salts as follows ("salts B"): 10 g of $MgSO_4 \cdot 7H_2O$, 0.5 g of $FeSO_4 \cdot$ $7H_2O$, 0.8 g of $MnSO_4 \cdot 4H_2O$, and 0.5 g of NaCl. Dissolve and make up salts B to 250 ml with distilled water, adding 1 drop of concentrated HCl to prevent formation of a precipitate. Both solutions are stable at room temperature, but are better refrigerated to inhibit mold growth in salts A and oxidation of Fe^{2+} in salts B.

Purine and pyrimidine bases. Dissolve xanthine and guanine by warming in dilute HCl; dissolve adenine and uracil in warm water. Combine them, and add water so that the concentration of each base in the mixture is 1 mg/ml. The solution is stable when refrigerated.

Vitamins. Combine all of the vitamins except biotin and folic acid in 10 times the amounts listed in Table 14. Prepare stock solutions of biotin and of folic acid separately at a concentration of 1 mg/ml. Biotin is solubilized by warming; folic acid is brought into solution by adding a small quantity of NH_4OH. Add 0.02 ml of the biotin solution and 0.2 ml of the folic acid solution to the combined vitamins, make up to

TABLE 4. *Compositions of undefined media (A) 25 through 37 for the bacteria listed in Table 1* [a]

Component	Amt (in units given in column 1)/liter for medium:												
	25A[b]	26A	27A	28A	29A[c,d]	30A	31A*	32A	33A*	34A[e]	35A[f]	36A[g]	37A[h]
Glucose[i] (g)				1.0			2.0			1.0	2.0	3.0	3.0
(NH$_4$)$_2$SO$_4$ (g)												0.45	4.0
Na$_2$HPO$_4$ (g)					2.5								
KH$_2$PO$_4$ (g)												0.9	2.7
MgSO$_4$·7H$_2$O (g)												0.02	0.2
MnSO$_4$·H$_2$O (g)												0.01	
NaCl (g)	5.0	5.0		10	30	5.0	5.0	5.0	5.0		1.0	0.9	
CaCl$_2$·2H$_2$O (g)												0.02	
Yeast extract (g)				5.0	3.0			3.0		2	5.0		
Peptone (g)	17							10	10	2			
Tryptone (g)		8.0		10			20						
Nutrient broth* (g)						8.0					16		5.0
Special media			Yes[j]						Yes[k]				
pH	7.1	7.0		7.0	6–9	7.0	7.3	7.0–8.0	7.4	6.8	7.0	6.7	6.6–6.9

[a] Prepared media that are available commercially are marked with an asterisk.

[b] Add (g/liter): lactose, 10; proteose peptone, 3.0; and bile salts, 1.5.

[c] Use filtered seawater for marine strains; it can be replaced by a commercial seawater substitute or by a mixture of the following salts (g/liter): NaCl, 25; MgSO$_4$·7H$_2$O, 5; CaCl$_2$·2H$_2$O, 1.0; and KCl, 1.0. Add sodium glycerophosphate (0.1 g/liter) and Tris buffer (Sigma) (1.0 g/liter).

[d] Proteose peptone (5 g/liter).

[e] Dissolve 6.05 g of Tris and 5.8 g of malic acid in water and adjust to pH 6.8 with NaOH.

[f] Add separately sterilized solutions of L-ascorbic acid (0.1 g) and L-cystine (0.1 g) to the other components just before use. Sterilize L-ascorbic acid by filtration and L-cystine either by filtration or by autoclaving.

[g] Add 200 ml of clarified rumen fluid (see below), 2 g of Trypticase (BBL), and 1.0 mg of hemin to the other components of the medium; adjust to pH 6.7, make up to 1 liter, and autoclave. Just before use, add sterile solutions of L-cysteine·HCl·H$_2$O (0.5 g) and Na$_2$CO$_3$ (4.0 g) to the autoclaved medium. Incubate the cultures under a partial CO_2 atmosphere. Clarified rumen fluid is prepared from rumen fluid freshly collected from a cow on an alfalfa hay-grain ration. Filter through gauze and allow to sediment; siphon off the clear liquid and centrifuge at 14,000 rpm for 20 min (21).

[h] Add 17.8 g of Na$_2$SO$_4$ to the medium.

[i] For most media it is recommended that glucose be sterilized (either by autoclaving or filtering through appropriate membranes) separately from the other components of the medium and then combined aseptically.

[j] To prepare chocolate agar, supplement Trypticase soy broth* with 3% (wt/vol) yeast extract and 1% agar. After sterilization, add defibrinated sheep blood (4% final concentration) and hold the medium at 70°C for about 15 min. Then cool to 50°C and add IsoVitaleX* (5 ml/liter of medium) and CaCl$_2$ (final concentration, 0.5 mM).

[k] Beef heart infusion agar: remove fat from 450 g of beef heart infused in water overnight at 4 to 6°C, add peptone and NaCl as indicated above, boil for 30 min, filter, combine with dissolved agar (2% in final medium), adjust to pH 7.4, and autoclave.

TABLE 5. *Compositions of undefined (A) and semidefined (B) media 38 through 50 for the bacteria listed in Table 1*

Component	Amt (in units given in column 1)/liter for medium:												
	38A[a]	39A[b]	40A[c]	41A[d]	42A[e]	43B[f]	44A[g]	45A[h]	46A[i]	47A[j]	48A[k]	49A	50A
Glucose[l] (g)	5.0	10	10			5.0				10	10		20
$(NH_4)_2SO_4$ (g)							0.5		2.89				
KNO_3 (g)	1.0												
Na_2HPO_4 (g)	5.0					2.25							
KH_2PO_4 (g)						0.75	0.5		1.0				
K_2HPO_4 (g)									1.0		2.5		
$MgSO_4 \cdot 7H_2O$ (g)						0.096	0.3		0.1	0.2			
$MnSO_4 \cdot H_2O$ (g)						0.001				0.05			
NaCl (g)		5.0							2.0				
$CaCl_2 \cdot 2H_2O$ (g)		15			0.074	0.002			0.1				
$FeSO_4 \cdot 7H_2O$ (mg)						2.0							
Trace elements						Yes[m]							
Amino acids						Yes[n]							
Vitamins						Yes[o]							
Beef extract (g)	3.0	1.0							10				
Yeast extract (g)	1.0	5.0	2.0	5.0	3.0		4.0	5.0		5.0	10		2.0
Acid-hydrolyzed casein (e.g., Casamino Acids) (g)						20							
Enzyme-hydrolyzed casein (e.g., Casitone) (g)											10		
Peptone (g)										10			
Tryptone (g)		5.0											
Special media						Yes[p]						Yes[q]	
pH	7.6	7.2–7.4	6.5–7.0	7.0	7.4	7.0	7.4	6.6	6–8	6.7	5.5	7.0	

[a] Add the following: cysteine·HCl·H_2O, 0.5 g; soluble starch, 2.0 g; proteose peptone, 10 g; and sterile serum, 50 ml. Incubate under a 90% N_2–10% CO_2 atmosphere.

[b] To the listed components add the following: cysteine·HCl·H_2O, 0.05 g; sodium thioglycolate, 2.0 g; sodium formaldehyde sulfoxylate, 1.0 g; proteose peptone, 10 g; and hemin, 0.002 g. For primary isolation, 5% CO_2 is essential.

[c] Add (per liter): cysteine·HCl·H_2O, 0.5 g; and Trypticase (BBL), 10 g. Incubate under 95% N_2–5% CO_2.

[d] Requires complex media. In addition to the components listed, the medium contains the following: butyrate, 1 to 3 g; oleic acid, 30 mg; infusion from 25 g of beef heart muscle; and sheep serum, 100 ml.

[e] Sodium glutamate (10 mg/liter), ribonucleic acid (250 mg/liter).

[f] Purine-pyrimidine supplement (per liter): 10 mg each of guanine, uracil, and hypoxanthine. Cultures on solid media are incubated in anaerobic jars flushed with 95% N_2–5% CO_2.

[g] Add (per liter): thioglycolic acid, 0.3 g; soluble starch, 20 g; and DL-sodium lactate, 13 g. Incubate under an atmosphere containing 5% or more CO_2, the remainder being H_2.

[h] Add (per liter): lactic acid (85%), 10 g; sodium thioglycolate, 0.75 g; Tween 80, 0.1 g; and Trypticase, 10 g. Adjust pH with solid K_2CO_3.

[i] Add 6 g of $NaHCO_3$ and rumen fluid to 30%. The atmosphere is 80% H_2–20% CO_2. Acetate can serve as carbon source. Some strains require 2-methyl-n-butyrate, a few amino acids, and an as yet unidentified vitamin-like growth factor present in rumen fluid.

[j] Add (per liter): ammonium citrate, 5.0 g; sodium acetate, 2.0 g; and Tween 80, 1 g.

[k] Add potassium acetate to a concentration of 0.25%.

[l] For most media it is recommended that glucose be sterilized (either by autoclaving or filtering through appropriate membranes) separately from other components of the medium and then combined aseptically.

[m] Add (per liter): $CoCl_2 \cdot 6H_2O$, $VSO_4 \cdot 7H_2O$, $ZnSO_4$, and $NaMoO_4 \cdot 2H_2O$, 1 mg each; $CuSO_4 \cdot 5H_2O$, 0.5 mg; and nitriloacetic acid, 20 mg.

[n] DL-Tryptophan (100 mg/liter), L-cysteine hydrochloride (350 mg/liter).

[o] Per liter of medium, add the following: pyridoxal hydrochloride and calcium pantothenate, 1 mg each; thiamine hydrochloride, niacin, and riboflavin, 50 μg each; p-aminobenzoic acid, 10 μg; biotin, 2 μg; folic acid and vitamin B_{12}, 1 μg each.

[p] Brain heart infusion broth (available commercially), 37 g/liter.

[q] Growth is abundant on clarified rumen fluid-cellobiose agar or broth. Clarified rumen fluid is prepared from rumen fluid freshly collected from a cow on an alfalfa hay-grain ration. Filter through gauze and allow to sediment; siphon off the clear liquid and centrifuge at 14,000 rpm for 20 min (21).

TABLE 6. *Compositions of undefined (A) and semidefined (B) media 51 through 63 for the bacteria listed in Table 1*

Component	Amt (in units given in column 1)/liter for medium:												
	51Bᵃ	52A	53Aᵇ	54Aᶜ	55Aᵈ	56A	57Aᵉ	58Aᶠ	59Aᵍ	60Aʰ	61Aⁱ	62Aʲ	63A
Glucose^k (g)	10	18		10	10		12	2.5		5.0		1.0	
(NH₄)₂SO₄ (g)						12.7				1.0			
KNO₃ (g)											3.0		
Na₂HPO₄ (g)	7.0					4.5	30						
KH₂PO₄ (g)	0.5	8.7	0.04			0.4	1.0		3.3	15			
K₂HPO₄ (g)	5.5	8.7	0.04						3.3		1.0	2.0	
MgSO₄·7H₂O (g)	0.1				1.5		0.2		0.4	0.2	0.5	0.2	
MnSO₄·H₂O (g)			0.008				0.002		0.02	0.0015			
NaCl (g)	2.0		0.08						0.2		0.5		
CaCl₂·2H₂O (g)			0.008							0.02			
FeSO₄·7H₂O (mg)									20	40	10	10	
Trace elements	Yesˡ												
Amino acids	Yesᵐ		Yesⁿ						Yesᵒ				
Vitamins	Yesᵖ						Yesq		Yesʳ	Yesˢ			
Beef extract (g)						5.0							
Yeast extract (g)		10			10		3.0	10	10	1.0		1.0	1.0–2.0
Acid-hydrolyzed casein (e.g., Casamino Acids) (g)	3.0						4.0		7.5			2.0	
Enzyme-hydrolyzed casein (e.g., Casitone) (g)										4.0			
Peptone (g)			20	5.0	10	5.0					20		
Special media		Yesᵗ											Yesᵘ
pH	7.4	7.4		7.2		8.8	7.8–8.1	7.0	5.4	7.2		6.8	

ᵃ Other components which were either essential or stimulatory (mg/liter): sodium acetate, 820; uracil, 112; xanthine, 152; and adenine, 135.

ᵇ Add 400 mg of NaHCO₃ per liter.

ᶜ Add 1.5 g of Na₂SO₄ per liter. Use tap water instead of distilled water. Glucose may be replaced by sodium lactate (3.5 g). In the presence of pyruvate as substrate, sulfate is not required for growth.

ᵈ Add 10 g of sodium acetate per liter.

ᵉ Add 1.0 g of sodium citrate, 0.3 g of sodium thioglycolate, 0.06 g of sodium oleate, and 3.0 g of saponin per liter of medium.

ᶠ Most of the lactic acid bacteria can be maintained by monthly transfer in stab culture using this medium solidified with 2% agar. All components are autoclaved together. Broth cultures should be supplemented with the inorganic salts listed in Table 14. Note that these are given as amount per liter of double-strength medium.

ᵍ Include the following components in the medium (per liter): sucrose, 10 g; potassium acetate, 1.5 g; and ascorbic acid, 0.5 g. This medium serves as the inoculum medium and for mass culture. When solidified with 1.5 to 2% agar, it is used to maintain the microorganism by transferring every 2 weeks in stab culture. All components are autoclaved together.

ʰ Also include (per liter of medium): cysteine·HCl, 1 g; soluble purified starch, 1 g; NaMoO₄·2H₂O, 1.5 mg; adenine sulfate, uracil, xanthine, and thymine, 20 mg each; coenzyme A, 100 μg; and oleic acid, 1 mg. The final medium supports good growth of *Actinomyces israelii* strains, as well as *A. bovis*, and also several species of *Corynebacterium*, *Lactobacillus bifidus*, and *Leptotrichia buccalis*. Suspend starch in a small amount of cold water, add to boiling water, and then add the other ingredients, except glucose, which is autoclaved separately. This medium is also used for maintaining the culture. Use 0.7% agar for stabs.

ⁱ Add 30 g of sucrose.

ʲ Also include (per liter of medium): ammonium hydrogen citrate, 1.0 g; and terregens factor, 100 μg.

ᵏ For most media it is recommended that glucose be sterilized (either by autoclaving or filtering through appropriate membranes) separately from other components of the medium and then combined aseptically.

ˡ Add 1 mg of ferric citrate per liter.

ᵐ Add 100 mg of L-cysteine hydrochloride and 50 mg of DL-tryptophan per liter.

ⁿ Add 0.5 g of cysteine·HCl per liter.

ᵒ Add 1 g of histidine·HCl·H₂O per liter.

ᵖ Add (per liter): nicotinic acid, 1 mg; thiamine hydrochloride, 1 mg; and biotin 1 μg.

q Add (per liter): riboflavin, 0.5 mg; thiamine, 1.0 mg; and calcium pantothenate, 1.0 mg.

ʳ Add 16.5 μg of pyridoxamine·HCl per liter.

ˢ Add (per liter): thiamine·HCl, 2 mg; pyridoxine·HCl, 1.0 mg; pyridoxamine·2HCl, 0.5 mg; pyridoxal·HCl, 0.5 mg; calcium pantothenate, 2.0 mg; riboflavin, 2.0 mg; nicotinic acid, 1.0 mg; nicotinamide, 1.0 mg; *p*-aminobenzoic acid, 100 μg; biotin, 50 μg; and folic acid, 50 μg.

ᵗ Brain heart infusion broth (commercially available) (37 g/liter).

ᵘ Brain heart infusion (commercially available), half strength (18.5 g/liter), and hemin (200 μg/liter).

TABLE 7. *Compositions of undefined media (A) 64 through 75 for the bacteria listed in Table 1*

Component	Amt (in units given in column 1)/liter for medium:											
	64A	65A^a	66A^b	67A	68A^c	69A^d	70A	71A^e	72A^f	73A	74A	75A
Glucose^g (g)	2.0–10	10	10		1.0		10	6.0	10		1.0	10
Glycerol (g)						88						
(NH₄)₂SO₄ (g)	0.3											0.2
Na₂HPO₄ (g)	5.0											
KH₂PO₄ (g)	1.0											3.0
K₂HPO₄ (g)									0.2			
MgSO₄·7H₂O (g)	0.1								0.3			0.5
NaCl (g)	4.0	5.0						2.5				
CaCl₂·2H₂O (g)	0.1											0.25
FeSO₄·7H₂O (mg)										5.0		
Amino acids								Yes^h				
Beef extract (g)		3.0				1.0	1.0					
Yeast extract (g)	0.1–1.0		5.0	3.0	1.0		1.0			0.1	10	1.0^i
Acid-hydrolyzed casein (e.g., Casamino Acids) (g)										0.1		
Enzyme-hydrolyzed casein (e.g., Casitone) (g)			5.0			5.0	2.0					
Peptone (g)		10										
Tryptone (g)											10	
Special media					Yes^j					Yes^k	Yes^l	
pH	7.0	7.0–7.2	7.4–7.5		6.8	6.5–7.0	7.3	7.4	7.0	7.6–7.8	7.8	1.0–2.0

^a Mix with 3 liters of horse serum; filter and sterilize by filtration.

^b Add soluble starch (2%) and CaCO₃ (0.1%).

^c Add soluble starch (0.1%), CaCO₃ (0.1%), and agar (1.5%).

^d Use tap water in place of distilled water.

^e Add (per liter): Na₂SO₃, 0.1 g; sodium thioglycolate, 0.5 g; Phytone (BBL), 3.0 g; and Trypticase (BBL), 17 g.

^f Add 0.2 g of CaCO₃ per liter.

^g For most media it is recommended that glucose be sterilized (either by autoclaving or filtering through appropriate membranes) separately from other components of the medium, and then combined aseptically.

^h Add 0.25 g of L-cystine per liter.

^i The active component of yeast extract has been isolated and partially characterized as one or more peptides of molecular weight of ~1,000 (201).

^j Rolled oats, 20 to 65 g; tap water, 1,000 ml. Cook to a thin gruel in a double boiler, filter through cheesecloth, and make up to 1 liter while still hot. Add 18 to 20 g of agar; adjust to pH 7.2 with NaOH (232).

^k Medium of Hayflick and Chanock (77): PPLO broth (Difco; a dehydrated medium which contains beef heart infusion + peptone + NaCl), 34 g/liter; pooled horse serum, 200 ml; 25% extract of fresh bakers' yeast, 100 ml; and agar (semisoft agar is preferable), 1%. A humid atmosphere is desirable.

^l Add (per liter of final medium): PPLO broth (Difco), 21 g; sorbitol, 70 g; fructose, 1 g; sucrose, 10 g; and horse serum, 200 ml. Five percent CO₂ in N₂ stimulates growth on solid media.

60 ml with distilled water, and warm to dissolve. Store this solution at 5°C and renew every 2 weeks.

Glucose, sodium acetate, and ammonium chloride. Weigh glucose, sodium acetate, and ammonium chloride into the medium as needed.

To prepare 500 ml of double-strength basal medium, weigh out 3.1 g of amino acid mixture (minus the weight contributed to that mixture by any amino acid omitted) and dissolve the mixture by warming in ~400 ml of water. Add 25 g of glucose, 20 g of sodium acetate, and 3 g of NH₄Cl. Add 6 ml of salts A, 5 ml of salts B, 10 ml of the stock solution of purine and pyrimidine bases, and 3.0 ml of the vitamin solution. Make up to 500 ml with distilled water. Adjust the medium to pH 6.8 (This and similar media

based on this method of analysis are also available commercially, e.g., from Difco Laboratories.)

Preparation of samples.

For accurate results, growth factors in samples to be assayed must be present in the same chemical form as the compound used as a standard. Thus, proteins should be carefully dried for weighing and then completely hydrolyzed. (Many peptides of a given amino acid can be absorbed and hydrolyzed intracellularly to supply a limiting amino acid; however, the quantitative response to a given amount of peptide-bound amino acid may not be identical to that of the free amino acid.) Any procedure that results in comple hydrolysis without significant

TABLE 8. *Compositions of defined media (C) 1 through 14 for the bacteria listed in Table 1*

Component	Amt (in units given in column 1)/liter for medium:													
	1C[a]	2C[b]	3C[c]	4C[d]	5C[e]	6C[f]	7C	8C	9C[g]	10C	11C[h]	12C[h]	13C	14C[i]
Glucose[j] (g)						2.0		1.0	1.0	1.0	0.2	2.0		
Glycerol (g)														0.2
NH$_4$Cl (g)	1.0	1.0	2.0	0.5–2.0	0.08	1.0					0.25	0.5	1.0	
NaNO$_3$ or KNO$_3$ (g)								0.5						
Na$_2$HPO$_4 \cdot$2H$_2$O (g)										0.04	5.24	1.74		0.53
K$_2$HPO$_4$ (g)						0.48		1.0		0.027	0.08[k]	0.08[k]	0.5	
KH$_2$PO$_4$ (g)	0.33	1.0	1.0	0.6				1.13		0.04	2.77	1.06		0.07
Tris buffer (g)						2.0								
NaCl (g)		10	10			24								
KCl (g)	0.33					0.7		0.5						
MnCl$_2 \cdot$4H$_2$O (mg)	0.03	0.005	0.005	22	0.01			2.0	5.0	100				
MgSO$_4 \cdot$7H$_2$O (g)	0.62	1.29	2.6	0.25	0.01	8.0	1.0	0.5	0.075	0.2			0.2	0.15
CaCl$_2 \cdot$2H$_2$O (g)	0.33	0.1	0.2	0.094	0.05	0.5		0.2	0.05	0.05	0.1		0.1	0.004
FeSO$_4 \cdot$7H$_2$O (mg)	2.0	0.5	0.5	15	3.5			20	10[l]					
Trace elements	Yes[m]	Yes[n]	Yes[n]	Yes[o]	Yes[p]	Yes[q]			Yes[r]					Yes[s]
Amino acids							Yes[t]			Yes[u]				
Vitamins			Yes[v]	Yes[w]						Yes[x]	Yes[y]	Yes[z]	Yes[aa]	Yes[bb]
Tween 60 (mg)														200
Tween 80 (mg)														50
pH	6.5–7.5	7.3	6.5–7.6	6.8	7.0	7–9	7.2–7.8	7.2	6.0–7.5	6–7	7.0	7.0	6.3–6.6	7.4–7.6

[a] Additional components: 0.2 g of malic acid and 0.005 g of ethylenediaminetetraacetic acid (EDTA), pH 6.8; solutions containing 1.5 g of $NaHCO_3$ and 1.0 g of $Na_2S \cdot 9H_2O$, separately sterilized by filtration.

[b] Separately sterilize by filtration solutions containing 20 g of $NaHCO_3$ and 1.0 g of $Na_2S \cdot 9H_2O$, and add to the sterile medium.

[c] Add to the other components 0.2 g of EDTA and filter-sterilized solutions of 12.76 g of Na_2CO_3 and 0.1 g of $Na_2S \cdot 9H_2O$.

[d] Add 0.025 g of EDTA. Malic acid, 2 to 6 g, contributes to the carbon pool.

[e] Addition of 0.5 g of sodium acetate per liter improves cell yield. Also add 2 g of $NaHCO_3$, 42 mg of $CaSO_4$, and 0.001 g of EDTA per liter.

[f] For *Flexithrix dorotheae*, glucose as a carbon source may be replaced by galactose or sucrose (1 g/liter); ammonium chloride as nitrogen source may be replaced by sodium glutamate or $NaNO_3$ (10 g/liter). Add sodium glycerophosphate (0.1 g/liter).

[g] Sodium glutamate (0.5 g/liter) and aspartic acid (0.5 g/liter) serve as nitrogen source.

[h] Medium contains 20 ml of Hutner's vitamin-free mineral base per liter. Hutner's vitamin-free mineral base as modified by Cohen-Bazire et al. (35) has the following composition: nitrilotriacetic acid (dissolved and neutralized with KOH), 10 g; $MgSO_4 \cdot 7H_2O$, 14.45 g; $CaCl_2 \cdot 2H_2O$, 3.33 g; $(NH_4)_6Mo_7O_{24} \cdot 24H_2O$, 9.25 mg; $FeSO_4 \cdot 7H_2O$, 99 mg; metals 44 (see below), 50 ml; distilled water to 1 liter; pH adjusted to 6.6 to 6.8. Metals 44 contains, in milligrams per 100 ml: ethylenediaminetetraacetic acid, 250; $ZnSO_4 \cdot 7H_2O$, 1,095; $FeSO_4 \cdot 7H_2O$, 500; $MnSO_4 \cdot H_2O$, 154; $CuSO_4 \cdot 5H_2O$, 39.2; $Co(NO_3)_2 \cdot 6H_2O$, 24.8; and $Na_2B_4O_7 \cdot 10H_2O$, 17.7; add a few drops of 6 N H_2SO_4 to retard precipitation.

[i] Add (g/liter): asparagine, 0.5; and EDTA, 0.01.

[j] For most of these media, glucose is sterilized separately (usually by autoclaving) and added aseptically to a sterile solution of the other components.

[k] Autoclave separately and then add aseptically to the autoclaved solution of remaining components.

[l] Separately sterilize by filtration.

[m] Add (mg/liter): $ZnSO_4 \cdot 7H_2O$, 0.1; H_3BO_3, 0.3; $CoCl_2 \cdot 6H_2O$, 0.2; $CuCl_2 \cdot 2H_2O$, 0.01; $NiCl_2 \cdot 6H_2O$, 0.02; and $Na_2MoO_4 \cdot 2H_2O$, 0.03.

[n] Add (mg/liter): $ZnSO_4 \cdot 7H_2O$, 0.1; H_3BO_3, 0.1; $CoCl_2 \cdot 6H_2O$, 0.05; and $CuCl_2 \cdot 2H_2O$, 0.005.

[o] Add (mg/liter): $ZnSO_4 \cdot 7H_2O$, 2.4; H_3BO_3, 28; $CuCl_2 \cdot 2H_2O$, 0.4; and $Na_2MoO_4 \cdot 2H_2O$, 7.5.

[p] Add (per liter): $ZnSO_4 \cdot 7H_2O$, 0.05 mg; H_3BO_3, 0.05 mg; $CoCl_2 \cdot 6H_2O$, 0.005 mg; $CuCl_2 \cdot 2H_2O$, 0.025 µg; and $Na_2MoO_4 \cdot 2H_2O$, 0.005 mg.

[q] Add 1 ml of the following trace element solution per liter of medium: soluble salts of B, Fe, Mn (each as element, 0.5 mg/liter); soluble salts of Co, Cu, Mo, Zn (each as element, 0.01 mg/liter).

[r] Milligrams per liter: $ZnSO_4 \cdot 7H_2O$, 0.1; H_3BO_3, 0.1; $CuCl_2 \cdot 2H_2O$, 0.1; $Na_2MoO_4 \cdot 2H_2O$, 0.05.

[s] $(NH_4)_2Fe(SO_4)_2 \cdot 6H_2O$, 6 mg/liter; $CaCO_3$, 4 mg/liter.

[t] Amino acids (g/liter): phenylalanine (1.0), leucine (1.0), isoleucine (0.5), valine (0.1), and methionine (0.05) are required; glycine (0.05), proline (0.5), asparagine (0.44), alanine (1.0), lysine (0.25), threonine (0.1), and tryptophan (1.0) are stimulatory. The following are added to the medium even though they have no observable effect on growth: histidine (0.05), arginine (0.1), and serine (0.1).

[u] Any one of the following can serve as nitrogen source: leucine, proline, asparagine, alanine, tryptophan, arginine, aspartic acid, and cystine (supplied at 1 g/liter). Although good growth is supported by the individual amino acids, it is not as rapid or as extensive as that obtained with a mixture of amino acids in the form of Casamino Acids. The microorganism can also use inorganic nitrogen if glucose is replaced by sucrose, glycerol, or succinate as carbon source.

[v] Vitamin B_{12} (cobalamine) is required; it is supplied at a concentration of 0.005 mg/liter.

[w] Biotin is required and is supplied at 15 µg/liter.

[x] Vitamin B_{12} (0.005 mg/liter) is required; biotin and thiamine hydrochloride (supplied in the medium at 0.005 and 0.1 mg/liter, respectively) are required by some strains.

[y] Biotin (0.005 mg/liter) reportedly is required by some strains of *Sphaerotilus natans*.

[z] None of the isolated strains grow in a medium free from vitamins; biotin, thiamine, and vitamin B_{12} are either essential or stimulatory for most strains. The vitamin supplement contains, per liter: biotin, 2 mg; folic acid, 2 mg; pyridoxine hydrochloride, 10 mg; riboflavin, 5 mg; thiamine hydrochloride, 5 mg; nicotinamide, 5 mg; calcium pantothenate, 5 mg; vitamin B_{12}, 0.1 mg; and p-aminobenzoic acid, 0.1 mg. Add 10 ml of this solution, sterilized by autoclaving or membrane filtration, per liter of medium.

[aa] Add biotin, 0.2 µg/liter.

[bb] Add thiamine hydrochloride and vitamin B_{12} (each separately sterilized by filtration) at a concentration of 1.0 and 0.003 mg/liter, respectively, after sterilization of medium.

TABLE 9. *Compositions of defined media (C) 15 through 27 for the bacteria listed in Table 1*

Component	Amt (in units given in column 1)/liter for medium:												
	15C[a]	16C	17C	18C	19C	20C	21C	22C	23C	24C	25C[b]	26C	27C[c]
Glucose[d] (g)	10	5.0	15	20	25	20	1–3	10	10	5		3.0[e]	4.0
Glycerol (g)											11.7		
NH_4Cl (g)	1.0	1.0					1.0		1.0	1.0		3.0	
$NaNO_3$ or KNO_3 (g)								0.6					
$Na_2HPO_4 \cdot 2H_2O$ (g)	1.0						3.49	0.45					
K_2HPO_4 (g)			1.0	0.8	1.74	1.0				7.0		3.0	500
KH_2PO_4 (g)	0.25	2.0		0.2			2.77		2.0	3.0			
NaCl (g)									5				
$MnCl_2 \cdot 4H_2O$ (mg)	10		0.005	16.9									
$MgSO_4 \cdot 7H_2O$ (g)	0.01	0.2	0.5	0.5	0.62	0.2		0.1	5.0	0.01	0.09		10
$CaCl_2 \cdot 2H_2O$ (g)								0.2					
$FeSO_4 \cdot 7H_2O$ (mg)	10		50	50	13.9	5.0		10			0.45		
Trace elements		Yes[f]	Yes[g]	Yes[h]		Yes[i]	Yes[j]		Yes[k]				
Amino acids	Yes[l]				Yes[m]					Yes[n]	Yes[o]		
Casein hydrolysate (acid) (g)		Yes[p]											
Purines and pyrimidines	Yes[q]												
Vitamins	Yes[r]	Yes[s]			Yes[t]		Yes[u]	Yes[v]		Yes[w]			
Tween 80 (mg)											600		
pH	7.4	6.8	7.2	6.9	6.6–7.4	5.5–9.0	6.8	7.2–7.5	6.8	7.0–7.2	7.2	7.0	7.0

[a] One percent crystalline serum albumin is added as a detoxicant.

[b] Add 45 mg of K_2SO_4 per liter.

[c] Dissolve the components (in addition to those listed, include 100 g of citric acid·H_2O and 175 g of $NaHNH_4PO_4 \cdot 4H_2O$) successively into 670 ml of distilled water at 45°C. Aseptically add glucose (sterilized separately) to the sterile medium just before use. This medium is a 50-fold concentrate and must be diluted appropriately before use. The addition of 5 ml of chloroform as a preservative permits storage in a stoppered container at room temperature.

[d] For most of these media, glucose is sterilized separately (usually by autoclaving) and added aseptically to a sterile solution of the other components.

[e] Other carbohydrates, glycerol, and citric acid are utilized as well.

[f] Trace elements (µg/liter): B (H_3BO_3), 5; Ca ($CaCO_3$), 100; Cu ($CuSO_4 \cdot 5H_2O$), 10; Fe [$FeSO_4(NH_4)_2SO_4 \cdot 6H_2O$], 100 to 500; I (KI), 1.0; Mn ($MnSO_4 \cdot H_2O$) 10 to 20; Mo (MoO_3), 10; Zn ($ZnSO_4 \cdot 7H_2O$), 50.

[g] Trace of molybdate (0.005 g of $Na_2MoO_4 \cdot 2H_2O$ per liter) is required for nitrogen fixation.

[h] For *Beijerinckia indica* molybdate is required for nitrogen fixation (see footnote g above) but is not replaceable by vanadate. Some strains also require Cu and Zn; these are supplied as $ZnSO_4 \cdot 7H_2O$ (0.005 g/liter) and $CuSO_4 \cdot 5H_2O$ (0.004 g/liter).

[i] Ca^{2+} is added as $CaSO_4$ (0.2 g/liter); also include 0.5 mg of $Na_2MoO_4 \cdot 2H_2O$ per liter.

[j] Hutner's vitamin-free mineral base, 20 ml/liter (see footnote h of Table 8).

[k] All species of *Aeromonas* grow on minimal medium plus a mixture of L-arginine, asparagine, leucine, and methionine.

[l] Arginine, aspartic acid, cysteine, glutamic acid, histidine, leucine, lysine, methionine, phenylalanine, threonine, tryptophan, and valine are essential for growth; serine and tyrosine are stimulatory; and alanine, hydroxyproline, and glycine have no effect on growth. All are supplied in the medium at a concentration of 0.001 M. Proline can replace glutamic acid. Glutamine (1.02 g/liter), separately sterilized by filtration, is added to the sterile medium just before use.

[m] Amino acids essential for optimal yields (g/liter): D-alanine 1.2; L-arginine·HCl, 0.77; L-cysteine, 0.1; L-glutamic acid, 3.9; L-lysine·HCl, 0.77; DL-methionine, 0.94.

[n] Add the following amino acids (mg/liter of medium): threonine, 276; proline, 100; leucine, 426; histidine, 126; glycine, 176; arginine, 160; glutamic acid, 200; valine, 185; tyrosine, 35. The first six amino acids listed are essential. The remainder are stimulatory.

[o] Add (per liter): DL-alanine, 1.4 g; DL-tryptophan, 216 mg; and DL-threonine, L-lysine, DL-methionine, DL-histidine, DL-phenylalanine, L-leucine, DL-isoleucine, DL-valine, and L-arginine, 60 mg each.

[p] Addition of 0.5% casein hydrolysate or 0.1% glutamic acid stimulates growth.

[q] Cytosine or uracil at 10 mg/liter of medium satisfies the requirement; adenine (10 mg/liter) is stimulatory.

[r] The vitamin requirements have not been determined. Vitamins are supplied at the following concentrations per liter: calcium pantothenate, 1 mg; *p*-aminobenzoic acid, 10 mg; biotin, 10 µg; choline, 50 mg; folic acid, 100 µg; inositol, 5 mg; nicotinic acid, 1.0 mg; pyridoxal, 5 mg; pyridoxamine, 5 mg; pyridoxine, 5 mg; riboflavin, 1 mg; thiamine, 1 mg; and ascorbic acid, 1.23 g.

[s] Biotin, 1 µg/liter.

[t] Nicotinic acid (2.0 mg/liter) and thiamine·HCl (0.15 mg/liter). For *Brucella abortus* biotin is essential. Five percent carbon dioxide in air satisfies most dependent strains such as *B. abortus*.

[u] Supplement with the following vitamins (mg/liter): biotin, 0.01; nicotinic acid, 1.0; thiamine hydrochloride, 0.5.

[v] Biotin (0.5 µg/liter), thiamine hydrochloride (100 µg/liter).

[w] Add (µg/liter): calcium pantothenate, 10; biotin, 1; nicotinamide, 10; thiamine hydrochloride, 10; and pyridoxine hydrochloride, 20. Pantothenate, nicotinamide, and thiamine are essential; biotin and pyridoxine are stimulatory.

TABLE 10. *Compositions of defined media (C) 28 through 40 for the bacteria listed in Table 1*

Components	Amt (in units given in column 1)/liter for medium:												
	28C[a]	29C[b]	30C[c]	31C[d]	32C[e]	33C[f]	34C[g]	35C[h]	36C[i]	37C[j]	38C[k]	39C	40C[l]
Glucose[m] (g)	1.0	12.5		2.5	3.0	2.0	5.0			5.0	5.0	5.0[n]	
Glycerol (g)										0.25	6.3		
Sodium acetate (g)	6.0									3.4			
NH4Cl (g)			0.5			1.0	0.9	4.0	3.0	2.2	0.4		
Na2HPO4·2H2O (g)											1.3	2.25	0.6
K2HPO4 (g)	1.56			8.7					0.65	3.48			
KH2PO4 (g)	0.14		5.0		6.8	0.01	0.9	2.7	0.35	2.72	0.17	0.75	0.2
Tris buffer (g)					6.05								
NaCl (g)			5.3	2.5	0.02	0.05	0.9	—[o]		5.8	5.8		0.1
KCl (g)						0.05					0.19		
MnCl2·4H2O (mg)	0.36				20	4.2		1.0	25		1.5	1.0	7.6
MgSO4·7H2O (g)	0.01	1.6	0.04		0.04	0.238	0.02	0.2	0.02	0.41	0.62	0.096	0.1
CaCl2·2H2O (g)	0.006					0.001	0.02	0.01	0.01	0.37	0.074	0.002	
FeSO4·7H2O (mg)	0.1				20	5.0	10	1.0	20	4.0	2.8	2.0	12.5
Trace elements	Yes[p]	Yes[q]	Yes[r]	Yes[s]		Yes[t]	Yes[u]					Yes[v]	
Amino acids	Yes[w]	Yes[x]	Yes[y]	Yes[z]	Yes[aa]	Yes[bb]				Yes[cc]	Yes[dd]	Yes[ee]	Yes[ff]
Casein hydrolysate (acid) (g)													5.0[gg]
Purines and pyrimidines	Yes[hh]									Yes[ii]		Yes[jj]	Yes[kk]
Vitamins	Yes[ll]	Yes[mm]	Yes[nn]			Yes[oo]	Yes[pp]		Yes[qq]	Yes[rr]		Yes[ss]	Yes[tt]
Tween 80 (ml)										5			
pH	7.8	7.5	7.6	7.5	6.6–6.9	6.8	6.7	6.8–6.9	7.0	7.45	7.4 ± 0.1	7.0	6.8[uu]

[a] After sterilization, supplement the medium with aseptic additions of 500 μg of putrescine and 100 μg of filter-sterilized nicotinamide adenine dinucleotide.

[b] Dissolve hematin (50 μg) and oleic acid (10 μg) in 0.001 N NaOH and autoclave. Hematin can be replaced in the growth medium by any one of several compounds which decompose H_2O_2. It is not required for anaerobic growth. Add 10 g of lactic acid (85%). Glucose alone inhibits aerobic growth, but glucose plus another carbon source, such as lactic acid or sucrose, permits satisfactory growth.

[c] Add 4.4 g of sodium lactate (70%) per liter.

[d] Medium is sterilized by filtration.

[e] Hematin, blood, or incubation under reduced atmospheric pressure is required when the medium is used as a solid medium.

[f] Add 5.8 g of malic acid, and adjust the medium to pH 6.8 with NaOH.

[g] To 1 liter of medium add 1 mg of hemin and volatile fatty acids at the indicated concentrations: acetic acid, 2.8×10^{-2} M; propionic acid, 9×10^{-3} M; n-butyric acid, 4.5×10^{-3} M; isobutyric acid, 9×10^{-4} M, n-valeric acid, 9×10^{-4} M; isovaleric acid, 9×10^{-4} M; and DL-methylbutyric acid, 9×10^{-4} M. Finally, add 4 g of Na_2CO_3 and 0.25 g of cysteine·HCl (both separately sterilized by filtration) just before use. Cultures of *Bacteroides ruminicola* are incubated in an O_2-free CO_2 atmosphere.

[h] Choline chloride (5 g/liter) is used as carbon source in this medium. Other carbon sources: lactate, malate, pyruvate. Add 2.6 g of Na_2SO_4 and 1 mg of Na_2MoO_4 per liter. $Na_2S\cdot9H_2O$ (0.05%) or sodium hydrosulfite (0.0035%) is used to remove oxygen from the culture medium. CO_2 (included in the medium as $NaHCO_3$, 0.3 to 0.5%) and yeast extract (0.1 to 0.2%) are stimulatory.

[i] Sodium pyruvate, 3 g/liter, is the carbon source. Carbon sources are limited to Krebs cycle intermediates and close derivatives.

[j] Add (per liter of medium): sodium lactate, 250 mg; oxalacetic acid, 200 mg; polyvinyl alcohol, 5 mg; spermine tetrahydrochloride, 87 mg; hemin, 1 mg; 2,2′,2″-nitrilotriethanol, 4% (vol/vol); nicotinamide adenine dinucleotide, 2 mg; thiamine pyrophosphate chloride, 0.46 mg; NaHCO3, 42 mg; and K2SO4, 1 g. The oxalacetic acid, nicotinamide adenine dinucleotide, and thiamine pyrophosphate chloride should be separately sterilized by filtration before they are added.

[k] Add (per liter): lactic acid, 5 g; citric acid·H2O, 0.42 g; and EDTA, 0.004 g.

[l] Add (per liter): lactic acid (85%), 10 ml; sodium glycolate, 0.75 g; and putrescine, 4.5 mg. Only freshly constituted media are used. Inoculations are generally made immediately on cooling.

[m] For most of these media, glucose is sterilized separately (usually by autoclaving) and added aseptically to a sterile solution of the other components.

[n] In this case, autoclave glucose with other components of the medium; the derivative products from medium-autoclaved glucose (not glucose itself) are stimulatory or essential.

[o] For marine strains, supplement the medium with 2.5% NaCl.

[p] Add (per liter): ZnSO4·7H2O, 0.44 mg; CoCl2·6H2O, 0.4 mg; and CuCl2·2H2O, 0.27 mg.

[q] Dissolve MgSO4·7H2O (amount listed in Table 10) and 80 mg of FeSO4(NH4)2SO4·6H2O in 8.7×10^{-6} M citric acid and autoclave.

[r] Dissolve 29 mg of Fe(NH4)2(SO4)2·6H2O in 0.02 M HCl and filter sterilize.

[s] Add 1 ml of a solution containing 5% MgSO4, 0.5% MnCl2·4H2O, 0.5% FeCl3, and 0.4% nitriloacetic acid per liter.

[t] Add (mg/liter): ZnSO4·7H2O, 7.2; CuSO4·H2O, 1.4; and CoSO4·5H2O, 1.4.

[u] Add 1 mg of CoCl2·6H2O per liter.

[v] Add (per liter): CoCl2·6H2O, VSO4·7H2O, ZnSO4, and NaMoO4·2H2O, 1 mg each; CuSO4·5H2O, 0.5 mg; and nitriloacetic acid, 20 mg.

[w] Add (per liter): DL-alanine, DL-aspartic acid, and L-glutamic acid, 1,000 mg each; L-arginine·HCl and L-lysine·HCl·H2O, 200 mg each; and L-histidine, L-leucine, L-isoleucine, L-methionine, L-phenylalanine, L-proline, L-threonine, L-tyrosine, L-valine, L-tryptophan, L-cysteine, L-serine, and L-glycine, 100 mg each.

[x] Add (g/liter): glycine, 0.1; DL-alanine, 0.1; DL-valine, 0.15; DL-leucine, 0.15; DL-isoleucine, 0.1; DL-aspartic acid, 0.2; DL-glutamic acid, 0.2; L-proline, 0.2; DL-phenylalanine, 0.1; DL-tyrosine, 0.1; L-arginine, 0.04; DL-histidine, 0.03; DL-serine, 0.2; DL-threonine, 0.2; DL-methionine, 0.2; and L-cystine, 0.4. Dissolve the first 12 amino acids in 0.033 M phosphate buffer and sterilize by autoclaving; dissolve serine, threonine, and methionine in water and filter sterilize; dissolve L-cystine in HCl and filter sterilize.

TABLE 10—*Continued*

[y] Any one of the following amino acids at a concentration of 0.62 or 0.36 M can serve as carbon and energy source or nitrogen source: glutamic acid, aspartic acid, serine, proline, α-alanine, asparagine, glutamine, and threonine. However, growth is better on the mixture or an equivalent quantity of Casamino Acids. Glucose, maltose, or sucrose can also serve as a carbon source.

[z] Add 2.5 g of L-asparagine per liter.

[aa] Add (mg/liter): DL-alanine, 37; DL-aspartic acid, 120; L-arginine·HCl, 90; L-cysteine·HCl (autoclave separately and then add aseptically to the autoclaved solution of remaining components), 13; L-glutamic acid, 642; glycine, 10; L-histidine·HCl·H$_2$O, 68; DL-isoleucine, 94; DL-leucine, 100; L-lysine·HCl, 155; DL-methionine, 65; DL-phenylalanine, 78; L-proline, 175; DL-serine, 100; DL-threonine, 70; DL-tryptophan, 31; L-tyrosine, 110; and DL-valine, 160.

[bb] Add each of the following amino acids at 60 mg/liter: alanine, arginine, aspartic acid, cysteine, glutamic acid; glycine, histidine, isoleucine, leucine, lysine, methionine, ornithine, hydroxyproline, phenylalanine, proline, serine, threonine, tryptophan, tyrosine, and valine. Replacement of these amino acids with hydrolyzed casein produces a similar rate and extent of growth. If the amino acids are omitted, the NH$_4$Cl concentration is increased to 2 g/liter.

[cc] Add (mg/liter): L-aspartic acid, 500; L-glutamic acid, 1,300; L-arginine·HCl, 150; glycine, 25; L-serine, 50; L-leucine, 90; L-isoleucine, 30; L-valine, 60; tyrosine, 70; L-lysine·HCl, 50; L-proline, 50; L-tryptophan, 80; L-threonine, 50; L-phenylalanine, 25; L-asparagine, 25; cysteine·HCl·H$_2$O (separately sterilized by filtration), 61.25; cystine, 36; L-glutamine (separately sterilized by filtration), 50; L-methionine, 14.9; glutathione (reduced), 46; and histidine, 16.5.

[dd] Add (per liter): L-arginine·HCl, 0.1 g; L-cysteine·HCl·H$_2$O, 10 mg; glycine, 150 mg; glutamic acid, 1.18 g; and serine, 21 mg.

[ee] Add (mg/liter): DL-tryptophan, 120; sodium glutamate, 5,000; L-tyrosine, 150; DL-phenylalanine, 250; L-arginine·HCl, 400; L-histidine·HCl·H$_2$O, 300; glycine, 200; DL-valine, 400; DL-serine, 2,000; DL-isoleucine, 125; and L-cysteine·HCl, 350.

[ff] Add (mg/liter): L-cysteine·HCl, 12.5; DL-tryptophan, 50; L-tyrosine, 20; L-proline, 10; DL-phenylalanine, 20; and DL-histidine, 10.

[gg] Although acid-hydrolyzed casein contains several minor components in addition to amino acids and therefore the medium is not "defined" in the strictest sense, it is labeled "C" because the casein has been purified by adsorption on Norit A at two different H$^+$ ion concentrations and some of the known components associated with casein (e.g., polyamines, vitamins) have been supplied in the medium.

[hh] Add 10 mg each of guanine, adenine, and uracil per liter.

[ii] Add 3.2 mg of hypoxanthine and 8 mg of uracil per liter.

[jj] Add 10 mg each of guanine, uracil, and hypoxanthine per liter.

[kk] Add (mg/liter): hypoxanthine, 30; and uracil, 2.

[ll] Add (per liter of medium): inositol, 20 mg; folic acid, 10 μg; p-aminobenzoic acid, 1 μg; riboflavin, 100 μg; calcium pantothenate, 1 mg; biotin 1 μg; thiamine hydrochloride, 100 μg; nicotinic acid, 500 μg; nicotinamide, 500 μg; and pyridoxine hydrochloride, 2 mg.

[mm] Add (per liter of medium): calcium pantothenate, 9.5 mg; nicotinamide, 2.4 mg; and thiamine hydrochloride, 3.3 mg (all filter sterilized).

[nn] Add 2.4 mg of nicotinic acid per liter.

[oo] Add 50 μg of calcium pantothenate per liter.

[pp] Add (mg/liter): riboflavin, 2; calcium pantothenate, 2; biotin, 0.05; nicotinic acid, 2; thiamine hydrochloride, 2; vitamin B$_{12}$, 0.02; DL-thioctic acid, 0.05; pyridoxal, 2.0; p-aminobenzoic acid, 0.1; and folic acid, 0.05.

[qq] Add (per liter): biotin, 1 μg; and thiamine hydrochloride, 1 mg. Pantothenate is markedly stimulatory.

[rr] Add (per liter): thiamine hydrochloride, 2 mg; calcium pantothenate, 2 mg; choline chloride, 1.4 mg; *myo*-inositol, 3.6 mg; and biotin, 0.7 mg.

[ss] Add (per liter of medium): pyridoxal hydrochloride and calcium pantothenate, 1 mg each; thiamine hydrochloride, niacin, and riboflavin, 50 μg each; p-aminobenzoic acid, 10 μg; biotin, 2 μg; and folic acid and vitamin B$_{12}$, 1 μg each.

[tt] Add (per liter of medium): pyridoxal hydrochloride and thiamine hydrochloride, 2 mg each; calcium pantothenate and niacin, 200 μg each; riboflavin, 50 μg; p-aminobenzoic acid and folic acid, 1 μg each; and biotin, 10 μg.

[uu] Adjust pH to 6.8 with solid K$_2$CO$_3$.

destruction of the amino acid to be determined may be used.

For determining arginine, dry the protein samples by heating for 3 h at 60°C under 20 mmHg (2.7 kPa). Carefully weigh about a 200-mg sample, place in a 25 by 200 mm Pyrex tube, add 10 ml of 4 N HCl, seal the tube, and autoclave for 8 h at 15 lb/in^2. When cool, open the tube, neutralize to pH 6.8, and dilute to a convenient volume. Some amino acids (tryptophan and tyrosine to some extent, and small amounts of several other amino acids) are destroyed by acid hydrolysis. Modified procedures should be used in these cases.

The amino acid which will serve as the standard (in this case, arginine) is likewise dried (2 h at 60°C under 20 mmHg) and carefully weighed and dissolved.

Assay Procedure.

The inoculum is prepared as follows: make up "inoculum tubes" by supplementing the complete double-strength medium (Table 14) with 0.2% yeast extract, dilute with an equal volume of distilled water, dispense 10-ml aliquots of the medium into culture tubes, cap, and sterilize by autoclaving. Several of these inoculum tubes can be made up at a time and stored at 5°C. The day before the assay is to be performed, use an inoculating needle to transfer some bacteria aseptically from the agar stab culture to an inoculum tube and incubate at 37°C for 15 to 18 h. Immediately before use, collect the cells by centrifugation, decant the supernatant medium, and suspend the cells in 10 ml of sterile water. Centrifuge again, and resuspend the pellet of

TABLE 11. *Compositions of defined media (C) 41 through 53 for the bacteria listed in Table 1*

Component	Amt (in units given in column 1)/liter for medium:												
	41C[a]	42C[b]	43C	44C[c]	45C	46C[d]	47C[e]	48C	49C[f]	50C[g]	51C	52C[h]	53C[i]
Glucose[j] (g)								10	10		10		20
Sodium acetate (g)									2.0		20	1.6	
NH₄Cl (g)			0.8	0.52	0.3		0.5	1.0		0.5	2.0	0.4	
Na₂HPO₄·2H₂O (g)						1.0							2.5
K₂HPO₄ (g)		0.002	0.016	0.28		0.6	3.48	4.0	2.0		4.0		
KH₂PO₄ (g)	0.15				3.0		2.7		2.0			0.9	2.5
NaCl (g)										0.04	0.9		
MnCl₂·4H₂O (mg)		0.066	0.2				1.0	10	0.05	8.0	0.03	10	20
MgSO₄·7H₂O (g)	0.15	0.1	0.2	0.25	0.5		0.01	0.2	0.2	2.0	0.16	0.01	0.4
CaCl₂·2H₂O (g)		0.006	0.02	0.07	0.25		0.01			0.01		0.01	
FeSO₄·7H₂O (mg)	30			10			2.0	10	50	4.0	8.0		10
Trace elements		Yes[k]	Yes[l]	Yes[m]	Yes[m]		Yes[n]			Yes[o]		Yes[p]	
Amino acids								Yes[q]	Yes[r]		Yes[s]		Yes[t]
Casein hydrolysate (acid) (g)								5.0[u]				2.0[v]	
Purines and pyrimidines								Yes[w]	Yes[x]		Yes[y]		
Vitamins								Yes[z]	Yes[aa]	Yes[bb]	Yes[cc]	Yes[dd]	Yes[ee]
Tween 80 (ml)										1			
pH	7.5–8.0	7.5–8.0	7.5–7.8	2.5–3.2	4.8–5.0	7.0–7.2	6.8–7.4	7.0–7.2	6.7	7.0	5.5	6.5	1–9.8

[a] Fixation of atmospheric CO_2 provides source of carbon, and oxidation of nitrite to nitrate furnishes energy. The medium contains, in addition to the salts listed, 2 g of $NaNO_2$ and 0.5 g of $CaCO_3$ per liter; tap water or seawater is used in place of distilled water.

[b] No growth in freshwater even if NaCl is included; 70 to 100% seawater, 1 mM in $NaNO_2$ plus inorganic salts listed, permits optimal growth. Atmospheric CO_2 is the carbon source.

[c] Aerate the medium intermittently with air containing 5% CO_2.

[d] Potassium thiocyanate (0.2 g/liter) is the source of energy, nitrogen, and part of carbon. *Thiobacillus thioparus* does not grow on complex media such as nutrient broth, peptone, yeast extract, or casein hydrolysate.

[e] After sterilization, add the following aseptically to the medium: 200 to 400 mg of $Na_2S·9H_2O$, 168 mg of $NaHCO_3$, and 410 mg of sodium acetate or 320 mg of methyl alcohol, in a total volume of 1 liter.

[f] Add 2 g of ammonium citrate per liter.

[g] Add 1.0 g of potassium glycerophosphate and 0.5 g of ethylenediaminetetraacetic acid per liter. L-Glutamate, pyruvic acid, acetate, or lactate is used as carbon source, but a mixture of 5 g of L-glutamic acid and 10 g of DL-lactate per liter gives the best growth.

[h] Include also (per liter): cellobiose, 3.0 g; acetate, 20 mmol; isovalerate, 1 mmol; isobutyrate, 1 mmol. Autoclave separately under N_2 solutions containing 0.5 g of $Na_2S·9H_2O$ and 4 g of Na_2CO_3, and add to the medium just prior to inoculation.

[i] A relatively heavy and active inoculum is necessary to obtain growth on this medium.

[j] For most of these media, glucose is sterilized separately (usually by autoclaving) and added aseptically to a sterile solution of the other components.

[k] Add (per liter): chelated iron, 1 mg; $Na_2MoO_4·2H_2O$, 30 µg; $CoCl_2·6H_2O$, 0.6 µg; $CuSO_4·5H_2O$, 6.0 µg; and $ZnSO_4·7H_2O$, 30 µg.

[l] Add (per liter): chelated iron, 1 mg; $Na_2MoO_4·2H_2O$, 100 µg; $CoCl_2·6H_2O$, 2.0 µg; $ZnSO_4·7H_2O$, 100 µg; and $CuSO_4·5H_2O$, 20 µg.

[m] Add 1.56 g of elemental sulfur per liter, sterilized separately by autoclaving.

[n] Add 1.0 mg of $Na_2MoO_4·2H_2O$ per liter.

[o] Add (mg/liter): Mo (as NH_4 molybdate), 0.8; Zn (as $ZnSO_4·7H_2O$), 10; B (as H_3BO_3), 4.0; Cu (as $CuSO_4·5H_2O$), 1.0; and Co (as $CoSO_4·7H_2O$), 1.0.

[p] Add 4 mg of $CoCl_2·6H_2O$ per liter.

[q] Add 100 mg each of L-cystine and L-tryptophan per liter.

[r] Add (mg/liter): L-alanine, 300: L-arginine, 400; L-aspartic acid, 750; L-cysteine, 200; L-glutamic acid, 750; glycine, 200; L-histidine, 500; DL-isoleucine, 200; L-leucine, 200; L-lysine, 800; DL-methionine, 150; DL-phenylalanine, 200; L-proline, 500; DL-serine, 400; DL-threonine, 200; L-tryptophan, 200; L-tyrosine, 100; and L-valine, 300.

[s] Add 200 mg of each of the following per liter: L-tryptophan, L-cysteine, DL-threonine, DL-methionine, DL-valine, DL-leucine, DL-isoleucine, DL-lysine, DL-phenylalanine, L-tyrosine, L-proline, glycine, DL-alanine, L-arginine, DL-serine, L-glutamic acid, and L-aspartic acid.

[t] Add 0.2 g of each of the following per liter: DL-arginine, DL-valine, L-glutamic acid, DL-serine, DL-histidine, DL-isoleucine, DL-leucine, L-tyrosine, DL-methionine, DL-tryptophan, and DL-phenylalanine.

[u] Casein hydrolysate can be replaced by 18 amino acids without affecting the rate or extent of growth.

[v] Casein hydrolysate can be deleted from the medium without decreasing the extent of growth, demonstrating that ammonium salts serve as sole nitrogen source.

[w] Add 5 mg each of adenine, guanine, uracil, and xanthine per liter. Either guanine or xanthine satisfies the purine requirement; adenine is satisfactory for some strains.

[x] Add 5 mg each of guanine, adenine, uracil, and xanthine per liter.

[y] Add 5 mg each of xanthine, guanine, adenine, thymine, and uracil per liter.

[z] Add (per liter): biotin, 100 µg; folic acid, 10 µg; nicotinic acid, 2 mg; calcium pantothenate, 1 mg; pyridoxine hydrochloride, 2 mg; riboflavin, 1 mg; and thiamine hydrochloride, 1 mg. Pantothenate, nicotinic acid, and biotin are either required or markedly stimulatory to all strains.

TABLE 11—*Continued*

[aa] Add (per liter): pyridoxal, 5 mg; nicotinic acid, 1 mg; calcium pantothenate, 1 mg; riboflavin, 1 mg; thiamine, 1 mg; vitamin B_{12}, 1 µg; biotin, 10 µg; p-aminobenzoic acid, 5 µg; and folic acid, 10 µg.

[bb] Add 10 µg of biotin per liter. Biotin requirement is satisfied by biocytin or dethiobiotin or by 1,000 times the amount of Tween 80 or sodium oleate.

[cc] Add (per liter): nicotinic acid, 5 mg; calcium pantothenate, 1 mg; pyridoxine, 1 mg; biotin, 100 µg; and folinic acid-SF (leucovorin), 0.01 mg. Mevalonic acid and riboflavin are required by some strains of *Pediococcus*.

[dd] Add (per liter): thiamine·HCl, 2 mg; calcium pantothenate, 2 mg; riboflavin, 2 mg; nicotinamide, 2 mg; pyridoxamine· 2HCl, 1.0 mg; pyridoxal·HCl, 1 mg; pyridoxine·HCl, 1.0 mg; p-aminobenzoic acid, 0.1 mg; biotin, 0.05 mg; folic acid, 0.05 mg; and cobalamine 0.005 mg.

[ee] Add 2 mg of nicotinic acid and 0.5 mg of biotin per liter.

cells in 10 ml of sterile water. This provides the washed suspension of cells used to inoculate the assay tubes.

The assay is carried out in 18 by 150 mm lipless culture tubes individually supported in a metal rack. When growth response is based on acid production, the total volume may be as low as 2 ml, as recommended by Steele et al. (216), or as high as 10 ml. For turbidimetric measurements, the total volume is 10 ml. An assay based upon 10-ml volumes is described here.

The standard curve is obtained by using the following amounts of L-arginine·HCl per 10-ml total volume: 0, 20, 40, 60, 80, 100, 120, 160, and 200 µg. Add the samples to the assay tubes in graded amounts estimated to supply several concentrations providing 10 to 100 µg of arginine per 10-ml assay tube. Make up the volume of each assay tube to 5 ml with distilled water; then add 5 ml of double-strength medium. Cap the tubes, sterilize by autoclaving for 10 min at 15 lb/in², cool, and inoculate with a washed suspension of cells obtained as described above. The size of the inoculum is not a significant factor in amino acid assays as long as the same amount is added to each tube (1 drop is convenient).

For turbidimetric measurement of growth response, incubate the culture tubes for 20 h at 37°C. Determine cell growth photometrically at a wavelength of >600 nm to minimize interference from the color of the growth medium (using, for example, a Bausch & Lomb Spectronic 20 colorimeter). When using the 2-ml assay, the incubation period is extended to 72 h, and growth response is measured by titrating the acid produced in each assay tube with 0.02 N NaOH.

For calculating results, prepare a standard dose-response curve for arginine by plotting the extent of growth (as measured by the cell density at 20 h) or the titration values against the amount of standard arginine added. For each growth response produced by known aliquots of sample, read from the standard curve the amount of arginine that elicits the same growth

response. Only those values that fall on the rapidly ascending portion of the curve (where growth is linearly related to arginine concentration) should be used. Several concentrations of sample should fall within this range, and the individual values calculated for the arginine content of the original sample should show only random variations from the mean. Results may be reported as milligrams of arginine per gram of sample or on any other useful basis desired.

7.3.2. Vitamins

Assays for vitamins are performed like those for amino acids. Hydrolyzed vitamin-free casein (supplemented with amino acids that have been destroyed by acid hydrolysis or are present in suboptimal amounts in casein) may be used in place of the mixture of amino acids in the growth medium. Vitamins are required in much smaller amounts than are amino acids; e.g., the optimum biotin requirement of a typical assay bacterium, *Lactobacillus plantarum,* is 1.0 ng/10-ml culture. For this reason, it is necessary to thoroughly clean all glassware that will be used, particularly the culture tubes. This is generally good practice for all assays. Use of chromate-based cleaning solution is not recommended since it is difficult to remove the last traces from glassware even by repeated rinsings, and residual traces inhibit growth of most bacteria. We find the following preliminary procedure satisfactory: fill culture tubes with distilled water, autoclave for 10 min, empty, and dry. Culture tube caps are also a source of chemical contamination and must be washed before reuse.

Since vitamins vary greatly in stability and in the nature of their combined forms, no single procedure for liberating bound vitamins is suitable in all cases. Treatments based on the following procedures have been widely used (204): (i) hydrolyze with dilute acid, neutralize, bring to volume, and filter; or (ii) treat with a mixture of enzymes, autoclave to stop enzymatic action, bring to volume, and filter.

TABLE 12. *Compositions of defined (C) and semidefined (B) media 54 through 66 for the bacteria listed in Table 1*

Component	Amt (in units given in column 1)/liter for medium:												
	54C[a]	55C[b]	56C[c]	57C[d]	58C[e]	59C[f]	60C[g]	61C[h]	62C	63C[i]	64B[j]	65C[k]	66C[l]
Glucose[m] (g)	10	25	10	10	30	30	2.5^n	10.4	2.0		10	10	10
Sodium acetate (g)		20		10					10	5.0	10	6.0	1.67
NH$_4$Cl (g)	0.5	3.0								1.0			
Na$_2$HPO$_4$·2H$_2$O (g)	3.3										1.75		
K$_2$HPO$_4$ (g)		0.6	3.0	8.7	5.0	7.0^o	0.5	8.3	0.5	1.0		0.5	16.7
KH$_2$PO$_4$ (g)	1.0	0.6	3.0	8.7		1.5	0.5	1.6	0.5			0.5	16.7
NaCl (g)	1.0	0.01	0.01	0.01	1.0		0.001		0.01		0.01	0.01	0.01
KCl (g)										3.0			
MnCl$_2$·4H$_2$O (mg)		20	10	10	40	40	1.0		10		10	13	10
MgSO$_4$·7H$_2$O (g)	0.7	0.2	0.2	0.2	0.5	0.5	0.02	0.02	0.2	0.3	0.2	0.2	0.5
CaCl$_2$·2H$_2$O (g)					0.3	0.3				0.01			
FeSO$_4$·7H$_2$O (mg)	10	10	10	10	2.5	2.5	1.0	10	10	10	10	10	10
Trace elements					Yes[p]	Yes[p]	Yes[q]	Yes[r]					
Amino acids	Yes[s]	Yes[t]	Yes[u]	Yes[v]	Yes[w]		Yes[x]	Yes[y]	Yes[z]	Yes[aa]	Yes[bb]	Yes[cc]	Yes[dd]
Casein hydrolysate (acid) (g)			5.0[ee]								5.0[ff]	5.0[gg]	
Purines and pyrimidines		Yes[hh]		Yes[ii]				Yes[jj]	Yes[kk]		Yes[ll]	Yes[mm]	Yes[nn]
Vitamins	Yes[oo]	Yes[pp]	Yes[qq]	Yes[rr]	Yes[ss]		Yes[tt]	Yes[uu]	Yes[vv]	Yes[ww]	Yes[xx]	Yes[yy]	Yes[zz]
pH	7.1	6.8–7.0	6.8	7.0	7.0	7.6	7.0–7.4	7.4	6.8	7.8	6.5	7.0	6.0

[a] Although cell yield on the synthetic medium is similar to that on complex media, the production of enterotoxin is only one-seventh as much. Anaerobic growth requires the addition of purines and uracil or of acetate (10 mM) or pyruvate. Acetate can be replaced by one-tenth the amount of mevalonate.

[b] Autoclave glucose with other medium components. This medium supplemented with folinic acid (citrovorin) also supports good growth of some *Pediococcus* strains.

[c] Autoclave glucose with other medium components. Add 20 g of sodium citrate per liter.

[d] Add 0.1 g of L-asparagine. Addition of 0.1 g of ammonium acetate and 1 g of NaHCO$_3$ per liter improves growth.

[e] Several of the amino acids listed in medium 58C are required by *Bacillus cereus* and *B. brevis*, but no vitamins are required; *B. alvei* requires amino acids and thiamine; *B. circulans* and *B. coagulans* require amino acids and usually thiamine and biotin; *B. sphaericus* requires amino acids, thiamine, and biotin. *B. pasteurii* requires amino acids, thiamine, biotin, nicotinic acid, and ammonia (NH$_4$Cl, 1%). Growth of *B. pasteurii* is favored by a pH of 8.5 and a temperature of 28°C. *B. anthracis* also grows on the amino acid medium; some strains require thiamine and are stimulated by adenosine and/or adenylic acid. Most species of this genus are aerobic but some (e.g., *B. cereus*) can grow anaerobically also.

[f] *Bacillus subtilis, B. licheniformis,* and *B. megaterium* require no vitamins; *B. pumilus* and *B. polymyxa* grow on NH$_4^+$ salts as a source of N but require the addition of biotin to the medium; *B. macerans* grows on medium 59C if biotin and thiamine are supplied.

[g] Mercaptoacetate (0.05%) and cysteine (0.025%) are added as reducing agents. CO$_2$ has been found to be essential for rapid and regular growth. Some strains of *Clostridium botulinum* grow on the defined medium only when an enzyme digest of casein is added (5.0 g/liter) or when the medium contains the following casein-replacing factors: NaHCO$_3$ (200 mg/liter), yeast extract (5.0 g/liter), or oxalacetate, (5.0 g/liter). Medium 60C supports good growth of *C. parabotulinum* types A and B and also of *C. sporogenes*. Although the medium supports satisfactory growth, the toxin titer is only one-tenth as high as on brain heart infusion broth.

[h] Add 5 g of sodium thioglycolate per liter. Growth on medium 61C is not as good as on brain heart infusion broth.

[i] Add 0.15 g of nitriloacetic acid and 0.01 g of Na$_2$S·9H$_2$O per liter. Adjust to desired pH with Tris.

[j] Autoclave glucose with the other medium components. Add (per liter): oleic acid, 0.01 g; Tween 40, 1.0 g (serves as a detoxicant for oleic acid); and enzymatic casein digest, 25 mg. The last is partially replaceable by 10 mg of glutamine (separately sterilized by filtration).

[k] Autoclave glucose with the other medium components. Add (mg/liter): L-ascorbic acid, 500; L-glutamine (separately sterilized by filtration), 300; and spermine·HCl, 1.

[l] Autoclave glucose with the other medium components. Add (per liter): ascorbic acid, 500 mg; potassium acetate, 1.67 g; and Tween-oleate (10 ml of a solution containing 2 g of Tween 40, 20 mg of sodium oleate, and water to 20 ml).

[m] Unless indicated otherwise, glucose is sterilized separately by autoclaving and added aseptically to a sterile solution of the other components.

[n] Sterilize separately by filtration.

[o] (NH$_4$)$_2$HPO$_4$ is used instead of K$_2$HPO$_4$.

[p] Add 2.0 mg of ammonium molybdate per liter.

[q] Add (mg/liter): ZnSO$_4$·7H$_2$O, 8; CuSO$_4$·5H$_2$O, 4.5; CoCl$_2$, 9; CaCl$_2$·2H$_2$O, 6; Na$_2$B$_4$O$_7$·10H$_2$O, 14; and (NH$_4$)$_2$MoO$_4$, 2.5.

[r] Add (mg/liter): H$_3$BO$_3$, 0.114; CuSO$_4$·5H$_2$O, 0.31; Fe(NH$_2$)(SO$_4$)$_2$·6H$_2$O, 2.8; CoSO$_4$·7H$_2$O, 0.01; MnSO$_4$·H$_2$O, 0.12; (NH$_4$)$_6$Mo$_7$O$_{24}$·4H$_2$O, 0.07; and ZnSO$_4$·7H$_2$O, 1.6.

[s] Add 25 mg of each of the following per liter: glycine, valine, leucine, threonine, phenylalanine, tyrosine, cysteine, methionine, proline, arginine, histidine, and tryptophan.

[t] Add (mg/liter): DL-α-alanine, 200; L-arginine·HCl, 242; L-asparagine, 400; L-aspartic acid, 100; L-cysteine, 50; L-glutamic acid, 300; glycine, 100; L-histidine·HCl, 62; DL-isoleucine, 250; DL-leucine, 250; L-lysine·HCl, 250; DL-methionine, 100; DL-phenylalanine, 100; L-proline, 100; DL-serine, 50; DL-threonine, 200; DL-tryptophan, 40; L-tyrosine, 100; and DL-valine, 250.

[u] Add (mg/liter): L-asparagine, 100; L-cysteine, 400; L-tryptophan, 50; and glycine, 200.

TABLE 12—*continued*

^{v} Add the following at 0.1 g/liter: L-alanine, L-arginine, L-aspartic acid, L-asparagine, L-hydroxyproline, L-isoleucine, L-leucine, L-lysine·HCl, L-methionine, L-norleucine, L-phenylalanine, L-proline, L-serine, L-threonine, L-tryptophan, L-tyrosine, L-valine. Add the following at the indicated concentrations (g/liter): L-cysteine·HCl, 0.05; L-cystine, 0.05; L-glutamic acid, 0.5; glycine, 0.2; L-histidine·HCl, 0.2; and L-glutamine (separately sterilized by filtration), 0.02.

^{w} Fourteen-amino acid medium (mg/liter): DL-alanine, 80; DL-aspartic acid, 100; L-arginine, 160; L-glutamic acid, 500; glycine, 80; L-histidine, 160; L-leucine, 100; DL-methionine, 160; DL-phenylalanine, 200; DL-serine, 80; DL-tryptophan, 200; L-tyrosine, 140; L-cysteine, 250; and DL-valine, 320.

^{x} Add (g/liter): L-arginine, 3.0; DL-phenylalanine, 2.0; L-tyrosine, 0.25; DL-valine, 2.0; DL-leucine, 1.5; DL-isoleucine, 0.5; L-tryptophan, 0.05; DL-threonine, 1.0; DL-methionine, 0.6; L-proline, 0.45; L-hydroxyproline, 0.1; L-histidine, 0.2; DL-glutamic acid, 1.0; DL-aspartic acid, 0.9; DL-lysine, 1.2; DL-alanine, 0.42; DL-serine, 1.0; L-cysteine, 0.25; glycine, 0.1.

^{y} Add (per liter): L-tryptophan, 100 mg; glycine, 325 mg; DL-valine, 1.9 g; L-leucine, 1.0 g; DL-isoleucine, 1.45 g; DL-phenylalanine, 1.0 g; L-tyrosine, 13 mg; DL-serine, 200 mg; DL-threonine, 1.0 g; L-glutamic acid, 3.15 g; DL-methionine, 540 mg; L-arginine, 475 mg; L-histidine, 350 mg; DL-lysine, 2.2 g; and DL-aspartic acid, 2.0 g.

^{z} Add (per liter): DL-alanine, 1.0 g; DL-aspartic acid, 1.0 g; L-glutamic acid, 500 mg; L-arginine·HCl, 200 mg; L-lysine·HCl, 200 mg; L-histidine·HCl, 200 mg; DL-isoleucine, 400 mg; L-leucine, 200 mg; DL-methionine, 400 mg; L-phenylalanine, 200 mg; L-proline, 200 mg; DL-threonine, 400 mg; L-tyrosine, 200 mg; DL-valine, 400 mg; DL-tryptophan, 400 mg; DL-serine, 400 mg; glycine, 200 mg; and L-cysteine, 200 mg.

^{aa} Add 1 g of monosodium glutamate per liter.

^{bb} Add (g/liter): L-asparagine, 0.1; L-tryptophan, 0.05; L-cystine, 0.1; and L-cysteine, 0.1.

^{cc} Add (g/liter): L-cystine, 0.1; L-tryptophan, 0.1; L-cysteine (separately sterilized by filtration), 0.4; L-serine, 0.1; and L-asparagine, 0.25.

^{dd} Add (mg/liter): DL-alanine, 167; L-arginine, 167; L-asparagine, 84; L-cysteine·HCl, 83.5; L-glutamic acid, 167; glycine, 167; L-histidine·HCl, 500; DL-isoleucine, 167; DL-leucine, 167; L-lysine·HCl, 167; DL-methionine, 167; DL-phenylalanine, 167; L-proline, 167; DL-serine, 167; DL-threonine, 167; DL-tryptophan, 33; L-tyrosine, 83.5; and DL-valine, 167.

^{ee} May be replaced without loss in cell yield by a mixture of amino acids supplied at the concentrations at which they are present in casein (16).

^{ff} Prepared according to Snell and Wright (206) or obtainable commercially (NBC-acid hydrolyzed casein, vitamin-free, low salt). Replaceable by a mixture of amino acids (69).

^{gg} Replaceable by individual amino acids (69).

^{hh} Add 10 mg each of adenine sulfate·H₂O, guanine·HCl·2H₂O, uracil, and xanthine per liter.

^{ii} Add 0.01 g each of adenine, guanine, and uracil per liter.

^{jj} Add 10 mg each of uracil and adenine per liter.

^{kk} Add 10 mg each of adenine sulfate, guanine hydrochloride, uracil, and xanthine per liter.

^{ll} Add 0.01 g each of adenine and guanine and 0.06 g of uracil per liter.

^{mm} Add (per liter): adenine, guanine·HCl, and xanthine, 5 mg each; guanylic acid, 50 mg; and uracil, 15 mg.

^{nn} Add (per liter): hypoxanthine, 33 mg; uracil, 16.7 mg; guanine, 5 mg; and adenine, 83.5 mg.

^{oo} Add (per liter): thiamine hydrochloride, 0.033 mg; nicotinic acid, 1.2 mg; and biotin, 0.003 mg.

^{pp} Add (mg/liter): thiamine·HCl, 0.5; pyridoxine·HCl, 1.0; pyridoxamine·HCl, 0.3; pyridoxal·HCl, 0.3; calcium pantothenate, 0.5; riboflavin, 0.5; nicotinic acid, 1.0; p-aminobenzoic acid, 0.1; biotin, 0.001; and folic acid, 0.01.

^{qq} Add (μg/liter): p-aminobenzoic acid, 200; biotin, 2; calcium pantothenate, 400; folic acid, 10; nicotinic acid, 400; riboflavin, 400; thiamine hydrochloride, 200; and pyridoxal or pyridoxamine hydrochloride, 10.

^{rr} Add (mg/liter): folic acid, 5; biotin, 2.5; p-aminobenzoic acid, 100; thiamine, 500; riboflavin, 500; pyridoxal·HCl, 1,000; pyridoxamine, 1,000; calcium pantothenate, 500; nicotinic acid, 1,000.

^{ss} Add (μg/liter): biotin, 1; folic acid, 2; riboflavin, 100; and thiamine, nicotinic acid, pyridoxine hydrochloride, and calcium pantothenate, 500 each.

^{tt} Add (mg/liter): biotin, 0.0005; thiamine, 0.4; folic acid, 0.01; choline, 2.5; calcium pantothenate, 1.0; pyridoxine, 0.5; pyridoxamine, 0.5; pyridoxal·HCl, 0.5; pyridoxal 5'-phosphate, 0.5; nicotinic acid, 1.0; nicotinamide, 1.0; riboflavin, 0.5; inositol, 40; citrovorum factor (folinic acid-SF), 0.01; p-aminobenzoic acid, 0.01; and vitamin B₁₂, 0.05.

^{uuu} Add (per liter): biotin, 10 μg; thiamine, 1 mg; pyridoxamine, 2.0 mg; nicotinic acid, 2.0 mg; calcium pantothenate, 1.0 mg; inositol, 1.0 mg; and folic acid, 6.0 μg.

^{vv} Add (per liter): thiamine, 1 mg; riboflavin, 1 mg; folic acid, 10 μg; nicotinic acid, 2.5 mg; p-aminobenzoic acid, 2.5 mg; calcium pantothenate, 1 mg; folic acid, 250 μg; and pyridoxal, 5 mg.

^{ww} Add 50 μg of biotin and 200 μg of thiamine·HCl per liter.

^{xx} Add (mg/liter): pyridoxal·HCl, thiamine chloride, and p-aminobenzoic acid, 0.4 each; riboflavin and nicotinic acid, 0.8 each; folic acid, 0.016; biotin, 0.002; vitamin B₁₂, 0.005; ascorbic acid, 300; and pantethine, 0.03.

^{yy} Add (μg/liter): riboflavin, 400; nicotinic acid, 400; calcium pantothenate, 400; thiamine·HCl, 200; p-aminobenzoic acid, 200; folic acid, 10; pyridoxal·HCl, 10; and biotin, 2.

^{zz} Add (μg/liter): p-aminobenzoic acid, 200; biotin, 2; folic acid, 16.7; nicotinic acid, 1,670; riboflavin, 400; thiamine, 200; calcium pantothenate, 800; pyridoxamine·2HCl, 16.7.

TABLE 13. *Compositions of defined (C) and semidefined (B) media 67 through 79 for the bacteria listed in Table 1*

Component	Amt (in units given in column 1)/liter for medium:												
	67C	68C[a]	69C[b]	70C[c]	71C	72C	73C	74C[d]	75B	76C[e]	77C[f]	78B[g]	79B[h]
Glucose[i] (g)	10	1.0	5.0		5.0	10		20	1.0	10	10	7.5	
Glycerol (g)							88						
Sodium acetate (g)										0.3	8.0		0.05
NH_4Cl (g)		10	0.4		1.4	1.06		1.0	1.0	1.0		2.8	
$NaNO_3$ or KNO_3 (g)		0.5											
$Na_2HPO_4\cdot2H_2O$ (g)	8.2			30						6.5		0.1	0.14
K_2HPO_4 (g)	3.28	1.0	0.1			5.65	1.0	5.0			4.0		
KH_2PO_4 (g)					1.0	2.38				1.0	15		
Tris buffer (g)			1.0									6.0	
NaCl (g)		0.1			0.1						4.0	6.8	6.6
KCl (g)			0.5	0.4								0.4	0.4
$MnCl_2\cdot4H_2O$ (mg)				2.5	0.4	7.9		0.3		0.15	3.0		
$MgSO_4\cdot7H_2O$ (g)	0.2	0.2	1.0	1.0	0.5	1.0	0.1	0.05	0.02	0.2	0.16	0.2	0.2
$CaCl_2\cdot2H_2O$ (g)		0.1	0.11	0.2	0.1		0.01		0.005	0.02			0.2
$FeSO_4\cdot7H_2O$ (mg)		10		5.0		1.1		10	5.0	0.5	4.0	8.0	0.50
Trace elements			Yes[j]	Yes[k]	Yes[l]	Yes[m]	Yes[n]	Yes[o]		Yes[p]			Yes[q]
Amino acids	Yes[r]	Yes[s]	Yes[t]	Yes[u]			Yes[v]	Yes[w]	Yes[x]	Yes[y]	Yes[z]	Yes[aa]	Yes[bb]
Casein hydrolysate (acid) (g)											5.0[cc]		
Purines and pyrimidines			Yes[dd]							Yes[ee]	Yes[dd]	Yes[ff]	Yes[gg]
Vitamins	Yes[hh]	Yes[ii]	Yes[jj]	Yes[kk]						Yes[ll]	Yes[mm]	Yes[nn]	Yes[oo]
Tween 80 (mg)			500					5000	200				
pH	7.2	6.8		6.0–7.0	6.0–7.0	6.8–7.0	6.5–7.0		7.2	6.4–7.0	7.0–7.2	8.3	7.6–7.8

[a] Add 100 μg of terregens factor per liter.

[b] Add 200 mg of nitrilotriacetic acid and 4 g of sodium β-glycerophosphate per liter.

[c] Add (per liter): pimelic acid, 0.15 mg; ethyl alcohol, 7 ml; and D-lactic acid, 17.5 g. For *Corynebacterium diphtheriae*, the pimelic acid and β-alanine requirements can be satisfied by biotin and calcium pantothenate, respectively.

[d] The pH during growth must be carefully controlled by the addition of NaOH or by including in the medium lactic or fumaric acid at a concentration of 5 g/liter.

[e] Add (mg/liter): hemin, 0.2; pimelic acid, 0.1; oleic acid, 1; and glutathione, 500.

[f] Add 0.1 g of sodium thioglycolate per liter.

[g] Add 10 g of bovine serum albumin fraction V per liter (commercially available). All species of *Mycoplasma* examined except *M. laidlawii*, *M. granularum*, and *Mycoplasma* sp. strain S-743 require cholesterol for growth (171). Cholesterol is usually supplied at a concentration of 1 to 20 μg/liter dissolved in ethanol or Tween 80.

[h] Add (mg/liter): lipoprotein factor, 500; cholesteryl laurate, 10; and lecithin, 5.

[i] For most of these media, glucose is sterilized separately (usually by autoclaving) and added aseptically to a sterile solution of the other components.

[j] Add (mg/liter): Fe, 200 [as $FeSO_4(NH_4)_2SO_4\cdot6H_2O$]; Zn, 100 (as $ZnSO_4\cdot7H_2O$); Mn, 120 (as $MnSO_4\cdot4H_2O$); Cu, 8 (as $CuSO_4\cdot5H_2O$); Co, 10 (as $CoSO_4\cdot7H_2O$); B, 200 (as H_3BO_3); and Mo, 5 [as $(NH_4)_6Mo_7O_{24}\cdot4H_2O$].

[k] Add (mg/liter): $CuSO_4\cdot5H_2O$, 5; and $ZnCl_2$, 2.5.

[l] Add (μg/liter): boric acid, 500; $CuSO_4\cdot5H_2O$, 40; KI, 100; $FeCl_3\cdot6H_2O$, 200; and $(NH_4)_6Mo_7O_{24}\cdot4H_2O$, 200.

[m] Add (mg/liter): $CuSO_4\cdot5H_2O$, 6.4; and $ZnSO_4\cdot7H_2O$, 1.5.

[n] Add 10 mg of $ZnSO_4$ per liter.

[o] Add (mg/liter): $CuSO_4\cdot5H_2O$, 1; and $ZnSO_4$, 0.8.

[p] Add 0.15 mg of $NaMoO_4\cdot2H_2O$ per liter.

[q] Add ash equivalent to 5 g of peptone. Sn can partially replace ash.

[r] Add (mg/liter): L-cysteine·HCl, 100; L-leucine, 100; DL-isoleucine 200; DL-valine, 200; L-glutamine (separately sterilized by filtration), 600; DL-methionine, 200; L-histidine·HCl, 200; L-arginine·HCl, 200; and DL-tryptophan, 200.

[s] Add a mixture of amino acids based on the percentage composition of casein as given by Hawk, Oser, and Summerson (76).

[t] Add (g/liter): DL-alanine, 4; DL-aspartic acid, 1.6; L-glutamic acid, 2.0; L-arginine·HCl, 0.8; L-lysine·HCl, 0.8; L-histidine, 0.4; DL-isoleucine 0.8; L-leucine, 0.4; DL-methionine, 0.8; DL-phenylalanine, 0.8; L-proline, 0.4; DL-threonine, 0.8; L-tyrosine, 0.4; DL-valine, 0.8; DL-tryptophan, 0.8; L-cystine, 0.4; DL-serine, 0.8; and glycine, 0.4. In place of the amino acid mixture, an acid hydrolysate of casein can be used. A 10% solution of Casamino Acids is treated with charcoal and supplemented with (mg/liter): L-tryptophan, 2,000; L-phenylalanine, 400; L-tyrosine, 400; L-proline, 200; DL-histidine·HCl, 200; and L-cysteine, 1,000. Use at 5 mg/100 ml for final medium.

[u] Add (per liter): L-cysteine, 0.7 g; DL-valine, 2.0 g; L-methionine, 0.6 g; L-tyrosine, 0.5 g; L-proline, 0.75 g; L-aspartic acid, 5 g; L-glutamic acid, 7.5 g; and β-alanine, 2.3 mg.

[v] Add (g/liter): L-glutamic acid, 1.5; and L-arginine, 1.0.

[w] Add 0.5 g each of L-glutamic acid, L-arginine, L-proline, and L-histidine per liter.

[x] A mixture of the following (at concentrations of 0.7 g each per liter) stimulates the rate of growth: L-glutamic acid, L-asparagine, DL-α-alanine, L-glutamine, L-histidine, and DL-aspartic acid. A bovine albumin (fraction V) solution (separately sterilized by filtration) inactivated by heating to 56°C is added to the sterile medium to a concentration of 0.01%.

TABLE 13—*continued*

y Add (per liter): glycine, L-alanine, DL-serine, L-threonine, L-leucine, L-isoleucine, DL-valine, L-aspartic acid, L-asparagine, L-lysine, L-histidine, L-arginine, L-proline, L-hydroxyproline, L-phenylalanine, L-tyrosine, DL-methionine, L-cysteine·HCl, 100 mg each; L-tryptophan, 40 mg; and L-glutamic acid, 500 mg. Casein hydrolysate (4 g/liter) can be substituted for the amino acid mixture.

z Add 0.1 g each of L-cystine and L-tryptophan per liter.

aa Add (mg/liter): L-cystine, 30; DL-isoleucine, 600; L-glutamine, 100; L-asparagine, 200; glycine, 100; DL-alanine, 200; DL-phenylalanine, 400; L-tyrosine, 50; L-histidine·HCl, 150; DL-leucine, 800; L-arginine, 200; DL-methionine, 200; and DL-threonine, 400.

bb Add (mg/liter): L-arginine, 70; L-aspartic acid, 60; L-cysteine, 1; L-glutamic acid, 75; L-glutamine, 100; DL-isoleucine, 40; L-methionine, 15; DL-phenylalanine, 50; and DL-tryptophan, 20.

cc Casein hydrolysate can be replaced by amino acids. Although this permits continual serial transfer, the extent of growth is diminished for some strains.

dd Add adenine, guanine, uracil, and xanthine to give a final concentration in the medium of 0.5 mg of each per 100 ml.

ee Add 20 mg each of adenine sulfate, guanine·HCl, uracil, xanthine, and thymine per liter.

ff Add 100 mg each of adenosine, guanosine, and cytidine per liter.

gg Add (mg/liter): adenosine triphosphate, 1; guanine, 0.3; hypoxanthine, 0.3; ribonucleic acid, 5; and deoxyribonucleic acid, 5.

hh Add (μg/liter): riboflavin 1,000; biotin, 100; thiamine·HCl, 1,000; and DL-thioctic acid, 1.0.

ii Add (μg/liter): thiamine, 80; pyridoxal, 40; pyridoxamine, 4; pyridoxine, 4; *p*-aminobenzoic acid, 40; folic acid, 4; choline, 80; inositol, 2,000; biotin, 0.1; vitamin B_{12}, 0.4; and calcium pantothenate, 80.

jj Add (μg/liter): thiamine·HCl, 1,000; *p*-aminobenzoic acid, 500; nicotinic acid, 1,000; choline chloride, 10,000; inositol, 10,000; pyridoxine, 500; pyridoxal·HCl, 1,000; calcium pantothenate, 1,000; biotin, 20; folic acid, 10; vitamin B_{12}, 10; and riboflavin, 500. The vitamins are nonessential except for one strain, which requires nicotinic acid in the absence of amino acids.

kk Add 2.3 mg of nicotinic acid per liter.

ll Add (per liter): vitamin B_{12}, 1 μg; folic acid, 0.5 mg; *p*-aminobenzoic acid, 2 mg; thiamine·HCl, 2 mg; riboflavin, 2 mg; nicotinic acid, 1 mg; nicotinamide, 1 mg; pyridoxal·HCl, 1 mg; pyridoxamine·2HCl, 1 mg; calcium pantothenate, 2 mg; inositol, 2 mg; DL-thioctic acid, 0.1 mg; and biotin, 0.1 mg. (Biotin, inositol, nicotinic acid, pyridoxal, and riboflavin are required.)

mm Add (per liter): biotin, 1.0 μg; and calcium pantothenate, thiamine, and *p*-aminobenzoic acid, 1 mg each.

nn Add (mg/liter): nicotinic acid, 2.5; riboflavin, 2.5; folinic acid, 1.0; pyridoxine·HCl, 2.5; pyridoxal·HCl, 2.5; and thiamine, 2.5.

oo Add (mg/liter): choline, 0.5; biotin, 0.01; folic acid, 0.01; calcium pantothenate, 0.01; pyridoxine, 0.05, and thiamine, 0.01.

TABLE 14. *Composition of a defined, complete basal medium for Streptococcus equinus*

Component	Amt/liter of double-strength medium
Amino acids:	
DL-α-Alanine	400 mg
L-Arginine·HCl	484 mg
L-Asparagine	800 mg
L-Aspartic acid	200 mg
L-Cysteine	100 mg
L-Glutamic acid	600 mg
Glycine	200 mg
L-Histidine·HCl	124 mg
DL-Isoleucine	500 mg
DL-Leucine	500 mg
DL-Lysine·HCl	500 mg
DL-Methionine	200 mg
DL-Phenylalanine	200 mg
L-Proline	200 mg
DL-Serine	100 mg
DL-Threonine	400 mg
DL-Tryptophan	80 mg
L-Tyrosine	200 mg
DL-Valine	500 mg
Glucose	50 g
Sodium acetate	40 g
Ammonium chloride	6 g
Inorganic salts:	
KH_2PO_4	1.2 g
K_2HPO_4	1.2 g
$MgSO_4·7H_2O$	400 mg
$FeSO_4·7H_2O$	20 mg
$MnSO_4·4H_2O$	40 mg
NaCl	20 mg
Purine and pyrimidine bases:	
Adenine sulfate·H_2O	20 mg
Guanine·HCl·H_2O	20 mg
Uracil	20 mg
Xanthine	20 mg
Vitamins:	
Thiamine·HCl	1.0 mg
Pyridoxine·HCl	2.0 mg
Pyridoxamine·HCl	0.6 mg
Pyridoxal·HCl	0.6 mg
Calcium DL-pantothenate	1.0 mg
Riboflavin	1.0 mg
Nicotinic acid	2.0 mg
p-Aminobenzoic acid	0.2 mg
Biotin	0.002 mg
Folic acid	0.02 mg
Distilled water to	1,000 ml

7.4. LITERATURE CITED

1. **Aaronson, S.** 1955. J. Bacteriol. **69**:67–69.
2. **Allison, M. J., M. P. Bryant, and R. N. Doetsch.** 1958. Science **128**:474–475.
3. **Allison, M. J., M. P. Bryant, and R. N. Doetsch.** 1962. J. Bacteriol. **83**:523–532.
4. **American Type Culture Collection.** 1978. Catalogue of strains I, 13th ed. American Type Culture Collection, Rockville, Md.
5. **Aristovskaya, T. V.** 1961. Dokl. Akad. Nauk. USSR **136**:954–957 (Transl. p. 113).
6. **Baird-Parker, A. C.** 1957. Nature (London) **180**:1056–1057.
7. **Baird-Parker, A. C.** 1960. J. Gen. Microbiol. **22**:458–469.
8. **Baldacci, E., and R. Locci.** 1974. *In* R. E. Buchanan and N. E. Gibbons (ed.), Bergey's manual of determinative bacteriology, 8th ed., p. 829–831. The Williams & Wilkins Co., Baltimore.
9. **Baron, L. S., W. F. Carey, and W. M. Spilman.** 1959. Proc. Natl. Acad. Sci. U.S.A. **45**:976–984.
10. **Baumann, P., M. Doudoroff, and R. Y. Stanier.** 1968. J. Bacteriol. **95**:58–73.
11. **Baumann, P., M. Doudoroff, and R. Y. Stanier.** 1968. J. Bacteriol. **95**:1520–1541.
12. **Becking, H. J.** 1961. Plant Soil **14**:49–81, 297–322.
13. **Becking, H. J.** 1974. *In* R. E. Buchanan and N. E. Gibbons (ed.), Bergey's manual of determinative bacteriology, 8th ed., p. 258–259. The Williams & Wilkins Co., Baltimore.
14. **Belaich, J. P., and J. P. Senez.** 1965. J. Bacteriol. **89**:1195–1200.
15. **Bergan, T., and B. Hovig.** 1968. Acta Pathol. Microbiol. Scand. **74**:421–430.
16. **Block, R. J., and D. Bolling.** 1945. The amino acid composition of proteins and foods, p. 485–492. Charles C Thomas, Publisher, Springfield, Ill.
17. **Bose, S. K.** 1963. *In* H. Gest, A. San Pietro, and L. P. Vernon (ed.), Bacterial photosynthesis, p. 501–510. The Antioch Press, Yellow Springs, Ohio.
18. **Bowdre, J. H., N. R. Krieg, P. S. Hoffman, and R. M. Smibert.** 1976. Appl. Environ. Microbiol. **31**:127–133.
19. **Brewer, J. H.** 1942. Science **95**:587.
20. **Bryant, M. P., and I. M. Robinson.** 1962. J. Bacteriol. **84**:605–614.
21. **Bryant, M. P., I. M. Robinson, and H. Chu.** 1959. J. Dairy Sci. **42**:1831–1847.
22. **Bryant, M. P., and N. Small.** 1956. J. Bacteriol. **72**:16–21.
23. **Bryant, M. P., and N. Small.** 1956. J. Bacteriol. **72**:22–26.
24. **Bryant, M. P., N. Small, C. Bouma, and H. Chu.** 1958. J. Bacteriol. **76**:15–23.
25. **Burton, M. O., and A. G. Lockhead.** 1953. Can. J. Bot. **31**:145–151.
26. **Callaham, L. T., III, and S. H. Richardson.** 1973. Infect. Immun. **7**:567–572.
27. **Campbell, L. L., Jr., H. A. Frank, and E. R. Hall.** 1957. J. Bacteriol. **73**:516–521.
28. **Canale-Parola, E., and R. S. Wolfe.** 1960. J. Bacteriol. **79**:857–859; 860–862.
29. **Caraway, B. H., and N. R. Krieg.** 1972. Can. J. Microbiol. **18**:1749–1759.
30. **Catlin, B. W.** 1973. J. Infect. Dis. **128**:178–194.
31. **Catlin, W.** 1960. J. Bacteriol. **79**:579–590.
32. **Catlin, W.** 1962. J. Bacteriol. **83**:470–474.
33. **Chang, G. W., and E. E. Snell.** 1968. Biochemistry **7**:2005–2012.

34. **Christie, A. O., and J. W. Porteous.** 1962. J. Gen. Microbiol. **28:**443–454; 455–460.

35. **Cohen-Bazire, G., W. R. Sistrom, and R. Y. Stanier.** 1957. J. Cell. Comp. Physiol. **49:**25–68.

36. **Cole, R. M., J. G. Tully, T. J. Popkin, and J. M. Bove.** 1973. J. Bacteriol. **115:**367–386.

37. **Couch, J. N.** 1950. J. Elisha Mitchell Sci. Soc. **66:**87–92.

38. **Couch, J. N.** 1955. J. Elisha Mitchell Sci. Soc. **71:**148–155.

39. **Couch, J. N.** 1963. J. Elisha Mitchell Sci. Soc. **79:**53–70.

40. **Craig, J. A., and E. E. Snell.** 1951. J. Bacteriol. **61:**283–291.

41. **Cross, T.** 1968. J. Appl. Bacteriol. **31:**36–53.

42. **Cross, T.** 1974. In R. E. Buchanan and N. E. Gibbons (ed.), Bergey's manual of determinative bacteriology, 8th ed., p. 859. The Williams & Wilkins Co., Baltimore.

43. **Darland, G., T. D. Brock, W. Samsonoff, and S. F. Conti.** 1970. Science **170:**1416–1418.

44. **Davis, B. R., and J. M. Woodward.** 1957. Can. J. Microbiol. **3:**591–597.

45. **DeLey, J.** 1961. J. Gen. Microbiol. **24:**31–50.

46. **DeLey, J., and J. Frateur.** 1974. In R. E. Buchanan and N. E. Gibbons (ed.), Bergey's manual of determinative bacteriology, 8th ed., p. 277. The Williams & Wilkins Co., Baltimore.

47. **Delwiche, E. A.** 1949. J. Bacteriol. **58:**395–398.

48. **DeMoss, R. O., and M. E. Happel.** 1959. J. Bacteriol. **77:**137–141.

49. **Difco.** 1962. Difco manual, 9th ed., p. 116, 121, 271, 276. Difco Laboratories, Detroit.

50. **Dondero, N. C., R. A. Phillips, and H. Heukelekian.** 1961. Appl. Microbiol. **9:**219–227.

51. **Dreizen, S., J. J. Mosny, E. J. Gilley, and T. D. Spies.** 1954. J. Dent. Res. **33:**339–345.

52. **Duchow, E., and H. C. Douglas.** 1949. J. Bacteriol. **58:**409–416.

53. **Dworkin, M.** 1962. J. Bacteriol. **84:**250–257.

54. **Dworkin, M., and J. W. Foster.** 1956. J. Bacteriol. **72:**646–659.

55. **Ehrlich, J., D. Gottlieb, P. R. Burkholder, L. E. Anderson, and T. G. Pridham.** 1948. J. Bacteriol. **56:**467–477.

56. **Englesberg, E., and J. B. Levy.** 1954. J. Bacteriol. **67:**438–449.

57. **Friedman, M. E., and W. G. Roessler.** 1961. J. Bacteriol. **83:**528–533.

58. **Frobisher, M.** 1968. Fundamentals of microbiology, p. 538. The W. B. Saunders Co., Philadelphia.

59. **Fuller, R. C.** 1963. In H. Gest, A. San Pietro, and L. P. Vernon (ed.), Bacterial photosynthesis, p. 507–509. The Antioch Press, Yellow Springs, Ohio.

60. **Gardner, G. A.** 1969. J. Appl. Bacteriol. **32:**371–380.

61. **Gardner, J. F., and J. Lascelles.** 1962. J. Gen. Microbiol. **29:**157–164.

62. **Garvie, E. T.** 1967. J. Gen. Microbiol. **48:**439–447.

63. **Georg, L. K., and J. M. Brown.** 1967. Int. J. Syst. Bacteriol. **17:**79–88.

64. **Georg, L. K., G. W. Roberstad, and S. A. Brinkman.** 1964. J. Bacteriol. **88:**477–490.

65. **Gerhardt, P.** 1958. Bacteriol. Rev. **22:**81–98.

66. **Gordon, M. A., and M. R. Edwards.** 1963. J. Bacteriol. **86:**1101–1115.

67. **Gould, R. G.** 1944. J. Biol. Chem. **153:**143–150.

68. **Griffin, P. J., and E. Racker.** 1956. J. Bacteriol. **71:**717–721.

69. **Guirard, B. M.** 1976. In G. D. Fasman (ed.), Handbook of biochemistry and molecular biology, 3rd ed., vol. 2, p. 629. CRC Press, Cleveland.

70. **Guirard, B. M., and E. E. Snell.** 1962. In I. C. Gunsalus and R. Y. Stanier (ed.), The bacteria, vol. 4, p. 33–93. Academic Press, Inc., New York.

71. **Guirard, B. M., and E. E. Snell.** 1964. J. Bacteriol. **87:**370–376.

72. **Gutierrez, J., R. E. Davis, I. L. Lindahl, and E. J. Warwick.** 1959. Appl. Microbiol. **7:**16–22.

73. **Hammerstrom, R. A., K. D. Claus, J. W. Coghlan, and R. H. McBee.** 1955. Arch. Biochem. Biophys. **56:**123–129.

74. **Hanert, H.** 1968. Arch. Mikrobiol. **60:**348–376.

75. **Happold, F. C., K. I. Johnstone, H. J. Rogers, and J. B. Yovatt.** 1954. J. Gen. Microbiol. **10:** 261–266.

76. **Hawk, P. B., B. L. Oser, and W. H. Summerson.** 1951. Practical physiological chemistry, 12th ed., p. 109. The Blakiston Co., New York.

77. **Hayflick, L., and R. M. Chanock.** 1965. Bacteriol. Rev. **29:**185–221.

78. **Haywood, H. R., and T. C. Stadtman.** 1959. J. Bacteriol. **78:**557–561.

79. **Heilman, F. R.** 1941. J. Infect. Dis. **69:**32–44; 45–51.

80. **Hemphill, H. E., and S. A. Zahler.** 1968. J. Bacteriol. **95:**1011–1017.

81. **Henderson, L. M., and E. E. Snell.** 1948. J. Biol. Chem. **172:**15–29.

82. **Hendrie, M. S., W. Hodgkiss, and J. W. Shewan.** 1970. J. Gen. Microbiol. **64:**151–169.

83. **Henssen, A.** 1957. Arch. Mikrobiol. **26:**373–414.

84. **Herbst, E. J., and E. E. Snell.** 1948. J. Biol. Chem. **176:**989–990.

85. **Hespell, R. B., and E. Canale-Parola.** 1970. J. Bacteriol. **103:**216–226.

86. **Hirsch, P.** 1974. In R. E. Buchanan and N. E. Gibbons (ed.), Bergey's manual of determinative bacteriology, 8th ed., p. 149–151. The Williams & Wilkins Co., Baltimore.

87. **Hofstad, T.** 1967. Acta. Pathol. Microbiol. Scand. **69:**543–548.

88. **Holt, S. C., S. F. Conti, and R. C. Fuller.** 1966. J. Bacteriol. **91:**311–323.

89. **Howell, A., and L. Pine.** 1956. J. Bacteriol. **71:** 47–53.

90. **Hutner, S. H.** 1942. J. Bacteriol. **43:**629–640.

91. **Hutner, S. H., L. Provasoli, A. Schatz, and C. P. Haskins.** 1950. Proc. Am. Philos. Soc. **94:** 152–170.

92. **Jackins, H. C., and H. A. Barker.** 1951. J. Bacteriol. **61:**101–114.

93. **Janke, A.** 1950. Arch. Mikrobiol. **15:**116–118.

94. **Jensen, E. M., and H. W. Seeley.** 1954. J. Bacteriol. **67**:484-488.
95. **Jensen, H. L.** 1932. Proc. Linn. Soc. N.S.W. **57**: 173-180.
96. **Jensen, H. L., E. J. Peterson, P. K. De, and R. Bhattacharya.** 1960. Arch. Mikrobiol. **36**: 182-195.
97. **Johnson, R. C., and J. B. Wilson.** 1960. J. Bacteriol. **80**:406-411.
98. **Jones, H. E., and R. W. A. Park.** 1967. J. Gen. Microbiol. **47**:369-378.
99. **Jones, J. B., and T. C. Stadtman.** 1977. J. Bacteriol. **130**:1404-1406.
100. **Jordan, R. M. M.** 1952. Br. J. Exp. Pathol. **33**: 27-35; 36-54.
101. **Kane, W. D.** 1966. J. Elisha Mitchell Sci. Soc. **82**:220-230.
102. **Kasai, G. J.** 1965. J. Dent. Res. **44**:1015-1022.
103. **Katznelson, H., and A. G. Lockhead.** 1952. J. Bacteriol. **64**:97-103.
104. **Kauffmann, F.** 1966. The bacteriology of Enterobacteriaceae, p. 311-332. The Williams & Wilkins Co., Baltimore.
105. **Kaufman, L., and J. C. Humphries.** 1958. Appl. Microbiol. **6**:311-315.
105a. **Kavanaugh, F.** 1963, 1972. Analytical microbiology, vol. 1 and 2. Academic Press, Inc., New York.
106. **Kellogg, D. S., W. L. Peacock, Jr., W. E. Deacon, L. Brown, and C. I. Pirkle.** 1963. J. Bacteriol. **85**:1274-1279.
107. **Kihara, H., and E. E. Snell.** 1960. J. Biol. Chem. **235**:1409-1414.
108. **Kindler, S. H., J. Mager, and N. Grossowicz.** 1956. J. Gen. Microbiol. **15**:386-393.
109. **King, B. M., and D. L. Alder.** 1964. Am. J. Clin. Pathol. **41**:230-232.
110. **Kitahara, K., and C. L. Lai.** 1967. J. Gen. Appl. Microbiol. **13**:197-203.
110a. **Kitahara, K., and J. Suzuki.** 1963. J. Gen. Appl. Microbiol. **9**:59-71.
111. **Knight, B. C. J. G., and H. Proom.** 1950. J. Gen. Microbiol. **4**:508-538.
112. **Knight, S. G.** 1951. In C. H. Werkman and P. W. Wilson (ed.), Bacterial metabolism, p. 500-516. Academic Press, Inc., New York.
113. **Kondrat'eva, E. N., E. N. Krasil'nikova, and L. M. Novikova.** 1968. Mikrobiologiya **37**: 417-427.
114. **Koser, S.** 1968. Vitamin requirements of yeast and bacteria. Charles C Thomas, Publisher, Springfield, Ill.
115. **Kowallik, U., and E. G. Pringsheim.** 1966. Am. J. Bot. **53**:801-806.
116. **Krasil'nikov, N. A., and N. S. Agre.** 1964. Microbiology (USSR) **33**:829-835.
117. **Krieg, N. R.** 1974. In R. E. Buchanan and N. E. Gibbons (ed.), Bergey's manual of determinative bacteriology, 8th ed., p. 198-200. The Williams & Wilkins Co., Baltimore.
118. **Kubica, G.-P., T. H. Kim, and F. P. Dunbar.** 1972. Int. J. Syst. Bacteriol. **22**:99-106.
119. **Kulp, W. L., and D. G. Borden.** 1942. J. Bacteriol. **44**:673-685.
120. **Lankford, C. E., and P. K. Skaggs.** 1946. Arch. Biochem. **9**:265-283.
121. **Lankford, C. E., and E. E. Snell.** 1943. J. Bacteriol. **45**:410-411.
122. **La Page, S. P., J. E. Shelton, and T. G. Mitchell.** 1970. In J. R. Norris and D. W. Ribbons (ed.), Methods in microbiology, vol. 3A, p. 1-133. Academic Press, Inc., New York.
123. **Larsen, H.** 1952. J. Bacteriol. **64**:187-196.
124. **Larsen, H.** 1963. In H. Gest, A. San Pietro, and L. P. Vernon (ed.), Bacterial photosynthesis, p. 509. The Antioch Press, Yellow Springs, Ohio.
125. **Lechevalier, H. A., M. Solotorovsky, and C. I. McDurmont.** 1961. J. Gen. Microbiol. **25**: 11-18.
126. **Lechevalier, M. P., H. H. Lechevalier, and P. E. Holbert.** 1968. Ann. Inst. Pasteur Paris **114**: 277-285.
127. **Lennette, E. H., A. Balows, W. J. Hausler, Jr., and J. P. Truant (ed.).** 1980. Manual of clinical microbiology, 3rd ed., p. 965-999. American Society for Microbiology, Washington, D.C.
128. **Lewin, R. A.** 1962. Can. J. Microbiol. **8**:555-563.
129. **Lewin, R. A.** 1970. Can. J. Microbiol. **16**:511-515.
130. **Lewin, R. A., and D. M. Lounsberry.** 1969. J. Gen. Microbiol. **58**:145-157.
131. **Lubin, M.** 1964. In J. H. Hoffman (ed.), The cellular function of membrane transport, p. 193-209. Prentice-Hall, Inc., Englewood Cliffs, N.J.
132. **Luedemann, G. M.** 1968. J. Bacteriol. **96**:1848-1858.
133. **Luria, S. E., and J. W. Burrows.** 1957. J. Bacteriol. **74**:461.
134. **MacLeod, R. A., and E. E. Snell.** 1950. J. Bacteriol. **59**:783-792.
135. **Mager, J., S. H. Kindler, and N. Grossowicz.** 1954. J. Gen. Microbiol. **10**:130-141.
136. **Mah, R. A., D. Y. C. Fung, and S. A. Morse.** 1967. Appl. Microbiol. **15**:866-870.
137. **Marshak, A.** 1951. J. Bacteriol. **61**:1-16.
138. **Matsumae, A. M., M. Obtani, H. Takeshima, and T. Hata.** 1968. J. Antibiot. **21**:616-625.
139. **McCleskey, C. S.** 1952. J. Bacteriol. **64**:140-141.
140. **McCurdy, H. D.** 1964. Can. J. Microbiol. **10**:935-936.
141. **McCurdy, H. D.** 1969. Can. J. Microbiol. **15**: 1453-1461.
142. **McCurdy, H. D.** 1974. In R. E. Buchanan and N. E. Gibbons (ed.), Bergey's manual of determinative bacteriology, 8th ed., p. 87-88. The Williams & Wilkins Co., Baltimore.
143. **Mickelson, M. N.** 1964. J. Bacteriol. **88**:158-164.
144. **Midgley, J., S. P. Lapage, B. A. G. Jenkins, G. I. Barrow, M. E. Roberts, and A. G. Buck.** 1970. J. Med. Microbiol. **3**:91-98.
145. **Mildvan, A. S.** 1970. In P. D. Boyer (ed.), The enzymes, student edition, vol. 2, p. 445-536. Academic Press, Inc., New York.
146. **Miller, T. L., and J. B. Evans.** 1970. J. Gen. Microbiol. **61**:131-135.
147. **Mitchell, T. G., M. S. Hendrie, and J. M. Shewan.** 1969. J. Appl. Bacteriol. **32**:40-50.
147a. **Moss, C. W., V. R. Dowell, Jr., D. Farshtchi,**

L. J. Ranier, and W. B. Cherry. 1969. J. Bacteriol. 97:561–570.

148. Moss, C. W., V. R. Dowell, Jr., V. J. Lewis, and M. A. Schekter. 1967. J. Bacteriol. 94: 1300–1305.

149. Mueller, J. H. 1938. J. Bacteriol. 36:499–515.

150. Mulder, E. G. 1964. J. Appl. Bacteriol. 27:151–173.

151. Mulder, E. G., and W. L. vanLeen. 1974. In R. E. Buchanan and N. E. Gibbons (ed.), Bergey's manual of determinative bacteriology, 8th ed., p. 133. The Williams & Wilkins Co., Baltimore.

152. Myers, W. F., L. D. Cutler, and C. L. Wisseman, Jr. 1969. J. Bacteriol. 97:663–666.

153. Norris, D. O. 1958. Nature (London) 182:734–735.

154. Ormerod, J. G., K. S. Ormerod, and H. Gest. 1961. Arch. Biochem. Biophys. 94:449–463.

155. Oxender, D. L. 1972. Annu. Rev. Biochem. 41: 777–814.

156. Pan, S. F., R. Yee, and H. M. Gezon. 1957. J. Bacteriol. 73:402–409.

157. Perry, J. J., and J. B. Evans. 1966. J. Bacteriol. 91:33–38.

158. Pfennig, N., and K. D. Lippert. 1966. Arch. Mikrobiol. 55:245–256.

159. Phillips, J. E. 1974. In R. E. Buchanan and N. E. Gibbons (ed.), Bergey's manual of determinative bacteriology, 8th ed., p. 373–374. The Williams & Wilkins Co., Baltimore.

160. Pine, L., and S. J. Watson. 1959. J. Lab. Clin. Med. 54:107–114.

161. Pittman, K. A., and M. P. Bryant. 1964. J. Bacteriol. 88:401–410.

162. Poindexter, J. S. 1964. Bacteriol. Rev. 28:231–295.

163. Postgate, J. R. 1956. J. Gen. Microbiol. 14:545–572.

164. Postgate, J. R., and L. L. Campbell. 1963. J. Bacteriol. 86:274–279.

165. Pratt, D. 1952. J. Bacteriol. 64:651–658; 659–665.

166. Pridham, T. G., and D. Gottlieb. 1948. J. Bacteriol. 56:107–114.

167. Puttlitz, D. H., and H. W. Seeley, Jr. 1968. J. Bacteriol. 96:931–938.

168. Rabinowitz, J. C., and E. E. Snell. 1947. J. Biol. Chem. 169:631–642.

169. Razin, S. 1978. Microbiol. Rev. 42:414–470.

170. Razin, S., and A. Cohen. 1963. J. Gen. Microbiol. 30:141–154.

171. Razin, S., and J. G. Tully. 1970. J. Bacteriol. 102:306–310.

172. Rechcigl, M., Jr. (ed.). 1978. CRC handbook series in nutrition and food, sect. G, vol. 3, p. 83–281. CRC Press, Cleveland.

173. Reich, C. V., and J. H. Hanks. 1964. J. Bacteriol. 87:1317–1320.

174. Reichenback, H. 1970. Arch. Mikrobiol. 70:119–138.

175. Reichenback, H., and M. Dworkin. 1969. J. Gen. Microbiol. 58:3–14.

176. Reyn, A. 1974. In R. E. Buchanan and N. E. Gibbons (ed.), Bergey's manual of determinative bacteriology, 8th ed., p. 432–433. The Williams & Wilkins Co., Baltimore.

177. Rockenmacker, M., H. A. James, and S. S. Elberg. 1952. J. Bacteriol. 63:785–794.

178. Rode, L. J., G. Ogelsby, and V. T. Schuhardt. 1950. J. Bacteriol. 60:661–668.

179. Rogosa, M. 1964. J. Bacteriol. 87:162 170.

180. Rogosa, M. 1964. J. Bacteriol. 87:574–587.

181. Rogosa, M. 1969. J. Bacteriol. 98:756–766.

182. Rogosa, M. 1971. Int. Syst. Bacteriol. 21:187–189; 231–233.

183. Roth, J. R. 1970. Methods Enzymol. 17A:6.

184. Rothwell, F. M. 1957. Mycologia 49:68–72.

185. Rowatt, E. 1957. J. Gen Microbiol. 17:297–326.

186. Sanders, T. H., K. Higuchi, and C. R. Brewer. 1953. J. Bacteriol. 66:294–299.

187. Sanderson, K. E. 1967. Bacteriol. Rev. 31:354–372.

188. Schubert, R. H. W. 1974. In R. E. Buchanan and N. E. Gibbons (ed.), Bergey's manual of determinative bacteriology, 8th ed., p. 345–346. The Williams & Wilkins Co., Baltimore.

189. Scotten, H. L., and J. L. Stokes. 1962. Arch. Microbiol. 42:353–368.

190. Sebald, M., and M. Vernon. 1963. Ann. Inst. Pasteur Paris 105:897–910.

191. Sebrell, W. H., and R. S. Harris (ed.). 1954–1972. The vitamins, 2nd ed., vol. 2, 4, 5, 7. Academic Press, Inc., New York.

192. Shenberg, E. 1967. J. Bacteriol. 93:1598–1606.

193. Shivvers, D. W., and T. D. Brock. 1973. J. Bacteriol. 114:706–710.

194. Sijpesteijn, A. K., and G. Fahraeus. 1949. J. Gen. Microbiol. 3:224–234.

195. Skeggs, H. R., L. D. Wright, E. L. Cresson, G. D. E. MacRae, C. H. Hoffman, D. E. Wolf, and K. Folkers. 1956. J. Bacteriol. 72: 519–524.

196. Skerman, T. M., and D. J. J. Jayne-Williams. 1966. J. Appl. Bacteriol. 29:167–178.

197. Slack, J. M. 1974. In R. E. Buchanan and N. E. Gibbons (ed.), Bergey's manual of determinative bacteriology, 8th ed., p. 671–673. The Williams & Wilkins Co., Baltimore.

198. Slotnick, I. J., and M. Dougherty. 1964. Antonie van Leeuwenhoek J. Microbiol. Serol. 30: 261–272.

199. Slotnick, I. J., and M. Dougherty. 1965. Antonie van Leeuwenhoek J. Microbiol. Serol. 31: 355–360.

200. Smith, C. F. 1955. Proc. Soc. Exp. Biol. Med. 88: 628–631.

201. Smith, P. F., T. A. Langworthy, and M. R. Smith. 1975. J. Bacteriol. 124:884–892.

202. Smith, P. H., and R. E. Hungate. 1958. J. Bacteriol. 75:713–718.

203. Sneath, P. H. A. 1960. Iowa State J. Sci. 34: 243–500.

204. Snell, E. E. 1950. In P. Gyorgy (ed.), Vitamin methods, vol. 1, p. 327–505. Academic Press, Inc., New York.

205. Snell, E. E. 1957. In J. H. Yoe and H. J. Koch, Jr. (ed.), Trace analysis, p. 547–574. John Wiley & Sons, Inc., New York.

206. Snell, E. E., and L. D. Wright. 1941. J. Biol. Chem. 139:675–686.

207. Sperry, J. F., and T. D. Wilkins. 1976. J.

Bacteriol. **127**:780–784.

208. **Stadtman, E. R., T. C. Stadtman, I. Pastan, and L. DS. Smith.** 1972. J. Bacteriol. **110**: 758–760.
209. **Stadtman, T. C., and H. A. Barker.** 1951. J. Bacteriol. **61**:67–80.
210. **Stadtman, T. C., and H. A. Barker.** 1951. J. Bacteriol. **62**:269–280.
211. **Staley, J. T.** 1968. J. Bacteriol. **95**:1921–1942.
212. **Stanier, R. Y.** 1942. Bacteriol. Rev. **6**:143–196.
213. **Stanier, R. Y., N. J. Palleroni, and M. Doudoroff.** 1966. J. Gen. Microbiol. **43**:159–271.
214. **Starr, M. P.** 1946. J. Bacteriol. **51**:131–143.
215. **Starr, M. P.** 1946. J. Bacteriol. **52**:187–194.
216. **Steele, B. F., H. E. Sauberlich, M. S. Reynolds, and C. A. Baumann.** 1949. J. Biol. Chem. **177**:533–551.
217. **Steinman, H. G., H. Eagle, and V. I. Oyama.** 1952. J. Bacteriol. **64**:265–269.
218. **Steinman, H. G., H. Eagle, and V. I. Oyama.** 1953. J. Biol. Chem. **200**:775–785.
219. **Stokes, J. L.** 1954. J. Bacteriol. **67**:278–291.
220. **Szaniszlo, P. J.** 1968. J. Elisha Mitchell Sci. Soc. **84**:24–26.
221. **Tamura, G.** 1956. J. Gen. Appl. Microbiol. **2**: 431–434.
222. **Tamura, G.** 1957. Bull. Agric. Chem. Soc. Jpn. **21**:202.
223. **Tehan, Y. T.** 1953. Proc. Linn. Soc. N.S.W. **78**: 85–89.
224. **Thiele, H. H.** 1968. Arch. Mikrobiol. **60**:124–138.
225. **Thiemann, J. E., and G. Beretta.** 1968. Arch. Mikrobiol. **62**:157–166.
226. **Thiemann, J. E., H. Pagani, and G. Beretta.** 1967. Arch. Mikrobiol. **58**:42–52.
227. **Thompson, R. S., and E. R. Leadbetter.** 1963. Arch. Mikrobiol. **45**:27–32.
228. **Tsukiura, H., M. Okanishi, H. Koshiyama, T. Ohmori, T. Miyaki, and H. Kawaguchi.** 1964. J. Antibiot. Ser. A **17**:223–229.
229. **Van Veen, W. L., E. G. Mulder, and M. H. Deinema.** 1978. Microbiol. Rev. **42**:329–356.
230. **Vogel, H. J., and D. M. Bonner.** 1956. J. Biol.

Chem. **218**:97.

231. **Vogler, K. G., and W. W. Umbreit.** 1941. Soil Sci. **51**:331–337.
232. **Waksman, S. A.** 1961. The Actinomycetes, vol. 2, p. 328–334. The Williams & Wilkins Co., Baltimore.
233. **Waksman, S. A., W. W. Umbreit, and T. C. Cordon.** 1939. Soil Sci. **47**:37–61.
234. **Waldie, E. L.** 1945. Iowa State Coll. J. Sci. **19**: 435–484.
235. **Watson, S. W.** 1971. Arch. Mikrobiol. **75**:179–188.
236. **Watson, S. W.** 1971. Int. J. Syst. Bacteriol. **21**: 254–270.
237. **Watson, S. W., L. B. Graham, C. C. Remsen, and F. W. Valois.** 1971. Arch. Mikrobiol. **76**: 183–203.
238. **Watson, S. W., and J. B. Waterbury.** 1971. Arch. Mikrobiol. **77**:203–230.
239. **Weeks, O. B.** 1969. J. Appl. Bacteriol. **32**:13–18.
240. **Weinman, D.** 1968. *In* D. Weinman and M. Ristic (ed.), Infectious blood diseases of man and animals, vol. 2, p. 3–24. Academic Press, Inc., New York.
241. **Welshimer, H. J.** 1963. J. Bacteriol. **85**:1156–1159.
242. **Werkman, C. H., and G. F. Gillen.** 1932. J. Bacteriol. **23**:167–182.
243. **White, D. C., G. Leidy, J. D. Jamison, and R. E. Shope.** 1964. J. Exp. Med. **120**:1–12.
244. **Wilson, G. S., and A. A. Miles.** 1964. Topley and Wilson's principles of bacteriology and immunology, 5th ed., vol. 1, p. 849–865. The Williams & Wilkins Co., Baltimore.
245. **Winogradsky, H.** 1974. *In* R. E. Buchanan and N. E. Gibbons (ed.), Bergey's manual of determinative bacteriology, 8th ed., p. 454–455. The Williams & Wilkins Co., Baltimore.
246. **Zavarzin, G., and R. Legunkova.** 1959. J. Gen. Microbiol. **21**:186–190.
247. **Zimmerman, K., K. B. Rogers, J. Frazer, and J. M. H. Boyce.** 1968. J. Pathol. Bacteriol. **96**: 413–419.

Chapter 8

Enrichment and Isolation

NOEL R. KRIEG

One rarely encounters pure cultures of bacteria in natural habitats, yet much of the present knowledge of the properties of bacteria and of the interactions occurring among bacteria is based on studies with pure cultures. Consequently, one often needs to be able to isolate into pure culture the various kinds of bacteria that coexist in a habitat. Enrichment techniques are a major step in making it possible to obtain pure cultures. Enrichments also make it possible to assess the differential effects of environmental factors imposed on mixed microbial populations, and they permit the selection of organisms capable of attacking or degrading particular substrates or of thriving under unusual conditions.

For successful isolation of a given organism into pure culture, the organism generally must comprise a sufficiently high proportion of the mixed population. Isolation is most easily accomplished when the organism is the numerically dominant member of the population. Enrichment methods are designed to achieve an increase in the relative numbers of a particular organism by favoring its growth, survival, or its spatial separation from other members of the population. **Physical methods** may make use of such conditions as growth temperature, heat treatment, sonic oscillation, or ultraviolet irradiation to kill or inhibit the rest of the population. They may also take advantage of some physical property of the desired organism, such as its size or motility, that can allow the organism to be preferentially separated from the rest of the population. **Chemical methods** may employ toxic agents to kill or inhibit the rest of the population without affecting the desired organism; alternatively, they may provide nutrient sources that can be used preferentially by a particular component of the mixed population. **Biological methods** may make use of specific hosts for selective growth of a particular organism, or they may take advantage of some pathogenic property, such as invasiveness, which the rest of the population does not possess. In many enrichment procedures a number of physical, chemical, or biological methods are used in combination to achieve a maximum effect.

Most enrichments are carried out in closed systems, such as ordinary batch cultures in a flask or tube, where the concentrations of nutrients and metabolic products in the culture vessel continually change during bacterial growth. Open systems have also been used for enrichment. For example, the use of a chemostat enables one to provide a constant environment for cultivation of bacterial cells by continuously supplying a growth-limiting nutrient and continually removing metabolic products. By altering the dilution rate for a chemostat, one can control the concentration of growth-limiting nutrient. This in turn can differentially affect the growth rate of the various organisms in a mixed culture, making it possible for one or another member of the mixed population to become predominant.

One usually isolates bacteria from enrichment cultures by spatially separating the organisms in or on a solid medium and subsequently allowing them to grow into colonies. Dilution to extinction can be used for organisms that cannot grow on solid media, by allowing separation of cells into individual tubes of a liquid medium. Because ordinary isolation methods do not absolutely ensure purity, one may sometimes wish to employ somewhat more difficult methods whereby an individual bacterial cell from a mixed population can be spatially isolated under a microscope before being cultured into a clone.

In this chapter it is the intention to exemplify the multiplicity, and in many cases the considerable ingenuity, of the enrichment methods that have been used for bacteria by presenting specific examples. In many cases several different approaches are combined for the enrichment of a given bacterium. Although emphasis is given to the practical aspects of enrichment and isolation, the general references (1–4) given at the end of the chapter can provide further information concerning the underlying principles. For enrichment and isolation of mutant cells, see 13.4–13.6.

8.1. BIOPHYSICAL ENRICHMENT

8.1.1. Low-Temperature Incubation for Psychrophiles and Psychrotrophs (52)

Low temperature will retard growth of bacteria other than psychrophiles and psychrotrophs, so incubate enrichment cultures at 0 to 5°C before attempting isolation. For instance, Inoue (31) was able to isolate psychrophilic bacteria from Antarctic soil by spreading dilutions over plates of a glucose-yeast extract-peptone agar (with the temperature maintained throughout at <5°C) and incubating for 14 to 24 days. In this manner nine strains of bacteria were obtained that had maximum growth temperatures of ca. 20°C, i.e., that were obligately psychrophilic (32).

8.1.2. Low-Temperature Incubation for *Listeria monocytogenes* (39)

It is difficult to isolate *Listeria* from clinical specimens containing large numbers of other bacteria (e.g., specimens from the cervix, vagina, meconium, feces, nasopharynx, or tissues), and the following enrichment procedure is often helpful. Place specimen swabs into screw-capped tubes containing 5 ml of listeria broth (8.5.25); for fecal or tissue specimens, prepare 10% (vol/vol) suspensions in listeria broth. Store the broths at 4°C for up to 3 months, and attempt isolation at weekly intervals for the first 4 weeks and at monthly intervals thereafter. The cold storage enriches for *Listeria*. This has been variously attributed to an ability to survive longer than many other bacteria at 4°C (84), an ability to multiply at 4°C (67), or release of the organisms from their intracellular location (15). After cold enrichment, transfer 0.2 ml of the broth suspension to 5 ml of KCNS-listeria broth (add 3.75% potassium thiocyanate to listeria broth prior to sterilization) and incubate at 25°C for 48 h. For isolation, remove a loopful from just below the surface and streak a plate of McBride listeria agar (8.5.29).

8.1.3. High-Temperature Incubation for Thermophiles (6)

The growth of organisms other than thermophiles is inhibited by a high incubation temperature, so incubate enrichments for thermophiles at an appropriately high temperature. For milk thermophiles, for example, plate serial dilutions of milk in standard methods agar (8.5.48) and incubate the cultures at 55°C.

8.1.4. High-Temperature Incubation for *Thermus aquaticus* (11)

T. aquaticus has an optimum temperature of 70 to 72°C and cannot grow below 40°C or above 79°C. Add samples of hot tap water, microbial mats or water from hot springs, or water from thermally polluted rivers, to 10-ml portions of basal salts medium (8.5.7) containing 0.1% pancreatic digest of casein and 0.1% yeast extract. Incubate without agitation at 70 to 75°C. Look for visible turbidity within 1 to 2 days. When the turbidity is heavy, the color of the microbial mass is usually yellow or orange. Examine the culture microscopically to see if short and long filaments, and also some spheroplasts, are present. Isolate pure cultures by steaking plates of the same medium solidified with 3% agar. Look for compact, spreading, yellow colonies that contain filamentous organisms.

8.1.5. Heating for Thermodurics (6)

Pasteurize a sample of raw milk by heating it to 62.8°C and holding it at this temperature for 30 min. Cool, add dilutions to standard methods agar (8.5.48), and incubate the plates at 32°C.

8.1.6. Heating for Mesophilic Sporeformers

Heat water samples or soil suspensions at 80°C for 10 min; then streak onto plating media. For *Bacillus* species, nutrient agar is usually suitable; for *Clostridium* species, streak a roll tube of prereduced chopped meat agar (20.3.10) or streak a plate of freshly prepared agar medium and incubate it in an anaerobic jar.

If small numbers of spores are present in the samples, one may be able to increase the numbers by adding the sample to a suitable sporulation medium and incubating the culture for various periods (2 to 21 days) prior to the heat treatment. For most *Bacillus* species, a suitable sporulation medium is nutrient broth (20.3.25) containing 0.5% yeast extract, 7×10^{-4} M $CaCl_2$, 1×10^{-3} M $MgCl_2$, and 5×10^{-5} M $MnCl_2$. For most *Clostridium* species, no one medium is optimum for production of spores, but chopped meat medium (20.3.10) with and without glucose often supports sporulation (27).

Another approach is to add some of the sample in which spores are suspected to a suitable growth medium, heat at 80°C for 10 min, and then incubate the broth for a day or so to permit growth to occur before streaking the culture onto solid media. For *Bacillus* species, nutrient broth containing 0.5% yeast extract is usually a suitable growth medium; for *Clostridium* species, prereduced starch broth (PY broth + 1.0% soluble starch; 20.3.28) is often satisfactory.

The heat treatment employed may need to be modified for some types of sporeformers, since the endospores of some bacterial strains may not be as heat resistant as others. One may wish to try a lower temperature, e.g., 75 or 70°C, if satisfactory results are not obtained at the usual temperature of 80°C.

8.1.7. Heating for Thermophilic Sporeformers (6)

Thermophilic sporeformers may occur in various sweetening agents used in ice cream. Prepare a 20% (wt/vol or vol/vol) solution of the sweetening agent (beet or cane sugar, lactose, cerelose, invert syrup, corn syrup, corn sugar, maple syrup, liquid sugar, or honey) and heat to 100°C for 5 min. Dilute with sterile water to give a final concentration of 13.3%.

Thermophilic, aerobic flat-sour organisms.

Plate 2-ml portions of the boiled sugar solution in glucose-tryptone agar (8.5.20). Incubate in a humid chamber at 55°C for 48 h. Characteristic surface colonies are round, 2 to 5 mm in diameter, possess a typical opaque central spot, and are surrounded by a yellow halo in the medium. Subsurface colonies are compact and may be pinpoint in size; these should be subcultured and streaked onto glucose-tryptone agar.

Thermophilic, anaerobic hydrogen sulfide producers.

Add samples of the boiled sugar solution to deep tubes of sulfite agar (1% tryptone or Trypticase, 0.1% sodium sulfite, 2% agar). The tubes of medium should be melted and cooled to 55°C prior to inoculation. Allow the tube contents to solidify, and incubate at 55°C for 72 h. Sulfide producers will form blackened spherical areas in the medium.

8.1.8. Heating for *Bacillus fastidiosus* (41)

Add 0.2 g of soil to 50 ml of uric acid medium (1.0% uric acid and 0.1% K_2HPO_4 in distilled water). Incubate with shaking for 48 h, heat at 80°C for 15 min, and plate onto solidified uric acid medium. In this procedure advantage is taken not only of heat resistance of *B. fastidiosus* but also of its ability to use a carbon and energy source not often used by other bacteria.

8.1.9. Heating for *Bacillus pasteurii* (82)

In addition to making use of the heat resistance of the spores of the urea-decomposing species *B. pasteurii*, this method takes advantage of the ability of the organisms to grow at a pH high enough to discourage most other bacteria (indeed, the organisms *require* a pH >8.0 in order to grow!). After heating soil suspensions to select for spores, incubate the samples at 30°C in an alkaline enrichment medium consisting of 1% yeast extract and 5% urea (sterilized by filtration). After growth occurs, streak the culture onto Wiley and Stokes' alkaline medium (8.5.58) to obtain isolated colonies.

8.1.10. Swarming Motility of *Clostridium tetani* (71)

Inoculate the specimen (soil, animal feces, clinical material) onto a small area of a freshly prepared plate of blood agar (20.3.7). Incubate the plate at 37°C in an anaerobic jar for 1 day, and examine the agar surface carefully for evidence of swarming (a thin film of growth that has spread outward from the inoculated area). It may be helpful to scrape the surface of the medium with a needle to verify the occurrence of swarming. Suspend some growth from the edge of the swarming area in broth, and streak onto solid media containing 5% agar in order to obtain isolated colonies.

8.1.11. Swimming Motility of Treponemes (24–26, 61–63)

Cut a well into the center of a suitable agar medium (8.5.42) contained in either petri dishes or beakers. The well should be at least 7 mm deep and from 2 to 10 mm in diameter and should not be cut to the bottom of the plate. Inoculate the well with the specimen (samples from the oral cavity, intestinal contents, or feces). Inoculate large wells (10-mm diameter) with up to 0.2 ml of sample; inoculate small wells (2-mm diameter) by stabbing ca. 2 mm obliquely into one side of the well. Do not allow any of the sample to be deposited on the surface of the agar. Immediately place the plate in an anaerobic jar (6.6.3) and incubate at 37°C for 4 to 7 days. In contrast to most other bacteria, treponemes can migrate through agar media. Look for treponemal growth occurring as a "haze" in the medium at some distance from the well. Subculture from the outermost portions of this hazy region into a suitable prereduced semisolid medium (e.g., broth containing 0.15% agar).

Since more than one kind of treponeme may be present, streak the subculture onto a roll tube of prereduced medium to obtain isolated colonies. These appear as hazy, whitish, dense areas in the medium.

8.1.12. Swimming Motility of *Spirillum volutans* (30, 60)

Obtain mixed cultures of the giant microaerophilic bacterium *S. volutans* by preparing a hay infusion with stagnant pond water. After a surface scum develops, examine samples taken just beneath the scum. Look for very large spirilla (1.4 to 1.7 μm in diameter and up to 60 μm in length) with bipolar flagellar fascicles that are clearly visible by dark-field microscopy. Enrich the culture by inoculating some of the hay infusion into Pringsheim's soil medium (8.5.41) and incubating at room temperature. Even with this enrichment *S. volutans* will be vastly outnumbered by other bacteria. However, *S. volutans* can be isolated by using a capillary tube procedure first devised by Giesberger and later applied successful to *S. volutans* (60). Soften the center of a short section of a sterile, cotton-plugged piece of 5-mm glass tubing in a flame. Pinch it with square-ended forceps until it is almost closed. Reheat the flattened portion, and draw it out rapidly to form a long capillary tube 15 to 30 cm long and 0.1 to 0.3 mm wide, oval in cross section. Seal the ends of the capillary in a flame. Break the capillary near one end with sterile forceps, and draw up 10 to 20 cm of sterile Pringsheim's soil medium (supernatant). Then dip the capillary into the enrichment culture and draw up another 2 to 4 cm, making sure that no air space occurs between the sterile medium and the culture. Seal the tip of the capillary, leaving a small air space. Mount the capillary on the stage of a 100× microscope. *S. volutans* will often be able to swim faster than the other bacteria in the enrichment culture and thus can reach the distal end of the capillary first. As soon as some spirilla reach the distal end, break the capillary behind them, expel the spirilla into a tube of semisolid CHSS medium (8.5.15), and incubate at 30°C. Confirm purity of cultures by phase-contrast microscopy.

8.1.13. Gliding Motility of Cytophagas and Flexibacters

Spread dilutions of the specimen (soil, water, or animal dung which has been in contact with the soil) onto agar media that have a low nutrient concentration (e.g., 0.1% tryptone or Trypticase, or one-tenth-strength nutrient agar). Another procedure is to smear terrestrial plant leaf material, algal fronds, or marine plants on nutrient-poor media. Cytophagas and flexibacters

have the ability to migrate on the surface of solid media; their colonies can be recognized as thin, often nearly translucent colonies with finger-like projections, which develop far beyond the streak or deposition line. Subculture from these colonies. The incorporation of penicillin G (15 U/ml) and chloramphenicol (5 μg/ml) into the agar media often helps to suppress the growth of other bacteria (79).

8.1.14. Gliding Motility of *Beggiatoa* (19, 64)

Beggiatoa forms colorless filaments containing cells in chains. The filaments exhibit gliding motility on surfaces. *Beggiatoa* occurs in aerobic freshwater or marine environments rich in H_2S, and the cells can oxidize the sulfide to elemental sulfur which is deposited intracellularly.

For enrichment, first prepare extracted hay by the following procedure. Cut dried hay into small pieces, and extract by boiling the hay in a large volume of water. Change the water three times during the extraction. The final wash should have an amber color. Drain the hay and place it on trays at 37°C to dry.

The enrichment medium consists of 0.8% (wt/vol) of the dried, extracted hay in tap water, distributed in 70-ml volumes in 125-ml cotton-stoppered Erlenmeyer flasks. Sterilize by autoclaving. Inoculate the flasks with 5-ml portions of mud containing decaying plant materials, as from small ponds, lakes, or streams. Incubate the flasks at ca. 25°C for 10 days.

Positive enrichments are indicated by a strong odor of H_2S and the development of a white film on the surface of the medium and on the submerged upper walls of the flasks. Examine the film for the characteristic filaments of *Beggiatoa*. Wash portions of the surface film several times in sterile tap water and place on the surface of a sterile agar medium consisting of 1% agar and 0.2% beef extract. Incubate at 28°C until some of the filaments of *Beggiatoa* have migrated to the periphery of the agar surface and away from contaminants. Cut out agar blocks containing single, isolated filaments, and place them filament side down on fresh plates of agar medium. After incubation, again select isolated filaments. Repeat the selection procedure until the cultures appear to be free from contaminants. Test for contaminants by inoculating a variety of bacteriological media.

8.1.15. Filterability of Treponemes (27, 70)

Prepare a petri dish of some suitable agar medium (8.5.42). Place a membrane filter (pore size, 0.15 μm) on the surface of the agar. Place an O-ring (25 to 30 mm in diameter) that has been lightly coated with vacuum grease on top of the filter. Place several drops of the diluted specimen onto the center of the membrane filter. Incubate the plate in an anaerobic jar at 37°C for 1 to 2 weeks. Treponemes are small enough to migrate through the pores of the filter and penetrate the underlying agar, where they grow as a "haze" in the medium. Remove the O-ring and membrane filter from the agar surface, remove a plug of agar from the hazy region with a Pasteur pipette, and examine it by dark-field microscopy for treponemes and contaminants. Subculture and purify as described in 8.1.11. The inclusion of polymyxin B (800 U/ml) and nalidixic acid (800 U/ml) in the agar medium often helps to suppress the growth of contaminants.

8.1.16. Filterability of *Campylobacter fetus* (68, 69)

Dilute 5 g of fecal or intestinal material (from cattle, sheep, pigs or birds) with 50 ml of nutrient broth, and filter through several layers of cheesecloth to remove large particles. Pass the filtrate through a membrane filter (pore size, 0.65 μm), using a prefilter of loosely packed glass wool. Soak sterile cotton swabs with the filtrate, and streak heavily onto plates of brucella agar (8.5.14). Incubate plates under microaerobic conditions (5% O_2, 10% CO_2, 85% N_2) for 4 to 5 days at 37°C. The inclusion of antibiotics (bacitracin, 2 U/ml, and novobiocin, 2 μg/ml) in the medium is often helpful in suppressing the growth of other organisms.

8.1.17. Filterability of *Aquaspirillum gracile* (12)

Use a membrane filter with a pore size of 0.45 μm, and place it on the surface of the isolation agar (8.5.1.). Deposit 0.05 ml of pond or stream water in the center of the filter. Incubate the plates for 1.5 to 2.0 h at room temperature; then remove the filter and continue the incubation for 3 days or longer. *A. gracile* has a cell diameter of 0.2 to 0.3 μm and is small enough to penetrate the pores of the filter into the underlying agar. Look for spreading, semitransparent areas of growth within the agar medium, and subculture from these areas. Other bacteria of small size (small vibrios, cocci, or short rods) may also pass through the filter to form small colonies on the surface of the isolation medium; these colonies can be easily distinguished from the subsurface, spreading, semitransparent growth typical of *A. gracile*.

8.1.18. Visible Illumination for Unicellular Cyanobacteria (59, 73)

For freshwater cyanobacteria (blue-green algae), inoculate samples from ponds, streams, or reservoirs into tubes of BG-11 medium (8.5.8). For cyanobacteria from marine environments, use MN medium (8.5.32); some grow poorly in this medium, so ASN-III medium (8.5.2) may occasionally be preferable. All glassware should be washed by hand and then rinsed successively in tap water, concentrated nitric acid, and deionized water. Incubate enrichment cultures in an illuminated water bath at 35°C; this temperature will inhibit the growth of most eucaryotic algae. Use a light intensity of 2,000 to 3,000 lx from white fluorescent or incandescent sources; some cyanobacteria may require lower intensities (500 lx or less). Examine the enrichment cultures periodically until there is evidence of development of cyanobacteria. Then streak onto media solidified with 2% bacteriological-grade agar. Incubate plates under illumination (<500 lx) at 25°C in air or in an atmosphere slightly enriched with CO_2; clear plastic boxes such as vegetable crispers make good chambers to help prevent evaporation. Use a dissecting microscope to detect the compact, deeply pigmented colonies of nonmotile cyanobacteria. Restreaking several times may be necessary to eliminate bacterial contaminants. In the case of motile cyanobacteria, enrich by placing a small patch of culture material at one side of a petri dish of agar medium. Illuminate the dish from the opposite side. The cyanobacteria will respond phototactically by gliding across the agar toward the region of higher light intensity. When some of the organisms have reached the opposite side, subculture them to a new plate and repeat the procedure until bacterial contaminants have been eliminated. This technique has also been used effectively with some filamentous cyanobacteria (73).

For slow-growing cyanobacteria such as members of the *Pleurocapsales*, primary cultures are best established by direct isolation of colonies on solid media, rather than by preliminary cultivation in a liquid medium (80). During transport of rock chips, mollusk shells, or macroalgae from intertidal zones to the laboratory, keep samples in closed bottles or tubes containing a damp piece of filter paper; do not submerge the samples in seawater, as this promotes development of contaminants. In the laboratory suspend material scraped from the natural substrates in sterile liquid medium, and streak several plates directly from the suspension in addition to preparing a liquid enrichment culture. If the suspension contains many contaminants, wash it repeatedly in sterile medium by low-speed centrifugation to reduce the level of contamination before streaking plates.

Confirm the purity of cultures of cyanobacteria by microscopic observation and also by inoculating complex media and incubating in the dark at 30°C.

8.1.19. Visible Illumination for *Rhodospirillaceae* (76)

Although the members of the *Rhodospirillaceae* have been termed "non-sulfur purple bacteria," a number of species can in fact use sulfide as an electron donor for growth, but only when the sulfide is maintained at low, nontoxic concentrations. Consequently, the organisms are ordinarily cultured with organic electron donors. Depending on the carbon source and the vitamins provided, rather specific enrichment for particular species is possible. For enrichment, only substrates that cannot be fermented by nonphototrophic organisms are provided in the enrichment medium.

Members of the *Rhodospirillaceae* can be readily isolated from freshwater and marine sediments and less often from field, lawn, or garden soils. Use the basal enrichment medium (8.5.4) and supplement it according to the instructions in Table 1. Place 0.1 g of the specimen into a screw-capped tube, and fill the tube completely with the enrichment medium. Tighten the cap so there is no air space at the top, and incubate the culture at ca. 25°C. Illuminate the culture continuously with incandescent light (not fluorescent light); use a 50- or 75-W lamp at a distance of 40 to 60 cm. Make sure the cultures do not become heated from the lamp. Look for development of turbidity with a brown, yellow, or pink tinge in 3 to 7 days. Transfer a drop of culture to a second tube of medium (see Table 1) for a secondary enrichment. For purification, use an agar medium containing 1% yeast extract or peptone, 0.2% sodium malate, and 1.5% agar. Use the shake tube method (8.4.2), or use pour plates and incubate in an anaerobic jar. Illuminate the cultures during incubation. Many of the

Rhodospirillaceae will also grow in the dark under an air atmosphere or under microaerophilic conditions, although the colonies are less highly pigmented than when grown anaerobically in the light.

8.1.20. Visible Illumination for *Chromatiaceae* and *Chlorobiaceae*

The sulfur phototrophs belonging to the *Chromatiaceae* and *Chlorobiaceae* are not as easy to cultivate as the non-sulfur forms, although massive developments ("blooms") of them are often readily visible to the eye, particularly in marine or brackish environments. Unless one has inocula from such blooms, a useful preliminary enrichment is to set up a Winogradsky column. This method provides anaerobic conditions and a long-lasting supply of H_2S, and successive blooms of the sulfur phototrophs (and many other microbes as well) usually result.

To prepare a Winogradsky column, obtain mud from freshwater, brackish, or marine environments (for example, mud from the edge of a freshwater pond, stream, or lake, or from a salt marsh). Mix three parts of mud with one part of $CaSO_4 \cdot H_2O$. Add some insoluble organic material such as finely shredded filter paper or small pieces of roots from aquatic plants. If paper is used, also add a small amount of NH_4MgPO_4. Pour the mixture into a tall glass cylinder (at least 5 cm in diameter) to a height of at least 15 cm, stirring to avoid air pockets. Fill the cylinder with water (if marine organisms are sought, use seawater). Incubate the cylinder in the dark for 2 to 3 days to minimize the development of oxygenic photosynthetic organisms. Expose the cylinder to incandescent light or to diffuse daylight at 18 to 25°C during subsequent incubation.

TABLE 1. *Substrates and vitamins to be added to the basal enrichment medium for Rhodospirillaceae*[a]

Species to be enriched	Substrate		Vitamin
	Primary enrichment	Secondary enrichment	
Rhodospirillum rubrum	Ethanol or acetate, 2 g/liter	L-Alanine, 3 g/liter	Biotin, 20 µg/liter
Rhodopseudomonas capsulata	Propionate, 2 g/liter	Same as for primary	Thiamine, 1 mg/liter; some strains also need biotin and niacin
Rhodopseudomonas gelatinosa	Isopropanol, 2 g/liter	Same as for primary	Biotin, 20 µg/liter; thiamine, 1 mg/liter; some strains also need pantothenate
Rhodopseudomonas spheroides	Ethanol or acetate, 2 g/liter	Tartrate, 3 g/liter	Biotin, 20 µg/liter; thiamine, 1 mg/liter; niacin, 1 mg/liter
Rhodopseudomonas palustris	Benzoate, 0.5 g/liter	Thiosulfate, 2 g/liter	p-Aminobenzoic acid, 200 µg/liter; some strains also require biotin
Rhodospirillum fulvum	Benzoate, 0.5 g/liter	Pelargonate, 0.3 g/liter	p-Aminobenzoic acid, 200 µg/liter
Rhodopseudomonas acidophila	Succinate, 2 g/liter; also, adjust pH of the medium to 5.1	Same as for primary	None
Rhodospirillum tenue	Pelargonate, 0.3 g/liter	Same as for primary	None

[a] From van Niel (76). For basal enrichment medium, see 8.5.4.

Anaerobic decomposition of the organic material in the column (with concomitant production of CO_2, alcohols, fatty acids, hydroxy acids, organic acids, and amines) and the formation of H_2S from the $CaSO_4$ will provide an appropriate array of micro-habitats in which the sulfur phototrophs can thrive and form distinctive purple, red, or green patches on the sides of the glass column or layers or bands in the water column above the sediment-water interface.

Use Pasteur pipettes to obtain organisms from the glass surfaces or from the distinctive layers for microscopic examination, isolation, or further enrichment in liquid cultures. Alternatively, one can sequentially remove portions of the sediment with a spoon or spatula to expose the various zones of growth.

Enrichment can also be accomplished in defined liquid media, and one can select for specific sulfur phototrophs by varying the cultural conditions with respect to the wavelength and intensity of the illumination used, the temperature of incubation, and the type and concentration of the electron donor. For a clear, detailed exposition of the many intricacies of such enrichment and the subsequent isolation of sulfur phototrophs into pure culture, see the very useful essay by Van Niel (76).

8.1.21. Ultraviolet Irradiation for Cyanobacteria

It is often difficult to obtain cultures of cyanobacteria (blue-green algae) free from bacterial contaminants, which frequently penetrate and live in the gelatinous sheaths that surround the cells and filaments of cyanobacteria. Gerloff et al. (22) were able to free cyanobacterial cultures from contaminating bacteria by the following method. Place a dilute suspension of cyanobacteria in a quartz chamber and irradiate with 275-nm ultraviolet light from a quartz-jacketed mercury vapor lamp. Agitate the suspension by continuous stirring during the irradiation. With the proper exposure time it is possible to kill the contaminating bacteria and yet recover viable cyanobacteria. At periodic intervals during the irradiation, remove samples and prepare a large number of dilution cultures from each sample. In the dilution cultures that show growth of the cyanobacteria, test for bacterial contamination by microscopic examination and by inoculating a variety of bacteriological media.

One disadvantage of this method is that the final pure culture may contain cyanobacteria in which mutations have occurred because of the ultraviolet light treatment.

8.1.22. Ultrasonication for *Sporocytophaga myxococcoides*

Enrichment of the microcyst-forming cellulolytic cytophaga *S. myxococcoides* takes advantage of a sonication-resistant form in the organism's life cycle (E. R. Leadbetter, Bacteriol. Proc., p. 42, 1963). Place 100 ml of *Sporocytophaga* medium (8.5.47) into a 500-ml Erlenmeyer flask and inoculate with ca. 0.1 g of soil, mud, or plant material. After incubation at 30°C for 7 to 10 days with moderate agitation, remove 5 ml of culture, subject to sonic oscillation for 15 to 30 s, and then inoculate a secondary enrichment flask. After ca. 5 to 7 days, look for a distinctive yellow hue in the flask, and use phase-contrast microscopy to look for both microcysts and vegetative cells on and around the cellulose fibers. Subject a portion of this culture to sonic oscillation, and then use the pour-plate method to obtain isolated colonies as described in part 8.2.25.

8.1.23. Antiserum Agglutination

The use of antiserum may enrich for the minority organism in a two-membered population. In a study of a mixed culture in which two species of marine spirilla occurred—a large and a small organism—Linn and Krieg (45) found the smaller organism to be greatly predominant. This made it impossible to obtain isolated colonies of the larger organism by plating; moreover, none of a variety of selection methods was applicable. To enrich for the larger spirillum, they isolated the smaller organism and used it to immunize a rabbit. When the resulting antiserum was added to the mixed culture, the smaller spirilla agglutinated and settled to the bottom of the tube. The supernatant, now containing a high proportion of the large spirilla, was used to obtain isolated colonies by plating.

8.2. BIOCHEMICAL ENRICHMENT

8.2.1. Alkali Treatment for *Mycobacterium tuberculosis* (20)

Add 10 ml of the suspected sputum sample to 10 ml of *N*-acetyl-L-cysteine-sodium hydroxide solution (8.5.36) in a sterile, disposable 50-ml screw-capped centrifuge tube. Tighten the screw cap and mix well in a Vortex mixer for 5 to 20 s. Allow to stand at room temperature for 15 min. The function of the *N*-acetyl-L-cysteine is to act as a mucolytic agent; it converts the thick sputum to a thin, watery consistency. The function of the sodium hydroxide is to destroy many of the contaminants present in the sputum, whereas *M. tuberculosis* is relatively resistant to the alkaline treatment. Fill the centrifuge tube to within 1.3 cm of the top with sterile 0.067 M phosphate buffer to neutralize the action of the sodium hydroxide. Centrifuge at 1,800 to 2,400 × g for 15 min to concentrate the mycobacteria, and decant the supernatant. Add 1 ml of sterile 0.2% bovine serum albumin to the sediment and gently mix. Use this suspension to inoculate suitable culture media (e.g., 8.5.26).

8.2.2. Alkaline pH Incubation for Vibrios (37)

To enrich for *Vibrio cholerae* and *V. parahaemolyticus*, directly streak the clinical specimen (fluid stool or rectal swab) lightly onto a nonselective agar medium such as nutrient agar (20.3.25) and heavily onto a selective agar medium such as TCBS agar (8.5.50). Also inoculate swabs of stool material into an enrichment broth (1% peptone, pH 8.4 to 8.5; in the case of *V. parahaemolyticus*, add 3.0% NaCl to the peptone broth). The high pH of the TCBS agar and the enrichment broth inhibit the growth of most contaminants. In the case of *V. cholerae*, incubate the enrichment broth for 6 to 8 h at 35°C; then streak nonselective and selective agar. For *V. parahaemolyticus*, incubate the enrichment broth overnight; then streak onto TCBS agar. On TCBS agar colonies of *V. cholerae* are yellow (sucrose fermenting) and oxidase positive; colonies of *V. parahaemolyticus* are bluish green and oxidase positive.

8.2.3. Acidic pH Incubation for Lactobacilli (46)

For lactobacilli in cheddar cheese, streak dilutions of the cheese on modified Rogosa's medium (8.5.33). This medium has a pH of 5.35 and has an acetic/acetate buffer system. At this pH lactobacilli such as *L. casei* and *L. plantarum* will form colonies, but the common dairy organism *Streptococcus lactis* will not.

8.2.4. Acidic pH Incubation for *Thiobacillus thiooxidans* (78)

Inoculate shallow layers of *T. thiooxidans* medium (8.5.53) with samples from soil, mud, or water (marine mud is the most reliable source). Look for a drop in pH to 2.0 after 3 to 4 days or more, which virtually assures the predominance of this acid-tolerant organism. Purify by streaking on the solidified medium.

8.2.5. Tellurite Inhibition for Corynebacteria and Certain Streptococci

Potassium tellurite inhibits gram-negative bacteria and most gram-positive bacteria when used at a suitable concentration. To select for

corynebacteria such as *C. diphtheriae*, use potassium tellurite at a concentration of 0.0375%, as in cystine tellurite blood agar (8.5.16). Colonies of corynebacteria are gray or black as a result of reduction of the tellurite. For selection of certain streptococci (*S. mitis*, *S. salivarius*, and enterococci), use potassium tellurite at a concentration of 0.001%, as in mitis-salivarius agar (8.5.31). *S. mitis* forms tiny blue colonies, *S. salivarius* forms larger blue "gum-drop" colonies, and enterococci form small blue-black colonies.

8.2.6. Thallium Inhibition for Mycoplasmas and Enterococci (10, 38)

Use thallous acetate at a concentration of 0.032%, as in E agar (8.5.17) and E broth (8.5.18), to select for mycoplasmas from the respiratory tract. Extract specimens collected on swabs into 2 ml of soybean-casein digest broth (8.5.45) containing 0.5% bovine serum albumin. Inoculate 0.1-ml amounts of the suspension into biphasic E medium (8.5.17) and also onto plates of E agar. Incubate the cultures at 37°C, with the plates being incubated aerobically in sealed containers. Examine the plates at intervals up to 30 days by means of a dissecting microscope (20 to 60× magnification) for the appearance of minute colonies (10 to 100 μm) with a typical "fried-egg" appearance. In the case of the biphasic cultures, examine these microscopically by looking through the side of the tube for "spherules" (fluid medium colonies); also observe the cultures for a decrease in pH (yellowing of the phenol red indicator). Inoculate E agar plates from the biphasic medium to obtain isolated colonies.

Thallous acetate has also been used at a concentration of 0.1%, as in thallous acetate agar (8.5.51), to select for enterococci.

8.2.7. Selenite Inhibition for Salmonellae

Use sodium hydrogen selenite at a concentration of 0.4%, as in selenite F broth (8.5.44), to temporarily suppress the growth of coliforms while allowing salmonellae to grow. In addition to directly streaking stool samples onto selective and nonselective agar media, inoculate selenite F broth heavily (ca. 1 g or 1 ml of sample in 8 to 10 ml of broth). Incubate this enrichment broth at 35 to 37°C for 12 to 16 h; then streak onto the plating media.

8.2.8. Phenylethanol Inhibition for Streptococci and Staphylococci

Use the reagent in agar media at a concentration of 0.25%, as in phenylethyl alcohol agar (8.5.40), to inhibit the growth of gram-negative bacteria, particularly *Proteus*, when these occur in mixed culture with gram-positive cocci. For example, one application is for the isolation of coagulase-positive staphylococci from a stool specimen.

8.2.9. Triphenylmethane Dye Inhibition for Gram-Negative Bacteria and Mycobacteria

Gram-positive bacteria are generally inhibited by lower concentrations of triphenylmethane dyes (such as crystal violet, basic fuchsin, brilliant green, and malachite green) than are gram-negative bacteria. For example, malachite green in media at a concentration of 1:4,000,000 will inhibit the growth of *Bacillus subtilis* and at 1:1,000,000 will inhibit the growth of staphylococci; yet concentrations of 1:30,000 to 1:40,000 are required to inhibit *Escherichia coli* or *Salmonella typhi* (21). Brilliant green is used in media for the confirmed test for coliforms (brilliant green lactose bile broth, 8.5.13) and in several selective media for *Enterobacteriaceae* such as salmonella-shigella agar (8.5.43) and brilliant green agar (8.5.12). Crystal violet is also used for selection of *Enterobacteriaceae*, as in violet red bile agar (8.5.56) and MacConkey agar (8.5.27). Mycobacteria are very resistant to dyes; malachite green is often incorporated into media used for isolation of *Mycobacterium tuberculosis* to inhibit the growth of contaminants as, for example, in Lowenstein-Jensen medium (8.5.26).

8.2.10. Salt Inhibition for Staphylococci

Use 7.5% NaCl in media to select for staphylococci; e.g., see mannitol salt agar (8.5.28). Growth of many other genera of bacteria is suppressed at this concentration of salt.

8.2.11. Salt Inhibition for Halobacteria (23)

The red halophilic rods (*Halobacterium*) and cocci (*Halococcus*) occur in heavily salted proteinaceous materials (such as salted fish), in salterns, and in the Dead Sea and other highly saline lakes. They can often be isolated from samples of solar salt. They require a high concentration of NaCl for growth (ca. 25% NaCl) and cannot grow with less than ca. 15% NaCl. They are killed by even short exposures to salt concentrations less than ca. 15%. At concentrations of NaCl of 20% or greater, only halophiles grow, thus making their selection from natural sources a simple matter. An example of a suitable enrichment medium is given in part 8.5.21. Incubate inoculated flasks at 37°C with agitation, and subsequently obtain isolated colonies by streaking from the enrichment culture onto plates of medium solidified with 2% agar. Incubate the plates at 37°C for 3 to 14 days in plastic bags to prevent excessive drying.

8.2.12. Bile Inhibition for Enteric Bacteria

Bile or bile salts are often incorporated into culture media as selective agents for intestinal bacteria. There are some exceptions to this selectivity; for example, *Yersinia pestis* can grow in pure bile even though it is not an intestinal organism (53). However, the rule holds sufficiently well to make bile an important selective agent for gram-negative enteric rods and for enterococci. For selection of members of the family *Enterobacteriaceae*, bile or bile salts are incorporated into such media as MacConkey agar (8.5.27), salmonella-shigella agar (8.5.43), violet red bile agar (8.5.56), and many others. For enterococci, bile esculin agar (8.5.9) has proved to be an excellent selective and differential medium (20).

8.2.13. Penicillin Inhibition for Mycoplasmas

Because mycoplasmas lack cell walls, they are resistant to very high concentrations of penicillin, which inhibit most other bacteria. Use the antibiotic at a concentration of 194 U/ml in such isolation media as E agar (8.5.17).

8.2.14. Penicillin Inhibition for *Bordetella pertussis*

Obtain the clinical specimen by means of a nasopharyngeal swab and suspend it in broth. Streak the suspension onto plates of Bordet-Gengou agar (8.5.11) containing penicillin (0.5 U/ml). The penicillin helps to suppress the growth of normal flora while permitting *B. pertussis* to form characteristic pearl-gray ("mercury-drop") colonies in ca. 4 days.

8.2.15. Multiple Antibiotics for *Campylobacter fetus* (68, 69)

Incorporate bacitracin (2 U/ml) and novobiocin (2 µg/ml) into brucella agar (8.5.14) to aid in the selection of *C. fetus* from feces and intestinal contents of cattle, sheep, pigs, and birds. This may be used in conjunction with the filtration method previously described for *C. fetus* (8.1.16).

8.2.16. Multiple Antibiotics for Neisseriae

Incorporate vancomycin, colistin, and nystatin into Thayer-Martin agar (8.5.52) to aid in the selection of *Neisseria meningitidis* from nasopharyngeal samples or of *N. gonorrhoeae* from urethral exudates, cervical swabs, etc. The antibiotics help to suppress the growth of normal flora.

8.2.17. Formaldehyde Treatment for *Gallionella ferruginea* (54)

The microaerophilic chemolithotroph *G. ferruginea* occurs in cold ferrous iron-bearing waters and forms distinctive twisted stalks encrusted with iron hydroxide. Add 0.5 ml of Formalin (40% formaldehyde) to a 150-ml rectangular, screw-capped milk-dilution bottle containing 10 ml of ferrous sulfide agar and 100 ml of fluid medium (see modified Wolfe's medium, 8.5.34). Centrifuge samples of *Gallionella* from a natural source at 3,000 × *g* for 3 min. Transfer 1 to 5 ml of sediment to the bottle of medium. Incubate at 25°C for 1 to 2 days. The Formalin will kill contaminating bacteria but not *Gallionella*. Transfer 1-ml portions of the Formalin-treated culture to fresh bottles of medium without Formalin. Incubate cultures at 25°C and transfer every 2 to 3 weeks. The organisms grow as a fluffy mat on the surface of the ferrous sulfide agar; use a portion of this mat for subcultures. Test the purity of the cultures by microscopic examination and by inoculating a variety of other types of bacteriological media (there should be no growth).

8.2.18. Dilute Medium for Caulobacters (57)

Caulobacters are stalked bacteria that can grow at levels of nutrients that do not support good growth of many contaminants. To samples of water from ponds, streams, or lakes, or to samples of tap water, add 0.01% peptone and incubate aerobically at 20 to 25°C in bottles or flasks loosely covered with paper or aluminum foil. Examine the surface film daily by phase-contrast microscopy. When stalked bacteria occur in a relative proportion of ca. 1 in 10 or 20 cells (usually about 4 days), streak the surface film onto plates of tap water agar containing 0.05% peptone and incubate at 30°C. The low peptone level allows the caulobacters to form tiny colonies but does not allow heavy overgrowth by other bacteria. After ca. 4 days or more of incubation, pick microcolonies of caulobacters under a dissecting microscope and transfer as patches to a richer medium (0.2% peptone, 0.1% yeast extract, 0.02% MgSO$_4$· 7H$_2$O, and 1.0% agar). After 2 days of incubation, prepare wet mounts from the patches to detect caulobacters. Obtain isolated colonies by streaking growth from the patches onto fresh plates.

In the case of marine caulobacters, add 0.01% peptone to samples of stored seawater and incubate at 13°C for ca. 7 days. When microscopic observation indicates the development of a suitable proportion of stalked bacteria, streak the surface film onto plates of seawater agar containing 0.05% peptone and incubate at 25°C.

Pick microcolonies to fresh plates as patches, and later purify by streaking.

8.2.19. Dilute Medium for Aquatic Spirilla (83)

In dilute media, aerobic chemoheterotrophic spirilla (the genera *Aquaspirillum* and *Oceanospirillum*) can often compete successfully with other bacteria for the nutrients present. For freshwater spirilla, add 1% peptone or yeast autolysate to samples of source water (e.g., water from stagnant ponds), and incubate at room temperature for ca. 7 days or until spirilla become numerous. Then add part of this initial culture to an equal part of the source water and sterilize by autoclaving. Inoculate this mixture from the unsterilized portion of the original culture. After incubation and further development of the spirilla, dilute a portion of the second culture with more source water, sterilize the mixture, and inoculate it from the unsterilized portion. Continue to deplete the nutrients in this manner until the spirilla predominate. Obtain colonies by streaking onto plates of MPSS agar (8.5.35).

Another approach depends on low levels of nitrogen sources (83). Supplement samples of source water with 1% calcium malate or lactate, and incubate at room temperature for ca. 1 week. Make a serial transfer into sterile source water containing 1% of the carbon source, and incubate. Continue in this manner (three or four serial transfers) until the spirilla predominate. In this method it is important not to add nitrogen sources such as NH$_4$Cl in order to prevent overgrowth of the spirilla by contaminants.

In the case of marine spirilla, mix the seawater sample with an equal volume of Giesberger's base medium (8.5.19) supplemented with 1% calcium lactate (83). After incubation, remove a portion of the culture, mix with an equal volume of Giesberger's medium containing lactate but lacking the NH$_4$Cl. Sterilize by autoclaving; then inoculate the mixture from the unsterilized portion of the original culture and incubate. Continue to successively deplete the nitrogen content by repeating this procedure until spirilla become the predominating organisms. Obtain isolated colonies by streaking MPSS agar (8.5.35).

Another way to enrich for spirilla is to use a continuous culture system that provides low levels of nutrients. Jannasch (34) found that, in chemostat experiments using a mixture of a marine spirillum and a pseudomonad, the growth rate of the spirillum exceeded that of the pseudomonad when the dilution rate of the chemostat was decreased to the point where the limiting carbon and energy source (lactate) fell below 10 mg/liter. In other experiments the growth rate of the spirillum exceeded that of *Escherichia coli* when the lactate concentration fell below 5 mg/liter (35). In similar experiments using a freshwater spirillum and a pseudomonad, Matin and Veldkamp (49) found that when the lactate concentration fell below ca. 0.09 mg/liter the pseudomonad was eliminated from the chemostat as a nongrowing population. The more efficient scavenging ability of the spirillum for lactate may be attributable to a lower K_m and a higher V_{max} of the transport system for lactate, and also to a higher surface-to-volume ratio for the spirillum (49).

Under starvation conditions spirilla appear to have a survival advantage (48). This may be related to the ability of spirilla to form intracellular reserves of poly-β-hydroxybutyrate under conditions of prior growth with limiting levels of carbon and energy sources. The role of poly-β-hydroxybutyrate in bacterial survival has also been reported for other types of bacteria (16). Consequently, starvation conditions should be considered as a possible way to select for bacteria that form this polymer.

8.2.20. Dilute Medium for *Sphaerotilus* (8)

The sheathed bacteria of the genus *Sphaerotilus* occur in streams contaminated with sewage or organic matter and form slimy "tassels" attached to submerged surfaces. They can also be isolated from rivers, open drains, or ditches where there is no initial evidence of their presence. The enrichment procedure takes advantage of the ability of *Sphaerotilus* to grow at very low nutrient concentrations. To 50-ml volumes of *Sphaerotilus* medium (8.5.46) contained in French square bottles, add 25-ml volumes of the water sample, or 1-, 5-, and 10-ml portions of settled sewage or the settled liquor from various stages of sewage treatment. Incubate at 22 to 25°C for 5 days. Examine microscopically for evidence of filamentous growth daily after the 2nd day. Obtain pure cultures by picking a filament from the enrichment broth and streaking it onto plates of a solid medium (0.05% meat extract and 1.5% agar). Incubate plates for 24 h at 25°C, and examine under a dissecting microscope for the typical curling filaments of *Sphaerotilus*. Transfer isolates to a Trypticase-glycerol broth (Trypticase [BBL Microbiology Systems, Cockeysville, Md.], 5 g; glycerol, 5 g; distilled water, 1,000 ml; pH 7.0 to 7.2) and incubate for up to 2 weeks. *Sphaerotilus* forms a heavy surface pellicle in 2 to 3 days, and the underlying broth remains clear. If turbidity develops, contamination has occurred. Even without devel-

opment of turbidity, one should reisolate *Sphaerotilus* from the surface pellicle by an additional streaking onto meat extract agar to ensure purity.

To confirm that the isolated organisms form a sheath, place a small piece of slime growth on a slide in a drop of water, apply a cover slip, and press down on the cover slip with blotting paper. Place a very small drop of 1% crystal violet solution at the edge of the cover slip so that it will flow into the preparation by capillary action. After 30 s, press again with blotting paper to remove excess dye and observe with a bright-phase oil-immersion lens. Both the cells and the sheath should be clearly visible.

8.2.21. Tryptophan Substrate for Pseudomonads (44)

To enrich for pseudomonads capable of using tryptophan as a sole carbon and nitrogen source, inoculate a 250-ml Erlenmeyer flask containing 40 ml of tryptophan medium (8.5.54) with ca. 0.1 g of soil. Incubate with shaking at 25°C for 5 to 7 days. Transfer 0.1 ml to a second flask of medium and incubate for 2 to 3 days. After a further serial transfer, obtain pure cultures by streaking onto tryptophan medium solidified with 15 g of agar per liter. After 1 to 3 days of incubation, subculture isolated colonies to agar slants.

8.2.22. N₂ Substrate for *Azospirillum* and *Azotobacter*

Azospirillum is a microaerophilic nitrogen fixer associated with the roots of a variety of plants; it also occurs in soil (17, 18). Place washed root pieces 5 to 8 mm long, macerated with a forceps, into nitrogen-free semisolid medium (Nfb medium, 8.5.37). Alternatively, inoculate the medium with a loopful of soil. Incubate without agitation for 40 h at 32°C; then test the enrichment culture for acetylene-reducing activity (see 20.2.7). Be careful not to disturb the dense subsurface pellicle that forms in the medium, as this may stop nitrogenase activity. If the culture reduces acetylene, enrich further by a serial transfer to fresh Nfb medium. Examine microscopically for plump, curved, motile rods, ca. 1 μm in width, filled with intracellular granules (of poly-β-hydroxybutyrate). Then streak onto plates of Nfb medium (solidified with 1.5% agar) containing 20 mg of yeast extract per liter. After 1 week, look for small, white, dense colonies and transfer to Nfb semisolid medium. For final purification streak the Nfb culture onto BMS agar (8.5.10). Look for development of typical pink, often wrinkled colonies.

Unlike *Azospirillum*, which is an obligate microaerophile under nitrogen-fixing conditions, *Azotobacter* has mechanisms to protect its oxygen-labile nitrogenase from oxygen and can fix nitrogen aerobically. Inoculate 0.1 g of soil into 100 ml of nitrogen-free *Azotobacter* medium (8.5.3) contained in a 1-liter flask. Incubate at 30°C with agitation. Observe the culture microscopically at periodic intervals for the development of large, ovoid cells, 2 μm or more in diameter. Prepare a secondary enrichment culture; then purify by obtaining isolated colonies on nitrogen-free agar medium (*Azotobacter* medium solidified with 1.5% agar).

8.2.23. Methanol Substrate for Hyphomicrobia (9)

Members of the genus *Hyphomicrobium* have the ability to use one-carbon compounds such as methanol or methylamine as sole carbon sources. Add the inoculum (5.0 ml of pond or ditch water, or 0.3 g of mud or soil—all preferably with a low organic content) to stoppered bottles (75- to 125-ml capacity), and add nonsterile *Hyphomicrobium* medium (8.5.22) through which nitrogen has been bubbled to provide anaerobic conditions. Fill the bottles completely with the medium. Incubate in the dark at 30°C. Be sure the bottles remain completely filled by adding fresh medium if necessary. Hyphomicrobia generally develop in ca. 8 days. Monitor their development by phase-contrast microscopy: look for rod-shaped cells with pointed ends, oval or egg- or bean-shaped forms, which produce filamentous outgrowths (hyphae) that vary in length and may show branching. Prepare a secondary enrichment, this time using sterile medium. Finally, streak onto solidified medium (8.5.22) and incubate aerobically to obtain isolated colonies.

8.2.24. H₂ Substrate for *Aquaspirillum autotrophicum* (7)

Filter water from a eutrophic lake through membrane filters; then place the filters on the surface of mineral agar plates (8.5.30). Incubate under an atmosphere of 60% H_2–30% air–10% CO_2 at 30°C. Obtain pure cultures by repeated streaking on mineral agar plates. Look for spirilla 0.6 to 0.8 μm wide with bipolar tufts of flagella. Confirm hydrogen autotrophy by demonstrating that both H_2 and CO_2 are required for growth.

8.2.25. Cellulose Substrate for Cytophagas

Enrich for cellulose-decomposing cytophagas by placing a piece of Whatman no. 1 filter paper on the surface of a basal mineral agar (8.5.6). Place particles of soil or plant materials on the filter paper, and incubate the plates at room temperature. Look for the development of yellow, orange, or pink pigmentation around the particles, together with cellulolysis; also look for slender, flexible cytophaga cells by phase-contrast microscopy. For isolation of cytophagas that are not obligate cellulose decomposers, streak onto tryptone agar. For obligate cellulose decomposers, obtain isolated colonies by the pour-plate method, using the basal mineral agar containing a thick suspension of finely divided cellulose (20.1.13). The agar medium should be sufficiently firm (0.85 to 1.0% agar) to prevent the motility of flagellated bacteria, but not so firm as to prevent the gliding motility characteristics of the cytophagas. It is sometimes helpful to use the inoculated cellulose agar as a thin overlay on plates of the basal mineral agar; zones of clearing due to cellulolysis may be more easily discernible by this technique.

8.2.26. Agar Substrate for *Cytophaga fermentans* and *C. salmonicolor* (77)

The facultatively anaerobic marine organisms *C. fermentans* and *C. salmonicolor* are able to hydrolyze and ferment agar; this trait is a valuable selective feature that can be used for their enrichment. Fill glass-stoppered bottles or screw-capped tubes to the top with nonsterile Veldkamp's medium (8.5.55). Inoculate with marine mud from areas with decaying algae, stopper the bottles, and incubate them in the dark at 30°C. Look for development of turbidity, gas formation, and a drop in pH in 3 to 7 days. Isolate colonies by the shake tube method, using sterile Veldkamp's medium containing 2% agar. Look for the development of colonies which on microscopic examination exhibit the flexing movements characteristic of cytophagas. Subculture the colonies to media containing 1% agar and 1% yeast extract to demonstrate softening or liquefaction of the agar.

8.2.27. Lactate Substrate for Propionibacteria

The ability of propionibacteria to ferment lactate provides a basis for selection of these organisms. Fill a screw-capped culture tube (25-ml capacity) with a freshly boiled and cooled medium consisting of 4% sodium lactate and 1% yeast extract. Add ca. 0.2 g of $CaCO_3$ to the tube and inoculate with a small piece of Swiss-type cheese. Tighten the screw cap and incubate the tube at 30°C. Look for the development of a reddish-brown turbidity in ca. 5 to 7 days. Obtain isolated colonies by streaking the culture onto sterile lactate-yeast extract agar and incubating anaerobically in a CO_2-enriched atmosphere, such as that provided by a GasPak (BBL).

8.2.28. Ammonium or Nitrite Substrates for Nitrifying Bacteria (50, 58)

Use the basal medium described in part 8.5.38. For ammonium oxidizers, which oxidize ammonium to nitrite, supplement the medium with 0.5 g of $(NH_4)SO_4$ per liter; for nitrite oxidizers, which oxidize nitrite to nitrate, use 0.5 g of $NaNO_2$ per liter. Prepare a 1:10 dilution of soil, and inoculate each flask of the enrichment medium with 1.0 ml of the dilution. Incubate at 28°C. At weekly intervals, test for disappearance of the ammonium or nitrite by removing samples of the enrichment culture to a spot plate. For ammonium, test 3-drop samples with Nessler's solution (20.4.10); an orange or yellow color indicates ammonium. For nitrite, add 3 drops of solution A and 3 drops of solution B (20.4.12) to 0.5-ml samples; a red color indicates nitrite. Compare the intensity of the colors with those obtained with a set of dilutions from standard solutions of ammonium or nitrite, and replace the amount of substrate lost from the enrichment culture by addition of fresh substrate from a sterile stock solution. If no nitrite is present in the enrichment culture, test for the formation of nitrate by adding a small amount of zinc dust after the nitrite test reagents; the zinc will reduce any nitrate present to nitrite, which will then yield a red color. Continue to incubate the enrichment culture and replace ammonium or nitrite at weekly intervals, until a population of nitrifiers is built up as indicated by microscopic observation.

The isolation of nitrifiers into pure culture is *extremely* difficult. Completely inorganic media must be used, and the colonies formed are tiny, ca. 100 μm in diameter. Moreover, nitrifiers grow very slowly and are thus often overgrown by contaminants. Using sterile enrichment medium as a diluent, prepare a decimal dilution series from the enrichment culture. Place 1.0-ml samples of each dilution into petri dishes; then add silica gel medium prepared with double-strength enrichment medium as described in 9.1.3. Mix the culture dilution with the medium immediately, and allow the plates to solidify. Incubate the plates in a humid atmosphere at 28°C, and examine periodically under a microscope for the development of tiny colonies. With a Pasteur pipette freshly drawn out in a flame to a fine tip, pick well-isolated colonies and subculture into sterile medium. Determine the ability of the subcultures to use ammonium or nitrite, and also test for contaminants by streaking samples onto organic media such as nutrient agar (no growth should occur). Purify any apparently pure cultures in silica gel medium several more times, each time picking well-isolated colonies under the microscope, testing for the ability of subcultures to use ammonia or nitrite, and also testing for the presence of contaminants.

8.2.29. Nitrate plus Organic Acid Substrates for Pseudomonads

Some members of the aerobic genus *Pseudomonas* can be enriched and isolated by taking advantage of their ability to use nitrate as a source of cellular nitrogen, together with their ability to use the salts of various organic acids as carbon and energy sources (74). Add 0.1 g of soil or mud, or 0.1 ml of pond or river water, to 3 ml of succinate-salts medium (8.5.49). Incubate at 30°C to enrich for members of the fluorescent group of pseudomonads, such as *P. putida*, *P. fluorescens*, or *P. aeruginosa*. Incubate at 41°C to select for *P. aeruginosa* (*P. putida* and *P. fluorescens* cannot grow at this temperature). Enrich for *P. acidovorans* by substituting glycolate, muconate, or norleucine for succinate in the enrichment medium and incubating at 30°C.

Because some pseudomonads not only can use oxygen as a terminal electron acceptor but also can use nitrate, one can often select for such strains by increasing the nitrate concentration of the enrichment medium to 1.0% and by filling screw-capped tubes completely with the medium to establish an oxygen-limiting condition, which favors nitrate respiration.

After preparing secondary enrichment cultures of pseudomonads, obtain isolated colonies by streaking agar plates (use the appropriate enrichment media solidified with 15 g of agar per liter). Incubate the plates aerobically at 30°C.

8.2.30. Sulfate plus Organic Acid Substrates for Sulfate Reducers

Members of the anaerobic genus *Desulfovibrio*, and most of the members of *Desulfotomaculum*, are able to oxidize lactate in the presence of sulfate, with the latter being reduced to H_2S. Inoculate screw-capped tubes or stoppered bottles with soil, mud, water, or fecal material, and fill the vessels completely with lactate medium (8.5.24). Incubate at 30°C. A blackening of the medium indicates sulfate reduction. Prepare a secondary enrichment before streaking onto solid media. In the case of *Desulfotomaculum*, one can select for this sporeformer by heating enrichment cultures at 70°C for 10 min before streaking solid media. Obtain isolated colonies of the sulfate reducers by streaking plates of Iverson's medium (8.5.23) and incubating them under a hydrogen atmosphere for 7 to 10 days at 30°C.

To enrich for *Desulfotomaculum acetoxidans* (81), which cannot use lactate but can use acetate, use the enrichment medium of Widdel and

Pfennig (8.5.57), which contains acetate as the oxidizable substrate. Look for development of motile, straight or slightly curved rods 1.0 to 1.5 μm wide and 3.5 to 9.0 μm long and which contain spores and also bright, refractile areas by phase-contrast microscopy. Use the shake tube method to obtain isolated colonies. After 3 weeks of incubation, pick colonies and heat them at 70°C for 10 min; *D. acetoxidans* is a spore-former and can be selected by this heat treatment.

8.2.31. Elemental Sulfur Substrate for *Desulfuromonas acetoxidans* (56)

D. acetoxidans obtains energy for growth by anaerobic sulfur respiration, with acetate, ethanol, or propanol serving as carbon and energy sources. Add samples of anaerobic, sulfide-containing water or mud from freshwater or marine sources to sterile screw-cap bottles. Fill the bottles with Pfennig and Biebl's medium (8.5.39), leaving a small air bubble. Tighten the caps and incubate the bottles at 28°C for 1 week with agitation (e.g., a rotary shaker at 150 rpm). The action of the glass beads in the medium will cause a gradual grinding of the sulfur to a very fine suspension. Positive enrichments can be recognized by a strong odor of H_2S. After two serial transfers in Pfennig and Biebl's medium supplemented with a vitamin solution (8.5.57), allow the sulfur to settle and examine the supernatant for a faint turbidity of cells. Observe microscopically for small rods, 0.4 to 0.7 μm by 1 to 4 μm, some of which may be motile. Obtain isolated colonies by the shake tube procedure. To get a fine, homogeneous distribution of sulfur in the agar tubes, add 3 drops of an autoclaved polysulfide solution (10 g of $Na_2S \cdot 9H_2O$ and 3 g of sulfur flower dissolved in 15 ml of distilled water) per 50 ml of the agar medium. Look for development of pink to ochre-colored colonies.

Because the sulfide formed by *D. acetoxidans* eventually causes inhibition of growth (no more than 0.1% H_2S can be tolerated), an alternative enrichment method has been devised in which the sulfur medium is inoculated not only with the mud or water sample but also with a pure culture of a green sulfur bacterium (of the family *Chlorobiaceae*). The latter organism continuously consumes the H_2S formed by *D. acetoxidans* by re-oxidizing it to elemental sulfur, allowing fast-growing and highly enriched cultures which can be directly used to isolate pure cultures (56).

8.2.32. Toluene Substrate for Toluene Oxidizers (14)

To enrich for toluene-oxidizing bacteria, use cotton-stoppered flasks of basal inorganic me-dium A or B (8.5.5). Inoculate with a small quantity of moist soil previously treated with toluene vapor for several days, or use fresh soil. Incubate the flasks at 25 to 30°C for 1 to 3 weeks in a closed chamber containing a beaker of water saturated with toluene. When growth occurs, obtain isolated colonies by streaking onto solidified medium and incubating the plates at 25 to 30°C in a toluene-containing atmosphere.

8.2.33. Bacterial Cell Substrate for Myxococci

Myxococcus species are able to lyse the cells of other bacteria by means of bacteriolytic enzymes and to use the compounds liberated from the bacteria for growth. One can take advantage of this in isolating myxococci from samples of soil, water, or plant material (47, 55, 66). Obtain bacterial cells for use as substrate (*Enterobacter aerogenes* is suitable and convenient) by removing an entire 4-mm colony from the surface of an agar medium, or by centrifuging the bacteria from a broth culture and washing them several times. Make a streak or smear of the bacteria ca. 1 cm wide and 4 cm long on the surface of a plate of water-agar medium (1.5% agar in distilled water). The smear of cells should be sufficiently thick so as to be barely visible to the eye. At one end of the smear place two or three particles of soil or bits of plant material (e.g., bark or leaf). After 2 to 3 days and at daily intervals thereafter, examine the plates with a dissecting microscope for evidence of dissolution of the bacterial smear near the added particles. Also look for development of fruiting bodies, usually yellow, orange, or pink, on and at the sides of the smear. Transfer the fruiting bodies found the greatest distance from the smear to an agar medium such as 0.2% tryptone or Casitone (Difco Laboratories, Detroit, Mich.) and 1.5% agar; crush each fruiting body in a drop of sterile water between two slides before streaking, in order to liberate the myxospores.

The incorporation of cycloheximide (25 μg/ml) in both the non-nutritive agar and the subsequent plating media will retard the growth of fungi, thereby aiding the isolation of the myxococci (55).

8.3. BIOLOGICAL ENRICHMENT

8.3.1 Bacterial Parasitism by Bdellovibrios (75)

The tiny vibrios called bdellovibrios are capable of attaching to a wide variety of gram-negative bacteria, penetrating the cell wall, and multiplying within the periplasmic space, with consequent lysis of the host bacteria. The method for isolating bdellovibrios resembles that used for bacteriophages in many respects.

Suspend 500 g of a soil sample in 500 ml of tap water and shake vigorously for 1 h. Centrifuge the suspension for 5 min at 500 × g to remove the larger particles. Pass the supernatant through membrane filters of decreasing pore size: 3.0, 1.2, 0.8, 0.65, and 0.45 μm. Mix 0.5 ml of the final filtrate with 0.5 ml of a suspension (ca. 5 × 10^{10} cells/ml) of the host bacteria (*Enterobacter aerogenes* or *Pseudomonas fluorescens*). Add the mixture to 4 ml of molten semisolid YP medium (8.5.60), mix, and pour over the surface of a plate of solid YP medium. After overnight incubation, examine the plates for plaques (areas of lysis). If plaques form within 24 h, they are attributable to bacteriophages rather than to bdellovibrios. Mark such plaques so that they will not be confused with plaques formed by bdellovibrios, which take at least 2 days to appear. Cut out plaques suspected to be caused by *Bdellovibrio*, suspend them in YP solution, and prepare a dilution series to be applied to lawns of host bacteria to obtain plaques that are well isolated. Examine one of the plaques by phase-contrast microscopy; look for tiny, highly motile vibrios ca. 0.3 μm in width. Suspend material from a plaque in YP broth, pass it through a 0.45-μm filter, dilute the filtrate, and plate it onto lawns of host bacteria. After three successive plaque isolations, one may regard the bdellovibrio strains as representing the descendants of a single bdellovibrio cell.

8.3.2. Plant Symbiosis by Rhizobia (5)

The nodules found on the roots of legumes represent a natural enrichment system for symbiotic nitrogen-fixing bacteria of the genus *Rhizobium*. Obtain nodulated roots of alfalfa or red clover and wash the soil from them. Remove a nodule from the root, leaving a small portion of the root attached to the nodule. Use a camel-hair brush to remove any soil still adhering to the nodule while holding the nodule under running water. Submerge the nodule in a 1:1,000 solution of $HgCl_2$ for 3 to 6 min; move the nodule around occasionally with sterile forceps. Transfer the nodule to 75% ethanol and agitate it in the solution for several minutes. Then remove it to sterile water and agitate it for several minutes. Add 1 ml of sterile water to each of six sterile petri dishes. Transfer the nodule to the first dish and crush it with sterile forceps. Mix the exudate with the water. Transfer one to two loopfuls of the suspension to the sterile water in the second dish and mix. Continue to serially dilute in this manner for the remaining dishes. To each of the dilutions, add molten yeast extract mannitol agar (8.5.59) at 45°C. Incubate the solidified

plates at room temperature and subculture from well-isolated colonies.

8.3.3. Animal Parasitism by *Streptococcus pneumoniae* (20)

By inoculating a host animal with a mixed culture containing a pathogen, one can select for the latter. The pathogen will predominate in the infected animal, often occurring in pure culture in the blood and tissues. Nonpathogenic contaminants are inhibited or destroyed by the defense mechanisms of the animal.

For instance, if a mouse is injected intraperitoneally with 1 ml of emulsified sputum containing *S. pneumoniae* and other bacteria, one can obtain a pure culture of the pneumococci 4 to 6 h later by inserting a sterile, sharp-tipped capillary pipette through the abdomen and collecting some of the peritoneal fluid.

8.3.4. Animal Infection by *Yersinia pestis*

In isolating the plague bacillus *Yersinia pestis* from a putrid cadaver, the problem is made difficult by the large number of putrefactive contaminants present. In this case one can rub the putrid material onto the shaven abdomen of a guinea pig. The small size of *Y. pestis* permits it to penetrate the minute abrasions of the shaven skin and infect the animal, whereas most of the contaminants are not pathogenic or are not of sufficiently small size to allow ready penetration of the skin. Pure cultures of *Y. pestis* can subsequently be recovered from the animals by culturing material from the buboes or spleen.

8.4. ISOLATION

Relatively few methods are available for the isolation of bacteria into pure cultures. Isolation is most commonly done by separating individual cells in or on a solidified nutrient medium, using either a streak-plate or pour-plate method. However, the obtaining of a single colony does not always ensure purity, since colonies can arise from aggregates of cells as well as from individual cells. In the case of slime producers, contaminants frequently adhere to the slime; in the case of *Bacillus* species or actinomycetes, contaminants may be enmeshed in the chains or filaments formed by these organisms. It is best to use nonselective media for purification because contaminants are more likely to grow and be detected on such media. Even with nonselective media one should not "pick" (subculture) colonies too soon, because slow-growing contaminants may not yet have made their presence known.

A pure culture should yield colonies that appear similar to one another, and microscopic observation of the culture should reveal cells that are reasonably similar to each other in appearance, particularly in regard to cell diameter and Gram reaction. There are, of course, some exceptions to these criteria: for example, colonies growing from a pure culture may exhibit smooth-rough (S-R) variation; coccoid bodies, cysts, and spores may occur in pure cultures of various organisms; and some organisms may show Gram variability. Nevertheless, the criteria are generally useful and apply in most cases.

8.4.1. Spatially Streaking or Spreading on Solid Medium

There are many methods for streaking plates of solid media ("streak plates"), but the one illustrated in Fig. 1 almost invariably yields well-isolated colonies, even when done by a novice. Alternatively, one can spread dilutions of a mixed culture onto the surface of plates of solid media (for details, see 11.2.1). In the case of anaerobes, plates streaked or spread under an air atmosphere can subsequently be incubated in an anaerobic jar. Solid media for anaerobes should be freshly prepared and streaked within 4 h to avoid accumulating too much dissolved oxygen.

Even so, it takes some time for an anaerobe jar to remove oxygen and establish anaerobic conditions; the use of "roll tubes" containing prereduced media eliminates this difficulty entirely (27–29). Such tubes are prepared by spinning sealed tubes of melted prereduced media so that the agar solidifies on the walls of the tubes as a thin layer. The method of streaking a roll tube is illustrated in Fig. 2. A roll tube can also be inoculated by adding a dilute suspension of cells and then rotating the tube to spread the cells over the surface. Also see 6.6.4.

8.4.2. Serially Diluting in Solidified Medium

The simplest method for preparing a "pour plate" is to inoculate a tube of sterile, melted agar medium (cooled to 45 to 50°C) with a loopful of the sample, mix, pour the inoculated medium into a petri dish, and allow it to solidify. However, one often needs to dilute the sample in order to obtain well-isolated colonies, and the best approach is to use a decimal series of dilutions of the sample. Add 1.0-ml portions of each dilution to petri dishes, add 15 to 20 ml of the melted agar medium, mix by rotating the dishes several times, and allow the plates to solidify. For anaerobes, dilutions of the sample can be mixed with the melted, cooled prereduced me-

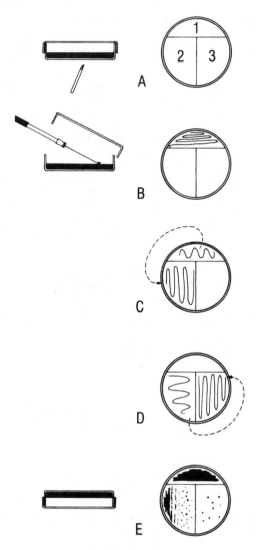

FIG. 1. *A useful streak-plate method for obtaining well-isolated colonies. (A) With a glass marker pencil draw a "T" on the bottom of the petri dish to divide the plate into three sections. (B) Streak a loopful of culture lightly back and forth on the surface of the agar over section 1 as shown. Raise the lid of the dish just enough to allow the streaking to be done, then replace it. Flame-sterilize the loop and allow it to cool (15 s). (C) Draw the loop over section 1 as shown, and immediately streak back and forth over section 2. Flame the needle and allow it to cool. (D) Draw the loop over section 2 as shown, and then streak back and forth over section 3. (E) Incubate the dish in an inverted position as shown to prevent drops of condensed water on lid from falling onto the agar surface. Section 1 will develop the heaviest amount of growth, while section 2 or 3 will usually have well-isolated colonies.*

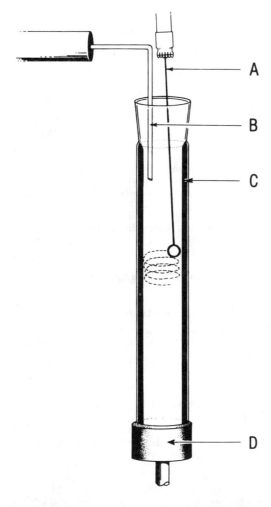

FIG. 2. *Streaking an anaerobic roll tube (27). (A) Loop needle (platinum or stainless steel; nichrome will cause oxidation of the medium). (B) Gassing cannula for continuous purging of the tube with oxygen-free gas. (C) Prereduced agar medium coating the inner wall of the tube. (D) Motor-driven tube holder for rotating roll tube during streaking. Insert the needle with a loopful of inoculum to bottom of tube, press the loop flat against the agar, and draw it upward. After streaking in this manner for one-fourth of the way up the tube, turn the loop so that it is perpendicular to the agar (as shown), and continue to streak upwards to the top. Remove the gas cannula, replace the rubber stopper in the tube, and incubate the culture in a vertical position.*

dium in roll tubes just before coating the walls of the tubes.

One disadvantage of diluting in agar media is that many of the isolated colonies are submerged in the agar and can be removed only by digging

them out with a sterile instrument or punching them out with a sterile Pasteur pipette. Another disadvantage is that the bacteria to be isolated must be able to withstand temporarily the 45 to 50°C temperature of the molten agar.

The "shake tube" method has often been used for isolation of anaerobic bacteria, especially in the case of phototrophic and sulfate-reducing bacteria (76). For this method, prepare a series of sterile tubes held in a water bath at 45°C and half fill them with molten agar medium at the same temperature. Inoculate the first tube of the series with a few drops of the mixed culture and mix gently with the medium. Then transfer one-tenth of the medium to the second tube of the series. Place the first tube in a vertical position, and immerse the bottom part in cold water to solidify the agar. Mix the contents of the second tube, transfer one-tenth of the medium to the third tube, and cool the second tube. Proceed in this manner for the remaining tubes of the series. Then overlay the solidified agar in each tube with a melted, sterile mixture (1:1) of paraffin and paraffin oil (mineral oil) to a depth of ca. 2 cm. This seals the medium from the air. During solidification of the wax, the plug may contract, resulting in an incomplete seal; in this case apply mild, local heating and also tap the tube to remove any air bubbles under the seal. After the bacteria have grown, select a tube that contains well-isolated colonies, and remove the agar from the tube by first melting and discarding the paraffin seal and then inserting a sterile capillary pipette between the glass wall of the tube and the agar. Push the tip of the pipette down to the bottom of the tube, and apply air pressure to push the column of agar out of the tube into a sterile dish. Dissect the agar to remove the desired colonies.

8.4.3. Serially Diluting to Extinction in Liquid Medium

This method is useful when the desired organism cannot grow on solid media. A prerequisite is that the organism desired must be the predominant member of the mixed population. Prepare a dilution of the mixed culture such that, when aliquots are added to a large number of tubes of growth medium, the mean number of bacteria inoculated per tube will be <0.05. In other words, if 100 ml of the dilution contained a total of 5 bacteria, and if 1.0-ml aliquots were inoculated into 100 tubes, the mean number of bacteria inoculated per tube would be 0.05. Upon incubation, most of the tubes would show no growth, but the few that did would be likely to have received only a single bacterium ($P =$

0.975). The smaller the mean number of bacteria inoculated per tube, the greater is the probability that the growth in a tube arose from a single bacterium. It is therefore imperative that most of the tubes inoculated exhibit no growth, so that the few tubes that do exhibit growth will have a high likelihood of having been inoculated with a single cell. For a discussion of the theoretical aspects of this method, see reference 51.

8.4.4. Isolating Single Cells

A review of the various methods used for single-cell isolation may be found in reference 36. When many isolations are required, the use of a micromanipulator is advisable. For occasional needs, the following procedure described by Lederberg (42) is useful.

On the back of a clean microscope slide draw a grid of 5-mm squares with India ink. Sterilize the face of the slide in a flame. After the slide cools, coat it with paraffin (mineral) oil to a depth of ca. 0.5 mm. (It is not necessary to sterilize the oil.) Heat 4-mm glass tubing in a flame and draw it out to form a capillary having a terminal diameter of ca. 0.1 mm. Attach rubber tubing to the opposite end. Dilute the culture to a density of 10^6 to 10^7 bacteria per ml in medium, and draw up the suspension into the capillary by applying suction to the rubber tubing. Deposit a drop of suspension from the capillary at the center of each square, under the oil. The drops will adhere to the glass and will flatten out to a diameter of 0.1 to 0.2 mm. Scan the flattened drops by phase-contrast or dark-field microscopy, and determine which drops contain only a single cell of the desired type of organism. Frequently, such a single cell can be recovered by repeatedly flushing the drop in and out of a capillary pipette containing sterile medium.

Another procedure is to add ca. 10^{-5} ml (10 μl) of sterile medium to the drop and incubate the slide in a container of oil until a clone develops; then remove some of the cells of the clone with a' capillary pipette. Since the growth conditions for the clone are semianaerobic, this procedure may not work for strictly aerobic organisms.

8.5. MEDIA AND REAGENTS

8.5.1. *Aquaspirillum gracile* Isolation Medium (12)

Peptone	5.0 g
Yeast extract	0.5 g
Tween 80 (sorbitan monooleate polyoxyethylene)	0.02 g
K₂HPO₄	0.1 g
Agar	10.0 g
Distilled water	1,000 ml

Adjust to pH 7.2. Boil to dissolve the agar. Sterilize at 121°C for 15 min.

8.5.2. ASN-III Medium (59)

NaCl	25.0 g
MgCl₂·6H₂O	2.0 g
KCl	0.5 g
NaNO₃	0.75 g
K₂HPO₂·3H₂O	0.02 g
MgSO₄·7H₂O	3.5 g
CaCl₂·2H₂O	0.5 g
Citric acid	0.003 g
Ferric ammonium citrate	0.003 g
Ethylenediaminetetraacetic acid (EDTA), disodium magnesium salt	0.0005 g
Na₂CO₃	0.02 g
Trace metal mix A5 (see BG-11 medium [8.5.8])	1.0 ml
Deionized water	1,000 ml

After autoclaving and cooling, pH of medium should be 7.5.

8.5.3. *Azotobacter* Medium

Mannitol	2.0 g
K₂HPO₄	0.5 g
MgSO₄·7H₂O	0.2 g
FeSO₄·7H₂O	0.1 g
Distilled water	1,000 ml

Adjust the pH to 7.3 to 7.6. Sterilize at 121°C for 15 min.

8.5.4. Basal Enrichment Medium for *Rhodospirillaceae* (76)

Solution A:

NaHCO₃	2.0 g
Distilled water	25.0 ml

Sterilize by filtration (positive pressure).

Solution B:

NH₄Cl	1.0 g
KH₂PO₄	0.5 g
MgCl₂	0.5 g
(NaCl, for organisms from brackish or marine environments)	(20–30 g)
Trace metal solution (see below)	1.0 ml
Distilled water	975 ml

Sterilize at 121°C for 15 min.

Solution C:

Na₂S·9H₂O	3.0 g
Distilled water	200 ml

Sterilize in a flask with a Teflon-covered magnetized stirring bar at 121°C for 15 min. When cool, add 1.5 ml of sterile 2 M H₂SO₄ with stirring.

To prepare medium, combine solutions A and B. Adjust the pH to 7 with sterile Na₃CO₃ and H₃PO₄ as required. At the time of inoculation of primary enrichment cultures, add 1 ml of solution C per 100 ml of medium. For subsequent transfers, when *Rhodospirillaceae* have established themselves, omit the Na_2S and substitute $MgSO_4$ for the $MgCl_2$ in the medium.

Trace element solution:

Ethylenediaminetetraacetic acid (EDTA), disodium salt	500 mg
$FeSO_4 \cdot 7H_2O$	200 mg
$ZnSO_4 \cdot 7H_2O$	10 mg
$MnCl_2 \cdot 4H_2O$	3 mg
H_3BO_3	30 mg
$CoCl_2 \cdot 6H_2O$	20 mg
$CuCl_2 \cdot 2H_2O$	1 mg
$NiCl_2 \cdot 6H_2O$	2 mg
$Na_2MoO_4 \cdot 2H_2O$	3 mg
Deionized water	1,000 ml

Dissolve the EDTA in a portion of the water. Separately dissolve the other ingredients in water and add them to the EDTA solution. Adjust the solution to ~pH 3 and bring to a final volume of 1,000 ml.

8.5.5. Basal Inorganic Media A and B (14)

Medium A.

Solution 1:

$(NH_4)_2SO_4$	1.2 g
$CaCl_2 \cdot 2H_2O$	0.1 g
$MgSO_4 \cdot 7H_2O$	0.1 g
Ferric citrate	0.002 g
Distilled water	1,000 ml

Sterilize at 121°C for 15 min.

Solution 2:

K_2HPO_4	0.2 g
KH_2PO_4	0.1 g
Distilled water	200 ml

Sterilize at 121°C for 15 min.

To prepare medium, combine solutions 1 and 2 aseptically. For a solid medium, use 20 g of agar per liter.

Medium B.

K_2HPO_4	0.8 g
KH_2PO_4	0.2 g
$CaSO_4 \cdot 2H_2O$	0.05 g
$MgSO_4 \cdot 7H_2O$	0.5 g
$FeSO_4 \cdot 7H_2O$	0.01 g
$(NH_4)_2SO_4$	1.0 g
Distilled water	1,000 ml

Sterilize at 121°C for 15 min. For a solid medium, use 20 g of agar per liter.

8.5.6. Basal Mineral Agar for Cellulolytic Cytophagas (72)

$(NH_4)_2SO_4$ or KNO_3	1.0 g
K_2HPO_4	1.0 g
$MgSO_4$	0.2 g

$CaCl_2$	0.1 g
$FeCl_3$	0.02 g
Agar	10.0 g
Tap water	1,000 ml

Adjust to pH 7.0 to 7.5. Boil to dissolve agar. Sterilize at 121°C for 15 min.

8.5.7. Basal Salts Medium for *Thermus* (11)

Nitrilotriacetic acid	100 µg
$CaSO_4 \cdot 2H_2O$	60 µg
$MgSO_4 \cdot 7H_2O$	100 µg
NaCl	8 µg
KNO_3	103 µg
$NaNO_3$	689 µg
Na_2HPO_4	111 µg
$FeCl_3$	2,800 ng
$MnSO_4 \cdot H_2O$	22 µg
$ZnSO_4 \cdot 7H_2O$	5 µg
H_3BO_3	5 µg
$CuSO_4$	160 ng
$Na_2MoO_4 \cdot 2H_2O$	250 ng
$CoCl_2 \cdot 6H_2O$	460 ng
Deionized water	1,000 ml

Adjust pH to 8.2 with NaOH.

8.5.8. BG-11 Medium (59)

$NaNO_3$	1.5 g
$K_2HPO_4 \cdot 3H_2O$	0.04 g
$MgSO_4 \cdot 7H_2O$	0.075 g
$CaCl_2 \cdot 2H_2O$	0.036 g
Citric acid	0.006 g
Ferric ammonium citrate	0.006 g
Ethylenediaminetetraacetic acid (EDTA), disodium magnesium salt	0.001 g
Na_2CO_3	0.02 g
Trace metal mix A5 (see below)	1.0 ml
Deionized water	1,000 ml

After autoclaving and cooling, pH of medium should be 7.4.

Trace metal mix A5:

H_3BO_3	2.86 mg/ml
$MnCl_2 \cdot 4H_2O$	1.81 mg/ml
$ZnSO_4 \cdot 7H_2O$	0.222 mg/ml
$Na_2MoO_4 \cdot 2H_2O$	0.39 mg/ml
$CuSO_4 \cdot 5H_2O$	0.079 mg/ml
$Co(NO_3)_2 \cdot 6H_2O$	0.0494 mg/ml

8.5.9. Bile Esculin Agar (BBL or Difco)

Solution A:

Beef extract	3.0 g
Peptone	5.0 g
Agar	15.0 g
Distilled water	400 ml

Solution B:

Oxgall	40.0 g
Distilled water	400 ml

Solution C:

Ferric citrate	0.5 g
Distilled water	100 ml

Combine the three solutions and heat to 100°C for 10 min. Sterilize at 121°C for 15 min. Cool to 50°C. Add aseptically 100 ml of a 1% solution of esculin (sterilized by filtration). Dispense into sterile tubes (for slants).

8.5.10. BMS Agar (Potato Agar) (17)

Washed, peeled, sliced potatoes	200 g
L-Malic acid	2.5 g
KOH	2.0 g
Raw cane sugar	2.5 g
Vitamin solution (see 8.5.37)	1.0 ml
Bromothymol blue (0.5% alcoholic solution)	2 drops
Agar	15.0 g
Distilled water	1,000 ml

Place the potatoes in a gauze bag. Boil in 1,000 ml of water for 30 min; then filter through cotton, and save the filtrate. Dissolve the malic acid in 50 ml of water and add 2 drops of bromothymol blue. Add KOH until the malic acid solution is green (pH 7.0). Add this solution together with the cane sugar, vitamins, and agar to the potato filtrate. Make up volume to 1,000 ml with distilled water. Boil to dissolve the agar. Sterilize at 121°C for 15 min.

8.5.11. Bordet-Gengou Agar, Modified (3)

Place 125 g of washed, peeled, and sliced potatoes in a gauze bag. Submerge in a mixture of 10.0 ml of glycerol and 500 ml of water. Boil until the potatoes are soft; then strain through the gauze into the water-glycerol. Allow the fluid to stand in a tall cylinder until the supernatant is relatively clear. Decant the supernatant and make up to 1,000 ml with distilled water. Add 5.6 g of NaCl. Heat. Add 22.5 g of agar, and dissolve by boiling with constant stirring. Sterilize at 121°C for 15 min and cool to 45°C. Add 200 ml of defibrinated sheep blood (see 20.3.7) aseptically and mix. Dispense into petri dishes. (Bordet-Gengou Agar Base is available from BBL or Difco.)

8.5.12. Brilliant Green Agar (BBL or Difco)

Pancreatic digest of casein USP	5.0 g
Peptic digest of animal tissue USP	5.0 g
Yeast extract	3.0 g
NaCl	5.0 g
Lactose	10.0 g
Sucrose	10.0 g
Phenol red	0.08 g
Brilliant green	0.0125 g
Agar	20.0 g
Distilled water	1,000 ml

Adjust pH to 6.9. Boil to dissolve agar. Sterilize at 121°C for 15 min.

8.5.13. Brilliant Green Lactose Bile Broth (available from BBL or Difco as Brilliant Green Bile 2%)

Solution A:

Peptone	10.0 g
Lactose	10.0 g
Distilled water	500 ml

Solution B:

Oxgall	20.0 g
Distilled water	200 ml

Mix solutions A and B. Make up to 975 ml with distilled water. Adjust the pH to 7.4. Add 13.3 ml of a 0.1% aqueous solution of brilliant green. Make final volume to 1,000 ml with distilled water. Dispense into tubes containing inverted gas vials and sterilize at 121°C for 15 min.

8.5.14. Brucella Agar (BBL or Difco)

Pancreatic digest of casein	10.0 g
Peptic digest of animal tissue	10.0 g
Glucose	1.0 g
Yeast autolysate	2.0 g
NaCl	5.0 g
Sodium bisulfite	0.1 g
Agar	15.0 g
Distilled water	1,000 ml

Adjust pH to 7.0. Boil to dissolve agar. Sterilize at 121°C for 15 min.

8.5.15. CHSS Medium for *Spirillum volutans*

Acid-hydrolyzed casein, vitamin-free, salt-free (ICN Nutritional Biochemicals, Cleveland, Ohio)	2.5 g
Succinic acid (free acid)	1.0 g
$(NH_4)_2SO_4$	1.0 g
$MgSO_4 \cdot 7H_2O$	1.0 g
NaCl	0.1 g
$FeCl_3 \cdot 6H_2O$	0.002 g
$MnSO_4 \cdot H_2O$	0.002 g
Distilled water	1,000 ml

Adjust pH to 7.0 with 2 N KOH. For semisolid medium add 1.5 g of agar and boil to dissolve the agar. Sterilize at 121°C for 20 min.

8.5.16. Cystine Tellurite Blood Agar (3)

Heart infusion agar (Blood Agar Base, 20.3.31)	500 ml
Agar	2.5 g

Adjust pH to 7.4. Boil to dissolve agar. Sterilize at 121°C for 15 min. Cool to 56°C. Add the following ingredients aseptically:

Defibrinated rabbit or sheep blood (20.3.7)	25 ml
0.3% potassium tellurite solution (sterilized by autoclaving)	75 ml
L-Cystine, powder	22 mg

Stir while dispensing medium into petri dishes to keep the cystine suspended.

8.5.17. E Agar (3)

Papaic digest of soy meal USP	20.0 g
NaCl	5.0 g
Agar	10.0 g
Deionized water	1,000 ml

Heat to dissolve ingredients. Cool and adjust pH to 7.4 with NaOH. Dispense and sterilize at 121°C for 15 min. Cool to 50°C. To 65 ml of solution, add aseptically 10 ml of yeast dialysate (see below), 25 ml of horse serum, 2 ml of penicillin (10,000 U/ml), and 1 ml of 3.3% thallous acetate (sterilized by filtration). Dispense 5-ml amounts into 10 by 35 mm petri dishes, and incubate overnight at room temperature before use.

Yeast dialysate:
Suspend 450 g of active dried yeast in 1,250 ml of distilled water at 40°C. Heat at 121°C for 5 min. Dialyze against 1 liter of distilled water at 4°C for 2 days. Discard the dialysis sac and its contents. Sterilize the dialysate at 121°C for 15 min. Store in a freezer.

Biphasic medium.

For biphasic medium, aseptically dispense 3-ml amounts of E agar into sterile 16 by 125 mm screw-capped tubes. After the medium solidifies, overlay it with 3 ml of E broth (see 8.5.18). Store at room temperature.

8.5.18. E Broth (3)

Papaic digest of soy meal USP	20.0 g
NaCl	5.0 g
Glucose	10.0 g
Phenol red, 2% aqueous solution	2.0 ml
Deionized water	1,000 ml

Adjust pH to 7.6. Dispense and sterilize at 121°C for 15 min. After the broth cools, add the same supplements as for E agar (see 8.5.17).

8.5.19. Giesberger Base Medium (83)

NH$_4$Cl	1.0 g
K$_2$HPO$_4$	0.5 g
MgSO$_4$	0.5 g
Distilled water (see note)	1,000 ml

Adjust to pH 7. Sterilize at 121°C for 15 min. (*Note*: For marine organisms, substitute seawater for the distilled water.)

8.5.20. Glucose-Tryptone Agar (6)

Tryptone (Difco) or Trypticase (BBL)	10.0 g
Glucose	5.0 g
Agar	15.0 g
Bromocresol purple	0.04 g
Distilled water	1,000 ml

Adjust pH to 6.7. Boil to dissolve agar. Sterilize at 121°C for 15 min.

8.5.21. Halophile Medium (65)

Casamino Acids (Difco)	7.5 g
Yeast extract	10.0 g
Trisodium citrate	3.0 g
KCl	2.0 g
MgSO$_4$·7H$_2$O	20.0 g
FeCl$_2$	0.023 g
NaCl	250.0 g
Distilled water	1,000 ml

Dissolve the solutes in 800 ml of the distilled water, and adjust the pH to 7.5 to 7.8 with 1 N KOH. Autoclave the medium at 120°C for 5 min; then filter to remove the precipitate. Adjust the pH to 7.4 with 1 N HCl, and make the medium up to 1,000 ml. Sterilize by autoclaving. For a solid medium, add 20 g of agar per liter, boil to dissolve the agar, and then autoclave. Dispense agar media into petri dishes at 60 to 70°C to prevent premature solidification.

8.5.22. *Hyphomicrobium* Medium (9)

K$_2$HPO$_4$	1.74 g
NaH$_2$PO$_4$·H$_2$O	1.38 g
(NH$_4$)$_2$SO$_4$	0.5 g
MgSO$_4$·7H$_2$O	0.2 g
CaCl$_2$·2H$_2$O	0.025 mg
FeCl$_2$·4H$_2$O	3.5 mg
Methanol	5.0 ml
KNO$_3$	5.0 g
Trace element solution (see below)	0.5 ml
Deionized water	1,000 ml

Adjust pH to 7.0 with NaOH. Do not sterilize when using the medium for enrichment. For sterile medium for purification, sterilize at 121°C for 15 min prior to adding the methanol. For solid medium to be used for isolation, add 15 g of agar per liter and omit the KNO$_3$; sterilize at 121°C for 15 min prior to adding the methanol. Remove the oxygen from all liquid media by bubbling nitrogen through them before use; the KNO$_3$ in these media serves as the electron acceptor under anaerobic conditions.

Trace element solution:

ZnSO$_4$·7H$_2$O	50 mg
MnCl$_2$·4H$_2$O	400 mg
CoCl$_2$·6H$_2$O	1 mg

$CuSO_4 \cdot 5H_2O$	0.4 mg
H_3BO_3	2,000 mg
$Na_2MoO_4 \cdot 2H_2O$	500 mg
Deionized water	1,000 ml

8.5.23. Iverson Medium (33)

Trypticase soy agar (dehydrated, BBL)	40.0 g
Agar	5.0 g
Sodium lactate, 0.4% solution	600 ml
$MgSO_4 \cdot 7H_2O$	2.0 g
Ferrous ammonium sulfate	0.5 g
Distilled water	400 ml

Adjust pH to 7.2 to 7.4. Boil to dissolve the agar. Sterilize at 121°C for 15 min.

8.5.24. Lactate Medium

Yeast extract	1.0 g
Sodium lactate	4.0 g
NH_4Cl	0.5 g
K_2HPO_4	1.0 g
$MgSO_4 \cdot 7H_2O$	0.2 g
$CaCl_2 \cdot 2H_2O$	0.1 g
$FeSO_4 \cdot 7H_2O$	0.1 g
Na_2SO_4	0.5 g
(NaCl, for marine organisms)	(20–30 g)

Sterilize at 121°C for 15 min.

8.5.25. Listeria Broth (3) (available from Difco as Tryptose Phosphate Broth)

Peptone, Tryptose (Difco), or Biosate (BBL)	20.0 g
Glucose (dextrose)	2.0 g
NaCl	5.0 g
Na_2HPO_4	2.5 g
Distilled water	1,000 ml

Adjust pH to 7.3. Sterilize at 121°C for 15 min.

8.5.26. Lowenstein-Jensen Medium (BBL or Difco)

KH_2PO_4	2.4 g
$MgSO_4 \cdot 7H_2O$	0.24 g
Magnesium citrate	0.60 g
Asparagine	3.6 g
Potato flour	30.0 g
Glycerol	12.0 ml
Distilled water	600 ml
Homogenized whole eggs	1,000 ml
Malachite green, 2% aqueous solution	200 ml

Dissolve the salts and asparagine in the water. Add the glycerol and potato flour, and autoclave at 121°C for 30 min. Cleanse whole eggs, not more than 1 week old, by scrubbing with 5% soap solution. Allow to stand for 30 min in soap solution; then rinse thoroughly in cold running water. Immerse the eggs in 70% ethanol for 15 min, remove, and break into a sterile flask. Ho-

mogenize by shaking with sterile glass beads. Filter through four layers of sterile gauze. Add 1 liter of the homogenized eggs to the flask of cooled potato-salt mixture. Add the malachite green. Mix well and dispense into sterile, screw-capped 20 by 150 ml tubes, 6 to 8 ml per tube. Slant the tubes and inspissate them at 85°C for 50 min. Incubate for 48 h at 37°C to check sterility. Store the tubes in a refrigerator with their caps tightly sealed.

8.5.27. MacConkey Agar (BBL or Difco)

Peptone (Difco) or Gelysate (BBL)	17.0 g
Proteose Peptone (Difco) or Polypeptone (BBL)	3.0 g
Lactose	10.0 g
Bile salts mixture	1.5 g
NaCl	5.0 g
Neutral red	0.03 g
Crystal violet	0.001 g
Agar	13.5 g
Distilled water	1,000 ml

Adjust pH to 7.1. Boil to dissolve agar. Sterilize at 121°C for 15 min.

8.5.28. Mannitol Salt Agar (BBL or Difco)

Beef extract	1.0 g
Peptone (Difco) or Polypeptone (BBL)	10.0 g
NaCl	75.0 g
Mannitol	10.0 g
Phenol red	0.025 g
Agar	15.0 g
Distilled water	1,000 ml

Adjust pH to 7.4. Boil to dissolve agar. Sterilize at 121°C for 15 min. Cool to 55°C and pour into petri dishes.

8.5.29. McBride Listeria Medium (3) (Difco)

Peptone, Tryptose (Difco) or Biosate (BBL)	10.0 g
Beef extract	3.0 g
NaCl	5.0 g
Glycine anhydride	10.0 g
LiCl	0.5 g
Phenylethanol	2.5 g
Agar	15.0 g
Distilled water	1,000 ml

Adjust pH to 7.3. Boil to dissolve agar. Sterilize at 121°C for 15 min. Cool to 45 to 50°C, and add sterile defibrinated sheep blood to a final concentration of 5%.

8.5.30. Mineral Agar (7)

$Na_2HPO_4 \cdot 12H_2O$	9.0 g
KH_2PO_4	1.5 g
$MgSO_4 \cdot 7H_2O$	0.2 g
NH_4Cl	1.0 g

Ferric ammonium citrate	0.005 g
CaCl$_2 \cdot$2H$_2$O	0.010 g
Trace elements solution (see below)	3.0 ml
Double-distilled water	1,000 ml

Adjust pH to 7.1. Sterilize at 121°C for 15 min. For a solid medium, add 17 g of agar per liter, boil to dissolve the agar, and sterilize by autoclaving.

After the medium has cooled to 45 to 50°C, add sufficient NaHCO$_3$ solution (sterilized by filtration) to give a final concentration of 0.5 g of NaHCO$_3$ per liter.

Trace elements solution:

ZnSO$_4 \cdot$7H$_2$O	10 mg
MnCl$_2 \cdot$4H$_2$O	3 mg
H$_3$BO$_3$	30 mg
CoCl$_2 \cdot$6H$_2$O	20 mg
CuCl$_2 \cdot$6H$_2$O	0.79 mg
NiCl$_2 \cdot$6H$_2$O	2 mg
Na$_2$MoO$_4 \cdot$2H$_2$O	3 mg
Double-distilled water	1,000 ml

8.5.31. Mitis-Salivarius Agar (13) (Difco)

Tryptone (Difco)	10.0 g
Proteose Peptone No. 3 (Difco) ...	5.0 g
Proteose Peptone (Difco)	5.0 g
Glucose	1.0 g
Sucrose	50.0 g
K$_2$HPO$_4$	4.0 g
Trypan blue	0.075 g
Crystal violet	0.0008 g
Agar	15.0 g
Distilled water	1,000 ml

Adjust pH. Boil to dissolve agar. Sterilize at 121°C for 15 min. Cool to 50 to 55°C. Add 1.0 ml of 0.1% potassium tellurite (sterilized by filtration), mix, and dispense into petri dishes.

8.5.32. MN Medium (59)

NaNO$_3$	0.75 g
K$_2$HPO$_4 \cdot$3H$_2$O	0.02 g
MgSO$_4 \cdot$7H$_2$O	0.038 g
CaCl$_2 \cdot$2H$_2$O	0.018 g
Citric acid	0.003 g
Ferric ammonium citrate	0.003 g
Ethylenediaminetetraacetic acid (EDTA), disodium magnesium salt	0.0005 g
Na$_2$CO$_3$	0.02 g
Trace metal mix A5 (see BG-11 medium, 8.5.8)	1.0 ml
Seawater	750 ml
Deionized water	250 ml

After autoclaving and cooling, pH of medium should be 8.3.

8.5.33. Modified Rogosa Medium (46)

Solution A:

MgSO$_4 \cdot$7H$_2$O	11.5 g
MnSO$_4 \cdot$4H$_2$O	2.8 g
FeSO$_4 \cdot$7H$_2$O	0.08 g
Distilled water	100 ml

Solution B:

Yeast extract	6.0 g
Diammonium hydrogen citrate	2.4 g
KH$_2$PO$_4$	7.2 g
Glucose	24.0 g
Tween 80	1.2 g
Distilled water	100 ml

Solution C:

Add 6.0 ml of solution A to 100 ml of solution B. Heat gently until the ingredients are dissolved. Then add 60 ml of 4 M sodium acetate-acetic acid buffer (pH 5.37), and make up to a final volume of 200 ml with distilled water. Final pH, 5.0.

Solution D:

Separated raw milk (adjusted to pH 8.5)	1,000 ml
Trypsin	5 g
Chloroform	10 ml

Incubate at 37°C for 24 h, steam for 20 min, filter while hot, and adjust the pH to 6.65 with glacial acetic acid (ca. 0.5 ml per liter).

To prepare complete medium, add 19 g of agar to 700 ml of solution D, and dissolve by autoclaving at 121°C for 20 min. While hot, add 185 ml of solution C (previously warmed to 50°C). Make up to 1,000 ml with hot solution D. A sample diluted with 3 parts of warm water should have a pH of 5.35 ± 0.05 at 30°C. Dispense the medium in 10-ml quantities into tubes and store in a refrigerator without sterilizing. When preparing plates, melt the medium with as little heating as possible to avoid darkening and the formation of a precipitate.

8.5.34. Modified Wolfe Medium (40, 54)

Preparation of ferrous sulfide agar.

React equimolar quantities of ferrous ammonium sulfate and sodium sulfide in boiling distilled water. Allow the precipitate to settle in a completely filled and stoppered volumetric flask. Decant the supernatant, add boiling distilled water, and invert the flask to suspend the ferrous sulfide. Allow to settle, and decant the supernatant, add more boiling water, etc. Wash the ferrous sulfide four times in this manner; then store it in the volumetric flask completely filled to the top with water and stoppered. The ferrous sulfide can be removed from the bottom of the flask with a pipette.

Add 15 g of purified agar to 100-ml portions of the ferrous sulfide precipitate, boil to dissolve the agar, and sterilize at 121°C for 20 min. Pipette 10-ml portions into sterile screw-capped, rectangular 150-ml milk-dilution bottles, and allow to solidify.

Preparation of liquid overlay.

Cover the agar in each bottle with 50 ml of the following overlay (100 ml in the case of Formalin enrichment):

NH_4Cl	1.0 g
K_2HPO_4	0.5 g
$MgSO_4 \cdot 7H_2O$	0.2 g
$CaCl_2$	0.1 g
Bromothymol blue	0.016 g
Bromocresol purple	0.004 g
Distilled water	1,000 ml

Sterilize at 121°C for 20 min. When cool, add medium aseptically to the bottles containing ferrous sulfide agar. Bubble CO_2 into the liquid overlay for 15 s or until the pH is lowered to 6.0.

8.5.35. MPSS Agar

Peptone (Difco)	5.0 g
Succinic acid (free acid)	1.0 g
$(NH_4)_2SO_4$	1.0 g
$MgSO_4 \cdot 7H_2O$	1.0 g
$FeCl_3 \cdot 6H_2O$	0.002 g
$MnSO_4 \cdot H_2O$	0.002 g
Agar	15.0 g
Distilled water (see note)	1,000 ml

Adjust to pH 7.0 with KOH. Boil to dissolve the agar. Sterilize at 121°C for 15 min. (*Note:* For marine organisms, substitute seawater for the distilled water.)

8.5.36. N-Acetyl-L-Cysteine–Sodium Hydroxide Reagent (20)

Combine equal volumes of 4% NaOH and 2.94% sodium citrate·2H₂O. Dissolve 0.5% *N*-acetyl-L-cysteine powder in this solution. Use within 24 h.

8.5.37. Nfb Medium (17)

L-Malic acid	5.0 g
K_2HPO_4	0.5 g
$MgSO_4 \cdot 7H_2O$	0.2 g
NaCl	0.1 g
$CaCl_2$	0.02 g
Trace metal solution (see below)	2.0 ml
Bromothymol blue (5% alcoholic solution)	2.0 ml
Fe EDTA (1.64% solution)	4.0 ml
Vitamin solution (see below)	1.0 ml
KOH	4.0 g
Agar	1.75 g
Distilled water	1,000 ml

Adjust pH to 6.8. Boil to dissolve agar. Sterilize

at 121°C for 15 min. Cool to 45 to 50°C, and dispense 4-ml amounts into 6-ml serum vials with rubber diaphragms.

Trace metal solution.

$Na_2MoO_4 \cdot 2H_2O$	0.2 g
$MnSO_4 \cdot H_2O$	0.235 g
H_3BO_3	0.28 g
$CuSO_4 \cdot 5H_2O$	0.008 g
$ZnSO_4 \cdot 7H_2O$	0.024 g
Distilled water	200 ml

Vitamin solution:

Biotin	10 mg
Pyridoxine	20 mg
Distilled water	100 ml

8.5.38. Nitrification Medium (58)

Na_2HPO_4	13.5 g
KH_2PO_4	0.7 g
$MgSO_4 \cdot 7H_2O$	0.1 g
$NaHCO_3$	0.5 g
$FeCl_3 \cdot 6H_2O$	0.014 g
$CaCl_2 \cdot 2H_2O$	0.18 g
Distilled water	1,000 ml

Place 75-ml amounts of medium into 250-ml Erlenmeyer flasks, and sterilize at 121°C for 15 min. For nitrite oxidizers add 0.5 g of $NaNO_2$ per liter prior to sterilization. For ammonium oxidizers, sterilize a stock solution of $(NH_4)_2SO_4$ separately from the basal medium and add aseptically to give a final concentration of 0.5 g per liter.

8.5.39. Pfennig and Biebl Sulfur Medium (56)

KH_2PO_4	1.0 g
NH_4Cl	0.3 g
$MgSO_4 \cdot 7H_2O$	1.0 g
$MgCl_2 \cdot 6H_2O$	2.0 g
(NaCl, for marine organisms)	(20.0 g)
$CaCl_2 \cdot 2H_2O$	0.1 g
Trace element solution (see below)	10.0 ml
2 M H_2SO_4 solution	2.0 ml
Distilled water	1,000 ml

Sterilize by autoclaving. When cool, add the following components from sterile stock solutions:

Sodium acetate	0.5 g
$NaHCO_3$ (sterilized by filtration under positive pressure)	4.0 g
$Na_2S \cdot 9H_2O$	0.3 g
Biotin	20 µg

Adjust the pH to 7.8. Grind highly purified sulfur flower in a mortar together with distilled water, and sterilize at 112 to 115°C for 30 min; decant the excess water. For every 50 ml of medium, add a pea-sized amount of the sulfur. Also add several sterile glass beads to each bottle of medium.

Trace element solution:

Ethylenediaminetetraacetic acid, disodium salt	500 mg
$FeSO_4 \cdot 7H_2O$	200 mg
$ZnSO_4 \cdot 7H_2O$	10 mg
$MnCl_2 \cdot 4H_2O$	3 mg
H_3BO_3	30 mg
$CoCl_2 \cdot 6H_2O$	20 mg
$CuCl_2 \cdot 2H_2O$	1 mg
$NiCl_2 \cdot 6H_2O$	2 mg
$Na_2MoO_4 \cdot 2H_2O$	3 mg
Deionized water	1,000 ml

8.5.40. Phenylethyl Alcohol Agar (BBL or Difco)

Pancreatic digest of casein USP	15.0 g
Papaic digest of soya meal USP	5.0 g
NaCl	5.0 g
Phenylethyl alcohol	2.5 g
Agar	15.0 g
Distilled water	1,000 ml

Adjust pH to 7.3. Boil to dissolve the agar, and sterilize at 118°C for 15 min.

8.5.41. Pringsheim Soil Medium (60)

Place one wheat or barley grain in a large test tube, and cover with 3 to 4 cm of garden soil. Fill the tube almost to the top with tap water. Sterilize the medium at 121°C for 30 min.

8.5.42. RGCA-SC Medium (27)

Glucose	0.0248 g
Cellobiose	0.0248 g
Soluble starch	0.05 g
$(NH_4)_2SO_4$	0.1 g
Resazurin solution (0.025%)	0.4 ml
Distilled water	20 ml
Salts solution (see 20.3.28)	50 ml
Rumen fluid (Randolph Biologicals, Houston, Tex.)	30 ml
Cysteine hydrochloride	0.05 g
Hemin solution (see 20.3.28)	1.0 ml
Vitamin K_1 solution (see 20.3.28)	0.02 ml

Prepare in a manner similar to that described for PY broth (see 20.3.28). Dispense 10-ml amounts into tubes containing 0.2 g of agar, and autoclave. To prepare the final medium, melt two tubes and cool to 50°C. To each tube add aseptically:

Sterile inactivated rabbit serum (heated at 60°C for 4 h)	1.5 ml
Cocarboxylase solution, 0.025% (sterilized by filtration)	0.2 ml
Sterile prereduced PY broth (see 20.3.28)	2.0 ml

Mix, and pour the contents of both tubes into a

petri plate. Allow to solidify. Use immediately or store in an anaerobic jar until needed.

8.5.43. Salmonella-Shigella (SS) Agar (BBL or Difco)

Beef extract	5.0 g
Peptone	5.0 g
Lactose	10.0 g
Bile salts mixture	8.5 g
Sodium citrate	8.5 g
Sodium thiosulfate	8.5 g
Ferric citrate	1.0 g
Brilliant green	0.33 g
Neutral red	0.025 g
Agar	13.5 g
Distilled water	1,000 ml

Adjust pH to 7.0. Heat to boiling to dissolve the agar. Do not sterilize by autoclaving. Cool to 42 to 45°C and dispense into petri dishes.

8.5.44. Selenite F Broth (43) (BBL or Difco)

Polypeptone (BBL) or Tryptone (Difco)	5.0 g
Lactose	4.0 g
Na_2HPO_4	10.0 g
Sodium hydrogen selenite	4.0 g
Distilled water	1,000 ml

Adjust pH to 7.0. Use immediately without sterilization, or place tubes in flowing steam for 30 min and store until needed.

8.5.45. Soybean-Casein Digest Broth (also known as Trypticase Soy Broth [BBL] or Tryptic Soy Broth [Difco])

Pancreatic digest of casein USP	17.0 g
Papaic digest of soy meal USP	3.0 g
NaCl	5.0 g
K_2HPO_4	2.5 g
Glucose	2.5 g
Distilled water	1,000 ml

Adjust pH to 7.3. Sterilize at 118 to 121°C for 15 min.

8.5.46. *Sphaerotilus* Medium (8)

Sodium lactate	100 mg
NH_4Cl	1.7 mg
KH_2PO_4	8.5 mg
K_2HPO_4	21.5 mg
$Na_2HPO_4 \cdot 7H_2O$	34.4 mg
$MgSO_4 \cdot 7H_2O$	22.5 mg
$CaCl_2$	27.5 mg
$FeCl_3 \cdot 6H_2O$	0.25 mg
Distilled water	1,000 ml

Adjust pH, if necessary, to 7.1 to 7.2. Dispense 50-ml volumes into French square bottles. Sterilize at 116°C for 15 min.

8.5.47. *Sporocytophaga* Medium

Whatman Chromedia 11	10 g
KNO₃	0.5 g
MgSO₄·7H₂O	0.2 g
CaCl₂·2H₂O	0.1 g
FeCl₃	0.02 g
Distilled water	1,000 ml

8.5.48. Standard Methods Agar (6) (BBL or Difco)

Pancreatic digest of casein, USP	5.0 g
Yeast extract	2.5 g
Glucose	1.0 g
Agar	15.0 g
Distilled water	1,000 ml

Adjust pH to 7.0. Boil to dissolve agar. Sterilize at 121°C for 15 min.

8.5.49. Succinate-Salts Medium

Sodium succinate	4.0 g
KNO₃	0.5 g
K₂HPO₄	0.5 g
MgSO₄·7H₂O	0.2 g
CaCl₂·2H₂O	0.1 g
FeSO₄·7H₂O	0.2 g

Adjust to pH 7.0. Sterilize at 121°C for 15 min.

8.5.50. TCBS Agar (3) (BBL)

Sodium thiosulfate	10.0 g
Sodium citrate	10.0 g
Oxgall	5.0 g
Sodium cholate	3.0 g
Sucrose	20.0 g
Pancreatic digest of casein USP	5.0 g
Peptic digest of animal tissue USP	5.0 g
Yeast extract	5.0 g
NaCl	10.0 g
Iron citrate	1.0 g
Thymol blue	0.04 g
Bromothymol blue	0.04 g
Agar	14.0 g
Distilled water	1,000 ml

Heat with agitation and boil for 1 min. Cool to 45 to 50°C, and dispense into plates. Do not autoclave. Final pH is 8.6.

8.5.51. Thallous Acetate Agar

Thallous acetate	1.0 g
Peptone	10.0 g
Yeast extract	10.0 g
Glucose	10.0 g
Agar	13.0 g
Distilled water	1,000 ml

Adjust pH to 6.0. Boil to dissolve the agar, and sterilize at 118°C for 15 min. Cool to 45 to 50°C, and add aseptically 10 ml of triphenyl tetrazolium chloride (1% aqueous solution, sterilized by filtration). Dispense into petri dishes.

8.5.52. Thayer-Martin Agar

This medium is made most conveniently by using commercial concentrates and solutions or by using prepared media (BBL or Difco).

Solution A (double-strength G C Agar Base):

Pancreatic digest of casein USP	15.0 g
Peptic digest of animal tissue USP	15.0 g
Cornstarch	2.0 g
K₂HPO₄	8.0 g
KH₂PO₄	2.0 g
NaCl	10.0 g
Agar (see note)	20.0 g
Distilled water	1,000 ml

Boil to dissolve agar. Sterilize at 121°C for 15 min. Cool to 50°C. *Note:* Finegold et al. (20) recommend increasing the agar to 30 g.

Solution B:

Hemoglobin, dry	20.0 g
Distilled water	1,000 ml

Add dry powder gradually to a little water to make a smooth paste; then gradually add the rest of the water. Sterilize at 121°C for 15 min.

Solution C:

Vitamin B₁₂	0.010 g
L-Glutamine	10.0 g
Adenine	1.0 g
Guanine hydrochloride	0.03 g
p-Aminobenzoic acid	0.013 g
L-Cystine	1.10 g
Glucose	100.0 g
Nicotinamide adenine dinucleotide (NAD)	0.250 g
Thiamine pyrophosphate (cocarboxylase)	0.100 g
Ferric nitrate	0.020 g
Thiamine hydrochloride	0.003 g
Cysteine hydrochloride	25.900 g
Distilled water	1,000 ml

Sterilize by filtration.

Solution D:

Vancomycin	30.0 mg
Colistin	75.0 mg
Nystatin	125,000 U
Distilled water	100 ml

Sterilize by filtration.

Aseptically combine 1,000 ml of solution A, 1,000 ml of solution B, 20 ml of solution C, and 20 ml of solution D. Dispense into petri dishes.

8.5.53. *Thiobacillus thiooxidans* Medium (78)

Na₂S₂O₃·5H₂O	10.0 g
KH₂PO₄	4.0 g
K₂HPO₄	4.0 g
MgSO₄·7H₂O	0.8 g
NH₄Cl	0.4 g
Trace metal solution (see below)	10.0 ml
Deionized water	1,000 ml

Sterilize at 121°C for 15 min. Adjust to pH 3.5 to 4.0 with sterile H_2SO_4. For a solid medium, add 15 g of agar per liter before adjusting the pH. Boil to dissolve the agar, sterilize by autoclaving, and cool to 45 to 50°C. Adjust pH to 3.5 to 4.0 with sterile H_2SO_4.

Trace metal solution:

Ethylenediaminetetraacetic acid (EDTA)	50.0 g
$ZnSO_4 \cdot 7H_2O$	22.0 g
$CaCl_2$	5.54 g
$MnCl_2 \cdot 4H_2O$	5.06 g
$FeSO_4 \cdot 7H_2O$	4.99 g
$(NH_4)_6Mo_7O_{24} \cdot 4H_2O$	1.10 g
$CuSO_4 \cdot 5H_2O$	1.57 g
$CoCl_2 \cdot 6H_2O$	1.61 g
Distilled water	1,000 ml

Adjust to pH 6.0 with KOH.

8.5.54. Tryptophan Medium (44)

$MgSO_4 \cdot 7H_2O$	0.2 g
K_2HPO_4	1.0 g
$MnCl_2 \cdot 4H_2O$	0.002 g
$FeSO_4 \cdot 7H_2O$	0.05 g
$CaCl_2$	0.02 g
$NaMoO_4 \cdot 2H_2O$	0.001 g
L-Tryptophan	1.0 g
Distilled water	1,000 ml

Adjust pH to 6.8 to 7.0. Sterilize by autoclaving.

8.5.55. Veldkamp Medium (77)

NaCl	30.0 g
KH_2PO_4	1.0 g
NH_4Cl	1.0 g
$MgCl_2 \cdot 6H_2O$	0.5 g
$CaCl_2$	0.04 g
$NaHCO_3$	5.0 g
$Na_2S \cdot 9H_2O$	0.1 g
Ferric citrate, 0.004 M solution	5.0 ml
Trace element solution (see below)	2.0 ml
Powdered agar	5.0 g
Yeast extract	0.3 g
Distilled water	1,000 ml

Adjust to pH 7.0. Use without sterilization for primary enrichment. For a solid medium to be used for pour plates, prepare the medium at double strength with the omission of the powdered agar, and adjust the pH. Sterilize by filtration (positive pressure). Prepare a 4% solution of agar in freshly distilled boiling water, and sterilize by briefly autoclaving. Bring both the double-strength broth and the molten agar to 45 to 50°C, and combine equal volumes aseptically.

Trace element solution:

H_3BO_3	2.8 g
$MnSO_4 \cdot H_2O$	2.1 g
$Cu(NO_3)_2 \cdot 3H_2O$	0.2 g
$Na_2MoO_4 \cdot 2H_2O$	0.75 g

$CoCl_2 \cdot 6H_2O$	0.2 g
$Zn(NO_3)_2 \cdot 6H_2O$	0.25 g
Deionized water	1,000 ml

8.5.56. Violet Red Bile Agar (6) (BBL or Difco)

Yeast extract	3.0 g
Peptone (Difco) or Gelysate (BBL)	7.0 g
Bile salts mixture	1.5 g
Lactose	10.0 g
NaCl	5.0 g
Neutral red	0.03 g
Crystal violet	0.002 g
Agar	15.0 g
Distilled water	1,000 ml

Adjust to pH 7.4. Boil to dissolve agar. Cool to ca. 45°C, and use for pour plates. After the medium has solidified, overlay it with more medium to prevent surface growth and spreading of colonies.

8.5.57. Widdel and Pfennig Medium (81)

Sodium acetate	1.23 g
Na_2SO_4	2.84 g
KH_2PO_4	0.68 g
$MgCl_2$	0.19 g
NH_4Cl	0.32 g
$CaCl_2$	0.07 g
Trace element solution (see below)	10 ml
Distilled water	1,000 ml

Sterilize at 121°C for 15 min. Then add the following ingredients from sterile stock solutions:

$FeCl_2$ (from acidified stock solution)	0.00025 g
$NaHCO_3$ (sterilized by positive pressure filtration)	1.68 g
Na_2S (sterilized by filtration)	0.117 g
Vitamin solution (sterilized by filtration, see below)	5.0 ml
Vitamin B_{12} (sterilized by filtration)	20 µg

Adjust the pH of the medium to 7.1 with sterile H_3PO_4. After inoculating media, add 0.0315 g of $Na_2S_2O_4 \cdot 2H_2O$ per liter from a freshly prepared stock solution sterilized by filtration. For marine organisms, add 20 g of NaCl and 1.14 g of $MgCl_2$ per liter of medium.

Trace element solution:

$ZnSO_4 \cdot 7H_2O$	10 mg
$MnCl_2 \cdot 4H_2O$	3 mg
H_3BO_3	30 mg
$CoCl_2 \cdot 6H_2O$	20 mg
$CuCl_2 \cdot 2H_2O$	1 mg
$NiCl_2 \cdot 6H_2O$	2 mg
$Na_2MoO_4 \cdot 2H_2O$	3 mg
Deionized water	1,000 ml

Vitamin solution:

Biotin	0.2 mg
Niacin	2.0 mg

Thiamine	1.0 mg
p-Aminobenzoic acid	1.0 mg
Panthothenic acid	0.5 mg
Pyridoxamine HCl	5.0 mg
Distilled water	100 ml

Dissolve ingredients. Sterilize by filtration. Store at 4°C.

8.5.58. Wiley and Stokes Alkaline Medium (82)

Yeast extract	20.0 g
(NH₄)₂SO₄	10.0 g
Tris(hydroxymethyl)aminomethane (Tris buffer)	15.7 g
Distilled water	1,000 ml

Prepare stock solutions of each ingredient separately and sterilize by autoclaving. (The pH of the Tris buffer solution should be 9.0.) Combine the ingredients aseptically from the sterile stock solutions. The final pH should be 8.7. For a solid medium, incorporate 15 g of agar per liter.

8.5.59. Yeast Extract Mannitol Agar

Mannitol	10.0 g
K₂HPO₄	0.5 g
MgSO₄·7H₂O	0.2 g
NaCl	0.1 g
CaCO₃	3.0 g
Yeast extract	0.2 g
Agar	15.0 g
Distilled water	1,000 ml

Boil to dissolve the agar. Sterilize at 121°C for 15 min.

8.5.60. YP Medium (75)

Yeast extract	3.0 g
Peptone	0.6 g
Distilled water	1,000 ml

Adjust pH to 7.2. Sterilize at 121°C for 15 min. For semisolid YP medium to be used as an overlay, add 6.0 g of agar per liter; add 19.0 g of agar per liter for a solid medium. Boil to dissolve the agar; then sterilize by autoclaving.

8.6. LITERATURE CITED

8.6.1. General References

1. **Aaronson, S.** 1970. Experimental microbial ecology. Academic Press, Inc., New York.
   Contains a wealth of detailed methods for enrichment and isolation of a great variety of bacteria.
2. **Collins, V. G.** 1969. Isolation, cultivation and maintenance of autotrophs, p. 1–52. *In* J. R. Norris and D. W. Ribbons (ed.), Methods in microbiology, vol. 3B. Academic Press, Inc., New York.
   A comprehensive treatment of the principles and techniques for enrichment and isolation of photo- and chemo-autotrophs.

3. **Lennette, E. H., A. Balows, W. J. Hausler, Jr., and J. P. Truant (ed.).** 1980. Manual of clinical microbiology, 3rd ed. American Society for Microbiology, Washington, D.C.
   Principles and methodology for enrichment and isolation of pathogenic bacteria, rickettsias, viruses, and fungi.
4. **Veldkamp, H.** 1970. Enrichment cultures of prokaryotic organisms, p. 305–361. *In* J. R. Norris and D. W. Ribbons (ed.), Methods in microbiology, vol. 3A. Academic Press, Inc., New York.
   Emphasizes the theoretical aspects of enrichment cultures and presents methods for the enrichment of specific organisms.

8.6.2. Specific References

5. **Allen, O. A.** 1957. Experiments in soil bacteriology, 3rd ed. Burgess Publishing Co., Minneapolis.
6. **American Public Health Association.** 1960. Standard methods for the examination of dairy products, 11th ed. American Public Health Association, New York.
7. **Aragno, M., and H. G. Schlegel.** 1978. *Aquaspirillum autotrophicum*, a new species of hydrogen-oxidizing, facultatively autotrophic bacteria. Int. J. Syst. Bacteriol. **28**:112–116.
8. **Armbruster, E. H.** 1969. Improved technique for isolation and identification of *Sphaerotilus*. Appl. Microbiol. **17**:320–321.
9. **Attwood, M. M., and W. Harder.** 1962. A rapid and specific enrichment procedure for *Hyphomicrobium* spp. Antonie van Leeuwenhoek J. Microbiol. Serol. **38**:369–378.
10. **Barnes, E. M.** 1956. Methods for the isolation of faecal streptococci (Lancefield Group D) from bacon factories. J. Appl. Bacteriol. **19**:193–203.
11. **Brock, T. D., and H. Freeze.** 1969. *Thermus aquaticus* gen. n. and sp. n., non-sporulating extreme thermophile. J. Bacteriol. **98**:289–297.
12. **Canale-Parola, E., S. L. Rosenthal, and D. G. Kupfer.** 1966. Morphological and physiological characteristics of *Spirillum gracile* sp. n. Antonie van Leeuwenhoek J. Microbiol. Serol. **32**: 113–124.
13. **Chapman, G. H.** 1944. The isolation of streptococci from mixed cultures. J. Bacteriol. **48**:113–114.
14. **Claus, D., and N. Walker.** 1964. The decomposition of toluene by soil bacteria. J. Gen. Microbiol. **36**:107–122.
15. **Davis, B. D., R. Dulbecco, H. N. Eisen, H. S. Ginsberg, W. B. Wood, and M. McCarty.** 1973. Microbiology, 2nd ed. Harper & Row, Hagerstown, Md.
16. **Dawes, E. A., and P. J. Senior.** 1973. The role and regulation of energy reserve polymers in micro-organisms. Adv. Microb. Physiol. **10**:135–266.
17. **Döbereiner, J., and V. L. D. Baldani.** 1979. Selective infection of maize roots by streptomycin-resistant *Azospirillum lipoferum* and other bacteria. Can. J. Microbiol. **25**:1264–1269.
18. **Döbereiner, J., I. E. Marriel, and M. Nery.** 1976. Ecological distribution of *Spirillum lipo-*

ferum Beijerinck. Can. J. Microbiol. **22**:1464–1473.

19. **Faust, L., and R. S. Wolfe.** 1961. Enrichment and cultivation of *Beggiatoa alba*. J. Bacteriol. **81**:99–106.

20. **Finegold, S. M., W. J. Martin, and E. G. Scott.** 1978. Bailey and Scott's diagnostic microbiology, 5th ed. C. V. Mosby Co., St. Louis.

21. **Freeman, B. A.** 1977. Burrows' textbook of microbiology, 21st ed. The W. B. Saunders Co., Philadelphia.

22. **Gerloff, G. C., G. P. Fitzgerald, and F. Skoog.** 1950. The isolation, purification and culture of blue-green algae. Am. J. Bot. **37**:216–218.

23. **Gibbons, N. E.** 1969. Isolation, growth and requirements of halophilic bacteria, p. 169–183. *In* J. R. Norris and D. W. Ribbons (ed.), Methods in microbiology, vol. 3B. Academic Press, Inc., New York.

24. **Hampp, E. G.** 1957. Isolation and identification of spirochaetes obtained from unexposed canals of pulp-involved teeth. Oral Surg. Oral Med. Oral Pathol. **10**:1100–1104.

25. **Hanson, A. W.** 1970. Isolation of spirochaetes from primates and other mammalian species. Br. J. Vener. Dis. **46**:303–306.

26. **Hanson, A. W., and G. R. Cannefax.** 1964. Isolation of *Borrelia refringens* in pure culture from patients with condylomata acuminata. J. Bacteriol. **88**:111–113.

27. **Holdeman, L. V., E. P. Cato, and W. E. C. Moore (ed.).** 1977. Anaerobe laboratory manual, 4th ed. Virginia Polytechnic Institute and State University, Blacksburg.

28. **Hungate, R. E.** 1950. The anaerobic cellulolytic bacteria. Bacteriol. Rev. **14**:1–49.

29. **Hungate, R. E.** 1969. A roll tube method for cultivation of strict anaerobes, p. 117–132. *In* J. R. Norris and D. W. Ribbons (ed.), Methods in microbiology, vol. 3B. Academic Press, Inc., New York.

30. **Hylemon, P. B., J. S. Wells, Jr., J. H. Bowdre, T. O. MacAdoo, and N. R. Krieg.** 1973. Designation of *Spirillum volutans* Ehrenberg 1832 as type species of the genus *Spirillum* Ehrenberg 1832 and designation of the neotype strain of *S. volutans*. Int. J. Syst. Bacteriol. **23**:20–27.

31. **Inoue, K.** 1976. Quantitative ecology of microorganisms of Syowa Station in Antarctica and isolation of psychrophiles. J. Gen. Appl. Microbiol. **22**:153–150.

32. **Inoue, K., and K. Komagata.** 1976. Taxonomic study on obligately psychrophilic bacteria isolated from Antarctica. J. Gen. Appl. Microbiol. **22**:165–176.

33. **Iverson, W. P.** 1966. Growth of *Desulfovibrio* on the surface of agar media. Appl. Microbiol. **14**:529–534.

34. **Jannasch, H.** 1967. Enrichments of aquatic spirilla in continuous culture. Arch. Mikrobiol. **59**:165–173.

35. **Jannasch, H.** 1968. Competitive elimination of *Enterobacteriaceae* from seawater. Appl. Microbiol. **16**:1616–1618.

36. **Johnstone, K. I.** 1969. The isolation and cultivation of single organisms, p. 455–471. *In* J. R. Norris and D. W. Ribbons (ed.), Methods in microbiology, vol. 1. Academic Press, Inc., New York.

37. **Joklik, W. G., and H. P. Willett.** 1976. Zinsser microbiology, 16th ed. Appleton-Century Crofts, New York.

38. **Kenny, G. E.** 1974. *Mycoplasma*, p. 333–337. *In* E. H. Lennette, E. H. Spaulding, and J. P. Truant (ed.), Manual of clinical microbiology, 2nd ed. American Society for Microbiology, Washington, D.C.

39. **Killinger, A. H.** 1974. *Listeria monocytogenes*, p. 135–139. *In* E. H. Lennette, E. H. Spaulding, and J. P. Truant (ed.), Manual of clinical microbiology, 2nd ed. American Society for Microbiology, Washington, D.C.

40. **Kucera, S., and R. S. Wolfe.** 1957. A selective enrichment medium for *Gallionella ferruginea*. J. Bacteriol. **74**:344–349.

41. **Leadbetter, E. R., and S. C. Holt.** 1968. The fine structure of *Bacillus fastidiosus*. J. Gen. Microbiol. **52**:299–307.

42. **Lederberg, J.** 1954. A simple method for isolating individual microbes. J. Bacteriol. **68**:258–259.

43. **Leifson, E.** 1936. New selenite enrichment media for the isolation of typhoid and paratyphoid (*Salmonella*) bacilli. Am. J. Hyg. **24**:423–432.

44. **Lichstein, H. C., and E. L. Oginsky.** 1965. Experimental microbial physiology. W. H. Freeman and Co., San Francisco.

45. **Linn, D. M., and N. R. Krieg.** 1978. Occurrence of two organisms in the type strain of *Spirillum lunatum*: rejection of the name *Spirillum lunatum* and characterization of *Oceanospirillum maris* subsp. *williamsiae* and an unclassified vibrioid bacterium. Int. J. Syst. Bacteriol. **28**:132–138.

46. **Mabbitt, L. A., and M. Zielinska.** 1956. The use of a selective medium for the enumeration of lactobacilli in cheddar cheese. J. Appl. Bacteriol. **19**:95–101.

47. **McCurdy, H. J., Jr.** 1963. A method for the isolation of myxobacteria in pure culture. Can. J. Microbiol. **9**:282–285.

48. **Matin, A., C. Veldhuis, V. Stegeman, and M. Veenhuis.** 1979. Selective advantage of a *Spirillum* sp. in a carbon-limited environment. Accumulation of poly-β-hydroxybutyric acid and its role in starvation. J. Gen. Microbiol. **112**:349–355.

49. **Matin, A., and H. Veldkamp.** 1978. Physiological basis of the selective advantage of a *Spirillum* sp. in a carbon-limited environment. J. Gen. Microbiol. **105**:187–197.

50. **Meiklejohn, J.** 1950. The isolation of *Nitrosomonas europaea* in pure culture. J. Gen. Microbiol. **4**:185–191.

51. **Meynell, G. G., and E. Meynell.** 1965. Theory and practice in experimental bacteriology. Cambridge University Press, London.

52. **Morita, R. Y.** 1975. Psychrophilic bacteria. Bacteriol. Rev. **39**:144–167.

53. **Myrvik, Q. N., N. N. Pearsall, and R. S. Weiser.** 1974. Fundamentals of medical bacteriology and mycology. Lea & Febiger, Philadelphia.

54. **Nunley, J. W., and N. R. Krieg.** 1968. Isolation of *Gallionella ferruginea* by use of formalin. Can. J. Microbiol. **14**:385–389.

55. **Peterson, J. E.** 1969. Isolation, cultivation and maintenance of the myxobacteria, p. 185 210. *In* J. R. Norris and D. W. Ribbons (ed.), Methods in microbiology, vol. 3B. Academic Press, Inc., New York.

56. **Pfennig, N., and H. Biebl.** 1976. *Desulfuromonas acetoxidans* gen. nov. and sp. nov., a new anaerobic, sulfur-reducing, acetate-oxidizing bacterium. Arch. Microbiol. **110**:3–12.

57. **Poindexter, J. S.** 1964. Biological properties and classification of the *Caulobacter* group. Bacteriol. Rev. **28**:231–295.

58. **Pramer, D. A., and E. L. Schmidt.** 1964. Experimental soil microbiology. Burgess Publishing Co., Minneapolis.

59. **Rippka, R., J. Deruelles, J. B. Waterbury, M. Herdman, and R. Y. Stanier.** 1979. Generic assignments, strain histories and properties of pure cultures of cyanobacteria. J. Gen. Microbiol. **111**:1–61.

60. **Rittenberg, B. T., and S. C. Rittenberg.** 1962. The growth of *Spirillum volutans* in mixed and pure cultures. Arch. Mikrobiol. **42**:138–153.

61. **Rosebury, T.** 1962. Microorganisms indigenous to man. McGraw-Hill Book Co., New York.

62. **Rosebury, T., and G. Foley.** 1942. Isolation and pure cultivation of the smaller mouth spirochaetes by an improved method. Proc. Soc. Exp. Biol. Med. **47**:368–374.

63. **Rosebury, T., J. B. McDonald, S. A. Ellison, and S. G. Engel.** 1951. Media and methods for separation and cultivation of oral spirochaetes. Oral Surg. Oral Med. Oral Pathol. **4**:68–85.

64. **Scotten, H. L., and J. L. Stokes.** 1962. Isolation and properties of *Beggiatoa*. Arch. Mikrobiol. **42**:353–368.

65. **Sehgal, S. N., and N. E. Gibbons.** 1960. Effect of some metal ions on the growth of *Halobacterium cutrirubrum*. Can. J. Microbiol. **6**:165–169.

66. **Singh, B.** 1947. Myxobacteria in soils and composts: their distribution, number and lytic action on bacteria. J. Gen. Microbiol. **1**:1–10.

67. **Slack, J. M., and I. S. Snyder.** 1978. Bacteria and human disease. Year Book Medical Publishers, Chicago.

68. **Smibert, R. M.** 1965. *Vibrio fetus* var. *intestinalis* isolated from fecal and intestinal contents of clinically normal sheep: isolation of microaerophilic vibrios. Am. J. Vet. Res. **26**:315–319.

69. **Smibert, R. M.** 1969. *Vibrio fetus* var. *intestinalis* isolated from the intestinal contents of birds. Am. J. Vet. Res. **30**:1437–1442.

70. **Smibert, R. M., and R. L. Claterbaugh, Jr.** 1972. A chemically-defined medium for *Treponema* strain PR-7 isolated from the intestine of a pig with swine dysentery. Can. J. Microbiol. **18**:1073–1078.

71. **Smith, L. D., and V. R. Dowell.** 1974. *Clostridium*, p. 376–380. *In* E. H. Lennette, E. H. Spaulding, and J. P. Truant (ed.), Manual of clinical microbiology, 2nd ed. American Society for Microbiology, Washington, D.C.

72. **Stanier, R. Y.** 1942. The *Cytophaga* group: a contribution to the biology of myxobacteria. Bacteriol. Rev. **6**:143–196.

73. **Stanier, R. Y., R. Kunisawa, M. Mandel, and G. Cohen-Bazire.** 1971. Purification and properties of unicellular blue-green algae (order Chroococcales). Bacteriol. Rev. **35**:171–205.

74. **Stanier, R. Y., N. J. Palleroni, and M. Doudoroff.** 1966. The aerobic pseudomonads: a taxonomic study. J. Gen. Microbiol. **43**:159–271.

75. **Stolp, H., and M. P. Starr.** 1963. *Bdellovibrio bacteriovorus* gen. et sp. n., a predatory, ectoparasitic, and bacteriolytic microorganism. Antonie van Leeuwenhoek J. Microbiol. Serol. **29**:217–248.

76. **Van Niel, C. B.** 1971. Techniques for the enrichment, isolation, and maintenance of the photosynthetic bacteria. Methods Enzymol. **23**:3–28.

77. **Veldkamp, H.** 1961. A study of two marine agar-decomposing, facultatively anaerobic myxobacteria. J. Gen. Microbiol. **26**:331–342.

78. **Vishniac, W., and M. Santer.** 1957. The thiobacilli. Bacteriol. Rev. **21**:195–213.

79. **Warke, G. M., and S. A. Dhala.** 1968. Use of inhibitors for selective isolation and enumeration of Cytophagas from natural substrates. J. Gen. Microbiol. **51**:43–48.

80. **Waterbury, J. B., and R. Y. Stanier.** 1978. Patterns of growth and development in pleurocapsalean cyanobacteria. Microbiol. Rev. **42**:2–44.

81. **Widdel, F., and N. Pfennig.** 1977. A new anaerobic, sporing, acetate-oxidizing sulfate-reducing bacterium, *Desulfotomaculum* (emend.) *acetoxidans*. Arch. Microbiol. **112**:119–122.

82. **Wiley, W. R., and J. L. Stokes.** 1962. Requirement of an alkaline pH and ammonia for substrate oxidation by *Bacillus pasteurii*. J. Bacteriol. **84**:730–734.

83. **Williams, M. A., and S. C. Rittenberg.** 1957. A taxonomic study of the genus *Spirillum* Ehrenberg. Int. Bull. Bacteriol. Nomencl. Taxon. **7**:49–111.

84. **Wilson, G. S., and A. A. Miles.** 1964. Topley and Wilson's principles of bacteriology and immunity, 5th ed. The Williams & Wilkins Co., Baltimore.

Chapter 9

Solid Culture

NOEL R. KRIEG AND PHILIPP GERHARDT

Media prepared in the solid state, in the form of firm gels, have been used in bacteriology since adopted by Robert Koch. The most important uses of solidified media stem from their enabling separated colonies to arise from individual cells in a population diluted into or on a solidified medium. Thus, the "streak plate" is a simple but effective technique for isolating pure cultures of bacteria (8.4.1), and the "pour plate," "spread plate," and "layered plate" are similarly valuable for enumerating viable bacteria (11.2). Other techniques that rely on solid culture are multiple-point inoculation with velveteen (13.6.2 and 20.2.5) and the auxanographic method (20.2.6).

The purpose of this chapter is to describe the nature of the main solidifying agents and their uses in bacteriological media for several solid and semisolid culture techniques not described elsewhere in this manual. The general subject has been reviewed by Codner (6).

9.1. SOLIDIFYING AGENTS

9.1.1. Agar

Agar is extracted from certain red marine algae and is the most commonly used solidifying agent for bacteriological media. Agar consists of two polysaccharides, agarose and agaropectin, with the former comprising about 70% of the mixture. When first extracted, agar is contaminated by algal cell debris and various impurities, most of which must be removed before the product is suitable for bacteriological purposes. A summary of commercial extraction and purification procedures can be found in reference 4.

For further information about the chemical composition and structure of agar, see references 4 and 17.

The most important properties of agar for bacteriological work are as follows: agar is not enzymatically degraded by most bacterial species; agar gels are stable up to 65°C or higher, yet molten agar does not gel until cooled to ca. 40°C; and agar gels have a high degree of transparency.

Commercial agar occurs in various grades, but for most purposes "bacteriological" grade is satisfactory. "Special" and "purified" grades contain decreased levels of impurities, are more suitable for electrophoretic and serological applications, and may sometimes be more suitable for nutritional studies. A laboratory procedure for washing bacteriological-grade agar free from many impurities is as follows (12). Soak granular agar in about 10 volumes of distilled water for several hours and filter; do this 10 times over a period of 2 days. Soak the resulting agar with an equal volume of 95% ethanol for 12 h at room temperature and filter; resoak the agar in fresh ethanol for 4 h and filter again. Add the agar to boiling 95% ethanol, bring the alcohol again to a boil, and then filter. Spread out the washed agar to dry at room temperature.

To prepare an agar-solidified medium, first adjust the pH of the basal liquid medium to the desired value and then add the granular agar. Bacteriological or higher grades of agar will not alter the pH of the medium appreciably. For solid medium, use 15 to 20 g of agar per liter; for semisolid medium, use 1.5 to 4.0 g per liter, depending on the consistency required. Different brands and grades of agar may require different concentrations to achieve a particular degree of firmness, and the instructions of the manufacturer should always be consulted.

After adding the agar to a liquid basal medium, heat the mixture to boiling to dissolve the agar completely. If heated over a flame or on a hot plate, stir the medium constantly during heating to prevent the agar from settling to the bottom of the pan where it can caramelize and char; then bring the medium to a rolling boil for a minute or so, being careful to avoid having it foam up or so over the edge of the pan. Alternatively, the agar can be dissolved by placing the mixture in a steamer or microwave oven for an appropriate length of time. After the agar is melted, mix to ensure uniformity in concentration throughout the medium. Dispense the molten medium into tubes, flasks, or bottles and sterilize in an autoclave.

Note: When an agar medium having a pH of 6.0 or less is required, one must initially prepare and sterilize the medium at a pH value greater than 6.0; otherwise, the agar will be hydrolyzed during heating and will fail to solidify when the medium is later cooled. Once the medium has been sterilized and cooled to 45 to 50°C, add sufficient sterile acid aseptically to achieve the final pH value desired. It is useful to prepare an extra portion of medium to experiment with in order to determine the correct amount of acid to add to the main batch.

If one wants tubes of slanted medium, place the hot tubes from the autoclave in a tilted position and allow them to cool and solidify. For preparing petri dishes of medium, first cool the molten medium to 45 to 50°C for ca. 10 min in a water bath, and then dispense the medium aseptically into the sterile dishes (usually 15 to 20 ml per dish). Cover the plates with newspaper or other insulation to prevent precipitation of moisture underneath the tops of the petri dishes. After the agar solidifies, drops of moisture (water of syneresis) may form on the surface. This moisture should be evaporated before the plates are inoculated. Usually, storage of the plates in an inverted position overnight at room temperature will result in sufficient drying of the agar. For more rapid drying, invert the plates on the shelf of a 45°C gravity convection incubator, and adjust the agar-containing halves of the plates so that they are slightly ajar. Incubate in this manner until the water of syneresis disappears. For some procedures, such as in the use of velveteen replicators, plates that have been dried more extensively may be required.

For storage of agar plates under conditions that prevent severe drying, place the plates in closed polyethylene bags (the bags in which plastic petri dishes are packaged make excellent storage bags for media). Plates can be stored in this way at room temperature for 4 to 5 weeks.

Store all culture media, liquid and solid, in a dark cabinet or cupboard if possible. Storage under illumination, especially sunlight (as near a window), may result in photochemical generation of hydrogen peroxide or other toxic forms of oxygen which can render the media inhibitory for the growth of bacteria (16, 20). Microaerophiles appear to be particularly reluctant to grow on media that have been subjected to illumination (3, 10).

9.1.2. Carrageenan

Carrageenan, long used in the food and dairy industry, has only recently been employed as a solidifying agent for bacteriological media. Also known as "Irish moss" or "vegetable gelatin," it is extracted from certain red marine algae. There are several types of carrageenan: kappa, lambda, mu, and iota (17). The potassium salt of kappa carrageenan is capable of forming rigid transparent gels which can be an effective substitute for agar in many bacteriological media (11, 21).

Carrageenan is considerably less expensive than agar. Like agar gels, carrageenan gels can be used for streaking or spreading inocula. Carrageenan is not degraded by most species of bacteria, and gels stable to temperatures of 60°C can be prepared. However, carrageenan does have certain limitations. The high temperature at which liquid carageenan media must be dispensed into petri dishes (55 to 60°C) probably precludes, for most species, the use of carrageenan for enumerating viable bacteria by the method of incorporating dilutions of bacterial suspensions into a molten medium before it solidifies.

Variability may exist between lots of carrageenan (21). Some investigators have found carrageenan to be unsuitable for semisolid media, whereas others have found it to be satisfactory (11). In contrast to agar, carrageenan may cause alterations in the pH of some media during preparation and sterilization; consequently, one should determine for any given kind of medium whether and to what magnitude such pH changes occur. It may be necessary to compensate for such changes by preparing the medium at a different initial pH value or by increasing the buffering capacity of the medium. Media containing a phosphate buffering system seem to be particularly prone to exhibit a large decrease in pH (11).

The preparation of carrageenan media is similar to that for agar media. Carrageenan Type I (Sigma Chemical Co., St. Louis, Mo.), a commercial grade that consists predominantly of kappa carrageenan and a lesser amount of lambda carrageenan, appears to be satisfactory for general purposes. For gels stable during incubation at 45°C, use 2.0% carrageenan; for gels stable at 60°C, use 2.4% carrageenan (11). After adding carrageenan to the basal liquid medium, boil the medium to dissolve the carrageenan completely and then sterilize by autoclaving. Cool the sterile medium to 55 to 60°C, dispense into petri dishes, and allow to solidify. It is difficult to dispense carrageenan media by pipetting; the media should be dispensed by pouring from a bottle or flask. Remove water of syneresis in the same manner as for agar gels.

9.1.3. Silica Gel

Media solidified with the inorganic agent silica gel provide a means for solid culture of autotrophic bacteria in the complete absence of organic substances. Moreover, by supplementing such inorganic media with various organic compounds, one can study the ability of heterotrophic bacteria to use such compounds as sole carbon sources. Vitamin requirements can also be determined by the use of silica gel media.

A variety of methods are available for preparing silica gel media (4). The method of Funk and Krulwich (7) is simple and reliable, and yields clear gels that are firm enough to be streaked lightly with an inoculating needle or to be used for spreading an inoculum with a glass rod. Prepare the following solutions:

Double-strength liquid nutrient medium. Prepare as usual but in double strength, and sterilize by autoclaving.

Potassium silicate solution. Add 10 g of powdered silica gel (certified grade 923, 100 to 200 mesh, Fisher Scientific Co., Pittsburgh, Pa.) or silicic acid (reagent grade, J. T. Baker Chemical Co., Phillipsburg, N.J.) to 100 ml of 7% (wt/vol) aqueous KOH, and dissolve by heating. Dispense 20-ml amounts into flasks, and sterilize in an autoclave.

Phosphoric acid solution. Prepare a 20% solution of o-phosphoric acid (85%, certified grade, Fisher Scientific Co.).

Add 20 ml of the sterile double-strength liquid medium to 20 ml of the sterile potassium silicate solution. Rapidly add a measured amount of the phosphoric acid solution sufficient (as determined from prior test, approximately 4 ml) to provide a pH of 7.0. Mix the solutions and immediately pour into two petri dishes. The medium will begin to solidify in 1 min and will become firm in 15 min. Remove water of syneresis by allowing it to evaporate in an incubator. After inoculating the plates, incubate in a moist atmosphere to prevent drying and cracking of the gels.

9.2. SOLID CULTURE TECHNIQUES

9.2.1. Solid-Surface Mass Culture

The maximum density of bacterial cells is obtained by colonial growth on a solid surface; only the interstitial (intercellular) space is occupied by liquid, although this may represent 20 to 30% of the total volume (27% for close-packed spheres, regardless of their size). If the inoculum is highly diluted, each cell develops into a discrete colony; such a procedure may be desirable even for mass culture if uniformity of colony characteristics is desired prerequisite to harvesting of the solid culture (for example, if one wants only a variant that produces "smooth" colonies). More frequently, the inoculum is not diluted and is spread evenly over the surface so that confluent solid growth results. The solid surface usually is that of an agar or otherwise solidified medium. However, a number of techniques have been described in which colonial growth is obtained on a membrane over a reservoir of liquid medium (14).

Mass culture of bacteria on a solid surface has certain advantages over that submerged in a liquid medium (see Chapter 10):

1. Solid cultures are already concentrated, so one does not need to use a centrifuge or other means for harvesting the cells. Instead, one pipettes a small amount of sterile saline or buffer onto the surface of the agar cultures and then suspends the growth in the fluid by means of an appropriately bent sterile glass rod. Large crops of cells can be obtained in this manner by harvesting growth from the surface of agar media contained in large numbers of petri dishes, in large flat culture bottles (e.g., Roux bottles), or in large flat dishes (e.g., covered household Pyrex baking dishes). The avoidance or minimization of centrifugation (which produces aerosols) may be particularly useful in obtaining masses of pathogenic or otherwise harmful bacteria.

2. Solid cultures are relatively free from macromolecular components and completely free from particulate components of the nutrient medium because these tend to be held within the agar gel. Furthermore, solid cultures are also relatively free from small molecular nutrients and their own metabolite products because these tend to be diluted into the greatly larger volume of the medium. Consequently, solid cultures may be particularly useful for preparing antigens or

for other purposes where cell purity is important.

3. Solid cultures may yield results that are otherwise unobtainable. For example, fruiting bodies of myxobacteria and endospores of certain *Bacillus* species are only or better produced from growth on solid medium. The reasons for such results include the ability to retain physical associations among cells, the maximum oxygen supply from the atmosphere, and the diffusion and dilution away from the immediate cell environment of metabolite products (e.g., acids) that otherwise would inhibit growth or development.

On the other hand, solid culture of bacteria has certain disadvantages compared with liquid culture:

1. Solid culture is limited in the scale to which it can be increased (e.g., one can effectively produce gram amounts of cells, decigram amounts become difficult, and centigram or kilogram amounts are impossible in the laboratory).

2. Solid cultures are not homogeneous in physiological properties of the cells. For example, the cells at the top surface of aerobic solid growth (which is likely to be 1,000 cells deep) will be nutrient starved but oxygen rich, whereas the opposite situation will prevail at the bottom surface of the growth.

3. Solid cultures often yield a small amount of cells from a given amount of medium.

9.2.2. Biphasic Mass Culture

The advantages of solid-surface culture can be retained and the disadvantages obviated by use of a liquid/solid culture technique called biphasic culture. Both techniques essentially represent a form of dialysis or diffusion culture in which the usual membrane is represented instead by an interface as the diffusion barrier separating the culture from a nutrient and product reservoir. The principles and development of interfacial dialysis culture in context with membrane dialysis culture have been reviewed by Schultz and Gerhardt (14), and membrane dialysis culture techniques are described in the succeeding chapter on liquid culture (10.3.4). A biphasic system for concentrated bacterial culture was systematically studied by Tyrrell, MacDonald, and Gerhardt (19).

The biphasic technique consists simply of a thick layer of solidified nutrient medium overlaid with a thin layer of nutrient broth (Fig. 1). To prepare such a system, partially fill the container with hot medium containing 2 to 3% agar. After the agar base is solidified, overlay it aseptically with a small volume of broth, inoculate, and incubate. Clamp the container on a shaking

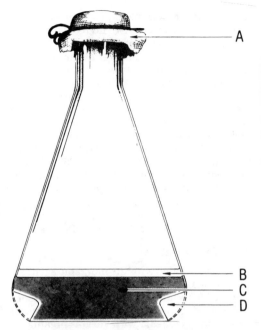

FIG. 1. *Biphasic system for concentrated cultures. (A) Cotton and gauze pad to provide adequate supply of air (11.1.2). (B) Liquid overlay of broth medium or of water diffusate. (C) Solid base of agar medium. (D) Flask indentations to hold the agar base and prevent its breakup if incubated on a shaking machine.*

machine to provide aeration and agitation of the broth during incubation, if the bacteria are aerobic. The culture is confined to the liquid overlay but has diffusional access to the reservoir of nutrients in the solidified base, and consequently becomes densely concentrated. Populations in excess of 10^{11} cells per ml can be obtained in this way, and the technique is applicable apparently to any type of bacterium.

An Erlenmeyer flask is a convenient container, but indentations at the base are helpful to hold the agar if placed on a shaking machine; rectangular containers also are useful in this way. Also, use a greater percentage of agar if breakup occurs during shaking. Since the movement of nutrients is dependent on diffusion, the agar base should be limited to about 5 cm in depth. The ratio of solid to liquid should be at least 4 and less than 10, depending on whether the yield or concentration of cells is more important.

If all of the medium components are incorporated into the agar base, the overlay can be distilled water. After an overnight equilibration period, the resulting clear diffusate in the overlay can be inoculated. For bacteria such as gonococci, which normally must be grown in a turbid medium enriched with blood and starch and which yield sparse populations, a relatively clean and dense population of cells can be harvested from such an overlay (8).

Another example of a biphasic culture system is that devised by Castaneda (5) for the isolation of *Brucella* from blood (Fig. 2). However, the principle of dialysis is involved to only a minor extent. To prepare such a system, first sterilize the 3% agar medium in a cotton-stoppered rectangular bottle; then place the bottle on its side so that the molten medium will solidify as a layer on the wall of the bottle. Stand the bottle upright and add sterile liquid medium aseptically. Seal the bottle with a rubber diaphragm. For some bacteria, such as *Brucella abortus*, it may be necessary to replace the air in the bottle with an atmosphere containing 10% carbon dioxide. To use the bottle, inject a blood sample through the rubber diaphragm by means of a syringe. Incubate the bottle in an upright position. At 2-day intervals tilt the bottle so that the liquid medium wets the entire surface of the agar and drains back down to the bottom. Bacteria present in the liquid medium will thus inoculate the agar surface and develop into visible colonies. The method has several advantages over the use of broth. (i) It is unnecessary to open the bottle repeatedly to withdraw samples from microscopic observation or subcultur-

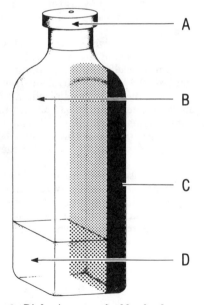

FIG. 2. *Biphasic system for blood cultures on both solid and liquid media. (A) Serum bottle stopper through which blood sample is injected. (B) Atmosphere containing air and 10% carbon dioxide. (C) Agar-solidified culture medium. (D) Liquid culture medium.*

ing in order to detect bacterial growth. This minimizes the chance of laboratory infections with highly infectious organisms such as *Brucella* and also minimizes the chance of contaminating the culture system. (ii) Organisms that grow poorly in broth and develop only a slight turbidity may be able to grow better on the surface of the agar. (iii) Growth of bacteria in the presence of colored or turbid substances in broth can be easily detected by observing the development of colonies on the agar surface. A recent study by Hall et al. (9) indicates the usefulness of such a culture system for detecting the growth of a large variety of pathogenic bacteria.

9.2.3. Bioautography

Bioautography is a version of paper chromatography in which one employs the growth of bacteria as a highly sensitive indicator for locating the position of certain compounds on a paper chromatogram. The method has the advantage of specifically detecting compounds with biological activity, which chemical or radioisotope detection systems lack. The method is particularly applicable to locating the position of growth factors on chromatograms of culture supernatants or cell extracts when the concentration of the growth factors is so low as to preclude the use of ordinary detection systems. For example, the location of as little as 5 to 10 ng of folic acid can be determined on a chromatogram by use of bioautography. For a good example of bioautography applied to detection of growth factors in cell extracts, see reference 15. Bioautography is also widely used by the pharmaceutical industry for detection of antibiotic agents on paper chromatograms (22).

For detection of growth factors, prepare an agar medium containing all of the nutrients and growth factors required to support dense growth of the indicator strain, except for the particular growth factor to be tested. Inoculate the sterile molten medium at 45 to 50°C with a washed suspension of the indicator organism. Pour this "seeded" agar medium into a sterile, covered dish large enough to accommodate the paper chromatogram (a Pyrex baking dish is often suitable). After the medium has solidified, place the dried paper chromatogram face down on the agar surface, taking care to avoid air spaces between the agar and the paper. After incubating the dish, remove the chromatogram and look for areas of dense growth in the agar that indicate the location of the growth factor. Flooding the agar surface with water often helps to increase the visibility of these areas.

For such bioautography to be successful, it is

usually helpful to grow the inoculum for the agar in a broth containing a minimum concentration of the particular growth factor to be tested; this is to avoid accumulation of excess growth factor within the bacteria, which might allow them to grow non-specifically throughout the seeded agar medium. To achieve maximum sensitivity of detection, it is helpful to use a high concentration of washed cells to seed the agar, so that easily visible areas of dense growth will occur where the growth factor is located on the chromatogram. One should perform preliminary experiments to determine the most suitable concentration of the inoculum. Contaminating bacteria that may be present on the paper chromatogram usually do not interfere with the experiments, since their growth occurs as discrete colonies which are not easily confused with the more dense, diffuse growth of the indicator organisms within the agar. However, one should attempt to keep contamination to a minimum. For example, store the dried chromatograms in sterile envelopes until needed, and apply the chromatograms to seeded agar in a room or cabinet where there are no dust-carrying air currents.

For bioautography of antibiotic agents, the agar medium should be nutritionally complete in order to allow the indicator organism to grow throughout the dish except at those regions of the chromatogram where the antibiotic substance is located. Because of the high potency of some antibiotic substances, it may sometimes be advisable to leave the paper chromatogram on the agar surface for only a short time (e.g., 10 min) and then remove it before incubating the dish. If the chromatogram is left on the agar surface, the zones of growth inhibition may be too large to be clearly resolved.

9.3. SEMISOLID CULTURE TECHNIQUES

Semisolid media contain a low concentration of gelling agent and have a soft, jelly-like consistency. Such media are useful for cultivation of microaerophilic bacteria and also for studying various aspects of motility and chemotaxis.

9.3.1. Microaerophiles

Microaerophilic bacteria have a strictly respiratory type of metabolism, with oxygen being used as the terminal electron acceptor; yet they cannot grow under an atmosphere containing the level of oxygen present in air (21% oxygen). For example, *Campylobacter fetus* generally grows best under a 6% oxygen atmosphere. Also, certain nitrogen-fixing bacteria that have a respiratory type of metabolism can grow as aerobes

when supplied with a source of fixed nitrogen (such as ammonium sulfate), but can only grow as microaerophiles in nitrogen-deficient media. For example, *Azospirillum* grows best under a 1% oxygen atmosphere under nitrogen-fixing conditions. This is because of the inactivation of the nitrogenase complex by excess oxygen.

One can grow microaerophiles in liquid media or on solid media by supplying a suitable low level of oxygen, but this is not necessary if one uses semisolid media containing 0.1 to 0.4% agar. The gelling agent stratifies the medium so that convection currents cannot mix the oxygen-rich upper layers of the medium with the underlying layers. The only way oxygen can penetrate into the depths of a tube of semisolid medium is by diffusion. Thus, a tube of semisolid medium provides an oxygen gradient, with the surface of the medium being the most highly oxygenated region. After inoculation of the medium by stabbing deeply with a loop, microaerophiles begin to grow at some distance below the surface where the oxygen level is most suitable. Growth begins in the form of a thin disk that is visible several millimeters or centimeters below the surface; as the growth increases, the disk becomes more dense and migrates closer to the surface. Eventually, the cells may form a very dense, diffuse layer just below the surface. In contrast, anaerobes begin to grow in the lower levels of a semisolid medium, but may also eventually grow throughout as they develop their own anaerobic conditions.

One can easily cultivate the microaerophile *Campylobacter fetus* in tubes of semisolid brucella medium (brucella broth containing 0.15 to 0.3% agar; see 8.5.14). The nitrogen fixer *Azospirillum* will grow well in a semisolid nitrogen-deficient malate medium (8.5.37 or 20.3.4), and large crops of these cells for physiological studies can be obtained by using 1-liter Roux bottles that contain a layer of the semisolid medium (13).

If tubes of semisolid media are stored for long periods, oxygen may eventually diffuse deeply into the media. If one places such tubes in a boiling-water bath for several minutes and then allows the tubes to cool and resolidify, the oxygen gradient can be restored.

9.3.2. Motility

One can also use semisolid media to determine whether a bacterial strain is motile. Inoculate a tube of motility medium (broth containing 0.4% agar) with a straight needle to one-half the depth of the tube. During growth, motile bacteria will migrate from the line of inoculation to form a diffuse turbidity in the surrounding medium;

nonmotile bacteria will grow only along the line of inoculation.

A modification of this method is to place a piece of open-ended glass tubing in the semisolid medium prior to sterilization (Fig. 3). Inoculate the medium within the glass tubing. If the organism is motile, it will migrate downward, emerge from the bottom of the tubing, and then migrate upward to the surface of the medium where it will grow extensively. This method is also helpful for selecting highly motile organisms for use in preparing H-antigens for immunization.

9.3.3. Chemotaxis

Semisolid media are also useful in studies of chemotaxis. For example, Adler (1) used a semisolid medium containing an oxidizable carbon and energy course to investigate positive chemotaxis in *Escherichia coli*. When a petri dish of the medium was inoculated heavily at its center, the bacteria migrated outward in the form of an expanding ring of cells that consumed all of the oxidizable substrate as it moved. When two oxidizable carbon sources were provided in the medium, two rings formed which migrated at different rates. In contrast to capillary tube methods for studying chemotaxis, the supply of oxygen on the agar plate was never exhausted; thus, the migrating rings formed only in response to self-created gradients of the substrate and not in response to self-created oxygen gradients.

Negative chemotaxis also has been studied by using semisolid media (18). *E. coli* cells were suspended in the medium at a concentration sufficient to give visible turbidity, and the inoculated medium was dispensed into a petri dish. After the medium gelled, a plug of hard agar (2% agar) containing the suspected repellent compound was inserted into the medium. If the compound was a repellent, the surrounding bacteria migrated away from the plug and left a clear zone that became visible in ca. 30 min.

The use of plates of semisolid media also makes it possible to select nonchemotactic mutants (2, 18). In the case of positive chemotaxis, nonchemotactic mutants (and also nonmotile mutants) fail to respond to a self-created substrate gradient and remain near the center of the petri plate, where they can be removed for subculturing and purification. In the case of negative chemotaxis, the nonchemotactic mutants can be isolated from the clear zone surrounding the hard agar plug.

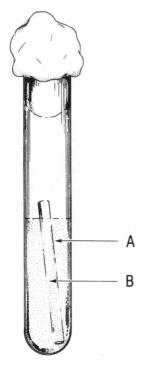

FIG. 3. *Semisolid medium (A) into which a piece of glass tubing (B) had been placed prior to autoclaving. The medium inside the glass tubing is inoculated. Motile bacteria will migrate downward inside the tubing, emerge from the bottom of the tubing, and then migrate upwards to the surface of the medium, where they will grow extensively.*

9.4. LITERATURE CITED

1. **Adler, J.** 1966. Chemotaxis in bacteria. Science **153:**708–716.
2. **Armstrong, J. B., J. Adler, and M. M. Dahl.** 1967. Non-chemotactic mutants of *Escherichia coli.* J. Bacteriol. **93:**390–398.
3. **Bowdre, J. H., and N. R. Krieg.** 1974. Water quality monitoring: bacteria as indicators. Va. Polytech. Inst. State Univ. Water Resour. Cent. Bull. 69.
4. **Bridson, E. Y., and A. Brecker.** 1970. Design and formulation of culture media, p. 229–295. *In* J. R. Norris and D. W. Ribbons (ed.), Methods in microbiology, vol. 3A. Academic Press, Inc., New York.
5. **Castaneda, M. R.** 1947. A practical method for routine blood cultures in brucellosis. Proc. Soc. Exp. Biol. Med. **64:**114–115.
6. **Codner, R. C.** 1969. Solid and solidified growth media in microbiology, p. 427–454. *In* J. R. Norris and D. W. Ribbons (ed.), Methods in microbiology, vol. 1. Academic Press, Inc., New York.
7. **Funk, H. B., and T. A. Krulwich.** 1964. Preparation of clear silica gels that can be streaked. J. Bacteriol. **88:**1200–1201.
8. **Gerhardt, P., and C. G. Hedén.** 1960. Concentrated culture of gonococci in clear liquid medium. Proc. Soc. Exp. Biol. Med. **105:**49–51.

9. **Hall, M. M., C. A. Mueske, D. M. Ilstrup, and J. A. Washington II.** 1979. Evaluation of a biphasic medium for blood cultures. J. Clin. Microbiol. **10:**673–676.

10. **Hoffman, P. H., H. A. George, N. R. Krieg, and R. M. Smibert.** 1979. Studies of the microaerophilic nature of *Campylobacter fetus* subspecies *jejuni*. II. Role of exogenous superoxide anions and hydrogen peroxide. Can. J. Microbiol. **25:**8–16.

11. **Lines, A. D.** 1977. Value of the K^+ salt of carrageenan as an agar substitute in routine bacteriological media. Appl. Environ. Microbiol. **34:**637–639.

12. **Meynell, G. G., and E. Meynell.** 1965. Theory and practice in experimental bacteriology. Cambridge University Press, London.

13. **Okon, Y., S. L. Albrecht, and R. H. Burris.** 1976. Carbon and ammonia metabolism of *Spirillum lipoferum*. J. Bacteriol. **128:**592–597.

14. **Schultz, J. S., and P. Gerhardt.** 1969. Dialysis culture of microorganisms: design, theory and results. Bacteriol. Rev. **33:**1–47.

15. **Sirotnak, F. M., G. J. Donati, and D. J. Hutchison.** 1963. Folic acid derivatives synthesized during growth of *Diplococcus pneumoniae*. J. Bacteriol. **85:**658–665.

16. **Sneath, P. H. A.** 1955. Failure of *Chromobacterium violaceum* to grow on nutrient agar, attributed to hydrogen peroxide. J. Gen. Microbiol. **13:**1.

17. **Towle, G. A., and R. L. Whistler.** 1973. Hemicellulose and gums, p. 198–248. *In* L. P. Miller (ed.), Phytochemistry, vol. 1. Van Nostrand Reinhold Co., New York.

18. **Tso, W., and J. Adler.** 1974. Negative chemotaxis in *Escherichia coli*. J. Bacteriol. **118:**560–576.

19. **Tyrrell, E. A., R. E. MacDonald, and P. Gerhardt.** 1958. Biphasic system for growing bacteria in concentrated culture. J. Bacteriol. **75:**1–4.

20. **Waterworth, P. M.** 1969. The action of light on culture media. J. Clin. Pathol. **22:**273–277.

21. **Watson, N., and D. Apiron.** 1976. Substitute for agar in solid media for common usages in microbiology. Appl. Environ. Microbiol. **31:**509–513.

22. **Weinstein, M. J., and G. H. Wagman (ed.).** 1978. Antibiotics: isolation, separation and purification. Elsevier Scientific Publishing Co., New York.

Chapter 10

Liquid Culture

STEPHEN W. DREW

Bacterial cell cultivation is the process of increasing the bulk concentrations of some or all of the components of a population and is usually defined implicitly by the types of measurements used to monitor the process. The measurements most often used reflect an increase in cell numbers or cell mass. Occasionally, growth is monitored by measurement of the synthesis of a macromolecule when an increase in protein, ribonucleic acid, deoxyribonucleic acid, lipid, or other macromolecule is coincident with an increase in cell mass and cell numbers (**balanced growth**). However, one or more of these growth parameters may be independent of the others (**unbalanced growth**), such as in the early and late stages of the batch growth cycle when cell mass is independent of cell numbers. In the following chapter of this manual, Koch discusses the definition of bacterial growth, the techniques for measurement of growth, and the assumptions required for the use of each method.

In this chapter are described techniques for the cultivation of bacterial cells in liquid culture ranging in volume from 10 ml to 100 liters. The techniques provide for the cultivation of a suspension of a single type of bacterium in aqueous medium with varying degrees of control over the physical conditions. The techniques are applicable with some modification to the mixed cultivation of more than one type of bacterium, but a discussion of this exciting area of bacterial cell culture is outside the scope of this book.

The techniques that are used to measure growth obscure the fact that all bacterial cultures are grossly heterogeneous and really monitor average values that describe the growth of a population rather than of individual cells. Although all members of a particular population may be genetically identical, the individual members will vary with respect to doubling time, age, composition, metabolic characteristics, and size. The magnitudes of these variations are influenced by the environment and can often be minimized by careful design and development of the system for cultivation. Most of the systems described here are designed to minimize heterogeneity within the population of a pure culture.

In such systems, the growth behavior of a bacterial population can be predicted by the simple relationship

$$\frac{dx}{dt} = \mu x \qquad (1)$$

where dx is the increase in amount of biomass,

dt is the time interval, x is the amount of biomass, and μ is the **specific growth rate**, representing the rate of growth per unit amount of biomass and having dimensions of reciprocal time (1/t).

If μ is constant, integration of equation 1 shows that x will increase exponentially with time as follows:

$$\ln \frac{x}{x_0} = \mu t \qquad (2)$$

where ln is the natural logarithm (to the base e = 2.303), and x_0 is the amount of biomass when $t = 0$.

It follows by rearrangement of equation 2 that the final biomass concentration, x, is

$$x = x_0 \, e^{\mu t} \qquad (3)$$

Bacterial growth which follows this relationship is called **logarithmic or exponential growth.**

The relationship between the biomass doubling time (t_d) and the specific growth rate is found by letting $x = 2x_0$ at $t = t_d$ in equation 2 and solving the equation as follows:

$$t_d = \frac{\ln 2}{\mu} = \frac{0.693}{\mu} \qquad (4)$$

Equations 1 to 4 predict the growth of bacteria in simple systems where the factors influencing growth are constant, but do not allow prediction of deviation from constant growth (**steady-state**) conditions. The techniques for continuous cultivation described below very nearly establish steady-state conditions of theoretically infinite duration, whereas the techniques for batch cultivation allow significant changes in the environment during the time course of cultivation. In batch culture, equations 1 to 4 will apply without adjustment in the value of μ only during that portion of the growth cycle in which the changes in the growth environment have no influence on population growth (i.e., during the exponential growth phase).

10.1. BATCH SYSTEMS

A batch culture system is one in which nothing (with the frequent exception of the gas phase) is added to or removed from an environment after inoculating a medium of appropriate composition with living cells (a **closed system**). It follows, therefore, that a batch system can support cell multiplication for only a limited time and with progressive changes in the original medium and environment.

Normal growth cycle.

Figure 1 shows an idealized normal growth cycle for a simple, homogeneous, batch culture of bacteria. Growth proceeds through a **lag phase**, during which cell numbers do not increase, and into a **growth phase**, which usually is characterized by an exponential increase and follows the relationships of equations 1 to 4. Ultimately, changes in the chemical or physical environment result in a phase of no net increase in numbers, the **maximum stationary phase**. Cells in stationary phase still require an energy source for the maintenance of viability. The availability of an energy source in a batch culture is limited by definition, and hence a **death phase** follows, which is often characterized by an exponential decrease in the number of living cells.

The lag phase may be brought about by the

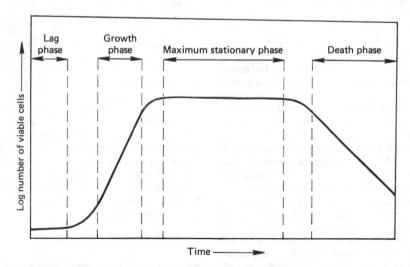

FIG. 1. *Idealized normal growth cycle for a bacterial population in a batch culture system.*

shock of rapid change in culture environment. In fresh medium, the length of the lag phase depends upon the size of the inoculum, the age of the inoculum, and the changes in nutrient composition and concentration experienced by the cells. A small inoculum volume transferred to a large volume of fresh medium may result in outward diffusion of vitamins, cofactors, and ions which are required for many intracellular enzyme activities. If cells are inoculated from a rich medium to a minimum medium, lag time may be affected by inoculum size as a result of carry-over of trace nutrients from the original medium.

The age of the inoculum will influence the lag in a fresh medium as a result of toxic materials accumulated and essential nutrients depleted within the cells during their prior growth. Both a positive and a negative effect on the length of lag phase in fresh medium can occur as a function of inoculum age. In general, increasing inoculum age lengthens the lag phase when cells are inoculated from a nutritionally simple medium into a richer one. A plot of the relationship between inoculum age and length of the lag phase for inoculation from rich to simple medium may show a definite minimum point because of the trade-off between nutrient and toxic product buildup.

Finally, changes in nutrient composition and concentration between the inoculum culture and the fresh medium may trigger the control and regulation of enzyme activities within the cells or morphological differentiation, such as spore germination. If the cells are transferred from a simple medium to a richer medium, both time and nutrients will be expended to allow an increase in enzymes essential for metabolism. When cells are inoculated from a rich medium to a medium of lower nutrient level, the cells may resume exponential growth immediately but at a slower rate.

The fact that constant exponential growth can occur for even a limited time in batch culture shows that growth rate can be virtually unaffected by changes in substrate concentrations over wide ranges. Under these conditions the culture is said to be in balanced growth (see above) and can be described by a single numerical value, μ. Eventually, the culture will deviate from constant exponential growth and can no longer be described only by the value of μ, even though it is possible to calculate this value for the case of nutrient substrate limitation (equation 5). Monod (44) described the relationship between substrate concentration and bacterial growth in simple systems at steady state as

$$\mu = \frac{\mu_{max} S}{(K_s + S)} \qquad (5)$$

where μ is the specific growth rate, μ_{max} is the maximum value of μ obtained when $S \gg K_s$, K_s is the saturation constant equivalent to a Michaelis-Menten constant, and S is the instantaneous or steady-state substrate concentration.

Several alternative models exist for growth response to substrate concentration (44). Equation 5 may be used to describe growth response to substrate limitation only under conditions of steady state with uncomplicated bacterial systems. When the culture ceases exponential growth as a result of substrate limitation, conditions are no longer steady state for cell mass accumulation or substrate concentration, and equation 1 must be used with equations 5 and 6 to adequately model the response of the culture to diminishing substrate. Simultaneous solution of equations 1, 5, and 6 yields a somewhat bulky model for batch growth, which is at best a rough estimation for very simple systems (44).

The maximum population in a batch culture can be estimated from experimental data relating the increase in cell number or mass to the corresponding decrease in substrate concentration:

$$\frac{(x - x_0)}{(S_0 - S)} = Y \qquad (6)$$

where x and S are the cell and substrate concentrations at time t, x_0 and S_0 are the cell and substrate concentrations at an earlier time, t_0, and Y is the overall yield coefficient. Equation 6 accurately describes the relationship between cell concentration and substrate concentration during exponential growth. If exponential growth continues unabated until the stationary phase is reached and substrate is completely consumed during exponential growth, then the maximum cell number or concentration will be given by equation 6. This estimation assumes that the yield coefficient is constant throughout the growth cycle and neglects substrate consumption during lag and stationary phases. In fact, the yield coefficient cannot be assumed constant for other than constant exponential growth conditions at a single specific growth rate. It follows that the prediction of a maximum population from equation 6 will lead to an overestimation. The concept of yield coefficient is discussed in greater detail later in this chapter (10.5). Prediction of the time required to attain maximum population density in batch culture also requires simplifying assumptions (2, 4).

Aberrant growth cycles.

The growth cycle (curve) shown in Fig. 1 is idealized and assumes constant morphology and asynchronous binary fission. An increase in biomass with concomitant increase in cell number and continuous utilization of substrate result. Variations in the growth cycle are common. Morphological change in the culture (such as increases in opacity, cellular refractive index, individual cell size, or cell aggregation) can lead to apparent changes in the growth cycle if growth is determined by optical measurement. For example, stationary bacterial cells are often more transparent than those growing in exponential phase. As a result, a growth curve plotted as the logarithm of absorbance versus the time may show an apparent decrease in stationary-phase cell concentration compared to that attained at the end of exponential growth. In this case, the aberrant growth curve represents a change in the morphological characteristics of the cell rather than a change in the number of cells.

Growth curves plotting the log of cell number versus time occasionally show an unusually rapid increase in cell number just after the lag phase, which then settles into a slower rate of cell number accumulation. The unusual burst of growth may in fact be an indication of partial or complete culture synchrony. This **synchronized growth** may result from culture acclimation to a new nutritional environment or from spore germination. Culture synchrony degenerates rapidly, and asynchronous growth will usually dominate within two generations. However, special techniques have been devised to synchronize cell divisions in a growing population as a way to mimic individual cell growth (10.3.3).

Although most bacterial cultures reproduce by binary fission as individual cells, some reproduce as filamentous extensions (i.e., actinomycetes). If growth occurs primarily through extension of the tips of the hyphae, then the increase in cell mass with respect to time will be arithmetic.

Arithmetic (linear) growth also occurs when the supply of a critical nutrient is regulated by an arithmetic process (such as by dropwise addition or diffusion) and this process becomes limiting. For example, limited diffusion of air through the cotton plug in a test tube or of nutrients through a membrane in a dialysis culture (51) may cause a shift from exponential to arithmetic growth.

Filamentous microorganisms often grow as pellets in liquid culture; when this occurs, biomass increases more slowly than the classical exponential rate and is proportional to the cube of time. Microbial growth in pellets may be severely affected by diffusion of nutrients to the pellet and diffusion of metabolic products away from the pellet, a possibility which equation 5 ignores.

Bacterial cells in complex environments often metabolize usable substrates in a sequential manner. That is, the presence of certain substrates may lead to repression of the enzymes for metabolism of other substrates. In this instance, only when the concentration of the repressing substrate has been reduced through bacterial consumption can the enzymes for metabolism of other substrates be elaborated. The regulation of bacterial physiology leads to an aberrant growth cycle which shows one or more intermediate but transient stationary phases. This response to a changing environment is termed diauxie. A classical example of **diauxic growth** response is that of *Escherichia coli* growing in the presence of both glucose and lactose (Fig. 2). Rapid growth on glucose occurs first. At the point of glucose exhaustion, an inflection in the biomass curve occurs. (There may even be a decline in biomass.) A new enzyme system for metabolism of lactose is induced during this lag, and biomass accumulation at the expense of lactose continues.

10.1.1. Culture Tubes

The lipless Pyrex-glass culture tube (usually 16 by 150 mm in size) is the most convenient and widely used container for batch liquid culture of bacteria. Yet, for aerobic cultivation, test tubes usually provide only minimally effective conditions of oxygen supply. Fortunately, most bacteria are facultatively aerobic and most bacteriological uses of culture tubes do not require optimum growth conditions.

To improve aerobiosis in a culture tube, increase the surface-to-volume ratio of the liquid medium by reducing the volume and slanting the tubes, or preferably mount them on a rotary shaking machine to induce a vortex. Also increase the availability of air by using a small and loosely packed cotton plug, a Morton-type cap (Scientific Products), or a gas-permeable mem-

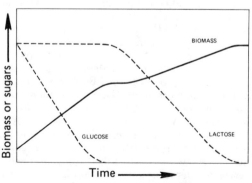

FIG. 2. *Idealized diauxic growth of a bacterial population (biomass) in a batch culture system with two usable substrates (glucose and lactose).*

brane closure. The latter is not commonly used but is effective because it is greatly permeable to gas exchange yet almost impermeable to water vapor (consequently, the medium does not evaporate). Membrane closures are available commercially as a Morton-type cap but with a thin polycarbonate membrane built into the top (Kimble P. M. Cap Closure; Scientific Products).

Conditions for optimization of aerobiosis in laboratory culture containers are described more completely in the following part on shake flask systems (10.1.2.). Conditions that ensure strict anaerobic conditions for tube cultures are described elsewhere in this manual (6.6).

10.1.2. Shake Flasks

An Erlenmeyer flask (100- to 2,000-ml capacity) commonly is used as a container for producing masses of cells in the laboratory or as the first stage in scale-up study of an industrial process involving batch liquid culture of bacteria. Consequently, maximally effective conditions of air (oxygen) supply are sought for obligately or facultatively aerobic types of bacteria, whereas exclusion of air is prerequisite for anaerobes.

Aerobic culture.

Aerobic flask culture must be carried out with shaking of the flask to facilitate mass transfer of oxygen (but also other gases and nutrients) at two levels: from the gas phase across the gas-liquid interface into the liquid phase, and from the liquid phase across the liquid-cell interface into the cell. Two major factors influence the ability of flask culture techniques to meet the oxygen requirements of the cells: gas exchange through the flask closure, and the attainable liquid surface area available for oxygen transport from the gas phase.

Closure design for culture flasks must allow adequate gas exchange between the external environment and the flask interior, yet maintain asepsis. Although nonabsorbent cotton plugs wrapped in gauze or cheesecloth are often adequate for low-density aerobic cultivation of bacteria, cotton plugs wrapped too tightly or oversized for the flask neck can lead to oxygen limitation in high-density bacterial cultures (58). Instead, prepare a cotton-and-gauze pad (see Fig. 10), which allows better oxygen transfer than a cotton plug, as described below:

1. Layer one thickness of nonabsorbent sheet cotton between two layers of surgical gauze.

2. Cut the cotton and gauze pad into squares sized to wrap over the top of the flasks.

3. Place the square pad over the top of a flask and secure the pad over the lip of the flask with a metal spring or two (for insurance) rubber bands. The metal springs can be prepared from heavy piano wire bent to form by local shop personnel.

A Morton-type cap also provides an adequate aseptic seal and good gas exchange, but may occasionally spin off a flask at high rotational speed. Foam plastic plugs also are effective aseptic closures for shake flask cultivation and provide good gas exchange, but the plugs must be sized to fit snugly into the neck of the flask without severe compression. Furthermore, foam plugs may contain plasticizers or other toxic chemicals which must be thoroughly washed from the plug prior to its use as a closure for biological culture. Caps and plugs can be purchased from various supply houses (e.g., Arthur H. Thomas, Scientific Products, Bellco Glass, Inc.).

The rate at which dissolved oxygen is consumed in a liquid culture of bacteria is determined by the density and growth rate. The demand for dissolved oxygen is created by cell consumption and is met by continuous diffusion of oxygen from the gas phase to the liquid phase. The contact surface area between gas and liquid often controls the flux or volumetric rate of oxygen transfer. High interfacial surface area results in high volumetric rates of oxygen transfer and therefore allows rapid growth of high-density cultures without oxygen limitation. Interfacial surface area can be maximized by maintaining low ratios of liquid volume to flask volume and by shaking the flasks. Oxygen limitation due to limited diffusion can usually be avoided by preparing flasks with liquid volumes of no more than 20% of flask volumes and by shaking the flasks at 200 to 350 rpm on a 2.5-cm throw rotary shaker or on a 150- to 250-strokes per min reciprocating shaker.

During incubation of the inoculated flasks at an appropriate temperature, rotary shaking should cause the liquid meniscus to rise to approximately two-thirds of the flask height. This condition can be met for 250-ml baffled Erlenmeyer flasks containing 50 ml of medium by rotation at approximately 220 rpm on a 2.5-cm throw. Reciprocal shaking, such as is found in most water-bath shakers, should be sufficiently vigorous to cause significant breaking (turbulence) of the liquid wave as it moves from side to side in the flask (23).

Baffled flasks have indentations in the bottom surface, which act to break vortex formation in flasks on a rotary shaking machine. Baffled flasks with optimum design and positioning of the baffles are available commercially (Bellco Glass Co. and other glassware distributors).

The following procedure illustrates shake flask cultivation of aerobic bacteria using the gauze pads:

1. Fill a 250-ml Erlenmeyer flask with 50 ml of appropriate liquid medium.
2. Prepare the gauze pad as described above.
3. Fit the pad on the flask, secure it with a band or spring, and cover it with aluminum foil.
4. Press the foil around the cover and the flask neck, to prevent wetting from condensation within the autoclave.
5. Repeat this procedure for other flasks, and sterilize the flasks in an autoclave, as described in Chapter 23.
6. Just prior to inoculation, remove the aluminum covers from the flasks.
7. Carefully remove the rubber bands or retaining spring without disturbing the pad. (Pads of proper thickness will retain their caplike appearance after autoclaving and will shield the mouth and neck of the flask.)
8. Lift the gauze cap from the flask neck, and hold the cap upright to prevent airborne contamination of the sterile underside.
9. Immediately place the neck of the flask into a gas flame, and briefly rotate the flask to evaporate condensation inside the neck.
10. Remove the flask from the flame, aseptically transfer inoculum into the flask, and replace the pad quickly.
11. Secure the pad with rubber bands or metal spring.

The procedure described above utilizes a flask closure that will allow very rapid exchange of gas across the closure while maintaining aseptic conditions. Flaming of the flask neck may be eliminated to allow somewhat more rapid inoculation of flasks and will, with practice, often provide sufficient maintenance of asepsis.

Anaerobic culture.

Erlenmeyer flasks, or flasks or bottles of any other convenient shape, are used to cultivate anaerobic bacteria in an intermediate size scale. Two main principles prevail: the nutrient medium must be prereduced to an Eh of -150 mV or lower, and air must be removed from the immediate environment of the culture. The special procedures for handling anaerobic bacteria are described elsewhere in this manual (6.6).

A few practical tips apply to mass cultivation of anaerobes. Fill the container nearly full to minimize the effect of the overlying gas phase. Use as large an inoculum as possible distributed from the bottom upward in the column of prereduced medium. Displace the gas phase with oxygen-free nitrogen or argon, plus 5 to 10% CO_2, and allow growth to begin with the medium

quiescent. Once growth has progressed actively throughout the medium, provide some agitation by means of a magnetic stirring bar or by purging nitrogen or natural gas through a tube with its outlet at the bottom of the container. Use a water trap to permit the escape of introduced and metabolic gases without the entrance of air. CAUTION: Some anaerobic bacteria produce hydrogen or methane gas, which may cause flammable or even explosive conditions; and most anaerobes produce CO_2 and other gases, which will increase the internal pressure dangerously if adequate venting is not provided.

10.1.3. Carboys

Carboys have traditionally been used to cultivate multi-liter quantities of bacteria in liquid culture. However, there are several factors which severely limit the usefulness of this technique. Handling and manipulating large glass carboys during medium preparation, sterilization, and inoculation is hazardous. Shaking of filled glass carboys for aerobic cultivation is risky. Plastic carboys lessen these concerns. However, dispersion in any carboy is so poor, even with stirrers, that adequate oxygenation is unlikely. Although preparation of media in carboys is often necessary for large-scale work, cultivation in carboys should be avoided.

10.1.4. Fermentors

The fermentors and procedures described below are oriented primarily toward cultivation of aerobic bacteria but are applicable, with little modification, to cultivation of anaerobes also. Fermentor design aspects discussed here were developed to minimize population heterogeneity. Other designs, such as plug-flow, closed-loop, and tubular fermentors, offer exciting research opportunities in the control of population heterogeneity, but are beyond the scope of this manual. A guide to the literature on fermentor design and operation is provided in Table 1.

Fermentors or stirred tank reactors for cultivation of bacteria are designed to provide a self-contained environment. The operational volume typically ranges from 500 ml to thousands of liters, with typical laboratory units having operating volumes of 1 to 20 liters. Since they are designed to minimize thermal and chemical concentration gradients in the culture environment, fermentors are always agitated (stirred) in a manner that yields rapid and complete mixing of the liquid phase. Inadequate agitation will allow development of regions of poor mixing ("dead spots") and will result in uncontrolled local concentration gradients. The physiological result of poor mixing is the cultivation of a highly heterogeneous cell population. For example, in-

TABLE 1. *Guide to the literature on fermentor design and operation*

Topic	References
Fermentor design	
General construction	2, 4, 9, 18, 22, 27, 29, 45
Operational mode:	
Batch	32, 36
Fed batch	16, 32
Continuous	18, 43
Dialysis	51
Fermentor operation	
Theory and control logic	2, 3, 17, 33, 35, 43, 47, 51–53
Physiological effects:	
Batch and fed batch	2, 3, 5, 16, 24, 32, 38, 40
Continuous culture	1, 2, 23–25, 37, 41, 54
Dialysis	14, 51
Immobilized cell	6, 7, 13

adequate agitation of aerobic cultures can result in insufficient contact between the oxygen-containing phase and the liquid cultivation medium. As a result, some regions of the fermentor may be nearly anaerobic. If the organism being cultivated is facultative, the cell population will have a wide range of respiratory characteristics and the cell yield may be reduced.

Although it is certainly possible to build an adequate fermentor in a machine shop, the necessary attention to detail of design for mixing, aeration, sterilization, and maintenance of asepsis warrant the purchase of commercially available units (12, 20, 31, 39). Commercially available fermentors (Table 2) are capable of adequate aeration and agitation of relatively dense cultures. These tanks are usually equipped with a multi-port head plate through which pH probes, dissolved oxygen probes, and sampling or feed lines can be introduced into the fermentor. Table 2 also includes fermentation hardware suppliers.

Aeration and agitation.

The minimum agitation speed for adequate oxygen transfer is subject to change as the culture density (cell concentration) or operating conditions (volumetric air flow rate, temperature, and fermentor liquid volume) change. Since the minimum agitation speed for adequate oxygen transfer is quite sensitive to environmental parameters, it is not convenient to control a fermentation on the basis of this parameter. Rather, the operator should adjust the agitation speed such that the dissolved oxygen concentration never falls below 10 to 15% of the initial saturated dissolved oxygen concentration. Main-

tenance of dissolved oxygen concentrations above 10% of the saturation value at ambient temperatures will allow growth of most bacterial cultures under conditions in which oxygen is not the growth-limiting substrate. The fundamental theory behind this general rule for adequate agitation in sparged systems is based upon classical Monod kinetics for microbial growth in response to rate-limiting substrates (1).

Agitation requirements for adequate oxygen transfer are almost always well in excess of the agitation requirement for mixing of highly soluble nutrients. This rule also applies to agitation conditions established for optimum dissolution of other sparingly soluble substrates, such as immiscible hydrocarbons and steroids. In the

TABLE 2. *Manufacturers of instrumentation and equipment for laboratory-scale fermentors*[a]

Item	Manufacturer
Air sterilization filters	Dollinger Corp. Michigan Dynamics Millipore Corp.
Dissolved oxygen monitors and controllers	Sensorlabs, Division of Instrumentation Laboratories, Inc. Beckman Instruments, Inc. Chemapec, Inc. New Brunswick Scientific Co.
Effluent gas analyzers	Beckman Instruments, Inc. Infrared Industries, Inc. Leeds and Northrup Co. Mine Safety Appliances Co. Perkin-Elmer Corp.
Exhaust air incinerators	Air Resources, Inc. Trane Thermal Co.
Fermentor systems	B. Braun Melsungen AG Chemapec, Inc. LKB Inc. L. E. Marubushi Co., Ltd. New Brunswick Scientific Co. The VirTis Co.
Foam controllers	Chemapec, Inc. Chemineer, Inc. Mixing Equipment Co., Inc. New Brunswick Scientific Co.
Metering pumps and controllers	Cole-Palmer Instrument Co. Manostat Corp.
pH monitors and controllers	Beckman Instruments, Inc. Leeds and Northrup Co. The London Co.
Temperature controllers	Beckman Instruments, Inc. Leeds and Northrup Co. Taylor Instrument Co.

[a] For the addresses of manufacturers, see 10.6.

case of oxygen transfer, both agitation speed and volumetric air flow rate can be varied to establish the desired oxygen concentration in the fermentation medium. Instrumentation for dissolved oxygen monitoring and for control of air flow rate is useful for cultivation of aerobic microorganisms and allows the operator to control dissolved oxygen concentration as an independent fermentation parameter (20).

Temperature control.

Most commercial fermentors for laboratory-scale operation are equipped with internal heat exchangers. The temperature of the circulated water (or other coolant liquid) is controlled by a temperature control unit to maintain a constant temperature in the fermentation medium. Commercial units are either equipped with a standard temperature control unit or offer these devices as ancillary equipment. In some cases, the fermentor is submerged in a constant-temperature bath as a means of temperature control; this technique is effective for control of temperature when accuracy is required.

Air sterilization.

The air or gas supply to sparged fermentors must be sterilized prior to injection into the fermentation vessel. This can be accomplished most easily by physical removal of airborne microorganisms using a fibrous-medium filter. Most commercially available laboratory fermentors utilize fiber glass-packed pipe filters, although some utilize membrane filters. In either case, the filtration medium must be changed regularly to ensure adequate performance. A supplementary or back-up air-sterilization filter may be needed for long-duration cultivation as a precaution against fouling of the duty filter.

An air-sterilization filter can be created by using a 20-cm section of 2.5-cm (inside diameter) stainless-steel pipe threaded at both ends to mate with threaded pipe caps or reducers. The pipe caps can be drilled and nipples can be threaded into the holes to allow connection of the device to rubber tubing. The thoroughly cleaned stainless-steel pipe body should be loosely packed with approximately 90 to 150 kg/m^3 of glass wool, and the clean caps fitted with connector nipples should be screwed firmly into place on each end of the filter. Teflon pipe-sealing tape applied to the threads of the pipe body prior to attachment of the pipe caps will ensure a gas-tight seal.

The air-sterilization filter should be sterilized before its initial use by wrapping it in aluminum foil and heating at 170 to 190°C in an oven for 2 h. After the filter has cooled, make sure that the caps are securely in place and connect the filter to the air supply line. Rubber tubing may be used to connect the outlet of the filter to the fermentor system. The air filter can be used two or three times before repacking becomes necessary. This type of air filter also is useful for in-line sterilization of the oxygen-free gases supplied to anaerobic cultures. A more detailed description of fibrous air filter design is given in reference 32.

Foam control.

Most agitated and aerated cultures will produce relatively stable foam at some point in the culture cycle. If foams are allowed to develop unchecked, they may wet the air filters and lead to back-contamination of the culture as well as to the nuisance of foam spillage. Foams may be controlled by mechanical foam breakers or by addition of chemical antifoam agents. Many commercial fermentors are equipped with mechanical foam breakers as standard or optional equipment. Current designs for mechanical foam breakers are adequate for all but a very few culture conditions.

Chemical antifoam agents provide a less expensive means of foam control where intermittent or light use is expected (11). However, antifoam agents must be added to the fermentation medium and therefore contribute to the overall medium composition. Some antifoam agents such as vegetable (corn oil, cotton seed oil) and animal (lard oil) ones may be metabolized by the culture and therefore contribute to the carbon-substrate pool (28). On the other hand, nonmetabolizable antifoam agents, such as the silicone antifoams (available from Dow Corning Co., General Electric Corp., and Union Carbide), may be toxic at high concentrations. However, their action as surfactant materials requires their use only in very small concentrations. Toxicity should be determined on an individual basis. Antifoam agents can be added directly to the medium prior to sterilization, or they can be added to the fermentation system through a feed line in the head plate (the antifoam agent, its reservoir, and the feed line must be sterilized prior to use). Automatic control units for antifoam addition are available commercially (New Brunswick Scientific Co., The VirTis Co., Chemapec Inc.).

Sampling and inoculation.

Preparation and operation of the fermentor requires that all points of entry or withdrawal from the system be designed for aseptic operation to prevent contamination of the fermentor from external sources. Most laboratory fermentors have a sample line that penetrates the fermentor head plate and extends to within a few centimeters of the bottom of the fermentor vessel. Follow the manufacturer's directions for sampling and inoculation carefully. Note that positive pressure on the fermentor must always be maintained.

Anaerobic operation.

The primary concern in system design for anaerobic cultivation in fermentors is exclusion of oxygen. Fill the fermentor as much as practical, and sparge with oxygen-free gas during the early hours of cultivation to maintain a slight positive pressure. Use large inocula (10 to 20% of total operating volume) to allow rapid establishment of the culture and to reduce the sensitivity of the system to leaks of oxygen from the external environment. If pH is to be controlled or nutrients are to be added in a continuous or semicontinuous fashion, keep the flexible tubing connecting the reservoirs to the fermentor as short as possible, since oxygen will penetrate most rubber tubing. Tygon tubing is relatively oxygen impermeable but softens during autoclaving and is not recommended unless it can be wired in place. Finally, fit the gas exit of the fermentor with a water-gas trap to prevent back diffusion of oxygen into the fermentor head space.

A simple water trap can be created by placing a water-filled graduated cylinder or other glass container in an inverted position in a water bath. Place the gas exit line from the fermentor under the inverted container so that the escaping gas will collect within the vessel. Again, the connecting tubing should be oxygen impermeable or as short as possible to minimize oxygen diffusion through the walls of the tubing. As gas collects within the container, the water will be displaced but will maintain a very effective seal against atmospheric oxygen. The water-bath container should be sufficiently large so that the displaced water will not cause the bath to overflow. When one container is full, transfer a second, water-filled container from its storage position in the water bath to the gas collection position. If graduated cylinders are used as the collection vessels, the volume of gas can be easily determined. A water seal is seldom necessary when active sparging of the fermentor by oxygen-free gas is in operation. However, once the anaerobic culture has become established in a growth pattern, the sparging of oxygen-free gas through the fermentor will probably not be necessary. If gas sparging is stopped, the water-gas trap must be installed immediately.

In all cases, prereduce the medium by heating within the fermentor, or charge prereduced medium into a fermentor that has been sparged with oxygen-free gas. Prepare and maintain acid, base, or nutrient solutions in an oxygen-deficient condition for addition to the fermentor during operation. Absolute exclusion of oxygen from acid, base, or other solutions to be added to the fermentor in small quantities is not necessary since the reducing characteristics of an active anaerobic culture will adequately cope with very slight additions of oxygen through the feed systems.

10.2. CONTINUOUS SYSTEMS

Continuous cultivation differs from batch cultivation in that a fresh supply of nutrients is added continuously at the same rate that medium is withdrawn from the culture (an **open system**). If the culture is well mixed, a sample representative of both the population and the substrate concentrations within the fermentor will be withdrawn with the fermentation broth. The technique of continuous cultivation theoretically allows continuous exponential growth of the culture in a system requiring constant addition of fresh medium and withdrawal of culture broth so that the culture volume remains constant with time. In its broadest sense, continuous cultivation does not require a constant cell concentration (steady state). However, most literature accounts of quantitative study of continuous cultivation have dealt with systems of fixed biomass concentration. The principles developed below deal only with steady-state continuous cultivation.

The term **steady state** is often applied to continuous cultivation and means literally that no change in status occurs during the time span studied. In reality, this definition is too broad since practical application of continuous cultivation theory often results in changes in some parameters while others remain constant. A continuous culture is therefore defined at steady state by an invariant biomass concentration with respect to the time span of observation. In contrast, a batch culture may have a steady-state dissolved oxygen concentration maintained by constant replacement but is not a continuous culture because biomass concentration changes with time.

Continuous cultivation at steady state is possible only when all factors contributing to the accumulation of biomass are exactly balanced by all factors contributing to the loss of biomass from the system. This is shown by the following general material balance on the bacterial cells:

$$\begin{matrix} \text{cells added} \\ \text{to the sys-} \\ \text{tem} \end{matrix} - \begin{matrix} \text{cells removed} \\ \text{from the system} \end{matrix} + \begin{matrix} \text{cells produced} \\ \text{through growth} \end{matrix}$$

$$- \begin{matrix} \text{cells consumed} \\ \text{through death} \end{matrix} = \begin{matrix} \text{cells accumulated} \\ \text{within the system} \end{matrix}$$

This equation is shown mathematically as follows:

$$\frac{FX_0}{V} - \frac{FX}{V} + \mu X$$
$$- \alpha X = \frac{dX}{dt} \qquad (7)$$

where F is the medium flow rate to and from the fermentor (liters per hour), V is the liquid volume within the fermentor (liters), X_0 and X are the cell masses (grams per liter) in the feed and fermentor, respectively, μ is the specific growth rate, α is the specific death rate (hours^{-1}), and dX/dt is the rate of change in cell mass (grams per liter-hour).

At steady state, $dX/dt = 0$. The volume of a true continuous culture is fixed in theory and undergoes negligible variation in practice. Therefore, the flow rates to and from the fermentor must be identical. Finally, the specific death rate is almost always much less than the specific growth rate, so the death term may be ignored. Exceptions to this rule may occur at very low growth rates, in the presence of toxic substances, or when cultivation occurs at conditions of extreme biophysical environment.

If the feed to the fermentor is sterile, $X_0 = 0$, and if the death term, αX, can be ignored, then at steady state:

$$\mu = \frac{F}{V} = D \qquad (8)$$

That is, the specific rate of growth of the population within the fermentor is determined by the dilution rate, D, where $D = F/V$.

Many types of continuous cultures are possible. References 2 and 4 describe several different types of continuous cultivation in detail. The following discussion is limited to the two major types of continuous culture: the "**chemostat**" achieves steady state by controlling the availability of a growth-limiting substrate; the "**turbidostat**" achieves steady state by actually removing the cell mass and replacing it with fresh medium at the same rate as cell growth. Table 3 compares the chemostat and turbidostat with respect to various operating parameters. The chemostat mode for continuous cultivation allows precise control over the growth-limiting condition (nutrient limitation) in contrast to turbidostat operation. However, turbidostats are better suited to continuous cultivation studies of growth at or near the maximum specific growth rate, μ_{max}.

Luedeking (38) presented a very useful discussion of continuous culture theory and practice. The article describes graphical design and analysis of continuous culture systems which avoid the requirement for an accurate model of growth response to changing environmental properties (such as the Monod model). The empirical approach to continuous cultivation based on batch data has its pitfalls, which were discussed by Luedeking (38) and by others (6–8); however, the approach is still useful for obtaining approximate design criteria.

Deviations from simple chemostat theory abound, but most are the result of: (i) product formation, which may require a sophisticated model of yield as a function of μ; (ii) imperfect mixing in the fermentor, which may allow a stable dilution rate that is larger than the critical dilution rate; (iii) wall growth, which has an effect similar to imperfect mixing, but more pronounced; and (iv) idiosyncrasies of physiological response. Pirt (4) discussed the first three of these deviations; Wang and Sinskey (59) discussed some aspects of the last deviation. Other aspects of the idiosyncratic physiological response (nutritional step-up and step-down responses) are discussed in a subsequent part of this chapter (10.5.).

10.2.1. Chemostat

Theory.

Chemostat cultivation is based on the realization that microbial growth in its simplest form can be described as a collection of reaction steps in which the rate of growth of a culture will be determined by the slowest step of nutrient metabolism. Although a growing culture requires the metabolism of many different nutrients, the growth rate of an ideal culture at any given instant in time will be determined by the rate of metabolism of a single nutrient. Monod (45, 46) found that growth rate dependence on substrate concentration could be predicted by an equation whose form is essentially a Michaelis-Menten type function, as follows:

$$\mu = \mu_{max}(S/[K_s + S]) \qquad (9)$$

where μ is the specific growth rate of the culture and is equal to $(\ln 2)/t_d$ for bacterial cultures in exponential growth, μ_{max} is the maximum obtainable specific growth rate in the medium concerned, S is the medium concentration of the substrate being studied, and K_s is the saturation constant for the substrate under study and is numerically equal to the growth-limiting substrate concentration at one-half the maximum specific growth rate. Monod's equation is most easily applied for experimental conditions of steady state. For our purposes, a steady state exists when the substrate concentration does not fluctuate with time and the resulting microbial

TABLE 3. *Comparison of chemostat and turbidostat*

Operating parameter	Chemostat	Turbidostat
Operation at or near maximum specific growth rate	Unstable	Stable, very nearly steady state
Operation at low specific growth rates	Stable steady states	Unstable, transient with pulsatile response
Dilution rate equals specific growth rate	Only at steady state	At all times
Cell concentration at constant specific growth rate depends on	Substrate concentration in the feed	Substrate concentration in the feed
Dilution rate	Predetermined	Controlled as a function of cell mass
Substrate concentration for steady-state operation	Requires a single limiting substrate	All substrates may be present in excess

population in the continuous culture vessel is constant with time. When a chemostat is operated with a sterile feed ($X_0 = 0$) and without recycle, the specific growth rate is numerically equal to the dilution rate (equation 8). This identity is forced by control of limiting nutrient availability through fresh medium addition.

A limiting nutrient balance can be written for a chemostat:

$$\text{input} - \text{output} - \text{consumed} = \frac{\text{accumulation}}{}$$

The mathematical expression, similar to equation 7 for cell mass, is as follows:

$$DS_0 - DS - \frac{\mu X}{Y_{x/s}} = \frac{dS}{dt} \quad (10)$$

where D is the dilution rate ($D = F/V$), S_0 and S are the limiting substrate concentrations in the feed and fermentor, respectively, μ is the specific growth rate of the culture in the fermentor, X is the dry cell mass in the fermentor, $Y_{x/s}$ is the overall yield coefficient (cells formed/substrate consumed), and dS/dt is the rate of change of substrate concentration in the fermentor. The overall yield term is a composite that includes contributions for both growth and maintenance (see part 10.5). No products other than cells are assumed. If product formation occurs, an additional consumption term must be added. At steady state, equation 10 becomes

$$D(S_0 - S) = \frac{\mu X}{Y_{x/s}} \quad (11)$$

Substitution of equation 8 into equation 11 gives

$$X = Y_{x/s}(S_0 - S) \quad (12)$$

Note that equation 12 was presented earlier without derivation for use with batch systems (equation 6). The overall growth yield is assumed to be dependent on only the limiting nutrient and independent of specific growth rate. Exceptions to these assumptions are discussed

in part 10.5. Equations 8 and 12 are the steady-state equations for continuous cultivation.

A model expressing the specific growth rate as a function of substrate concentration must be assumed before biomass concentration, substrate concentration, and specific growth rate can be related to define a stable set of operating conditions. Equation 5 provides an adequate model for many situations. Substitution of equation 8 into equation 5 yields

$$D = \frac{D_c S}{K_s + S} \quad (13)$$

where D_c is the critical dilution rate corresponding to the maximum specific growth rate, μ_{max}. Operation of the chemostat at dilution rates above D_c will result in complete wash-out of the culture.

Equation 13 can be rearranged to give

$$S = \frac{DK_s}{D_c - D} \quad (14)$$

Substitution of equation 14 into equation 12 results in

$$X = Y_{x/s}\left(S_0 - \frac{DK_s}{D_c - D}\right) \quad (15)$$

Equation 15 relates the steady-state biomass concentration to the dilution rate. Figure 3 depicts the generalized system response to dilution rate predicted by equations 14 and 15.

The Monod model (equation 5) is one of several models relating growth rate and substrate concentration. The assumptions implicit with this model are quite simple and do not adequately describe all systems. Excellent discussions of the fundamental theory of continuous culture have been presented by Tempest (56) and others (2, 32).

Apparatus.

Most of the equipment described above for batch cultivation of microorganisms in stirred

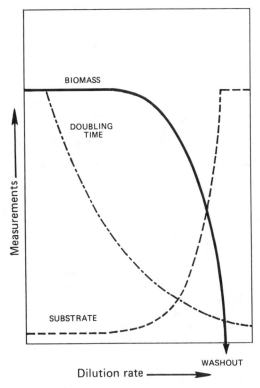

FIG. 3. *Generalized effects of changes in the dilution rate on culture variables in a chemostat system for continuous cultivation, where the dilution rate approaches μ_{max} at washout.*

tanks can be used directly for continuous cultivation. Modification of the stirred tank may be necessary to allow continuous withdrawal of culture broth; a dual pump system (one for fresh medium addition and one for culture broth withdrawal) used with a level controller can allow continuous culture operation by withdrawing culture broth through the sample port. Pump and level control systems for continuous culture are available from the major manufacturers of fermentation equipment (New Brunswick Scientific Co., The VirTis Co., and Chemapec Inc.).

Although the dual pump and level controller system for medium addition and broth withdrawal is the preferred method for continuous cultivation in multi-liter fermentors, it may be more convenient to use gravity for overflow withdrawal of culture broth in sub-liter ones. Special culture vessels manufactured with a side arm for liquid withdrawal can be created by modifying standard glass or stainless-steel vessels. Special care must be taken to provide adequate surface mixing on a level with the withdrawal port. Adequate surface mixing can be accomplished by insertion of a liquid draft tube

(21) or by installing a turbine paddle with a 5° forward pitch on the agitation shaft at the level of the withdrawal port. Bellco Glass, Inc., manufactures an all-glass continuous culture vessel with gravity overflow (catalog no. 1970). This vessel is adequate for sub-liter cultivation of low-density bacterial cultures. The manufacturers listed in Table 2 market considerably more versatile continuous culture units capable of meeting the aeration and mixing requirements of even the most dense of bacterial cultures; however, these units also require withdrawal of culture fluids by pumping.

Medium flow-rate determination.

Careful and precise control of medium flow rate into and out of the constant-volume fermentor is absolutely necessary for establishment of steady-state conditions for growth. The flow rate of fresh medium into the fermentor will determine the culture density and specific growth rate in chemostat-type continuous cultivation. The medium flow rate should be carefully set and periodically checked during the fermentation. Figure 4 illustrates a simple device for the determination of medium flow rate. Fill the calibrated pipette (or burette) with a measured volume of fluid from the reservoir by briefly closing pinch clamp B, and close pinch clamp A until ready for determination of the flow rate. Measure the flow rate by opening pinch clamp A while simultaneously closing pinch clamp C and measuring the time necessary for withdrawal of a measured volume from the

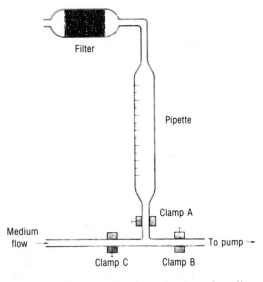

FIG. 4. *Apparatus for determination of medium flow rate.*

pipette. Open pinch clamp C and close pinch clamp A to resume normal operation. The dilution rate (or specific growth rate for single culture vessels without recycle) is then calculated by dividing the flow rate by the fluid volume in the fermentor.

Preventing back contamination of the medium reservoir.

The aeration and agitation conditions of continuous culture result in production of an aerosol containing large numbers of the cultured microorganism. As a result, the medium feed line is subject to back-contamination and must be fitted with a medium "break" tube which acts as an aseptic seal between the culture vessel and the medium reservoir. Figure 5 illustrates the basic principle of such a tube (21). The lower section provides a slow steady flow of sterile air to prevent aerosol contamination of the medium feed line while the upper chamber provides a secondary break in the liquid line. This device should be used for fed-batch cultivation as well as continuous cultivation.

Medium and product reservoirs.

Continuous cultivation requires the use of rather large quantities of medium and produces equally large volumes of product broth. Table 4 presents the liquid volume of fresh medium necessary for 20 h of continuous operation at different rates of dilution. The corresponding mass doubling times are listed for convenience. It is advisable to prepare enough medium for 20 h of operation to allow flexibility in scheduling of reservoir transfer. Since most laboratory-scale operations will require autoclaving of the medium as the means of sterilization, it will probably be most convenient to prepare the required volumes of medium in carboys of appropriate size. (*CAUTION*: Exercise extreme care in handling large volumes of hot liquids. Use autoclavable plastic carboys rather than glass ones.) For example, a 14-liter fermentor with a 10-liter liquid working volume maintained so that the bacterial culture doubles every hour (dilution rate equal to 0.69 h^{-1}) would require 138 liters of fresh medium per 20-h interval. The same culturing conditions but with a working liquid volume of 1 liter would require 13.8 liters of medium.

Small volumes of medium may be easily autoclaved and stored in standard 20-liter carboys. The preparation of medium in small volumes (10 to 14 liters) will allow adequate heat sterilization without excessive cooking. Polypropylene or polycarbonate bottles can be purchased from most of the major supply houses and are well suited to sterile medium storage. Figure 6 illus-

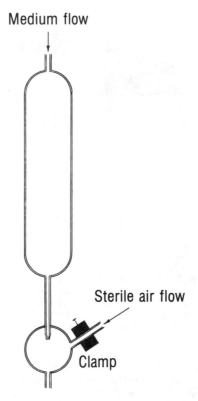

FIG. 5. *Apparatus to prevent back contamination of a medium reservoir for a continuous or fed-batch culture system.*

TABLE 4. *Medium requirements for continuous cultivation at various dilution rates*[a]

Dilution rate[b] (h^{-1})	Corresponding mass doubling time		Medium vol per fermentor vol required for 20-h continuous operation (ml/ml)
	h	min	
2.77	0.25	15	55.4
1.39	0.50	30	27.8
0.92	0.75	45	18.4
0.69	1.00	60	13.8
0.46	1.50	90	9.2
0.35	2.00	120	7.0
0.17	4.00	240	3.4
0.09	8.00	480	1.8

[a] A bacterial culture maintained in 500 ml of liquid working volume at a mass doubling rate of 30 min ($D = 1.39$ h^{-1}) will require 13.9 liters of fresh medium per 20-h period.

[b] Dilution rate calculated for a single vessel without recycle.

trates a typical reservoir configuration for upright sterilization in an autoclave. The medium volume in culture reservoirs should not exceed 75% of the total reservoir volume to minimize the chance of boil-over during autoclaving. Fur-

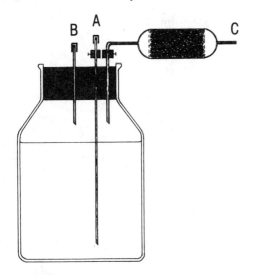

FIG. 6. *Medium reservoir for upright sterilization in an autoclave. Lines A and B are fitted with gravity caps for aseptic closure, lines A and C are fitted with pinch clamps, and line C is fitted with an air filter. During use, line A is connected to the fermentor by a "quick connector" (see Fig. 7).*

thermore, it is essential that the gas space in the reservoir be vented during autoclaving. After sterilization, several carboys can be connected in series to increase the reservoir volume without changing the reservoir; this is accomplished by connecting line A of the first reservoir to line B of the adjacent reservoir (Fig. 6).

Aseptic connections.

Among reservoirs and between the reservoirs and the fermentor, use the aseptic "quick connect" shown in Fig. 7. It consists of two sections of stainless-steel tubing sized so that one of the stainless-steel tubing sections will easily slip into the other and such that the gum rubber connecting lines fit securely over the free ends of both of the stainless tubing sections. The individual tubing sections should be wrapped with gauze or aluminum foil prior to autoclaving. Aseptic connection is made by carefully removing the aluminum foil from the male and female sections to be joined, flaming each section, and then inserting the smaller tube into the larger until the smaller-diameter tube makes a secure fit with the tubing connected to the larger diameter metal tube. The smaller metal tubing must be somewhat longer than the larger metal tubing section to allow complete penetration and seat in the rubber tubing line. The seal may be secured by wiring it in place or using a small hose clamp. The "quick connect" should be held so that the larger female tube is placed with its

open end pointing downward. The quick connect may also be rewrapped in sterile covering to prevent ingress of dust and airborne organisms between the tubes. These precautions will allow opening of the connection followed by flame sterilization and reconnection. The connector can be opened to allow changing of the reservoir by reversing the procedure, being sure to flame the smaller tubing prior to reconnection with the next sterile line.

Sterilization of medium.

Standard procedures for sterilization of liquid media are described elsewhere in this manual (Chapter 23). Filter sterilization may be necessary to prevent destruction of heat-labile medium components. However, most simple media for cultivation of bacteria can be autoclaved prior to use, but larger volumes of medium require longer autoclaving times. Vessels containing approximately 10 liters of liquid should be autoclaved at 121°C for 30 to 90 min, depending upon medium constituents. Media containing solids such as cornmeal flour or soy flour may require up to 90 min for complete sterilization of a 10-liter volume. However, media containing only dissolved components will usually be sterilized after 30 min at 121°C. All vessels should be vented during autoclaving to allow equilibration of pressure.

Take special care to ensure establishment and maintenance of the chemically reduced state when autoclaving large volumes of prereduced medium for the cultivation of anaerobic bacteria. Vented vessels containing prereduced media should be connected to an oxygen-free gas sup-

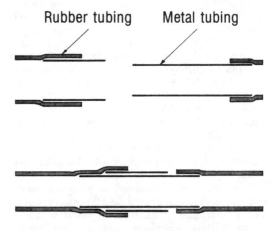

FIG. 7. *Connector for quick and aseptic connections of tubing among and between reservoirs for continuous culture systems. Top: disassembled connector parts. Bottom: assembled connector.*

ply immediately upon removal from the autoclave. As the steam in the gas head space of the vessel begins to condense, a partial vacuum will form. Sparging with oxygen-free gas will allow release of the vacuum formed during cooling without contamination of the medium by oxygen in the air.

Start-up.

Continuous cultivation is always preceded by transient batch cultivation during which time the cell mass accumulates at the expense of the substrate. Start the continuous culture while the batch cultivation is in the constant exponential phase of growth; this will minimize oscillations due to nutritional step-up and avoid inadvertent wash-out due to physiological lag. Start the dilution at a rate less than the desired operational dilution rate and then increase to the operational dilution rate within one reactor residence time; this will minimize oscillations from toxic substrates such as phenol. Chemostat response to toxic substrates is discussed in more detail by Pirt (4).

Sampling procedure.

Sampling is best accomplished with an independent sample line system. Alternatively, samples may be taken from the effluent line although this latter procedure is not recommended when small sample size is necessitated, since the lumen of the effluent tube may become coated with adherent bacteria which may break away and result in non-representative samples. Follow the fermentor manufacturer's instruction for sampling during continuous cultivation. If an external sampling system is used in place of sample collection from the effluent line, the sample size must be kept below 10% of the reactor working volume so that steady-state conditions will not be greatly disturbed. Finally, the sample line must be out-purged of its entire contents before samples are collected for analysis. Some fermentors are equipped with a sterile air backpurge which displaces the sample-line contents back into the fermentor after a sample is taken. If the latter system is used, purging prior to sample collection is not necessary.

10.2.2. Turbidostat

The turbidostat is the simplest of the well-mixed continuous cultivation systems (61). Cell concentration is monitored and maintained constant by adjusting the feed rate of fresh nutrients. In contrast to the chemostat, a biomass concentration is chosen by the operator, and the dilution rate (and therefore the substrate concentrations) is adjusted to maintain the predetermined biomass level. It follows that sub-

strate(s) need not be present in limiting amounts but, rather, will usually be present in excess. Operation is therefore most stable when the specific growth rate of the culture is near the maximum value, μ_{max}, for the particular medium in use. The system may then be operated over a wide range of biomass concentrations near the critical dilution rate so long as all medium components are in excess. These are precisely the conditions where the chemostat is least stable. Also, unlike the chemostat, the dilution rate will always equal the specific growth rate, whether or not the system is at steady state, if cell mass can be controlled with precision.

However, the precision with which the cell mass can be controlled has historically been the weakest aspect of turbidostat cultivation. Most of the older methods of monitoring population density relied on optical monitoring with some type of photoelectric sensor (scattered or transmitted light). These methods suffered from interference by foam and bubbles produced during aeration, wall growth on the fermentor and, even more critically, on the optical device. Bubble interference can be eliminated by using an external flow-through optical cell; however, foaming continues to be a problem. Wall growth on optical surfaces can be partially remedied by wiping the surface, but the many designs for this have proved cumbersome and failure prone. New techniques with fiber optics offer some promise but are mostly unproven.

The term turbidostat has come to include any technique that holds the cell density constant and includes monitoring techniques based on cellular metabolism such as the "pH-stat" (48) and "CO₂-stat." Since certain metabolic functions (such as oxygen uptake, carbon dioxide evolution, and, in some cases, pH change) are intimately linked to cell growth rate and ultimately to the specific growth rate, these parameters may be used as control variables for medium replacement.

Theoretically, any metabolic parameter that can be measured can be used as the control variable for turbidostat cultivation. Practically, only those parameters that are tightly linked to cell growth (i.e., that have small time delays in culture response) are useful control variables. The advent of sophisticated computer-sensor-fermentor control systems allows a powerful new level of use for turbidostat theory, although this is generally not recognized.

10.3. SPECIAL SYSTEMS

10.3.1. Controlled Batch Culture (pH)

Controlled batch culture requires instrumentation that will monitor an environmental parameter and trigger the addition of medium to the fermentor so that the indicator parameter is maintained at a steady value. An example of environmental parameter control in a fermentation system is the maintenance of pH at a constant value by the automatic addition of acid or base. The basic components of a pH control system are the pH electrode(s) with its shielded connector cable, the pH meter, an endpoint titrator, an acid or base reservoir with flexible tubing for connection to the fermentor, and a simple peristaltic pump or solenoid valve. Although separate reference and pH electrodes can be used, it is usually more convenient to use a steam-sterilizable combination electrode to minimize space requirements in a small laboratory-scale fermentor. All components of the fermentor system, including pH and dissolved oxygen electrodes, must be designed for steam sterilization or be compatible with other sterilization techniques, such as ethylene oxide or formaldehyde-steam sparging.

The Ingold-type electrode is particularly reliable and may be repeatedly autoclaved. The pH meter and pH titrator must be interconnected and therefore should probably be purchased from the same manufacturer to ensure equipment match. Although a peristaltic pump can allow foolproof addition of acid or base against relatively high fermentor back pressure, a simple tubing-pinch solenoid valve for control of gravity feed of acid or base to the fermentor is usually sufficient for laboratory-scale operation. Radiometer/Copenhagen manufactures a very reliable pH measurement and titration system that is easily adaptable to fermentor operation. The Radiometer pH control system and Ingold electrodes are available from The London Co. Dissolved oxygen control systems are available from all of the major fermentor manufacturers and usually allow control of both agitation and airflow rate in a sequential manner. The dissolved oxygen electrode available from Instrumentation Laboratories, Inc., is particularly reliable and can be steam sterilized many times before membrane replacement is necessary.

Further to pH control, see 6.1 and 16.2.1.

10.3.2. Fed-Batch Culture

In a batch culture, the transition from exponential growth to stationary phase may occur for a variety of reasons, including depletion of an important nutrient or build-up of a toxic metabolite in the medium. When the transition results from nutrient depletion, growth will continue if fresh medium is added. The medium addition rate and the culture volume in a closed system would have to be increased exponentially in order to maintain a constant rate of exponen-

tial culture growth. This technique for growth maintenance is called "fed-batch cultivation" (19). Periodic removal of culture broth to allow additional feeding is called "extended-batch cultivation" or "repeated fed-batch cultivation." Continuous culture differs from batch, fed-batch, or extended-batch cultivation in that a supply of medium is added at the same rate that culture is withdrawn.

The technique of fed-batch culture may be used to supply large quantities of a potentially toxic substrate while maintaining a low concentration of the substrate in the medium. Fed-batch operation of a fermentor is a compromise between ordinary batch and continuous operation. In some cases, the fed-batch mode will allow significant improvements in cell-mass or product productivity over an ordinary batch or a continuous operation. Fed-batch operation on the laboratory scale can be easily accomplished by gravity feed addition to the fermentor of primary or secondary substrates.

True fed-batch operation requires that the volume of the liquid medium in the fermentor increase during the fermentation. This requirement places an upper limit on the culture time based on feed rate and also leads to changing conditions of aeration and agitation effectiveness. Furthermore, it is possible to back-contaminate the feed reservoir from the fermentor; for this reason, it is desirable to use a medium breaker (Fig. 5).

One of the greatest utilities of fed-batch and extended-batch cultivation is the ability to control transient changes between stable growth rates. This aspect has not been extensively studied (4), but may hold the key to improved industrial processes such as antibiotic biosynthesis.

10.3.3. Synchronous Batch Culture

Biochemical events associated with specific times in the individual cell cycle can be studied in a population of cells by batch cultivation in a synchronous manner. Culture synchrony occurs when all of the cells divide at nearly the same instant. In such cases, a plot of the logarithm of cell number versus cultivation time will resemble a stairstep rather than a straight line. Although development of culture synchrony is relatively straightforward, maintenance of synchrony over long periods of time is a difficult task requiring precise control.

Culture synchrony can be achieved by periodically varying a critical environmental condition. The technique forces the synchronization of cell multiplication by interrupting, promoting, or retarding metabolic function in a cyclic manner. The population will gradually synchronize its response to these periodic disturbances of metabolic activity and will ultimately synchronize its growth pattern. The technique, however, requires severe disturbance of normal metabolic activity and is therefore of limited use in the study of growth cycle-linked bacterial physiology. One exception to this limitation is bacterial spore germination as a means of initiating culture synchrony. Step or pulse changes in the culture environment can trigger spore germination and may actually closely model natural occurrences. However, population homogeneity (100% sporulation or germination is virtually impossible to achieve) requires some form of physical preselection.

A generally useful method of synchronization is based on physical selection of a homogeneous fraction from a heterogeneous population. This approach is often termed **selection synchrony** and avoids most of the problems of metabolic disturbance during synchronization by physically selecting cells that are in similar states of the cell growth cycle.

The data of Kubitschek (34) and Poole (49) show that the cell volume of *Escherichia coli* increases linearly during the cell cycle while its cell mass increases exponentially. The observation that cell volume is smallest just after cell division suggests that centrifugation (particularly in a density gradient) might allow recovery of a population of new daughter cells from an asynchronous culture. However, the observation that volume increase is linear while mass increase is exponential means that those cells which are just ready to divide or those cells which have just divided will have the greatest cell density in spite of cell-size differences. Density centrifugation will result in cell fractions in which the most dense fraction contains both young (daughter) cells and mature (ready to divide) cells, while the least dense fraction will contain a homogeneous population of cells that have progressed through a common fraction of their cell cycles.

Selection-synchrony through density centrifugation cannot supply new daughter cells for direct study of cell cycle-linked physiology because these fractions will always be contaminated with mature cells ready to divide. However, density centrifugation can supply an adequate inoculum for synchronous culture growth. Cell cycle-linked physiology may then be studied by direct sampling of the synchronous culture. The procedure described below is presented as a general guideline for development of synchronous cultures and is based upon the technique described by Mitchison and Vincent (43). This technique may be used as an initial guide for development of a synchronous cultivation technique specifically designed for the organism in use (37).

Selection-synchrony procedure.

Delay synchronization experiments until reproducible batch cultivation conditions can be established, including determination of asynchronous culture kinetics. When these prerequisites are met, establish batch cultivation conditions so that two to five cell mass doublings during logarithmic growth will occur.

1. Prepare 500 ml of sterilized medium in several culture vessels (250-ml Erlenmeyer flasks may be convenient). Inoculate half of the flasks and incubate these under appropriate conditions. Store the remaining sterile flasks under identical conditions.

2. Harvest the cells from the batch cultivation at mid-exponential growth phase. Rapidly cool the culture to 0 to 4°C by swirling the flasks in an ice bath. Harvest the cells by refrigerated centrifugation at $10,000 \times g$ for 10 min.

3. Resuspend the sedimented cells in 2 ml of appropriate ice-cold buffer (0.1 M potassium phosphate at pH 7.0) by vigorous agitation with a Vortex mixer. If the cells tend to aggregate, use mild sonication or mild homogenization by a blender or tissue grinder to prepare a suspension of discrete cells. For cultures that are particularly difficult to suspend, use 0.01% (wt/vol) Tween 80 in the suspending buffer before mechanical or sonic treatment.

4. Rapidly, but carefully, layer the suspended cells on a sterile, precooled density gradient prepared from Ficoll, sucrose, colloidal silica sol, or other appropriate material as dictated by the cell system. To prepare an exponential sucrose gradient in a discontinuous manner, layer sterile ice-cold solutions of increasing sucrose concentration into presterilized, precooled (0°C) centrifuge tubes. For example, place 10 ml of 35% (wt/vol) sucrose in phosphate buffer into the bottom of a sterile centrifuge tube. Sequentially layer 10 ml each of sucrose-buffer solutions containing 26.5, 25.5, 24.5, 22.0, 19.0, and 15.0% (wt/vol), respectively, onto the 35% sucrose cushion. (The buffers may all be prepared from the 35% sucrose stock solution.) Filter sterilize the sucrose solutions prior to use in forming the gradient. All solutions must be ice cold and preaerated (for aerobic cells).

5. Carefully centrifuge the cell-charged, sealed centrifuge tubes in a precooled centrifuge held at $2,500 \times g$ for 15 to 20 min at 0°C. For more precise separations, adjust the time and speed of centrifugation so that the optically dense band of bacteria moves no more than two-thirds of the way down the centrifuge tube; speeds and times will vary somewhat depending

on the culture being handled.

6. Inoculate prewarmed flasks of growth medium with 0.5 ml directly from the lightest-density fraction of cells.

7. Carefully monitor the optical density of the newly inoculated culture flasks for a definite step-increase in optical density, an indication of synchronous cultivation.

Culture synchrony should be maintained for two to three cycles; study over longer periods will require re-establishment of synchrony through the procedure described above. Continuous density gradient zonal centrifugation can allow large-scale preparation of inocula for synchronous cultivation of microorganisms and offers the intriguing possibility of "continuous" synchronization through timed inoculation of a turbidimetrically controlled continuous culture (turbidostat). Figure 8 shows a typical growth curve for a selection-synchronized population of bacteria, showing decay of synchrony after two cycles.

10.3.4. Dialysis Batch and Continuous Culture (by P. Gerhardt)

Dialysis is a process for separation of solute molecules by means of their unequal diffusion

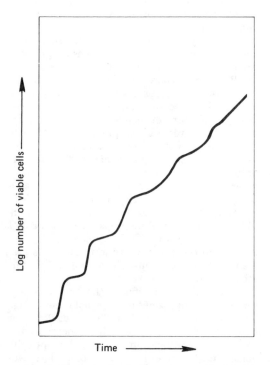

FIG. 8. *Typical growth curve for a selection-synchronized population of bacteria.*

through a semipermeable membrane because of a concentration gradient. The process is applied to the growth and maintenance of living cells by a technique called dialysis (or diffusion or perfusion) culture.

There are other processes that involve membranes (e.g., electrodialysis, ultrafiltration, microfiltration, and reverse osmosis). However, dialysis is most easily applied to the production of cells and their metabolites since membranes clog less during dialysis than during other processes. To employ the technique of dialysis culture, a membrane is introduced between a culture chamber and a dialysate reservoir. For dialysis culture to be effective, the volume of the dialysate must be large compared to that of the culture, or else the dialysate must be replenishable. Further, the permeability and area of the membrane must be sufficient to permit useful diffusion in rate and amount. Such a system of dialysis culture can be operated in vitro or in vivo and batchwise, continuously, or in a combination of these modes.

Three main types of membranes, differing in porosity, are applicable to dialysis culture. **Dialysis membranes** have a nominal pore size in the order of 10 nm in diameter, so they exclude cells and macromolecules but pass small molecules such as the nutrients for bacterial growth. **Filter membranes** (or membrane filters) have a nominal porosity in the order of 100 nm, so they also exclude cells but pass low-molecular-weight molecules. Dialysis membrane tubing in various sizes is manufactured from regenerated cellulose by the Visking process (Union Carbide Corp.). Sheets of membrane are commercially available in a wide range of porosities and materials (Gelman Sciences, Inc., Millipore Corp., Nuclepore Corp.). Factors to consider for bacteriological use include autoclavability, pore size, inertness, and permeability (which is governed by porosity, void space, and thickness). **Solution transport membranes** have no pores and pass gases because of their solubility in the membrane material itself (e.g., rubber or polycarbonate).

An interface between two physically different phases may also be used to separate the culture and dialysate (**interface dialysis culture**), such as in a solid/liquid biphasic system (9.2.2).

Specific advantages and reasons for using dialysis for bacterial culture include: (i) prolongation of the exponential growth phase in the batch cycle, which allows the attainment of high densities of viable cells; (ii) extension of the stationary phase, thus permitting increased production of metabolites associated with this phase; (iii) relief of product inhibition control by removal or dilution of metabolite products, thus enabling their greater production in batch or continuous culture; (iv) establishment of a steady-state population with mainly maintenance metabolism, thus immobilizing the cells for prolonged production of a metabolite product; (v) production of metabolites free from cells and conversely production of cells free from medium macromolecules; (vi) means to study and recover a cell population placed in an in situ or in vivo environment such as in an ecological or animal system; and (vii) the capability for study of interactions between separated populations of cells.

The history of dialysis culture can be traced back to 1896 when Metchnikoff inserted a collodion sac containing cholera bacteria into the peritoneum of a live guinea pig to learn whether the bacteria produced a diffusible toxin. The history, generalized design, mathematical theory, and practical applications of dialysis culture were brought together in a comprehensive review article by Schultz and Gerhardt in 1969 (51). Recently, Abbott (7) reviewed the fermentative uses of dialysis culture for cell immobilization (10.3.5), and Stieber (Ph.D. thesis, Michigan State University, East Lansing, 1979) provided a comprehensive updating review of the entire subject.

In vivo systems.

The term "diffusion chamber" is often used to describe a small dialysis culture unit that can be implanted within a living experimental animal (e.g., in the peritoneum or rumen or beneath the skin). Among a number of systems, the best are made with a nondegradable filter membrane sealed on both sides of an inert plastic cylinder (like a drum) through which sampling access can be provided. Bacteria introduced into diluent within the chamber grow entirely on diffusible nutrients from the host. Such systems have been used in bacteriology to study immune reactions and the growth of fastidious pathogens (e.g., *Mycobacterium leprae*, *Treponema pallidum*, and *Neisseria gonorrhoeae*).

Ex vivo systems.

The carotid-jugular blood stream of a large experimental animal (e.g., a goat) is surgically externalized and circulated across one side of a membrane, on the other side of which is contained a bacterial culture. Such an ex vivo hemodialysis culture system enables the bacteria to grow entirely on the nutrients of the blood, yet separate from the macromolecular and cellular defense mechanisms, if a dialysis type of membrane of appropriate molecular exclusion is employed. In one such system, the bacteria are contained in an ordinary laboratory fermentor

and the culture is continuously circulated through the jacket of a hollow-fiber type of artificial-kidney hemodialyzer, which is connected by tubing to the blood stream of the restrained animal (50). In another such system, the bacteria are contained in a small (3 ml) clear-plastic (polycarbonate) chamber over a membrane, on the opposite side of which is a passage for the blood stream; the entire unit is mounted on the neck of the ambulatory animal. The latter system was used successfully with many facultatively aerobic, but no obligately anaerobic, bacteria and was designed to study host-parasite reactions of fastidious pathogenic bacteria (26).

In vitro systems.

Dialysis culture is often attempted in the laboratory by suspending a membrane sac containing the culture in a tube, flask, or carboy of medium. The simplest such arrangement is to employ a length of dialysis membrane tubing that is intussuscepted so as to form a double-walled tube, in the annular space of which the culture is contained (Fig. 9).

The advantages of shake flasks (10.1.2.) are

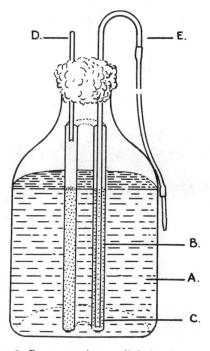

FIG. 9. *Representative sac-dialysis culture system. A tube of dialysis membrane is intussuscepted so as to form a double-walled sac (C), the annular space of which contains culture (B). The medium (A) is contained in a carboy. There also is a tube for sampling and inoculating (D) and a harvesting siphon (E). Reproduced from Sterne and Wentzel (53).*

combined with those of dialysis culture in a unit assembled from flanged Pyrex glass pipe (Fig. 10). This and other flanged flask designs (51) enable the use of sheet membrane of any type.

Scale-up of dialysis culture to laboratory and larger sizes of fermentors is possible by carrying out growth in a separate culture circuit that is connected with a separate dialysate circuit by means of an intermediate dialyzer (22). Such a dialyzer-dialysis culture system, depicted schematically in Fig. 11, can be operated batchwise in both circuits (Fig. 11a), continuously in both circuits (Fig. 11b), or in two combinations of these modes (Fig. 11c and d).

The key to effective operation of such systems is a suitably designed dialyzer. The best principle of design resembles a plate-and-frame filter press and contains molded silicone-rubber separators that promote turbulent flow of the culture and dialysate on the opposing sides of the membranes (51). Various types of artificial-kidney hemodialyzers also have been employed.

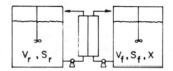

a) BATCH   RESERVOIR AND FERMENTOR

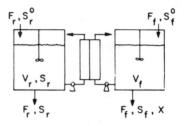

b) CONTINUOUS RESERVOIR AND FERMENTOR

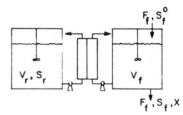

c) BATCH RESERVOIR , CONTINUOUS FERMENTOR

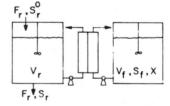

d) CONTINUOUS RESERVOIR , BATCH FERMENTOR

FIG. 11. *Four modes of operating a dialysis culture. Although a dialyzer-dialysis system is diagrammed, the principles are intrinsically applicable to any design. The symbol F is the flow rate, S is the substrate concentration, V is the volume, and X is the cell concentration into, within, and out from the reservoir and fermentor vessels. Reproduced, with revisions, from Schultz and Gerhardt (51).*

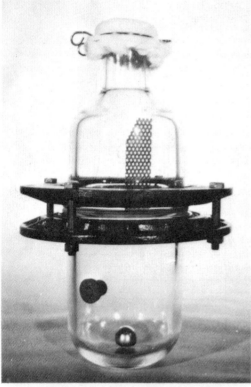

FIG. 10. *Dialysis culture flask, designed to be held on a carriage mounted on a rotary shaking machine. The bottom compartment is filled with sterile medium which is stirred by the rotating ball. The top compartment holds the culture, which is turbulently aerated by swirling and baffling. Reproduced from Gerhardt and Gallup (25).*

The design of the system can be optimized for a given culture situation. For a situation in which growth is limited by a toxic metabolite product (e.g., lactic acid), a completely continuous system is operated most effectively as shown schematically in Fig. 12. The substrate (S_f^o) in relatively high concentration is fed directly into the fermentor and the usual reservoir vessel for dialysate is eliminated. Instead, the dialysate circuit consists only of the tubing, pump, and dialysate side of the dialyzer and contains a relatively small volume (V_d). Only water is fed into

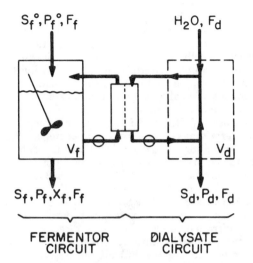

FIG. 12. *Schematic of fermentor-fed dialysis continuous culture system in which a stream of water is used in the dialysate circuit to maximize the withdrawal of a toxic metabolite product. Reproduced from Coulman et al. (15).*

the dialysate circuit at a relatively high flow rate (F_d). This system provides the greatest concentration gradient possible for dialysis. This operational mode is illustrated by Stieber et al. (54) to maximize the conversion of concentrated (25%) whey lactose into lactic acid.

Mathematical modeling and computer simulation.

The foregoing developments in system design and practical uses of dialysis continuous culture (fermentation) also exemplify the power of mathematical modeling and computer simulation for predicting the results to be expected in a bacterial fermentation process. Laboratory experiments need be conducted only to validate the theoretical predictions by using a relatively limited number of changes in experimental conditions at preselected critical points. The experimental results in turn are used to establish growth constants and to indicate the need for additional terms in the equations. By this process of successive theoretical prediction and experimental validation, the model becomes increasingly accurate and useful to predict fermentation process management.

An example of such a combined theoretical and experimental approach to a bacterial process is provided by the results of Stieber et al. (15, 54, 55) for the ammonium-lactate fermentation. A set of material-balance and rate-relationship equations is developed for substrate, product, and cell mass in the fermentor and dialysate circuits. These equations are com-

bined, the variables are defined in dimensionless parameters, and the time derivatives are set at zero to obtain a generalized solution for the steady state which consists of five quadratic equations. These are programmed on a digital computer using selected real values for the various terms. A typical comparison of simulated predictions with experimental results (Fig. 13) shows the close correlation that can be attained.

Such a methodology of theoretical prediction and experimental validation is commonly used in the field of industrial fermentations but much less so in other fields of bacteriology. This powerful method could be much more widely employed in the science, e.g., to predict the outcome of infectious diseases.

10.3.5. Immobilized-Cell Reactors (by P. Gerhardt)

In most bacterial culture methods, maximum multiplication of cells is the objective, and the

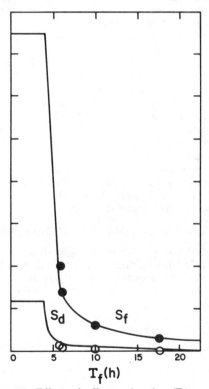

FIG. 13. *Effects of cell retention time (T_f) on residual lactose in the fermentor circuit (S_f) and the dialysate circuit (S_d) during dialysis continuous fermentation. The curves were plotted by computer from the mathematical model, and the points were plotted from experimental data to demonstrate the close fit between the experimental results and the simulated predictions. Reproduced from Stieber and Gerhardt (55).*

medium nutrients are mainly consumed for growth. In an industrial fermentation process, however, maximum production of a metabolite compound is the objective, and the medium nutrients should be mainly consumed for products rather than growth. During the past decade, a great deal of research has been directed toward the use of bacterial enzymes or cells that are "immobilized" within a reactor in order to maximize the conversion of a substrate into a useful product.

Cell immobilization is considered by Abbott (6) "as a physical confinement or localization of a microorganism that permits the economical reuse of the microorganisms." This admittedly operational definition can be replaced by a physiological one describing a reactor that maximizes maintenance metabolism and minimizes growth metabolism of a microbial population.

The principle is exemplified by the old "quick vinegar" process in which acetic acid bacteria are immobilized as a colonial film on a column of wood shavings, through which an ethanol solution is trickled downward while air passes upward, yielding an acetic acid solution. Once the bacterial film is established, little further growth occurs, and the bacterial maintenance energy is derived from the oxidation of ethanol to acetate. The process can be operated continuously over periods of weeks or months.

Physical adherence or chemical bonding to a solid carrier remains as one of the general methods for cell immobilization. Cell pellets or flocs are formed in other methods.

Detailed description of methods for immobilized-cell reactors is perhaps beyond the scope of this book, and the reader is referred to recent reviews on the subject (6, 7, 13).

10.4. HARVESTING AND CLEANING

The choice of a process for separation of bacterial cells from culture broth will be dictated by the ultimate disposition of the fractions. If cells are to be used for further study, cake filtration may not be desirable owing to contamination of the cells by a filter-aid material. Similarly, recovery of the culture broth for further study by batch centrifugation may require discarding a portion of the broth to avoid resuspension of the cell pellet during removal of the supernatant.

10.4.1. Filtration

Cake filtration.

Two types of filtration are usually encountered in recovery of bacterial cells from culture broths: cake and membrane filtration. The former utilizes a filtration medium that is highly porous. Often, the filtration medium is a com-

posite of a filter support (often filter paper or cloth of relatively large pore diameter) with a relatively thick overlay of suspendible filter aid such as cellulose, diatomaceous earth, kieselguhr, or other material that can be freely dispersed in aqueous media to form a pad whose porosity is small enough to retain bacteria (57). The use of filter aid increases the effective surface area of a filter and increases the amount of bacteria that can be removed by a single filter. Filter aid may be added directly to the culture broth or prelayered on the filter support. The practice of adding filter aid, such as cellulose, to the culture broth can allow filtration of as much as 10 times the culture broth that can be treated by prelayering the same quantity of cellulose on the filtration support. However, direct addition of filter aid to the culture broth prior to filtration greatly increases the difficulty of recovering the microbial cells for later study.

To assemble a simple, but effective, cake filtration system, fit a Büchner funnel to a standard filtration flask with a side arm. Place a piece of highly porous filter paper, such as Whatman no. 1, on the funnel so that the paper covers the holes in the funnel plate. Cover the filter paper with a layer of calcified diatomaceous earth 2 to 5 mm deep by suspending the appropriate amount of filter aid in water and gently, but uniformly, pouring this suspension onto the filter paper while vacuum is applied. Since the thick filter aid has uniform filtration characteristics, gently scrape the cake surface during filtration to remove bacteria as they accumulate; this allows prolonged filtration without having to change the filter cake.

Cake filtration of bacterial cultures can remove most of the organisms from the culture broth, leaving a highly clarified filtrate. However, this technique is not a reliable method of filter sterilizing liquid media. Cells recovered from the filtration cake will usually be contaminated with large amounts of filter aid which may interfere with further studies on the cells themselves.

Membrane filtration.

Unlike cake filtration, membrane filtration does not rely upon the depth or tortuosity of pores in the filtration medium for physical removal of particles from the fluid. Rather, pore diameter at the surface of the membrane is controlled to within very narrow tolerances. The technique of membrane filtration is based on exclusion of the bacteria at the surface of the filtration medium. Pressure drop across the membrane is often high, and practical operation of a membrane filtration system requires pressurized feed to the filter. Membrane filtration apparatus and membranes of various pore sizes are commercially available (Millipore Corp., Nuclepore Corp., Gelman Sciences, Inc.).

The choice of a specific membrane will depend upon the culture characteristics and ultimate recovery of cell or filtrate fractions. Since an average bacterium is roughly 1.5 μm by 0.5 μm, absolute removal of bacteria for the purpose of filter sterilization will require use of a microporous membrane filter whose mean pore diameter is less than 0.5 μm. Often, membranes of 0.22 $\pm$ 0.02 μm are used in filter-sterilization procedures. Although this type of membrane will be adequate for most sterilization procedures, bacteria are not rigid and therefore could theoretically pass. It follows that solutions which have been filter sterilized through 0.22-μm membranes should be periodically checked to ensure sterility.

The use of microporous membrane filters with a mean pore diameter of 0.45 $\pm$ 0.04 μm will generally allow rapid and efficient filtration of bacterial cultures but will not generally ensure sterility of the filtrate. Furthermore, filtration membranes and apparatus for filter sterilization must be sterilizable by autoclaving or other procedures. Pressure filtration with nitrogen as the pressurizing gas is the preferred method of harvesting bacterial cells and preparing sterile solutions by filtration. Pressure filtration with nitrogen minimizes or eliminates oxidation of the filtrate by maintaining positive pressure with an inert gas within the filtration device. If cells are to be aseptically recovered, the nitrogen gas must be filter sterilized. Specific operational pressures and equipment size must be determined by consultation with the manufacturers.

Additional reading on cake and membrane filtration can be found in references 47, 58, and 60.

10.4.2. Centrifugation

Recovery of bacterial cells by centrifugation usually results in less complete separation of the culture broth and the cells than can be achieved by filtration. However, this technique allows recovery of broth and cells free from contamination by extraneous materials such as filter aids and is the method most often used for laboratory-scale harvesting. Centrifugation is not an adequate method for preparation of sterile solutions, however.

The properties which govern the rate of settling of cells during centrifugation are the viscosity of the fluid through which centrifugal settling occurs, the size of the particles (diameter), and the density difference between the suspending fluid and the cells. Although these factors interact to dictate specific centrifugation

times for pellet formation, most bacteria can be removed from relatively nonviscous culture broths (such as nutrient broth) by centrifugation at 10,000 $\times$ g for 10 to 15 min. Since the packing characteristics of bacteria are dependent upon cell shape and cultivation conditions, it may be necessary to adjust the centrifugal force (rotational speed of the rotor) or centrifugation time so that the bacterial pellet will be less subject to resuspension during removal of the centrifuged tubes and withdrawal of the supernatant.

Centrifugation equipment and operation.

Laboratory centrifuges are available for either batch or continuous operation, and often a single device can be equipped for both types of operation. Many rotor designs are available, but all operate on the same principle, i.e., controlled application of centrifugal force through rotation of the centrifuge head. Since the average particle diameter of a bacterial culture is quite small, centrifugal recovery of cells and broth will usually require high-speed rotation. High-speed centrifuges inevitably generate large amounts of heat and should therefore be equipped with a self-contained refrigeration system or be amenable to operation in a cold room. Furthermore, since centrifuge heads are available in many sizes, the user should be aware that centrifugal force (usually expressed as multiples of the gravitational force, g) together with centrifugation time should be used rather than rotational speed and time in designing and reporting centrifugation procedures. Specific details of rotor speed and equivalent centrifugal force are available from the major manufacturers (DuPont Instruments, Ivan Sorvall, Inc., International Equipment Co., Beckman Instruments, Inc.).

Large volumes of culture medium may be treated by continuous centrifugation. The DuPont-Sorvall refrigerated centrifuges may be equipped with a special rotor and delivery system to convert them to continuous operation, which is quite adequate for most bacterial cultures but will not adequately treat highly flocculated or fungal cultures, which tend to plug the feed and distributor lines.

The Sharples centrifuges have long narrow spindle-like rotors through which the culture is passed in an upward flow. The high speed of spindle rotation causes the cells to settle on the walls of the centrifuge vessel. This type of centrifuge offers very rapid processing capability and the potential for differential centrifugation of particles of different size or density. Models are available with either finite microbial sludge capacity or continuous microbial sludge discharge.

Procedure.

An example of a centrifugation procedure is illustrated below. The procedure allows aseptic preparation of washed cells. The procedure may be simplified if asepsis is not required.

1. Prepare an appropriate diluent for washing of the cells recovered by centrifugation (25.1). Prepare enough diluent so that four to six times the culture volume is available during the washing step. Be sure that the ionic strength and pH of the diluent closely approximate those of the culture medium. Sterilize the washing diluent and cool it to 0 to 4°C prior to use.

2. Precool the appropriate centrifuge rotor by storage in a refrigerator. *CAUTION*: Do not place a room-temperature rotor onto the centrifuge spindle and then allow the centrifuge refrigeration system to cool the head. If this is done, the centrifuge rotor will contract and seize the spindle, making it very difficult to remove at a later time.

3. Sterilize several capped centrifuge tubes (polypropylene and polycarbonate tubes are autoclavable) and cool before use. Wrap aluminum foil around the cap and lip of each centrifuge tube prior to autoclaving to maintain sterility of the centrifuge tube lip during handling.

4. Determine the volume of the centrifuge tubes in use and aseptically transfer two-thirds of that volume of bacterial culture into each centrifuge tube. Be sure that the pairs of centrifuge tubes weigh exactly the same. If the last centrifuge tube contains less than the others, balance it against a separate centrifuge tube containing enough distilled water to allow the balance of tube weights. Be sure that the sterile caps are securely in place on each centrifuge tube.

5. Place the centrifuge tubes into the rotor so that each tube is placed in the position opposite its balanced partner.

6. Carry out centrifugation under appropriate conditions following the manufacturer's procedures. Centrifugation at 10,000 × g for 10 to 15 min will suffice for most bacterial cultures. Some centrifuge systems have a special protective rotor head plate that must be secured in place prior to centrifugation. Removal of this head plate after the rotor has come to a gentle stop presents the potential to jar the rotor, an action that can resuspend the bacterial pellet. The head plate should be removed carefully while bracing the rotor to prevent rotor movement on the spindle.

7. Carefully remove each centrifuge tube, noting the position of the bacterial pellet at the bottom of the tube. Carry out aseptic transfer of the supernatant to a sterile reservoir by decanting or pipetting. It will probably be impossible to prevent partial resuspension of the bacterial pellet during this procedure. Bacterial resuspension (contamination of the supernatant) can be minimized by allowing some of the supernatant to remain in the centrifuge tube with the cells.

8. Aseptically fill each centrifuge tube containing a bacterial pellet with two-thirds of the tube volume of precooled sterile washing diluent. Replace and secure the centrifuge caps and resuspend the bacterial pellets by agitation on a Vortex mixer. Repeat steps 4 through 6 (substituting washing diluent for culture) for a minimum of four washings.

9. After the last wash water has been aseptically removed, resuspend the bacteria in a small amount of sterile diluent. This suspension may then be used for further study.

The number of washings required in either the filtration or centrifugation technique will depend upon the culture conditions, volume of wash water, and volume of wet cell mass. If the wet cell mass occupies roughly one-tenth of the volume of the total culture suspension, four washings with volumes equal to the original culture suspension volume will result in approximately 99.99% removal of soluble contaminants. It is important that multiple washes rather than a single wash whose volume equals that of the total multiple washes be used since each sequential washing reduces contaminant level by roughly a factor of 10 for the conditions described above.

10.5. GROWTH-YIELD CALCULATIONS

The concept of growth yield is developed from a material balance of a stable cultivation system. Monod originally defined the yield constant, Y, in terms of mass units:

$$Y = \frac{\text{grams of cells formed}}{\text{grams of substrate consumed}}$$

The yield on substrate is often indicated by $Y_{x/s}$, indicating a ratio of the respective concentration. The convention of mass ratios for expressing yield of cells on the basis of carbon substrate consumption is fairly well accepted (10, 42). Yields can be easily calculated by determining the production of dry cell mass over a given time period of the cultivation and dividing by the mass of carbon substrate consumed during the same time period.

If yield is to be expressed on the basis of oxygen consumed or adenosine triphosphate (ATP) produced during growth, the cell yield can be defined as follows:

$$Y_{x/\text{ATP}} = \frac{\text{grams of dry cells formed}}{\text{moles of ATP formed}}$$

$$Y_{x/O_2} = \frac{\text{grams of dry cells formed}}{\text{moles of oxygen consumed}}$$

Under many conditions, the yield with respect to ATP synthesis or oxygen consumption is relatively constant for many organisms. For example, the yield of cells per mole of ATP synthesized under conditions of energy substrate limitation and fairly rapid growth rate is approximately 10 ± 2 g of dry cells formed per mol of ATP.

However, the yield of cells per mole of ATP is not constant for all bacteria and is generally variable if the energy substrate does not limit growth or if the growth rate of the culture is significantly slower than the maximum culture growth rate. Occasionally, carbon and energy sources will be consumed not only for cellular growth but also for product formation. Furthermore, energy substrates are involved in cell maintenance as well as cell growth. Since cells can use energy substrates for endogenous respiration without growth, maintenance energy requirements will contribute to the consumption of energy substrates without a concomitant increase in dry cell weight. Although the influence of endogenous metabolism on the calculation of cell growth yield is minor when the culture growth rate is near its maximum value with the energy substrate as the limiting nutrient, consumption of energy substrates for maintenance can be quite significant at growth rates less than the maximum or under conditions of extreme biophysical growth environment. Since meaningful calculation of yield values requires careful and precise control of the cultivation conditions, yield data are best obtained from chemostat systems. In this way, the growth-limiting substrate and the specific rate of growth, μ, can be specified and controlled.

Calculation of the overall growth yield from continuous culture data assumes steady-state substrate and cell concentration and, therefore, is most easily accomplished in a chemostat. The medium for chemostat operation must be designed so that a single nutrient limits the rate of cell growth. This can be easily accomplished by ensuring that all nutrients, other than the one on which yields will be based, are in excess. An insight into this design of medium can be obtained from an elemental analysis of the bacteria to be grown in the chemostat. The medium must supply all of the components of the cell and can therefore be based upon elemental analysis. The ultimate level of growth will be determined by the concentration of the limiting nutrient. Since cell concentration in a chemostat will automatically adjust to match the rate of limiting nutrient addition, the ratios of medium components are more important than their absolute amounts. Media for determination of overall growth yield should therefore be designed such that the ratios of elements supplied by the medium to that element chosen as the limiting nutrient are considerably greater than predicted by taking the ratio of elements from elemental analysis data.

The calculations described below require the establishment of steady-state conditions in terms of cell concentration and substrate concentration. Make sure that this condition is met by monitoring both cell concentration and substrate concentration over a period of time long enough to allow displacement of one reactor volume of medium. Whenever conditions are changed (such as increasing or decreasing the rate of medium addition to a chemostat), the operator must allow sufficient time for a return to steady state. This can be ensured by allowing at least four turnovers of fermentor liquid volume. Since a continuous reactor operates on a dilution principle, a step or pulse change in operating conditions would cause deviation from the preexistent steady state which would approach a new steady-state value at a rate predicted by the dilution rate of the culture. That is, four mean residence times would be necessary for a close asymptotic approach to the new steady-state equilibrium value. For example, a 2-liter chemostat in which the rate of medium addition and withdrawal is 0.5 liter per h is operating at a dilution rate of $F/V = D = 0.25$ h^{-1}. The amount of time necessary for four turnovers of medium volume is calculated by dividing four by the dilution rate. In this case, the reactor medium volume will have been completely replaced four times after 16 h (4/0.25 h^{-1}). That is, at least 16 h must elapse after changing the growth conditions of a continuous culture prior to achievement of steady-state conditions.

In reality, the amount of time necessary to obtain a close asymptotic approach to the new steady-state value may be dependent upon the way in which operating conditions have been changed. If the change in operating conditions involves a nutritional step-down, for instance, four mean residence times should be adequate. If, however, the change involves a nutritional step-up under conditions of rapid growth, the system may require more than four mean residence times to achieve essential steady state. In the latter case the rapidly growing cells may require elaboration of new enzymes for metabolism of the added nutrients.

Once steady-state continuous cultivation is achieved, the overall growth yield is determined as shown below:

$$Y_{x/s} = \frac{(X - X_0)}{(S_0 - S)}, \frac{\text{g of dry cells produced}}{\text{g of substrate used}}$$

where X is grams of dry cells per liter of effluent medium, X_0 is grams of dry cells per liter of influent medium (usually zero), S is grams of substrate per liter of effluent medium, and S_0 is grams of substrate per influent medium. If the medium added to the fermentor is sterile, the value of X_0 is zero and the yield becomes $Y_{x/s} = X/(S_0 - S)$.

Determine the dry cell weight by quantitatively filtering a precisely measured volume of culture effluent through a preweighed 0.22-μm membrane filter. Wash the collected cells with buffer or distilled water to remove excess medium. Dry the membrane filter and cells to constant weight in an oven whose temperature is not greater than 105°C. Lower oven temperatures may be required to prevent extensive browning and therefore weight loss of the sample. *CAUTION:* Many filters lose weight to a variable extent after washing and drying. Therefore, several samples should be filtered and weighed, and several filters should be washed, dried, and weighed as controls. It is more accurate to centrifuge and wash cells as described above (part 10.4.2) from a larger volume of medium, resuspend in a small volume of water, dry, and weigh (see part 25.2.2).

Overall growth yield is related to maintenance and growth requirements for limiting substrates which act as energy sources according to the following equation:

$$\frac{1}{Y_{x/s}} = \frac{m}{\mu} + \frac{1}{Y_G}$$

where μ is specific growth rate (hours^{-1}), m is specific rate of substrate uptake for cellular maintenance (hours^{-1}), $Y_{x/s}$ is overall yield, and Y_G is growth specific yield. The values of m and Y_G can be estimated by plotting $1/Y_{x/s}$ versus $1/\mu$ on rectangular graph paper. (Remember that $\mu = D$ for simple continuous culture without cell recycle.) If the data form a straight-line relationship, the intercept will be the value of $1/Y_G$ and the slope of the line will be the value of m.

Although yields are conventionally expressed in terms of mass of cells formed per mass or moles of substrate used, this mixture of terms is often confusing. Herbert (30) has suggested that yields be expressed in terms of gram-atoms of cellular carbon formed per gram-atom of element consumed from the limiting substrate. Adoption of this convention would standardize reporting procedures and eliminate much of the confusion in interpreting yield data. Expression of growth in terms of gram-atoms of cellular carbon formed does not require a determination of the complete elementary composition of the cells. Rather, only the carbon content of the cells need be determined to allow use of this convention. Table 5 presents a summary of terms related to yield calculation.

TABLE 5. *Summary of terms for growth-yield calculations*

Symbol[a]	Definition
$Y_{x/s}$, Y_s, Y	Overall cellular yield coefficient, $Y_{x/s}$ = grams of cell formed per gram of nutrient consumed
$Y_{x/m}$	Molar yield coefficient, $Y_{x/m}$ = grams of cell formed per mole of nutrient consumed
$Y_{x/s(MAX)}$, Y_{MAX}	Theoretical or maximum yield coefficient, Y_{MAX} = grams of cell formed per gram of nutrient consumed
Y_G	Growth specific yield coefficient, Y_G = grams of cell formed per gram of nutrient consumed for the production of cells alone
Y_{x/O_2}, Y_{O_2}	Overall cellular yield coefficient, Y_{x/O_2} = grams of cell formed per mole of oxygen (O_2) consumed
$Y_{x/ATP}$, Y_{ATP}	Overall cellular yield coefficient, $Y_{x/ATP}$ = grams of product chemical formed per mole of ATP formed
$Y_{p/s}$, Y_p	Overall product yield coefficient, $Y_{p/s}$ = grams of product chemical produced per gram of substrate consumed
$Y_{p/x}$	Specific product yield coefficient, $Y_{p/x}$ = grams of product chemical produced per gram of cell formed
$Y_{p/s(MAX)}$, $Y_{p(MAX)}$	Theoretical or maximum product conversion yield coefficient, $Y_{p/s(MAX)}$ = grams of product per gram of substrate converted to product
μ	Specific growth rate (grams of cells formed per grams of cells per hour)
m	Maintenance coefficient (grams of substrate consumed per gram per cell·hours

[a] The symbol listed first is the preferred designation.

10.6. COMMERCIAL SOURCES

Aaron Equipment Co., Div. Areco Inc., 9301 W. Bernice St., Schiller Park, IL 60176

Air Resources, Inc., 800 E. Northwest Highway, Palatine, IL 60067

Alloy Crafts Co., Subsidiary of Lox Equipment Co., RR # Box 2C-A, Delphi, IN 46923

Artisan Industries, Inc., 73 Pond St., Waltham, MA 02154

B. Braun Melsungen AG, P.O.B. 346, D-3508 Melsungen, West Germany

Beckman Instruments, Inc., 2500 Harbor Blvd., Fullerton, CA 92634

Bellco Glass, Inc., 340 Edrudo Rd., Vineland, NJ 08360

Brighton Corp., 11861 Mosteller Rd., Cincinnati, OH 45241

Chemapec, Inc., 230-C Crossway Park Dr., Woodbury, NY 11797

Chemineer Agitators, P.O. Box 1123, Dayton, OH 45401

Cole-Palmer Instrument Co., 7425 N. Oak Park Ave., Chicago, IL 60648

Crepaco, Inc., 8303 W. Higgins Rd., Chicago, IL 60631

Dollinger Corp., P.O. Box 23200, Rochester, NY 14692

Dow Corning Corp., P.O. Box 1767, Midland, MI 48640

DuPont Co., Instrument Products, Scientific and Process Div., Concord Plaza, Quillen Bldg., Wilmington, DE 19898

Fisher Scientific Co., 711 Forbes Ave., Pittsburgh, PA 15219

Gelman Sciences, Inc., 600 S. Wagner Rd., Ann Arbor, MI 48106

General Electric Co. (Silicone Products Dept.), Waterford, NY 12188

Glitsch Inc., 4900 Singleton Blvd., P.O. Box 226227, Dallas, TX 75266

Infrared Industries, Inc., Box 989, Santa Barbara, CA 93102

Koch Engineering Co., Inc., 4111 S. 37 St. N., Wichita, KS 67220

Leeds and Northrup Co., Sumneytown Pike, North Wales, PA 19454

LKB Instruments, Inc., 12221 Parklawn Dr., Rockville, MD 20852

The London Co., 811 Sharon Dr., Cleveland, OH 44145

Manostat Corp., 519 Eighth Ave., New York, NY 10018

L. E. Marubushi, Co., Ltd., Koki Bldg., 8-7, 2-Chome, Kaji-cho, Chiyoda, Tokyo, Japan

Michigan Dynamics, Division of AMBAC Industries Inc., 32400 Ford Rd., Garden City, MI 48135

Millipore Corp., Ashby Rd., Bedford, MA 01730

Mine Safety Appliances Co., 600 Penn Center Blvd., Pittsburgh, PA 15235

Mixing Equipment Co., Inc., Unit of General Signal, 177 Mt. Read Blvd., Rochester, NY 14603

New Brunswick Scientific Co., Inc., 44 Talmadge Rd., P.O. Box 986, Edison, NJ 08811

Nuclepore Corp., 7035 Commerce Circle, Pleasanton, CA 94566

Perkin-Elmer Corp., Maine Ave., Norwalk, CT 06856

Perry Products Corp., Mount Laurel Rd., Hanesport, NJ 08036

The Pfaudler Co., The Sybron Corp., 1000 West Ave., Rochester, NY 14603

Scientific Products, Division of American Hospital Supply, 1430 Waukegan Rd., McGaw Park, IL 60085

Sensorlabs, Division of Instrumentation Laboratories, Inc., 113 Hartwell, Lexington, MA 02173

Sharples-Stokes, Division of Pennwalt Corp., 955 Mearens Rd., Warminster, PA 18974

Taylor Instrument Co., 95 Ames St., Rochester, NY 14601

Trane Thermal Co., 250 Brook Rd., Conshohocken, PA 19428

Union Carbide Corp., 270 Park Ave., New York, NY 10017

The VirTis Co., Route 208, Gardiner, NY 12525

10.7. LITERATURE CITED

10.7..1 General References

1. **Aiba, S., A. E. Humphrey, and N. F. Millis.** 1973. Biochemical engineering, 2nd ed. Academic Press, Inc., New York.
   This text develops the fundamental bases for many of the cultivation and harvesting techniques presented here.

2. **Bailey, J. E., and D. F. Ollis.** 1977. Biochemical engineering fundamentals. McGraw-Hill Book Co., New York.
   This text presents an excellent discussion of microbial growth kinetics and can be easily read by microbiologists with little or no engineering background.

3. **Holdeman, L. V., E. P. Cato, and W. E. C. Moore.** 1977. Anaerobe laboratory manual, 4th ed. Virginia Polytechnic Institute and State University, Blacksburg.
   This is an extraordinarily detailed manual for laboratory study of anaerobic bacteriology and is hailed by many as the definitive source of anaerobic laboratory procedure.

4. **Pirt, J. S.** 1975. Principles of microbe and cell cultivation. John Wiley & Sons, Inc., New York
   This text is one of the best works available on the theory of cell cultivation.

5. **Seeley, H. W., and P. J. Van Demark.** 1972. Microbes in action, 2nd ed. W. H. Freeman and Co., San Francisco.

10.7.2 Specific Articles

6. **Abbott, B. J.** 1977. Annu. Rep. Ferm. Proc. **1:** 205–233.

7. **Abbott, B. J.** 1978. Annu. Rep. Ferm. Proc. **2:**91–124.

8. **Abbott, B. J., A. I. Laskin, and C. J. McCoy.** 1974. Appl. Microbiol. **28:**58–63.

9. **Borkowski, J. D., and M. J. Johnson.** 1967. Biotechnol. Bioeng. **9:**635–639.

10. **Boyles, D. T.** 1978. Biotechnol. Bioeng. **20:**1101–1104.

11. **Bryant, J.** 1970. *In* J. R. Norris and D. W. Ribbons (ed.), Methods in microbiology, vol. 2, p. 187–203. Academic Press, Inc., New York.

12. **Cameron, J., and E. I. Godfrey.** 1969. Biotechnol. Bioeng. **11:**967–985.

13. **Chibata, I.** 1977. Adv. Appl. Microbiol. **22:**1–27.

14. **Collins, E. B., and A. W. Tillion.** 1976. J. Dairy Sci. **60:**387–393.
15. **Coulman, G. A., R. W. Stieber, and P. Gerhardt.** 1977. Appl. Environ. Microbiol. **34:**725–732.
16. **Dietrich, G. G., R. J. Watson, and G. J. Silverman.** 1972. Appl. Microbiol. **24:**561–566.
17. **Difco.** 1953. Difco manual of dehydrated culture media and reagents for microbiological and clinical laboratory procedures, 9th ed. Difco Laboratories, Detroit, Mich.
18. **Dowell, V. R., and T. M. Hawkins.** 1974. Laboratory methods in anaerobic bacteriology. Center for Disease Control Laboratory Manual. U.S. Government Printing Office, Washington, D.C.
19. **Dunn, I. J., and M. R. Mor.** 1975. Biotechnol. Bioeng. **18:**1805–1822.
20. **Elsworth, R.** 1972. Chem. Eng. (London) **258:**63–71.
21. **Evans, C. G. T., D. Herbert, and D. W. Tempest.** 1970. *In* J. R. Norris and D. W. Ribbons (ed.), Methods in microbiology, vol. 2, p. 277–327. Academic Press, Inc., New York.
22. **Evans, J. B.** 1975. J. Gen. Microbiol. **91:**188–190.
23. **Freedman, D.** 1970. *In* J. R. Norris and D. W. Ribbons (ed.), Methods in microbiology, vol. 2, p. 175–185. Academic Press, Inc., New York.
24. **Gallup, D. M., and P. Gerhardt.** 1963. Appl. Microbiol. **11:**506–512.
25. **Gerhardt, P., and D. M. Gallup.** 1963. J. Bacteriol. **86:**919–929.
26. **Gerhardt, P., J. M. Quarles, T. C. Beaman, and R. C. Belding.** 1977. J. Infect. Dis. **135:**42–50.
27. **Grant, G. A., Y. W. Han, and A. W. Anderson.** 1978. Appl. Environ. Microbiol. **35:**549–553.
28. **Hall, M. J., and C. Ratledge.** 1977. Appl. Environ. Microbiol. **33:**577–584.
29. **Harrison, D. E. F.** 1972. J. Appl. Chem. Biotechnol. **22:**417–400.
30. **Herbert, D.** 1976. *In* A. C. R. Dean, D. C. Ellwood, C. G. T. Evans, and J. Melling (ed.), Continuous culture 6: applications and new fields, p. 1–30. Ellis Horwood Ltd., Chichester.
31. **Hishinuma, K., Y. Iida, K. Kobayashi, and S. Ikeda.** 1972. *In* G. Terui (ed.), Proceedings of the 4th International Fermentation Symposium: Fermentation Technology Today, p. 155–161.
32. **Humphrey, A. E.** 1977. *In* J. J. McKetta and W. A. Cunningham (ed.), Encyclopedia of chemical process design, vol. 4, p. 359–394. Marcel Dekker, Inc., New York.
33. **Khaergaard, L.** 1978. Biotechnol. Bioeng. **20:**1691–1694.
34. **Kubitschek, H. E.** 1968. Biophys. J. **8:**792–804.
35. **Lien, H. C., B. J. Chem, and C. C. Creagan.** 1977. Biotechnol. Bioeng. **19:**425–433.
36. **Linek, V. M. Sobotka, and A. Prokop.** 1973. Biotechnol. Bioeng. Symp. **4:**429–453.
37. **Lloyd, D. L., John C. Edwards, and A. H. Chagla.** 1975. J. Gen. Microbiol. **88:**153–158.
38. **Luedeking, R.** 1976. *In* N. Blackebrough (ed.), Biochemical and biological engineering science, vol. 1, p. 181–243. Academic Press, Inc., New York.
39. **Margaritis, A., and J. Zajic.** 1978. Biotechnol. Bioeng. **20:**939–1001.
40. **Martin, D. F., E. C. Kutt, and Y. S. Kim.** 1974. Environ. Lett. **7:**39–46.
41. **Melling, J.** 1976. *In* A. Wiseman (ed.), Topics in enzyme and fermentation biotechnology, p. 10–39. Ellis Horwood Ltd. (John Wiley & Sons Inc., distributors), Chichester.
42. **Mickelson, M. N.** 1971. J. Bacteriol. **190:**96–105.
43. **Mitchison, J. M., and W. S. Vincent.** 1965. Nature (London) **205:**987–989.
44. **Monod, J.** 1942. Recherches sur le croissance des cultures bacteriennes, 2nd ed. Hermann, Paris.
45. **Monod, J.** 1949. Annu. Rev. Microbiol. **3:**371–394.
46. **Monod, J.** 1950. Ann. Inst. Pasteur Paris **79:**390–410.
47. **Mulvany, J. G.** 1969. *In* J. R. Norris and D. W. Ribbons (ed.), Methods in microbiology, vol 1, p. 205–253. Academic Press, Inc., New York.
48. **Oltmann, L. F., G. S. Schoenmaker, W. N. M. Reunders, and A. H. Stauthamer.** 1978. Biotechnol. Bioeng. **20:**921–925.
49. **Poole, R. K.** 1977. J. Gen. Microbiol. **98:**177–186.
50. **Quarles, J. M., R. C. Belding, T. C. Beaman, and P. Gerhardt.** 1974. Infect. Immun. **9:**550–558.
51. **Schultz, J. S., and P. Gerhardt.** 1969. Bacteriol. Rev. **33:**1–47.
52. **Scott, D. C., and C. W. Hancher.** 1976. Biotechnol. Bioeng. **18:**1393–1403.
53. **Sterne, M., and I. M. Wentzel.** 1950. J. Immunol. **65:**175–183.
54. **Stieber, R. W., G. A. Coulman, and P. Gerhardt.** 1977. Appl. Environ. Microbiol. **34:**733–739.
55. **Stieber, R. W., and P. Gerhardt.** 1979. Appl. Environ. Microbiol. **37:**487–495.
56. **Tempest, D. W.** 1970. *In* J. R. Norris and D. W. Ribbons (ed.), Methods in microbiology, vol. 2, p. 259–276. Academic Press, Inc., New York.
57. **Thomson, R. O., and W. H. Foster.** 1970. *In* J. R. Norris and D. W. Ribbons (ed.), Methods in microbiology, vol. 2, p. 377–405. Academic Press, Inc., New York.
58. **Van Suijdam, J. C., N. W. F. Kossen, and A. C. Joha.** 1978. Biotechnol. Bioeng. **20:**1695–1709.
59. **Wang, D. I. C., and A. J. Sinskey.** 1970. Adv. Appl. Microbiol. **12:**121–152.
60. **Watson, T. G.** 1969. J. Gen. Microbiol. **59:**83–89.
61. **Watson, T. G.** 1972. J. Appl. Chem. Biotechnol. **22:**229–243.

Chapter 11

Growth Measurement

ARTHUR L. KOCH

In order to measure bacterial growth, a precise definition is needed. Probably the most basic definition of growth is based on the ability to multiply the number of individuals, i.e., to initiate and complete cell division. This definition implies monitoring the increase in total number of discrete bacterial particles. There are two basic ways to do this: by microscopic enumeration of the particles in a fixed volume, and by electronic enumeration of the particles passing through an orifice.

Since some cells may be dead or dying, this definition of growth may be different from one based on the detection of viable organisms. In the long run, the increase in the number of organisms capable of indefinite growth is the only important consideration. This is the reason that colony counting and most probable number methods of measurement are so important.

Viable counting methods, which seem so natural to a bacteriologist, are really quite special in that cultures are diluted so that individual organisms cannot interact. For example, these methods cannot in principle be applied to sexually reproductive organisms requiring male-female interaction. Even when applied to procar-

yotes, there are many special restrictions and limitations.

Macromolecular synthesis and increased capability for synthesis of cell components is the obvious basis for the measurement of growth by the bacterial physiologist, the biochemist, and the molecular biologist. From their point of view, cell division is an essential but minor process that seldom limits growth and what usually limits growth is the ability of enzymatic systems to rapidly utilize resources to form biomass.

Balanced growth.

These several viewpoints on growth become synonymous under a single circumstance: that is, when growth is balanced (13). An asynchronous culture can be said to be in balanced growth when every extensive property increases exponentially with time. "Extensive property" is a term from physical chemistry and refers to those properties of the system that increase when there is more of it in the system. Thus, mass and cell number per culture of fixed volume are extensive properties of a culture, but temperature or ratios, such as deoxyribonucleic acid (DNA) per cell, are not. The application of this

179

physical-chemical principle to bacteriology lies in the thought that, if a culture is grown for a long enough duration where growth is sparse enough not to alter the environment significantly, sooner or later the bacteria will achieve the same balance and adjustment to this constant environment no matter what their state was initially. Once balanced growth has been achieved, if the conditions remain the same, the culture will remain in balanced growth indefinitely (until it is altered by mutation and selection). The above is dogma. Nobody could possible have measured everything about a culture, particularly when that culture is maintained at infinite dilution. But it is not so difficult to study cultures that are substantially in balanced growth, if we substitute less inclusive criteria.

Imagine a culture in balanced growth. Imagine any extensive property such as biomass, DNA, ribonucleic acid (RNA), or surface area per milliliter of culture, and let us call this property X. The rate of formation of X will be proportional to the amount of biomass and therefore of X; thus:

$$\frac{dX}{dt} = \lambda X$$

where λ is a proportionality constant. Through the laws of calculus, this equation can be integrated and a boundary condition ($X = X_0$ when $t = 0$) can be imposed, yielding:

$$X = X_0 e^{\lambda t}$$

That is, for every substance X, the increase of X will be exponential in time. If the ratio of one substance to every other is to remain constant, the same proportionality constant must apply for every choice of X. This common value of λ is called the **specific growth rate**. Not only will any substance chosen to be X increase exponentially, but any combination of substances or the rate of change of any substance will also increase exponentially with the same specific growth rate, λ. Practically, this allows the rate of oxygen uptake or metabolite production to be used as an index of growth.

Intensive properties such as λ and the ratio of the concentrations of different substances will necessarily remain constant under balanced growth conditions. In addition, the cell size distribution will stay constant. An average cell will have a constant rate of carrying out every cellular process, and newly arisen daughter cells will have a constant probability of being able to form a colony (the percentage of nonviable cells will remain constant). Therefore, under these special conditions, no matter what measure of growth is used (whether it is particle counting,

colony formation, or chemical determination of a cell substance), the same specific growth rate will be obtained and that rate will be constant through time.

Such halcyon conditions in principle occur when bacteria are cultivated under long-term continuous culture. Balanced growth should arise both for growth limited by the rate of nutrient addition under chemostat conditions and for unlimited growth under turbidostat conditions where the culture is mechanically diluted to maintain a constant amount of biomass. It is also believed to be achieved by repeated dilution under batch growth (39, 49).

Variation at different growth rates.

Organisms respond to environmental conditions, both physical and chemical, by altering their own composition. These changes have been well documented for certain enteric bacteria (31, 39), but may occur with all procaryotes. In general, favorable growth conditions mean faster growth which requires a higher concentration of ribosomes and associated proteins. In terms of gross composition, this means a higher RNA content. Also, under favorable growth conditions the cells can lay down reserve material such as glycogen and poly-β-hydroxybutyric acid. The cells also become bigger. Teleologically, this is because they do not need as high a surface-to-volume ratio to pump poorer, scarcer, or fewer kinds of resources into the cell. These changes in composition lead to the possible pitfall of falsely relating one measure of growth to another. This can only be done when growth occurs in a single medium under conditions where the composition is not altered by the bacteria themselves.

Unbalanced growth.

During the culture cycle resulting from diluting a stationary culture into fresh medium, the properties of organisms change drastically. Once again, this is well documented for certain enteric bacteria (1, 11) but probably applies throughout the procaryotic kingdom. In some sense, the culture cycle is an artifact and depends on the medium and the age and condition of the inoculum, but in all cases the cycle has relevance to growth measurements. The phenomena exhibited during the cycle are particularly relevant in ecological studies, where the conditions under which the organisms grow are important and to a large degree uncontrollable. The changes in characteristics are also involved in response to the fluctuations in natural conditions (31, 34). Briefly, when a stationary culture is diluted into rich medium, macromolecular synthesis accelerates. The components of the protein-synthesizing system (i.e., ribosomal proteins, and ribosomal RNA) are made first. Only after considerable macromolecular growth has ensued does cell division take place. During this phase, the average size of cells increases greatly. When the capacity of the medium to support rapid growth is exceeded, cellular processes continue more slowly and result in small, RNA-deficient cells. Finally, the bacteria die; that is, there are a decreasing number of colony-forming units (16).

Probably most workers using the techniques from this manual will employ intentional perturbations of growth. These perturbations will result from nutritional shifts or deficiencies, the use of growth inhibitors (particularly antibiotics), and the effects of radiations and other extreme physical conditions (e.g., high or low temperature and osmotic pressure). The effect on growth can be quite complex, and the worker must be both cautious and critical.

Pitfalls.

There are four classes of pitfalls. The first major pitfall is the general tendency toward clumping and filamentous growth under mildly toxic conditions for bacteria that ordinarily divide regularly. The second pitfall is the variability of viability of injured bacteria under different culture conditions. Repair processes may permit the recovery of viable cells under some, but not other, conditions (8). A third major pitfall is the possible development of resistant stages. Bacteria well known to form resistant stages pose no problem since the controls and measurements to correct for such forms are well known. But when resistant stages are not suspected, error can arise. The fourth pitfall pertains to the way in which the inoculum is exposed to the new environment. Different results may be obtained if the concentration of an agent is raised gradually, if it is raised discontinuously, or if a high concentration is temporarily presented and then removed or lowered. The cell concentration at the time of challenge can be critical. Bacteria may have special, sometimes inducible or sometimes unknown, ways to protect themselves against toxic agents. The protection may be dependent on the number of organisms cooperating in detoxification. Rapid growth itself can provide protection against slowly penetrating toxic agents.

The above series of phenomena can be illuminated with a single example from work on the interaction of rifampin and *Escherichia coli* (36). Doses of rifampin can be detoxified by a sufficient number of cells, and therefore higher doses of drug can be tolerated when the cells have previously been gradually exposed to the antibiotic. Independent of this phenomenon, fast-growing bacteria are more resistant than slow-growing ones. This observation can largely be explained by the shorter time available for the drug to penetrate relative to the growth rate of more rapidly growing bacteria. For example, a dose of antibiotic that is subsequently diluted to a low dose blocks growth under conditions where the low dose by itself would not inhibit growth. Thus, when growth is slowed by the brief high dose, the inhibition remains permanent because now sufficient time is available to allow penetration of enough drug to maintain the blockage of growth.

Mycelial habit of growth.

The evaluation of mycelial growth is much more difficult than is that of "well-behaved" organisms that engage in binary fission and then promptly separate. The problems and methods for colonial growth were discussed by Calam (12) and Koch (33) and will be briefly dealt with here. Filtering and weighing filamentous cells, with or without drying, is much more convenient than with smaller nonfilamentous cells. In addition, one can follow the increase in size of a mycelial colony of filamentous cells. In the extreme case, a meter-long tube containing a layer of nutrient agar is inoculated at one end, and the mycelial mass grows along the tube and may even be continued into another tube, and then another, indefinitely (47). Such colonial growth is one dimensional, whereas growth of colonies on agar surfaces is two dimensional. However, many mycelial bacteria grow in shake cultures as three-dimensional colonies.

The rate of increase of size of the colony in one, two, or three dimensions depends on the rate of elongation of the terminal hyphae that happen to be growing perpendicular to the surface of the colony, but the mobilization of resources into the mycelial mass depends on the surface area of the colony. Therefore, with shake cultures particularly, the results depend on the nature and size of inoculated fragments; with large fragments, growth becomes limited by diffusion of nutrients at an earlier stage into the mycelial mass. Thus, the major pitfall with the mycelial habit of growth is that the growth is likely to deviate from exponential growth and depend on the geometry of growth and the nature of the inoculum. In shake cultures, the apparent growth may depend on the shaking speed, which affects the tendency of the mycelia to break into smaller pieces.

Cell differentiation.

The change of enteric bacteria from large, RNA-rich forms to small, RNA-poor forms on the entry to stationary phase has many of the aspects of differentiation. Bacteriology, however, has much clearer examples of cell differentiation in the cases of transition of rod to coccus and of vegetative bacillus to endospore, and in the formation of exospores, cysts, and buds. The tendency to form filaments in certain circumstances can also be considered a differentiation. These changes and their reversal pose potential pitfalls for all approaches to growth measurement.

Adsorption on surfaces.

Many bacteria can adapt or can mutate to achieve a high avidity for surfaces, including glass. Many experiments with chemostat culture have failed to achieve their primary goal because the organisms adhere to the vessel walls. It is for these reasons that plastic or Teflon should be used whenever there is long-term contact of the organism with a culture vessel (31). Other approaches to minimizing the effects of growth on vessel walls include the use of fluorocarbons, the use of large culture volumes in large containers, the employment of violent agitation, and the frequent subculture of the bacteria into fresh glassware. In addition, the use of detergents, silicone coatings, and a high-ionic-strength medium can to some degree alleviate or lessen this problem.

This problem is particularly pertinent during dilution of the culture for measurement by highly sensitive means, such as microscopic and plate counts. It is, therefore, given further consideration in the discussion of those methods (11.1.1, 11.2.1). Also see part 25.1.

Heterogeneous environment.

If the problem at hand is to measure growth in natural conditions, tremendous difficulties must be overcome (9). In nature, growth almost always involves mixed cultures with bacteria attached to each other or to particles of dirt, leaves, or fronds. Four kinds of approaches have been used to deal with the measurement problem in these cases. First, $^{14}CO_2$ fixation has been employed (53) to monitor biomass increase. Second, tritiated thymidine autoradiography has been used to identify cells engaged in DNA replication (10). Third, antibodies tagged with fluorescent labels have been used to scan soil samples and identify particular organisms (48). Fourth, sophisticated mass spectrometry of the gases emanating from natural soil samples has been used to provide information about the growth of organisms within the sample (41). Further discussion is beyond the scope of this manual.

Principles of bacterial growth are also discussed in the preceding chapter (Chapter 10) and in references 1–7.

11.1. DIRECT COUNTS

11.1.1. Microscopic Enumeration

Microscopic enumeration is a commonly used technique that is quick and cheap and uses equipment readily available in the bacteriological laboratory. However, it is subject to gross errors. These problems were overcome to a large degree in the work of Norris and Powell (42), but the specialized techniques and apparatus required have prevented wide use of their procedures.

The major difficulty in direct microscopic enumeration is the reproducibility of filling the chamber with fluid (42). In the technique recommended below, the thickness of the fluid filling is measured with a good microscope by focusing on bacteria attached to the top and bottom interfaces and measuring the distance between them with the micrometer scale on the microscope. The horizontal dimensions between the scribe marks in commercial chambers are quite accurate (42) and cause no problem.

The second major problem is the adsorption of bacteria on the surfaces of glassware, including pipettes. This problem can be decreased by carrying out dilutions in high-ionic-strength medium instead of water, or by using a solution of formaldehyde that has been neutralized with K_2HPO_4 together with a trace of anionic detergent (42). The advantage of formaldehyde is that it stops growth and motility. An advantage of the K_2HPO_4-detergent solution is that it prevents aggregation; a possible disadvantage is that the detergent may lyse certain organisms even though it is only used in minute amounts. Alternatively, one can use plastic containers and plastic pipette tips for the dilution; to a large degree, this also prevents the loss by adsorption. In the procedure described below 0.1 N HCl is used as the final diluent. The sample is quickly delivered to a glass counting chamber with a plastic-tipped pipette, and the cells attach to the surfaces of the counting chamber.

The Hawksley counting chamber (A70 Helber; Hawksley, Ltd., Lansing, England) is recommended in preference to the Petroff-Hausser counting chamber (Arthur H. Thomas Co., Philadelphia, Pa.). The former's prime advantage is that its optical path with an ordinary cover slip is short enough so that the chamber can be used under an oil-immersion objective. While most counts are done under a high-dry objective, it is sometimes necessary to use the oil-immersion one either because there are too many cells or because they tend to clump. The Hawksley chamber consists of a slide with a circular well 20 μm deep. In the bottom of the well, scribe marks define areas of 50 μm by 50 μm.

One major advantage of microscopic examination is that one can gain additional information about the size and morphology of the objects counted. Oil immersion makes counting more tedious, but critical distinctions can be made. Alternative procedures are supplied by the manufacturers of chambers and in numerous laboratory manuals of microbiology courses. Additional discussion was presented by Meynell and Meynell (5) and Postgate (46). For a discussion of dilution techniques and statistical considerations, see part 11.5.

Procedure.

1. Clean the chamber and the cover slip with water containing a small amount of anionic detergent; rinse with water and then alcohol, blot, and let air dry.

2. Make preliminary estimation of the concentration of cells. Proceed to the next step if the concentration is less than 3×10^8 cells per ml. Otherwise, make a primary dilution of the cell suspension in Norris-Powell diluent prepared as follows. Add 5 ml of Formalin (40% formaldehyde) to 1 liter of water. Adjust to pH 7.2 to 7.4 (indicator paper is sufficiently accurate) by adding solid disodium hydrogen phosphate; the amount of the phosphate needed will vary with the amount of formic acid in the Formalin. Add a little sodium dodecyl sulfate (a small spatula tip full is usually sufficient), and repeat until bubbles do not break immediately when air is passed through the solution with a Pasteur pipette.

3. Carry out a final single dilution of the sample in a ratio of at least 1:1 with 0.1 N HCl. This kills the bacteria and gives their surface a net positive charge so they will not aggregate but will adsorb onto glass.

4. Immediately fill the Hawksley chamber with approximately 5 μl of the diluted sample, using a Pipetteman, Eppendorf, Centaur, or other plastic-tipped pipette. Let the chamber rest in an inverted position for 1 to 2 min, right it, and allow it to rest for an additional 3 to 5 min.

5. Examine with a good phase-contrast microscope under a high-dry or oil-immersion objective. Most of the bacteria will have attached to the bottom interface; a few cells will not have attached and will exhibit Brownian motion, and a few cells will be attached on the top interface.

6. Focus on those organisms that have attached on the bottom of the chamber, and read the markings on the dial of the focusing knob. On many microscopes of high quality, this dial reads directly in micrometers. Next, focus on organisms on the upper interface. Note down the distance between the top and bottom, and augment the difference by one bacterial diameter. This will quite accurately measure the chamber thickness, which is nominally 20 μm. When gaining familiarity with the technique, make depth measurements in several well-separated regions of the chamber to find how uniform the depth of the filling is.

7. Count the cells lying within small squares. Optimally, the number in each small square should be in the range of 5 to 15. Score the cells that cross the boundaries of a square if they are on the upper or right side but not if they are on the lower or left side. A hand tally counter is convenient to count the cells within a square. A second hand tally is convenient to keep track of the number of squares. Some workers use a hand tally to count the tens and mentally count the digits. At least 600 total organisms should be counted for accurate work (see below), but this need not be done with a single filling. Some authors recommend multiple fillings of the chamber and thus averaging the variability of the fillings. However, this is not necessary with the method described here, because the thickness of the filling is measured each time. It is best to count squares chosen in a systematic fashion, such as the four corner squares and the major diagonal squares. This prevents counting the same square twice and averages a possible geometric gradient of cells in the chamber.

8. Calculate the number of cells by the following formula:

$$\frac{\text{total bacteria counted}}{\begin{array}{c}\text{number of small squares counted}\\ \times \text{ thickness (in } \mu\text{m)}\end{array}} \times \text{dilution factor} \times 4 \times 10^8$$

If the cover slip is precisely positioned, the volume of the solution on top of a small square is $50 \times 50 \times 20$ $\mu\text{m}^3 = 5 \times 10^4$ $\mu\text{m}^3 = 5 \times 10^{-8}$ ml. The reciprocal of this, 2×10^7/ml, is the usual factor in the formula quoted in the instructions supplied by the manufacturer. The formula given above reduces to this, if the thickness of the chamber is 20 μm. For the procedure to be successful, only a few cells need be attached to each surface for the thickness measurement. It is most convenient, however, if most cells are on one surface for counting. Then the final act of counting a small square is focusing through the suspension to count the cells not attached and the ones on the interface that has fewer cells. One then tallies the square, refocuses on the original surface, and moves on.

11.1.2. Electronic Enumeration

The Coulter Counter (Coulter Electronics, Hialeah, Fla.), its commercial competitors, and particularly the laboratory-built versions have been important in the development of bacteriology in the last quarter century. Such instruments are practical and routinely used in clinical hematology. They are also very useful in enumeration of nonfilamentous yeasts and protozoa, but are not applicable to mycelial or filamentous growth. Although use of these counters has led to important conceptions in bacteriology, the technique is very difficult to apply in a valid way to small rod-shaped bacteria. Attempts to improve and validate the technique for use in bacteriology have been at the research level and have involved people with backgrounds in physics or engineering. Evidence of its difficulties is the fact that many of the people who helped develop the technique no longer use it. This manual, therefore, can only present the principles, mention the difficulties and the attempted solutions to these difficulties, and direct the reader to published literature. Then the reader will be able to consider the applicability of the technique to a specific bacteriological problem.

The principle of electronic enumeration is as follows. The diluted bacterial suspension is forced to flow through a very small orifice connecting two fluid compartments. Electrodes in each measure the electrical resistance of the system. Even though the medium conducts electricity readily, the orifice is so small that its electrical resistance is very high whereas the electrical resistance of the rest of the electrical path is negligible by comparison. When a particle is carried through the orifice, the resistance further increases since the conductivity of bacteria is less than that of the medium. This change in resistance is sensed by a measuring circuit and converted into a voltage or current pulse, which is then shaped electronically. The pulses may be counted by an electronics circuit similar to that used in counting radioactivity, with very small pulses eliminated by a discriminator circuit. The high pulses may be eliminated by an upper discriminator, if it is thought that they are due to dirt or other irrelevant particles. In advanced models, the pulses may be analyzed by size and stored in a multichannel analyzer; later, the data may be recovered and plotted in a histogram, and the numbers, mean size, and standard deviation may be calculated. All the data may be collected and the discrimination against pulses that are too high or too low may be carried out as the data are analyzed. The instruments need some method of forcing an accurately known volume through the orifice during the counting period. This is usually done by displacing the fluid in contact with a mercury column past triggering electrodes that conduct effectively through the mercury but not through the diluent medium.

There are three major problem areas. First, some bacteria are very small (less than $0.4 \ \mu m^3$), and the resistance pulses produced as they pass through the orifice are comparable to the noise generated by the turbulence that develops in the fluid flowing through the orifice. One can set the discriminator dial on the instrument to reject the turbulence noise, but one loses sample information as well, particularly about newly divided cells. One can run blanks and subtract the blank values, but blanks are particularly variable for the size of resistance pulses corresponding to small cell sizes. In addition, patterns of turbulence can establish themselves, remain for a while, and then be replaced with other patterns. Finally, the overall error increases when the blank has statistical variation (see 11.5.5). There is no problem if one only considers bacteria growing in rich medium where even newly formed cells are larger than the pulses produced by the turbulence or if one restricts interest to the relatively few large bacteria such as *Azotobacter agilis*, which are several times the size of *E. coli*.

The second major difficulty results from the failure of a cell to separate promptly from its sister after it is formed by cell division. This, and the tendency to form filaments, can be minimized by careful choice of the organism and the conditions. Although such behavior is rare, it is clear that the choice of *E. coli* for physiological and genetic studies has been influenced to a large degree by the relatively small extent to which it aggregates, forms chains and filaments, or remains attached to its sister cell. One can use various physical techniques such as mild ultrasound treatment or vigorous blending in a Vortex mixer to try to disperse the aggregates, break up filaments, and separate sister-sister pairs. However, the Coulter Counter is no better than the oil-immersion microscope in providing evidence that all the particles are single cells and that cell destruction is negligible.

There is a related problem of coincidence (see 11.5.5), i.e., the passage of more than one particle through the orifice in a short enough time so that a single larger particle is registered by the electronics. This problem can be dealt with (see below) by varying the dilution so that the probability of coincidence is altered. This is an especially vexing problem when cell size distributions are being measured.

The third major problem with the use of the

Coulter Counter for bacteriology is the clogging of the orifice. The resistance change is a smaller proportion of the total resistance across the orifice when the orifice diameter is larger. Consequently, for small rod-shaped bacteria such as *E. coli*, orifices with diameters in the range of 12 to 30 μm must be used. The exact choice is determined by the trade off of increased signal against increased noise and chance of becoming clogged. One way to prevent clogging is by ultrafiltration of all reagents. Alternatively, the diluent can be allowed to settle for a long time in a siphon bottle so that the particulate-free solution can be withdrawn. Choosing solvents that do not tend to generate particulate matter can also be of help. Kubitschek (38) recommended 0.1 N HCl for this reason; it is entirely volatile and does not leach materials out of the glass which later may form precipitates. But the problems in practice are severe and become worse in some of the modifications needed to size the cells more accurately.

The Coulter Counter has influenced the study of bacterial growth almost more because of its ability in principle to measure the cell size distribution of bacterial cultures than because of its ability to enumerate them. In fact, it is very difficult to measure cell size accurately because of the nature of the resistance pulse generated by a particle passing through an orifice. Attempts have been made to overcome this difficulty by using a relatively long pore (100 μm) (37). While this partially solves the physical problem, the flow slows and increases the chance of clogging. A second approach involves special hydrodynamic focusing of solutions in such a way that the bacterial cells very nearly pass down the center of the pore surrounded by fluid containing no particles (50, 52, 55).

Additional information about this technique can be found in references 2, 17, and 38 and in the instruction manuals of the instruments.

11.2. COLONY COUNTS

Bacteriology really became an experimental science when Herr Dr. Robert Koch listened to Frau Fannie Hesse and developed the agar plate. This allowed not only the cloning of pure strains but also the enumeration of colonies arising from individual "viable cells." Many variations have been used: (i) **pour plates**, in which the sample is pipetted into the empty sterile petri dish, molten but cool (45°C) agar is poured on top, and the contents are swirled and then allowed to harden; (ii) **spread plates**, in which a small volume of diluted culture (0.1 ml) is pipetted directly on solidified agar in a petri plate, and then the culture is distributed with a glass or Teflon spreader; (iii) **thin-layer plates**, for which the diluted culture is added to a tube containing a small volume (2.5 to 3.5 ml) of molten soft (0.6 to 0.75%) agar, and this is poured into a sterile agar plate and the overlay is allowed to harden; (iv) **layered plates**, which are like the thin-layer plates except that an additional layer of sterile agar is poured or pipetted onto the newly congealed agar so that all colonies are subsurface; and (v) **membrane filter methods**, in which the diluted culture is filtered onto an appropriate membrane filter (carefully prewashed) and then placed on an agar nutrient plate or onto blotter pads with concentrated nutrient medium.

The pour-plate methods have variations in which the cells are grown in roller tubes or microtubes (44) and are examined with microscopes when the colonies are small (45). There are many individual variations of techniques, sometimes resulting from historical accidents, but also the result of the special bacteriological circumstances. Automation of colony counting has put additional special restrictions on techniques, but allows petri plates to be counted rapidly without operator error.

Two methods will be described. The first is the spread plate in which all colonies are surface colonies. It is chosen for presentation since surface colonies are required to produce the proper color responses with many indicator agars. In many cases, different colors are given from subsurface colonies because the oxygenation is different, and, therefore, the acid production and reducing potential are different than on the surface. The second method is the layered plate. It is very useful because all colonies are subsurface and, therefore, much smaller and compact. They can be intensely colored. Many more colonies may be present and yet the coincidence by fusion of colonies is small; this means that several thousand colonies per plate can be used to give meaningful results. This approach is recommended because the main difficulty with usual colony counting methods is the lack of dynamic range. Rules have been issued that between 30 and 300 colonies are required. The lower limit is set by statistical accuracy, and the upper limit is set by coincidence limitations. This 10-fold range is inconvenient for many purposes, because in many cases one cannot guess the number within a factor of 10 when one carries out the dilution. The extra care needed to prepare the overlay plates is justified because it allows one to count in the larger range of 30 to 2,000. A second major advantage is flexibility for nutrient supplementation. Minimum agar can be used to pour many petri plates for indefinite storage.

Stock supplies of the minimum soft agar can be kept on hand. Then 10- to 50-fold excesses of needed special nutrients can be added to the aliquots used for the molten top agar. In some cases dyes and chromogenic substrates, to allow screening of the colonies, can be added as needed to the soft agar.

Several articles and books have been devoted to attempts to speed and automate growth measurement (2, 23). This is a field that is in so much flux that further discussion is not pertinent here.

Certain organisms are very sensitive to substances present in the agar. Meynell and Meynell (5) presented an excellent discussion of these problems. Injured organisms may have additional special requirements, and the entire 26th symposium of the Society for General Microbiology (19) was devoted to these problems. It remains for the reader to decide whether to use conditions that favor high or low survival. Genetically defective organisms pose their own individualistic problems that can be research problems on their own, e.g., the ability of various repair mutants to form countable colonies (14).

The problem of quantitating the number of organisms in cultures of strict anaerobes is dealt with in references 15, 24, and 25.

11.2.1. Spread Plates

Prepare suitable agar medium in ordinary glass or plastic 9-cm petri plates. Various procedures can be adopted to pour and dry plates. The following procedure is recommended:

1. Place the covered container with molten agar, after removal from the autoclave and after any addition of thermolabile substances (sugars, dyes, antibiotics, chromogenic substance), into a dishpan containing hot tap water (45 to 50°C). The agar cools quickly but uniformly throughout the vessel and reaches a temperature plateau above the solidification point of the agar. For 1,000 ml of agar, 30 min is sufficient. With this procedure there is no gelling around the edges of the container. The agar is now sufficiently cooled that little condensation forms on the undersurface of the petri dish lid.

2. Pour the petri plates with a lighted Bunsen burner nearby. If bubbles occur on the agar surface, the flame can be momentarily directed downward on them until they burst. With 9-cm petri plates anywhere from 15 to 20 ml of agar is satisfactory. This corresponds to a thickness ranging from 0.24 to 0.32 cm. The thicker the plates are poured, the less contrast is made by the colonies against their background. Too thin plates may result in small colonies, in a reduced number of colonies over the plate, or in thin spots, because some nutrient is limiting or the plate becomes locally dried.

3. Dry the plates at room temperature overnight or for 24 h, depending on the relative humidity.

4. Store indefinitely at 4°C in closed plastic containers.

The spreading procedure is as follows:

1. Prepare a suitable dilution of the culture based on all information and hunches at your disposal. Calculate to get 100 to 200 colonies, but use the results of plates containing between 30 and 300. If the organism forms only small colonies, up to 500 may be counted. Make the dilutions with a solution that does not favor adsorption to glass, if ordinary glassware and pipette are to be used. High ionic strength, pH between 4 and 5, and the presence of small amounts of anionic detergents (if not toxic) are helpful in this regard. Alternatively, deal with presterilized plastic vessels and pipette tips for the Pipetteman-type device.

2. The actual dilutions can be carried out in many ways. Historically, large volumes of diluent (99 ml) and 1-ml samples were used. As pipette and volumetric apparatuses have been improved, smaller volumes have been employed, economizing on reagents and mixing time. Modern plastic-tipped semiautomatic pipettes allow very small samples of bacterial culture to be used, but then the problem is proper sterilization. For many purposes, it is suggested that only 0.1-ml serological pipettes be used to deliver a 0.1-ml volume of culture to 0.9 or 4.9 ml of diluent contained in 13 by 100 mm tubes or 9.9 ml of diluent in 16 by 150 mm tubes. The dilution factors are about optimum, for ease of making an accurate dilution and ease of mixing adequately.

3. Pipette 0.1 ml of the final dilution on the agar surface of the petri plate. Form two or three free-falling drops on the surface; then blow the remaining fluid on the surface. If using a pipetting device, push to the second stop of the pipettor. The cells may have a tendency to become immediately attached in situ, so do not delay.

4. Sterilize a spreader (Fig. 1) by dipping it in alcohol, shaking off the excess alcohol, and flaming. A spreader prepared from a Giant Gem paper clip (Nestling) is recommended. Bend it out, being careful to keep the longest straight part of the wire unbent. Then place a length of no. 16 shrinkable Teflon tubing (Small Parts, Inc., Miami, Fla.) over the straight part, and gently heat to shrink it onto the wire. The part serving as a handle can be bent to your convenience. This spreader has the advantage of low heat capacity, cools quickly compared with solid glass rods, and has much less affinity for bacterial cells. Also recommended is one that can be quickly made from a Pasteur pipette.

FIG. 1. *Spreaders. The one at the top is bent from a paper clip and fitted into a length of Teflon tubing. The one at the bottom is made by fusing the end of a Pasteur pipette and then bending it in two places to define the spreading surface. The handles of both are bent to the convenience of the user. Both spreaders have low heat capacity. The spreader at the top absorbs fewer bacteria because of its Teflon construction.*

5. Spread the plate. Try to achieve a uniform coverage as close to the edges as possible. The major difficulty with this method is learning the technique for uniform distribution of the culture. Therefore, after the plates have been incubated, examine the distribution of colonies on all of the plates to learn how uniform your spreading technique has become. Those that have a larger number of colonies are especially useful in this regard, even if they have too many colonies to count. Also note whether the number of colonies is higher in the vicinity of the original droplets. If so, then the technique must be improved. If drops of water remain on the lid of the petri dish, a vigorous shake can remove the water (to the floor). The plates should be dry enough so that the 0.1 ml delivered from the pipette is absorbed in 15 to 20 s by the agar.

6. Incubate the inverted plates in a constant-temperature room or chamber whose temperature regulation is good (6.3). Incubating in closed containers is advantageous, but do not overfill the containers. If so done, an increased time is needed for the temperature of the plates to equilibrate to the temperature of the incubator.

This will be especially important when utilizing temperature-sensitive mutants. Storing in closed containers also avoids the effects of any noxious gases which may be present in general-purpose constant-temperature rooms (e.g., acetic acid fumes from gel destaining). Opaque containers protect against inactivation of colored drugs and dyes by keeping out light. Finally, in closed containers the plates do not dry out and can be used to look for slow- or late-developing colonies. Another point of concern has to do with CO_2. Even organisms not usually isolated in a high CO_2 atmosphere may have a CO_2 requirement. This may be particularly evident when single cells are spread at high dilution and incubated in normal laboratory air. Some of these considerations may be of small importance in any particular case, but should be kept in mind.

7. Observe plates before they have fully developed, mature colonies. Many times, one can see that too many colonies are developing. It may be possible to make fresh dilutions, or to count the very small colonies under a dissecting scope, to obtain reliable information with less coincidence correction.

8. Count the colonies. Depending on the circumstance, various types of illumination are advantageous and may not be obtainable with the commercial colony counters. Experiment with various types of magnifying glasses that can be worn or clipped onto your own glasses. Also try various lamps that have a magnifying lens as an integral part. The colonies may be enumerated by marking the bottom of the petri plate with a pen, or they may be counted by hand with an electronic counter, hand tally, or television-based scanning equipment. One technique that clearly marks the individual colonies is to stick the point of a colored pencil into the colony. Not all brands of colored pencils transfer color to the agar; Eberhard Faber Mongol colored pencils do this well, but some colors work better than others (e.g., French Green no. 898). Several colors can be used for differential counting. One can speed the counting process by using a hand tally to record the tens of colonies and mentally keeping track of the units. If using the electronic scanning counters, careful attention must be made to be sure that false-positive counts are not registered due to dirt and imperfections in the plastic or glass petri dishes (which would not cause difficulty to a human eye). Dyes can be incorporated into the agar to increase the contrast needed to avoid these errors during the electronic scan.

11.2.2. Layered Plates

1. Prepare the petri plates with a base of agar medium. The usual medium containing 1.5%

agar and a base approximately 0.4 cm thick are satisfactory. Pour the plates on a level surface. Check your work area with a level very carefully; if not true, use any shim material (metal strips, wood strips, pieces of paper) to level a piece of plate glass, and pour the three layers of agar while the plate rests on this level surface.

Prepare small test tubes (13 by 100 mm) with 2.5 to 3 ml of soft agar (0.7%). It is convenient to prepare stock bottles with this strength agar in your basal medium. Melt the agar in these bottles. (A microwave oven is the quick way, but experience is necessary to find the settings that melt but do not explosively boil the agar.) Additions of nutrients, dyes, and inhibitors are made at this point, and then aliquots are pipetted into Wasserman tubes (13 by 100 mm) previously placed in a heating block at 45°C. This pipetting can be done while the medium is still very hot, decreasing the chance of contamination. The agar will remain liquid at 45°C for several hours. It remains liquid for a very long time at 50°C, but this temperature is more prone to cause some killing.

2. Dilute the sample as in the previous procedure (11.2.1).

3. Bring the petri plates from storage to at least 25°C, or better to 30°C.

4. Pipette 0.1 ml of the diluted sample onto the lip of the Wasserman tube. Pipette on an identifiable side of the tube (e.g., the side with a trademark, or a side that has a frosted spot or an identifying mark made with a marking pen).

5. Immediately pour the soft agar out of the tube onto the agar plate. Pour it over the side on which the sample has been pipetted. This will wash all of the organisms to the plate in such a way that they will be quantitatively transferred and uniformly mixed with the rest of the soft agar on the petri plate. They will not have a chance to become adsorbed to the glass of the test tube or to localized spots on the basal level of agar.

6. Tilt and swirl the petri plate so that the melted agar covers the surface.

7. Place the petri plate on the level work area to congeal.

8. Carry out the platings needed for the remainder of the experiment.

9. As convenient, pipette 2.5 to 3 ml of soft agar onto the congealed surface, or pour a tube with 2.5 to 3 ml of soft agar on each seeded and congealed plate, and distribute by rocking. Then let this final layer congeal on the level surface.

10. Incubate the plates in either orientation, as contamination and degree of dryness are much less critical with this technique than with spread plates.

11. Examine and count the colonies. The sub- surface colonies are compact with well-defined edges and are smaller than the usual surface colonies. This makes them a little more difficult to count, but the magnifying glass of colony counters or the readily available magnifying glasses with head bands help. These glasses have prisms built in that reduce eye strain.

The extra trouble involved in the overlay technique is worthwhile when many colonies are present. The colonies should be uniformly distributed, and this can be checked by visual inspection. One can count a fraction of a plate and prorate the count. If necessary, a low-power dissecting binocular scope allows one to virtually eliminate coincidence counts because one can visualize the colonies in three dimensions. It is laborious to count under such conditions, but it can save an experiment that has been very carefully carried out from being incomplete and possibly inconclusive.

11.3. MOST PROBABLE NUMBERS

The concentration of viable cells can be estimated by the most probable number (MPN) method, which involves the mathematical inference of the viable count from the fraction of cultures that fail to show growth in a series of tubes containing a suitable medium.

This method consists of making several replicate dilutions in a growth medium and recording the fraction of tubes showing bacterial growth. The tubes exhibiting no growth presumably failed to receive even a single organism that was capable of growth. Since the distribution of such particles must follow a Poisson distribution (see below), the mean number at this dilution can be calculated from the formula $P_0 = e^{-m}$, where m is the mean number and P_0 is the ratio of the number of tubes with no growth to the total number of tubes. The mean number is then simply multiplied by the dilution factor and by the volume inoculated into the growth tube to yield the viable count of the original culture. The MPN method is a very inefficient method from the point of view of statistics, because each tube corresponds to a small fraction of the surface of a petri dish. Consequently, many tubes must be used or the worker must be prepared to settle for a very approximate answer.

When is the MPN method of advantage? Firstly, it can be used if there is no way to culture the bacteria on solidified media. Secondly, it is preferred if the kinetics of growth are highly variable. Suppose some cells grow immediately and rapidly, and end up making a large colony on solid agar that spreads over and obscures colonies of the organism of interest that form later. The small colony formers may be more numerous but unmeasurable on plates because of fewer but highly motile or rapidly growing bacteria. Thirdly, if other organisms not of interest are present in the sample and no selective method is available, the method has utility when the bacterium of interest produces some detectable product (e.g., a colored material, specific virus, or antibiotic). Then, even though any contaminating organisms may overgrow the culture, the bacterium in question can be estimated by the fraction of the tubes that fail to produce the characteristic product. Fourthly, if agar and other solidifying materials have some factors (such as heavy metals) that may alter the reliability of the count or interfere with the object of the experimental plan, the MPN method can be used.

Modern developments in laboratory techniques can be used to speed the execution of the MPN method. Machines are available to fill the wells of plastic trays that have as many as 144 depressions. Scanning devices designed for other purposes can be used to aid in counting the number of wells with no growth. Similarly, automatic and semiautomatic pipettes can be used to fill small test tubes. Because these procedures make it possible to examine many more cultures, the classical tables of fixed numbers of tubes and fixed dilution series should be abandoned. Now a different approach and method of calculation will be needed.

Although statistically inefficient, the MPN method used at a single level of dilution is most accurate when the mean number of bacterial cells capable of indefinite growth is 1.59 per tube (18). This will result in 20.8% of the tubes remaining sterile (Fig. 2). If the expected number is known quite accurately, all the available tubes should be seeded with a dilution that is expected to have the value 1.59. The accuracy falls off rapidly as one deviates from this optimum, particularly outside of the range of 1 to 2.5 cells per tube. Thus, the dilutions 10-fold away from the optimum contribute almost no information about the mean number of cells. Consequently, if the viable count is known within a factor of 5-fold, the optimum strategy is to use all the allotted growth tubes at the dilution expected to have 1.59 cells, assuming that the viable count is in the middle of the possible range.

Usually, prior knowledge is not available and growth tubes at several dilutions are needed. One approach is to make the dilution levels 10-fold or more different. For this approach, discard the data from dilutions where the percentage of sterile tube lies outside of 8 to 36%, or outside of the range of 5 to 50% if a larger range of error is acceptable. The error from the dilution within the range can then be read or interpolated from

Fig. 2. This method is simple, but wasteful. Consequently, statistical methods have been developed to combine data from different dilution levels when a specified number of tubes is run at each level.

Table 1 gives goodness-of-fit calculations for the best estimate of the MPN. It applies when data from 10 cultures at each of three successive 10-fold dilutions are employed. Of the 1,000 possible outcomes of such tests, Halvorson and Ziegler (21) listed the 210 cases that could happen with P values of greater than 0.01% (see footnote to Table 1 for explanation and definition of terms). Finney (18) gave a good exposition of the statistical aspects and the approaches to calculate the MPN of replicates at each dilution. He pointed out that there can be a definite advantage in computation when carrying out error calculations if the dilution series extends over a range such that, in at least one dilution,

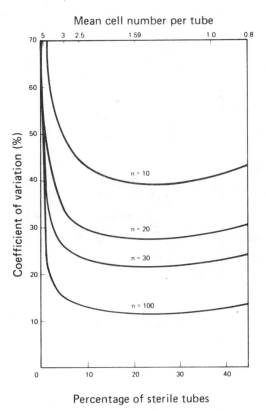

FIG. 2. Errors (coefficient of variation) that arise in using the method of most probable numbers (MPN) as a function of the percentage of sterile tubes (bottom abscissa) and the mean number of cells per tube (top abscissa); the respective optima occur at 20.8 and 1.59. A varying number of tubes (n) are prepared at a single dilution level.

TABLE 1. *Probabilities for method of most probable number (reproduced from Halvorson and Ziegler [21])[a]*

	Code		X	P		Code		X	P		Code		X	P
								Probability						
10	10	10	100.0	100.0	10	10	0	2.40	3.4	10	9	0	1.70	5.30
10	10	9	23.0	38.7	10	9	9	6.07	0.01	10	8	7	3.10	0.01
10	10	8	16.2	30.2	10	9	8	5.26	0.04	10	8	6	2.78	0.14
10	10	7	12.0	26.2	10	9	7	4.58	0.27	10	8	5	2.49	0.56
10	10	6	9.18	25.0	10	9	6	3.98	0.85	10	8	4	2.21	1.64
10	10	5	7.02	24.2	10	9	5	3.46	2.25	10	8	3	1.96	4.50
10	10	4	5.42	23.4	10	9	4	2.98	4.80	10	8	2	1.71	8.46
10	10	3	4.28	21.7	10	9	3	2.63	8.57	10	8	1	1.50	10.83
10	10	2	3.49	17.3	10	9	2	2.28	11.77	10	8	0	1.30	7.21
10	10	1	2.75	10.2	10	9	1	1.97	11.02	10	7	6	2.19	0.01
10	7	5	1.95	0.20	10	6	3	1.25	0.59	10	4	5	1.073	0.01
10	7	4	1.74	0.80	10	6	2	1.09	2.73	10	4	4	0.943	0.07
10	7	3	1.53	2.57	10	6	1	0.933	8.42	10	4	3	0.818	0.48
10	7	2	1.33	6.17	10	6	0	0.792	10.73	10	4	2	0.700	2.26
10	7	1	1.16	10.01	10	5	5	1.30	0.02	10	4	1	0.589	7.97
10	7	0	1.01	9.33	10	5	4	1.15	0.14	10	4	0	0.493	14.99
10	6	6	1.75	0.01	10	5	3	1.02	0.91	10	3	4	0.773	0.02
10	6	5	1.57	0.08	10	5	2	0.872	3.30	10	3	3	0.662	0.24
10	6	4	1.41	0.11	10	5	1	0.742	8.94	10	3	2	0.561	1.60
10	6	3	1.25	0.59	10	5	0	0.622	12.68	10	3	1	0.474	6.93
10	3	0	0.399	17.67	10	0	2	0.314	0.15	9	5	1	0.372	0.86
10	2	4	0.631	0.01	10	0	1	0.268	1.44	9	5	0	0.334	1.78
10	2	3	0.534	0.12	10	0	0	0.231	7.95	9	4	3	0.408	0.02
10	2	2	0.456	0.99	9	8	0	0.499	0.03	9	4	2	0.365	0.22
10	2	1	0.388	5.30	9	7	1	0.488	0.02	9	4	1	0.324	1.13
10	2	0	0.329	18.14	9	7	0	0.435	0.17	9	4	0	0.290	3.58
10	1	3	0.442	0.04	9	6	2	0.474	0.06	9	3	3	0.362	0.04
10	1	2	0.376	0.60	9	6	1	0.425	0.23	9	3	2	0.324	0.26
10	1	1	0.317	3.71	9	6	0	0.381	0.56	9	3	1	0.288	1.77
10	1	0	0.275	15.14	9	5	2	0.416	0.10	9	3	0	0.255	6.23
9	2	3	0.316	0.03	8	6	0	0.270	0.11	8	2	2	0.210	0.14
9	2	2	0.284	0.26	8	5	1	0.267	0.11	8	2	1	0.188	1.24
9	2	1	0.253	2.19	8	5	0	0.242	0.46	8	2	0	0.169	7.12
9	2	0	0.223	9.53	8	4	2	0.266	0.04	8	1	2	0.187	0.11
9	1	2	0.249	0.19	8	4	1	0.240	0.33	8	1	1	0.166	1.46
9	1	1	0.221	1.84	8	4	0	0.217	1.47	8	1	0	0.147	9.31
9	1	0	0.193	9.03	8	3	2	0.239	0.08	8	0	2	0.166	0.06
9	0	2	0.217	0.07	8	3	1	0.214	0.77	8	0	1	0.146	0.88
9	0	1	0.191	0.79	8	3	0	0.193	3.73	8	0	0	0.128	6.41
9	0	0	0.164	5.02	8	2	3	0.233	0.01	7	6	0	0.212	0.02
7	5	1	0.209	0.08	7	2	0	0.133	5.49	6	4	0	0.139	0.32
7	5	0	0.191	0.16	7	1	2	0.149	0.08	6	3	2	0.153	0.01
7	4	2	0.208	0.01	7	1	1	0.132	1.10	6	3	1	0.138	0.20
7	4	1	0.188	0.12	7	1	0	0.116	8.98	6	3	0	0.123	1.35
7	4	0	0.171	0.68	7	0	2	0.132	0.04	6	2	2	0.137	0.03
7	3	2	0.188	0.03	7	0	1	0.116	0.80	6	2	1	0.122	0.45
7	3	1	0.169	0.35	7	0	0	0.101	7.92	6	2	0	0.107	4.12
7	3	0	0.152	2.44	6	5	0	0.155	0.08	6	1	2	0.121	0.06
7	2	2	0.167	0.04	6	4	2	0.171	0.00	6	1	1	0.106	0.87
7	2	1	0.150	0.78	6	4	1	0.155	0.04	6	1	0	0.092	8.73
6	0	2	0.106	0.04	5	2	0	0.086	3.36	4	3	0	0.080	0.43
6	0	1	0.092	0.83	5	1	2	0.099	0.03	4	2	1	0.080	0.16
6	0	0	0.078	9.77	5	1	1	0.086	0.58	4	2	0	0.068	2.16

TABLE 1—*Continued*

	Code		X	P		Code		X	P		Code		X	P
				Probability										
5	5	0	0.128	0.04	5	1	0	0.073	8.28	4	1	2	0.080	0.01
5	4	1	0.127	'0.02	5	0	2	0.085	0.01	4	1	1	0.068	0.50
5	4	0	0.114	0.14	5	0	1	0.072	0.79	4	1	0	0.056	7.21
5	3	1	0.113	0.08	5	0	0	0.060	12.10	4	0	2	0.067	0.02
5	3	0	0.100	0.79	4	5	0	0.106	0.02	4	0	1	0.056	0.76
5	2	2	0.113	0.01	4	4	0	0.093	0.06	4	0	0	0.045	14.62
5	2	1	0.099	0.28	4	3	1	0.092	0.03	3	4	0	0.076	0.04
3	3	2	0.086	0.00	3	0	0	0.032	18.64	1	2	1	0.038	0.02
3	3	1	0.075	0.02	2	4	0	0.062	0.02	1	2	0	0.029	0.53
3	3	0	0.064	0.23	2	3	0	0.051	0.11	1	1	1	0.028	0.12
3	2	1	0.064	0.08	2	2	0	0.041	3.95	1	1	0	0.019	4.76
3	2	0	0.053	1.53	2	1	1	0.040	0.02	1	0	1	0.019	0.34
3	1	2	0.064	0.01	2	1	0	0.030	6.01	1	0	0	0.110	34.58
3	1	1	0.053	0.31	2	0	2	0.040	0.01	0	2	0	0.018	0.21
3	1	0	0.043	7.00	2	0	1	0.030	0.41	0	1	1	0.018	0.04
3	0	2	0.052	0.02	2	0	0	0.020	24.13	0	1	0	0.009	3.05
3	0	1	0.042	0.51	1	3	0	0.038	0.04	0	0	1	0.009	0.26

[a] This table indicates the number of bacteria per milliliter based on the number of tubes showing growth which have been inoculated with dilutions as outlined. Ten tubes are to be inoculated with 10 ml each, 10 tubes with 1 ml each, and 10 tubes with 0.1 ml each. The code number is made up of the number of tubes showing growth in each case, the first number of the code representing the number of tubes showing growth that are inoculated with 10 ml each, the second those inoculated with 1 ml each, and the last those inoculated with 0.1 ml each. The column labeled X then gives the most probable number of bacteria per milliliter in the material used for inoculation. The column labeled P gives the percentage of times that the code would be obtained if an infinite number of determinations were made of a solution containing the number of organisms indicated by X.

Examples: Suppose that a culture is inoculated into broth in a series of 10 dilutions in steps of 10. Assume that 9 of the tubes inoculated with 10^{-7} ml, that 3 of the tubes inoculated with 10^{-8} ml, and that none of the tubes inoculated with 10^{-9} ml show growth. The code will then be 9 3 0. Referring to the table, X is found to be 0.255. This means that the most probable number of bacteria in the 10^{-8} ml dilution was 0.255 bacteria per ml, or that the most probable number in the original solution was 0.255 times 10^8 or 25,500,000. If this experiment were repeated an infinite number of times on a solution containing this number, this result would be obtained 6.23% of the time.

every tube shows growth and such that, in another dilution, every tube shows no growth. Even then the proper statistical calculations are cumbersome.

With modern programmable calculators, it is easy to do away with tables and to calculate the MPN for any arrangement of tubes and dilutions, as in the example shown in Table 2. Even for a standard series of dilutions (such as used for Table 1), this method is of value when outcomes (codes) not listed in the published tables are obtained.

For the case of 10 tubes at each of three 10-fold dilutions, although 30 tubes are employed, the accuracy is less than if all 30 tubes had been used at a single dilution (as given by the curve method $n = 30$ in Fig. 2). However, the accuracy is approximately shown by the line marked $n = 20$.

11.4. LIGHT SCATTERING

Light-scattering methods are the techniques most generally used to follow the growth of pure cultures. They can be very powerful and very useful, but they can lead to erroneous results. The major advantages are that they can be performed quickly and nondestructively. However, they may give information about a quantity not of primary interest to the investigator. They mainly give information about macromolecular content (dry weight) and not about the number of cells. The physics and mathematics of light scattering are complex; still, without difficult physics, an elementary consideration can give most of the needed answers.

The basic principle is that of Huygens. Electromagnetic radiation interacts with the electronic charges in all matter. When the light energy cannot be absorbed, a light quantum of the same energy (color) must be re-radiated. This light can emerge in any direction. This means that all atoms in a physical body serve as secondary sources of light. Those photons that happen to go in the direction of the original wave will stay in phase, but those that go in other directions will differ in phase, at an observation point, from the incident wave arriving directly from the light source. The photon also may differ in phase, depending on distance and

TABLE 2. *Calculation of most probable number with programmable pocket calculator*

General formula: $\sum \dfrac{\alpha_i \rho_i}{1 - e^{-\alpha_i x}} = \Sigma \alpha_i \eta_i$

Symbol	Value(s) in example
x = most probable number	?
α_i = volume tested	10, 1.0, 0.1
η_i = number of total tests	10, 10, 10
$\Sigma \alpha_i \eta_i$	$10 \times 10 + 10 \times 1 +$ $10 \times 0.1 = 111$ ml
ρ_i = number showing growth	8, 5, 1
$\alpha_i \rho_i$	80, 5, 0.1

Specific formula: $\dfrac{80}{1 - e^{-10x}} + \dfrac{5}{1 - e^{-x}} + \dfrac{0.1}{1 - e^{-0.1x}} = 111$ ml = total volume tested

Set up a programmable calculator to add $\dfrac{Y}{1 - e^{-X}}$ to a register, where Y and X are the contents of the working memories.

A program for Hewlett-Packard model 33E is as follows: 32, 15-1, 1, 41, 32, 71, 51, 13–00. Then to initialize, key R/S.

Guess a value of x (e.g., 0.3). Clear registers,
Key 80, enter 3.0, key R/S,
Key 5, enter 0.3, key R/S,
Key 0.1, enter 0.03, key R/S.
    Results 106.87. Record. Clear.
Choose a new value of x—say, 0.25.
Key 80, enter 2.5, key R/S,
Key 5, enter 0.25, key R/S,
Key 0.1, enter 0.025, key R/S.
    Results 113.81. Record. Compare.

These results bracket the 111 ml tested, so interpolate or try an intermediate value searching for a value of x that yields 111 ml for the sum of the three terms. Either way, a value for x of 0.27 is obtained which checks with the value read from Table 1.

direction of the emitting atom from photons emitted from other points of the physical body. Light going through matter is slowed by these interactions proportionally to the index of refraction.

It is the above circumstance that controls how light is bent and focused in large bodies such as prisms, lenses, raindrops, or a pane of glass. In the last case, light scattered in every direction but straight ahead cancels out, leaving a beam of light going in the original direction but slightly retarded relative to a light ray not going through the pane.

11.4.1. Turbidimetry

Bacterial suspensions are in between the size limits of atoms and objects like window panes. Consequently, most of the scattered light is directed almost but not quite in the same direction as was the incident beam (Fig. 3). The light scattered from an atom or a very small particle

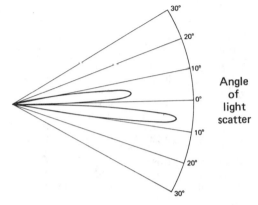

Angle of light scatter

FIG. 3. *Light-scattering patterns from randomly oriented bacteria showing the angular distribution of light scattered by a beam traversing from left to right through a suspension of cells with the size and physical properties of Escherichia coli grown in minimal medium. The lower pattern shows the distribution if the cells are spherical. The upper pattern shows the distribution if the cells are ellipsoidal with an axial ratio of 4:1.*

depends inversely on the fourth power of the wavelength of light. Small particles, therefore, scatter blue light more strongly and transmit red light more efficiently. This is the reason that the sky appears blue and that smoke and many virus suspensions are bluish. For a pane of glass, there is no light left uncancelled in any but the original direction and the light retains its original hue. For bacteria, the light scattered away from the original direction most nearly approximates an inverse second power (28). Therefore, while the sky appears blue and a pane of glass is transparent, suspensions of bacteria (like a cloud) are intermediate and appear white but not transparent. Because bacterial suspensions appear turbid or cloudy, instruments to measure the phenomenon are called turbidimeters (L. *turbidus* = confused).

The common practice in bacteriology is to use any available colorimeter or spectrophotometer to measure turbidity. Ideally, such instruments measure the primary beam of light that passes into the sample *without deviation* to reach the photocell (Fig. 4A). Usually, the measurement is made of the light intensity relative to that which reaches the photocell when the suspending medium has replaced the cell suspension. From what has been said, an ideal photometer must be designed with a narrow beam so that only the light scatter in the forward direction reaches the photocell. That is, the instrument must have well-collimated optics. Such an ideal instrument gives larger apparent absorbance values than simple instruments with poorly collimated op-

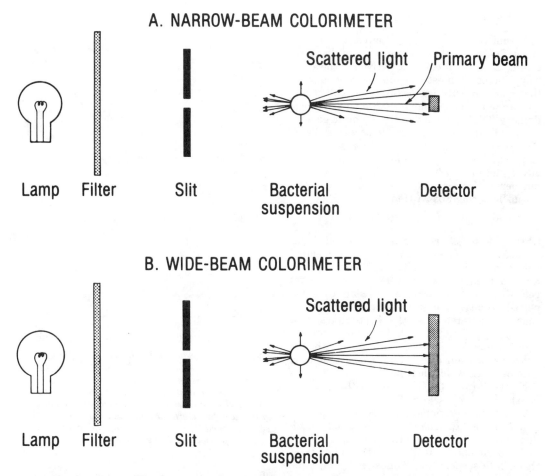

A. NARROW-BEAM COLORIMETER

Scattered light   Primary beam

Lamp   Filter   Slit   Bacterial   Detector
suspension

B. WIDE-BEAM COLORIMETER

Scattered light

Lamp   Filter   Slit   Bacterial   Detector
suspension

FIG. 4. *Schematic designs for filter-type "colorimeter" instruments. (A) Narrow-beam instrument; almost all the scattered light deviates enough to escape falling on the photodetector. (B) Wide-beam instrument; a variable fraction of the scattered light falls on the detector and thus the sensitivity is lowered.*

tics, since in the latter a large percentage of the light scattered by the suspension is intercepted by the phototube (Fig. 4B). Therefore, the measuring system responds as if there were less light scattered than is actually deviated from the forward direction.

The larger the number of bacteria in the light path is, the less the intensity of the light that gets through the sample. At low turbidities this is a simple geometric relationship, since the intensity of the unscattered light decreases exponentially as the number of bacteria increases. The geometric relationship between turbidity and bacterial numbers can be deduced by considering a suspension of bacteria which reduces the light intensity to one-tenth of the original intensity. Now consider two identical suspensions in two cuvettes in a row such that the light beam goes through both suspensions. By how

much would the light intensity be reduced after it goes through both tubes? The answer is 0.1 × 0.1, or 0.01 of the intensity of the incident beam. In the ideal case the same relationship holds when the concentration of bacteria is doubled and only one tube is used. When this is expressed mathematically, the intensity of the unscattered light (I) is equal to the intensity of the incident light (I_0) multiplied by $10^{-W/W_{10}}$, where W_{10} is that concentration of bacteria which gives a 10-fold decrease in the light intensity. That is:

$$I = I_0 \cdot 10^{-W/W_{10}}$$

Thus, if $W = W_{10}$, then I would be 0.1 of I_0. If this concentration of bacteria were doubled, then I would be 0.01 of I_0. By taking logarithms of both sides of the equation, this equation can be rewritten:

$$-\log I/I_0 = \log I_0/I = W/W_{10}$$

A similar law, the Beer-Lambert law (see 16.1.1), works for the absorption of light by colored samples. It could be derived in the same way. Therefore, most instruments have a scale that reads directly $\log I_0/I$. This is called **absorbance** (A) or, on older instruments, **optical density** (OD). Optical density is a more general term and can be used for turbid as well as light-absorbing solutions, but absorbance is the term employed here. From the above equation, ideally, a plot of A of bacterial cultures versus cell numbers yields a straight line going through the origin. This could be written:

$$A = K \cdot W$$

where K is the slope (and is $1/W_{10}$) and A has replaced $\log I_0/I$. As shown in Fig. 5, the actual absorbance becomes increasingly less than the formula predicts (see also 16.1.1). More light gets through the turbid suspension than expected because light scattered from one bacterium is rescattered by another bacterium so that the light is redirected back into the phototube, and because the bacteria interfere with the Brownian motion of one another so that they become more evenly distributed and scatter less light away from the beam (like the pane of glass).

These considerations allow one to grasp some practical implications about bacterial turbidimetry. The results of both theoretical (28, 29) and experimental (20, 28, 30, 31) studies show that dilute suspensions of most bacteria, independently of cell size, have nearly the same absorbance per unit of the dry weight concentrations. However, very different absorbancies are found per particle or per colony-forming unit with different sizes of bacterial cells. An approximate rule has been proposed that the dry weight concentration is directly proportional to the absorbance (28). This rule applies to both cocci and rods, and is a first approximation to a more precise rule, i.e., that the light scattered out of the primary beam is proportional to the four-thirds power of the average volume of cells in the culture. Objects smaller than bacteria, such as suspensions of viruses, will not obey these rules well. The rules also fail to apply when the suspensions of bacteria have a bluish cast. Larger objects (yeasts, and filaments and aggregates of bacteria) may still appear cloudy and not necessarily colored, but may not obey the rule. Where the rule applies, the proportionality constant relating the absorbancy measurement to the dry weight concentration is the same for any good, well-collimated photometer. In such instruments the phototube is at a considerable

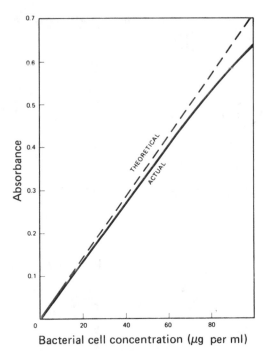

FIG. 5. *Absorbance (at 420 nm) as a function of bacterial cell concentration. The dashed line shows the theoretical relationship. The solid line shows an actual result obtained with 15 dilutions of a bacterial culture, the concentrations of which were measured by dry-weight determinations and the plot of which was obtained by the best-fitting quadratic curve forced to pass through the origin.*

distance and/or the instruments are so well collimated that a narrow beam passes through and only a narrow beam is intercepted by the detector. The Zeiss and Cary instruments possess this design character. Other instruments can be assessed by examination of their blueprints, by examination of their optical design, or by comparison of the same suspension of cells within a short period in a Zeiss or Cary instrument and in the spectrophotometer under consideration.

If the instrument or the size of organism does not meet these criteria, valid measurements in many cases can be made by establishing a standard curve on either a relative or an absolute basis. Difficulty arises only as physical or biological conditions vary with the samples to be compared, since then the experimenter cannot be sure that the same standard curve applies.

The above considerations apply only to dilute suspensions. Deviation from Beer's law must be applied when the absorbance exceeds about 0.3 when light of 500 nm or shorter is used (see example in Fig. 5). A good discussion of this question was given by Kavenagh (26). Failure to take this deviation into consideration is the most common difficulty encountered in the microbiological literature that involves turbidimetry. One way to avoid this problem is to restrict culture densities, or dilute cultures for measurement, below the 0.3 absorbance level. However, the photometric measurements at this level are not accurate, and an additional dilution with the accompanying pipetting error may be introduced. For such small absorbance measurements, it is also critical that the cuvettes be well matched and scrupulously clean. In addition, precise placement of the cuvette in the spectrophotometer is needed. It is usually better to make measurements at higher densities and correct for deviations from Beer's law.

Procedure.

The important point about turbidimetric methods is that there is no set procedure. One uses the equipment at hand and seeks to obtain useful results. The most costly of spectrophotometers increase the spectrophotometric precision and the stability and accuracy of wavelength settings. But the main advantage of such instruments for the present purpose is that they have a larger dynamic range. Lower absorbancies have validity when measured on good equipment because of electronic stability and increased reproducibility in positioning the cuvettes, and because the cuvettes have plane parallel faces and are of accurate dimension. Higher absorbancies, for which very little light reaches the phototube, have validity when measured on such equipment because there is less stray light. Moreover, the electronics can be compensated by a dark-current adjustment, and they may be read more accurately.

The more important distinction among various types of equipment for turbidity measurements is between single- and double-beam photometers. The critical test of the instrument is to obscure the phototube with the shutter or to place an opaque object in the light path and follow the kinetics of decrease in transmission. In cheaper instruments, there will be a very quick fall which is followed by a much slower fall that may continue for a very long time. Similarly, when the shutter is subsequently opened or the opaque object is removed from poorer instruments, after the transmission has risen precipitously it then rises further very slowly and never achieves a truly stable value. These drifts result from changes in phototube response due to memory of prior illumination conditions. Single-phototube, double-beam instruments avoid this because the same photodetector alternately measures the blank and the

sample. However, simple single-beam instruments can be used with very little error if the phototube is alternately exposed to blanks, sample, and dark-current measurements in a routine standardized way with fixed timing.

There is a second reason for adopting a routine timing procedure for measuring cell turbidity, even with the best spectrophotometer. This is because turbidity fluctuates depending on the alignment of rod-shaped particles. The longer the rods, the more severe the alignment will be in recently mixed cultures. For organisms with moderate ratios of length to width, such as *E. coli*, one approach is to mix the culture in the cuvette, place it in the spectrophotometer, wait a definite time (e.g., 45 s), and then record the reading. Waiting for a long time in hopes of constancy is not recommended. This is because the cuvette will heat up from the light source, creating convection currents that cause partial vertical alignment of rod-shaped organisms. Another approach for rod-shaped organisms is to make the measurements in a slowly flowing system. Then rod-shaped organisms will be oriented by the laminar flow (22). The orientation is greater nearer the walls than the center because the gradient of velocity is greater there. Of course, the procedure must be calibrated under the same floating conditions with organisms of the same axial ratio.

Calibration.

The procedure adopted for a given culture and instrument must be calibrated in some way to allow for comparison with other work. It is not valid to report turbidity measurements in absorbancy units at a specified wavelength, for reasons to be itemized below. On the other hand, if a narrow-beam spectrophotometer is used, if the cells have a size range from 0.4 to 2 μm^3, and if reporting the measurement in terms of dry weight concentration is acceptable, then the conversion factors and deviations from Beer's law given below and elsewhere (30, 32) can be used.

To calibrate any adopted procedure, prepare a concentrated suspension of the organism grown under the appropriate conditions. Carry out a series of dilutions of the well-mixed suspension, and make measurements as quickly as possible so that growth need not be stopped or interfered with. Alternatively, place the concentrated suspension in an ice bath and make dilutions and measurements, or introduce an agent such as chloramphenicol to stop protein synthesis. These are three ways to minimize biomass changes while measurements are made. In any case, make a repeat dilution from the stock of the first dilution after all other measurements have been made. The first and last must check. Plot the data as they are obtained on a working graph similar to Fig. 5 except with an abscissa expressed as the percent concentration relative to the undiluted stock. A feeling for the shape of the curve is achieved with this working graph. Inspect the graph after each point is added. This will allow the detection of observational errors. Repeat dilutions when an error is suspected. Make additional dilutions to define those parts of the curve where there is more curvature. Repeating a dilution can give an immediate estimation of error, but an additional intermediate dilution defines the curvature better. There need be no significant pipetting errors since the stock can be adjusted such that dilutions need not be high and relatively large volumes of solution can be pipetted into volumetric glassware. This working graph is sufficient to permit accurate estimates of growth of a culture, since only the relative change in biomass with time is needed for calculating specific growth rates and doubling times. However, for many other uses an absolute instead of a relative measure is needed. In addition, if biomass is measured on an absolute basis, the work is reportable in a way that allows accurate replication in other laboratories. Therefore, it is suggested that calibration be done on an absolute basis.

For absolute calibration it is necessary to measure accurately the stock suspension content of some other measure of bacterial biomass or numbers. This may be done by dry weight determinations (40; also, part 25.2.2) or by chemical determinations of a nucleic acid (17.5), protein (17.6), or total nitrogen (17.6.9.). This must be done with adequate replication and quality control. Use many independent dilutions for viable counts, and use many independent fillings and adequate counts in the counting chamber for microscopic counts. Accuracy at this stage is to be emphasized since the value obtained will pervade future work.

Prepare a graph using all the data that relate meter readings on the photometric instruments to cell number or mass, such as shown in Fig. 5. If computer facilities are available, calculate the best linear, quadratic, or cubic fit to the data. Programs available at most computing centers carry out an F-test to define which of the types of curves fit the data most efficiently. The best-fitting curve can then be calculated by simple programs with whatever computer facilities are available. Programmable pocket calculators are adequate.

In work from this laboratory, we did these things using the narrow-beam Cary and Zeiss spectrophotometers. We also used and have tested a range of commercial instruments with

bacterial cells of various sizes (30). Over the size range from 0.4 to 2 μm^3, the results were very nearly the same when expressed on a dry weight basis. Since the same curve applied to all the bacteria tested, the conversion formulas no doubt will apply to other uncolored procaryotes whose mean size is within these limits. When the absorbance range is limited to the region below 1.1, a quadratic which passed through the origin fitted the data adequately: $A_\lambda = a_\lambda W - b_\lambda W^2$, where W is the dry weight concentration of bacteria, and a_λ and b_λ are constants that apply at a given wavelength λ. The empirical equations for two convenient wavelengths are

$$A_{420} = 7.114 \times 10^{-3}\, W - 7.702 \times 10^{-6}\, W^2$$
$$A_{660} = 2.742 \times 10^{-3}\, W - 0.138 \times 10^{-6}\, W^2$$

The coefficient of the first-power W (a_λ) decreases with increasing wavelength. The coefficient can be calculated at other wavelengths since it depends inversely on the wavelength raised to approximately the second power, actually the 2.11 power. Consequently, a_λ can be calculated at other intermediate wavelengths from

$$a_\lambda = 2{,}439/\lambda^{2.11}$$

There is no rule to estimate b_λ at other wavelengths, although Kavenagh (26) made measurements of this quantity at a number of wavelengths. This constant depends more critically on shape, internal heterogeneity, design of the photometer, etc. In fact, the quadratic fit is only marginally satisfactory at higher turbidities. For this and for photometric reasons, it is best to work at absorbancies less than 1. Increasing the wavelength decreases b_λ, it decreases multiple scattering of light into the detector, and it improves linearity.

For routine calculation of W, the equations given above can be solved to yield in the units of micrograms of dry weight per milliliter:

$$W = 461.81\, (1 - \sqrt{1 - 0.60881\, A_{420}})$$

and

$$W = 9929\, (1 - \sqrt{1 - 0.07347\, A_{660}})$$

The latter expression can be expanded with no loss of accuracy to:

$$W = 364.74\, A_{660} + 6.7\, A_{660}^2$$

All three expressions can be used for absorbancies up to 1.

In the absence of internal pigments, the choice of wavelengths should be based on the following considerations. The lower the wavelength, the larger the turbidity. Therefore, lower concentrations can be directly measured. Conversely, low sensitivity is desirable for some purposes because more dense suspensions are of interest and an extra step of dilution can be avoided. An additional reason for using higher wavelengths is that many growth media contain substances that absorb near the blue end of the visible spectrum and, therefore, appear yellow or brown. Then 660 nm is a desirable wavelength because the medium blank is not much different from a water blank. Any higher wavelengths would be inconvenient because a different phototube would be required in many instruments. Chemical changes due to bacterial action, air oxidation, or variation in autoclaving alter the light transmission in complex media at shorter wavelengths. In any case, it is important when using a new medium that the supernatant of a culture of high density be compared with the original growth medium to see whether pigments are introduced or removed by bacterial growth.

Photometric methods can be used with colored organisms. The worker may be able to choose a suitable wavelength to avoid absorption bands. Alternatively, choose a wavelength where the absorption by cellular pigments is maximum, and report the biomass using the pigment as an indicator. The latter procedure has the difficulty that in many cases the pigment content varies with culture conditions.

If a narrow-beam spectrophotometer is not available for routine use, the above formula for dry weight can be of use if there is temporary access to a narrow-beam instrument. Simply carry out the calibration procedure, but instead of the other methods listed above, measure the turbidity of a known dilution on a narrow-beam instrument and use the formulas given above.

With wide-beam instruments the measured turbidity is less than in the narrow-beam instruments when a cuvette of the same size and shape is used; therefore, the sensitivity is lower (see above). This is a minor factor in most work. Of more significance is the fact that the calibration may change every time the phototube or the light bulb is changed; in some instruments, simply removing and replacing the housing compartment for the measuring cuvette causes measurable changes. Frequent restandardization of the wide-beam instruments against a narrow-beam instrument is recommended. As an example, the reading on an instrument of the Beckman-DU type exhibits dramatic changes in sensitivity as thermospacers are inserted around the cuvette compartment (30). This is not unexpected, since the thermospacers lengthen the working distance and make the instrument approach more closely the quality of a narrow-beam instrument with its greater sensitivity.

Special quality control needs to be used with spectrophotometers utilizing gratings to disperse the spectrum. As the grating replicas age, they become warped, and an increased amount of stray light hits the phototube.

Klett-Summerson colorimeter.

Of the many colorimeters and relatively inexpensive spectrophotometers, the Klett-Summerson colorimeter must be singled out for special comment. It was one of the first photoelectric instruments, it is rugged and stable, and it has been used by many workers for more than 40 years. Its stability arises because it is a double-beam instrument so that fluctuations in the light source are cancelled out and because of a simple CdS sensor which, while not sensitive, produces enough current so that a sensitive galvanometer suffices. The detector is far enough from the cuvettes so that less of the scattered light is trapped than in some other instruments.

It is calibrated in a different scale than more modern instruments: an absorbance of 1 on other instruments corresponds to 500 on the Klett scale. It reads so sluggishly that the turbidity of rod-shaped bacteria becomes stable by the time the dial can be properly adjusted; by then, the cells have achieved random orientation. Take care in reading the instrument dial and adjusting the dial to achieve a null reading on the galvanometer. Stand in a reproducible position so that parallax error is avoided. The major fault of this instrument is that there is no dark-current adjustment. Light can pass around the contents of the cuvette by being scattered on instrument surfaces and reflected through the glass walls of the cuvette. Test this (and other instruments without dark-current adjustments) by filling a cuvette with a highly absorbing material with an essentially infinite optical density (0% transmittance). In the case of the Klett colorimeter, the reading should be beyond the printed scale of the instrument and in fact should be right on the vertical line that forms the left-hand side of the letter R in COPYRIGHT. Frequently, the Klett colorimeter is out of calibration, and then there is an adventitious instrument error which creates additional deviations to the Beer's law relationship. Special adjustment, repair, or replacement of the light source is needed. The problem can be partially alleviated, in some cases, by reblackening the internal surfaces.

11.4.2 Nephelometry

While most of the light scattered is nearly in the forward direction, instruments that measure the light scattered at 90° from the primary beam have been used to measure bacterial concentrations. Instruments designed to make right-angle measurements of light scattering are called nephelometers. Bacterial concentration may be measured with this principle in a spectrofluorometer when the dials are set so that the excitation monochromator and the emission monochromator are at the same wavelength.

Nephelometric measurements can be ultrasensitive, so that very small numbers of organisms can be measured. There are two difficulties as far as routine measurements are concerned. The first is caused by false signals from particulate impurities in the medium; this can be partially overcome by ultrafiltration of all reagents. The second problem is that it is necessary to standardize the instrument repeatedly during a series of measurements. A different kind of standardization is needed with a nephelometer or spectrofluorometer than with a spectrophotometer. With these, the operator sets the dark-current adjustment so that the scale reads zero with a reagent blank. Then the gain is set to give a fixed response from a standard scattering material. That is, one needs to run a one-point standard curve routinely. Since a suspension of bacteria would not be stable, a secondary standard such as a piece of opal glass must be used. One can also use a suspension of glass beads, such as those used in rupturing bacteria and other biological material. A tube can be prepared and sealed and used to calibrate the instrument. Then one can prepare a standard curve relating meter readings after standardization to bacterial concentration. The standard curves are not linear at higher bacterial densities.

Although various specialized instruments have been made and used over the years, these instruments are of limited value because most of the light that is scattered from particles in the size range of bacteria is nearly in the forward direction (Fig. 3). If this light is to be sensed, the instrument must be capable of orientation to within a very few degrees of the direction of the incident light without intercepting undeviated light from the primary beam. This means that very well-collimated light is necessary. Today, this is readily achieved with a laser beam and instruments are made that use a low-angle range between 2 and 12° from the beam, but none is commercially available at a reasonable price. Measurements at higher angles contain information about the amount of cell material, but they also have information about internal structure and distribution of material within the cell. If it is known which factors are important and whether elongated particles are oriented or randomly distributed in space, then light-scattering measurement over a range of angles can give information about: the state of aggregation of the protoplasm, such as whether the ribosomes are in polysomes or as monosomes within the living cell (6); thickness of the cell envelope (27, 29, 35, 54); and the distribution of cell mass from the center of the cell (27, 54).

Wyatt (54) has developed an apparatus to measure the angular dependency of the light-scattering signal from 30 to 150°C. In this range, different organisms, different treatments of a culture, and cultures in different phases of growth give characteristic patterns. These may be of use as a fingerprint technique for the diagnosis and study of drug action. For the reasons stated above, the light-scattering signals in the high-angle range cannot be used reliably for growth measurement because they are sensitive to details of subcellular structure.

Light-scattering measurements made over a range of angles from 0 to 30° are relatively independent of these three factors, but are more critically dependent on overall size. Measurement in this range could nondestructively yield the biomass concentration, the number concentration, the average axial ratio, and a measure of average biomass distribution around the center of the cells for bacteria in balanced growth.

11.5 STATISTICS AND CALCULATIONS

Several aspects relevant to the measurement of bacterial growth almost never get included in courses in basic microbiology or in statistics taken by undergraduate majors in bacteriology. For the former, this is because the needed material is a little beyond the scope of the courses, and for the latter, this is because the counts of colonies or particles is an all-or-none event and of lesser utility for most types of users of statistics. The following description is intended to make these mathematical tools readily available.

11.5.1 Population Distributions

The **binomial distribution** describes the chance of occurrence of two alternative events. For example, it provides the answer to the question, "What is the chance of having 5 boys in a family of 9, assuming that births of boys represent 0.56 of the total live births?" The numerical answer is $P_5 = 0.2600$. The relevant formula is as follows:

$$P_r = \frac{n!}{(n-r)!r!} \cdot p^r \cdot (1-p)^{n-r}$$

where p is the chance of a specified response on a single try, n is the total number of trials, and

r is the number of specified responses and would vary from 0 to n. In different families of 9 children, use of the formula shows that more families would have 5 boys than any other number. For example $P_4 = 0.2044$, $P_5 = 0.2600$, and $P_6 = 0.2207$. The plot of P_r against r is the distribution histogram. The binomial distribution provides a way to estimate the mean and standard deviation of p from experimental data. Assuming that there are data on only a single family with 9 children and that there are by chance 5 boys in that family, statistical theory shows that the best estimate of p is r/n. Numerically for this example, p is $5/9 = 0.5556$ and its standard deviation is $\sqrt{p\,(1-p)/n} = \sqrt{(5/9)\cdot(4/9)/9} = 0.1656$.

This result is not very reliable because the coefficient of variation (c.v.) is nearly 30% (c.v. $= 0.1656 \times 100\%/0.5556 = 29.81\%$). The precision in estimating p could be improved by looking for a family which had many more children and applying the same formula. It would not only be easier, but better, to pool the census data of a number of families. Suppose 50,000 boys are counted out of 90,000 children in a large pool of families. Then the same formula yields $p = 0.5556$, but now the c.v. will be 0.2981%. Not only is this more precise because large numbers are counted, but also many families, for genetic and sociological reasons, may have different values of p. *It is desirable to estimate the value of p that applies generally to the entire population.*

The formula for the distribution is hard to calculate when the numbers are large. An important contribution of Gauss was to rewrite the binomial distribution for this large-number case. Thus, the **Gaussian distribution** applies as a generalization of the binomial distribution for the case in which the numbers involved are so large that they can be treated as a continuous distribution instead of one with discrete variables. The variables of the Gaussian distribution replacing n and p are the population mean (m) and standard deviation (σ). The formula then becomes:

$$P_x = \frac{1}{\sqrt{2\pi}\sigma} e^{-(x-m)^2/2\sigma^2}$$

In this formula the continuous quantity (x) replaces the positive integer variable (r) as the measurement of response. This distribution, like the binomial, also can be mathematically manipulated so that one can go from data to estimations of these two parameters. The Gaussian distribution is also called the normal distribution, partly because it is symmetrical about the mean.

For data which follow a Gaussian distribution, the estimate of the mean is called $\bar{x}$ and is given by $\bar{x} = \Sigma x/n$. The estimate of the standard deviation is called s and is given by:

$$\sqrt{\frac{\Sigma x^2 - (\Sigma x)^2/n}{n-1}}$$

The c.v. is given by $s/\bar{x}$.

The other limiting distribution of the binomial is the **Poisson distribution**. It applies for the case where n is very large and p is very small, but the product $n \cdot p$ is finite. The best estimate of $n \cdot p$ is N, the observed number of total specified responses. This distribution would be useful if, for example, boys occurred very rarely (say, 1 in 10,000) but families were large (say, 100,000 children). Then, an average family would contain $N = 10$ boys and the standard deviation would be:

$$n\sqrt{p(1-p)/n} = \sqrt{n(N/n)\,(1-N/n)}$$
$$= \sqrt{10,000\cdot(10/10,000)\,(1-10/10,000)}$$
$$= 3.1621$$

The simplification of Poisson was to assume $N/n \ll 1$. Then the formula for the binomial distribution simplifies to a one-parameter distribution

$$P_r = \frac{e^{-m}\cdot m^r}{m!}$$

whose best estimate of the mean is $m = N$ and the best estimate of the standard deviation is $\sqrt{N}$. Note that this is not much different than with the binomial distribution, since $\sqrt{10} = 3.1623$ is not much different than 3.1621. Note that again n and p are replaced by different symbols, in this case by a single one, m. The important point is that *the count of the number of discrete objects provides not only the best estimate of the mean value* (i.e., $N = \bar{x} \simeq m$), *but also an estimate of the precision of the estimate* ($\sqrt{N} = s \simeq \sigma$).

Consequently, in the enumeration of objects, it does not matter how they are subdivided. Two replicate plates with 200 colonies on each are no better or worse from the point of view of Poisson statistics than is one plate with 400 colonies. In both cases the standard deviation (s) of the measurement is $\sqrt{400} = 20$, and the c.v. is $\sqrt{400}/400 = 5\%$. Therefore, to get the best estimate from a group of plates from the same or different dilutions of the same sample, simply add up the total counts on all the plates and divide it by the total volume of the original solution. The standard deviation is the square root of the total count divided by the plated volume of solution.

As an example, imagine that duplicate plates were made at dilutions of both 10^{-5} and 10^{-6} with counts of 534 and 580 and of 32 and 60, respectively. The total count is 1,206. If 0.1 ml of these dilutions were plated, then 2.2×10^{-6} ml was the total volume of original culture used to make the four plates. Therefore, the best estimate of the concentration is $1,206/2.2 \times 10^{-6} = 5.46 \times 10^8$/ml. The standard deviation is $\sqrt{1,206}/2.2 \times 10^{-6} = \pm 0.16 \times 10^8$/ml.

Justification for treating the results at the two different dilutions separately and not pooling them, as done above, depends on other kinds of errors being larger. Then, by comparing the results at different levels of dilution, some estimate is obtained of the variability due to the additional pipetting operation and to other sources of error that are not included in the calculation of the Poisson sampling error. While this variability is of interest, it could be more directly measured by carrying out independent dilutions from the same cell suspension. As an example, imagine that 0.1 ml were plated on a single plate from each of a dozen independent dilutions of 10^{-5}-fold and that the following set of colony counts were obtained: 534, 580, 760, 643, 565, 498, 573, 476, 555, 634, 514. The sum is 7,026 and the Poisson standard deviation is $\sqrt{7,026} = 83.8$. The mean of the numbers is 585.5 and the standard deviation by the Gaussian formula is ±81.2. Calculation of the bacterial count of the original suspension together with the two different estimates of error yields the following:

$$\frac{7,026}{1.2 \times 10^{-5} \text{ ml}}$$

$$= \frac{585.5}{0.1 \times 10^{-5} \text{ ml}} = 5.85 \times 10^8/\text{ml}$$

$$\pm \frac{81.2}{0.1 \times 10^{-5} \text{ ml}} = \pm 0.81 \times 10^8/\text{ml}$$

(Gaussian error)

$$\pm \frac{83.8}{1.2 \times 10^{-5} \text{ ml}} = \pm 0.07 \times 10^8/\text{ml}$$

(Poisson error)

The comparison of these two estimates of error suggests that considerable error is due to sources other than random sampling. Attempts should be made to find and reduce these sources of error. Until that is done, it is necessary to make many independent dilutions and use Gaussian statistics, and the Poisson error is irrelevant.

This same point can be made in another way from this example. Imagine that only one plate had been made, say the first one, in which case

only the Poisson error would be available for consideration. The count then would be $5.34 \times 10^8 \pm 0.23 \times 10^8$, and the real error would be fourfold underestimated. It is therefore cautioned not to rely on Poisson statistics until their use has been justified for the conditions. Instead, make a comparison on at least several occasions with a Gaussian statistic measurement of error as indicated above.

11.5.2 Statistical Tests

Much of the statistics taught in elementary courses is concerned with whether a body of data is consistent with an hypothesis. Usually the **probability** (P) that the observed deviations from the hypothesis could occur by chance is computed. If P is small, the hypothesis could still be true, although improbable. These statistical tests are generally made on the assumption that the data follow a Gaussian distribution. In many cases in bacteriology, this assumption should be questioned, but the statistical tools, when appropriate, can be very useful.

The **standard deviation** has been defined above. This is frequently confused with another term, the **standard error**, also called **standard deviation of the mean**. The standard deviation measures the deviation of an individual measurement from the mean of many measurements. The standard error measures the mean of all the data observed from the mean of a hypothetical data base containing an infinite number of observations, and is a measure of how close the average is to the true value.

The only test mentioned in the text below is Student's *t* test. This applies to the difference in the means of two groups. The difference is divided by the standard error of the combined data, and this ratio is compared to values given in tables to generate P values. Use of the tables requires a knowledge of the number of measurements and whether the deviations can occur on both sides or only one side of the mean. If P is very small, then the hypothesis that the two populations were identical may be rejected.

In recent years, the **analysis of variance** (ANOVA), which is a subbranch of statistics, has been elaborated so that it now can be applied to many problems and replaces many of the more specialized techniques previously used. This means that it requires work, but much less work than previously, to learn to use statistical methods where they are appropriate in bacteriology.

11.5.3 Error Propagation

The accuracy of an estimate depends on the accuracy of its component measurements. The

Poisson error of a colony count and the error of the dilution procedure both contribute to the error in the estimated concentration of organisms of the original undiluted suspension. Generally, additional errors can only further blur or make results less precise. Even though errors in one part may compensate for error in another part of an estimate, on the average they will make them larger. When errors in one measurement are independent of (uncorrelated with) errors of another measurement, the overall error can be calculated by two rules for "Propagation of Errors":

1. If two quantities (x and y) are to be added or subtracted, then the standard deviation (s) of the combined quantities is

$$s_{x+y} = s_{x-y} = \sqrt{s_x^2 + s_y^2}$$

2. If two quantities are to be multiplied or divided, then the coefficient of variation (c.v.) of the combined quantities is

$$\text{c.v.}_{x \cdot y} = \text{c.v.}_{x/y} = \sqrt{\text{c.v.}_x^2 + \text{c.v.}_y^2}$$

As an example, apply the second rule to estimate the overall error in a single plate count containing colonies from the series of 10^5-fold dilutions. Assume that the dilutions were performed in five steps of 10-fold each and that the pipetting error of a single 10-fold dilution has a c.v. of 0.02. Then the overall error of the five dilution steps is $\sqrt{5} \times 0.02$. This result is obtained by the repeated use of the second rule. It then must be combined with the Poisson error. Since the best estimate of the Poisson error c.v. is $1/\sqrt{585.5}$, then the overall c.v. is as follows:

c.v. $= \sqrt{1/585 + 5\,(0.02)^2}$

$= \sqrt{0.001709 + 0.00200} = 6.1\%$

This 6.1% error is composed of a Poisson counting error of $4.1\% = 100/\sqrt{585.5}$ and an error due to the cumulative pipetting errors of $\sqrt{5} \cdot 2\% = 4.47\%$ The rule to combine them gives a value smaller than their sum $(4.1 + 4.47 = 8.57\%)$ but larger than the largest component error.

Two important experimental considerations derive from this example. First, *there is no reason for increasing the accuracy of one part of an experiment unless other sources of error comparable to it are also increased in accuracy.* Second, *if an operation is to be done many times, it is worthwhile to devise a way to do it accurately and then not carry out elaborate statistical calculations.* In the previous section, the pipetting error was neglected because it was assumed that pipetting can be and was done accurately. This is a reasonable thing to do if the c.v. of this error is less than one-half of the

Poisson counting error. Imagine that each pipetting operation had been carried out with an accuracy of 1% instead of 2%. Then the overall pipetting error would have been $\sqrt{5} \cdot 1\% = 2.23\%$ and the overall total c.v. consequently would have been $\sqrt{4.1\%^2 + 2.23\%^2} = 4.7\%$, only a little bit larger than the Poisson counting error by itself.

Similar logic follows for cases in which blank values and background values are to be subtracted and in which the first rule for propagation of errors applies, or in which the instrument or procedure is standardized and the second rule applies. In the measurement of controls used repeatedly, errors should be reduced by repetitions or by more accurate measurement than for experimental values so as not to contribute significantly to the overall error.

11.5.4. Ratio Accuracy

There is a very powerful and general method applicable to experimental conditions varying between large natural ecosystems at one extreme and a drop of culture on an electron microscope grid at the other. This method is to add a known number of reference particles, which may be bacteria, ferritin particles, polystyrene spheres, abortively transduced bacteria, plasmids, viruses, etc. After mixing takes place, samples are taken and the ratio of the number of bacterial cells of interest to the number of reference particles is determined by appropriate means. The method may be illustrated for the case of microscopic smears containing a class of recognizable organism of unknown number and reference polystyrene beads. The smear was prepared from a known volume of cellular suspension and a known volume of suspension of beads of known concentration. Multiply the concentration of the known beads by the ratio of the counts of the unknown cells relative to those of the reference beads to calculate the concentration of unknown cells. The second rule for the propagation of error applies in this case. If the concentration of the reference particles is known without error in the original stock solution, then the coefficient of variation (c.v.) of the unknown particles is given by

$$\text{c.v.} = \sqrt{1/N_u + 1/N_r}$$

where N_u and N_r are the counts of the unknown and reference, respectively.

To minimize the number of total counts following the first argument in the previous action, N_u should be about equal to N_r. Then the c.v. will be about $\sqrt{2} = 1.4$-fold larger than if an infinite number of reference cells (or unknown cells) were counted.

11.5.5. Coincidence Correction

Coincidence corrections need to be applied when too many colonies are on a plate, or too many cells are on a square of a counting chamber, or if too many radioactive decays are recorded by a radioactivity counter. For the case of a colony count, assume that, if two cells initially are closer together than a distance r, they will be counted as a single colony. Let N_t be the true count and N_a be the actual count, and assume that the radius of the petri dish is R. Consider a single cell; the chance that another cell is within a distance r is $N_t \pi r^2 / \pi R^2$; thus, the count is decreased by $N_t r^2 / R^2$. If there are N_a colonies, the number lost will be $N_a N_t r^2 / R^2$. Therefore:

$$N_t = N_a + N_a N_t r^2 / R^2 = N_a (1 + cN_t)$$

where $c = r^2 / R^2$. It is usually convenient to substitute N_a for N_t on the far right-hand side when the correction is small. From this formula it is clear why a fourfold reduction in colony size reduces the coincidence correction at a given count by 16-fold. This is the basis of the overlay plate described above.

11.5.6. Exponential Growth Calculations

Under constant conditions after a long enough time when cell-cell interaction is small, growth measured in any manner is expected to proceed according to $X = X_0 e^{\lambda t}$. This can be written in any of the following equivalent ways:

$$\ln X = \ln X_0 + \lambda t$$
$$\log X = \log X_0 + \lambda t / 2.303$$

or

$$X = X_0 2^{t/T_2}$$

In the last equation, T_2 is the **doubling time** and can be calculated from the following:

$$T = (\ln 2)/\lambda = 0.6931/\lambda$$

Many symbols other than λ have been used for the **specific growth rate** (also called the growth rate constant), including a, k, and μ. Knowing these other symbols is important because many papers use the symbols without defining them. The most confusion arises with μ, which designates the specific growth rate in the literature on continuous culture but is used to measure the number of doublings per hour in the literature on cell physiology. The latter usage differs from the former usage by a factor of $\ln 2 = 0.6931$. The symbol of choice is λ because it is little used for other purposes in microbiology. Any time unit could be used, but per hour appears to be nearly standard. The doubling time (T_2) is reported in either minutes or hours.

11.5.7. Plotting and Fitting Exponential Data

There are several alternative ways to fit data to the exponential growth model that are equally valid but differ in their precision and in the additional information given to the experimenter.

The most simple conceptually is to look up the natural (base e = base 2.718) logarithms of the concentration of cells (or the dry weight of biomass, or other measurement of an extensive property of the bacteria) and plot them on ordinary arithmetic graph paper against the time the measurements were made. Then draw a straight line through the data points as close to the points as possible, and determine its slope by the rise-over-fall method. If the time scale is in hours, then the slope is in units of $(\text{hours})^{-1}$. At this point ask two questions: How appropriate is a straight line to the data? Is the line drawn a good summary of the data?

Common (base 10) logarithms may be used, in which case the slope must be multiplied by $2.303 = \ln 10$. Base 2 logarithms also may be used (tables are given in reference 5), in which case multiply by $0.6931 = \ln 2$. There is an advantage in using base 2 logarithms; this is because the reciprocal of the slope is the doubling time. Whatever the base of logarithm used, plot both the characteristic and mantissa numbers on the arithmetic scale of the graph paper. Do not make the all-too-frequent blunder of mixing them (e.g., $\log_{10}$ of 4×10^8 cells = 8.6 $\neq$ 8.4)!

It usually is more convenient to use semilogarithmic graph paper. Such paper that has six divisions between darker lines on the arithmetic scale and two cycles on the logarithmic scale is recommended (Keuffel and Esser Co., no. 46-4850). Define the abscissa (X-axis) according to hours or days on the major lines so that an even number of minutes or hours is represented by the minor lines. Mark each of the three unit labels with the appropriate powers of 10 on the ordinate (Y-axis). The printed scales may be multiplied by a constant, but it is invalid to add or subtract a constant from the logarithmic scale. Find the point corresponding to the amount of biomass, cell numbers, or other measure of extent of growth on the Y-axis and mark the point exactly. It is useful to make a small point marker for exactness and surround it with a larger circle for better visualization. It also is useful, when light-scattering methods are used, to plot each point as soon as obtained. This frequently will show when errors have been made, whether they be biological, instrumental, or arithmetic. If the error is detected immediately, then one has the opportunity to restart a culture, remeasure the culture, or replot a point. Once all the data have been plotted, draw the best straight line to fit the points.

The mathematical procedure that does this best is called the **least-squares fit**, which minimizes the square of the vertical distance of all points to the proposed line. This can be approximated visually by mentally noting the distance from those few points that are farthest from the position of a proposed straight line by moving a transparent ruler. By readjusting the ruler and remembering that the distances are to be squared, one can do a quite accurate job of drawing a line very nearly that which would be generated by the mathematical least-squares procedure.

Figure 6 shows an actual growth curve with an "eye-balled line" which is essentially the same as the computer-fitted line. This example is drawn from a carefully executed experiment with accurate data over a time permitting a 30-fold increase in cell mass. Note that the data points fit close to the line. This implies that no overt error was made during the experiment. This example was chosen to show how sensitive growth rate measurements are to the accuracy of the data and to show the way the line is drawn. The dashed line is almost as good a representation of the data and is indistinguishable from the computed line over the range from 10 to 20 μg (dry weight)/ml, but corresponds to a 2.7% difference in λ or T_2.

Note that there are many kinds of semilogarithmic graph paper classified by how many cycles (powers of 10) they span. The fewer the number of cycles, the more spread out the points are and the easier it is to define the line accurately. Note also that there is only one type of semilogarithmic paper. The same paper is to be used whether the user chooses to work and think with natural, common, or base 2 logarithms.

Use the graph to find the doubling time, since this is the measure of growth rate which has the most intuitive appeal to bacteriologists (later calculate the specific growth rate, if desired). The doubling time is easily found by measuring the time for the bacterial concentration to change twofold. Put a tick mark on the graph where the line crosses some major division on the logarithmic scale of the paper; make another on the division twofold higher or lower. Read both times off the abscissa and subtract them to

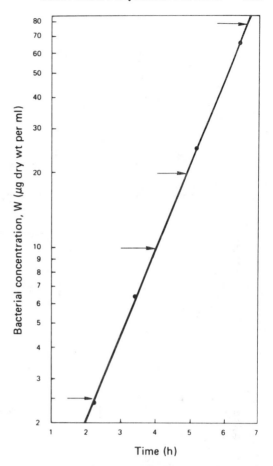

FIG. 6. *Growth curve plotted on semilogarithmic paper. See text for conditions. The horizontal arrows are drawn to the curve at values of bacterial concentrations that are convenient powers of 2 apart, which facilitates estimating the doubling time.*

obtain the doubling time. Alternatively, better precision can be obtained by using fourfold, eightfold, or 2^N multiples of mass and then dividing the time difference by 2, 3, or N to correct the resulting time difference to one doubling time. The specific growth rate or other measures of growth rate can then be calculated.

11.5.7. Least-Squares Fitting

The least-squares procedure is more precise for fitting data, and there is no reason today not to make the calculation this way since pocket calculators with linear regression capability are readily available at low cost. Table 3 shows the data for Fig. 6 treated on a Hewlett-Packard pocket calculator with least-squares capability. By simply following the instruction manual, any

TABLE 3. *Fitting the exponential growth model*

Data

Data were obtained from a subculture of *E. coli* ML308 growing in 0.2% glucose-M9 medium at 28°C:

As recorded		As converted	
Time (h:min)	A_{420}	Time (min)	W (µg/ml)
2:13	0.017	133	2.39
3:23	0.045	203	6.36
3:57	0.071	237	10.09
5:10	0.174	310	25.14
6:23	0.439	383	66.50

Graphical estimation

From Fig. 6 (which is a plot of data in column 1 versus data in column 4), determine the time at any value of W, and also at twice that value (e.g., at $W = 10$ and 20):

Time	W
4:00	10
4:52	20

The time difference = 52 min; T_2 = 52 min; and the ratio = 2.

Alternatively, determine the time at any value of W and at 2^N times that value (e.g., at $W = 2.5$ and for $N = 5$ at $2.5 \times 2^5 = 80$):

Time	W
2:15	2.5
6:38	80

The time difference = 4:23 = 263 min; $T_2 = \dfrac{263}{5} = 52.6$ min.

Pocket calculator computation

Obtain the regression of the logarithms of data in column 4 against data in column 3 with the following steps:

1. Clear memory
2. Key 2.39, ln, key 133, Σ+
3. Key 6.36, ln, key 203, Σ+
4. Key 10.09, ln, key 237, Σ+
5. Key 25.14, ln, key 310, Σ+
6. Key 66.50, ln, key 383, Σ+
7. Key linear regression. Read ln $W_0 = -0.85456$.
8. Key Y ⇄ X. Read $\lambda = +0.01321$/min.
9. These results can be expressed as ln $W = -0.85456 + 0.013213t$, or $W = 0.4255\ e^{+0.013213t}$, when t is expressed in minutes.
10. To reexpress the specific growth rate on a per hour basis, multiply by 60 to give $\lambda = 0.793$/h.
11. To reexpress in terms of doubling time, divide by ln 2 to give $T_2 = 52.47$ min.
12. For an error calculation (see text), $T_2 = 52.47 \pm 0.65$ min, and c.v. = 1.24%.

calculator with linear regression functions can be used.

The first step is to accumulate the sums, sums of squares, and sums of cross products. Take care to calculate the natural logarithm (ln) of the cell count (or other measure of cell substance) and to use it as the Y variable. Key the time in for the X variable. Care must be taken to convert the time to decimal units before executing the cumulation procedure (Σ+). Some calculators convert hours and minutes into decimal hours; thus, if 10.30 h is keyed in and this is thought of as 10 h and 30 min, then a function available on some calculators converts this to decimal time 10.50 which is then suitable for regression. Be careful if time goes past noon (or midnight) to add 12 h to the time. Alternatively, one can calculate elapsed time from the start of the growth run or have used an elapsed time counter that reads in decimal time during the execution of the experiment. Many clocks available for laboratory use read in minutes and decimals of minutes. The clock can also be watched and measurements can be made at the hour, the half hour, or at any regular time interval.

It is important to use time as the X variable. This is because the formulas used in the linear regression have been derived by assuming that all error in the measurement is error in the Y variable and that there is none in the X variable. In the case of growth measurements, the time can be precisely defined (if necessary, to the second), and the major error is in the measurement of the number of cells or other indices of biomass.

The calculator derives the best estimate of the intercept and slope by the formulas applicable to the linear equation $Y = a_0 + a_1X$:

$$a_0 = \frac{(\Sigma X^2)(\Sigma Y) - (\Sigma X)(\Sigma XY)}{n(\Sigma X^2) - (\Sigma X)^2}$$

$$a_1 = \frac{n(\Sigma XY) - (\Sigma X)(\Sigma Y)}{n(\Sigma X^2) - (\Sigma X)^2}$$

where a_0 is the estimate of the value of Y when X is zero, and a_1 is the slope of the regression. For the case of exponential growth, the value of a_0 is the natural logarithm of the initial cell density. This is of lesser interest than the slope (a_1), which is the best estimate of the specific growth rate (λ).

A very real problem arises with the usual use of a linear regression (in its unweighted form). To see this more clearly, examine the hypothetical situation where the problem does not arise. Imagine that growth is measured by viable counts and dilutions were made at every time point in such a way that precisely 400 colonies developed on each plate. In this case the coefficient of variation is $\sqrt{400}/400 = 5\%$ and is constant throughout the growth period, even though growth may have increased the number of cells per unit volume of culture many orders of magnitude during the experiment. This means that each value of ln N is known with a standard deviation of ln $(1.05) = -0.05$ above and ln $(0.95) = -0.05$ below the standard mean value, and is thus constant throughout the experiment. Constancy of the error in the Y variable was another of the basic assumptions made in deriving the least-square unweighted linear regression formulas.

Now imagine that growth is followed turbidimetrically and without diluting samples for measurement. At low turbidities the coefficient of variation is high, because of difficulty of matching and retaining balance of the blank. At an apparent absorbance of 0.4 the relative error is at a minimum; at higher densities the photometric precision falls and the scales cannot be accurately read. Consequently, the coefficient of variation rises. There are additional errors involved in correction at high density for deviations from Beer's law and for problems resulting from stray light.

For these reasons, after logarithms have been taken, the standard deviation is not constant but falls to a minimum at intermediate densities and then becomes larger. Therefore, the data at intermediate densities have higher accuracy, but do not determine the slope of the line set by the experimental data as well as points at the low and high end that span a longer stretch of the line. One can overcome this problem and use the data in a balanced way by carrying out a weighted regression. To do so, one must estimate the error of each datum in one or another way. Ordinarily this is a great deal of experimental and computational work. Alternatively, one can make turbidimetric measurements with techniques that are very precise so that the errors are very small. Additionally, one can take more data points and exercise more care in regions at the low and high ends of the curves.

If the data are treated by computer or hand calculator, no arithmetic mistakes are likely to be made. But human copying error can easily invalidate the results. For this reason, it is recommended strongly that a graph should be made so that the results of the computation can be compared and qualitatively checked with the graphical results. Alternatively, the error of the slope should be calculated. The error of the slope of a linear regression is given by:

$$s_{a_1} = s_\lambda = \sqrt{\frac{(\Sigma Y^2 - a_0\,\Sigma Y - a_1\Sigma XY)n}{(n-2)(\Sigma X^2 - (\Sigma X)^2/n)}}$$

Except for ΣY^2, most of the summations in this formula were also needed by the formula for calculating a_0 and a_1. Fortunately, ΣY^2 is needed for calculating correlation coefficients and for the standard deviation of the Y-values. For this reason, most modern pocket calculators calculate everything needed for the error calculation when the data pair is entered with the $\Sigma+$ key. Many older machines do not. With ΣY^2 accumulated, it is no great difficulty to program the programmable versions of the hand calculators to calculate the error of the slope. The error given in Table 3 was calculated this way. One could also calculate errors of the intercept, the mean of X and of Y, and also calculate the confidence interval. But the error of the specific growth rate is the crucial value and worth the effort to calculate. The standard error of the doubling time can be calculated to sufficient accuracy by assuming that λ and T_2 ($= \ln 2/\lambda$) have the same coefficient of variation. Algebraic manipulation shows that the standard deviation of the doubling time is the standard deviation of the specific growth rate multiplied by ln 2 and divided by λ^2:

$$s_{T_2} = \frac{s_\lambda \ln 2}{\lambda^2}$$

In our laboratory, we have programmed a minicomputer to take growth data and compute λ and T_2. The values of the coefficient of variation, which are also computed, have turned out to be extremely useful, particularly in detecting gross errors, whether due to copying errors or forgetting to replace the sparger in the culture after taking a sample. It serves the same role in actually executing a growth experiment as making a working plot on semilogarithmic graph paper while the experiment is in progress. The error in the slope of the regression has some of the same properties as does the Poissonian $\sqrt{N}$ as an estimate of the error of a count. It tells about the internal accuracy of the measurement, but does not assess external sources of variation.

The specific growth rate can be measured today with a precision within a run that is better than the reproducibility of the specific growth rate from run to run or of the reproducibility of subcultures of bacteria from day to day. The sources of this variation are unknown at present, but this is an important field of research for the future. However, these run-to-run variations must be taken into consideration for the practical purposes of making and using growth rate measurement to study aspects of bacterial biology.

Most often, the reason for measuring growth rate is to measure the effect of some treatment. This requires comparison of growth with and without treatment. Ultimately, comparison requires replicate control cultures and replicate experimental cultures. Then, the standard t test of the specific growth rates of the two groups suffices to test the probability that the groups are not different. Other statistical procedures should be used when it is evident that there is a difference and the effect of the treatment is to be quantitated (see any statistics text, e.g., reference 51). The real value of the calculation of s_λ is in allowing the experimenter to replicate the treatments and the controls less often, once the relationship of the internally measured error to the external measured error has been measured often enough to instill confidence. If they are always comparable when examined, then single growth runs have more significance—when backed up with the error of the slope.

Another way to increase the precision of the measurement of the effect of a treatment when the variation from subculture to subculture is important is the "subject as its own control" method. Start by following the growth of an untreated culture. When the growth rates have been determined, treat the culture and follow growth to remeasure the specific growth rate. This procedure isolates the treatment as the sole variable between the two halves of the growth run. Run control experiments in which the treatment is only a sham treatment.

11.5.9. Nonexponential Growth

Except under turbidostat and chemostat conditions, growth does not stay balanced (if ever achieved) for very long. Even under continuous culture conditions there can be long-term changes that cause deviations from a constant state of balanced growth. So in all cases, the assumption of an exponential model is a limited one from the point of view of the biological growth process, independent of the difficulties and imprecision of making measurements. It becomes necessary, therefore, to limit the range of measurement. For batch cultures, this means excluding low- and high-density data, or conducting the experimental measurements only over a limited range of growth density. Both of these approaches are attempts to obtain balanced growth by restricting the part of the total growth process considered. All such choices and judgments are difficult, and are impossible to justify from a statistical point of view. It is possible to refine growth measurements to the point where the constancy of growth can be assessed. This has been attempted by utilizing a computer-linked double-beam spectrometer and a long-path aerating cuvette to achieve accuracy and high sensitivity (39). With this equipment and under optimum conditions, the measurements are so numerous and accurate that the standard deviation is less than 1% of the specific growth rate in a measurement period of only 200 s. Certain conditions are achievable (very rich media, very low cell density) where growth is constant within this limit, but under other conditions the growth rate fluctuates in a way that depends on the previous history.

11.6. LITERATURE CITED

11.6.1. General References

1. **Dawson, P. S. S. (ed.).** 1974. Microbial growth. Halsted Press, New York.
   A collection of papers that dominated the study of physiology of bacterial growth.
2. **Gall, L. S., and W. A. Curby.** 1979. Instrumental systems for microbiological analysis of body fluids. CRC Press, West Palm Beach, Fla.
3. **Kavenagh, F. (ed.).** Analytical microbiology, vol. 1, 1963, and vol. 2, 1972. Academic Press, Inc., New York.
4. **Meadows, P., and S. J. Pirt (ed.).** 1969. Microbial growth. Symposium of the Society for General Microbiology, vol. 19. Cambridge University Press, London.
5. **Meynell, G. G., and E. Meynell.** 1970. Theory

and practice in experimental bacteriology, 2nd ed. Cambridge Press, Cambridge. *Excellent, critical discussion of techniques and methods.*

6. **Miller, J. H.** 1972. Experiments in molecular genetics. Cold Spring Harbor Laboratory, Cold Spring Harbor, N.Y.
7. **Pirt, S. J.** 1975. Principles of microbe and cell cultivation. Blackwell Scientific Publications, Oxford.

11.6.2. Specific References

8. **Bridges, B. A.** 1976. Symp. Soc. Gen. Microbiol. **26**:183-208.
9. **Brock, T. D.** 1967. Science **155**:81-83.
10. **Brock, T. D.** 1971. Bacteriol. Rev. **35**:39-58.
11. **Buchanan, R. E.** 1918. J. Infect. Dis. **23**:109-125. *This classic has been reprinted in reference 1.*
12. **Calam, C. T.** 1969. *In* J. R. Norris and D. W. Ribbons (ed.), Methods in microbiology, vol. 1, p. 567-592. Academic Press, Inc., New York.
13. **Campbell, A.** 1957. Bacteriol. Rev. **21**:263-272.
14. **Capaldo, F. N., and S. D. Barbour.** 1975. *In* P. C. Hanawalt and R. B. Setlow (ed.), Molecular mechanisms for repair of DNA, part A, p. 405-418. Plenum Publishing Corp., New York.
15. **Cato, E. P., C. S. Cummins, L. V. Holdeman, J. L. Johnson, W. E. C. Moore, R. M. Smibert, and K. D. S. Smith.** 1970. Outline of clinical methods in anaerobic bacteriology. Virginia Polytechnic University, Blacksburg.
16. **Clifton, C. E.** 1966. J. Bacteriol. **92**:905-912.
17. **Drake, J. F., and H. M. Tsuchiya.** 1973. Appl. Microbiol. **26**:9-13.
18. **Finney, D. J.** 1964. Statistical method in biological assay, 2nd ed., p. 570-586. Hafner Publishing Co., New York.
19. **Gray, T. R. G., and J. R. Postgate.** 1976. Symp. Soc. Gen. Microbiol. **26**:1-43.
20. **Günther, H. H., and F. Bergter.** 1971. Z. Allg. Mikrobiol. **11**:191-197.
21. **Halvorson, H. O., and N. R. Ziegler.** 1933. J. Bacteriol. **25**:101-121.
22. **Haney, T. A., J. R. Gerke, and J. F. Pagaon.** 1963. *In* F. Kavenagh (ed.), Analytical microbiology, p. 219-247. Academic Press, Inc., New York.
23. **Hedén, C.-G., and T. Illéni.** 1975. Automation in microbiology and immunology. John Wiley & Sons, Inc., New York.
24. **Holdeman, L. V., E. P. Cato, and W. E. C. Moore (ed.).** 1977. Anaerobe laboratory manual, 4th ed. Virginia Polytechnic Institute and State University, Blacksburg.
25. **Hungate, R. E.** 1971. *In* J. R. Norris and D. W. Ribbons (ed.), Methods in microbiology, vol. 3, p. 117-132. Academic Press, Inc., New York.
26. **Kavenagh, F.** 1972. *In* F. Kavenagh (ed.), Analytical microbiology, p. 43-121. Academic Press, Inc., New York.

27. **Kerker, M., D. D. Coke, H. Chew, and P. J. McNulty.** 1978. J. Opt. Soc. Am. **68**:592-601.
28. **Koch, A. L.** 1961. Biochim. Biophys. Acta **51**:429-441.
29. **Koch, A. L.** 1968. J. Theor. Biol. **18**:133-156.
30. **Koch, A. L.** 1970. Anal. Biochem. **38**:252-259.
31. **Koch, A. L.** 1971. Adv. Microb. Physiol. **6**:147-217.
32. **Koch, A. L.** 1975. J. Bacteriol. **124**:435-444.
33. **Koch, A. L.** 1975. J. Gen. Microbiol. **89**:209-216.
34. **Koch, A. L.** 1976. Perspect. Biol. Med. **20**:44-63.
35. **Koch, A. L., and E. Ehrenfeld.** 1968. Biochim. Biophys. Acta **169**:44-57.
36. **Koch, A. L., and G. H. Gross.** 1979. Antimicrob. Agents Chemother. **15**:220-228.
37. **Kubitschek, H. E.** 1964. Rev. Sci. Instrum. **35**: 1598-1599.
38. **Kubitschek, H. E.** 1969. *In* J. R. Norris and D. W. Ribbons (ed.), Methods in microbiology, vol. 1, p. 593-610. Academic Press, Inc., New York.
39. **Maaløe, O., and N. O. Kjeldgaard.** 1966. Control of macromolecular synthesis. W. A. Benjamin, Inc., New York.
40. **Mallette, M. F.** 1969. *In* J. R. Norris and D. W. Ribbons (ed.), Methods in microbiology, vol. 1, p. 521-566. Academic Press, Inc., New York.
41. **Mitruka, B. M.** 1976. Methods of detection and identification of bacteria. CRC Press, Cleveland.
42. **Norris, K. P., and E. O. Powell.** 1961. J. R. Microsc. Soc. **80**:107-119.
43. **Novick, A., and L. Szilard.** 1951. Proc. Natl. Acad. Sci. U.S.A. **36**:708-719.
44. **Perfil'ev, B. V., and D. R. Gabe.** 1969. Capillary methods of investigation of microorganisms. University of Toronto Press, Toronto.
45. **Postgate, J. R.** 1967. Adv. Microb. Physiol. **1**:2-23.
46. **Postgate, J. R.** 1969. *In* J. R. Norris and D. W. Ribbons (ed.), Methods in microbiology, vol. 1, p. 611-628. Academic Press, Inc., New York.
47. **Ryan, F. J., G. W. Beadle, and E. L. Tatum.** 1943. Am. J. Bot. **30**:789-799.
48. **Schmidt, E. L., R. O. Bantkole, and B. B. Bohlool.** 1968. J. Bacteriol. **95**:1987-1992.
49. **Shehata, T. E., and A. G. Marr.** 1971. J. Bacteriol. **107**:210-216.
50. **Shuler, M. L., R. Aris, and H. M. Tsuchiya.** 1972. Appl. Microbiol. **24**:384-388.
51. **Sokal, R. R., and F. J. Rohlf.** 1969. Biometry. W. H. Freeman & Co., San Francisco.
52. **Spielman, L., and S. L. Goren.** 1968. J. Colloid Interface Sci. **26**:175-182.
53. **Vollenweider, R. A.** 1974. A manual on methods for measuring primary production in aquatic environments. IBP Handbook No. 12. Blackwell Scientific Publications, Oxford.
54. **Wyatt, P. J.** 1973. *In* J. R. Norris and D. W. Ribbons (ed.), Methods in microbiology, vol. 8, p. 183-263. Academic Press, Inc., New York.
55. **Zimmerman, U., J. Schultz, and G. Pilwat.** 1973. Biophys. J. **13**:1013.

Chapter 12

Preservation

ROBERT L. GHERNA

Most bacteriology laboratories maintain stock cultures for educational, research, bioassay, industrial, and other purposes. The proper preservation of cultures is extremely important and should be (but often is not) given the same attention as the standardization of equipment and the selection of chemical compounds. Numerous programs have been hampered by the loss or variation of a stock culture due to improper preservation.

The primary aim of culture preservation is to maintain the organism alive, uncontaminated, and without variation or mutation, that is, in a condition which is as close as possible to the original isolate. Many methods have been employed to preserve bacteria, but not all species respond in a similar manner to a given method. The availability of equipment, storage space, and skilled labor often dictates the method employed.

Record keeping is an important and often-neglected aspect of stock-culture maintenance. Too often, isolates are not given strain designations, or if they are, the designations are not documented in strain data files. A strain designation (e.g., by number or letter) refers to a specific strain with distinct phenotypic characteristics which should be recorded. The documentation should also include the isolation source and the investigator who isolated the strain, since the same number may be assigned to other strains by different workers. Some strain variations which have been attributed to preservation conditions have turned out to be due to poor strain documentation, e.g., the wrong number was noted on the data card or on the containers.

12.1. SHORT-TERM METHODS

12.1.1 Subculturing

The traditional method of preserving bacterial cultures is through periodic transfer to fresh media. The interval between subcultures varies with the organism, the medium employed, and the external conditions. Some bacteria must be transferred every other day, others only after several weeks or months. Three conditions must be determined when using this method for the preservation of cultures: (i) suitable maintenance medium, (ii) ideal storage temperature, and (iii) the frequency between transfers.

Maintenance medium.

Minimal media (13) are preferred because they lower the metabolic rate of the organism and thus prolong the period between transfers. Some bacteria, however, require complex media for growth, or their retention of specific physiological properties necessitates the presence of complex compounds in the medium. When a complex medium is used, more frequent transfers may be necessary as a result of accelerated growth or metabolic end product accumulation.

Storage.

The simplest method of storage is at room temperature in a test tube rack or in a specially constructed storage box containing shelves with holes for test tubes. Cultures stored in this fashion require constant care since they tend to dry quickly and, unless the laboratory has a controlled environment, are subject to temperature fluctuations.

To minimize the dehydration process, use screw caps with rubber liners, wrap the top of tubes in Parafilm (VWR Scientific, Inc.; no. 52858-000), or place the tubes in a plastic bag (VWR Scientific, Inc.; no. 11215-790). To reduce the metabolic rate of the organism, place the culture in a refrigerator at about 5 to 8°C. The majority of bacteria can be kept for 3 to 5 months between transfers by using these precautions.

Transfer schedule

The subculturing interval is determined by experience. Keep subculturing to a minimum in order to avoid the selection of variants. Maintain duplicate tubes as a precaution against loss. Examine the cultures for purity after each transfer, and perform an abbreviated characterization check periodically to monitor any changes in phenotypic characteristics. Do not select single colonies in transferring cultures since the chances of selecting a mutant are greater when this technique is used.

The major disadvantages of the serial transfer technique are the risks of contamination, transposition of strain numbers or designations (mislabeling), selection of variants or mutants, the possible loss of culture, and the required storage space.

12.1.2. Immersing in Mineral Oil

Many bacterial species can be successfully preserved for months or years by the relatively simple and cheap method of immersing in sterile medicinal-grade mineral oil (paraffin oil of specific gravity of 0.865 to 0.890 is satisfactory). Contamination of cultures when this method is used is often due to improperly sterilized mineral oil.

Sterilize the oil by heating in an oven at 170°C for 1 to 2 h; autoclaving is not recommended (19).

Grow the culture in the appropriate medium as an agar slant or stab or as a broth culture. After suitable growth occurs, add the sterile mineral oil aseptically to a depth of at least 2 cm (a slant must be entirely covered) in order to prevent dehydration and reduce metabolic activity and growth of the culture (5, 11).

Store the oil-covered cultures in an upright position at refrigerator temperature. Perform viability tests periodically to determine whether a culture is deteriorating.

The culture under oil can be transferred with an inoculating needle to fresh medium, and the subsequent growth can be overlaid with sterile oil. Take care when flaming the needle since the oil will splatter and contaminate the surrounding area and personnel. Keep the original culture for at least several weeks to enable recovery of the culture in the event that the subculture is contaminated or shows aberrant characteristics.

The disadvantages of this method are the same as those of ordinary subculturing (12.1.1). In addition, the oil is messy.

12.1.3. Ordinary Freezing

The preservation of bacteria in the freezing compartment of a refrigerator or an ordinary freezer with a temperature range of 0 to −20°C produces variable results, and its success depends on the bacterial species. In general, it is not recommended for preservation because of the damage to cells by eutectic concentrated electrolyte solutions—NaCl has a eutectic temperature of −20°C (14, 17). Some bacteria can be kept for 6 months to 2 years by this method, however.

12.1.4. Drying

Most cultures die if left to dry in the laboratory. However, some cultures, especially sporeformers, can be preserved for years by drying on a suitable menstruum (3).

Soil.

Sporeforming bacteria have been successfully preserved for years by use of a mixture of sterile, air-dried soil (10). Sterilize the soil by autoclaving for several hours on two successive days. Inoculate a spore suspension (1 ml) into tubes containing the sterile soil and allow to stand at room temperature until visibly dry. Close the tube with a sterile rubber stopper and store at refrigerator temperature.

Paper.

A relatively simple and inexpensive method of preserving bacteria employs drying on sterile filter-paper strips or disks. The technique is ideally suited for the maintenance of quality control cultures. Many disks containing the same culture can be stored in a single test tube or screw-capped vial. A single disk can be removed aseptically with sterile forceps as needed and inoculated into a suitable broth. Members of the *Enterobacteriaceae* as well as a variety of other bacteria are successfully preserved for several years in this manner.

The procedure for this method is as follows. Saturate the sterile paper with a bacterial suspension containing 10^8 cells or more per ml, and dry in air or vacuum (2). Dry in vacuum for greater survival. Store the strips or disks sealed in tubes in desiccators or between two layers of sterile clear plastic (7). Store the desiccators in a refrigerator, because this extends the shelf life of the culture.

Gelatin.

Many heterotrophic bacteria can be preserved in dried gelatin drops or disks (24). The storage temperature is important for successful preservation. In general, $-20°C$ is superior to $+4°C$ or room temperature.

Prepare the gelatin cultures as follows. Grow the culture in the appropriate medium and harvest by aseptic centrifugation. Resuspend the pellet in a small amount of broth, and inoculate a tube containing 2.0 to 5.0 ml of melted nutrient gelatin held at $30°C$ to yield a density of 10^8 to 10^{10} cells per ml. Using a sterile Pasteur pipette or syringe, place a drop of the bacterial suspension on the bottom of a sterile plastic petri plate. Place the petri plate in a desiccator containing phosphorus pentoxide and evacuate with a vacuum pump. After the drops have dried, aseptically transfer them to sterile screw-capped tubes and store at refrigerator temperature. Propagate the cultures by aseptically transferring a gelatin drop into a tube containing a suitable medium.

12.2. LONG-TERM METHODS

12.2.1 Freeze-Drying (Lyophilization)

Freeze-drying or lyophilization is one of the most economical and effective methods for long-term preservation of bacteria and other microorganisms. Many physiologically diverse bacterial species and bacteriophages have been successfully preserved by this technique and have remained viable for over 30 years. The method enables large numbers of vials to be produced, and the small size of the vial facilitates storage. While the freeze-drying procedure is relatively simple, the theoretical aspects are complex. The reader is referred to several excellent reviews for an in-depth discussion (1, 21, 22).

Freeze-drying involves the removal of water from frozen bacterial suspensions by sublimation under reduced pressure; that is, the water is evaporated without going through a liquid phase. The dried cells can be stored for long periods if kept away from oxygen, moisture, and light. They can, at any time, be easily rehydrated and restored to their previous state.

Lyophilization can be performed in several ways for which various types of apparatus have been devised. The simplest forms consist of a desiccator which can be cooled to allow the cell suspension to remain frozen while it is attached to a vacuum pump. Details of other procedures and equipment are given by Wickerham and Andreasen (26) and Haynes et al. (12).

Two of the most common methods employed are centrifugal freeze-drying and prefreezing. For a detailed description of centrifugal freeze-drying, see Muggleton (18) and LaPage et al. (2).

Equipment.

Freeze-drying equipment varies from a simple and relatively inexpensive desiccator-vacuum pump system to a complex commercial freeze-dryer which costs thousands of dollars. Reproducibility and shelf life will depend on the system employed. In general, however, excellent results can be obtained with a simple system consisting of a high-vacuum pump, a condenser, and a chamber or manifold (Fig. 1 and 2).

Vials.

Although dimensions and shapes of vials are variable, there are two basic designs used for freeze-drying. When the cells are to be freeze-dried in a chamber (Fig. 1), double vials are recommended (Fig. 3). Such vials are easier to handle, less susceptible to contamination, and safer for preserving pathogens than are single manifold vials (25). The latter are dried and sealed while directly attached to a manifold (Fig. 2).

For double vials, use a glass (soft-glass) shell vial (VWR Scientific no. 27921-015; 11.5 mm by 35 mm) for the inner vial, and prepare and sterilize it as described below. Prepare the outer (soft-glass) vial (VWR Scientific no. 27921-015; 14.25 mm by 25 mm) by covering the bottom with silica gel granules (Fisher Scientific Co., grade 42, 6–16 mesh, Tel-Tale brand) and adding a small wad of cotton to cushion the inner vial. Heat the outer vials at $100°C$ overnight before use. The silica gel should be dark blue in color after heating and should remain blue during sealing and storage of the cultures.

Two styles of vials are used in the single-vial manifold procedure. For shell freezing of cultures, use a 1-ml bulb or "tear-drop" (Kontes, 8-mm outer diameter) vial. Use a 1-ml tubular type ("Edwards") vial (Bellco Glass, Inc., 8-mm outer diameter) when a pellet is desired.

Lightly plug the inner vials of the double-vial system, and the manifold vials, with absorbent cotton, and sterilize in a hot-air oven at $170°C$ for 2 h. Some cotton contains residual oils and salts, which are released during heat sterilization and will leave a slight film on the inner walls. This film is toxic to some bacterial species. An alternative method of sterilization is by autoclaving for 1 h with the vials on their sides to allow steam penetration.

Vials must be properly labeled with ink which does not come off easily. A labeling machine

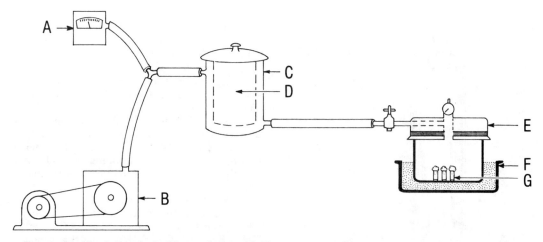

FIG. 1. *Double-vial method of freeze-drying. (A) Vacuum gauge; (B) vacuum pump; (C) VirTis condenser; (D) reservoir filled with dry ice and ethyl Cellosolve; (E) Atmo-vac plate; (F) stainless-steel pan filled with crushed dry ice and ethyl Cellosolve; (G) specimen*

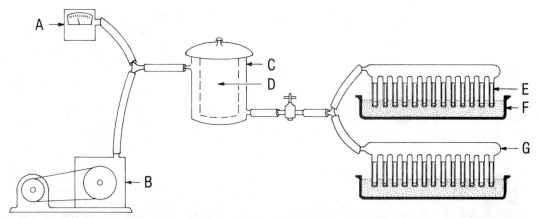

FIG. 2. *Manifold method of freeze-drying. (A) Vacuum gauge; (B) vacuum pump; (C) VirTis condenser; (D) reservoir filled with dry ice and ethyl Cellosolve; (E) specimen vial; (F) stainless-steel pan filled with crushed dry ice and ethyl Cellosolve; (G) manifold.*

(Markem Machine Co., model 135A) with specially formulated ink (Markem Machine Co., no. 7224) is recommended. This ink will withstand ultralow temperatures and brief exposure to solvents such as alcohol. After imprinting the letters and numbers on glass vials, heat them at 160°C for 20 min in a hot-air oven to set the ink.

Culture preparation.

Successful freeze-drying depends on using healthy cells grown under optimum conditions. Grow a sufficient number of cells to provide a suspension of at least 10^8 cells per ml. Harvest the culture at the time of maximum stability and viability, in the late logarithmic or early stationary phase.

Cryoprotective agents.

Prepare the cells for lyophilization by suspending them in a cryoprotective agent. The American Type Culture Collection has experienced considerable success in long-term preservation of physiologically diverse bacteria by employing either 20% skim milk for the double-vial method or a 24% sucrose solution diluted equally with growth medium (12% sucrose, final concentration) for the single-vial manifold procedure. Others have used 10% dextran, horse serum, inositol, and other chemical agents (8, 14, 20) as cryoprotective agents.

The following procedures are used for the double-vial method. Prepare a 20% (wt/vol) solution of skim milk (Difco 0032) and sterilize in

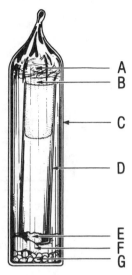

FIG. 3. *Double vial with dried specimen. (A) Glass-fiber wad; (B) cotton plug; (C) outer vial; (D) inner vial; (E) freeze-dried culture; (F) cotton cushion; (G) silica gel.*

small volumes (5 ml) at 116°C for 20 min. Avoid overheating; it can cause caramelization of the milk. For cultures grown on agar surfaces, harvest by aseptically washing the growth off with the 20% skim milk solution. Harvest broth cultures by aseptic centrifugation, and suspend the pellet with sterile skim milk to yield a cell suspension containing at least 10^6 cells per ml. Use the same procedure when employing sucrose as the cryoprotective agent.

Dispense 0.2 ml of the cell suspension into each vial, and trim the cotton plugs with scissors. Dispense as soon as the cell suspensions are prepared. The interval between dispensing and the freeze-drying process should be kept to a minimum to avoid possible alteration of the culture.

In the single-vial manifold method, dispense 0.1 ml of the bacterial suspension into each vial. Push the cotton plug about 1.3 cm below the rim of the vial, and flame the top to eliminate protruding cotton fibers.

Double-vial method.

Figure 1 shows a typical chamber freeze-drying system. All vacuum lines utilize Tygon tubing (⅜-inch [0.9 cm] bore, ⅞-inch [2.2-cm] wall thickness). The system is monitored by a thermistor vacuum gauge (Consolidated Vacuum Corp., model GT-340A) and should measure below 30 μm of mercury. When the vacuum sensor is placed between the condenser and the

vacuum pump, the pressure should remain constant. If the sensor is positioned between the chamber holding the vials and the condenser, it will sense the change in vapor pressure in the system and will register an increase in pressure as drying occurs. Upon completion of drying, however, the pressure should return to below 30 μm of mercury.

Place the filled "inner" vials upright in a stainless-steel pan. For efficient drying, use only a single layer of vials. Freeze the cells by placing the stainless-steel pan containing the vials for 1 h on the bottom of a mechanical freezer maintained at −60 to −65°C.

Prepare a moisture trap by trimming a block of dry ice to fit the center well of a condenser (The VirTis Co., Inc.), and add ethyl Cellosolve (Fisher Scientific Co., E-180). Close the stopcock in the vacuum line between the Atmo-vac plate (Refrigeration for Science, Inc.) and the condenser, and evacuate the system below 30 μm of mercury.

At the end of the hour, place the stainless-steel pan containing the vials in a larger shallow pan containing sufficient crushed ice to cover the bottom and surround the sides of the stainless-steel container. Let it stand for 2 to 3 h. Take care to avoid getting dry ice into the pan containing the vials, as this will interfere with the evacuation of the system. Attach the Atmo-vac plate immediately to the pan and hold firmly in place while opening the stopcock in the vacuum line. Monitor the stopcock to ensure that no pinhole leaks occur. Drying proceeds under vacuum of 20 to 30 μm of mercury for 18 h. For convenience, a run can be started late in the afternoon and allowed to proceed overnight.

Remove the outer vials containing the silica gel from the oven, and place in a dry box (maintained below 10% relative humidity) and allow to cool. Close the stopcock between the Atmo-vac plate and condenser. Connect the outlet on top of the Atmo-vac plate to a column of silica gel located inside the dry box, using rubber tubing. Open the valve on the Atmo-vac plate to admit dry air from the silica gel column to the stainless-steel pan. Transfer the vials to the dry box, and insert each vial into a soft-glass outer vial. Add a 0.6-cm wad of glass-fiber paper (Whatman, Inc.) above the cotton-plugged inner vial and tap down (the lines in Fig. 3 for the glass fiber depict creases, not layers).

Remove the double vials from the dry box and heat the outer vial just above the glass-fiber paper with an air/gas torch. Rotate the vial until the glass begins to constrict. Pull the vial slowly, using forceps, until the constriction is a narrow capillary.

Allow the vials to cool, and then attach them to a vacuum manifold (Fig. 2), using single-hole rubber stoppers which fit the open end of the vials. Evacuate to less than 50 μm of mercury and seal with a double-flame air/gas torch at the capillary constriction.

Manifold method (single-vial).

Attach the filled ampoules to sterile 2.5-cm pieces of amber latex intravenous tubing (American Hospital Supply Corp., nonpowdered, 17610-063; $\frac{3}{16}$ by $\frac{1}{16}$ inches [0.47 by 0.16 cm]) on the manifold (Fig. 2), using a rubber tubing stretcher. When the "Edwards"-type vials are employed, freeze the cell suspension by direct immersion in a dry ice-ethyl Cellosolve bath and attach to the manifold. Immerse the material in the bulb-type ampoules in the dry ice-ethyl Cellosolve bath, and rotate to shell freeze before placing the ampoules on the manifold.

Prepare a moisture trap as described for the double-vial freeze-drying method. The ampoules attached to the manifold remain in the dry ice-Cellosolve bath during the freeze-drying process. Dry with a vacuum below 30 μm of mercury for 18 h. During the freeze-drying cycle, the bath will warm to ambient temperature.

After completion of the drying process, seal the vials below the cotton plug, using a dual-tipped air/gas torch and continuously moving the flame up and down the ampoule within a 2.5-cm area. Allow the vials to cool.

Storage.

Freeze-dried cultures, both the double and single vial, are stored below 5°C. Extended shelf life has been obtained when cultures are stored at −30 or −69°C in a Revco freezer. Room-temperature storage of freeze-dried cultures should be avoided.

Recovery

Open double vials by heating the tapered end of the outer container vigorously in a Bunsen burner flame, and quickly add a drop or two of water to crack the hot glass. Remove the broken tip of the container carefully (preferably in a hood) with a sharp blow of a forceps, and withdraw the fiber-glass plug and inner vial. Clean the cotton plug of the inner vial to eliminate extraneous fibers which may contaminate the culture.

Open single vials by first scoring the ampoule with a triangular file approximately 2.5 cm from the tip. Disinfect the vial with a piece of gauze dampened with 70% alcohol. Wrap sterile gauze around the vial and break at the scored area. Flame the opened end gently, prior to rehydra-

tion of the culture. Open the vials in a laminar flow hood. Vials containing pathogenic bacteria should be opened only in a closed safety cabinet. Some investigators employ a tungsten needle, heated in an oxygen gas flame to white heat, to poke a small hole in the tip of the ampoule. This allows air to enter the vial more slowly and sterilizes the area around the hole. The culture is rehydrated by adding the broth with a sterile syringe and needle, and is removed with the same syringe.

Rehydrate freeze-dried cultures immediately after opening by adding 0.3 to 0.4 ml of a suitable sterile broth to the contents of the vial: mix well so that the pellet dissolves completely, and transfer to a test tube containing 5 ml of the rehydrating broth. After thorough mixing, transfer 0.2 ml of the rehydrated culture to an agar slant or a semisolid medium of the same composition. Pre-lyophilization and post-lyophilization purity checks must always be done. To accomplish this, the pre-lyophilization cell suspension or the reconstituted culture is serially diluted and streaked on solid media. Incubate the tubes and plates at the optimum growth temperature, and subculture onto fresh medium as soon as growth appears and purity is ascertained. Freeze-dried cultures often exhibit a prolonged lag period and should be allowed to incubate for an additional period before the culture is considered dead.

Monitoring viability.

Although freeze-drying has facilitated the long-term preservation of bacteria, viability checks must be done before and after freeze-drying in order to determine the effectiveness of the process. In addition, periodic viability tests must be done to ascertain the shelf life of the cultures. Characterization tests should also be performed on cultures which have been lyophilized to determine whether any changes have occurred as a result of the freeze-drying procedures or during storage.

12.2.2. Ultra-Freezing

Long-term preservation of bacterial species not amenable to freeze-drying has been achieved through storage in the frozen state at the temperature of liquid nitrogen (−196°C) or in the vapor phase (−150°C, well-insulated tanks). The American Type Culture Collection has successfully preserved many fastidious bacteria without the loss of phenotypic properties for over 15 years in this manner.

Long-term preservation of bacterial cultures also can be achieved by using ultra-low temperature mechanical freezers (e.g., Revco, Inc.) with a temperature of −70°C. This method of preserving cultures is fairly successful for a large variety of bacteria. Precautions must be taken, however, for electrical shutdowns or compressor malfunction. Adequate back-up systems such as alarms, back-up freezers, or an electrical generator will help to prevent the loss of a valuable collection. Commercial freezers (e.g., Revco, Inc.) are now equipped with battery-operated alarm systems as standard equipment.

Additional alarm systems are recommended, such as the sound/off power-temperature monitor (Arthur H. Thomas Co., no. 4052-W20) which has a temperature range of −75°C to 200°C (±0.3°C). The unit is battery powered and can be mounted on a wall.

Equipment.

Liquid nitrogen has been considered an expensive method for the long-term preservation of microorganisms. However, the recent history of the successful storage of physiologically diverse bacteria by this method, along with a decreased need for handling, makes liquid nitrogen feasible to use, especially when one considers the cost of labor. The cost of liquid nitrogen varies from 20 to 50 cents per liter.

A variety of liquid nitrogen refrigerators are now commercially available, with a wide assortment of features and storage capacity (e.g., MVE, Cryogenics, Linde, Union Carbide Corp.). The sizes range from 10 to 1,000 liters, permitting storage of 300 to 40,000 ampoules.

Vials.

Various kinds of liquid-nitrogen vials are available commercially. Several types which have been used successfully are heavy-walled borosilicate vials (Wheaton Scientific, Cryule no. 12483-unscored and no. 12523-prescored) and special wide-mouth vials (Wheaton, Gold Band Cell and Tissue Cryule no. 12742). These vials have a capacity of 1.2 ml and an indentation groove to facilitate opening. In addition, presterilized polypropylene tubes (38 mm by 12.5 mm) with screw caps and silicone washers (Vangard International) can be used in place of the glass ampoules.

Cryoprotective agents.

Cryoprotective compounds fall into two types: agents such as glycerol and dimethyl sulfoxide (DMSO), which readily pass through the cell membrane and appear to provide both intracellular and extracellular protection against freezing; and agents such as sucrose, lactose, glucose, mannitol, sorbitol, dextran, polyvinylpyrrolidone, and polyglycol, which appear to exert their protective effect external to the cell membrane. The former type has proven to be more effective, and glycerol and DMSO appear to be equally effective in preserving a wide range of bacteria. The exact choice of cryoprotective agent, however, depends on the bacterial species. A tolerance test must be done when freezing new species to ascertain whether the cryoprotective agent is toxic or beneficial.

Glycerol and DMSO are routinely employed at a concentration of 10% (vol/vol) and 5% (vol/vol), respectively, in an appropriate growth medium. Glycerol is sterilized by autoclaving at 15 lb/in^2 for 15 min. DMSO is filter sterilized by using 03 porosity Selas filter candles, collected in 10- to 15-ml quantities in sterile test tubes, and then stored in the frozen state at 5°C (DMSO freezes at 18°C). Because of the accumulation of oxidative breakdown products, an opened bottle of DMSO should not be used for more than 1 month.

Culture preparation.

The physiological condition of the culture plays an important part in the survival of the bacterium to liquid nitrogen freezing. In general, use actively growing cells at the mid- to late-logarithmic phase of growth (23).

Grow the cells in an appropriate medium. For broth cultures, harvest by aseptic centrifugation, and resuspend the pellet with sterile fresh medium containing either 10% (vol/vol) glycerol (prepared by adding 20% glycerol to an equal volume of sterile broth) or 5% (vol/vol) DMSO (prepared by adding the appropriate amount of 100% DMSO to the sterile broth). For agar cultures, wash the growth from the agar surface with sterile broth containing the suitable cryoprotective agent. Dispense 0.4 ml of the cell suspension containing at least 10^8 cells per ml into each sterile, prelabeled vial.

If the strain is pathogenic, trim the cotton plug level with the top of the ampoule. Store in unsealed ampoules in the vapor phase above the liquid nitrogen to avoid accidents during the sealing process and possible explosion (due to leakage of liquid nitrogen through undetected pinholes and rapid expansion of the gas at ambient temperature) during the thawing process.

For nonpathogenic bacteria, precool the ampoules for a minimum of 30 min at 4°C before sealing. Then heat the ampoules above the prescored area by rotating in the torch flame for several seconds to remove moisture. This procedure avoids pressure inside the vial, which could result in bubbling during sealing. Hold the ampoule at an angle to prevent the culture suspension from contacting the hot glass, and partially withdraw the cotton plug but stop just short of complete removal. Seal the vials, preferably with a semiautomatic sealer equipped with an oxygen gas torch (e.g., Kahlenberg-Globe Equipment Co.) which is designed to pull and seal, thus minimizing the formation of pinhole leaks.

Carefully remove the hot ampoule with a forceps and place in a test tube rack immersed in a cool-water bath deep enough to immerse the vial to a depth of 0.6 cm. Avoid getting water on the hot portion of the vial; otherwise, the glass will crack. When the ampoules have cooled, they are ready for freezing. If this process is performed satisfactorily, the temperature of the specimen should not rise above 25°C. If a semiautomatic sealer is unavailable, the vials can be sealed with a gas torch by hand, but caution must be used to determine the presence of pinhole leaks. After sealing, allow the glass ampoules to stand for approximately 30 min at

5°C in a 0.05% methylene blue solution. Improperly sealed vials are detected by penetration of the dye into the vial.

Freezing.

Much has been written about the effect on survival exerted by the cooling rate during freezing (4, 15). In general, best results in the survival and recovery of bacterial cultures have been obtained using a slow cooling rate (e.g., 1°C per min).

The procedures for this method are as follows:

1. Place the filled ampoules onto aluminum canes (Nasco no. A545); insert the canes into boxes (Murray & Heister, Inc., open-end cartons with rectangular holes, 2½ by 2½ by 11 inches [6.4 by 6.4 by 27.9 cm]).

2. Place the boxes into the freezing chamber of a programmed freezer (Union Carbide Corp.,

TABLE 1. *Expected shelf life of representative bacteria preserved by various methods*

Genus	Serial transfer[a]	Mineral oil	Sterile soil	Deep freeze	Freeze-dried	Liquid nitrogen
Acetobacter	1–2 mo	1 yr		1–3 yr	>30 yr	>30 yr
Achromobacter	1 mo	1–2 yr		1–3 yr	>30 yr	>30 yr
Acinetobacter	weekly				>30 yr	>30 yr
Actinobacillus	weekly	2–3 yr			>30 yr	>30 yr
Actinomyces	1 mo		1–2 yr	2–3 yr	>30 yr	>30 yr
Agrobacterium	1–2 mo	1–2 yr	1–2 yr		>30 yr	>30 yr
Arthrobacter	1–2 mo			1–2 yr	>30 yr	>30 yr
Bacillus	2–12 mo	1 yr	1–2 yr	2–3 yr	>30 yr	>30 yr
Bacteroides	weekly			1 yr	>30 yr	>30 yr
Bifidobacterium	weekly				>30 yr	>30 yr
Chromatium	1 mo				6 yr	>10 yr
Clostridium	6–12 mo	1–2 yr		2–3 yr	>30 yr	>30 yr
Corynebacterium	1–2 mo	1 yr		1–2 yr	>30 yr	>30 yr
Enterobacter	1–4 mo	1–2 yr			>30 yr	>30 yr
Escherichia	1–4 mo	1–2 yr			>30 yr	>30 yr
Erwinia	1–4 mo	1–2 yr			>30 yr	>30 yr
Flavobacterium	1 mo	2 yr			>30 yr	>30 yr
Gluconobacter	1 mo				>30 yr	>30 yr
Haemophilus	weekly	1 mo (37°C)			>30 yr	>30 yr
Klebsiella	1–4 mo	1 yr		1–2 yr	>30 yr	>30 yr
Lactobacillus	weekly				>30 yr	>30 yr
Methanobacterium	1 mo				>10 yr	>10 yr
Methanomonas	1 mo				>10 yr	>10 yr
Micromonospora	1 mo		1–2 yr	1 yr	>30 yr	>30 yr
Neisseria	1 mo				>30 yr	>30 yr
(*N. gonorrhoeae*)	weekly	1 mo (37°C)			>30 yr	>30 yr
(*N. meningitidis*)	weekly	1 mo (37°C)			>30 yr	>30 yr
Nocardia	1–4 mo	1 yr		1–2 yr	>30 yr	>30 yr
Proteus	1–2 mo	1 yr		1–2 yr	>30 yr	>30 yr
Pseudomonas	1–3 mo				>30 yr	>30 yr
Spirillum	weekly	6 mo		1 yr	>30 yr	>30 yr
Staphylococcus	1–2 mo			1 yr	>30 yr	>30 yr
Streptococcus	1–2 mo	1 yr			>30 yr	>30 yr
Streptomyces	1–8 mo	1–2 yr	2–3 yr	1–3 yr	>30 yr	>30 yr
Xanthomonas				1–2 yr	>30 yr	>30 yr

[a] Transfer schedule depends on media used. The times listed are approximations, and species variation occurs within genus.

Linde BF3-2) which has an adjustable cooling rate.

3. Freeze the cells at a controlled rate of 1°C per min to −30°C and than at a more rapid drop of 15 to 30°C per min to −150°C.

4. After this temperature is achieved, transfer the ampoules to a liquid nitrogen tank and store immersed in the liquid phase at −196°C or above in the vapor phase (−150°C).

If a programmable freezer is not available, slow cooling can be achieved by either of the following procedures.

1. Place filled vials or plastic ampoules in a stainless-steel pan on the bottom of a Revco freezer at −60°C for 1 h, and then plunge them into a liquid nitrogen bath for 5 min. The rate of cooling to −60°C using this method is approximtely 1.5°C per min.

2. Canale-Parola (6) recommended immersing the sealed ampoules (placed on a cane) in 95% ethyl alcohol contained in a graduated cylinder. The cylinder containing the ampoules is placed in a Revco freezer set at −85°C and allowed to reach that temperature before the canes and ampoules are placed in liquid nitrogen.

Thawing

Goos et al. (9) and others (14, 16, 17) found that rapid warming resulted in the greatest recovery of fungal spores and other microorganisms from the frozen state. This has also been the experience with bacterial cells frozen at the American Type Culture Collection.

To recover frozen cultures, rapidly thaw them with moderate agitation in a 37°C water bath until all the ice melts. This usually takes 40 to 60 s for glass ampoules and 60 to 120 s for the polypropylene ones. Immediately after thawing, remove the ampoule from the water bath and wipe with 70% ethanol to disinfect. Open the ampoule and aseptically transfer the culture to fresh medium. *CAUTION*: Wear protective gloves and a face shield when handling frozen glass ampoules, as they can explode upon removal from liquid nitrogen.

Check the viability of the culture to determine the effectiveness of the procedure for a given species.

The expected shelf life of various bacteria is shown in Table 1.

12.3. COMMERCIAL SOURCES

American Hospital Supply Corp., 1450 Waukegan Rd., McGaw Park, IL 60085

Bellco Glass, Inc., 340 Edrudo Rd., P.O. Box 'B,' Vineland, NJ 08360

Consolidated Vacuum Corp., 1775 Mt. Read Blvd., Rochester, NY 14603

Difco Laboratories, P.O. Box 1058A, Detroit, MI 48232

Fisher Scientific Co., 711 Forbes Ave., Pittsburgh, PA 15219

Kahlenberg-Globe Equipment Co., Sarasota, FL 33577

Kontos, Spruce St., P.O. Box 739, Vineland, NJ 03860

Markem Machine Co., 150 Congress St., Box 480, Keene, NH 03431

Minnesota Valley Engineering, Inc., New Prague, MN 56071

Murray & Heister, Inc., 10738 Tucker St., Beltsville, MD 20705

Nasco, 901 Janesville Ave., Fort Atkinson, WI 53538

Refrigeration for Science, Inc., 3441 Fifth St., Oceanside, NY 11572

Revco, Inc., 1100 Memorial Drive, West Columbia, SC 29169

Arthur H. Thomas Co., P.O. Box 779, Philadelphia, PA 19105

Union Carbide Corp., Linde Division, 270 Park Ave., New York, NY 10017

Vangard International, 1111-A Green Grove Rd., Neptune, NJ 07753

The VirTis Co., Inc., Gardiner, NY 12525

VWR Scientific, Inc., 3202 Race St., Philadelphia, PA 19104

Whatman, Inc., 9 Bridewell Place, Clifton, NH 07014

Wheaton Scientific, 1000 N. 10th St., Millville, NJ 08332

12.4. LITERATURE CITED

12.4.1. General References

1. **Cabasso, V. J., and R. H. Regamy (ed.).** 1977. International Symposium on Freeze-Drying of Biological Products, p. 1–398. S. Karger AG, Basel.
   A collection of papers on the theoretical aspects, available equipment, and future research in freeze-drying.

2. **Lapage, S. P., J. E. Shelton, T. G. Mitchell, and A. R. Mackenzie.** 1970. Culture collections and the preservation of bacteria, p. 135–228. *In* J. R. Norris and D. W. Ribbons (ed.), Methods in microbiology, vol. 3A. Academic Press, London.
   This paper discusses the activities of maintaining a large service culture collection, and also describes the use of centrifugal freeze-drying.

3. **Onions, A. H. S.** 1971. Preservation of fungi, p. 113–159. *In* C. Booth (ed.), Methods in microbiology, vol. 4. Academic Press, London.
   This chapter discusses the preservation of fungi using various methods such as serial transfer, freeze-drying and freezing, etc.

4. **Rinfret, A. P., and B. LaSalle (ed.).** 1975. Round Table Conference on the Cryogenic Preservation of Cell Cultures, p. 1–78. National Academy of Sciences, Washington, D.C.
   This report contains eight papers discussing various aspects of cell preservation with particular emphasis on liquid nitrogen storage.

12.4.2. Specific Articles

5. **Buell, C. B., and W. H. Weston.** 1947. Am. J. Bot. **34:**555–561.

6. **Canale-Parola, E.** 1973. *In* J. R. Norris and D. W. Ribbons (ed.), Methods in microbiology, vol. 8, p. 61–73. Academic Press, London.
7. **Coe, A. W., and S. P. Clark.** 1966. Mon. Bull Min. Health Public Health Lab. Serv. **25**:97–100.
8. **Fry, R. M., and R. I. N. Greaves.** 1951. J. Hyg. **49**:220–246.
9. **Goos, R. D., E. E. Davis, and W. Butterfield.** 1967. Mycologia **59**:58–66.
10. **Gordon, R. E., and T. K. Rynearson.** 1963. *In* S. M. Martin (ed.), Culture collections: perspectives and problems, p. 118–128. University of Toronto Press, Toronto.
11. **Hartsell, S. E.** 1956. Appl. Microbiol. **4**:350–355.
12. **Haynes, W. C., L. J. Wickerham, and C. W. Hasseltine.** 1955. Appl. Microbiol. **3**:361–368.
13. **Kauffmann, F.** 1966. The bacteriology of *Enterobacteriaceae*, p. 1–368. The Williams and Wilkins Co., Baltimore.
14. **Mackenzie, A. P.** 1977. *In* V. J. Cabasso and R. H. Regamy (ed.), International Symposium on Freeze-Drying of Biological Products, p. 263–277. S. Karger, Basel.
15. **Mazur, P.** 1956. J. Gen. Physiol. **39**:869–888.
16. **Mazur, P.** 1966. In H. T. Meryman (ed.), Cryobiology, p. 213–315. Academic Press, London.
17. **Meryman, H. T.** 1966. In H. T. Meryman (ed.), Cryobiology, p. 1–114. Academic Press, London.
18. **Muggleton, P. W.** 1963. Prog. Ind. Microbiol. **4**: 191–214.
19. **Perkins, J. J.** 1976. Principles and methods of sterilization in health science, p. 286–292. Charles C Thomas, Publisher, Springfield, Ill.
20. **Redway, K. F., and S. P. Lapage.** 1974. Cryobiology **11**:73–75.
21. **Rey, L. (ed.).** 1964. Aspect theoriques et industriels de la lyophilisation, p. 199–234. Hermann, Paris.
22. **Rey, L. (ed.).** 1966. Lyophilisation: recherches et applications nouvelles, p. 1–244. Hermann, Paris.
23. **Speck, M. L., and R. A. Cowman.** 1970. *In* H. Iizuka and T. Hasegawa (ed.), Proceedings of the First International Conference on Culture Collections, p. 241–250. University of Tokyo Press, Tokyo.
24. **Stamp, L.** 1947. J. Gen. Microbiol. **1**:251–265.
25. **Weiss, F. A.** 1957. *In* Manual of microbiological methods, Committee on Bacteriological Technic, Society of American Bacteriologists, p. 99–119. McGraw-Hill Book Co., New York.
26. **Wickerham, L. J., and A. A. Andreasen.** 1942. Wallerstein Lab. Commun. **5**:165–169.

Section III

GENETICS

Introduction to Genetics

EUGENE W. NESTER

Bacterial genetics has advanced at a remarkable pace from its beginnings in the 1940s when Lederberg and Tatum first demonstrated genetic recombination in *Escherichia coli*. These advances have included a detailed understanding of the mechanisms by which genes undergo mutation, are transferred, and recombine in bacteria. It is not surprising that the birth and development of molecular biology have paralleled the developments in bacterial genetics: the tools and techniques developed by the geneticists have provided the materials for study by the molecular biologists. It is not coincidental that *E. coli* is by far the most well-understood organism at the molecular level: most molecular studies have taken advantage of and actually depend on the genetic manipulations that can be performed on this organism. The tools of the bacterial geneticist include well-characterized mutants, donor and recipient strains that can conjugate and undergo recombination, and genomes that can be isolated and then analyzed in genetic and biochemical terms.

The techniques for working with each of these tools in bacteriology are described in this section of the *Manual*. In chapter 13 there are described various techniques for inducing, selecting, and characterizing bacterial mutants of the two most frequently used bacteria in genetic analysis, *E. coli* and *Bacillus subtilis*. The same techniques, with variations in the details of the procedure, should allow an investigator to isolate mutants of most any bacterium.

Gene transfer has been described in relatively few bacteria, although the number continues to increase as the laboratory conditions for demonstrating the phenomenon are better understood. The three known systems of gene transfer (transformation, transduction, and conjugation) are described in chapter 14. In some instances, systems were chosen because they were easy to perform (e.g., transformation in *Acinetobacter calcoaceticus*). In others, selection was based on

the system being studied in many research laboratories (e.g., transduction in *E. coli* with bacteriophage lambda).

In chapter 15, on plasmids, there are described the techniques which are most useful for isolating and characterizing plasmid deoxyribonucleic acid (DNA) molecules. The described techniques are constantly being improved, as suggested by the fact that some are published here for the first time (e.g., the isolation of plasmid DNA of high molecular mass).

The authors of each of the chapters are experienced with most of the described techniques, and they provide useful hints and point out potential pitfalls. Altogether, the section should prove useful to those who perform experiments in bacterial genetics as well as those who employ bacteria as experimental tools in molecular biology.

For general background, the reader is referred to selected books on bacterial genetics (1-5).

"Recombinant DNA" methods provide a powerful new tool for bacterial genetics, with revolutionary implications. The rapidly changing methodology and federal restrictions, as well as the specialization, preclude the inclusion of these methods in this manual. The reader is referred to an entire volume of *Methods in Enzymology* on recombinant DNA (6).

LITERATURE CITED

1. **Broda, P.** 1979. Plasmids. W. H. Freeman and Co., San Francisco.
2. **Falkow, S.** 1975. Infectious multiple drug resistance. Pion Limited, London.
3. **Lewin, B.** 1974. Gene expression, vol. 1: Bacterial genomes. John Wiley & Sons, London.
4. **Lewin, B.** 1977. Gene expression, vol. 3: Plasmids and phages. John Wiley & Sons, London.
5. **Miller, J.** 1972. Experiments in molecular genetics. Cold Spring Harbor Laboratory, Cold Spring Harbor, N.Y.
6. **Wu, R. (ed.).** 1979. Recombinant DNA. Methods Enzymol. **68**:1–555.

Chapter 13

Gene Mutation

BRUCE C. CARLTON AND BARBARA J. BROWN

Mutation, the occurrence of a heritable change in the genetic material, is an important biological phenomenon. It is the ultimate source of all biological variation and, together with genetic transfer mechanisms, provides the genetic variability on which the forces of evolution operate. In a practical sense, mutation and the induction of new mutations with mutagens provide important tools for genetic and biochemical analysis. First, the occurrence of a mutation in a specific gene function allows both the identification of that gene and, together with genetic mapping analyses, the localization of the gene on the chromosome. Second, the analysis of mutants (strains that contain mutations) which are defective in different parts of a complex biochemical pathway or system can reveal the details of genetic and biochemical organization. Third, a knowledge of the mode of action of various mutagenic agents (mutagens) can help establish correlations between the mutagenic and cancer-inducing (carcinogenic) actions of a variety of environmental agents such as chemicals, radiations, and other physical agents.

The application of mutagenesis for genetic analysis of bacteria requires the coordination of several sequential experimental steps, as follows:

1. determination of the type of mutant desired for the specific research objective;
2. choice of the most appropriate mutagen for inducing the desired mutation;
3. allowance for expression of the new mutation (outgrowth);

4. enrichment of the desired mutant to enhance the probability of its recovery;
5. detection of the new mutant by appropriate direct or indirect selection procedures;
6. characterization of the mutant; and
7. mapping of the site of the new mutation, to localize its position on the bacterial genome.

Methods needed for the first six of these objectives are dealt with in parts 13.1 to 13.7 of this chapter; mapping procedures are covered in the next chapter. This chapter also contains two specialized methods, one for isolating mutations in localized segments of the bacterial genome (13.8.1) and another for assessing the mutagenicity of various carcinogenic agents (13.8.2). Media recipes are in part 13.9, and the sources of bacterial strains and special materials are in part 13.10. General references to mutagenesis and mutant isolation are listed in part 13.11.1 (as references 1 through 4), and specific references pertinent to individual procedures follow (13.11.2).

Detailed procedures for the isolation of a specific mutant type may vary considerably in different situations. The particular bacterial strain, conditions of growth (media, temperature, etc.), and concentrations of mutagens may be optimum for one bacterium but totally inadequate for another species or even for different strains of the same species. The procedures described below should, therefore, be viewed as a starting point, but with the realization that the conditions may have to be varied considerably for specific applications to different bacteria.

13.1. MUTANT DETERMINATION

When initiating a mutant search, first determine what type of mutant is needed for a specific objective. For example, if one wishes to obtain a mutant which carries a stable genetic marker that identifies that particular bacterial strain through a variety of experimental manipulations, then a nonreverting and tightly blocked mutant such as a **deletion mutant** should be sought. On the other hand, if one wishes to examine the relationship between the structure of a particular protein and its function, then mutants which have acquired base substitution **point mutations** would be the most advantageous. Such mutants allow for selection of functional revertants. Base substitution mutations will frequently be those of the **missense type**, in which a triplet base sequence coding for one amino acid is changed to one which codes for a different amino acid. As a consequence of the degeneracy patterns of the genetic code, the mutant amino acid often is similar to the paren-

tal type, thus producing a "leaky" phenotype in which the gene function (usually a protein) is only partially lost. Such strains are likely to revert spontaneously to the parental type, thus showing genetic instability as well as physiological leakiness. A significant fraction of base substitution mutations may be of the **nonsense type**, in which an amino acid-coding triplet is changed to one coding for no amino acid. This type of mutation leads to termination of synthesis of the protein product at that site, producing an incomplete fragment of the protein which is almost always completely nonfunctional. A nonsense mutant will, thus, be phenotypically non-leaky, although the potential for reversion is still present. **Frameshift mutations** arise by the insertion or deletion of one or more bases into the deoxyribonucleic acid (DNA). These events result in a shift in the reading frame of the coded information, leading to a sequence of altered amino acids in the mutant strain.

If one wishes to obtain a mutant whose genetic defect cannot be corrected by the addition of some nutritional supplement (e.g., defects in DNA- or ribonucleic acid [RNA]-replicating enzymes, defects in components of the protein-synthesizing machinery, etc.), then one must seek **conditional lethal mutants**. These mutants are not viable under one set of conditions, but are viable under other conditions; temperature-sensitive and suppressible nonsense mutants are examples. Table 1 summarizes the properties of various types of mutations and may be used as a general guide for determining what type may be most useful for a specific purpose.

13.2. MUTAGEN CHOICE

Choosing a mutagen for any particular application will depend primarily on two factors: the particular type of mutation desired (i.e., base substitution, deletion, or frameshift), and the relative effectiveness of a mutagenic agent in inducing that type of mutation. More than one type of mutation may be appropriate for a given purpose, and more than one type of mutagen may effectively induce the desired mutation.

Numerous studies on the mutagenic specificity of various radiations, chemicals, and physical agents have shown that, although many mutagens induce one primary type of DNA alteration, most also produce other types at low to moderate frequencies. For example, mutagens such as ethyl methane sulfonate (EMS), nitrosoguanidine (NTG), hydroxylamine, and nitrous acid are generally conceded to induce primarily GC→AT transitions in double-stranded DNA (10, 22, 23). Yet, all four of these also yield a small fraction of AT→GC transitions, as well as

TABLE 1. *Functional and genetic properties of bacterial mutations*[a]

Type of mutation	Nature of DNA alteration	Types of genes involved	Effect on gene function	Revertable	Suppressible by tRNA suppressors
Point					
Missense	Base substitutions	Protein coding	Variable	Yes	Yes
Nonsense	Base substitutions	Protein coding	Complete loss[b]	Yes	Yes
Frameshift	Base insertions or deletions	Protein coding	Complete loss[b]	Yes[c]	Yes
Other	Base substitutions	Regulatory elements; tRNA and rRNA genes	Variable	Yes	No
Deletion	Loss of a DNA segment	All types	Complete loss	No	No
Insertion	Incorporation of a phage genome or transposable element	All types	Complete loss	Yes	No

[a] tRNA, Transfer RNA; rRNA, ribosomal RNA.
[b] Unless very near the distal end of the gene.
[c] By second-site compensating mutations of the opposite type.

some transversions. **Transitions** are single-base substitutions in which a purine base (adenine or guanine) at one site in a DNA strand is changed to the other purine, or one pyrimidine (thymine or cytosine) is exchanged for the other pyrimidine base. **Transversions**, on the other hand, are base substitution mutations in which a purine base at a given site is exchanged for one of the two pyrimidine bases, or vice versa. Similarly, base analogs such as 2-aminopurine and 5-bromouracil primarily induce AT→GC transitions, although GC→AT transitions are also induced by these chemicals at a much lower frequency. This lack of absolute mutagen specificity is probably a function of not only the actual mechanisms by which the mutagens act, but also the local environment of the base which is mutated (i.e., the nature of neighboring bases or the secondary structure of the DNA). (See reference 23 for a discussion of these considerations.) As a consequence of these variable mutagenic specificities, one must be cautious in assuming that mutations induced by even the most specific of mutagens are a particular type; further tests of the mutants are necessary to confirm the assumption. Despite these uncertainties, one can nevertheless use the major specificity characteristics of a given mutagen with reasonable confidence that an array of newly induced mutants will, for the most part, reflect that particular type of mutation. Table 2 summarizes the properties of some common mutagens used in bacterial genetics.

In addition to greatly increasing the rate of mutation, many mutagens are also highly potent **carcinogens**—agents capable of inducing tumors in mammalian systems (19). The development of bacterial test systems by Ames and co-workers (7, 8) has provided a means of effectively correlating these two properties and has revealed that most carcinogens are also mutagens. The details of this test will be dealt with later in this chapter (see 13.8.2).

CAUTION: When using mutagens, great care should be exercised in their handling and disposal. Solutions of mutagens should *never* be mouth pipetted. When weighing or measuring these agents, disposable gloves should be worn. All manipulations involving powdered or volatile forms should be carried out in a hood to avoid breathing the vapors. Solutions of mutagens should not be discarded down the sink, but rather should be absorbed onto a suitable solid medium (such as vermiculite) to minimize spillage and disposed of in a sealed container like other hazardous wastes. If mutagens are routinely used in solid plating media or liquid growth media, the incubators in which the plates or flasks are stored should be vented to a fume hood or to the outside atmosphere.

13.2.1. Ultraviolet Light

Mutagenesis with ultraviolet light is one of the simplest and most convenient ways to obtain a wide variety of bacterial mutants. Once calibrated for optimum dosage, most ultraviolet light sources that emit energy in the short wavelength (about 253.7 nm) can be used.

Procedure.

1. Grow 50 ml of the parental bacterial strain in L broth (13.9.1) or nutrient broth (13.9.11) at 37°C to a cell density of about 2×10^8/ml.

2. Chill the culture on ice for a few minutes to prevent further growth, divide the culture

TABLE 2. *Properties of some commonly used mutagenic agents*

Mutagen type	Presumed mechanism of mutagenesis	Type of mutation produced	Relative effectiveness	Special advantage or disadvantage
Radiations				
X rays, thermal neutrons	Primarily breakage of chromosomes	Deletions, inversions	High	Equipment may be difficult to obtain
Ultraviolet	Pyrimidine dimerization	GC → AT transitions, transversions, deletions	Medium	Wide-spectrum mutagen, extent of killing must be carefully controlled, photoreactivation must be prevented
Chemicals				
Base analogs: 2-aminopurine, 5-bromouracil, 5-bromodeoxyuridine	Errors in DNA replication	Primarily AT → GC transitions	Low	Relatively inefficient
Hydroxylamine	Deamination of cytosine	GC → AT transitions	Low	Preferentially induces one-directional transitions
Nitrous acid	Deamination of cytosine and adenine	Bidirectional transitions, deletions	Medium	
NTG	Alkylation of bases at replication fork	Primarily GC → AT transitions	Very high	Highly mutagenic at low killing levels, induces multiple mutations in localized regions
EMS	Alkylation of guanine	Primarily GC → AT transition, some AT → GT transitions and transversions	Medium	Similar mutational effect to that of NTG, but less likely to yield multiple mutations
Acridine half-mustards (ICR compounds)	Intercalation between bases during replication	Frameshifts—base insertions and deletions	High	Compounds may be difficult to obtain
Acridine dyes, ethidium bromide	Intercalation between bases during replication	Frameshifts, loss of extrachromosomal elements	Low	Especially effective in curing cells of plasmids
Novobiocin	Blocks DNA replication	Curing of plasmids	High	Very effective in curing cells of plasmids

into 5-ml portions, and centrifuge for 5 min at 5,000 to 6,000 × *g* to pellet the cells.

3. Resuspend each cell pellet in 0.5 volume of sterile distilled water containing 0.1 M $MgSO_4$. Plate one of these samples at 10^{-6} and 10^{-7} final dilutions on L agar (13.9.2) for unirradiated controls.

4. Transfer the remaining undiluted cultures to sterile glass petri plates 100 mm in diameter. Prewarm a short-wavelength (253.7 nm) ultraviolet lamp for 15 to 20 min. Place the petri plate under the lamp and remove the top of the plate. Tilt the plate slightly from side to side to allow for uniform exposure. Samples should be irradiated over about a 10-fold time scale (e.g., 15 s to 2.5 min) so as to obtain a range of killing.

5. Immediately dilute each sample over several log ranges, and plate under dim light on L agar (13.9.2) or nutrient agar (13.9.12) for determination of survival rates.

6. Centrifuge the remainder of each of the undiluted irradiated samples, and resuspend the cells in 10 ml of L broth (13.9.1).

7. Grow cells overnight at 37°C in tubes or in flasks covered with aluminum foil or in a darkened room to prevent photoreactivation. To obtain a number of independent mutations, divide each irradiated sample into several tubes before incubation (see 13.3).

Suggestions and possible problems.

The efficiency of killing with ultraviolet irradiation varies tremendously with different ultraviolet sources and experimental conditions. Conduct a survival curve assay in order to standardize irradiation conditions. If the bacterial populations are too concentrated (above 5×10^8 cells/ml), shielding effects will give variable killing. If a dosimeter is available, an average dosage of about 500 ergs/mm^2 should yield an optimal result. A cell survival rate of 0.1 to 1% is ideal because of the relatively low mutagenic efficiency of ultraviolet. Also, minimize photoreactivation effects by carrying out both the actual irradiations and the subsequent outgrowths in the dark.

13.2.2. Nitrosoguanidine

The methylating compound N-methyl-N'-nitro-N-nitrosoguanidine (NTG) is one of the most potent mutagens yet discovered for bacteria. It induces primarily base transition mutations of the GC→AT type, although AT→GC transitions, transversions, and even frameshifts arise at low frequencies.

Procedure.

1. Grow 10 ml of a suitable bacterial strain to the mid-logarithmic stage (ca. 5×10^8 cells/ml) in L broth (13.9.1) or nutrient broth (13.9.11).
2. Centrifuge the culture for 5 min at 5,000 × g to pellet the cells, and resuspend the cell pellet in an equal volume of Tris-maleic acid buffer (13.9.4) at pH 6.0.
3. Add a freshly prepared solution of NTG (1 mg/ml in sterile water) to the culture to a final concentration of 100 μg/ml. Incubate at 37°C *without shaking* for 30 min.
4. Centrifuge the treated culture and resuspend the pellets in an equal volume of M56 minimal medium (13.9.8).
5. Recentrifuge the cells and again resuspend in M56 minimal medium (13.9.8). This procedure, devised by Adelberg et al. (5), yields a high degree of mutagenesis without the excessive cell killing which occurs when cells are treated in rich broth media (at least for *Escherichia coli*). It is particularly effective for the isolation of auxotrophic mutants (13.6.2).

Suggestions and possible problems.

CAUTION: NTG is not only a potent mutagen but also a carcinogen! Always employ all of the precautions outlined in part 13.2.

You may wish to determine a killing curve for NTG by plating 10^{-6} dilutions of both the untreated culture and the treated culture sampled at 5-min intervals during the NTG treatment (4). The conditions employed should produce no more than about 50% killing—higher concentrations will give more killing but may yield up to 40% auxotrophs among the survivors (5).

One of the potential difficulties in working with this mutagen is the likelihood of inducing multiple mutations in a small region of the bacterial chromosome (12). As a consequence, the mutants could have complex growth requirements due to mutations affecting more than one type of metabolic function and fail to revert to wild type at expected frequencies. It is thus wise to check NTG-induced mutants carefully for reversion before attempting to use them for further analysis (see 13.5.2).

13.2.3. Base Analogs

Base analogs such as 5-bromouracil and 2-aminopurine are believed to induce replication copy errors when incorporated into DNA. These base analogs are able to exist in two tautomeric forms, the normal keto or amino form and the rare enol or imino form. Each form has a different base-pairing specificity, and thus a shift into the rare tautomeric form can cause mispairing during DNA replication. Since mispairing errors can occur at two different stages (at the time of initial incorporation or during a subsequent DNA replication event), they can lead to bidirectional transition mutations, in which a GC base pair is changed to AT or vice versa. Despite this capability, both 5-bromouracil and 2-aminopurine are more effective in producing AT → GC transitions than the GC → AT type (10, 22). Neither of these agents is especially useful in forward mutational studies because of their fairly low mutagenic efficiency, presumably due to their inability to effectively compete with the internal nucleotide pools of their normal analogs. They are quite useful in reversion studies, however, where the level of sensitivity for detecting infrequent mutations is higher (see 13.5.2).

Procedure for 5-bromouracil.

To obtain significant mutagenicity with 5-bromouracil, it is necessary to depress the levels of internal thymine, the natural analog of 5-bromouracil. This can be accomplished by using a thymine-requiring mutant or by pretreating the cells with agents which inhibit thymine synthesis, thereby lowering intracellular pool levels of this base and its nucleotide and nucleoside derivatives. Thymine is normally synthesized from deoxyuridine, via the enzyme thymidylate synthetase. Growth of cells in the presence of sulfanilamide, an inhibitor of folic acid synthesis, leads to a depletion in methylation donor concentrations and hence a depression in the levels of thymine nucleosides and nucleotides.

1. Grow the bacterial strain overnight in minimal medium (13.9.8 or 13.9.15) if the strain is wild type for thymine synthesis or in minimal medium supplemented with thymine (20 μg/ml) if the strain is thymine requiring.
2. For *thymine-positive strains*, dilute 1:25 into sulfanilamide broth (13.9.5) containing 50 μg of 5-bromouracil per ml.
3. Allow to shake at 35 to 37°C for 5 to 6 h.
4. Centrifuge the culture (or filter through a membrane filter), and resuspend the cells in fresh medium appropriate for the enrichment and/or selection procedures required for the type of mutant desired.

2a. For *thymine-requiring strains*, centrifuge the overnight culture, resuspend cells in minimal medium (13.9.8 or 13.9.15) lacking thymine but containing 0.1% acid-hydrolyzed casein and 20 μg of L-tryptophan per ml.

3a. Allow cells to incubate with gentle shaking for 20 to 30 min to deplete any residual thymine from the medium.

4a. Add 5-bromouracil to 20-μg/ml final concentration, and continue shaking for about 1 h.

5a. Centrifuge or filter the culture to remove 5-bromouracil, and resuspend the cells in medium containing 20 μg of thymine per ml to prevent thymineless death during outgrowth and/or selection procedures.

Procedure for 2-aminopurine.

1. Grow a culture of the test bacterium overnight in L broth (13.9.1) or nutrient broth (13.9.11).

2. Dilute the culture to 10^2 to 10^3 cells per ml in fresh broth containing 500 μg of 2-aminopurine per ml.

3. Incubate the culture with shaking until it becomes visibly turbid. This long growth period allows for the outgrowth and expression of new mutations, and one may proceed directly to an enrichment procedure if desired. If many independent mutations are needed, divide the culture among several tubes at the time of 2-aminopurine addition and select only one mutant of any given type from each subculture.

13.2.4. Acridine Half-Mustards

Compounds of this general type (usually identified by number with an ICR prefix, which refers to their study as potential antitumor agents at the Institute for Cancer Research, Philadelphia, Pa.) act as mutagens by intercalating between the stacked bases of the DNA double helix, leading to insertion or deletion of nucleotides during subsequent replication. The resulting frameshift mutations cause a shift in the translated reading frame of the coded information in the messenger RNA transcript, thus leading to an altered sequence of amino acids past the point of the insertion or deletion. Usually, an out-of-phase nonsense codon is generated such that, phenotypically, the mutants closely resemble base substitution nonsense mutants in being totally defective for the mutated functions.

Procedure.

1. Dilute the actively growing culture to be mutagenized to about 10^4 cells/ml in minimal salts medium (13.9.8) containing the acridine half-mustard compound ICR-191 (available from Polysciences, Inc., Warrington, PA 18976) at 5 to 10 μg/ml.

2. Grow the culture to a turbidity of about 100 units of a Klett-Summerson colorimeter, red filter (see 11.4.1), or for about 10 to 12 generations, in the presence of the mutagen.

3. Plate out dilutions to select for mutants, or subject to enrichment procedures as required.

Suggestions and possible problems.

It is advisable before carrying out this type of mutagenesis to run growth tests of the bacterium with increasing concentrations of the mutagen. For mutagenesis, select a concentration which allows slow growth of the bacterium (division time approximately twice the normal); there is a fairly good correlation between growth inhibition and mutagenic efficiency. This preliminary screening is important when using bacteria that have various recombinational or repair system deficiencies, as they may be ultrasensitive to the mutagen. Also, these compounds may enhance the induction of mutations by light-induced reactions. If only frameshift mutations are desired, the mutational induction should be conducted in the dark.

13.2.5. Nitrous Acid

Nitrous acid is a chemical mutagen which induces primarily bidirectional transition mutations, due to its deaminating activity on cytosine and adenine. In addition, it induces deletions at a significant level by an unknown mechanism.

Procedure.

1. Grow 5 ml of a culture to be mutagenized overnight in L broth (13.9.1) or nutrient broth (13.9.11), and centrifuge at $5,000 \times g$ for 5 min to pellet the cells.

2. Wash the cell pellet in 5 ml of sterile 0.1 M sodium acetate buffer, pH 4.6 (13.9.6).

3. Prepare a fresh nitrous acid solution (0.05 M sodium nitrite in 0.1 M sodium acetate buffer, pH 4.6).

4. Resuspend the washed cells in 1.0 ml of the nitrous acid solution, and incubate the mixture for 10 to 20 min at 30 to 37°C without shaking.

5. Add 10 ml of a minimal medium (13.9.8 or 13.9.15) to stop the reaction.

6. Centrifuge the suspension at $5,000 \times g$ for 5 min to pellet the cells, and resuspend in a suitable medium for outgrowth and detection of the desired mutant type. Since cell killing by nitrous acid is substantial, determine the survival levels by plating approximate dilutions on nutrient agar (13.9.12) before and after treatment. Optimally, survival levels should be between 0.1 and 0.01%.

13.2.6 Alkylating Agents

Alkylating agents such as EMS, diethyl sulfate, methyl methanesulfonate, etc., act on DNA primarily by modification of purine bases, especially guanine. The commonest alkylating agent in use for mutagenic studies is EMS, which is quite specific in its action, producing primarily $GC \rightarrow AT$ transition mutations (10, 22).

Procedure.

1. Grow a culture of the bacterium to be mutagenized to the late logarithmic phase of growth in L broth (13.9.1) or nutrient broth (13.9.11).
2. Mix 1 volume of the culture with an equal volume of a freshly prepared stock solution of EMS (0.1 ml of EMS in 2.5 ml of minimal medium [13.9.8 or 13.9.15] warmed to 37°C).
3. Aerate the mixture on a shaking incubator at 37°C for 1 to 2 h.
4. Dilute the culture 10-fold into fresh minimal medium, and allow to grow out for several hours. Depending upon the type of mutant desired, either proceed to an enrichment procedure or plate out dilutions directly for detection of mutants.

13.2.7. Plasmid-Curing Agents

A variety of chemical agents, such as acridine dyes, ethidium bromide, sodium dodecyl sulfate, and novobiocin, as well as growth at elevated temperatures, are able to free or "cure" bacterial cells of plasmid DNA molecules (20). Plasmid molecules, which exist as autonomously replicating circular DNA duplexes, are eliminated by these agents either because of interference with their replication (acridines, ethidium bromide, and novobiocin) or by alterations of their membrane attachment sites (sodium dodecyl sulfate and elevated temperatures). In view of the importance of plasmid DNAs as carriers of genetic determinants for antibiotic resistances, heavy metal resistances, antibacterial agents, and complex metabolic functions, methods for producing cured variants are gaining in attention among bacterial geneticists.

Procedures.

A typical curing procedure is as follows:
1. From a logarithmically growing culture of the test bacterial strain inoculate 10^3 to 10^4 cells into a series of tubes containing nutrient broth (13.9.11), L broth (13.9.1), or other suitable medium containing several different concentrations of the curing agent.
2. Incubate the cultures overnight, at the usual growth temperature for the particular bacterium.

3. Select a culture which shows a just detectable increase in turbidity, and plate out dilutions on nutrient agar (13.9.12) or similar medium.
4. Test individual colonies for loss of a plasmid-coded function (i.e., by streaking on antibiotic agar [13.9.14] or overlaying with an indicator strain to test for bacteriocins or other extracellular products).

It should be noted that the most effective concentration of a curing agent may vary as much as 100- to 1,000-fold depending on the bacterium being treated. The efficiency of curing is strain dependent as well as agent dependent.

Elevated temperatures (5 to 7°C above normal growth temperature) can also be used as another procedure for curing.
1. Grow a small amount (5 to 10 ml) of the test bacterial strain in nutrient broth (13.9.11) or other suitable rich medium to late log phase at an elevated temperature (43°C for strains which normally grow at 37°C).
2. Dilute the culture 1:20 into fresh medium and allow the culture to regrow at the elevated temperature. Repeat the step if necessary.
3. Plate out appropriate dilutions of the culture on nutrient agar (13.9.12) or similar medium to obtain single colonies.
4. Test individual colonies for loss of a plasmid-coded function as in step 4 above.
5. Verify loss of plasmid by isolation and fractionation of plasmid DNA from the suspected cured strain (see Chapter 15 for plasmid DNA isolation procedures).

13.3. MUTANT EXPRESSION

Although bacteria are haploid in the sense that they can survive with a single copy of their chromosome, the dynamics of cell division and chromosome replication are such that under most growth conditions the average bacterial cell possesses 1.5 to 2 complete chromosomes. Thus, a newly induced mutation of a potentially recessive phenotype (e.g., an enzymatic function) may be masked by the presence of an unmutated chromosome in the mixed clones that could arise from cultures plated immediately after mutagen treatment. Therefore, it is wise to allow mutagenized cultures to undergo a period of outgrowth which will permit the segregation of newly mutated chromosomes from nonmutated chromosomes in the same cell. The length of the outgrowth period will vary with the division time of the particular bacterium under the growth conditions being employed, as well as with the growth habits of the bacterium. Species or strains which tend to grow in chains or clusters (e.g., bacilli, streptococci, staphylococci, etc.) will require a longer outgrowth period to segregate the new mutation than will strains which tend to grow as single cells. If the outgrowth period is too lengthy, the mutant cells will also undergo duplication, thus leading to the possibility of recovering several replicas of the same mutation. If a number of independent mutants of the same type are desired, it is best to divide the culture either at, or immediately after, the mutagen induction period and to select only one mutant from each subculture after the outgrowth period. As noted in the previous mutagenesis procedures, if the exposure time to the mutagen is sufficiently lengthy, outgrowth is accomplished simultaneously with the mutagenesis induction, thus obviating the need for any further outgrowth period.

13.4. MUTANT ENRICHMENT

The occurrence of mutations, even when induced by highly efficient mutagens, is usually a relatively infrequent event. Mutant frequencies in a bacterial population may be as low as 1 per 10^6 or 10^7 cells. Recovering a particular mutant from such a large background of nonmutant cells may be tedious, especially when indirect detection techniques must be employed. In such cases, an enrichment step is often employed before mutagenized cultures are screened.

The most common enrichment procedure utilizes the antibiotic penicillin (11, 18). When a culture containing both mutant and nonmutant cells is growing in a synthetic medium which does not allow the mutant cells to grow, the addition of penicillin selectively kills the nonmutant cells by interfering with the cell wall synthesis of the actively dividing cells. Under optimal conditions a 1,000-fold enrichment of mutants to nonmutants may be achieved.

13.4.1. Penicillin Enrichment

1. Grow a culture of the bacterium which has been previously mutagenized until it is in the early logarithmic phase of growth (optical density at 600 nm of about 0.1 or Klett reading at 660 nm of 15 to 20).

2. Add 10,000 U of penicillin G per ml and continue incubation of the culture, with shaking, until the turbidity increase levels off (usually 60 to 90 min).

3. Remove the penicillin, either by filtration through a sterile membrane filter (e.g., a 0.45-μm membrane filter) or, for those bacteria which tend to plug such filters, by inactivating the penicillin with penicillinase. Add 100 U of the enzyme per ml, and incubate for 30 min at the same temperature at which the cells were previously growing.

4. After centrifugation at 5,000 $\times$ g for 5 min (or filtration as noted above), wash the cells once with fresh culture medium and resuspend in the same medium before proceeding to the selection or screening step.

13.4.2. Cycloserine Enrichment

In addition to penicillin, the amino acid analog D-cycloserine has been reported to be effective in enriching for spontaneously arising mutants in various *Pseudomonas* species (21). When added in conjunction with penicillin at 100 μg/ml to growing cultures of *Pseudomonas putida*, it promotes selective lysis of the nonmutant cells, leading to an enrichment of mutant cells by a factor of 100 to 1,000 during a 1- to 2-h exposure. This double antibiotic treatment offers the advantage over using either penicillin or cycloserine alone in that it does not select for spontaneous mutants resistant to either agent. By employing several successive cycles of the combined enrichment procedure, it is possible to raise the relative proportions of mutant cells in a population by a factor of 10^6 or greater, thus permitting the recovery of highly infrequent mutant types.

13.5. MUTANT DETECTION (DIRECT)

Mutants which have *gained* an activity lacking in nonmutant cells can be detected by direct selection on appropriate media. Such mutants include cells acquiring resistance to various antibiotics, phages, or chemical inhibitors which would normally be either bactericidal or bacteriostatic to the nonmutant bacterial parent. In addition, mutants which have gained the ability to utilize certain atypical sources of carbon or nitrogen can also be isolated by direct selection. Because of the great resolving power of direct selection (i.e., the ability to detect those infrequent mutant cells against a very great background of nonmutant cells), it is usually unnecessary to employ any enrichment techniques. Furthermore, since the types of genetic functions amenable to direct selection are expected to be dominant to their normal counterparts in cells carrying more than one chromosome, there is normally no need for a period of outgrowth prior to plating the mutagenized cells on selective media.

One of the major difficulties encountered with direct selection is in choosing the optimal concentration of the selective agent. For example, in selecting mutants resistant to various antibiotics, it is important to choose a level of antibiotic which prevents all growth of wild-type cells but which allows the appearance of resistant mutants. The determination of the optimal concentration by trial and error approaches can be time-consuming and difficult. In the first example given below, the use of a simple gradient plate technique is described which can be modified for the detection of many different types of mutants described above. In the second example, a simple filter disk assay is described for the induction and direct selection of revertants of auxotrophic mutants.

13.5.1. Antibiotic   Resistance   (Gradient Plate

Antibiotic-resistant mutants are easily isolated, and the resistances can often serve as convenient genetic markers for use in characterizing bacterial strains. In this example, a procedure for obtaining spontaneously arising mutants is described (24).

Preparation of gradient plates.

1. Measure 10 ml of autoclaved nutrient agar (13.9.12) into sterile petri plates 100 mm in diameter, and allow to cool with one edge elevated so that the agar level just reaches the intersection of the bottom and side of the plate (Fig. 1).

2. After the agar has hardened, level the plates and add an additional 10 ml of antibiotic agar (13.9.14), which contains streptomycin sulfate (100 μg/ml), penicillin (500 μg/ml), or rifampin (50 μg/ml).

3. Allow the plates to cool on a level table, thus producing a concentration gradient of the antibiotic from 0 μg/ml at one edge to the maximal concentration (50 to 500 μg/ml) at the opposite edge.

4. Mark the edges to show which way the gradient is oriented. The plates should be prepared at least 24 h before use so that the antibiotic gradients will be established.

Mutant isolation procedure.

1. Grow a culture of a prototrophic strain of the bacterium of interest from a single colony to mid-logarithmic growth phase in L broth (13.9.1) or nutrient broth (13.9.11).

2. Spread 0.1-ml samples on the surfaces of antibiotic gradient plates (13.9.14).

3. Incubate the plates for 24 to 36 h.

4. Pick resistant colonies with a sterile loop and streak out on a second gradient plate in the direction of increasing antibiotic concentration to determine their approximate level of resistance.

5. After a second incubation, pick single colonies and test on nongradient antibiotic plates (13.9.14) containing several different concentrations of the various antibiotics to confirm the levels of resistance obtained. Alternatively, if

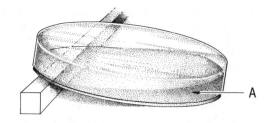

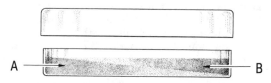

FIG. 1. *Preparation of gradient plate for isolation of antibiotic-resistant mutants. Top, step 1; bottom, step 2. (A) Nutrient agar; (B) antibiotic agar. The resulting antibiotic gradient extends from a low concentration on the left to high on the right of the plate.*

higher levels of resistance are desired, the strains isolated after the first cycle of selection may be replated on gradient plates containing 5 to 10 times steeper gradients of the antibiotic and the process repeated.

13.5.2. Reversion Analysis (Filter Disk)

Direct selection is widely employed to obtain revertants of previously isolated auxotrophic mutants. *Auxotrophs* represent a special class of conditional lethal mutants in that they cannot survive unless supplied with the metabolite (e.g., amino acid, vitamin, etc.) which they are unable to synthesize. If an auxotroph has arisen as a result of a base substitution mutation, it can be induced to "revert" (mutate back to its original type) by mutagens which reverse the original mutational event. In addition to these "true revertants," a variety of "phenotypic revertants" can arise by mutations at sites other than the original site of mutation. For example, frameshift mutants can often be reverted by a secondary compensating frameshift mutation located close to the original mutant site which restores the proper reading frame. Some base substitution mutants can revert phenotypically by a secondary mutation occurring elsewhere in the mutant gene, presumably because the secondary mutation partially compensates for the original mutation through interactions of the amino acids at the two altered sites in the protein. Base substitution mutants, particularly those of the nonsense type, can also be "reverted" by suppressor mutations occurring in various transfer RNA genes. These suppressor mutations allow the nonsense code word in a protein-coding gene to be translated by the altered suppressor transfer RNA, thus correcting, at least partially, the original mutation. In all of these instances, however, the "revertants" can be obtained by direct selection, simply by plating large numbers of the original mutant strain on a medium which lacks the required metabolite. In addition to the tremendous sensitivity of this procedure, which allows the detection of a single revertant against a background of 10^8 to 10^9 mutant cells, reversion analysis can also provide considerable information about the nature of the original mutation. By comparing the efficiency with which various mutagens reverse a particular mutation, one can often determine whether the original mutation was due to a base transition, a transversion, or an insertion or deletion (frameshift). The revertability of a mutation by any mutagen is regarded as evidence that the original mutation was a point mutation and not a deletion, inversion, or other chromosomal aberration.

The protocol described below utilizes a filter disk assay (6, 9) for detecting increased rates of reversion promoted by several commonly used chemical mutagens. Besides base analogs and hydroxylamine, one can examine the activities of an alkylating agent (EMS); the acridine half-mustard (ICR-191), which induces primarily base pair insertions and deletions (frameshifts); and NTG, which has a specificity spectrum similar to that of EMS.

Select several auxotrophic strains of the bacterium of interest—those requiring a specific amino acid for growth, for example. Typically, one might have a series of such mutants which were isolated following mutagenesis with several different mutagens and whose reversion properties are to be compared.

Procedure.

1. Grow 5 ml of each auxotrophic strain being tested to the late logarithmic stage of growth (ca. 10^9 cells/ml) in L broth (13.9.1) or nutrient broth (13.9.11) at 37°C.

2. Centrifuge the cells, and resuspend the cell pellets in the same volume of M56 minimal medium (13.9.8) or Davis minimal broth (13.9.15).

3. Spread 0.1-ml samples on the surface of thick (30 ml) minimal agar plates (13.9.10).

4. After the surfaces have been allowed to dry, use sterile forceps to equally space five 1-cm-diameter sterile filter paper disks.

5. Add to the surfaces of the disks 1 drop (ca. 50 μl) of one of the following: sterile water (a

control), 5-bromouracil (8 mg/ml), ICR-191 (1 mg/ml), NTG (1 mg/ml), hydroxylamine (17 mg/ml), and EMS (undiluted from the bottle).

6. Incubate the plates *right side up* at 37°C for 4 to 5 days or until colonies are readily evident. Induced reversion (mutation) will be revealed by rings of colonies around the disks—usually no colonies are observed immediately adjacent to the disks as a result of the killing effect of the mutagens. From the observed patterns and the principal mutagenic specificities of the test compounds, one can often determine the nature of the original mutations.

Characterization of revertants.

This semiquantitative analysis can be expanded to distinguish full revertants (i.e., to the original genotype/phenotype) from partial revertants and from strains containing unlinked suppressors. The latter two classes are almost always characterized by a very slow growth rate in the absence of the required supplement. Pick colonies of different sizes from the original disk reversion plate, purify by restreaking on minimal agar, grow in nutrient broth, and plate at appropriate dilutions on minimal agar. After 2 or 3 days of incubation at 37°C, determine the relative colony sizes. Partial revertants and suppressed strains will typically show considerably smaller colony sizes.

In some systems, one can also distinguish full revertants from partial revertants or suppressed strains by determining sensitivity to a metabolic inhibitor, such as an analog which inhibits growth by repressing the synthesis of messenger RNA coding for the various enzymes. In the tryptophan biosynthetic systems, for example, 5-methyltryptophan is an analog which exerts this effect. At low concentrations the inhibitor does not inhibit tryptophan synthesis in full revertants but it virtually shuts down tryptophan synthesis in strains producing limited amounts of tryptophan. Perform this test by spreading 10^6 to 10^7 cells of the strains to be tested on a minimal agar plate (13.9.10 or 13.9.15). Place a drop of 5-methyltryptophan solution (1 μg/ml) on a sterile disk as in the reversion assays. Partial revertants and suppressed strains will show a much larger zone of growth inhibition around the disk than full revertants. Alternatively, spread various revertant strains directly on minimal agar plates containing 0.01 μg of 5-methyltryptophan per ml, and compare their growth rates with that on minimal agar plates lacking the tryptophan analog.

A more difficult and time-consuming method of differentiating between full and partial revertants and suppressed strains involves assaying the enzymatic activities of the biochemical functions involved. If the function can be conveniently assayed, the enzyme level and stability from various revertants and their corresponding wild-type strain can be compared. This type of analysis can sometimes be misleading, however, due to regulatory effects caused by structural gene mutations influencing the levels of enzyme protein synthesized. For example, in the tryptophan system and other repressible enzyme systems, partial revertants and suppressed strains are usually derepressed relative to the wild type for the synthesis of enzymes of the pathway as a result of the fact that the end product is being synthesized in limited amounts, thus creating a partial starvation situation. A partial revertant strain may produce an enzyme of very low specific activity, but as a result of the derepression effect, the actual *amount* of enzyme relative to total protein may be even higher than in the wild type (see references 6 and 9 for examples).

13.6. MUTANT DETECTION (INDIRECT)

The majority of mutant types which arise cannot be detected by the direct selection methods described in the previous part. Any mutation which involves *loss* of a genetic function requires detection methods which can identify the very infrequent mutant type against the very large background of nonmutant cells. This group includes auxotrophic mutants, mutants involving carbohydrate fermentation, morphological variants of various types, and other conditional lethal mutants.

The most commonly employed indirect techniques for detecting such mutants are random colony screening and replica plating. Both techniques require the examination of a large number of bacterial colonies. They differ primarily in the way in which this is accomplished. Random screening involves the transfer of individual colonies one at a time from a medium on which the mutation is not expressed to a medium which reveals the mutational lesion. It is a time-consuming and tedious process, and for that reason is sometimes referred to as the "brute force" approach. One way to improve the likelihood of recovering a desired mutant by this process is to utilize a suitable chromogenic substrate in the plating medium. In this way, the mutant colonies can be distinguished from the nonmutants by a different color in, or adjacent to, the bacterial colonies. One example of this application, presented below, involves the detection of lactose-nonfermenting mutants by incorporating a tetrazolium dye into the plating medium (15).

13.6.1. Fermentation Mutants (Tetrazolium Method)

1. Grow an inoculum of an appropriate lactose-fermenting bacterium in L broth (13.9.1) or nutrient broth (13.9.11)
2. Mutagenize the culture and allow it to undergo a period of outgrowth as described in part 13.3.
3. Read the density of the culture on a Klett-Summerson colorimeter or other suitable turbidity-measuring instrument, and plate appropriate dilutions on lactose-tetrazolium plates (13.9.3) so as to obtain 50 to 100 colonies per plate.
4. Incubate until colonies are 2 to 3 mm in diameter. The lactose-positive parents are detected as white colonies while the lactose-negative variants are bright red. The basis for this differential staining is that lactose utilizers produce acid as a product of the fermentation, which in a weakly buffered medium lowers the pH and prevents reduction of the tetrazolium dye. The lactose-nonfermenting colonies have normal ability to reduce the tetrazolium to its red color form.

Suggestions and possible problems.

It is important that the colonies not be overcrowded on the plates because the release of acid by an excess of wild-type colonies will interfere with the reduction of the tetrazolium by the nonfermenting lactose-negative colonies. Dilutions should be adjusted so as to obtain no more than 100 colonies per plate. It is also important that a good quality of glucose-free lactose be used; otherwise, the results may be obscured by the leakiness often observed when lactose is contaminated with glucose (see list of reagents in part 13.10.2.).

This general type of screening assay can be used for any mutant for which there is a suitable chromogenic substrate available to distinguish mutants and nonmutants. Some examples are *p*-nitrophenyl phosphate for distinguishing phosphatase mutants, Giemsa or methyl green for detecting nuclease mutants and diazotized derivatives of casein, collagen, or albumin for proteolytic enzyme mutants. Although not a chromogenic indicator per se, the solubilization of gelatin from an opaque form to a transparent form on plates has long been used as a measure of proteolytic activity. Similarly, nuclease activity may be detected by the solubilization of RNA and DNA incorporated in an acid-insoluble form in agar plates.

13.6.2. Nutritional Auxotrophs (Replica Plating Method)

In 1952, Lederberg and Lederberg (17) devised

the replica plating technique for indirect selection of bacterial mutants. This technique, diagrammed in Fig. 2, permits the simultaneous transfer of a large number of colonies from one plating medium to another in a single operation by use of cotton velveteen cloth (13.9.16). Furthermore, the exact pattern of colonies on the master plate is reproduced on the replica plate, thus permitting the rapid examination of large numbers of individual colonies for mutant characteristics. A suitable dilution of mutagenized cells is first plated on the surface of a nonselective complete medium. After colonies form, they are replica plated to a selective or minimal medium. Colonies which grow on the supplemented but not on the minimal plate are then picked and tested for the desired mutation. By this technique one can readily examine several thousand colonies in a short time, without having to pick and transfer each one separately. It is a particularly useful technique for isolating auxotrophic mutants which grow on a complete medium but not on a minimal medium. By appropriate supplementation of the replica plate, one can identify auxotrophs of a specific type (e.g., single amino acid, vitamin, or nucleic acid precursor). When employed together with a highly efficient mutagen such as NTG, this technique can yield up to 20% total auxotrophs, or 1 to 2%

of a particular mutant type. This approach is illustrated by the following example for isolating various types of auxotrophic mutants. Also see 20.2.5.

Procedure.

1. Mutagenize the bacterium of interest, preferably with a mutagen having a wide spectrum of mutagenic effects (such as ultraviolet light).

2. Allow the mutagenized cells to grow out in nutrient broth (13.9.11) to express the mutations. If a number of independent mutations are desired, divide the culture into a number of subcultures before the outgrowth period.

3. Plate appropriate dilutions on the surface of nutrient agar (13.9.12) plates and incubate upside down at 37°C.

4. After colonies reach about 3 mm in diameter, replica plate to minimal agar (13.9.10 or 13.9.15) and to nutrient agar (13.9.12) plates. After the colonies have grown, identify those which grew on the nutrient agar but not on the minimal agar plates.

The procedure to this point (and as depicted in Fig. 2) identifies only total auxotrophs; it does not distinguish between those altered in different metabolic functions. To identify various types of auxotrophs more specifically, carry out the following additional steps.

a. Colonies grown on complete medium in master plate (Plate 1)

b. Colonies transferred to velveteen cloth

c. Master plate (Plate 1) removed

 on \ off /

d. Replica colonies transferred from velveteen cloth to minimal medium (Plate 2)

 on \ off /

e. Replica colonies transferred from velveteen cloth to complete medium (Plate 3)

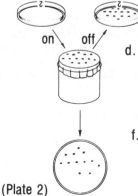 (Plate 2)

f. Time allowed for growth of colonies. Plates 2 and 3 then compared to identify the auxotrophic mutant colonies

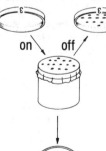 (Plate 3)

FIG. 2. *Replica plating method for screening auxotrophic mutants of a bacterium.*

5. Pick the identified auxotrophic colonies with sterile toothpicks, streak 0.6-cm wide patches onto fresh plates of nutrient agar (13.9.12), and allow to grow at 37°C. Each plate should accommodate 40 to 50 such patches.

6. Replica plate the patched clones to supplemented minimal agar plates (13.9.10 or 13.9.15) containing the nine combinations of metabolic pools shown in Table 3, as well as to unsupplemented minimal agar plates.

7. Determine on which group or groups of metabolic pools a given mutant will grow. From this information it should be possible to deduce the nature of the mutational defect. For example, a mutant which grows on pools 3 and 6 is most likely a cysteine-requiring mutant; one which grows on both 5 and 9 is probably an arginine auxotroph. Test each clone further by streaking it out on a minimal agar plate supplemented with the suspected growth requirement to verify the preliminary conclusions.

Suggestions and possible problems.

Some auxotrophs may respond to only a single metabolic pool or to none at all. The former may be a consequence of several factors: (i) two independent mutations, leading to unrelated growth requirements which happen to be included in the same metabolic pool; (ii) a mutation at an early step in a complex pathway (e.g., a polyaromatic acid-requiring strain, which would grow only on pool 8); (iii) inhibitory effects of other metabolites in the pool. Mutants which respond to none of the pools may be (i) mutants requiring some metabolite (e.g., a different vitamin) not tested or (ii) double mutants requiring a combination of two metabolites not represented in these pools. These mutants can be tested on plates containing a complex vitamin mixture, or on RNA hydrolysates or a complete mixture of amino acids (e.g., acid-hydrolyzed casein or Casamino Acids supplemented with L-tryptophan). If you isolate auxotrophs which fail to respond to any of the nine suggested metabolic pools, you may wish to try other compounds such as mixtures of vitamins, amino acids, or nucleic acid hydrolysates (16).

13.7. MUTANT CHARACTERIZATION

Once a mutant has been isolated, it is frequently desirable to characterize the nature of

the genetic defect by further biochemical or genetic analyses. For example, a nutritional requirement can result from mutations in several genes. Thus, an auxotrophic mutant unable to synthesize a specific amino acid, for example, may be blocked in any one of the several sequential enzymatic reactions in that synthetic pathway. Morphological mutants or those altered in macromolecular synthesis may be altered in even more diverse functions, such as those of assembly processes or complex regulatory functions.

While it is beyond the scope of this chapter to attempt a detailed description of the many possible approaches to mutant characterization, a few basic tests which can be applied are described below.

13.7.1. Nature of DNA Alteration

One can frequently obtain some idea of the nature of the change in the DNA by subjecting the mutant strain to reversion analysis, as described in part 13.5.2. If the mutant is induced to revert by any mutagen, it can be inferred that the mutation is a point mutation and not a deletion or other chromosomal aberration. If the reversion is enhanced by base analogs, by EMS, or by hydroxylamine, the preliminary conclusion is that the mutation must be a base substitution, probably of the transition type. If, instead, the reversion rate is enhanced by a mutagen such as ICR-191, then the mutation is most likely a frameshift. Accelerated reversion by agents such as ultraviolet light, but not by the other mutagens, would be indicative of a transversion mutation (but see caution note in 13.2). The frequency of spontaneous reversion is also an important parameter since it gives an estimate of the genetic *stability* of the mutation. This property can often be an important criterion in assessing the suitability of a given mutant for subsequent physiological or genetic studies.

13.7.2. Completeness of Genetic Block

A second criterion which influences the usefulness of a particular mutant is the extent to which the mutation inactivates the gene in which it has occurred. Leakiness frequently is a property of base substitution mutations, but much less so with frameshifts or deletions. The extent of leakiness is most easily assessed by

TABLE 3. *Combinations of supplements in minimal agar plates*[a]

Pool no.	1	2	3	4	5
6	Adenine	Guanine	Cysteine	Methionine	Vitamin B$_1$
7	Histidine	Leucine	Isoleucine	Valine	Lysine
8	Phenylalanine	Tyrosine	Tryptophan	Threonine	Proline
9	Glutamic acid	Serine	Alanine	Aspartic acid	Arginine

[a] All supplements are at a final concentration of 20 µg/ml except vitamin B$_1$, which is at 4 µg/ml.

plating the mutant cells at low density on the selective or nonpermissive medium and observing the extent of growth or other expression of that particular property. Leakiness can be a problem with mutants which are to be used for other physiological studies such as enzymatic assays or in complementation tests with other mutants.

13.7.3. Physiology of Genetic Block

For those mutants defective in one of several genes specifying related enzymes of a common biochemical pathway, the nature of the defect can often be elucidated by a combination of growth response and intermediate accumulation tests. In ideal situations, for any sequence of biochemical reactions $X \xrightarrow{1} A \xrightarrow{2} B \xrightarrow{3} C \xrightarrow{n} EP$, mutants blocked at any step in the sequence will respond to (grow on) any diffusible intermediate located *past* the point of the block. The intermediate(s) just *prior* to the point of the block will accumulate in the growth medium. Thus, for the hypothetical sequence given, a mutant in the gene which specifies enzyme 1 will accumulate intermediate X and will grow if supplied with A, B, or C (in addition to the end product [EP] of the reaction sequence). A mutant defective in enzyme 2 will accumulate intermediate A (and perhaps X) and will grow on B, C, or end product.

In actual practice, the characterization of mutants by accumulation and growth response tests is limited by several factors. First, many intermediates cannot be taken up by the cells. For example, normally phosphorylated intermediates cannot be taken up by cells and, thus, cannot be used as growth response intermediates (since usually these compounds are synthesized inside the cells). Second, some intermediates are chemically unstable and are rapidly degraded to other compounds or may be metabolized by other enzyme systems. Third, the usefulness of these tests is restricted by the availability of the appropriate intermediates and of simple tests for identifying them (e.g., colorimetric assays, etc.). Despite these limitations, accumulation and growth response tests can be extremely useful in distinguishing the different genetic and enzymatic functions in amino acid biosynthesis, purine and pyrimidine base synthesis, and vitamin synthesis.

13.8. SPECIAL APPLICATIONS

The induction and isolation of mutants are important primarily for the applications in which the mutants can be employed. Two special applications of mutagenesis have achieved considerable and widespread usefulness in recent years. These applications are the technique of localized mutagenesis, by which one can spe-cifically obtain mutations in a very restricted segment of the bacterial genome, and the Ames mutagenesis test, which allows one to estimate the potential carcinogenic (tumor-inducing) activity of an unknown substance by measuring its mutagenicity in a bacterial test system.

13.8.1. Localized Mutagenesis

The technique of localized mutagenesis, introduced in 1971 by Hong and Ames (13), provides a means for the isolation of new mutations in any small region of a bacterial chromosome. The mutations are induced in the short segments of bacterial DNA contained in transducing phage particles and are recovered in transductants by selection for a closely linked marker. In this manner new mutations can be selectively recovered for a region covering approximately 1% of the bacterial chromosome.

In the procedure to be described, a generalized transducing phage (P1) of *E. coli* is mutagenized with hydroxylamine, a chemical which produces GC → AT transition mutations. The mutagenized phage are used to transduce a recipient strain which carries a genetic defect in an early step in aromatic amino acid biosynthesis (*aroE*). The transductants are then examined for new mutations in either one of the closely linked drug resistance genes (streptomycin or spectinomycin) or in various ribosomal protein genes which can be isolated as temperature-sensitive conditionally lethal mutations (14).

Procedure for preparation of transducing phage.

1. Mix about 10^6 particles of P1*vir* phage with 0.5 ml of a mid-log-phase (1×10^8 to 2×10^8 cells/ml) L broth (13.9.1) culture of any wild-type *E. coli* strain (*aroE*$^+$) in a tube containing 3 ml of molten LC top agar (13.9.17).

2. Pour the mixture over the surface of a freshly prepared plate of LC agar (13.9.18) and incubate overnight at 37°C.

3. Scrape off the top agar layer with a sterilized spatula into a screw-cap tube containing 0.5 ml of chloroform.

4. Wash the remaining plate surface with 2 ml of L broth (13.9.1). Add the wash to the agar scrapings and shake vigorously.

5. Centrifuge at 5,000 to 6,000 × g in a benchtop centrifuge for 15 min. Remove the supernatant fraction with a Pasteur pipette and transfer to a clean sterile tube containing a few drops of chloroform.

Procedure for titration and mutagenesis of transducing phage.

1. Grow a culture of *E. coli* strain AB1157 to approximately 2×10^8 cells per ml in L broth

(13.9.1), centrifuge, and resuspend the cells in 0.5 volume of L broth.

2. Mix 0.1 ml of this cell suspension with 0.1 ml of various phage lysate dilutions (10^{-5}, 10^{-6}, 10^{-7}) in L broth. Incubate at 37°C for 20 min to allow phage adsorption.

3. Add 1.5 ml of LC top agar (13.9.17) at 55°C, mix by rotating the tube rapidly between the palms, and pour over the surface of a fresh LC agar plate (13.9.18).

4. Incubate the plates overnight at 37°C and count plaques the next day.

5. Concentrate the phage from the lysate by centrifugation at 48,000 × g for 3 h.

6. Discard the supernatant liquid and resuspend the pellet by allowing it to stand overnight in a small volume of L broth at 4°C.

7. The following day, remove the sediment by centrifugation at 15,000 × g for 10 min.

8. Retiter the phage as described in steps 2 through 4 above.

9. Mutagenize the phage by diluting the concentrated suspension (about 10^{11} plaque-forming units/ml) 10-fold in a solution of 0.45 M hydroxylamine, 2 mM ethylenediaminetetraacetate, and 10 mM $CaCl_2$, and incubate for 10 to 14 h at 37°C.

10. Centrifuge the mixture at 48,000 × g for 1 h to pellet the phage. Gently resuspend the pellet in 0.1 volume of L broth containing 2 mM ethylenediaminetetraacetate and 10 mM $CaCl_2$.

11. Centrifuge at 15,000 × g for 10 min to remove debris, and save the supernatant phage suspension. Under these conditions, the viability should be reduced by about 4 logs.

12. Retiter the phage as described above.

Transduction procedure.

1. To carry out the transduction (25), grow the recipient strain (*E. coli* AB2834 *aroE*) in L broth (13.9.1) at 37°C to about 1 × 10^8 cells/ml.

2. Centrifuge at 6,000 × g to pellet the cells and resuspend in 0.1 volume of L broth.

3. Mix 0.5 ml of the bacterial cell suspension (about 5 × 10^8 cells) with 0.5 ml of the mutagenized phage stock (~5 × 10^7 plaque-forming units/ml) and 0.5 ml of a mixture containing 0.015 M $CaCl_2$ and 0.03 M $MgSO_4$.

4. Incubate the mixture at 37°C for 20 min, centrifuge, and wash the cells twice with M56/2 buffer (13.9.9).

5. Resuspend the cells in 1 ml of M56/2 buffer, and spread 0.2-ml samples on minimal agar (13.9.10) plates supplemented with 0.2% glucose and 0.3% decolorized Casamino Acids. (Mix 3 g of Norit A with 20 g of vitamin-free Casamino Acids in 200 ml of water. Remove the charcoal by vacuum filtration. Repeat the procedure

twice with fresh Norit A. This gives a 10% solution of decolorized Casamino Acids.)

6. Incubate the plates at 30°C for about 18 h or until transductant colonies are just visible to the naked eye.

7. Replica plate to either spectinomycin or streptomycin plates (13.9.19) and to minimal agar plates (13.9.10) containing 0.3% Casamino Acids.

8. Incubate the minimal agar plates at 44°C and the antibiotic plates at 30°C. After incubation of the master and replica plates for an additional 24 h at their respective temperatures; superimpose the plates to score for drug-resistant isolates as well as colonies which failed to grow at 44°C.

9. Pick colonies of interest from the master plates, purify by single-colony isolation, and retest for the various selected properties.

Suggestions and possible problems.

P1 phage form very small plaques which are difficult to score on old plates. To minimize this problem, use only thick plates (30 to 35 ml of agar medium) poured no more than 8 to 12 h before use. Plaques larger than 1 mm in diameter are invariably contaminants (usually phage T1). One usually obtains 1 × 10^{10} to 5 × 10^{10} phage from a single plate lysate. Sufficient P1 for a localized mutagenesis experiment can be derived from 20 to 30 plates.

A transduction with 10^7 mutagenized P1 phage particles will normally yield between 200 and 2,000 transductants. The replica plating method will work only if there are fewer than 300 transductants per plate. Accordingly, it is suggested that a trial experiment be undertaken to determine the correct plating volumes. Approximately 5,000 to 10,000 transductants should be screened to ensure isolation of both drug-resistant and conditionally lethal mutants.

13.8.2. Mutagenicity of Carcinogens (Ames Test)

Chemical carcinogens are agents that induce tumor formation in laboratory animals with a high degree of predictability. Ames and co-workers (7, 8) have shown that most known carcinogens are also highly potent mutagens. Their test involves a simple plating assay which measures the induction of bacterial gene reversion by known carcinogens. Many of these agents require some type of activation before exerting their carcinogenic and mutagenic activities; in nature, this activation is usually carried out in the liver of the test animal. The bacterial reversion test assesses this requirement by incorporating post-mitochondrial extracts of mammalian liver into the assay mixture. About 85% of all known carcinogens are also mutagens, whereas less than 10% of non-carcinogenic compounds act as mutagens. This establishes a rather high correlation between these two properties and supports the idea that a significant proportion of mammalian cancers may arise as a consequence of induced somatic mutations.

The Ames test utilizes a series of histidine auxotrophs of *Salmonella typhimurium*. These strains carry specific mutations in one of the structural genes for histidine biosynthesis, including frameshift and base substitution mutations (see Table 4 for properties of strains). These strains also carry a defective excision repair system (*uvrB*), as well as a defect in the production of the lipopolysaccharide capsule which normally coats the outside of the bacterium (*rfa*). Other strains containing R factors (plasmids determining drug resistance) have recently been developed which further enhance the detection of mutagenicity for several classes of carcinogens.

The test is carried out by mixing the bacterial strains with a post-mitochondrial liver enzyme preparation together with the suspected mutagen in top agar on an appropriate minimal agar pour plate. Alternatively, a spot test can be performed with the bacteria and enzyme preparations poured into a top agar layer and the suspected carcinogen/mutagen added either in crystalline form or as a microdrop spotted on a sterile filter paper disk. Volatile compounds can be tested by incubating the bacterium-microsome-containing plates in a desiccator and introducing the compound to be tested into the atmosphere of the enclosed vessel. See reference 8 for additional experimental details.

Procedure for preparation of media.

1. Pour 30 ml of M56 minimal agar (13.9.10), containing 1.5% agar and 2% glucose, into 100 by 15 mm petri plates.

2. Prepare 100 ml of stock solution of top agar containing 0.65% agar and 0.5% NaCl (13.9.20). Add L-histidine·HCl and biotin to 0.05 mM final concentrations (10 ml of a stock solution which is 0.5 mM for both added to 100 ml of top agar) and maintain at 45°C until used. The small amount of histidine allows the bacterial cells to undergo several cell divisions, which enhances the mutagenic effects of the compounds to be tested.

Procedure for preparation of S-9 fractions.

Prepare the enzyme fractions (called S-9) as follows:

TABLE 4. *Salmonella strains used for mutagen testing*[a]

Strain	his muta- tion	Type of mutation	Other defects	R factor
TA 1535	G46	Base substitution	rfa, ΔuvrB	−
TA 1537	C3076	Frameshift	rfa, ΔuvrB	−
TA 98	D3052	Frameshift	rfa, ΔuvrB	+
TA 100	G46	Base substitution	rfa, ΔuvrB	+
TA 1538[b]	D3052	Frameshift	rfa, ΔuvrB	−
TA 1978[b]	D3052	Frameshift	rfa, uvr⁺	−

[a] Obtain from B. N. Ames, Department of Biochemistry, University of California, Berkeley, CA 94720.
[b] These two strains differ only in their excision repair capabilities. Comparison of the zones of growth inhibition by various compounds in the spot test gives a measure of the extent of killing due to DNA damage.

1. Induce the microsomal liver enzymes of male rats (Sprague-Dawley/Bio-1 strain) weighing about 200 g by injecting intraperitoneally 0.5 ml of a solution of Aroclor 1254 (a polychlorinated biphenyl) made up to 200 mg/ml in corn oil.

2. Feed the rats Purina Lab Chow and allow them to drink water freely for 5 days.

3. Withhold food for 12 h, and sacrifice the rats by stunning them on the head or by cervical dislocation followed by decapitation.

4. Remove and weigh the livers; rinse in sterile 0.1 M KCl, mince with sterile scissors, and homogenize them in a Potter-Elvehjem homogenizer with a Teflon pestle. Use 3 volumes of 0.15 M KCl per gram (wet weight) of liver as the homogenizing medium. Keep the preparations on ice throughout the procedure.

5. Centrifuge the homogenates for 10 min at 9,000 × g and distribute the supernatant (S-9) fraction in 1- to 2-ml portions in small plastic tubes; quick-freeze on dry ice, and store at −80°C until needed. Samples prepared in this manner should contain about 40 mg of protein per ml, and 1 ml should contain approximately the microsome content of 250 mg (wet weight) of liver.

If facilities and equipment are unavailable for preparing S-9 fractions as described above, such preparations are now apparently available commercially (see 13.10.2 for details). However, we have not had experience with such commercial preparations and cannot comment on their suitability.

Procedure for preparation of S-9 mix.

A tube of S-9 fraction is thawed at room temperature and kept on ice and is used for only 1 day's experiments; the remainder is discarded. The S-9 mix, used in the assay, contains, per ml: S-9 fraction, 0.04 to 0.1 ml; MgCl₂, 8 μmol; KCl,

33 μmol; glucose-6-phosphate, 5 μmol; nicotin-amide adenine dinucleotide phosphate (NADP), 4 μmol; and sodium phosphate buffer, pH 7.4, 100 μmol. Stock solutions of NADP (0.1 M) and glucose-6-phosphate (1 M) are prepared with sterile water and stored at −20°C. The stock salt solutions (0.4 M MgCl$_2$–1.65 M KCl) and phosphate buffer (0.2 M, pH 7.4) can likewise be prepared, autoclaved, and stored in a refrigerator. The S-9 mix is freshly prepared each day and may be kept for several hours on ice. Samples should be tested for bacterial contamination. If necessary, the S-9 mix can be filtered before use through a 0.45-μm membrane filter.

Procedure for preparation of test compounds.

Materials to be tested are prepared and stored in disposable glass screw-cap tubes with Teflon-lined caps. Water-insoluble compounds are usually dissolved in dimethyl sulfoxide. Other solvents which can be used are formamide, p-dioxane, ethanol, ethylene glycol dimethyl ether, or acetonitrile, depending upon the solubility properties of the substance being tested. Although many compounds can be stored at −20°C or in a refrigerator, some highly reactive agents such as alkylating agents should be freshly prepared at the time of testing.

Procedure for setting up the assay.

1. For the standard pour plate assay add 0.1 ml of an overnight Oxoid broth (13.9.7) culture of the bacterial tester strain to 2 ml of top agar (13.9.20) at 45°C. Also add 100 μl (or more, if insoluble) of the compound to be tested and 0.5 ml of the S-9 reaction mixture.
2. Mix the contents *immediately* by gently blending on a Vortex mixer or rotating the tube between the palms, and pour the contents over the surface of a minimal glucose agar plate (13.9.10).
3. Tilt the plate back and forth gently to evenly distribute the top agar layer. The entire operation should be carried out in 20 s or less.
4. Allow the plates to harden in the dark for a few minutes; then incubate upside down in the dark at 37°C.
5. After 2 days, score the histidine revertants (prototrophs) and confirm the presence of a slight background growth of bacteria on the plate due to the small amount of histidine added to the medium. To determine whether a particular compound requires activation for its mutagenic action, it is well to run samples with and without S-9 mix. In addition to the test plates, control plates of bacteria alone, bacteria plus known mutagens (including one which requires

activation), and bacteria plus S-9 (no mutagen) should also be prepared.

If mutagens are to be tested by a spot test, the mutagen is left out of the top agar mixture and is applied later as a few crystals or on a sterile 6-mm filter disk, as an aqueous, alcoholic, or dimethyl sulfoxide solution. With some compounds, this method is less effective, probably as a result of their adsorption to the filter. In some cases, the compounds can be spotted directly on the plates, although highly insoluble ones will still require much higher concentrations to give a positive result.

Suggestions and possible problems.

CAUTION: The test organism, *S. typhimurium*, is a class II human pathogen. It is one of the main causative agents of food poisoning and, if ingested, can bring about severe diarrhea, vomiting, and gastroenteritis. Although these laboratory strains are only slightly virulent, especially those lacking the external lipopolysaccharide capsule, they should still be handled very carefully. All pipettes should be cotton plugged, and mouth pipetting of the organism should be avoided. Cultures should be autoclaved before disposal. Laboratory workers taking antibiotics should be especially cautious in working with strains carrying R-determinant plasmids.

Also, many suspected carcinogens/mutagens are extremely toxic and should be handled with great care. Studies with such agents are best conducted in a laboratory separated from other general microbiological activities. A well-ventilated hood should be used for all weighing of hazardous materials and for spreading the top agar layers containing mutagens. Any incubator used for these experiments should be of the forced fan type, and the vented air should be taken into the hood or otherwise vented outside of the general laboratory air flow. Workers should wear laboratory coats and plastic gloves when handling these materials and should use respirators for working with volatile preparations.

In interpreting results, it should be realized that the spot test method, although useful for the general screening of a large number of suspected mutagens, is only qualitative. It is also of little use for chemicals such as many of the water-insoluble polycyclic hydrocarbons, which are unable to diffuse in the agar. Even with the more sensitive and quantitative overlay method, it is highly desirable to examine several different concentrations of the chemicals in question to determine whether the induction of revertants follows a linear response relationship to concentration. When there is a great amount of killing of the bacterial tester strains, as evidenced by a fainter than usual background growth of the bacterial lawn, it is necessary to be sure that the effect being noted is on the linear portion of the dose-response curve and not at a concentration which produces more killing relative to the mutagenic effect. The extent of cell killing may be determined by examining the differential effect of a given compound on a repair-defective (*uvrB*) versus a normal strain of the test organism. Another variable factor may lie in the S-9 fraction. The mutagenic effects may vary somewhat from one batch of S-9 to another, so it is well to spot test each new batch at several concentrations with several standard compounds, such as 2-aminofluorene, 7,12-dimethylbenzanthracene, and aflatoxin B1 (8).

In interpreting results, a compound is generally considered to be nonmutagenic if 1 mg tested by the pour plate assay fails to give a twofold elevation of reversion over the spontaneous background level. Some compounds may show very weak mutagenicity by the standard pour plate assay. The sensitivity of the assay can be enhanced for these compounds by carrying out a 30-min liquid preincubation prior to plating. For the preincubation, add together, *in the order given*, to sterile glass tubes: 0.1 ml of test compound solution, 0.5 ml of S-9 mix, and 0.1 ml of bacterial culture. Incubate at 30°C for 30 min (or at 37°C for 20 min) with gentle shaking. Add 2 ml of top agar (13.9.20), mix, and pour onto an M56 minimal agar plate (13.9.10). Incubate and score as for the standard assay procedure.

Occasionally, a high spontaneous reversion rate of the tester strains may obscure the results. If this problem arises, it usually is due to the presence of mutagens in the environment. Some batches of nutrient broth seem to contain such substances; for this reason, Oxoid broth (13.9.7) is recommended for growth of the tester strains. In addition, different batches of disposable plastic petri plates may contain significant levels of ethylene oxide, which is mutagenic. Petri plates are now available which are not sterilized with ethylene oxide (see 13.10.2).

13.9. RECIPES FOR MEDIA

13.9.1. L Broth (complex medium for rapid growth of *E. coli* and other *Enterobacteriaceae*)

Tryptone	10 g
Yeast extract	5 g
NaCl	0.5 g
Distilled water	1 liter

Add 1 M NaOH to adjust pH to 7.0. Autoclave for 15 min at 15 lb/in^2 (121°C). Add 10 ml of a sterile 20% dextrose solution after autoclaving.

13.9.2. L Agar (standard rich plating medium for many bacteria)

Add 15 g of agar per liter of L broth. Pour 20 to 25 ml into petri plates (100 by 15 mm).

13.9.3. Lactose-Tetrazolium Plates (for detection of lactose-nonfermenting mutants)

Difco Antibiotic Medium no. 2	25.5 g
Distilled water	950 ml
2,3,5-Triphenyltetrazolium chloride (Sigma Chemical Co.)	50 mg

Dissolve the Antibiotic Medium no. 2 first; then add the tetrazolium chloride and heat until it dissolves *completely*. Autoclave for 15 min at 121°C, and add 15 ml of filter-sterilized 20% lactose after autoclaving. Pour into 100 by 15 mm petri plates.

13.9.4. Tris-Maleic Acid Buffer (TM Buffer) (for mutagenesis with nitrosoguanidine)

2-Amino-2-(hydroxymethyl)-1,3-propanediol (Tris)	6 g
Maleic acid	5.8 g
$(NH_4)_2SO_4$	1.0 g
$FeSO_4 \cdot 7H_2O$	0.25 mg

Add distilled water to 1 liter, adjust pH (if necessary) to 6.0, and autoclave for 15 min at 121°C. After medium has cooled to about 50°C, aseptically add:

$MgSO_4 \cdot 7H_2O$	0.1 g
$Ca(NO_3)_2$	5 mg

13.9.5. Sulfanilamide Broth (used for 5-bromouracil mutagenesis)

M56 minimal salts medium (13.9.8) containing:

Vitamin-free Casamino Acids	0.1%
Xanthine	25 μg/ml
Uracil	2.5 μg/ml
5-Bromouracil	50 μg/ml
L-Tryptophan	20 μg/ml
Sulfanilamide	1 mg/ml

13.9.6. Sodium Acetate Buffer (for nitrous acid mutagenesis)

$NaC_2H_3O_2 \cdot 3H_2O$	13.6 g
Distilled water	1 liter

Adjust to pH 4.6 by adding concentrated acetic acid.

13.9.7. Oxoid Broth (for Ames test)

Oxoid Medium no. 2	25 g
Distilled water	1 liter

Autoclave at 121°C for 15 min to sterilize.

13.9.8. M56 Minimal Medium (standard liquid growth medium for *E. coli* and other bacteria not having fastidious growth requirements)

Na$_2$HPO$_4$·7H$_2$O	8.2 g
KH$_2$PO$_4$	2.7 g
(NH$_4$)$_2$SO$_4$	1.0 g
FeSO$_4$·7H$_2$O	0.25 mg

Add distilled water to 1 liter, adjust pH to 7.2, and autoclave for 15 min at 121°C (15 lb/in^2). After medium has cooled to about 50°C, aseptically add:

MgSO$_4$·7H$_2$O	0.1 g
Ca(NO$_3$)$_2$	5 mg

Supplement with 10 ml of a sterile 20% glucose solution for use as a growth medium.

13.9.9. M56/2 Medium (for localized mutagenesis)

Dilute M56 medium 1:1 with sterile distilled water.

13.9.10. M56 Minimal Agar (basic plating medium used for isolating nutritional auxotrophs; also for Ames test)

Prepare M56 minimal medium (13.9.8) and, after autoclaving, supplement with 10 ml of 20% glucose and appropriate amino acid and vitamin pools as required. Add 1 volume of supplemented M56 medium to 1 volume of 3.2% agar (32 g of agar per liter of water; autoclaved at 121°C, separately). Mix well, allow bubbles to dissipate, and pour into plates.

13.9.11. Nutrient Broth (general purpose medium)

Nutrient broth powder	8 g
Distilled water	1 liter

13.9.12. Nutrient Agar

Nutrient broth powder	8 g
Agar	15 g
Distilled water	1 liter

or

Nutrient agar	23 g
Distilled water	1 liter

Autoclave at 121°C (15 lb/in^2) for 15 min. Pour 20 to 25 ml into 100 by 15 mm petri plates.

13.9.13. Penassay Broth (rich growth medium commonly used for *Bacillus subtilis* and other bacteria requiring a buffered medium)

Difco Antibiotic Medium no. 3	17.5 g
Distilled water	1 liter

or

Beef extract	1.5 g
Yeast extract	1.5 g
Peptone	5.0 g
Dextrose	1.0 g
NaCl	3.5 g
KH$_2$PO$_4$	1.32 g
K$_2$HPO$_4$	3.68 g
Distilled water	1 liter

Autoclave at 121°C for 15 to 20 min.

13.9.14. Antibiotic Agar (for selection of antibiotic-resistant mutants by the gradient plate technique and for plasmid curing)

After autoclaving nutrient agar (13.9.12), add one of the following antibiotic solutions, sterilized by membrane filtration:

Streptomycin sulfate (Sigma)	100 µg/ml
Penicillin G (Sigma or Squibb)	500 µg/ml
Rifampin (Sigma)	50 µg/ml

13.9.15. Davis Minimal Broth or Agar (for *B. subtilis*)

In one flask (e.g., 2-liter Erlenmeyer):

K$_2$HPO$_4$	7 g
KH$_2$PO$_4$	2 g
Sodium citrate	0.5 g
MgSO$_4$·7H$_2$O	0.1 g
(NH$_4$)$_2$SO$_4$	1.0 g
Distilled water	500 ml

In a separate flask:

Agar (delete for use as broth)	15 g
Dextrose	5 g
Distilled water	500 ml

Autoclave both solutions at 121°C (15 lb/in^2) for 15 min, and allow to cool to about 55°C. Combine solutions, mix, and pour into 100 by 15 mm petri dishes.

13.9.16. Replicating Cloths (for isolation of nutritional auxotrophs)

Use ordinary cotton velveteen cloth, cut into pieces about 15 cm square. Wrap in packets of 10 to 20 in heavy paper. Autoclave for 10 to 15 min at 121°C on fast-exhaust and dry cycle.

13.9.17. LC Top Agar (for localized mutagenesis)

Tryptone	10 g
Yeast extract	5 g
NaCl	5 g
Agar	7 g
Distilled water	1 liter

Dissolve ingredients by heating, and dispense into bottles. Autoclave at 121°C for 20 to 25 min.

13.9.18. LC Agar Plates (for localized mutagenesis)

Tryptone	10 g
Yeast extract	5 g
NaCl	5 g
Agar	10 g
Distilled water	1 liter

Autoclave for 30 min at 121°C. Before pouring into plates, add, per liter of mixture, the following (each previously sterilized):

0.5 M CaCl$_2$	4 ml
20% glucose	5 ml
0.25% thymidine	40 ml

13.9.19. Spectinomycin or Streptomycin Agar (for localized mutagenesis)

Prepare L agar (13.9.2). After autoclaving, aseptically add spectinomycin (2 ml of a filter-sterilized 5% solution) or streptomycin sulfate (1 ml of a filter-sterilized 10% solution). Mix well and pour into plates.

13.9.20. Top Agar (for Ames test)

NaCl	5 g
Agar	6.5 g
Distilled water	1 liter

Dissolve by heating, and dispense 100-ml amounts into bottles. Autoclave for 15 min at 121°C, and cool to about 50°C. Add, after autoclaving, per 100 ml of top agar, 10 ml of a sterile stock mixture of:

L-Histidine·HCl (0.5 mM)	10.5 mg/100 ml
Biotin (0.5 mM)	12.2 mg/100 ml

13.10. SOURCES OF STRAINS AND SPECIAL MATERIALS

13.10.1. Strains

Escherichia coli strains. *E. coli* Genetic Stock Center, Dr. Barbara Bachmann, Department of Human Genetics, Yale University School of Medicine, 333 Cedar St., New Haven, CT 06510.

Bacillus subtilis strains. Curator, *Bacillus* Genetic Stock Center, Department of Microbiology, Ohio State University, Columbus, OH 43210.

Salmonella typhimurium strains for carcinogen-mutagen testing. Dr. Bruce N. Ames, Department of Biochemistry, University of California, Berkeley, Berkeley, CA 94720.

13.10.2. Media, Chemicals, and Reagents

Growth media. Agar, antibiotic media, nutrient broth, dextrose, etc.: Difco Laboratories, Detroit, MI 48232; BBL Microbiology Systems, Cockeysville, MD 21030.

Phosphate salts and other inorganic salts. Mallinckrodt Chemicals, St. Louis, MO 63147. Generally available through laboratory supply companies.

Amino acids, 2,3,5-triphenyltetrazolium chloride, 5-bromouracil, 2-aminopurine, hydroxylamine, 5-methyltryptophan, ethyl methane sulfonate, erythromycin, nitrosoguanidine, rifampin, penicillin G, streptomycin sulfate, daunomycin, glucose-6-phosphate, nicotinamide adenine dinucleotide phosphate, sulfanilamide: Sigma Chemical Co., P.O. Box 14508, St. Louis, MO 63178.

ICR-191 and ICR-170. Polysciences, Inc., Paul Valley Industrial Park, Warrington, PA 18976.

Lactose. Lactose CHR (glucose-free): U.S. Biochemical Corp., Cleveland, OH 44122.

Methyl methanesulfonate, ethyl methane sulfoxide, sodium nitrite. Eastman Organic Chemicals, Rochester, NY 14650.

Test mutagens. Benzo(α)pyrene, 2-aminofluorene, 4-nitro-*o*-phenylenediamine: Aldrich Chemical Co., 940 W. St. Paul Ave., Milwaukee, WI 53233. 9-Aminoacridine: Sigma Chemical Co., P.O. Box 14508, St. Louis, MO 63178.

Casamino Acids. Acid-hydrolyzed casein or Difco vitamin-free Casamino Acids: Difco Laboratories, Detroit, MI 48232. Casein hydrolysate-acid: U.S. Biochemical Corp., Cleveland, OH 44122.

Aroclor-1254. Analabs and Co., 80 Republic Drive, North Haven, CT 06473.

Dimethyl sulfoxide (spectral grade). Schwarz/Mann, Orangeburg, NY 10962.

Oxoid Medium no. 2. K. C. Biological, P.O. Box 5441, Lenaxa, KS 66215.

Non-ethylene oxide-sterilized petri dishes (for Ames test). Muta-Assay dishes, catalog no. 1028: Falcon Laboratories, 1950 Williams Dr., Oxnard, CA 93030 (805 485-8711).

Commercial S-9 preparations. Available from: Marketing Manager, Bionetics Laboratory Products, Litton Bionetics, Inc., 5516 Nicholson Lane, Kensington, MD 20795 (301/881-5600), or Meloy Laboratories, Inc., 6715 Electronic Dr., Springfield, VA 22151 (703/642-4800).

ACKNOWLEDGMENTS

We acknowledge the assistance of the following individuals in the preparation of this chapter: S. Kushner and W. S. Champney for advice in preparation of the localized mutagenesis procedures, Lynne Haroun for suggestions on the Ames test description, L. N. Ornston for his review of the entire chapter, and Debbie Chandler for numerous retypings of the entire manuscript.

13.11. LITERATURE CITED

13.11.1. General References

1. **Clowes, R. C., and W. Hayes.** 1968. Experiments in microbial genetics. John Wiley & Sons, Inc., New York.

    A collection of experiments with bacteria, bacteriophages, and the fungus Aspergillus, based upon a summer course taught at Hammersmith Hospital, London, England. The experiments are wide-ranging, from phage fluctuation tests to Aspergillus mapping. Presented in a step-by-step "cookbook" fashion.

2. **Drake, J. W.** 1970. The molecular basis of mutation. Holden-Day, San Francisco.

    One of the few comprehensive treatises on molecular mechanisms of mutation. Although several years old, still highly worthwhile for background reading on the subject of mutagenesis.

3. **Drake, J. W., and R. H. Baltz.** 1976. The biochemistry of mutagenesis. Annu. Rev. Biochem. **45:**11–37.

    An update of Drake's 1970 treatise, with special emphasis on various types of DNA misrepair mechanisms which may be mutagenic in their effects.

4. **Miller, J. H.** 1972. Experiments in molecular genetics. Cold Spring Harbor Laboratory, Cold Spring Harbor, N. Y.

    A compilation of rather sophisticated experiments in bacterial genetics, many of them based on the summer bacterial genetics courses conducted at Cold Spring Harbor. Although many of the experiments are specifically directed to the lactose operon of E. coli, the principles of mutagenic treatment, selection techniques, and mutant characterization are applicable to a broad variety of molecular genetics experiments.

13.11.2. Specific Articles

5. **Adelberg, E. A., M. Mandel, and G. C. C. Chen.** 1965. Optimal conditions for mutagenesis by N-methyl-N'-nitro-N-nitrosoguanidine in *Escherichia coli* K12. Biochem. Biophys. Res. Commun. **18:**788–795.

6. **Allen, M. K., and C. Yanofsky.** 1963. A biochemical and genetic study of reversion with the A-gene A-protein system of *Escherichia coli* tryptophan synthetase. Genetics **48:**1065–1083.

7. **Ames, B. N.** 1979. Identifying environmental chemicals causing mutations and cancer. Science **204:**587–593.

8. **Ames, B. N., J. McCann, and E. Yamasaki.** 1975. Methods for detecting carcinogens and mutagens with the Salmonella/mammalian-microsome mutagenicity test. Mutat. Res. **31:**347–

364.

9. **Carlton, B. C., and D. D. Whitt.** 1969. The isolation and genetic characterization of mutants of the tryptophan system of *Bacillus subtilis*. Genetics **62:**445–460.

10. **Coulondre, C., and J. H. Miller.** 1977. Genetic studies of the *lac* repressor. IV. Mutagenic specifity in the *lac* I gene of *Escherichia coli*. J. Mol. Biol. **117:**577–606.

11. **Davis, B. D.** 1948. Isolation of biochemically deficient mutants of bacteria by penicillin. J. Am. Chem. Soc. **70:**4267.

12. **Guerola, N., J. L. Ingraham, and E. Cerdá-Olmeda.** 1971. Induction of closely-linked multiple mutations by nitrosoguanidine. Nature (London) **230:**122–125.

13. **Hong, J., and B. N. Ames.** 1971. Localized mutagenesis of any specific small region of the bacterial chromosome. Proc. Natl. Acad. Sci. U.S.A. **68:**3158–3162.

14. **Kushner, S. R., V. F. Maples, and W. S. Champney.** 1977. Conditionally lethal ribosomal protein mutants: characterization of a locus required for modification of 50S subunit proteins. Proc. Natl. Acad. Sci. U.S.A. **74:**467–471.

15. **Lederberg, J.** 1948. Detection of fermentative variants with tetrazolium. J. Bacteriol. **56:**695.

16. **Lederberg, J.** 1950. Isolation and characterization of biochemical mutants of bacteria. Methods Med. Res. **3:**5–22.

17. **Lederberg, J., and E. M. Lederberg.** 1952. Replica plating and indirect selection of bacterial mutants. J. Bacteriol. **63:**399–406.

18. **Lederberg, J., and N. D. Zinder.** 1948. Concentration of biochemical mutants of bacteria with penicillin. J. Am. Chem. Soc. **70:**4267–4268.

19. **Miller, E. C., and J. A. Miller.** 1971. The mutagenicity of chemical carcinogens: correlations, problems and interpretations, p. 83–119. *In* A. Hollaender (ed.), Chemical mutagens, vol. 1, Principles and methods for their detection. Plenum Press, New York.

20. **Novick, R. P.** 1969. Extrachromosomal inheritance in bacteria. Bacteriol. Rev. **33:**210–263.

21. **Ornston, L. N., M. K. Ornston, and G. Chou.** 1969. Isolation of spontaneous mutant strains of *Pseudomonas putida*. Biochem. Biophys. Res. Commun. **36:**179–184.

22. **Osborn, M., S. Person, S. Phillips, and F. Funk.** 1967. A determination of mutagen specificity in bacteria using nonsense mutants of bacteriophage T4. J. Mol. Biol. **26:**437–447.

23. **Prakash, L. and F. Sherman.** 1973. Mutagenic specificity: reversion of iso-1-cytochrome *c* mutants of yeast. J. Mol. Biol. **79:**65–82.

24. **Szybalski, W.** 1952. Microbial selection. I. Gradient plate technique for study of bacterial resistance. Science **116:**46–48.

25. **Willetts, N. S., A. J. Clark, and K. B. Low.** 1969. Genetic location of certain mutations conferring recombination deficiency in *Escherichia coli*. J. Bacteriol. **97:**244–249.

Chapter 14

Gene Transfer

ROY CURTISS III

Three types of gene transfer are known in bacteria. **Transformation** is a gene transfer process in which deoxyribonucleic acid (DNA) from one bacterium, the donor, is taken up by another bacterium, the recipient. A recipient cell that now expresses a donor genetic trait is called a transformant. **Transduction** is a gene transfer process in which a bacterial virus propagating on one strain of bacterium, the donor, picks up genetic information and upon infection of another strain of bacterium, the recipient, sometimes causes a heritable change. A recipient cell that acquires a donor trait by this process is called a transductant. **Conjugation** is a gene transfer process in which one strain of bacterium, the donor, makes physical contact with another strain of bacterium, the recipient, and transfers genetic material. The recipient which acquires donor genetic information is called a transconjugant. Transformation, transduction, and conjugation, in conjunction with mutation, play an important role in allowing the evolution of new bacterial types. They are also extremely important processes in that they permit the bacterial geneticist to analyze the genetic and biochemical bases of bacterial functions and to establish principles of gene structure, function, and regulation as well as of the more complex processes of macromolecular synthesis, growth, and cell division.

14.1. TRANSFORMATION

Transformation is known to occur and has been well studied in certain strains of *Strepto-coccus pneumoniae, Haemophilus influenzae, Bacillus subtilis, Streptococcus sanguis, Neisseria gonorrhoeae*, and *Acinetobacter calcoaceticus* (25, 31, 36). The ability or inability to demonstrate transformation as a natural process in any given bacterial strain is dependent on the ability of the recipient strain to exhibit competence. **Competence** is defined as the ability of the recipient strain to transport DNA from the culture medium into the cell.

Transfection is a process analogous to transformation in which bacterial virus DNA is isolated and added to competent recipient cultures, which leads to production of phage particles. In instances in which the entire phage genome is not taken up by the competent recipient cell, it is necessary that several partial phage chromosomes be taken up and recombine in the recipient cell to result in an intact phage genome.

Successful development of a transformation system depends upon the use of high-molecular-weight, double-stranded DNA and the development of specific conditions to achieve maximum competence of the recipient cell population. The latter objective often depends on such factors as the stage in the growth cycle of the recipient population, the nutritive value of the culture medium, the degree of aerobiosis or anaerobiosis, the pH, the presence of appropriate concentrations of divalent cations, and, in most instances, the genotypic properties of the recipient. With regard to the last point, the structure of the cell surface, the production of competence factors, the ability to undergo autolysis, and the

243

presence or absence of prophages (genomes of temperate bacterial viruses replicating in synchrony with bacterial host genomes) are among the more important genotypic attributes influencing competence.

The study of transformation requires that the donor and recipient strains differ with respect to easily distinguishable genetic traits. Customarily, the recipient possesses mutations conferring inability to synthesize some metabolite or inability to utilize some carbon source, or the recipient possesses wild-type attributes of being sensitive to some antibiotic, drug, or phage. The donor parent, on the other hand, possesses the opposite allele which is most easily selected when transformants inherit that trait. Methods for the isolation and characterization of mutant strains of bacteria are described in Chapter 13.

14.1.1. Isolation of DNA

The optimum method for isolating high-molecular-weight DNA from a donor strain or transfecting DNA from a bacterial virus often depends on specific attributes of the bacterial species or of the bacterial virus. General principles and methods for disrupting gram-negative and gram-positive bacteria for the purpose of isolating DNA are given in Chapter 22. The most widely applicable method for the isolation and purification of high-molecular-weight DNA from microorganisms is that described by Marmur (29). This method is described in detail in part 22.2.1. In most instances, however, it is not essential to use highly purified preparations of donor DNA nor to remove ribonucleic acid (RNA) and other contaminating macromolecules; thus, treatment with ribonuclease can be omitted. Specialized methods for the isolation of plasmid DNA are described in Chapter 15.

A central feature of all methods of DNA isolation ensures that bacterial deoxyribonucleases are inactivated. Consequently, transforming DNA is generally suspended in a buffer that contains either citrate or sodium ethylenediaminetetraacetate (EDTA) to chelate divalent cations such as magnesium, which are essential cofactors for most deoxyribonucleases. The DNA preparations are usually stored in standard saline-citrate (0.15 M NaCl, 0.015 M trisodium citrate, pH 7.0) or, in the case of plasmid and phage DNAs where maintenance of supercoiled circular configurations may be important, in 10 mM tris(hydroxymethyl)aminomethane (Tris)–1 mM EDTA (pH 8.0) buffer.

14.1.2. Quantitation of DNA

For quantitation of purified DNA (free from RNA and other contaminating ultraviolet [UV]-absorbing compounds), the concentration of DNA can be determined spectrophotometrically. The absorbance in a cuvette with 1-cm light path at 260 nm divided by 20 gives the DNA concentration in milligrams per milliliter. Thus, 50 μg of DNA per ml has an A_{260} of 1.0. The measurement of absorbance at 234, 270, and 280 nm can be used in relation to the absorbance at 260 nm to indicate the degree of purity of the DNA preparation (22.4).

For quantitation of unpurified DNA, the colorimetric assay using diphenylamine can be used. This method is described in detail in part 22.4.3.

The concentration of DNA in a preparation to be used for transformation can also be determined by labeling the DNA to a known specific activity with radioactively labeled thymidine. This method requires that the donor strain possess a *thyA* mutation to block endogenous synthesis of thymidine monophosphate and preferably either a *deoB* or *deoC* mutation to prevent breakdown of deoxyribose 1-phosphate. The procedure is as follows:

1. Use a suitable minimal medium such as VB medium E (see 14.4.6) which can be supplemented with 0.5% vitamin-free Casamino Acids (14.4.8).

2. Add 5.0 μCi of either [^{14}C]thymidine or [^{3}H]thymidine and sufficient unlabeled thymidine to give a total thymidine concentration of 4.0 μg/ml.

3. Inoculate this medium with approximately 10^{6} *thyA deoB* cells per ml, and incubate the culture to allow seven or more generations of growth.

4. Isolate the DNA.

5. Place duplicate 25-μl samples of the DNA preparation on Whatman 3 MM filter disks which are mounted on stainless-steel pins. After 1 min or more, immerse the filters in 10% trichloroacetic acid (room temperature is adequate) in a beaker to precipitate the DNA. Subsequent washing of the filters is facilitated by placing them in a basket made by cutting out part of the top of a polyethylene bottle which has many holes in its bottom and sides. (A hot triangular file is a suitable instrument to use to make such holes.)

6. Wash the filters twice in 5% trichloroacetic acid and two or three times in ethanol by sequentially moving the "holey" basket to beakers containing each solution.

7. Dry the filters by placing the "holey" basket in a forced-air incubator, place them in a suitable scintillation fluid, and count them to obtain counts per minute in a scintillation counter (16.4.3).

8. Subtract background counts per minute, and calculate disintegrations per minute per mil-

liter of the DNA preparation by using the efficiency of the scintillation counter for the scintillation fluid used (16.4.3).

9. Calculate the concentration of DNA in micrograms per milliliter by using the specific activity in microcuries per microgram of the total labeled thymidine added to the culture, the mole percent of thymidine in the DNA, the determined value for disintegrations per minute per milliliter of the DNA preparation, and the constant 2.2×10^6 dpm per μCi.

14.1.3. Transformation in *Acinetobacter calcoaceticus*

A. calcoaceticus is a gram-negative, oxidase-negative, nonmotile, aerobic coccus which develops competence under various cultural conditions and in which transformation is independent of the presence of divalent cations (23). Thus, transformation of *A. calcoaceticus* is readily demonstrable in the laboratory. Sawula and Crawford (34) showed that the *trpA* and *trpB* genes, which encode the two subunits of tryptophan synthase, are closely linked and cotransformable, and they developed a method for scoring single and double transformants on the primary selection medium. The method is based on the observation that strains with the *trpA* mutation, which form large colonies on agar media with either 10 μg of L-tryptophan per ml or 40 μg of indole per ml, will form small colonies on media with only 5 μg of indole per ml. Thus, transformation of a recipient with a *trpB* mutation, which can only grow when supplied with 10 μg of L-tryptophan per ml, with DNA from a donor with a *trpA* mutation results in transformants and gives rise to both small and large colonies when plated on agar media containing 5 μg of indole per ml. The small colonies are due to cotransformation of the *trpA* mutation with the *trpB*$^+$ donor allele, whereas the large colonies arise from transformants that only inherit the donor *trpB*$^+$ allele.

Bacterial strains.

The donor and recipient strains were isolated from strain BD413, which is a microencapsulated mutant of the transformable wild-type *A. calcoaceticus* strain BD4 (23). The donor strain has the *trpA23* mutation and the recipient strain has the *trpB18* mutation (34). These strains are grown at 30 or 35°C and are stored on L-agar slants at 4°C. Fresh transfers should be made every 2 months.

Media.

A. calcoaceticus can be cultivated by use of a variety of complex media such as L broth and L agar, Penassay broth and agar, or brain heart infusion broth and agar. Various minimal media can also be used, although VB medium E is satisfactory. In using the *trpB* and *trpA* mutants, one can use an acid hydrolysate of casein such as Casamino Acids at 0.5% to stimulate more rapid growth or prototrophic recombinants while precluding growth of mutants, since acid hydrolysis destroys all tryptophan. See part 14.4 for methods of media preparation.

DNA isolation by method of Marmur (29).

DNA from the *trpA* donor strain can be isolated by the Marmur procedure (see 22.2.1) or more simply by one of the following procedures.

DNA isolation by method of Sawula and Crawford (34).

1. Grow donor cells with aeration at 30 to 35°C to early stationary phase in 10 ml of L broth (14.4.1). Aeration can be achieved by using either bubbler tubes (25 by 200 mm screw-cap tubes with a 4- to 5-mm-diameter glass rod encased in a rubber-tube sleeve placed through the cap and with a cotton plug at the top of the glass tube) connected to an aquarium pump or flasks agitated by a reciprocating or rotary shaker in a water bath.

2. Sediment cells by centrifugation and suspend in 2.4 ml of saline-EDTA buffer (0.15 M NaCl, 0.1 M EDTA, pH 8.0).

3. Add 100 μl of 6% sodium dodecyl sulfate and incubate the suspension at 37°C for 10 min to lyse the cells.

4. Clear the lysate by adding 50 μl of 5 M KCl, and after chilling on ice for 20 min, centrifuge at 39,000 $\times$ *g* for 20 min.

5. Heat the supernatant fluid (which contains the DNA) at 50 to 60°C for 5 min to kill any surviving cells.

6. Test the sterility of the DNA preparation by streaking a small amount on an L-agar (14.4.1) plate, which is incubated overnight at 30 to 35°C.

7. If desired, precipitate the DNA by adding twice the volume of cold ethanol, and wind it on a glass rod which has a small hook on one end (see 22.2.1). Dissolve the DNA in saline-citrate buffer.

8. Determine the DNA concentration by using the diphenylamine colorimetric assay (see 22.4.3) since the preparation will contain RNA and other UV-absorbing contaminants.

9. Store the DNA preparation at 4°C.

DNA isolation by simplified method of Juni (22).

1. Scrape a loopful of growth off either a slant or an agar plate, and suspend the cells in 0.5 ml

of a sterile solution containing 0.15 M NaCl, 0.015 M trisodium citrate, and 0.05% sodium dodecyl sulfate which is contained in a small test tube.

2. Heat the suspension at 60 to 65°C for 1 h.

3. The crude lysate is stored at 4°C and is stable with regard to transforming activity for at least 1 year.

Transformation procedure.

1. Grow the *trpB* recipient strain in L broth (14.4.1) with aeration at 35°C overnight. (Actually, *A. calcoaceticus* can grow at any temperature from 30 to 37°C, although slightly more rapid growth is achieved at 35°C.)

2. Dilute the overnight culture 1:10 into fresh prewarmed L broth, and aerate by either bubbling or shaking at 35°C for 2 h.

3. On each VB minimal agar (14.4.6) plate (supplemented with 0.5% Casamino Acids, 0.5% glucose, and 5 µg of indole per ml), spot 100 µl of the competent recipient culture and a varying amount of the donor DNA preparation (incremental amounts of 10 to 200 µl, or 20 to 500 mg).

4. Immediately mix the two preparations on the plate and distribute them evenly over the surface of the agar using a glass spreader which has been immersed in alcohol, flamed, and cooled on the agar surface.

5. Spot and spread 100 µl of the recipient cell population on one selective medium plate as a control, to verify the absence of *trpB*⁺ revertants or other contaminants.

6. Spot and spread 100 µl of the donor DNA preparation on one selective medium plate as an additional control, to verify that it is free from contaminating cells.

7. Add 10 µl of 100-µg/ml deoxyribonuclease in 0.2 M MgCl₂ to 100 µl of the donor DNA preparation (which is in saline-citrate buffer) about 2 to 3 min before mixing on a selective plate with 100 µl of competent recipient cells, as a control to verify that destruction of DNA eliminates transformant colonies.

8. Invert all plates and incubate at 35°C overnight or until colonies are clearly discernible.

9. Examine the plates and count the total numbers of small and large colonies on each plate without destroying the colonies.

10. Purify some of the small colony types to determine whether they have the properties expected of *trpA23 trpB*⁺ cotransformants (i.e., they still require either tryptophan or indole and form large colonies on media with 40 µg of indole per ml) and also purify some of the large colony types to determine whether their growth is independent of either indole or tryptophan. This can be done as follows:

a. Pick 10 to 20 individual colonies of each type (using either sterile toothpicks or a sterile needle) into 1 to 2 ml of buffered saline with gelatin (BSG) (14.4.7) in 13 by 100 mm test tubes, into a few drops of BSG contained in individual wells of a microtiter plate, or even into droplets of BSG contained on the inside of a plastic petri dish.

b. Streak these suspensions on selective agar plates in a manner to ensure obtaining isolated colonies. This is most routinely achieved by using a sterile needle if the colonies were suspended in very small volumes of BSG or by using a sterile loop with the excess fluid tapped off on the side of the test tube if the colonies were suspended in 1 to 2 ml of BSG. The selective agar plates can be marked into 4 to 10 equal-sized pie-shaped segments to allow purification of 4 to 10 isolates per plate.

c. Incubate these plates at 35°C for 1 to 2 days.

d. Pick individual isolated colonies from each original transformant into BSG, and streak these suspensions on minimal agar supplemented with 5 µg of indole per ml, 40 µg of indole per ml, and 10 µg of tryptophan per ml and on unsupplemented minimal agar. Incubate these plates at 35°C.

11. Analyze the results to determine whether the total number of transformants increases proportionally with increases in the amount of donor DNA used. This would be expected except at high concentrations of DNA when saturation may be achieved.

12. Calculate the frequency of cotransformation of the *trpA23* allele with *trpB*⁺ and determine whether this value is constant independent of DNA concentration.

Modifications and other transformation experiments with *A. calcoaceticus*.

The above-described procedure can be made quantitative by adding known quantities of donor DNA (1 to 500 ng/ml) to broth cultures of competent recipient cells at 1×10^8 to 2×10^8 cells per ml. After 10 to 15 min at 35°C, 10 µl of 0.5-mg/ml deoxyribonuclease suspended in 1 M MgCl₂ can be added per ml of donor DNA-recipient cell mixture to stop the reaction. Appropriate dilutions can be plated on selective media, and the log of the absolute titers of transformants can be plotted against the log of the DNA concentrations.

The stage in the growth cycle of the recipient culture for maximum expression of competence can also be determined by a slight modification of the simple plate transformation procedure or by the more quantitative method described in the preceding paragraph. In performing this type of experiment, it is important that the same concentration or number of recipient cells be used for each determination of the number of transformants. Precision in the design of this experiment is facilitated by first doing a growth curve for the recipient strain growing in L broth

(14.4.1) at 35°C with aeration. The general procedure for determining the effect of stage of growth on transformation frequency is as follows. Sediment the cells in a stationary aerated overnight culture by centrifugation and suspend them at the same density in fresh L broth. The titer of the suspension is determined by plating an appropriate dilution on L agar (14.4.1). The suspension is diluted 1:5, 1:10, 1:20, 1:40, and 1:80 into flasks containing L broth which are then incubated at 35°C with shaking. The 1:5 dilution is sampled immediately and after approximately 15 and 30 min for mixing with saturating concentrations of donor DNA to assess the competence of cells in the early-, mid- and late-lag phase. Samples from the cultures inoculated with the 1:10, 1:20, 1:40, and 1:80 dilutions can be taken when they achieve a cell density equal to that of the initial 1:5 dilution culture and will give values for various stages of exponential growth. The culture inoculated with the 1:80 dilution can be allowed to grow to higher densities to evaluate the competence of cells in the late-log, early-stationary, and stationary phases, but these samples will have to be diluted to the appropriate constant density before addition of DNA. Total viable recipient cell counts will have to be determined for each sample at the time donor DNA is added. By using these titers and the original inoculum dilutions, the growth curve can be graphed, and the frequency of transformants for each point can be plotted on the same

graph. Peak levels of competence should be noted for late-lag to early-log phase and for late-log to early-stationary phase.

Experiments to determine the absolute fraction of the recipient population competent to take up DNA involve either the use of radioactively labeled donor DNA and radioautographic analysis or else a mathematical analysis of data on frequencies of transformation of two unlinked donor markers singly and together as a function of donor DNA concentration. These methods are discussed in general terms by Hayes (18) and Notani and Setlow (31).

14.1.4. Transformation in *Escherichia coli*

Mandel and Higa (28) discovered that $CaCl_2$ treatment of *E. coli* cells, followed by incubation in the cold and then a heat shock, resulted in the uptake of DNA. This discovery led to a succession of refinements to artificially induce competence in various bacterial genera to allow transformation with chromosomal or plasmid DNA and transfection with bacteriophage DNA.

Bacterial strains.

Table 1 lists strains of *E. coli* K-12 that can be used as donors and recipients in transformation experiments, as well as those for use in transduction and conjugation experiments. In addition to genotypic differences between donor and recipient strains, it is frequently advantageous to include mutations in recipient strains

TABLE 1. *E. coli K-12 strains*[a]

Strain no.	Mating type	Genotype[b]	Equivalent designation
1	F⁻	pYA13 (pMB9::Tn3; Cel-*imm*⁺ *tet*⁺ *bla*⁺)/*supE42* λ⁻	χ2239
2	F⁻	*leu-5 proB59* Δ(*lacZOPI*)69 *tsx-98 supE42* Δ(*gal-uvrB*)40 λ⁻ *rpsL206 argH70*	χ2087
3	F⁻	*thr-16 tsx-63 purE41 supE42* λ⁻ Δ*trpE63 his-53 srl-2* Δ*thyA57 endA1 mtlA9 metE98 cycB2 cycA1*	χ1836
4	F⁻	*thr-1 ara-14 leu-6 proA2 lacY1 tsx-33 sup-37 galK2* λ⁻ *sbcB15 his-4 recB21 recC22 rpsL31 xyl-5 mtl-1 argE3 thi-1*	JC7623
5	F⁻	*supE42* λ⁻	χ289
6	F⁻	*galK2* λ⁻	W3102
7	F⁺	*lacZ81 supE42* λ⁻	χ226
8	F⁻	*proC63 supE42* λ⁻ *cycA1*	χ353
9	F′	F14 *lac*⁺/Δ(*proB-lac*)41 *supE42* λ⁻ *cycA1*	χ1301
10	F⁻	Δ(*proB-lac*)41 *supE42* λ⁻ *his-53 nalA95 xyl-14 cycB2 rpoB309 cycA1*	χ1997
11	R⁺	R100-1 (*drd-1 tet*⁺ *cat*⁺ *aadA*⁺ *sul*⁺ IncFII)/*tsx-63 supE42* λ⁻ *his-53 lysA32 xyl-14 arg-65*	χ1784
12	HfrOR4	λ⁻ O-*thr leu proA lac ... pyrB* F	χ433
13	F⁻	*leu-50 supE42* λ⁻ *rpsL206*	χ585
14	F⁻	*ara-14 leu-6 azi-6 tonA23 proA70 lacY1 tsx-67 purE42 supE44 galK2* λ⁻ *trpE38 nalA94 rpsL109 xyl-5 mtl-1 thi-1*	χ2634

[a] The strains listed have been selected because they possess the simplest constellation of suitable genotypic properties to illustrate the concepts and methodologies described in the text. These specific strains have not all been used in my laboratory or tested to conduct the procedures described in the text.

[b] Genetic nomenclature conforms to that of Demerec et al. (14) and uses notations and symbols for chromosomal and plasmid traits as described by Bachmann et al. (5) and Novick et al. (32), respectively.

that increase their transformability. In *E. coli* K-12, deletion of the *gal* operon (Δ*gal* mutations), such as in strain 2, eliminates galactose in the lipopolysaccharide core and substantially increases transformability of strains with plasmid DNA and transfection with lambda but not with T1 or T7 DNA. The *endA* mutation, such as in strain 3, abolishes the periplasmic endonuclease I and very much improves transformation with plasmid DNA, but has little effect on transformation with linear DNAs. Cosloy and Oishi (9) discovered that elimination of exonuclease V and exonuclease I in *recB recC sbcB* strains, such as strain 4, facilitates transformation of *E. coli* with chromosomal DNA and that these mutations also potentiate transfection with T1 and T7 DNA but not with lambda DNA. These mutations have little or no effect on transformation with plasmid DNA.

Media.

L broth and agar (14.4.1) or Penassay broth and agar (14.4.4) can be used as complex media. A diversity of minimal media can be employed, including M9 (4), VB medium E (37), and ML (10). Agar is added to 1.5% for solidified media. Amino acids, purines, pyrimidines, and vitamins are added to optimal concentrations as appropriate (Table 4) dependent upon the genotype of the recipient strain employed. Carbon sources are generally added to a 0.5% final concentration (Table 3), and antibiotics are added at concentrations that are dependent upon the level of resistance conferred by the drug-resistance mutation and/or gene (Table 5). See part 14.4 for media preparation methods.

DNA isolation.

Chromosomal DNA is most conveniently isolated from donor strains such as strain 1 (Table 1) by the method described by Marmur (29; see 22.2.1), although the final steps in the procedure to remove all traces of RNA are not necessary to obtain active transforming DNA preparations. pYA13 plasmid DNA can also be isolated from strain 1 and used for transformation. Isolate plasmid DNA by the methods described in Chapter 15.

Transformation procedure.

A large number of modifications of the original Mandel and Higa (28) procedure have been reported, but the method that seems to give the highest efficiencies of transformation or transfection with a diversity of phage, plasmid, and chromosomal DNAs and with a diversity of *E. coli* K-12 recipient strains has been worked out by Gurnam Gill and Laura Alexander in our

laboratory. It is a modification of a procedure originally used by Enea et al. (15). The procedure is as follows:

1. Grow a 5-ml overnight culture of the recipient strain in L broth (14.4.1) as a standing culture at 37°C. Standing overnight cultures (which have not been aerated by either bubbling or shaking) when subcultured the next day exhibit a much shorter lag phase prior to initiation of exponential growth than is the case with aerated overnight cultures which have already reached the stationary phase of growth.

2. Dilute this culture 1:100 into 40 ml of L broth and incubate with aeration by either bubbling or shaking at 37°C until the cells reach early log phase (about 1×10^8 to 2×10^8 cells per ml).

3. Chill this culture on ice and sediment the cells by centrifugation at $7,000 \times g$ at 4°C.

4. Gently suspend the cell pellet in 10 ml of ice-cold 10 mM NaCl–50 mM $MnCl_2$–10 mM sodium acetate buffer (pH 5.6).

5. Hold the cell suspension on ice for 20 min, and sediment the cells by centrifugation.

6. Gently suspend the cell pellet in 1 to 2 ml of ice-cold 75 mM $CaCl_2$–100 mM $MnCl_2$–10 mM sodium acetate buffer (pH 5.6). (Depending on the particular recipient strain and the type of DNA used, it may be advantageous to vary the $CaCl_2$ concentration from 25 to 100 mM and the $MnCl_2$ concentration from 50 to 200 mM.)

7. Add 200-μl samples of competent cells to chilled 13 by 100 mm test tubes.

8. Add 1 to 10 ng of plasmid or phage DNA or 100 to 1,000 ng of chromosomal DNA to each sample of competent cells. If the donor DNA is in a volume of less than 20 μl, the buffer used to suspend the DNA will have little effect on the result; however, if the DNA sample is larger than 20 μl, the DNA should be diluted into the $CaCl_2$–$MnCl_2$–sodium acetate buffer (pH 5.6) used to suspend the competent cells.

9. Hold the DNA-cell mixture on ice in the cold room for approximately 30 min.

10. Heat-shock the mixture at 30°C for 2.5 min. The rapidity of the temperature shift is important, but the actual temperature for the heat shock and the duration will vary depending on the properties of the recipient strain. For example, either 2 min at 37°C or 1 min at 42°C may be superior to 2.5 min at 30°C.

11. Following the heat shock, neutralize the mixture to pH 7.0 by addition of 3 μl of 2 M Tris buffer (pH 7.4).

12. If the donor marker being selected confers resistance to an antibiotic (especially one that is bactericidal), then dilute the mixture 1:10 into the growth medium and incubate at 37°C for 60 to 90 min to allow for expression of the drug

resistance marker.

13. Plate the mixture (with or without dilution, depending on the type of DNA used and on whether dilution was already made to permit expression) on medium to select transformants and on nonselective medium, after appropriate dilution to measure cell survival. Since the treatment of cells with $CaCl_2$ causes them to become fragile, it is important to use selective medium that has been freshly prepared and which is not too dry. It is also important not to spread the plates to dryness but, after distributing the diluted suspension over the surface of the agar, to allow it to dry into the agar by leaving the lid of the petri dish slightly ajar for a few minutes.

14. Incubate the plates at 37°C for 1 or 2 days as is appropriate, depending upon the selected marker and the growth rate of the recipient strain.

Applications.

Transformation is an important method for genetic analysis since naturally transformable species do not have conjugation systems and only a few have well-developed transductional systems. In terms of basic knowledge, transformation has provided information about the mechanism of genetic recombination (25, 31), and in *Bacillus subtilis* it has provided information about the origin and direction of chromosome replication (18). Data from homospecific and heterospecific transformation experiments have also been used to determine the degree of genetic relatedness of specific genetic information (31). These approaches augment studies of relatedness utilizing DNA reassociation kinetic analyses (see Chapter 22) and the newer recombinant DNA methodologies. Undoubtedly, an important use of transformation and transfection is in conjunction with recombinant DNA methodologies in which methods for the artificial induction of competence allow introduction of recombinant molecules into various bacterial species.

14.2. TRANSDUCTION

Bacterial viruses are classified as either virulent, in which case all infected cells die with liberation of new phage particles, or temperate, in which the infected bacteria either lyse with liberation of new phage progeny or become lysogenic (4). In the lysogenic state, the phage genome, now called a prophage, replicates in synchrony with the bacterial chromosome either by establishing a plasmid form or by integrating into the chromosome. Although virulent phages and virulent mutants of temperate phages are capable of transduction, temperate phages capable of establishing lysogeny probably account for most transduction-mediated gene flow among bacteria in nature and provide systems that are most easily studied.

Two types of transduction are known: specialized and generalized (18). In **specialized transduction**, only donor genetic markers adjacent to the integrated prophage are capable of being transduced. In **generalized transduction**, any donor trait, whether it be on a chromosome, plasmid, or prophage, is capable of being transduced. Because of these requirements, specialized transduction is observed only when the phage lysate is obtained either by spontaneous or induced liberation of phage from a lysogenized donor and never when the phage is propagated by the lytic productive cycle. On the other hand, with generalized transducing phages, donor genetic information can be picked up after induction of a lysogenized donor culture or by lytic productive infection of the donor culture. Specialized transducing phages contain segments of donor genetic formation that have replaced phage genome segments. Thus, these phages are recombinants containing both phage and host genetic information. Upon infection and lysogenization of a suitable recipient strain, specialized transducing phages establish a state of partial diploidy such that both the donor genetic information contained in the transducing phage and the homologous genetic information in the recipient chromosome persist. In forming the specialized transducing phage, if the donor genetic information replaces phage genetic information that is vital for the propagation of the phage, the resulting transducing phage becomes defective and is unable to multiply and propagate in the absence of an intact helper phage genome. In certain instances, however, the specialized transducing phage does not lose essential functions and is able to form plaques and to exhibit a normal productive lytic cycle.

In those systems best studied, the formation of generalized transducing phages involves the inadvertent packaging of donor host-genetic material in lieu of phage genetic information. Thus, those phage particles capable of transducing donor information are completely defective. A recipient cell infected with such a particle will only acquire the donor genetic material and no phage genetic information, provided that the recipient cell is not multiply infected with infective phage particles along with a transducing phage particle. Upon entry into the recipient cell, the donor genetic information (if it is chromosomal in origin) may integrate into and replace recipient genetic information or may persist without integration or replication to give rise to what is termed an abortive transductant. Generalized transducing phages can also pick up and transduce plasmid genetic information. Such plasmid elements, upon introduction into the recipient host, become established and commence to replicate in synchrony with the recipient chromosome.

The study of transduction requires that the donor and recipient differ with respect to appropriate and easily distinguishable genetic traits. Customarily, the recipient possesses mutations conferring inability to synthesize some metabolite or inability to utilize some carbon source, or the recipient possesses the wild-type attribute of being sensitive to some antibiotic, drug, or phage. The donor parent, on the other hand, possesses the opposite allele which is most easily selected when transductants inherit that trait.

General methods.

The preparation of specialized transducing lysates requires the use of a donor lysogenic for the transducing phage. The lysate can be prepared by inducing this lysogen with either UV or mitomycin C, depending upon the system, or even better by using a lysogen carrying a heat-inducible prophage in which lysogeny is stably maintained at 30°C but the lytic response is induced at 37°C (19). In preparing generalized transducing lysates, one usually infects the donor bacteria and obtains the phage lysate by growing the infected culture in a liquid medium or by plating with the soft agar overlay method to obtain a confluent plate lysate (2). In either case, one must grow the bacteria and infect them in a medium with the appropriate cation concentrations to maximize phage yields. Most transducing phage lysates are stored over chloroform, which also kills any uninfected or uninduced donor cells.

The conditions to ensure optimum infection of the recipient strain depend on the phage-host system being employed, but usually involve careful attention to the ionic environment and the means of growing the recipient. With most phage-host systems, phage are preadsorbed to the recipient cells prior to plating for transductants. Since most phage in a transducing lysate are infectious and able to kill recipient cells, methods are employed to circumvent infection and killing of transductant cells. This can be accomplished most readily by using a recipient which is lysogenic for the transducing phage since lysogenic cells are immune to superinfection with the homologous phage. Alternatively, one can use a multiplicity of less than one phage particle per recipient cell so as to minimize multiple infections and following preadsorption can add either antiserum against the phage or a chelator if the phage requires divalent cations

for adsorption. These latter two methods generally prevent infection of potential transductants by phage released from recipient cells that were productively infected.

Transductants are recovered and enumerated by plating on appropriate selective media. Such plating is delayed to allow phenotypic expression and segregation if the selected donor trait is recessive to the recipient allele and/or confers survivability to a bactericidal agent or compound.

14.2.1. Specialized Transduction (in *E. coli* with Bacteriophage Lambda)

Bacterial and phage strains.

Since lambda phage usually integrates into the *E. coli* chromosome between the *gal* and *bio* genes, it is these genetic traits that are normally transducible by lambda (4, 18, 22). Transduction can be demonstrated by using a nonlysogenic recipient strain that possesses a *gal* mutation such as strain 6 (Table 1) and a *gal*+ donor strain such as strain 5 (Table 1) that is lysogenized with λ *c*I857. λ *c*I857 has a mutation which causes the repressor protein necessary to establish and maintain lysogeny to be heat sensitive such that the bacterial strain remains lysogenic when grown at 30°C but is induced and undergoes lysis with liberation of phage when grown at 37°C.

Media.

Use Super broth (14.4.3) containing 10 mM Mg^{2+}, lambda agar or soft agar (14.4.2), EMB agar (14.4.5) containing 1% galactose (EMB + Gal) or containing no added carbohydrate (EMB0), VB minimal agar (14.4.6) containing 0.5% galactose (MA + Gal) (and any other nutritional supplements required for growth of the recipient used), and a 10 mM $MgCl_2$ (or $MgSO_4$) solution in water.

Preparation of transducing phage lysate.

A transducing phage lysate is most easily prepared by first lysogenizing the donor strain with λ *c*I857 and then inducing the lysogenized donor cells to undergo lysis. The procedure is as follows:

1. Grow a few milliliters of the donor culture in Super broth at 37°C to an approximate titer of 1×10^8 to 2×10^8 cells per ml.

2. Add approximately 200 µl of this culture to 2 ml of molten lambda soft agar which is held at 45°C, and immediately pour the contents onto the surface of a lambda agar plate. Rapidly tilt the plate back and forth to distribute the soft agar mixture uniformly over the plate and then

let it sit flat on the counter top at room temperature for several minutes until the soft agar layer solidifies. (The lambda agar plates should be equilibrated to room temperature or even pre-warmed, since use of "cold" plates will cause the soft agar to solidify before it can be evenly distributed on the surface of the plates.)

3. Place a drop of two of a λ cI857 lysate containing 10^9 to 10^{10} plaque-forming units (PFU) per ml onto the center of the plate. (If the λ lysate has been stored over $CHCl_3$, the residual $CHCl_3$ must be removed prior to use by incubating a sample of the lysate at 37°C with intermittent bubbling until $CHCl_3$ vapors are no longer detectable.) Leave the petri dish lid ajar until the spot has dried into the plate. Invert the plate and incubate at 30°C overnight. The next day there will be a confluent lawn of bacteria over the surface of the plate with a large plaque in the center caused by lysis of some of the cells infected with λ cI857. The center of this plaque will be somewhat turbid, however, because of the survival and growth of donor cells that have become lysogenized with λ cI857.

4. Use a sterile toothpick, wood applicator stick, or loop to scrape off some of the turbid growth from the center of the plaque to inoculate 10 ml of Super broth contained in a 125-ml sterile Erlenmeyer flask.

5. Incubate this culture at 30°C on a gyratory shaker to get good aeration. When the culture reaches approximately 2×10^8 cells per ml, shift the flask to 45°C for 1 to 2 min and then to 37°C. Continue aerated incubation for approximately 2 h, when lysis should be complete.

6. Add 1 drop of $CHCl_3$, shake vigorously to disperse, and continue shaking at 37°C for 10 to 15 min.

7. Remove bacterial debris from the lysate by centrifugation at about 6,000 to 7,000 $\times$ g for 5 to 10 min. Pour the supernatant fluid into a sterile bottle or screw-cap tube and add another drop of $CHCl_3$.

Determination of phage titer.

The λ cI857 lysate should have a titer of about 10^{10} PFU/ml, and the exact titer can be determined as follows:

1. Grow a 10-ml culture of either the nonlysogenic donor or recipient strain in Super broth with aeration by either bubbling or shaking at 37°C until a titer of 1×10^8 to 2×10^8 cells per ml is reached.

2. Sediment the cells in this culture by centrifugation at 6,000 to 7,000 $\times$ g for 5 to 10 min, and suspend the pellet in 10 mM $MgCl_2$.

3. Resediment the cells as a washing step, and suspend the pellet in 10 ml of 10 mM $MgCl_2$.

4. Starve this suspension for 60 min by shaking in a sterile Erlenmeyer flask at 37°C.

5. Distribute 1.9-ml amounts in sterile 13 by 100 mm test tubes and place these at 37°C.

6. Dilute the λ lysate serially using 10 mM $MgCl_2$.

7. Add 100 μl each of the 10^{-4}, 10^{-5}, 10^{-6}, and 10^{-7} dilutions to individual 1.9-ml samples of bacteria in the 13 by 100 mm test tubes at 37°C. Allow 10 to 15 min for preadsorption, and take 200 μl from each time and mix with 2 ml of molten lambda soft agar (equilibrated to 45°C). Immediately pour the contents on the surface of a lambda agar plate as described above.

8. After the soft agar has solidified, invert the plates and incubate at 37°C.

It is often wise to plate each dilution in duplicate in case one of the plates smears because of excess moisture. The most reliable titer is determined from assay plates that contain at least 50 and no more than 400 plaques. The titer will be equal to the mean plaque count per plate times 100 times the reciprocal of the initial dilution that gave rise to the plaque count. At the time of determining the transducing phage lysate titer, it is wise to take some of the lysate and streak on an EMB + Gal plate to verify that the lysate is free from all viable donor bacteria.

Specialized transduction procedure.

1. Grow the gal recipient in 10 ml of Super broth at 37°C with aeration by either bubbling or shaking to a titer of 2×10^8 to 4×10^8 cells per ml.

2. Sediment, wash, and suspend the cells in 10 ml of 10 mM $MgCl_2$; then starve the cells with aeration at 37°C for 60 min prior to infection.

3. Distribute 0.9-ml amounts of this culture in sterile 13 by 100 mm test tubes and incubate at 30°C.

4. Dilute the transducing phage lysate appropriately in 10 mM $MgCl_2$ so that addition of 100 μl to 0.9 ml of bacteria will result in multiplicities of 0.1, 1.0, and 3.0 phage per bacterium. Be sure to remove residual $CHCl_3$ from the lysate by blowing a stream of air through the lysate.

5. Add these amounts of lambda to individual 0.9-ml samples of recipient cells, and leave one 0.9-ml recipient culture uninfected to serve as a control to verify that the gal mutation in the recipient is stable and that the number of Gal^+ transductants is in excess of the number of Gal^+ revertants, should any appear.

6. Allow 10 to 15 min for preadsorption and then plate 100-μl amounts from each mixture by spreading on the surface of each of four MA + Gal plates. Incubate two plates from each infection at 30°C and two at 37°C.

7. Incubate the plates for 2 days; then count the number of Gal⁺ colonies.

Characterization of transductants.

In order to be characterized, transductants must first be picked and purified as described above. They can be purified on MA + Gal, EMB + Gal, or MacConkey + Gal plates which should, of course, be incubated at 30°C. As the last step, pick individual isolated colonies from each transductant into 2 ml of Super broth and grow to 1 × 10⁸ to 2 × 10⁸ cells per ml at 30°C. The Gal⁺ clones, if due to transduction, should have several properties. The procedure is as follows:

1. The transductants should be lysogenic for defective or nondefective derivatives of λ *cI857* and thus should be unable to grow at 37 or 42°C. This can be tested by spotting loopfuls of each culture on an EMB + Gal plate and incubating at 37 or 42°C. As a control, test for growth at 30°C as well.

2. The transductants should be immune to lambda, and this is most easily determined by cross-streak test. Approximately 50 μl of the λ *cI857* lysate is distributed evenly down the center of an EMB0 agar plate. After 5 to 10 min to allow the lysate to dry into the agar surface, full loopfuls of the individual Gal⁺ transductant cultures are streaked across the phage streak. For this test, the donor and recipient strains should be included as controls. Incubate plates at 30°C. The growth to one side of the λ *cI857* streak will serve as the uninfected control and will appear as white to pink on the EMB0 plates. Sensitivity to λ is revealed by the dark-red coloration of the surviving bacterial growth at the juncture of the λ *cI857* streak with the streak of a λ-sensitive strain. The change in color is due to acid production and/or liberation as a consequence of lysis of some of the bacteria. EMB0 is thus a very sensitive indicator of sensitivity of bacteria to temperate phages that do not lyse all infected cells.

3. The transductants should be partially diploid and heterozygous, containing both *gal*⁺ and *gal* genetic information. This is most easily determined by diluting a culture sufficiently so that when 100 μl is spread on the surface of an EMB + Gal plate, 100 to 300 colonies will result after incubation at 30°C. If the transductants are partially diploid and heterozygous, approximately 1 to 3% of the colonies will give a fermentation-negative appearance and be white to pink, and the remainder will give a fermentation-positive reaction and appear dark purple or black. The Gal⁻ types arise as a result of recombination leading to homozygosity for the *gal* mutation originally in the recipient.

4. Most Gal⁺ transductants arising from the low-multiplicity infections will only harbor the λ *gal* transducing phage genome, and some of these can be analyzed by incubating cultures at 37°C to see whether or not infectious lambda particles are liberated. This can be done by (i) diluting the cultures of purified transductants 1: 10 into 1 to 2 ml of Super broth in 13 by 100 mm test tubes, (ii) incubating these subcultures at 37°C for 2 h, (iii) adding a drop of CHCl₃ to kill surviving bacteria, (iv) spotting loopfuls of each "lysate" onto either an EMB0 or lambda agar plate that has been overlaid with 2 ml of lambda soft agar inoculated with 100 to 200 μl of a log-phase culture of a λ-sensitive strain, and (v) incubating the plates overnight at 37°C. More than likely, these Gal⁺ transductants will contain λ *dgal* particles and will not be able to yield plaque-forming particles to lyse the λ-sensitive indicator strain. This is because λ genes essential for the lytic response have been replaced by the *gal*⁺ genes.

5. Most Gal⁺ transductants formed in the higher-multiplicity infections will be double lysogens having a wild-type λ *cI857* prophage as well as λ *dgal* prophage. Incubation of these cultures at 37°C will lead to spontaneous lysis with liberation of a more or less equal mixture of λ *cI857* particles that will be able to form plaques and λ *dgal* particles. The plaque-forming ability of these lysates can be verified by the test outlined in the preceding paragraph. Such lysates have a high frequency of λ *gal* particles and can transduce 10 to 50% of the *gal* recipients to Gal⁺. This can be verified by repeating the transduction as outlined above (Specialized transduction procedure) with the modification that the infection mixture should be diluted (10⁻³ to 10⁻⁵) prior to plating on the MA + Gal plates. These have been called high-frequency transducing (HFT) lysates. This is in contrast to the low-frequency transducing (LFT) lysate formed after the λ *cI857* donor lysogen was initially heat induced at 37°C.

Applications.

Specialized transduction facilitates the genetic analyses of bacteria, including gene arrangement and regulation. *E. coli* has a number of phages like lambda that integrate into the chromosome at different locations and allow for specialized transduction of adjacent chromosome markers. Even more important, by using a donor host such as strain 2 (Table 1) that carrys a deletion for the normal integration site of the λ prophage, λ will integrate at low frequency into abnormal sites scattered throughout the chromosome. Induction of these lysogens will lead to formation of transducing phages carrying almost any gene desired (35). This approach can be used with other specialized transducing phages of *E. coli* as well as with those that propagate on other bacterial species. Many of the methods to accomplish these objectives as well as other uses of specialized transducing phages are described by Miller (2).

14.2.2. Generalized Transduction (in *E. coli* with Bacteriophage P1)

Bacterial and phage strains.

Bacteriophage P1 is a generalized transducing phage with an extremely broad host range. It is easy to isolate mutant derivatives of wild-type P1 that will plate with high efficiency on approximately 70% of *E. coli* strains obtained from clinical specimens and also to isolate derivatives that will readily infect bacterial strains that are members of the genera *Shigella*, *Salmonella*, *Enterobacter*, *Klebsiella*, and *Citrobacter* (17). Because of this attribute and the fact that P1 can transduce fragments of donor strain DNA having a mass of 60×10^6 daltons, it is a very useful transducing phage. On the other hand, compared to other frequently used coliphages, P1 adsorbs to bacteria more slowly, forms very small plaques, and yields lysates that are somewhat unstable unless completely purified from bacterial cell debris. P1 *kc* (24) is a P1 derivative that plates with high efficiency on *E. coli* K-12 and forms a relatively clear plaque that facilitates its assay. However, P1 *kc* is able to lysogenize bacterial hosts. P1 L4 (8) is a mutant derivative of P1 *kc* that possesses a mutation in the *cI* repressor gene and is thus unable to lysogenize.

Although a diversity of donor and recipient strains of *E. coli* K-12 can be used for P1-mediated transduction, it is convenient to illustrate the principles involved using a system that allows a simple demonstration of cotransduction. Thus the donor strain (strain 7, Table 1) should have a *lacZ* mutation conferring inability to synthesize β-galactosidase and thus inability to utilize lactose as a carbon source. The recipient (strain 8, Table 1) should have the genotype *lacZ⁺ proC* so that selection of *proC⁺* transductants will sometimes allow inheritance of the donor *lacZ* mutation by cotransduction with *proC⁺*. The two genes are approximately 30% cotransducible. Both P1 *kc* lysogenic and non-lysogenic derivatives of the recipient strain should be available.

Media.

Use L broth, L agar (1.2% agar), and L soft agar (0.65% agar) containing 2.5 mM CaCl₂ for growth of bacteria and P1 lysate preparation and assays; Penassay agar for bacterial assays; and VB medium E agar supplemented with 1% lactose, 0.4% disodium succinate, and 50 μg of triphenyltetrazolium chloride per ml for selection of transductants. On this last medium, *lac⁺ proC⁺* transductants will have white colonies and *lacZ proC⁺* transductants will be able to grow because of utilization of succinate as a carbon source and will appear as red colonies. The citrate in VB medium E will chelate the Ca^{2+} needed for P1 adsorption and will minimize P1 infection of bacteria on the selective medium. See part 14.4 for methods for preparation of media.

Determination of phage titer.

1. Add 1 ml of a standing overnight culture (see step 1 under Transformation procedure in part 14.1.4) of the P1-sensitive donor strain grown in L broth (without CaCl₂) to 9 ml of L broth containing 2.5 mM CaCl₂.

2. Aerate this culture in a bubbler tube or by shaking at 37°C for 60 to 90 min until the culture reaches a titer of approximately 2×10^8 cells per ml.

3. While the culture is growing, dilute a P1 L4 (or P1 *kc*) lysate (which should have a titer of approximately 10^{10} PFU/ml) 10^{-5}, 10^{-6}, and 10^{-7} in L broth + 2.5 mM CaCl₂.

4. Place 1.9-ml amounts of the grown P1-sensitive strain in 13 by 100 mm test tubes at 37°C.

5. After a few minutes for temperature equilibration, add 100 μl of each P1 dilution, and allow 20 or 30 min for adsorption of P1 to the bacterial cells.

6. Add 200 μl of each phage-bacteria mixture to 2 ml of L soft agar and immediately distribute on the surface of an L agar + 2.5 mM CaCl₂ plate. After 5 min, invert the plates and incubate at 37°C.

Plaques become visible after 6 h of incubation and are 0.5 to 1 mm in diameter. Plaque assay plates for P1 L4 can be incubated overnight prior to counting, but plaque assay plates for P1 *kc* should be read after about 6 h since continued growth of lysogenized cells in the center of P1 *kc* plaques causes them to become extremely turbid and difficult to visualize if plates are incubated overnight. The phage titer is equal to the plaque count times the reciprocal of the dilution times 100.

Preparation of transducing phage lysate.

1. Grow the desired bacterial donor strain as described above under Determination of phage titer (steps 1 and 2).

2. Dilute the P1 L4 (or P1 *kc*) lysate in L broth plus 2.5 mM CaCl₂ so that it contains between

6×10^7 and 8×10^7 PFU/ml, and add 100 μl to 1.9 ml of log-phase donor bacteria in 13 by 100 mm test tubes at 37°C.

3. Allow 20 to 30 min for P1 preadsorption.

4. Add 200 μl of the adsorption mixture per 2 ml of L soft agar containing 2.5 mM CaCl₂, and immediately distribute on an L agar + 2.5 mM CaCl₂ plate. Prepare five to eight plates in this way. (Each plate will be seeded with 6×10^5 to 8×10^5 P1 L4-infected [or P1 *kc*-infected] donor bacteria and an excess of uninfected donor bacteria.) After 5 min, invert the plates and incubate at 37°C for 6 h.

5. Add 3 to 5 ml of L broth containing 10 mM MgSO₄ (but no CaCl₂) per plate. Using a glass spreader, break up the soft agar layer and scrape into a 40- to 50-ml polypropylene (or another CHCl₃-resistant material) centrifuge tube. Add approximately 100 μl of CHCl₃ for each 4 ml of lysate, cap the centrifuge tube tightly, and blend vigorously on a Vortex mixer for 1 to 2 min. Let this tube sit for 2 h at room temperature or overnight in a refrigerator to permit the P1 to diffuse into the aqueous phase. Periodic blending can be done if desired.

6. Centrifuge the crude lysate at 5,000 to 6,000 $\times g$ for 10 to 15 min, and carefully decant the supernatant fluid into a bottle or screw-cap test tube. Add a few drops of CHCl₃ to this lysate and shake vigorously. Store lysate at 4°C.

The next day, titer the lysate as described above and test for sterility by streaking a small amount of lysate on the surface of an L or Penassay agar plate. These lysates are relatively stable but will drop one to two logs in titer over the period of 1 year when stored at 4°C. Complete stability of P1 can be achieved by banding the phage in CsCl and removing CsCl by dialysis against L broth containing 10 mM MgSO₄ (8).

Construction of P1 *kc* lysogens.

It is wise to use recipient strains that are lysogenic for P1 *kc* when the transduction experiment is performed to yield quantitative data on gene frequencies in donor cells, cotransduction, etc. This is neither necessary nor desirable when using transduction to construct strains. Lysogenic recipient strains are constructed as follows:

1. Grow the recipient strain as described above under Determination of phage titer (steps 1 and 2).

2. Add 0.9 ml of culture to a 13 by 100 mm test tube and equilibrate at 30°C for about 5 min.

3. Add 0.1 ml of a P1 *kc* lysate to achieve a multiplicity of about 5 PFU/bacterium. Be sure to remove traces of CHCl₃ if the P1 *kc* lysate is

not diluted more than 10^{-1} to prevent killing of recipient cells.

4. Incubate at 30°C for 2 to 3 h. With P1 infection, the lysogenic response is favored at low temperatures (20 to 30°C), and the lytic response is favored at high temperatures (37 to 42°C).

5. Streak out growth on a Penassay or L agar plate to obtain single colonies. Alternatively, plate an appropriate dilution to achieve single colonies.

6. Pick and purify 10 well-isolated colonies (nonmucoid and with regular smooth edges) on Penassay or L agar by the general methods described in step 10 under Transformation procedure in part 14.1.3.

7. Pick a well-isolated colony from each of the 10 original isolates into 2 ml of L broth containing 1 mM citrate (no CaCl₂) in 13 by 100 mm test tubes. Number the cultures from 1 to 10 and incubate at 42°C for 4 to 6 h, to about 2×10^8 cells per ml. Growth at 42°C increases the rate of spontaneous induction of P1 lysogens and the absence of Ca²⁺ permits the P1 phage liberated as a consequence of spontaneous induction to accumulate.

8. Test whether the putative P1 lysogens liberated P1 during growth at 42°C. This is done as follows:

a. Place 1 ml of each of the 10 cultures and 1 ml of a culture of the P1-sensitive starting strain (as a control) into numbered 13 by 100 mm test tubes at 42°C.

b. Add a drop of CHCl₃ to each culture, shake, and continue incubating for 15 to 20 min.

c. Make serial 10-fold dilutions of a P1 *kc* lysate from 10^{-1} to 10^{-8}.

d. Add 200 μl of a log-phase P1-sensitive strain grown in L broth containing 2.5 mM CaCl₂ to 2 ml of molten L soft agar containing 2.5 mM CaCl₂, and pour onto the surface of an EMB0 agar (containing 2.5 mM CaCl₂) plate. Let the soft agar layer solidify.

e. Spot full loopfuls of each of the 11 chloroformed cultures and the 8 serial P1 dilutions onto the surface of the EMB0 plate. By touching the side of the loop to the surface of the agar, the entire contents of the loop should be delivered. Leave the lid ajar until the spots have dried into the soft agar. Mark the bottom of the plate so that location of all spots is known.

f. It is best to leave the EMB0 plate right side up (i.e., uninverted) for incubation at 37 or 42°C overnight.

Areas of lysis (confluent or plaques) should be discernible for each chloroformed extract made from a P1 lysogen provided that sufficient P1 was liberated during growth at 42°C. The spots made from the P1 dilutions should indicate the sensitivity of the method.

9. Test whether the putative P1 lysogens are immune to P1. This is done as follows:

a. Distribute 50 to 100 μl of a P1 L4 lysate (ca. 10^{10} PFU/ml) down the center of an EMB0 agar (containing 2.5 mM CaCl$_2$) plate. Leave the lid ajar until the streak is dry. Mark the plate bottom to indicate the orientation and placement of streaks.

b. Streak full loopfuls of the 10 cultures of viable putative P1 lysogens and the culture of the viable P1-sensitive starting strain across the P1 L4 streak.

c. Let the streaks dry. Invert the plate and incubate overnight at 37°C.

A P1 L4 immune isolate that liberates P1 *kc* during growth at 42°C can be stocked as a P1 *kc* lysogen. P1 lysogens are somewhat unstable, presumably because the P1 prophage exists as a plasmid which is sometimes lost. Thus, P1 lysogens that have been subcultured over a period of time may require reisolation using the procedures and tests enumerated above. Growth of P1 lysogens in the presence of 2.5 mM CaCl$_2$ at 37°C will also permit spontaneously liberated P1 to infect nonlysogenic segregants and either lyse or relysogenize them.

Generalized transduction procedure.

When using the nonlysogenic recipient, it is customary to use a multiplicity of infection of 1 or slightly less so that cells infected with a transducing P1 particle will not also be infected with an infectious P1 particle and thereby be lost as a result of lysis. Since P1 does not adsorb very rapidly (only 30 to 50% of the phage will attach and inject within 20 to 30 min at 37°C), an initial multiplicity of 2 to 3 phage per bacterium is used to achieve an actual multiplicity of infection of 1. When using the lysogenic recipient, double infections will not decrease the recovery of transductants since the lysogenic cells are immune to superinfection with P1 L4. To illustrate these points, the P1 L4 lysate grown on the donor parent can be added to the nonlysogenic and lysogenic recipient strains at multiplicities of 0.5, 2.5, and 12.5 per bacterium. The highest multiplicity may be difficult to achieve unless the P1 L4 lysate grown on the donor strain has a titer of 10^{10} PFU/ml or higher. The transductions are conducted as follows:

1. Grow the P1 *kc* lysogenic and nonlysogenic *lacZ⁺ proC* recipient strains to 2 × 10^8 cells per ml as described above under Determination of phage titer in 14.2.2.

2. While the recipient strains are growing, place some of the P1 L4 lysate grown on the *lacZ proC⁺* donor at 37°C and aerate to remove all residual CHCl$_3$. Also, calculate the amount of diluted or undiluted lysate (usually 0.1 to 0.5 ml) to add to a selected volume of recipient cells at about 2 × 10^8/ml (usually 0.5 to 1.9 ml) to achieve a multiplicity of 12.5 PFU/bacterium. Then dilute the P1 L4 lysate in L broth containing 2.5 mM CaCl$_2$ so that the multiplicities of 0.5, 2.5, and 12.5 PFU/bacterium can be achieved by the addition of a constant volume of phage lysate to a constant volume of recipient cells.

3. Add the calculated volume of bacteria of each recipient strain to each of four 13 by 100 mm test tubes and place the eight cultures at 37°C. Determine the titers of the recipient cultures by plating appropriate dilutions on L or Penassay agar.

4. Add the calculated amounts of the donor P1 L4 lysate to three of the four cultures of each recipient strain and an equal volume of L broth containing 2.5 mM CaCl$_2$ to the fourth culture, which will serve as a control. It is best to stagger the infections at about 3-min intervals.

5. After 30 min for preadsorption, add 100 μl of 0.2 M sodium citrate to each phage-bacteria mixture to chelate the Ca^{2+} and stop further infection of P1.

6. Spread 100-μl amounts of undiluted and 10^{-1} diluted samples from each infected and uninfected culture per VB minimal agar plate (containing lactose, succinate, and triphenyltetrazolium chloride). Incubate these plates at 37°C for 2 days prior to counting.

7. Plate appropriate dilutions of the infected and uninfected cultures on L or Penassay agar to determine the viable titers. Incubate these plates overnight at 37°C.

8. Redetermine the titer of the P1 L4 lysate by the method described above under Determination of phage titer. This is especially important with newly made P1 lysates since the titer sometimes increases two- to threefold during the first week due to disaggregation by the action of CHCl$_3$ of P1 particles stuck together because of bacterial debris and/or residual agar remaining in the lysate.

9. Plate 100 μl of the P1 L4 lysate on the selective VB minimal agar to verify the absence of viable donor cells.

The number of transductants will be proportional to the titer of phage used when the P1 *kc* lysogenic recipient is used. However, when the nonlysogenic recipient is used, fewer transductants should be observed when multiplicities of infection are in excess of one than when multiplicities of infection are one or less. In either case, the frequency of cotransduction of *lacZ* with *proC⁺* should be approximately the same within the limits of statistical variation. The frequencies of killing of the nonlysogenic recipient for each infection can be calculated and used in conjunction with the Poisson distribution to estimate the actual multiplicities of infection (4).

Applications.

Generalized transduction with P1 has been very useful in unambiguously determining the order of genes on the bacterial chromosome. This has been accomplished by using appropriately marked strains and following inheritance of two or three nonselected alleles in conjunction with the selected marker and by performing the transductions in all possible reciprocal combinations for selected and unselected markers. It has also been possible to use P1-mediated transduction to construct strains of desired genotypes. A mutant allele of known properties can be cotransduced into a strain by selecting for inheritance of an adjacent wild-type marker. When cotransduction is not possible, mutant alleles can be introduced by P1 transduction and, after time for segregation and phenotypic expression, can be recovered after ampicillin-cycloserine enrichment. A newer and more elegant method is based on the use of transposable drug resistance elements (7) in which insertional inactivation by transposition of a drug resistance element into genes provides a means to select for or against inheritance of the induced mutations. Modifications of this methodology can also be employed to place a transposable drug resistance element adjacent to a wild-type or mutated allele of a given gene and to move that allele from one strain to another.

Another useful method was developed by Hong and Ames (20) to induce mutations in unknown genes that are closely linked to known genes. By treating a transducing phage lysate with nitrous acid, hydroxylamine, or nitrosoguanidine and then selecting transductants for inheritance of a known wild-type allele at 30°C, it is possible to isolate a high frequency of mutations causing thermosensitive defects in some nutritional or vital function in genes adjacent to the selected allele.

P1-mediated transduction can also be employed to determine whether a trait is chromosomally specified or plasmid specified. A P1 lysate propagated on the donor is irradiated with increasing doses of UV light, and these irradiated and unirradiated lysates are used to transduce a recipient strain with selection for inheritance of the donor trait in question. If the trait is specified by a chromosomal gene, small doses of UV will actually increase the total number of transductants since UV is recombinogenic, whereas if the trait is specified by a plasmid, the frequency of transductants inheriting the donor trait will decrease with increasing UV dose since UV-induced damage will preclude the replication of the plasmid after it is introduced into the recipient by transduction.

The availability of P1 derivatives that possess translocatable drug resistance elements permits positive selection of rare lysogens and facilitates the development of P1 transducing systems among other members of the Enterobacteriaceae (17). One can also select rare derivatives of P1 that are better able to propagate on a given host strain as well as mutant derivatives of a host strain that are better able to propagate P1.

Generalized transducing phages are known for a variety of bacterial genera. Probably the best studied system makes use of phage P22 of *Salmonella typhimurium* (39). This system has certain advantages over the P1 system of *E. coli* in that P22 is very easy to propagate in liquid lysates, forms large plaques, and yields lysates that are stable for years. Also important, this system allows analysis of mutant alleles for ability to complement each other by a process called abortive transduction (18, 33).

14.3 CONJUGATION

Conjugation is a cell-mediated type of gene transfer that requires that the donor parent possess a conjugative plasmid. This plasmid may either replicate in the cytoplasm in synchrony with the bacterial chromosome or be integrated into and replicate as a part of the bacterial chromosome. Conjugative plasmids possess *tra* genes that specify and/or control: (i) synthesis of appendages termed donor pili that are obligatory, at least for most conjugative plasmids, to allow donor cells to make contact with recipient cells; (ii) substances that minimize the occurrence of donor-donor matings; and (iii) conjugational transfer of plasmid or chromosomal DNA commencing from a definable transfer origin site on the conjugative plasmid molecule. Conjugative plasmids also possess genes to control (i) their vegetative replication, (ii) the number of copies of plasmid per chromosomal DNA equivalent, and (iii) incompatibility functions. Conjugative plasmids may or may not possess genes for other phenotypic attributes such as antibiotic resistance, heavy metal ion resistance, bacteriocin production, surface antigen production, and enterotoxin production. Conjugative plasmids seem to be ubiquitous in gram-negative bacteria and have recently been found in certain species of *Streptomyces* and *Streptococcus*.

The best studied conjugative plasmid is the fertility factor F of *E. coli* K-12, and this conjugative plasmid can exist either in an autonomous cytoplasmic state in an F^+ donor or in an integrated state in an Hfr donor. Recipient strains do not possess F and are termed F^-. F^+ donors transfer the F plasmid to essentially all F^- cells with which they mate, but are only able to transfer chromosomal genetic information at low frequency. All F$^+$ donor chromosomal markers are transferred at nearly equal frequency (ca. 10^{-5} for any donor marker per donor cell), and most of these transconjugants inherit F. Hfr donors transfer their chromosome in an oriented sequential manner from a genetically fixed origin dependent upon the site of F integration into the bacterial chromosome. Genetic markers near the origin of transfer of the Hfr chromosome are transferred at highest frequencies, with frequencies of transfer and inheritance decreasing proportionately as markers become more distal from the origin. This behavior is observed because of spontaneous random interruption of chromosome transfer. Most transconjugants inheriting Hfr chromosomal material remain F$^-$ unless the mating is of long enough duration to allow transfer of the entire Hfr chromosome, since parts of the integrated F plasmid are transferred as the terminal markers. The rate of Hfr chromosome transfer is constant at 37°C. By using a multiply mutant F$^-$ strain and periodically interrupting mating, it is possible to determine the length of mating time necessary to transfer any marker to the recipient and the interval of time between the transfer of sequentially transferred markers. In using Hfr matings at 37°C, it has been possible to determine that transfer of the entire E. coli K-12 chromosome would take 100 min and that approximately 26 × 10^6 daltons of DNA are transferred per minute. The F plasmid as well as other conjugative plasmids contain insertion sequences (IS), which are also present in the E. coli chromosome. It is recognition between these sequences followed by reciprocal recombination that allows F to insert in a defined manner into the E. coli chromosome (7). Some 25 to 30 F integration sites have been identified by genetic techniques, giving rise to Hfr donors with different origins and orientations of chromosome transfer (13, 27).

A more complete discussion of the process and mechanism of conjugation can be found in various reviews (3, 11, 12, 38) as well as in many standard textbooks and monographs (16, 18, 21, 25). Methods for performing conjugational crosses, in addition to those given below, can be found in the manuals by Clowes and Hayes (1) and Miller (2).

General methods.

Since the goal in conjugation experiments is to isolate transconjugants that have attributes from both donor and recipient parents, strains with readily selectable genetic markers must be used. In experiments in which the kinetics of conjugational transfer are being studied, it is important to be able to terminate the conjuga-

tional event at precise intervals of time and to preclude further matings on the selective medium, a situation which can arise when high densities of donor and recipient cells are plated. The simplest means to accomplish these goals is to use a recipient strain which is resistant to nalidixic acid because of a mutation in the nalA gene. Nalidixic acid inhibits conjugational DNA transfer, and its addition to a mating mixture with a nalidixic acid-sensitive donor and a nalidixic acid-resistant recipient immediately terminates transfer of either plasmid or chromosomal DNA. Recipients resistant to rifampin due to a mutation in the rpoB gene also have been used, but rifampin does not cause cessation of conjugational DNA transfer once it is in progress. Rifampin does terminate transcription of donor genetic information, which is highly advantageous when attempting to detect low-frequency transfer of plasmids specifying resistance to drugs such as ampicillin, since continued production of β-lactamase by the donor leads to inactivation of the ampicillin included in the selective medium and precludes detection of Ap^r transconjugants.

Most conjugative plasmids isolated from strains of bacteria in nature are repressed with regard to expression of the donor phenotype such that only 1 out of 1,000 to 10,000 cells harboring the conjugative plasmid is able to transfer it under optimum conditions. Since, in some instances, transient derepression occurs upon transfer to a recipient cell, the detection of conjugational plasmid transfer sometimes requires that matings be conducted over a period of several hours or longer and with periodic dilution of the mating mixture to maintain appropriate bacterial densities. Derepressed mutants of some repressed conjugative plasmids have been isolated such that essentially all donor cells harboring the conjugative plasmid are capable of acting as genetic donors. F is one of the best characterized derepressed conjugative plasmids.

Although some conjugative plasmids promote conjugational DNA transfer when matings are conducted in liquid media, others are more efficient at promoting transfer when the donor and recipient cells are either mated on agar plates or impinged on a membrane filter and the filter is placed on a nutritive agar medium. In the latter case, the nutrient medium is usually overlaid with nutrient soft agar, and the plates are incubated uninverted for a suitable period of time, after which the filter is moved to a selective agar plate that counterselects against the donor parent. Alternatively, the filter can be immersed in a buffered saline and the cells can be removed by intermittent blending in a Vortex mixer, after which suitable dilutions are plated on selective agar medium.

Another variable that must be considered in optimizing detection of conjugational transfer is temperature. Many conjugative plasmids found in members of the *Enterobacteriaceae* exhibit optimal transfer frequencies when matings are conducted at or near 37°C. With some of these, transfer is all but undetectable at 25°C, and leaving donor cultures at room temperature for a period of time prior to mixing with recipient cultures at 37°C can lead to a significant drop in the frequency of transfer. On the other hand, there are conjugative plasmids such as those in the IncH group that transfer optimally at 25°C and not at all at 37°C. In examining whether a trait might be transferred on a conjugative plasmid, it is wise to examine mating at 37°C versus 25°C.

Restriction is a process in which a microorganism recognizes entering DNA as either foreign or like self and if foreign causes its ultimate degradation. The inability to demonstrate conjugative plasmid transfer when conducting intraspecies, interspecies, or intergeneric matings may be associated with a restriction barrier being manifest in the recipient strain. Restrictionless (*hsdR* or *hsdS*) mutants are available in *E. coli* K-12, *S. typhimurium* LT2, and certain other well-characterized and frequently used bacterial species. Even in the absence of a recipient which is unable to carry out restriction, the potential problem can frequently be eliminated by heating the recipient parent prior to mating for 5 to 10 min at 50°C or 15 to 20 min at 45°C. Indeed, I have observed that growth of a large number of *E. coli* recipient strains at 42°C for five to seven generations prior to mating allowed conjugative plasmid transfer to over 95% of them.

14.3.1. Conjugative Plasmid Transfer (in *E. coli* F⁻ with F′ or R Plasmids).

Bacterial strains.

The study of conjugative plasmid transfer is facilitated if the plasmid specifies a readily selectable trait. F′ plasmids (which arise from Hfr donors) carry (in addition to F) chromosomal DNA sequences that were adjacent to the integrated F. A donor such as strain 9 (Table 1) which harbors an F′ *lac*⁺ plasmid can readily transfer the *lac*⁺ trait to an F⁻ such as strain 10 (Table 1), and Lac⁺ Nal^r transconjugants can be easily selected.

The transfer of conjugative R plasmids is also easily studied. The transfer of derepressed R-plasmid derivatives such as the R100 *drd-1* plasmid in strain 11 (Table 1) is easily detected since

R100 *drd-1* confers resistance to 25 to 50 μg of tetracycline per ml (Tc^r), 50 μg of chloramphenicol per ml (Cm^r), 50 μg of streptomycin per ml (Sm^r), 50 μg of spectinomycin per ml (Sp^r), and (on synthetic media) 25 to 40 μg of sulfanilamide per ml (Su^r). The donor harboring R100 *drd-1* should be sensitive to nalidixic acid and the recipient, such as strain 10 (Table 1), should not express resistance to any of the antibiotics for which R100 *drd-1* expresses resistance and should have a *nalA* mutation.

Media.

Either L or Penassay broth allows optimal growth and maximum expression of donor and recipient phenotypes. Penassay agar (with 8 g of NaCl per liter) can be used to titer donor and recipient cultures and, when supplemented with Tc, Cm, or Sm and Nal (see above and part 14.4, Table 4), to select R plasmid-containing transconjugants. VB medium E agar supplemented with 0.5% lactose, proline, and 50 μg of nalidixic acid per ml can be used to select Lac⁺ Nal^r transconjugants. See part 14.4 for methods for preparation of media.

Procedure for F′ transfer.

1. Inoculate 5 ml of Penassay broth contained in 16 by 125 (or 150) mm test tubes with loopfuls of growth off slants of each parent and incubate overnight at 37°C without aeration (i.e., as standing cultures).

2. Dilute these overnight cultures 1:100 or 1:200 into 10 to 20 ml of prewarmed Penassay broth. Aerate these cultures by either bubbling or shaking, and grow to a density of about 2 × 10⁸ cells per ml. It is advantageous for optimum expression of the donor phenotype to cease bubbling or shaking the donor culture 30 min prior to commencement of mating since nonaerated growth conditions favor synthesis of maximum numbers and lengths of donor pili specified by F. This, however, may not be the case for all conjugative plasmid types.

3. Place either flat-bottomed 125-ml micro-Fernbach flasks or 250-ml Erlenmeyer flasks in a 37°C water bath with lead collars around the necks of the flasks so that the water in the water bath surrounds the flasks to a level just below the caps. The lead collars can be cut in a doughnut shape out of 6 to 8 mm thick lead sheets. In the absence of lead collars, clamp the flasks so they will not float when immersed in the water bath. The immersion of flasks is important if the temperature of the mating culture is to be maintained at 37°C.

4. Add 9 ml of recipient culture to the mating flask to equilibrate, and determine its titer by plating a 10⁻⁶ dilution by spreading on Penassay

agar plates. Spread 100-μl amounts of the undiluted recipient culture on the selective agar medium to verify that the culture is free from contaminants and contains few, if any, revertants.

5. After 5 min, add 1 ml of donor culture to the mating flask and gently swirl to mix the donor and recipient cells. Titer the donor culture and plate 100 μl of the undiluted donor culture on selective medium. Note that the total volume of 10 ml gives a 2-mm depth in the mating flask, which allows sufficient oxygen diffusion to the mating cells and obviates any need for agitation by shaking of the mating mixtures. Although it is customary to use an excess of recipient cells in matings, different ratios ranging from 1:20 to 1:1 can be used by varying the volumes and/or densities of recipient and donor cultures added to the mating flask.

6. At 5-, 10-, or 15-min intervals up to 1 h, dilute the mating mixture and plate 100-μl samples (50 μl per plate) on selective agar medium. Total dilutions of 10^{-2} to 10^{-3} should be suitable up to 30 min of mating, and total dilutions of 10^{-3} and 10^{-4} should be suitable for longer durations of mating. Invert and incubate plates at 37°C.

Donor and recipient cultures grown in rich complex media give higher frequencies of transconjugants than those grown in synthetic media. However, the selection of transconjugants on synthetic media that are formed from broth-grown and mated parents constitutes a step-down growth condition that sometimes results in detection of fewer transconjugants than are actually formed. This potential difficulty can be circumvented by adding 10% Penassay broth (vol/vol) to the BSG diluent used to dilute the mating mixtures, since the small amount of broth added to the selective minimal agar medium will permit all Lac$^+$ Nal^r transconjugants to form colonies.

The frequency of transconjugants for each time of interruption is calculated by dividing the titer of transconjugants per milliliter of mating mixture by the titer of donor cells in the mating mixture at the commencement of mating.

Procedure for R-plasmid transfer.

Steps 1 to 5 are the same as described above for F′ transfer. If step 6 is performed as described above, then Tc^r Nal^r and Cm^r Nal^r transconjugants will be detected at higher frequencies than Sm^r Nal^r transconjugants, especially for the early times at which mating is interrupted, and the frequencies detected for all transconjugant classes will be less than the frequencies actually formed in the mating. These observations are due to the facts that (i) expression of R-plasmid-specified drug resistance requires transcription and translation of the R-plasmid genes in the recipient prior to drug exposure and (ii) tetracycline and chloramphenicol are bacteriostatic so that phenotypic expression of resistance to these drugs can sometimes be slowly expressed even in the presence of the drugs, whereas streptomycin is bactericidal and kills all transconjugants that have not yet expressed streptomycin resistance. These problems in phenotypic expression are even more pronounced when working with R plasmids that express resistance to either ampicillin or kanamycin.

To circumvent the above problems and to achieve an accurate measure of the rate and total frequency of R-plasmid transfer at each time of interruption of mating, place 100 μl of the mating mixture into 0.9 ml of Penassay broth containing 50 μg of nalidixic acid per ml in a 13 by 100 mm test tube at 37°C. Leave these tubes at 37°C for 20 to 30 min prior to further dilution and plating on selective media.

To verify that the various drug resistances are indeed transferred together on the same plasmid, pick 10 or so colonies of each transconjugant type with sterile toothpicks and streak onto the other types of selective media.

Applications.

The above protocols can be used to study conjugational transfer of a diversity of plasmid-mediated traits. Of course, detection of transfer of Col plasmids (3) and plasmids that specify traits such as antigen production and enterotoxins (16) requires special procedures since direct selection of the plasmid-mediated trait is not possible. Some conjugative plasmids are able to integrate into the bacterial chromosome to produce Hfr-type donors and upon excision rarely pick up a piece of chromosomal genetic information to generate R′, Col′, or F′ plasmids. Such plasmids can be used to determine dominance-recessiveness and complementation relationships between alleles and aspects of gene regulation. They also can be used to construct strains. These applications are described by Miller (2). Since these plasmids share homology with the chromosome, they can be used in lieu of Hfr-type donors to mobilize and transfer chromosomal genes in an oriented sequential manner (18). This can be particularly useful since the R′, Col′, and F′ plasmids can mobilize the chromosome of strains (or species) of bacteria in which Hfr donors have not been isolated and that are closely related to the strain which gave rise to the R′, Col′, or F′ plasmid.

14.3.2. Chromosome Transfer (in *E. coli* F⁻ with *E. coli* Hfr)

Bacterial strains.

Numerous Hfr strains with differing origins and directions of chromosome transfer have been isolated in *E. coli* (13, 27). The choice of a specific Hfr strain thus depends on the region of the bacterial chromosome to be studied with regard to the mapping of genetic loci. New Hfr donors can be readily isolated in different genetic backgrounds by using either a modified fluctuation test with an F⁺ strain (6) or integrative suppression in an F⁺ or R⁺ strain possessing a *dnaA*(Ts) mutation (30). The methods for conducting Hfr × F⁻ crosses can be illustrated by using an Hfr strain such as the prototrophic strain 12 (Table 1) that transfers its chromosome O-*thr leu proA lac . . . pyrB* F and is sensitive to nalidixic acid and streptomycin and an F⁻ strain such as strain 14 (Table 1) that possesses mutations conferring auxotrophy for leucine, proline, adenine, tryptophan, and thiamine and resistance to both nalidixic acid and streptomycin. Strain 14 also possesses other mutations, and inheritance of their wild-type alleles can be followed as unselected markers. The F⁻ strain 13 (Table 1) has fewer mutations than strain 14 and is more conveniently used to test the stability of the Hfr phenotype in strain 12.

Media.

Either L or Penassay broth allows optimum growth and maximum expression of donor and recipient phenotypes. Penassay agar (with 8 g of NaCl per liter) can be used to titer donor and recipient cultures. VB medium E minimal agar appropriately supplemented can be used to select transconjugants. For example, Leu⁺ Nal^r transconjugants in a cross between strains 12 and 14 would be selected on a medium containing 0.5% glucose and proline, adenine, tryptophan, thiamine, and nalidixic acid (see Tables 4 and 5 for concentrations of supplements). BSG can be used as the diluent but should be supplemented with 10% broth (vol/vol) when mating mixtures, which are in a complex medium, are diluted prior to plating on a synthetic medium. This small amount of broth ensures nutrients for recombination and phenotypic expression of donor traits and gives rise to maximal recovery of transconjugants. See part 14.4 for methods for preparation of media.

Testing stability of Hfr phenotype.

Since Hfr donors arise by integration of a conjugative plasmid into the bacterial chromosome, it follows that they can revert to a state in which the conjugative plasmid regains its cyto-plasmic state. Thus, it is important to verify that the majority of cells in an Hfr population do exhibit the Hfr phenotype prior to use of the strain in conjugation experiments. This can be done by either the replica plating method or the picking and streaking method.

Replica plating method:

1. Dilute the Hfr culture and plate on Penassay agar to achieve about 100 well-isolated colonies per plate. Incubate at 37°C for 12 to 16 h.

2. Spread 50 to 100 μl of a broth-grown log-phase culture of the F⁻ strain 13 (Table 1) on a minimal agar plate selective for Leu⁺ Str^r transconjugants.

3. Replica plate the colonies from the Hfr culture onto the minimal agar plate spread with the F⁻ strain. Mark the bottoms of both plates so that the orientation and placement are known. Incubate the minimal agar plate at 37°C for 24 to 48 h.

4. Confluent patches of recombinant growth should be evident for each colony from the Hfr culture that displays the Hfr phenotype and only 0, 1, or 2 recombinant colonies for a colony from the Hfr culture that has reverted to the F⁺ state.

5. If the culture contains an appreciable frequency of F⁺ revertants, the master Penassay agar plate can be used to pick a colony that displays the Hfr phenotype.

Picking and streaking method:

1. Plate the Hfr culture to get isolated colonies on a Penassay agar plate.

2. Pick 10 to 20 isolated colonies into individual 13 by 100 mm test tubes containing 2 ml of broth.

3. Incubate cultures at 37°C for 4 to 6 h to achieve densities of about 2×10^8 to 5×10^8 cells per ml. Also grow cultures of F⁻ and F⁺ strains such as strains 5 and 7 (Table 1) to be included as controls.

4. Place 50 to 100 μl of a broth-grown log-phase culture of the F⁻ strain 13 down the center of a minimal agar plate selective for Leu⁺ Str^r transconjugants. Leave the lid ajar until the streak has dried. (Two streaks can be accommodated per plate if desired.)

5. Immediately streak full loopfuls of each "Hfr" culture (being careful not to touch sides of tubes with the loop, which will cause loss of fluid and yield variable results) and the F⁻ and F⁺ cultures across the F⁻ streak. Let streaks dry and incubate the plate for 24 to 48 h at 37°C.

6. A culture giving the Hfr phenotype will give confluent growth of transconjugants at the juncture of the two streaks, whereas a culture that contains principally F⁺ cells will give only 5 to

10 isolated transconjugant colonies. An $F^- \times F^-$ cross streak should, of course, give no transconjugants.

Both of the above methods are based on the fact that, although the donor cells are ultimately killed by the streptomycin in the selective minimal agar, they are still able to transfer their chromosomes in the presence of streptomycin (albeit at a somewhat reduced frequency). Nalidixic acid cannot be used to counterselect against the donor in such "plate" matings since it causes an immediate cessation of genetic transfer by Nal^s donors.

Procedure for interrupted mating.

Use the reisolated Hfr, if necessary, and F^-, and follow the procedure outlined in steps 1 through 5 under Procedure for F′ transfer in part 14.3.1, except use minimal agar that is selective for Leu^+ Nal^r, Pro^+ Nal^r, Ade^+ Nal^r, and Trp^+ Nal^r transconjugants and the methods for dilution and plating that follow:

1. At appropriate intervals (every 2, 3, 5, or 10 min, but the more frequent sampling will necessitate use of two people to do plating), take 100 µl of the mating mixture and add to 0.9 ml of BSG containing 50 µg of nalidixic acid per ml in 13 by 100 mm test tubes at 37°C.

2. Incubate the 10^{-1} dilution tubes at 37°C for 30 to 40 min, but choose a time that is about 30 s after a time when the mating mixture is to be sampled since this will maximize the amount of time for making further dilutions and plating.

3. Plate either 50 µl on each of two selective agar plates or 100 µl on one selective agar plate of the appropriate dilution for each time of mating interruption. Table 2 gives the approximate final dilutions that should be plated on each selective agar medium for each time of mating interruption. These values are based on a donor to recipient cell ratio of 1:10, a total mating density of about 2×10^8 cells per ml, use of a pure Hfr culture, and use of conditions that optimize the Hfr phenotype and are optimal for Hfr chromosome transfer.

4. Incubate Penassay agar plates for titer determinations overnight at 37°C. If the Hfr was not reisolated prior to use, the Penassay plates can be used to determine the fraction of Hfr cells by one of the methods described above under Testing stability of Hfr phenotype.

5. Incubate minimal selective agar plates at 37°C for 2 days. Save plates and do not count colonies by destruction since they can be used to test inheritance of unselected markers.

6. Calculate transconjugant frequencies by dividing the titers of transconjugants of each type and for each time of interruption by the Hfr titer

TABLE 2. *Final dilutions for plating to recover various transconjugant classes in Hfr $\times$ F^- mating*[a]

Time of interruption (min)	Transconjugant class selected			
	Leu^+ Nal^r	Pro^+ Nal^r	Ade^+ Nal^r	Trp^+ Nal^r
5	10^{-2}	10^{-2}	—[b]	—[b]
10	10^{-2} 10^{-3}	10^{-2}	10^{-2}	—
15	10^{-3} 10^{-4}	10^{-2} 10^{-3}	10^{-2}	—
20	10^{-3} 10^{-4}	10^{-2} 10^{-3}	10^{-2} 10^{-3}	—
25	10^{-4} 10^{-5}	10^{-3} 10^{-4}	10^{-2} 10^{-3}	10^{-2}
30	10^{-4} 10^{-5c}	10^{-3} 10^{-4}	10^{-3} 10^{-4}	10^{-2}
35	10^{-4} 10^{-5}	10^{-3} 10^{-4}	10^{-3} 10^{-4}	10^{-2} 10^{-3}
40	10^{-4} 10^{-5}	10^{-4} 10^{-5c}	10^{-3} 10^{-4}	10^{-2} 10^{-3}
45	10^{-4} 10^{-5}	10^{-4} 10^{-5}	10^{-4c}	10^{-3} 10^{-4}
50	10^{-4} 10^{-5}	10^{-4} 10^{-5}	10^{-4}	10^{-3} 10^{-4}
55	10^{-4} 10^{-5}	10^{-4} 10^{-5}	10^{-4}	10^{-3} 10^{-4}
60	10^{-4} 10^{-5}	10^{-4} 10^{-5}	10^{-4}	10^{-3} 10^{-4}

[a] Based on results expected when strains 12 and 14 (Table 1) are grown and mated under optimal conditions.

[b] Plating is not necessary, since no transconjugants are expected for these times of interruption.

[c] Transconjugant titer should be nearly at the maximum expected frequency. Platings can be omitted for some or all of subsequent times of mating interruption.

in the mating mixture at the commencement of mating. The data can be plotted (arithmetic $\times$ arithmetic) to determine the time of entry of each selected Hfr marker.

Analysis of transconjugants for inheritance of unselected traits.

It is sometimes either not possible to directly select for inheritance of some donor markers or not feasible because of the number of markers that are subject to inheritance in a given mating. Thus, a more classical recombination analysis can be performed to evaluate the order of genetic loci. Customarily, this is accomplished by picking 100 to 200 colonies of a given transconjugant type with a needle or toothpicks into BSG and streaking each of these small suspensions on the same selective medium to obtain purified isolated colonies. These isolated colonies can then be grown as small cultures and used to test the genotypes by streaking on appropriate media or patched onto a suitable agar medium for subsequent analysis by replica plating. When using the latter method, the master plate should be selective for the transconjugant phenotype, and each test replica plate should evaluate one additional nutritional requirement or capability in addition to maintaining selection for the original transconjugant phenotype. This methodology precludes occasional ambiguity due to failure to completely purify the transconjugant type away from other transconjugant or parental types prior to patching on the master plate.

Since chromosome transfer by Hfr donors is spontaneously interrupted, the determination of an unambiguous gene order by recombinant analysis demands that recombinants inheriting a terminally transferred Hfr marker be tested for inheritance of proximally transferred markers. In the case of the cross between strains 12 and 14 (Table 1), 200 to 400 Trp$^+$ Nal^r transconjugants should be picked, purified, and tested for inheritance of the Hfr ara^+, leu^+, $tonA^+$, $proA^+$, lac^+, tsx^+, $purE^+$, and $galK^+$ markers. Testing for inheritance of these markers can be done as follows:

1. Patch 50 to 100 purified Trp$^+$ Nal^r transconjugants per minimal agar plate selective for Trp$^+$ Nal^r transconjugants. Incubate at 37°C overnight and use to replica plate sequentially to minimal agar that is selective for Ara$^+$ Trp$^+$ Nal^r, Leu$^+$ Trp$^+$ Nal^r, Pro$^+$ Trp$^+$ Nal^r, Lac$^+$ Trp$^+$ Nal^r, Ade$^+$ Trp$^+$ Nal^r, Gal$^+$ Trp$^+$ Nal^r, and Trp$^+$ Nal^r transconjugants. The last plate is included as a control to verify that cells were replicated to each plate.

2. Testing for resistance or sensitivity to T5 ($tonA$) and T6 (tsx) is most accurately done by cross-streaking broth cultures of purified Trp$^+$ Nal^r transconjugants against the respective phage on EMB0 plates (see step 2 under Characterization of transductants in part 14.2.1). This is laborious, however, and replica plating can sometimes give reasonably accurate results. For this purpose, the EMB0 plates are spread with 50 to 100 μl of a T5 or T6 lysate (ca. 10^{10} PFU/ml), and a separate velveteen is used to replica plate from the master plate to each of these since carry-over of phage from one plate to another will obscure the results. Even though the $tonA$ mutation also confers resistance to phage T1, this phage should not be used unless scrupulous care is taken and there is rigid adherence to all rules of aseptic technique. T1 is resistant to desiccation and has a very short latent period. It has been responsible for wiping out *E. coli* research in more than one laboratory.

After scoring the phenotype of all Trp$^+$ Nal^r transconjugants, the number in each recombinant class can be counted and used to calculate map distances between all markers and to determine the order of the markers on the Hfr chromosome.

Applications.

By using a diversity of Hfr parents with different origins and directions of chromosomal transfer in conjugation with the cross-streak method of mating, it is easy to initially identify the general location of the wild-type allele for a mutant allele isolated in an F$^-$ recipient (4). Interrupted matings with the appropriate Hfr can then be used to more precisely locate this locus with respect to other loci. P1-mediated generalized transduction may be necessary, however, to unambiguously locate this locus in relation to other closely linked loci since conjugation is often inadequate for this purpose.

14.4 MEDIA AND STRAINS

The media commonly used in gene transfer procedures are ordinary, commercially available ones. Only those media requiring modification and various supplements are described below. Sources of bacterial and phage strains also are listed.

14.4.1. L Broth and L Agar (24)

L broth contains 10 g of tryptone, 5 g of yeast extract, 5 g of NaCl, and 1 g of glucose per liter of water. The pH is adjusted to 7.0 by adding approximately 1.7 ml of 1 N NaOH prior to autoclaving. L agar is L broth containing 1.5% agar. When making confluent plate lysates for bacteriophage P1, agar is used at 1.2%. L soft agar is L broth containing 0.65% agar. For experiments with phages such as P1, CaCl$_2$ is added to L media to a final concentration of 2.5 mM. The CaCl$_2$ must be added to these media after they are autoclaved and allowed to cool to at least 70°C. Media to which CaCl$_2$ has been added cannot be reautoclaved since a precipitate will form. The CaCl$_2$ can be prepared as a sterile 1 M solution in water.

14.4.2 Lambda Agar

Attachment and injection of bacteriophage λ requires the presence of the maltose transport proteins on the *E. coli* cell surface, whose synthesis is inhibited by glucose due to catabolite repression. Thus, lambda agar cannot contain glucose. It contains 10 g of tryptone, 5 g of yeast extract, 5 g of NaCl, and 12 g of agar per liter of water. The pH is adjusted to 7.0 by adding 1.7 ml of 1 N NaOH prior to autoclaving. Lambda soft agar has the same ingredients but contains 0.65% agar and 10 mM Mg^{2+}.

14.4.3. Super Broth (26)

Preparation of high-titer lysates of λ and many other phages is best achieved by using a rich medium for growth of the host bacteria. Super broth (one such medium) contains 32 g of tryptone, 20 g of yeast extract, and 5 g of NaCl per liter of water. The pH is adjusted to 7.0 by adding about 5 ml of 1 N NaOH prior to autoclaving. For propagation of λ, add 10 mM MgSO$_4$ after autoclaving. The MgSO$_4$ can be prepared as a 1 M solution in water.

14.4.4. Penassay Agar

Commercial Penassay agar is satisfactory as a general complex plating medium, but adding 8 g of NaCl per liter very much improves its utility in many bacterial genetics experiments. Penassay broth as prepared commercially already contains monovalent cations.

14.4.5. EMB and MacConkey Agars

The EMB and MacConkey base agars commercially available are satisfactory fermentation indicator media when working with gram-negative enteric bacteria. However, since many of the *E. coli*, *Salmonella*, and other strains used in gene transfer experiments have mutations conferring auxotrophy for purines, pyrimidines, and/or vitamins, 5 g of yeast extract should be added per liter of medium. The addition of 5 g of NaCl per liter is also advisable if the media are to be used for testing strains of bacteria for sensitivity to phage. EMB and MacConkey agars are usually supplemented with sugars to a final concentration of 1% and are designated EMB+Gal, Mac+Lac, etc. EMB agar without added sugar is used for testing bacteria for sensitivity to phages and is designated EMB0 agar. Table 3 lists the concentrations for sterile stock solutions of the most commonly used sugars.

14.4.6. VB Minimal Medium E (37)

To prepare a 50× stock, dissolve successively in 670 ml of distilled or deionized water 10 g of $MgSO_4 \cdot 7H_2O$, 100 g of citric acid $\cdot H_2O$, 500 g of anhydrous K_2HPO_4, and 175 g of $NaNH_4PO_4 \cdot 4H_2O$. The final volume will be 1 liter. To prepare 1× VB minimal liquid medium E, add 20 ml of sterile 50× salts to about 960 ml of sterile distilled or deionized water so that, upon addition of sterile carbohydrate and other required supplements (see below), the final volume will be 1 liter. To prepare 1× VB minimal agar medium E, add 15 g of agar to about 960 ml of distilled or deionized water, autoclave, and,

TABLE 3. *Concentrations and preparation of sterile carbohydrate stock solutions*[a]

Carbohydrate	Concn of stock solution (%)
L-Arabinose	20
D-Galactose	30
D-Glucose	40
D-Glycerol[b]	40
Lactose	20
Maltose	30
D-Mannitol	10
Na₂ succinate	20
D-Xylose	30

[a] All concentrations are in weight per unit volume. Heating of distilled (or deionized) water facilitates solution of the carbohydrates, but these solutions should be allowed to cool to room temperature before they are brought to the correct final volume. All solutions should be sterilized by autoclaving (20 min for volumes of 250 ml or less) prior to use and should be stored at 4°C.

[b] Since glycerol is a liquid, add 31.7 ml/68.3 ml of water to achieve a 40% (wt/vol) solution.

when cooled to 60 to 70°C, add 20 ml of sterile 50× salts and appropriate sterile carbohydrate and other supplements. When triphenyltetrazolium chloride is used as a fermentation indicator, it should be added to the agar prior to autoclaving. The 1× minimal liquid (ML) and minimal agar (MA) media have a pH of 7.

14.4.7. Buffered Saline with Gelatin (BSG) (10; also see part 25.1)

BSG is a general-purpose diluent. The presence of gelatin stabilizes phage and fragile bacterial cells and improves the accuracy of dilutions. It contains 8.5 g of NaCl, 0.3 g of anhydrous KH_2PO_4, 0.6 g of anhydrous Na_2HPO_4, 10 ml of 1% gelatin, and 990 ml of water.

14.4.8. Supplements

Carbohydrate energy sources (Table 3) are customarily added to minimal media to a final concentration of 0.5%.

Casamino Acids or other acid hydrolysates of casein are often added to minimal media to a final concentration of 0.5%. Ten percent (wt/vol) sterile stock solutions can be stored at 4°C. The addition of amino acids to minimal media stimulates more rapid growth of bacteria and is particularly useful in conducting genetic experiments with mutants requiring tryptophan (which is destroyed during acid hydrolysis of proteins), purines, pyrimidines, and vitamins. In the last case, use vitamin-free amino acids.

The optimal concentrations for supplementing minimal media with amino acids, purines, pyrimidines, and vitamins to permit growth of strains with auxotrophic mutations must often be determined empirically. Factors of importance in deciding on concentrations to try include: amount of the compound in cellular matter, efficiency of transport of the compound, use of the compound for synthesis of other cellular constituents, and ability of the bacterium to degrade the compound. Table 4 lists the concentrations of amino acids, purines, pyrimidines, and vitamins in sterile stock solutions and also the levels for supplementation of minimal media to achieve maximal growth of mutant strains of *E. coli* and other gram-negative bacteria.

Strains of bacteria that are resistant to antibiotics due to either chromosomal mutations or the presence of R plasmids (Chapter 15) are frequently used in gene transfer experiments. Table 5 lists the concentrations of antibiotics for stock solutions and for supplementing media with those antibiotics commonly used in gene transfer experiments with gram-negative enteric bacteria. It should be noted that sterilization of antibiotic stock solutions is usually not necessary, provided that precautions are taken during

TABLE 4. *Amino acid, purine, pyrimidine, and vitamin concentrations for stock solutions and for supplementation of minimal media*[a]

Compound	Concn of stock solution (mg/ml)	Stock solution/liter of minimal media (ml)	Final concn in minimal media (μg/ml)[b]
Amino acids[c]			
DL-Alanine[d]	50.0	2.0	100
L-Arginine HCl	11.0	2.0	22
L-Asparagine	10.0	10.0	100
L-Aspartic acid	10.0	10.0	100
L-Cysteine HCl[e]	11.0	2.0	22
Glycine	50.0	2.0	100
L-Glutamic acid	10.0	10.0	100
L-Glutamine	20.0	5.0	100
L-Histidine HCl	11.0	2.0	22
L-Isoleucine	10.0	2.0	20
L-Leucine	1010	2.0	20
L-Lysine HCl	44.0	2.0	88
DL-Methionine	10.0	2.0	20
L-Phenylalanine	10.0[f]	2.0	20
L-Proline	15.0	2.0	30
DL-Serine	10.0	10.0	100
DL-Threonine	40.0	2.0	80
L-Tryptophan	4.0	5.0	20
L-Tyrosine	2.0[g]	10.0	20
DL-Valine	20.0	2.0	40
Purines and pyrimidines			
Adenine	2.0[h]	20.0	40
Thymidine	10.0	0.4 or 4.0	4 or 40[i]
Thymine	5.0	0.4 or 8.0	2 or 40[i]
Uracil	2.0	20.0	40
Vitamins			
Biotin	0.25	2.0	0.5
Niacin	0.5	2.0	1
Ca pantothenate	0.5	2.0	1
Pyridoxine HCl	0.5	2.0	1
Thiamine HCl	0.5	2.0	1

[a] All stock solutions are prepared in distilled or deionized water unless otherwise indicated and are stored at 4°C unless otherwise indicated. Sterilization can be either by autoclaving (15 min for volumes of 100 ml per bottle or less) or by filtration.

[b] Concentrations listed give optimum growth of *E. coli* auxotrophs. Somewhat different concentrations might be needed for mutant strains of other bacterial species.

[c] The L isomers have been listed when their price per gram is no more than twice the price of the DL racemic mixture.

[d] Both the D and L isomers are usable by bacteria which possess alanine racemase.

[e] Should be prepared freshly (at least every week), since cysteine is rapidly oxidized to cystine (which is very insoluble). Alternatively, the stock solution of cysteine can be overlaid with sterile mineral oil after autoclaving.

[f] Prepared in 0.001 N NaOH and stored at room temperature.

[g] Prepared in 0.01 N NaOH and stored at room temperature.

[h] Prepared in 0.03 N HCl.

[i] *thyA* mutants of enteric bacteria require relatively high concentrations of either thymine or thymidine. *thyA* strains with either *deoB* or *deoC* mutations (which eliminate functional deoxyribomutase and deoxyriboaldolase, respectively) cannot degrade deoxyribose 1-phosphate and thus grow with lower concentrations of either thymine or thymidine.

TABLE 5. *Antibiotic concentrations for stock solutions and medium supplementation*

Antibiotic	Concn of stock solution (mg/ml)	Final concn (μg/ml) in media for:	
		R plasmid-specified resistance	Resistance due to chromosomal mutation
Ampicillin	—[a]	25–50	10–20
Chloramphenicol	25[b]	25–50	—
Nalidixic acid	10[c]	—	50[d]
Rifampin	2[e]	—	50
Streptomycin	50[f]	25–50	200–400
Tetracycline	—[g]	25–50	—

[a] Ampicillin solutions are unstable. The powder is weighed out, suspended in sterile water, and added to media to the desired final concentration. Alternatively, stock solutions can be prepared and stored at −20°C. Corrections should be made for buffers or other ingredients to achieve the correct concentration of pure ampicillin.

[b] Suspended in 50% ethanol and stored at 4°C.

[c] Suspended in 0.1 N NaOH which converts the acid to the sodium salt. Stored at 4°C.

[d] Add at 75 μg/ml to EMB agar.

[e] Dissolved at 20 mg/ml in 100% methanol and then diluted to 2 mg/ml in 10% methanol with sterile water. Store at 4°C in the dark. Prepare freshly each week.

[f] Suspended in sterile water and stored at 4°C.

[g] Tetracycline solutions are unstable. The powder is weighed out, suspended in sterile water, and added to media to the desired final concentration. Alternatively, stock solutions can be prepared and stored in the dark at −20°C. Corrections should be made for the presence of high concentrations of ascorbic acid in most tetracycline preparations.

their preparation. These include use of weighing paper that has been taken from a closed box or envelope container (paper products as packaged are generally sterile), a sterile container (bottle or tube), and sterile water or alcohol for suspension. It should be noted that ampicillin, rifampin, and tetracycline are relatively unstable; therefore, media containing these antibiotics should be used within several days. Storing media with tetracycline and rifampin in the dark prolongs their useful life. Tetracycline decomposes to substances that are toxic to both Tc[s] and Tc[r] bacteria.

14.4.9. Preparation of Agar Media

Agar plates used to prepare or titer phage lysates must be poured on a level surface. Phage yield and plaque size are increased by using 35 to 45 ml of agar medium per plate. It is also important that plates for plaque assays be neither too dry nor too wet. A satisfactory procedure is to pour plates either 2 days prior to use and leave them at room temperature or the day before use and incubate them with lids closed at 37°C for about 12 h. When preparing lysates by the confluent plate method, agar plates should be poured the day before use and not dried at 37°C.

Minimal agar plates, which often need to be incubated for 2 days, should contain 25 to 30 ml of medium. Fermentation reactions on EMB and MacConkey agars are also improved if the plates contain at least 25 ml of media.

14.4.10. Storage of Agar Media

Most agar media, except those containing unstable antibiotics (see Table 5), can be placed in plastic bags and stored at 4°C for several months. It is wise to store media that contain dyes and other pigmented compounds in the dark. This is especially true for EMB agar which, because of photooxidation, becomes inhibitory to bacterial growth.

14.4.11. Sources of Bacterial and Phage Strains

E. coli strains. Coli Genetic Stock Center, Dr. Barbara Bachmann, Department of Human Genetics, Yale University School of Medicine, 333 Cedar Street, New Haven, CN 06510.

Other bacterial and phage strains. American Type Culture Collection, 12301 Parklawn Drive, Rockville, MD 20852.

14.5. LITERATURE CITED

14.5.1. General References

1. **Clowes, R. C., and W. Hayes.** 1968. Experiments in microbial genetics. Blackwell Scientific Publications, Oxford.
   Somewhat out of date but contains many useful methods and protocols for experiments in microbial genetics.
2. **Miller, J. H.** 1972. Experiments in molecular genetics. Cold Spring Harbor Laboratory, Cold Spring Harbor, N.Y.
   The most up to date and comprehensive compilation of methods and protocols for experiments in microbial genetics and molecular biology.

14.5.2. Specific References

3. **Achtman, M.** 1973. Curr. Top. Microbiol. Immunol. **60**:79–123.
4. **Adams, M. H.** 1959. Bacteriophages. Interscience Publishers, Inc., New York.
5. **Bachmann, B. J., K. B. Low, and A. L. Taylor.** 1976. Bacteriol. Rev. **40**:116–167.
6. **Berg, C. M., and R. Curtiss III.** 1967. Genetics **56**:503–525.
7. **Bukhari, A. I., J. A. Shapiro, and S. L. Adhya (ed.).** 1977. DNA insertion elements, plasmids, and episomes. Cold Spring Harbor Laboratory, Cold Spring Harbor, N.Y.
8. **Caro, L., and C. M. Berg.** 1971. Methods Enzymol. **21D**:444–458.
9. **Cosloy, S. D., and M. Oishi.** 1973. Proc. Natl. Acad. Sci. U.S.A. **70**:84–87.
10. **Curtiss, R., III.** 1965. J. Bacteriol. **89**:28–40.
11. **Curtiss, R., III.** 1969. Annu. Rev. Microbiol. **23**:69–136.
12. **Curtiss, R., III, R. G. Fenwick, Jr., R. Goldschmidt, and J. O. Falkingham III.** 1977. *In* S. Mitsuhashi (ed.), R factor drug resistance plasmid, p. 109–134. University of Tokyo Press, Tokyo.
13. **Curtiss, R., III, F. L. Macrina, and J. O. Falkingham III.** 1975. *In* R. C. King (ed.), Handbook of genetics, vol. 1, p. 115–133. Plenum Publishing Corp., New York
14 **Demerec, M., E. A. Adelberg, A. J. Clark, and P. E. Hartman.** 1966. Genetics **54**:61–76.
15. **Enea, V., G. F. Vovis, and N. D. Zinder.** 1975. J. Mol. Biol. **96**:495–509.
16. **Falkow, S.** 1975. Infectious multiple drug resistance. Pion Limited, London.
17. **Goldberg, R. B., R. A. Bender, and S. L. Streicher.** 1974. J. Bacteriol. **118**:810–814.
18. **Hayes, W.** 1968. The genetics of bacteria and their viruses, 2nd ed. John Wiley & Sons, Inc., New York.
19. **Hershey, A. D.** 1971. The bacteriophage lambda. Cold Spring Harbor Laboratory, Cold Spring Harbor, N.Y.
20. **Hong, J.-S., and B. N. Ames.** 1971. Proc. Natl. Acad. Sci. U.S.A. **68**:3158–3162.
21. **Jacob, F., and E. L. Wollman.** 1961. Sexuality and the genetics of bacteria. Academic Press, Inc., New York.
22. **Juni, E.** 1972. J. Bacteriol. **112**:917–931.
23. **Juni, E., and A. Janik.** 1969. J. Bacteriol. **98**:281–288.
24. **Lennox, E. S.** 1955. Virology **1**:190–206.
25. **Lewin, B.** 1977. Gene expression—3. Plasmids and phages. John Wiley & Sons, Inc., New York.
26. **Lodish, H. F.** 1970. J. Mol. Biol. **50**:689–702.
27. **Low, K. B.** 1972. Bacteriol. Rev. **36**:587–607.
28. **Mandel, M., and A. Higa.** 1970. J. Mol. Biol. **53**:159–162.
29. **Marmur, J.** 1961. J. Mol. Biol. **3**:208–218.
30. **Nishimura, Y., L. Caro, C. M. Berg, and Y. Hirota.** 1971. J. Mol. Biol. **55**:441–456.
31. **Notani, N. K., and J. K. Setlow.** 1974. Prog. Nucleic Acid Res. **14**:39–100.
32. **Novick, R. P., R. C. Clowes, S. N. Cohen, R. Curtiss III, N. Datta, and S. Falkow.** 1976. Bacteriol. Rev. **40**:168–189.
33. **Ozeki, H.** 1956. Carnegie Inst. Washington Publ. 612, p. 97–106.
34. **Sawula, R. V., and I. P. Crawford.** 1972. J. Bacteriol. **112**:797–805.
35. **Shimada, K., R. A. Weisberg, and M. E. Gottesman.** 1972. J. Mol. Biol. **63**:483–503.
36. **Tomasz, A.** 1969. Annu. Rev. Genet. **3**:217–232.
37. **Vogel, H. J., and D. M. Bonner.** 1956. J. Biol. Chem. **218**:97–106.
38. **Willetts, N. S.** 1977. *In* S. Mitsuhashi (ed.), R factor drug resistance plasmid, p. 89–107. University of Tokyo Press, Tokyo.
39. **Zinder, N. D. and J. Lederberg.** 1952. J. Bacteriol. **64**:679–699.

Chapter 15

Plasmids

JORGE H. CROSA and STANLEY FALKOW

The term plasmid originally was used by Lederberg (21) to describe all extrachromosomal hereditary determinants. Currently, the term is restricted to the autonomously replicating extrachromosomal deoxyribonucleic acid (DNA) of bacteria. While not essential for the survival of bacteria, plasmids may encode a wide variety of genetic determinants which permit their bacterial hosts to survive better in an adverse environment or to compete better with other microorganisms occupying the same ecological niche. Plasmids are found in a wide variety of bacteria, and it is as difficult to generalize about plasmids as it is to generalize about the bacteria that harbor them. The medical importance of plasmids that encode for antibiotic resistance, R plasmids, and those that contribute directly to microbial pathogenicity is well known. Plasmids are of equal importance, however, for the study of the structure and function of DNA. Most recently, plasmids have taken on paramount importance in recombinant DNA technology.

The methods for studying gene transfer by plasmids have been described in Chapter 14. This chapter deals with the isolation and characterization of plasmids, particularly the plasmids of gram-negative bacteria since they have been studied most extensively. Nevertheless, the methods can be successfully applied to the plasmids of any bacterium, so long as there is a suitable means for gently releasing the plasmids from the cells.

Many of the techniques currently employed for the isolation of plasmid DNA are based on its supercoiled covalently closed circular config-

uration. All of the techniques require some means for gently lysing bacterial cells so that the plasmid DNA is preserved intact and can be physically separated from the more massive chromosomal DNA. Although plasmids normally have a molecular mass between 0.5 and 100 megadaltons (10^6 daltons equals 1 megadalton [Mdal]), some large plasmids are in the 100- to 300-Mdal range (2, 12, 22, 24, 26, 27, 31). For comparison, *Escherichia coli* chromosomal DNA has a molecular mass between 2,200 and 3,000 megadaltons.

The initial characterization of a bacterial plasmid usually is at the genetic level. If a bacterial trait is suspected to be plasmid mediated, gene-transfer experiments will often document transmissibility of plasmid determinants independently of chromosomal determinants. Moreover, the elimination of a genetic trait by exposure of a bacterial population to "curing agents" such as acridine orange or ethidium bromide may strongly suggest the presence of a plasmid. In most cases, however, it is essential to document that a plasmid is present and unequivocally associated with the genetic trait in question.

If possible, it is best to transfer a plasmid by some genetic means into a bacterial host which is known to be devoid of plasmids. For many gram-negative bacteria, this is best achieved by using a well-characterized F⁻ strain of *E. coli* K-12, such as C600. Other equally well-characterized plasmid-free strains are known for other species (e.g., *Pseudomonas aeruginosa, Klebsiella pneumoniae, Serratia marcescens, Shigella flexneri, Salmonella typhi,* and representatives

266

of other genera). The advantage, of course, is that any single plasmid transferred to and subsequently isolated from such strains can be analyzed without fear of contamination by a host plasmid.

In some cases, genetic methods are not available so that one must directly examine a bacterium to determine its plasmid content. Such an analysis usually can be performed to determine simply whether a plasmid is or is not present. Subsequently, one can examine a strain "cured" of a trait by chemical means or spontaneously to determine whether a plasmid is lost concomitantly with a particular host-cell function.

In the clinical laboratory and in the recombinant DNA laboratory, it is often useful to screen a large number of isolates for their plasmid content. Clinically, the plasmid content of a cell can be a useful epidemiological marker. For the recombinant DNA laboratory, one often wishes to determine different classes of recombinant DNA.

15.1. ISOLATION OF PLASMID DNA

15.1.1. With Triton (in *E. coli* K-12)

The following procedure (4, 19) works well for isolating plasmid DNA from *E. coli* K-12 strains, if lysed by the nonionic detergent Triton X-100 (Beckman Instruments, Inc., Fullerton, CA 92634).

1. Grow the cells at 37°C either in a rich medium (e.g., brain heart infusion broth) or in a suitable minimal medium, with gentle aeration achieved by shaking a flask at a rate just sufficient to keep the surface of the medium in motion. The details that follow are for a 100-ml culture in a 250-ml Erlenmeyer flask, but the method can be scaled up or down proportionally. To ensure optimum lysis, harvest the cells in the mid-logarithmic phase of growth.

2. Harvest the cells by centrifugation in a 250-ml bottle (e.g., for 10 min at 12,100 × g at 5°C) in a Sorvall GS-A rotor, resuspend the pellet with 15 ml of TES buffer [0.05 M tris-(hydroxymethyl)aminomethane (Tris), 0.005 M ethylenediaminetetraacetate (EDTA), 0.05 M NaCl, pH 8.0], transfer to a 40-ml tube, and centrifuge for 10 min at 10,000 rpm at 5°C in a Sorvall SS-34 rotor. Resuspend the pellet with 1 ml of ice-cold 25% sucrose solution (in 0.05 M Tris and 0.001 M EDTA at pH 8.0), and place the tube in ice for 30 min.

3. Add lysozyme (0.2 ml of a 5-mg/ml solution in 0.25 M Tris at pH 8.0). Mix the contents by swirling the tube several times, and then place the tube in ice for 10 min.

4. Add 0.4 ml of 0.25 M EDTA at pH 8.0.

Swirl the tube and then place the tube in ice for another 10 min.

5. Add 1.6 ml of a Triton X-100 lytic mixture (1 ml of 10% Triton X-100 in 0.01 M Tris at pH 8.0, 25 ml of 0.25 M EDTA at pH 8.0, 5 ml of 1 M Tris at pH 8.0, 69 ml of water). After *very gentle* swirling to mix the contents thoroughly, place the tube in ice for 20 min.

6. Centrifuge at 35,000 × g at 5°C for 20 min in a Sorvall or Beckman J21 centrifuge (SS-34 or JA-20 rotor, respectively).

About 95% of the plasmid is separated from the bulk of the chromosomal DNA and cellular debris by this procedure. The plasmid-enriched supernatant fraction (3.2 ml) can be further purified and concentrated by centrifugation in a cesium chloride-ethidium bromide density gradient (see 15.3.3).

15.1.2. With Triton (in Other Gram-Negative Bacteria)

R. A. Quackenbush (personal communication) has introduced a modification of the foregoing procedure which improves the sharpness of the banding pattern in agarose gels (see 15.3.1) for lysates obtained from certain bacteria, e.g., *Pseudomonas aeruginosa*, *Serratia marcescens*, *Proteus retgeri*, and *Klebsiella pneumoniae*.

1. Grow 40 ml of culture with shaking in a 100-ml flask in brain heart infusion broth (Difco Laboratories, Detroit, Mich.).

2. Harvest the cells by centrifugation at 5,000 rpm for 10 min (Sorvall SS-34 type rotor).

3. Suspend the cell pellet in 5 ml of TES buffer (see 15.3.1).

4. Centrifuge the cells as before.

5. Suspend the cell pellet in 2 ml of a 25% sucrose solution (in 0.001 M EDTA and 0.05 M Tris at pH 8.0). Place the tube in ice for 20 min.

6. Add 0.4 ml of lysozyme (10 mg/ml in 0.25 M Tris at pH 8.0) to the suspension. Place the tube in ice for 20 min.

7. Add 0.8 ml of 0.5 M EDTA at pH 8.0 to the cell suspension.

8. Lyse the cells with 4.4 ml of Triton lytic mixture (see 15.3.1). Mix gently.

9. Heat the tube at 65°C for 20 min.

10. Remove the cellular debris by centrifugation at 27,200 × g for 40 min (Sorvall SS-34 rotor).

11. Adjust the solution to 0.5 M NaCl and 10% polyethylene glycol by use of stock solutions of 5 M NaCl and 40% polyethylene glycol (molecular weight, 1,000 to 6,000).

12. Store the tube at 4°C overnight.

13. Sediment the resulting precipitate by centrifugation at 3,000 × g for 10 min (Sorvall SS-

34 rotor), and resuspend the pellet in 1 to 2 ml of 0.25 M NaCl containing 0.001 M EDTA and 0.01 M Tris-hydrochloride at pH 8.0.

11. Precipitate the DNA by adding 2 volumes of 95% ethanol at −20°C, and let the tube stand overnight at −20°C.

15.1.3. With Sodium Dodecyl Sulfate

The following method described by Guerry et al. (13) is a modification of that described by Hirt (16). The method works well with bacteria that are lysed by the detergent sodium dodecyl sulfate (SDS). The method is based on the preferential precipitation of the high-molecular-weight chromosomal DNA by SDS in the presence of sodium chloride.

1. Inoculate 30 ml of brain heart infusion broth (Difco) in a 100-ml flask with a test bacterium, and grow overnight on a shaker in a water bath at 37°C.

2. Pellet the cells by centrifugation at 12,100 × g for 10 min at 4°C. Resuspend the pellet in 1.5 ml of 25% sucrose containing 0.05 M Tris and 0.001 M EDTA at pH 8.0.

3. Add 0.2 ml of lysozyme solution (10 mg of lysozyme per ml in 0.25 M Tris, pH 8.0) to the cell suspension and mix gently. Place the tube in ice for 15 min.

4. Add 0.1 ml of 0.25 M EDTA at pH 8.0, mix gently, and replace the tube in ice for 10 min.

5. Add 0.1 ml of 20% SDS solution. Mix the suspension gently and keep on ice for 10 min. During this time, the cells lyse and the solution becomes viscous.

6. Precipitate chromosomal DNA by adding sufficient 3 M NaCl to bring the final concentration to 1 M NaCl (usually about 0.9 ml of 3 M NaCl). Place the tube in ice for at least 2 h to allow complete precipitation. It is often convenient to place the tube in a refrigerator overnight.

7. Centrifuge the precipitated chromosomal DNA and any remaining cell debris at 17,000 rpm for 30 min at 4°C, and decant the supernatant (enriched for plasmid DNA) into a 15-ml Corex tube.

8. Remove the ribonucleic acid by the addition of 1 volume of distilled water and 4 μl of a 5-mg/ml ribonuclease solution in 0.15 M NaCl (the enzyme is heated at 100°C for 10 min prior to use to destroy deoxyribonucleases). Incubate the tube at 37°C for an additional 1 h.

9. Add an equal volume of Tris-saturated phenol to deproteinize the mixture. Shake the mixture vigorously and then centrifuge at 5,000 rpm for 20 min at 20°C. Remove the aqueous (upper) phase containing the DNA. Repeat this step once.

10. Transfer the aqueous phase to a 30-ml Corex tube, and add sufficient 3 M sodium acetate (usually about 0.6 ml) to make the final concentration 0.3 M.

11. Add 2 volumes of cold (−20°C) 95% ethanol. Mix the solution well and place at −20°C overnight.

12. Centrifuge the precipitated DNA at 12,100 × g for 20 min at −10°C, decant the ethanol thoroughly, and resuspend the DNA in 0.2 ml of 6 mM Tris buffer at pH 7.5.

The DNA can now be characterized by gel electrophoresis, density-gradient centrifugation, or homology determination detailed in part 15.3.

15.1.4. With Lysostaphin

The following procedure has been used successfully to obtain plasmid DNA from gram-positive bacteria and from gram-negative bacteria that are resistant to lysis by Triton or SDS.

1. Grow cells in 30 ml of brain heart infusion broth contained in a flask on a shaker, to the mid-logarithmic phase of growth. Pellet the cells by centrifugation as before, and suspend them in 1.5 ml of a solution that is 0.0075 M NaCl–0.050 M EDTA, pH 7.0.

2. Add lysostaphin (Sigma Chemical Co., St. Louis, MO 63178) to a final concentration of 15 μg/ml. (Double the enzyme concentration for *Staphylococcus epidermidis*.) Incubate the suspension at 37°C for 15 min with gentle agitation, and then place the flask on ice.

3. Add 1.5 volumes of a mixture containing 0.4% deoxycholate, 1% Brij-58, and 0.3 M EDTA (at pH 8.0) to achieve lysis. Mix the viscous contents of the tube gently, and put the tube on ice for 15 min.

4. Pellet the cellular debris by centrifugation at 23,000 × g for 20 min at 4°C (Sorvall SS-34 rotor), and decant the supernatant fluid into a 15-ml Corex tube.

5. Add 1 volume of distilled water and 4 μl of ribonuclease solution (1 mg/ml) to the supernatant fluid. Incubate the tube at 37°C for 1 h. Proceed with plasmid DNA purification as described in the SDS method (15.1.3).

15.1.5. With Penicillin

The following procedure can be used for obtaining plasmid DNA from gram-positive organisms and from gram-negative organisms that are resistant to Triton or SDS, but sensitive to penicillin. If the bacterium under study produces a β-lactamase, then a "penicillinase-resistant" antibiotic (e.g., a cephalosporin) may be effective. We are not certain whether other antimicrobial agents active against cell wall biosynthesis would prove successful, although it may be useful to investigate this possibility.

1. Grow bacteria to the mid-logarithmic phase of growth as described in part 15.1.4. Add 1 mg of penicillin G per ml, and incubate the culture at the optimum growth temperature for an additional 2 h. (It is wise to perform a preliminary experiment to determine the time and antibiotic concentration that gives optimum protoplast formation; the goal is to achieve this maximum without massive lysis. The addition of 1 M sucrose to the growth medium may be beneficial in achieving this goal.)

2. Harvest the penicillin-treated cells by centrifugation as before, and wash them twice with equal volumes of 0.01 M Tris-hydrochloride, pH 8.2.

3. Suspend the cell pellet in 0.25 volume of 0.02 M Tris at pH 8.2, 0.5 volume of 1 M sucrose, and 0.25 volume of lysozyme (4 mg/ml) in 0.02 M Tris at pH 8.2.

4. Incubate the cell suspension at 37°C for 1 h with shaking.

5. Centrifuge at 27,000 × g for 15 min.

6. Gently suspend the cells in 0.02 M Tris and 0.01 M EDTA at pH 8.2, taking care to ensure that the cells are homogeneously dispersed.

7. Add 0.1 volume of 10% SDS to the cells. Lysis should be complete within 10 min. Proceed from here as described in the SDS method (15.1.3).

15.2. ISOLATION OF PLASMID DNA OF LARGE MOLECULAR WEIGHT

The foregoing methods have a limitation in that they can be used only for isolating plasmid DNA with molecular weight no greater than 100×10^6. Large plasmids have been described in several bacterial systems, including the tumor-inducing plasmids of *Agrobacterium tumefaciens* (31), the H-incompatibility group of antibiotic resistance plasmids (12), the camphor degradative plasmid of *Pseudomonas putida* (2, 24), and the various F-prime factors (22, 26, 27). Recently, several methods have been developed which permit the isolation of these larger plasmids as well as the smaller plasmids (9, 14, 17, 30). In addition, several techniques have been developed which permit the use of very small working volumes or even material from a single colony (10).

The following method is one recently developed by Daniel Portnoy and Frank White (personal communication). It has the advantage of small working volumes, and it shares the best features of the previously published methods.

The method takes advantage of the resistance of plasmid DNA molecules to strand separation by alkali. Cell lysis also is carried out under alkaline conditions, which may account for the improved yields in plasmid DNA of very large

molecular weight (up to 350×10^6). This method is now employed by us for the routine screening of all bacterial cultures for all types of plasmids.

We have used this technique successfully for the isolation of plasmids from gram-negative bacteria (e.g., *Agrobacterium tumefaciens*, *Yersinia enterocolitica*, *Salmonella typhi*, *Klebsiella* sp., *E. coli*) and also gram-positive bacteria (e.g., *Staphylococcus* sp. and *Streptococcus* sp.). As in the other procedures, we noticed that it is necessary to carry out the washing steps in the presence of high salt concentrations (up to 1 M NaCl) for marine vibrios and other halophilic bacteria.

The plasmid DNA obtained by this method from even 2 ml of culture is relatively free from chromosomal DNA and, after a few additional steps, can be used directly for three restriction endonuclease analyses (see 15.3.1). The general and extended procedure is as follows:

1. Grow 2 ml of culture overnight in an appropriate medium (e.g., brain heart infusion broth) in 120-mm screw-cap tubes on a shaker. The cells may be used directly or can be employed to obtain logarithmically growing cells by diluting the overnight culture 1:20 into 2 ml of fresh medium and incubating the subculture for 2 to 3 h. Harvest the cells by centrifugation in a Sorvall table-top centrifuge at about 2,500 × g for 10 min.

2. Resuspend the cells in 2 ml of TE (0.05 M Tris-hydrochloride and 0.01 M EDTA at pH 8.0), recentrifuge, and resuspend the cell pellet thoroughly in 40 μl of TE.

3. Prepare lysis buffer (4% SDS in TE at pH 12.4) daily. It is important to determine the pH accurately (use a high-pH electrode) since values higher than 12.5 will irreversibly denature the plasmid DNA. Add 0.6 ml of the lysis buffer to a 1.5-ml centrifuge tube (e.g., Brinkmann Instruments Microtest tubes, catalog no. 22-36-911-1) and, with a Pasteur pipette, transfer the 40 μl of the cell suspension into the lysis buffer. Mix the suspension well, but avoid vigorous agitation or the use of mechanical mixing devices. (The culture could be grown directly in a 1.5-ml centrifuge tube to avoid the need for transfer in this step.)

4. Incubate the suspension at 37°C for 20 min to achieve full lysis of the cells.

5. Neutralize the solution by adding 30 μl of 2.0 M Tris at pH 7.0. Slowly invert the tube until a change in viscosity is noted.

6. Precipitate chromosomal DNA by adding sufficient NaCl (about 0.24 ml of 5 M NaCl) to bring the solution to 1 M. This step should be done quickly. For complete removal of the chromosomal DNA, put the tube on ice for 4 h. If one is screening a large number of cultures and

does not care about maximum plasmid purity, the time on ice can be shortened to 1 h.

7. Spin the tube in a microcentrifuge (e.g., Eppendorf model 5412) for 10 min to sediment debris, and then pour the supernatant fluid into another 1.5-ml centrifuge tube (do not attempt to obtain the small amount of fluid remaining in the bottom of the tube).

8. Add 0.55 ml of isopropanol to the supernatant fluid to precipitate the DNA. After mixing, place the tube at $-20°C$ for 30 min.

9. Centrifuge for 3 min in a microcentrifuge, pour off the supernatant fluid and invert the tube on a paper towel; then dry the tube under vacuum.

10. Resuspend the precipitate in 30 μl of TES (0.05 M Tris, 0.005 M EDTA, 0.05 M NaCl, pH 8.0). Allow the DNA to dissolve overnight at 4°C. Ordinarily, 10 μl of this solution will show readily visible DNA bands after gel electrophoresis through 0.7% agarose in Tris borate buffer for 3 h at 100 V.

Plasmid DNA obtained by this method should be further purified for restriction endonuclease analysis (see 15.3.2) by using the following additional steps:

11. Resuspend the pellet obtained in step 9 in 100 μl of 0.010 M Tris at pH 8.0. Add 100 μl of phenol equilibrated with 0.010 M Tris at pH 8.0. Mix well. Add 100 μl of chloroform.

12. Centrifuge for 30 s in a microcentrifuge to separate the aqueous phase from the phenol-chloroform phase. Remove the upper aqueous phase with a 100-μl micropipette, being careful to avoid the interface.

13. Precipitate the plasmid DNA with 2 volumes of ice-cold 95% ethanol and proceed as in step 9. Resuspend the precipitate in a suitable buffer for the restriction endonuclease reactions. The buffer depends on the particular restriction endonuclease used.

15.3. CHARACTERIZATION

15.3.1. Gel Electrophoresis (for Preliminary Study)

Simple adaptations of gel electrophoresis methods (10, 23) are suitable for the detection and preliminary characterization of plasmid DNA present in clinical isolates and laboratory strains of gram-negative and gram-positive bacteria. The isolation method using SDS (15.1.3) has been widely employed in conjunction with characterization by gel electrophoresis (26), but the molecular weights of the plasmids that can be detected range from 0.6×10^6 to 100×10^6 in partially purified cell lysates (23). Instead, we use the method described in part 15.2, which

permits the detection of plasmids with molecular weights ranging from 0.6×10^6 to 350×10^6.

When plasmid DNA is subjected to electrophoresis in gel, the migration of the different DNA species is related inversely to their molecular weights; i.e., the larger the molecular weight, the slower the rate of migration. All of the currently published procedures involve either a vertical or a horizontal slab of gel, a regulated power supply, and a short-wave ultraviolet light source. Even partially purified lysates (generally 10 to 20 μl of lysate is sufficient) can be subjected to electrophoresis.

For electrophoresis, use a standard vertical-slab gel apparatus (gel dimensions: 100 by 140 by 2.5 mm) with 12 slots (6.5 by 15 by 2.5 mm). The apparatus either may be purchased commercially (Aquebogue Machine and Repair Shop, P.O. Box 205, Aquebogue, NY 11931) or it may be fabricated. Gel concentrations vary from 0.25 to 1.5% agarose (Seakem type ME, Marine Colloids, Inc., Rockland, Maine) in a standard electrophoresis buffer (89 mM Tris, 2.5 mM disodium EDTA, and 8.9 mM boric acid). Electrophoresis is ordinarily carried out for 2 h at room temperature, 60 mA, and 120 V. The gel is then placed in a solution of ethidium bromide (0.4 to 1 $\mu g/ml$) and stained for 15 min. The DNA then can be visualized with short-wave ultraviolet light.

Figure 1 is a photograph of one such gel (0.7% agarose). Band CHR corresponds to chromosomal DNA present in the partially purified preparation. Column 1 shows, besides the band corresponding to chromosomal DNA (CHR), several bands corresponding to plasmids of different known molecular weight from different strains of bacteria that are used as standards. By plotting the $\log_{10}$ of the distance migrated from the origin by the plasmid DNA versus the $\log_{10}$ of the molecular mass of the plasmid DNA, a straight-line relationship is obtained from about 100 to 1 Mdal. Interpolation permits the determination of molecular weights for the plasmids from the various bacteria.

Plasmids of known molecular weight (Table 1) can be used as standards to determine the molecular weight of an unknown plasmid by agarose gel electrophoresis. In order to detect the large-molecular-weight plasmids, strains should be lysed by the method described in 15.2. It is advisable to distribute the plasmid DNA solutions obtained from each strain in 20-μl samples, which can then be stored at $-20°C$.

15.3.2. Gel Electrophoresis (with Restriction Endonuclease)

Restriction endonucleases catalyze double-stranded, staggered cleavages at specific recog-

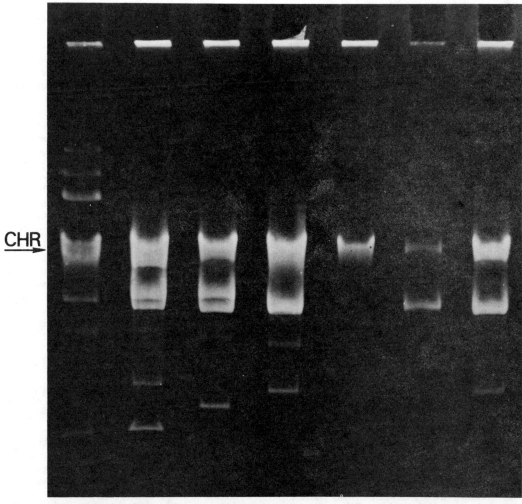

FIG. 1. *Agarose-gel electrophoresis of crude lysates obtained from different bacterial strains. Column 1 (left): standard plasmid DNAs ranging in molecular weights from 62×10^6 (uppermost band) to 1.8×10^6 (lowest band). Columns 2 to 7: plasmid contents of different bacterial strains. The chromosomal DNA is located in the band marked CHR. From Crosa et al. (8).*

nition sites in DNA (1, 29). For example, in the case of the enzyme *Eco*RI, the specific nucleotide sequence is

$$\downarrow$$
$$\text{GAATTC}$$
$$\text{CTTAAG}$$
$$\uparrow$$

(the position of the arrows specifies the cleavage site). Plasmid DNA (and all forms of DNA, for that matter) after endonuclease treatment have characteristic fragment patterns which depend upon the number and spacing of the specific recognition sites within the genome. The DNA fragments can be separated by electrophoresis in agarose gels. Comparison of the patterns ob-

tained for different plasmids can be used to determine qualitatively their degree of relatedness. The DNA fragments also can be transferred from the agarose gel to a strip of cellulose nitrate. After hybridization with a radioactive DNA "probe," the fragments containing sequences homologous to those present in the probe can be detected as sharp bands by autoradiography of the cellulose nitrate strip (28).

To date, approximately 150 separate restriction endonucleases have been described. A number of these enzymes can be purchased from commercial sources (Bethesda Research Laboratory, Rockville, MD 20850; Miles Laboratories, Inc., Elkhart, IN 46514; New England Biolabs, Beverly, MA 01915). Each enzyme requires

Bacterial source[a]	Mol wt of plasmid ($\times 10^6$)
Pseudomonas aeruginosa PA02 (pMG1)	312
P. aeruginosa PA02 (pMG5)	280
Escherichia coli DT78 (TP116)	143
E. coli DT41 (R27)	112
Salmonella typhimurium LT2	60
P. aeruginosa PA02 (RP1)	38
E. coli K-12 J5-3 (Sa)	26
E. coli K-12 W1485-1 (RSF2124)	7.4
E. coli K-12 W1485-1 (RSF1010)	5.5
E. coli JC411 (ColE1)	4.2
E. coli W1485-1 (PMB8)	1.8

[a] These strains can be obtained from Ester M. Lederberg, Plasmid Reference Center, Department of Medical Microbiology, Stanford University Medical School, Stanford, CA 94305 (telephone 415/321-2300).

specific reaction conditions which are supplied by the manufacturer. The usefulness of this method of analysis can be illustrated by the following example.

Crosa et al. (8) examined a number of ampicillin-resistant strains of *Shigella dysenteriae* type 1 which were isolated in different epidemics. The ampicillin-resistant determinant was mediated in all cases by a 5.5-Mdal plasmid. Figure 2 shows the banding patterns obtained after treatment with *Hinc*II restriction endonuclease. A similar 5.5-Mdal plasmid DNA was observed in all of the isolates. The results suggested that all of the plasmids were homologous.

The following procedure was used for *Hinc*II restriction endonuclease cleavage of the 5.5-Mdal plasmids:

1. Treat 20 µl of purified plasmid DNA (approximately 0.2 µg of DNA) in a buffer (containing 10 mM Tris-hydrochloride, 50 mM NaCl, 6 mM 2-mercaptoethanol, and 6 mM $MgCl_2$ at pH 7.0) with 2 U of *Hinc*II restriction endonuclease (New England Biolabs; 2,000 U/ml).

2. Incubate the reaction mixture at 37°C for 90 min.

3. Stop the reaction by adding 5 µl of an aqueous mixture containing 0.07% bromophenol blue, 7% SDS, and 33% glycerol.

4. Subject the total reaction mixture to electrophoresis in 0.7% agarose gel as described in part 15.3.1.

5. Compare the banding patterns after staining with ethidium bromide (15.3.1).

In some circumstances, the ability to extract fragments that have been separated on agarose gels is highly desirable. These fragments can be useful for physical and chemical studies as well as for fine-structure restriction analysis, DNA sequencing, heteroduplex analysis, and hybridization by the S1 endonuclease method (6) or by the "blotting" technique of Southern (28). The following technique (11) permits the extraction of a variety of physically intact, biologically active DNAs in higher yield directly from agarose gel bands. The procedure requires commercially available agarase (catalog no. 121811, Calbiochem, La Jolla, Calif.), an enzyme that digests agarose.

1. The endonuclease digestion is described above, and the agarose gel-ethidium bromide electrophoresis is described in part 15.3.1. DNA fragments separated on a 0.6% agarose gel (0.6-cm diameter by 11-cm length) are visualized by illumination with an ultraviolet lamp and are sliced out with a razor blade. Several gel slices (0.05 to 0.1 g each) containing the same restriction fragment can be processed simultaneously.

2. Crush a gel slice in 0.1 M Tris at pH 5.95 (0.1 ml per gel slice). Homogenize by forcing the mixture three times through a 1-ml plastic syringe (without the needle) into a Pyrex centrifuge tube.

3. Add agarase (50 µg per gel slice) dissolved at 1 mg/ml in 0.1 M Tris-hydrochloride at pH 5.95. Incubate at 37°C for 2 h.

4. Remove the agarose by centrifuging at 4°C for 30 min at 48,000 $\times$ *g*.

5. If necessary, the supernatant containing the DNA can be further concentrated by precipitation with 2 volumes of ethanol or by dialysis against polyethylene glycol.

6. Further purification can be achieved prior to the concentration step by passing the DNA supernatant through a 0.45-µm Swinnex filter (Millipore Corp., Bedford, Mass.).

Recovery of DNA following this procedure is about 75 to 85% for plasmid DNA of molecular weight between 7.4×10^6 and 60×10^6 and for *Eco*RI restriction fragments ranging in size from 0.75 to 13.9 Mdal. Agarase-extracted DNA has essentially the same transformation frequency (for *E. coli* cells) as DNA not exposed to agarase. DNA fragments extracted from gels have been successfully used in cloning experiments, as vehicles and/or passengers (11).

15.3.3. Density Gradient Centrifugation (with Ethidium Bromide)

Plasmid DNA molecules extracted from bacterial cells have been characterized as double stranded, covalently closed, circular, and with no free end of rotation. Ethidium bromide is one in a series of phenathridinium dyes that bind to DNA and ribonucleic acid and inhibit nucleic acid function (5, 15). Cells are lysed, and lysates are added to cesium chloride gradients contain-

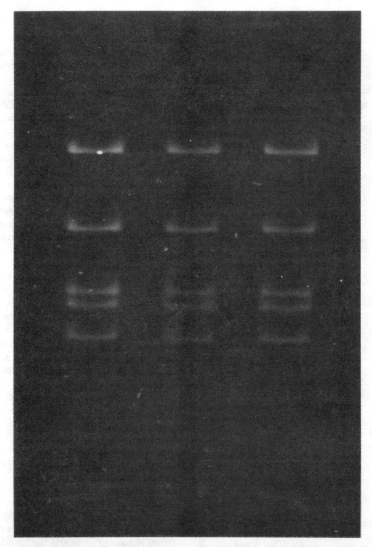

FIG. 2. *HincII restriction endonuclease patterns obtained with Ap^r plasmids from different enterobacteria. Reactions were carried out and electrophoresed as described in part 15.3.2. The gel was stained with ethidium bromide. The fluorescence produced by the intercalated dye when illuminated with an ultraviolet light source was photographed with a Polaroid film. From Crosa et al. (8).*

ing ethidium bromide. The DNA is sedimented to equilibrium by high-speed centrifugation. Ethidium bromide intercalates into the DNA, thus reducing its density (3, 25). A linear DNA molecule or an open circular (nicked) molecule of plasmid DNA does not have the same physical constraints that are imposed upon a covalently closed circular molecule of plasmid DNA. Consequently, the former types of DNA can bind significantly more ethidium bromide molecules and are rendered less dense than the latter type of DNA. This difference in binding permits the separation of the different forms of plasmid DNA. In gradients of cesium chloride, these forms can be visualized as discrete bands when the gradient is illuminated with ultraviolet light due to the fluorescence of the ethidium bromide intercalated in the DNA (Fig. 3).

The following procedure requires a preparative ultracentrifuge, hand-held ultraviolet light (Blak Ray, UVL-22; Ultraviolet Products, Inc., San Gabriel, Calif.) gradient dripper, and refractometer (Abbe 3L, Bausch & Lomb, Inc., Rochester, N.Y.). For example, the material obtained from the Triton isolation method (15.1.2) can be handled as follows:

1. To a Beckman polyallomer centrifuge tube (size: ⅝ by 3 inches [1.6 by 7.6 cm]), add about

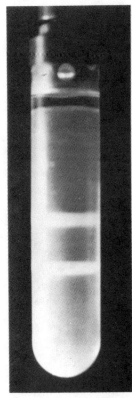

FIG. 3. *Cesium chloride-ethidium bromide density gradient centrifugation of a bacterial lysate obtained by using the Triton lysis procedure. Illumination with a long-wave ultraviolet light permits the visualization of two bands, due to the fluorescence of the ethidium bromide intercalated in the DNA. The upper band is chromosomal DNA, and the lower band is plasmid DNA.*

5 g of cesium chloride, 2 ml of TES buffer, and 3.2 ml of cell lysate.

2. Mix the contents of the tube until the cesium chloride is in solution, and then add 0.2 ml of ethidium bromide solution (10 mg/ml in TES buffer).

3. Adjust the refractive index of the solution to 1.3925 ± 0.001 g/cm^3 by adding TES buffer or solid cesium chloride. The refractive index is a suggested value and may vary for each instrument; it must be standardized by doing trial experiments. If necessary, fill the tubes with mineral oil.

4. Prepare the tubes for ultracentrifugation according to the centrifuge manufacturer's specifications, and place them in the appropriate rotor. Make certain the tubes are well balanced (gradients should have approximately the same height as well as the same weight to ensure proper balance).

Larger polyallomer tubes can be used according to the volume of lysate. For example, a scaled-up Triton isolation method can yield 20 ml of lysate which can be centrifuged in a Ti-60 Beckman polyallomer tube. Alternatively, about 10 ml of lysate can be spun in a 65-rotor Beckman polyallomer centrifuge tube. The different proportions of TES, cesium chloride, and ethidium bromide are as shown in Table 2.

5. Centrifuge the tubes at 40,000 rpm at 15°C for at least 44 h in a Beckman preparative ultracentrifuge (or its equivalent). After centrifugation, examine the gradients by illuminating the tubes with an ultraviolet light and looking for the presence of fluorescent bands due to the DNA-ethidium bromide complex. Collect the plasmid DNA (lower band in the gradient of Fig. 3) through the bottom of the tube using some type of gradient fractionator.

6. If a radioactive preparation of DNA was used, assay the fractions for radioactivity. A typical profile is illustrated in Fig. 4 and shows the clear distinction between the higher-density plasmid DNA fractions and the lower-density chromosomal DNA fractions.

7. Free the DNA of ethidium bromide by extracting the preparation at least four times with cesium chloride-saturated isopropanol.

8. Dialyze the DNA against an appropriate buffer to eliminate the cesium chloride.

9. Divide the purified plasmid DNA into different samples, and store at −20°C.

Cell lysis, followed by density gradient centrifugation, can be utilized to determine whether a particular bacterial strain contains plasmid DNA. In most laboratories this would not be the technique of choice since it requires expensive equipment, it is time-consuming, and only a relatively few cultures can be examined at one time. However, it is the best method for purifying sufficient quantities of plasmid DNA for further analysis.

Some tips about density-gradient centrifugation.

1. To totally eliminate chromosomal DNA contamination (upper band in gradient, Fig. 3), perform a second centrifugation of the collected plasmid DNA in a cesium chloride-ethidium bromide gradient. Collect the rebanded plasmid DNA dropwise from the bottom of the gradient tube. Alternatively, collect the plasmid DNA band by puncturing the side of the tube with a hypodermic needle.

2. Add the ethidium bromide after the CsCl, since the relaxation protein present in some plasmids which leads to nicking of the DNA can

TABLE 2. *Proportions of TES, CsCl, and ethidium bromide for use in density gradient centrifugation*

Sample	Rotor type	
	50	Ti-60
Lysate	3.2 ml	24 ml
TES	2.0 ml	3 ml
CsCl	4.9 g	23 g
Ethidium bromide (10 mg/ ml)	0.2 ml	1 ml

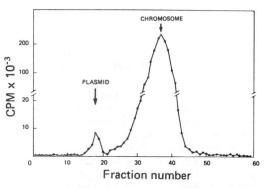

FIG. 4. *Determination of plasmid copy number. A plasmid-containing bacterial strain was grown in a minimal salts medium plus Casamino Acids and glucose. Labeling and lysis were carried out, and the lysate was centrifuged in the cesium chloride-ethidium bromide gradient as described in part 15.3.3. Seven-drop fractions were collected on microtiter trays and assayed for radioactivity.*

be activated by ethidium bromide (20). The high CsCl concentration (7 M) inactivates the relaxation complex (20) of most plasmids, and consequently higher yields of DNA can be obtained.

3. Even at a high salt concentration, visible light (in the presence of ethidium bromide) can induce nicking of plasmid DNA. Thus, all operations must be conducted under indirect illumination.

4. It is possible to obtain good banding and separation between plasmid DNA and chromosome DNA in an overnight run by centrifuging in a 65-type Beckman rotor at 45,000 rpm at 15°C. The recently developed vertical rotors offer an even more rapid means to achieve separation of plasmid DNA from chromosome DNA.

Determination of plasmid copy number.

The number of copies of a plasmid present per chromosome equivalent is a parameter that characterizes a plasmid and also gives information about the nature of its replication.

Plasmid copy number can be estimated by centrifuging a [³H]thymine-labeled total-cell lysate in an ethidium bromide-cesium chloride gradient. The ratio between the plasmid and chromosome peaks can be used for the calculation of copy numbers, as follows:

copy number
$$= \frac{\text{cpm of plasmid peak}}{\text{cpm of chromosome peak}} \times \frac{\text{MW}_c}{\text{MW}_p}$$

where MW_c and MW_p are the molecular mass weights of the chromosome and the plasmid, respectively.

If the bacterial strain requires thymine, use an adequate concentration of thymine for growth plus [³H]thymine (7 µCi/ml). For a strain that does not require thymine, use 1 µg of thymine, 250 µg of deoxyadenosine, and 7 µCi of [³H]-thymine per ml.

The procedure is as follows:

1. Grow a culture (1 ml) of a plasmid-containing strain at 37°C in the presence of 7 µCi of [³H]thymine per ml to a density of 2×10^8 cells per ml.

2. Centrifuge the cells, wash them in TE (0.01 M Tris-hydrochloride and 0.001 M EDTA at pH 8.1), and resuspend them in 400 µl of 0.5 M sucrose (ultrapure, ribonuclease-free sucrose in 50 mM Tris at pH 8.0).

3. Add 0.2 ml of 0.2 M EDTA at pH 7.8 to the cell suspension, and mix well.

4. Add 0.2 ml of 1% lysozyme (in 0.25 M Tris-hydrochloride at pH 8.00) to the cell suspension, and mix well. Shake gently, and place in ice for 15 min.

5. Add 1.2 ml of 1.2% Sarkosyl to the suspension, and then immediately add 0.1 ml of pronase (1 mg/ml) in 0.01 M Tris and 0.02 M EDTA at pH 8.1. Pronase should be first self-digested for 2 h at 37°C and then heated for 2 min at 80°C before using. Mix gently, and incubate at 37°C until the suspension clears completely.

6. Draw the DNA solution rapidly in and out of a 5-ml pipette (to shear the chromosomal DNA) until the solution is no longer viscous.

7. Centrifuge a sample containing about 500,000 cpm of radioactive DNA in a 5-ml ethidium bromide-cesium chloride gradient (the refractive index must be 0.0025 lower than the refractive index used for Triton lysate CsCl-ethidium bromide gradient). Also centrifuge a sample in a 5 to 20% linear sucrose gradient.

8. Collect the gradients on Whatman 3 MM filter disks.

9. Precipitate radioactive DNA with cold 5% trichloroacetic acid containing 50 µg of thymine per ml, wash with 95% ethanol, and dry. Determine radioactivity in a scintillation spectrometer.

Figure 4 shows a cesium chloride-ethidium

bromide gradient with the profile obtained for ^{3}H-labeled total-cell DNA from a plasmid-containing strain of *E. coli*.

Tips about determination of plasmid copy number.

1. Plasmid forms other than covalently closed circular (CCC) DNA are detected in the same cesium chloride-ethidium bromide gradient (or in sucrose gradients) by their banding characteristics. For example, dimers of two CCC DNA molecules will coband with CCC DNA in a CsCl-ethidium bromide gradient, but will sediment ahead of monomeric CCC DNA in a sucrose gradient. Dimers of one CCC and one open circular (OC) DNA molecule will band at a position intermediate between CCC DNA and OC DNA in the CsCl-ethidium bromide gradient. Dimers of two OC DNA molecules will coband with OC DNA in a CsCl-ethidium bromide gradient, but will band ahead of OC DNA in a sucrose gradient.

2. Sucrose gradients can be prepared with a gradient mixing chamber or, alternatively, by making a sucrose solution of an average concentration between the two desired extreme concentrations. Freezing of this solution at $-20°C$ and slow thawing at $4°C$ (overnight) permits the development of a gradient suitable for analytical purposes. For example, a frozen 12.5% sucrose solution after thawing will render a gradient that is approximately 5 to 20%.

15.3.4. Homology Determination (with Single-Strand Endonuclease)

The dissociation of double-stranded plasmid DNA followed by specific reannealing with single-stranded DNA from a homologous or heterologous source is the basis for all homology studies. Plasmid homo- and heteroduplexes can be analyzed with the single-strand specific endonuclease, "S1," of *Aspergillus oryzae*. The following procedure (6) permits an accurate and rapid determination of polynucleotide sequence relationships. It is particularly useful for surveys and for other investigations which require a large number of DNA-DNA hybridization assays.

Purified, radiolabeled plasmid DNA is used as the reference "probe" and is hybridized with purified total-cell DNA of plasmid-containing strains. DNA from a plasmidless strain serves as a control. DNA-DNA reassociation reactions assayed by the S1 endonuclease method are performed by incubating approximately 5,000 to 10,000 cpm (typically less than 0.001 μg of DNA) of sheared, denatured, purified plasmid DNA with 150 μg of unfractionated, total, sheared, denatured bacterial DNA preparations from

plasmid-containing and plasmidless bacteria in a total volume of 1 ml of 0.42 M NaCl. The DNA mixtures are incubated at a temperature which depends on the guanine plus cytosine (G+C) percentage of the plasmid DNA. The time of reassociation is such that essentially complete reassociation for the homologous reaction is achieved. The time required for a given plasmid to reassociate completely is a function of its molecular mass and the ionic concentration. The time of reassociation can be determined by calculating the $C_0t_{1/2}$. C_0 is the initial concentration of DNA; $t_{1/2}$ is the time for obtaining 50% reassociation of the DNA (at C_0 concentration) for a given temperature, salt concentration, and DNA fragment size. The empirical relationship (7) is as follows:

$$C_0t_{1/2} = \frac{\text{molecular weight of plasmid}}{3 \times 10^7}$$

The approximate time of incubation to get complete reassociation is equivalent to about 8 $C_0t_{1/2}$ in 0.42 M NaCl. Thus,

$$8\ C_0t_{1/2} = C_0t_x$$

where C_0 is the initial concentration of plasmid DNA (which can be calculated or estimated from the plasmid copy number) and t_x is the time of reassociation.

Under optimum conditions, less than 10% of the labeled plasmid DNA incubated alone should reassociate with itself, while more than 85% of the labeled plasmid DNA should reassociate with its homologous unlabeled DNA. The reassociation of labeled plasmid DNA with DNA of a plasmidless bacterial strain is also included in each experiment, and this value (about 10% or less) is subtracted from the values of all reactions.

Hybridization.

1. Reassociation mixture: 150 μg of unlabeled sonicated total-cell DNA, 5,000 to 10,000 cpm of labeled sonicated plasmid DNA, and sufficient NaCl to bring the solution to 0.42 M in a total volume of 1.0 ml.

2. Denature the DNA by boiling the solution for 10 min. Then place the DNA solution in ice.

3. Incubate the reassociation mixture at 55 to 70°C depending on the G+C percentage of the plasmid DNA.

[3H]thymine-labeled plasmid DNA is prepared by isolating CCC DNA by the Triton lysis technique (15.1.2) or it can also be labeled in vitro. The specific activity of the [3H]thymine plasmid DNA obtained by this method is usually about 10^6 cpm/μg. The plasmid DNA is subjected to sonic treatment to obtain fragments of an approximate molecular weight of 2.5×10^5.

Total cell DNA is prepared as described in Chapter 14. This DNA is also degraded by sonic treatment to an approximate molecular weight of 2.5×10^5.

S1 endonuclease reaction.

S1 endonuclease of very good activity can be purchased commercially (Sigma catalog no. N-5255).

1. Prepare stock solution of S1 reaction mixture (0.125 mM $ZnSO_4$, 87.5 mM NaCl, 37.5 mM sodium acetate buffer at pH 4.5, 25 μg of calf thymus DNA per ml). Store frozen at $-20°C$ in 0.8-ml samples. The reaction mixture will then contain 0.8 ml of the S1 reaction mixture and 0.2 ml of the reassociation mixture. The final reagent concentrations are 0.1 mM $ZnSO_4$, 150 mM NaCl (this includes the NaCl added with the 0.2 ml of the reassociation mixture), 30 mM sodium acetate buffer at pH 4.5, and 20 μg of calf thymus DNA per ml. For each reassociation reaction, two S1-treated and two untreated samples are prepared.

2. Add 187.5 U of S1 endonuclease to start the reaction. Incubate at 50°C for 20 min.

3. Stop the reaction by placing the tubes in an ice bath; add 50 μg of calf thymus DNA per ml as a carrier (any commercial, not very pure DNA can be used for this step), and add 0.3 ml of cold 20% trichloroacetic acid. Collect the tri-chloroacetic acid precipitate by vacuum filtration on a membrane filter (type HA, Millipore Corp.). Dry the filters at 70°C and determine the radioactivity by counting in a liquid scintillation spectrometer. Controls should include: native-labeled plasmid DNA, which will detect any double-stranded nuclease activity in the enzyme; denatured and reassociated labeled DNA alone, which is a control for the amount of self-reassociation of the labeled DNA under the conditions of reassociation; and a denatured labeled DNA, which is a control for the efficiency of the single-stranded activity of the S1-endonuclease preparations. Table 3 shows how to transform the raw radioactivity results into percentage of DNA homology. The results are expressed as the percentage of the untreated controls normalized to the values obtained for the homologous reaction.

15.3.5. Homology Determination (with Electron Microscopy)

If two partially complementary strands of plasmid DNA are allowed to renature, the resulting heteroduplex molecules can then be examined in an electron microscope. It is possible to map the regions of homology and nonhomology because single- and double-stranded nucleic acids are recognizably different if suitably prepared for electron microscope examination. Since the advent of analyses with restriction endonuclease and of methods for examining DNA homology in solution with radioactive

TABLE 3. *Determination of homology between heterologous plasmids by use of single-strand endonuclease*

Line	Sample	Counts/min after S1 nuclease[a]		Raw %	Avg %	Corrected and normalized % homology[b]
		Plus nuclease	No nuclease			
1	Heat-denatured ^{3}H-labeled plasmid DNA	20	821	2.4	2.3	
		18	832	2.2		
2	Heat-denatured and reassociated ^{3}H-labeled plasmid DNA	50	850	6.0	6.0	
		53	890	6.0		
3	Heat-denatured reassociated mixture of ^{3}H-labeled plasmid DNA and total cell DNA containing the same plasmid	790	850	92.9	94.8	100
		798	825	96.7		
4	Heat-denatured and reassociated mixture of ^{3}H-labeled plasmid DNA and total cell DNA containing a heterologous plasmid	45	820	5.5	6.3	<0.3
		63	881	7.1		
5	Heat-denatured and reassociated mixture of ^{3}H-labeled plasmid DNA and total cell DNA from a plasmidless derivative	60	860	7.0	7.0	<1
		62	880	7.0		

[a] The count-per-minute values were obtained by subtracting a background of 20 cpm.

[b] Values in the column were obtained by subtracting the value in line 2 from the respective values in lines 3, 4, and 5 and then normalizing to the corrected value in line 3.

probes, the electron microscope, heteroduplex methodology is not as widely used for the analysis of plasmid DNA. Although the technique requires considerable technical skill and specialized equipment, it is still highly recommended as a precise way to localize differences among plasmids and to better appreciate the organization of plasmid DNA.

In the standard Kleinschmidt technique (18), basic protein is attached to the DNA and is absorbed to a denatured protein monolayer at an air-water interface. Single-stranded DNA collapses and appears as a "bush" in the electron microscope. In the presence of formamide, however, the single strands are more extended and can be visualized together with double-stranded DNA. The addition of formamide thus allows the measurement of single-stranded regions in an otherwise double-stranded DNA molecule and is the basis for the heteroduplex formation analysis of plasmid DNA. The Kleinschmidt technique can be used very successfully to estimate molecular weights of plasmid DNA by measuring the contour length of well-spread, open circular or linear, double-stranded plasmid DNA molecules.

For heteroduplex experiments starting with CCC DNA, it is desirable to introduce a single-strand break so that, upon denaturation, an intact single-stranded linear molecule and a single-stranded circular molecule are generated. This can be achieved by first separating the CCC molecules and then lightly nicking them with deoxyribonuclease or with visible light in the presence of ethidium (3). Restriction endonuclease-cleaved CCC plasmid DNA also can be used successfully, so long as a single break is introduced. Denaturation is usually carried out by heating or by treatment with alkali. Generally, the method used to denature DNA must be vigorous enough to achieve complete strand separation, but not so vigorous as to degrade the DNA or to introduce further breakage. Renaturation in an aqueous solution requires a high salt concentration and heating to about 30°C below the midpoint (T_m) of thermal denaturation. These latter procedures can cause single-strand breaks. Thus, renaturation in a formamide solvent which effectively lowers the T_m is a better procedure. By using formamide, moderately dilute DNA solutions can be used, and the reactions can be carried out at room temperature or lower, and for longer times. Renaturation of approximately 50% of the DNA is optimum since the products of renaturation of a mixture of related DNAs consist of unrenatured single strands, homoduplexes, or heteroduplexes. Hence, the kinetics of renaturation after the 50 to 75% level is reached favors homoduplex for-

mation rather than the desired heteroduplexes. Figure 5 shows a typical heteroduplex between nicked DNAs obtained from a plasmid and its deletion mutant. Double-stranded and single-stranded regions can be easily recognized.

The basis of any electron microscope technique to analyze DNA is preparation of a monomolecular layer of the nucleic acid. To do this, a film of protein (cytochrome c) is floated onto an aqueous solution; the hypophase is used. Because the protein is surface denatured, it forms an insoluble film that can be considered a monomolecular layer. The DNA is absorbed onto this protein film and, as a consequence, is brought from a tridimensional configuration in solution to a bidimensional configuration on the film. The monolayer containing the DNA is then transferred onto a solid support film over a specimen grid for electron microscopy. Parlodion, a nitrocellulose derivative, can be used to provide the support film.

The specimen is shadowed with uranyl salts to improve contrast; metal shadowing also is used to increase contrast between double- and single-stranded DNA. (For principles of these methods, see Chapter 4.) Overshadowing should be avoided because irregularities in the support film and metal deposits in the background can trap additional metal, and the contrast between the object and background and between double- and single-stranded DNA is then reduced. Different procedures have been reported in the literature (18) for preparing the protein monolayer (e.g., a spreading procedure, a diffusion procedure, and a "one-step release" procedure). In the spreading procedure, a trough is filled with the hypophase and a solution containing both the protein and the nucleic acid is floated down a ramp (for instance, a glass slide) onto the surface of the hypophase.

The following procedure is recommended for determination of plasmid DNA homology by electron microscopy:

1. Place 0.1 μg of each of the two DNAs to be heteroduplexed in a small Eppendorf tube (West Coast Scientific Co., P. O. Box 2947, Rockridge Station, Oakland, CA 94618; catalog no. 3810). Denature the DNA by adding 0.25 ml of 0.1 M NaOH, 0.02 M disodium EDTA, and sufficient 10 N NaOH (1 to 5 μl) to bring the pH to 12.4 to 12.6. Allow the DNA to stand at room temperature for 10 min.

2. Neutralize the solution by adding 25 μl of 1.8 M Tris-hydrochloride–0.2 M Tris base and 0.25 ml of formamide (Mallinckrodt, 99%). The pH should be between 8.4 and 8.6. If not, add more Tris buffer (if the pH is higher than 8.6) or NaOH (if the pH is lower than 8.4); in either case, add 2 μl at a time.

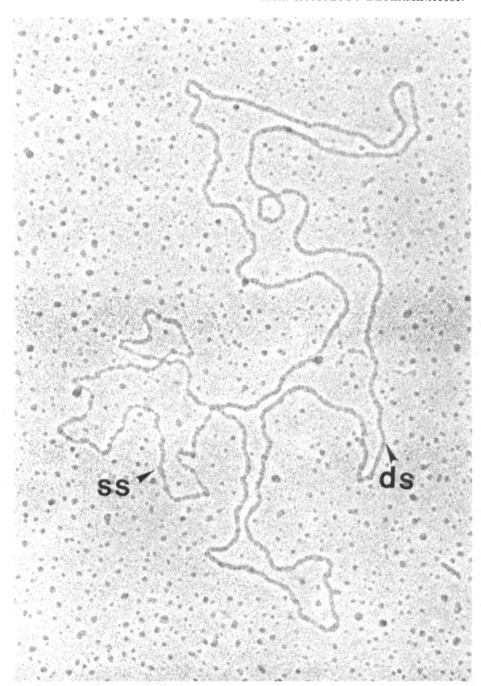

FIG. 5. *Heteroduplexing between the R-plasmid R6K and its deletion mutant RSF1040. The heteroduplex DNA molecules were obtained between the two X-ray-nicked plasmid DNAs by following the procedure described in part 15.3.5. The loop on the left is single-stranded (ss) R6K DNA which is deleted in RSF1040 DNA. The large loop on the right is double-stranded (ds) DNA which gives an account of the regions shared in common by both plasmids. The shared duplex region is about 70% of R6K DNA. From Crosa et al. (7).*

3. Allow the denatured plasmid DNA to reanneal for 1.5 h. To stop the renaturation, chill the solution to 0 to 4°C and dialyze it against 0.01 M Tris-hydrochloride and 0.001 M EDTA at pH 7 to 8.5 at 4°C for 2 h. The renaturation solution can be stored at 4°C and spread any time within

1 month, although DNA in the renaturation solution will continue to renature even at 4°C.

4. At this stage, it is convenient to have the trough and ramp ready. A small disposable plastic petri dish (such as a tissue culture dish, 50 mm in diameter) is a good trough, and glass microscope slides can be used as ramps. The slides should be acid cleaned, further washed with a detergent solution, and then thoroughly rinsed with distilled water. The trough should be rinsed with hypophase, which consists of a mixture of 45 ml of 0.011 M Tris-hydrochloride and 1.1 mM EDTA at pH 8.5 and 5 ml of 99% formamide (Mallinckrodt). Fill the trough with hypophase to give a convex meniscus.

5. Using clean forceps, thoroughly rinse the microscope slides with hypophase solution. Support the slide at a 30 to 45° angle on the inside edge of the dish.

6. Sprinkle a small amount of fine talcum powder about half way between the slide and the front of the dish. The talcum powder can be delivered by using a device consisting of a piece of a Pasteur pipette in which one of the ends is covered with tissue paper and the other end has an aspirator bulb. Alternatively, the talcum powder can be delivered with a camel-hair brush.

The talcum powder will serve as a measure of the location of the boundaries of the monolayer during spreading, and it provides some film compression by confining the edges of the monolayer.

7. In a 5-ml disposable beaker add 95 μl of the reannealed DNA solution and 5 μl of a freshly prepared solution (2 mg/ml) of cytochrome c (Sigma type II) in distilled water.

Pick up the DNA-cytochrome solution with a mechanical pipetting device fitted with a disposable 100-μl tip (the end of the plastic tip should be cut off).

8. Slowly deliver the DNA-cytochrome solution to the slide at about 0.5 cm above the meniscus. Maintain a continuous sheet of liquid between the pipette tip and the hypophase. Avoid drafts and breathing on the surface, and avoid moving the slide. The talcum should move abruptly away from the slide toward the edges of the dish as the protein monolayer is formed.

At this point, the distribution of the talcum powder is stationary and confined to the edges of the dish. If vortices are established, some talcum powder will swirl back toward the slide. The cytochrome film will be found in the areas free from talcum powder.

9. About 1 min after spreading, pick up a Parlodion-coated grid with a tweezer and gently lower the grid (Parlodion surface down) onto the monolayer; pause, and then rotate and lift the

grid at an angle between 20 and 30°. An adequate grid will be covered by a large droplet with a flat meniscus. Prepare two or three such grids so that several portions of the monolayer are sampled.

10. Stain the specimen by dipping the grid in 5×10^{-5} M uranyl acetate (see tips on methodology) for 30 s.

11. Dip the grid in isopentane for 10 s to facilitate drying.

12. Single- and double-stranded DNA should be visible in the electron microscope without shadowing, but shadowing improves discrimination. Afix the grids to a slide with double-sided tape. The grids are then rotary shadowed using 1.2 cm of Pt-Pd (80:20) 23-gauge wire (Ted Pella, Inc.) in a tungsten wire basket (E. F. Fullam, Inc.) at an angle of about 7°. The vacuum in the evaporation chamber of the shadowing apparatus must be below 6×10^{-5} mmHg before shadowing.

13. Grids are now ready to be examined in the electron microscope. About 20 different heteroduplex molecules give a statistically significant number from which the single- and double-stranded regions can be measured. It is convenient to include single-stranded ϕX phage and double-stranded DNA as markers. The calibration of the magnification used in the electron microscope can be done by using a standard grating replica available commercially (Ted Pella, Inc.).

Measurement of the DNA lengths from photographic negatives can be carried out by using a tracing assembly (Numonic or Hewlett-Packard electronic planimeter) or an overhead projection system (a photographic enlarger). Alternatively, a manual map-measuring device can be used but is much more laborious.

Tips about homology determination with electron microscopy.

1. Be careful about cleanliness. Surfactants such as detergents, oil vapors, and grease from fingerprints interfere with the formation and continuity of the protein in the monolayers. Surfactants must not be allowed to contaminate equipment, glassware, or solutions.

2. The Tris, uranyl acetate, disodium EDTA, and Parlodion should be stored in a desiccator. All reagents must be reagent grade. Filter sterilize the following solutions: 1 M Tris base, 0.1 M disodium EDTA, 1.8 M Tris-hydrochloride, 0.2 M Tris base, 0.1 M NaOH, and 0.02 M disodium EDTA. To prepare the stock uranyl acetate, make a 5×10^{-3} M solution of uranyl acetate in 10 ml of 0.05 M HCl, and filter sterilize it. For

staining, dilute the stock solution 1:100 in 90% ethanol. The stock keeps only 1 week, and the diluted staining solution keeps only for hours.

3. Formamide ($HCONH_2$) hydrolyzes to yield ammonium formate, which in turn produces ammonia and formic acid. The salt changes the ionic strength of the hyper- and hypophase solution, which can result in less than complete extension of single-stranded DNA. The ammonia can evaporate, resulting in a rapid lowering of the pH. As a consequence, aqueous formamide solutions should be used within 15 min of mixing. Mallinckrodt formamide (99%) can usually be used directly without purification; lower grades should be purified. Purity can be estimated by reading the optical density at 270 nm, which should be less than 0.2. Purification of formamide can be accomplished by recrystallization as follows: place formamide in a sealed bottle or flask at 0°C (sealing avoids absorption of water) until about one-third to one-half of the formamide has crystallized; the crystals can be recovered by filtration in a cold room or by centrifugation at 0°C.

4. Plastic petri dishes (50 mm in diameter) can be used for spreading. It is convenient to hold the petri dishes in place during spreading with double-stick masking tape.

5. Clean glass slides with and store in sulfuric-chromic acid mixture. Rinse them thoroughly with distilled or double-distilled water before use. After use, rinse them with water and return to the acid container.

6. To prepare Parlodion-coated grids, dissolve 60°C oven-dried Parlodion (pyroxylin, Mallinckrodt) in isoamylacetate (amylacetate can also be used) at a concentration of 3.5%. Mix but do not agitate to facilitate dissolution. Add molecular sieves to absorb water in the solution. Then proceed to coat the grids with the Parlodion in the following steps:

 a. Place a steel-wire screen on a stand at the bottom of a plugged Büchner funnel. Fill the funnel with double-distilled water. Deposit the grids on the screen. For general work, 200-mesh copper grids (E. F. Fullam, Inc.) are suitable. Either the dull or the shiny side can be up, but it is important to be consistent in order to know which side of the grid will have the sample.

 b. Clean the water surface by adding a few drops of the Parlodion solution and removing the film, after it dries, with paper.

 c. Form a film with 1 drop of the Parlodion solution. Cover the funnel with Saran wrap, and wait about 3 min until a shiny film with wide wrinkles on the edges is formed. Drain the water slowly, so that the film falls onto the grids. Wipe excess water from the back of the screen stand, and place it with the Parlodion-coated grids in a 60°C oven for 45 min. The film should appear silvery if properly prepared.

7. Some suppliers of electron microscope equipment are as follows:

E. F. Fullam, Inc., P.O. Box 444, Schenectady, NY 12301 (518/785-5533)

Numonics, P.O. Box 444, North Wales, PA 19454 (215/643-7410)

Ted Pella, Inc., P.O. Box 510, Tustin, CA 92680 (714/557-9434)

15.4. LITERATURE CITED

1. **Boyer, H. W.** 1974. Restriction and modification of DNA: enzyme substrates. Fed. Proc. 33:1125–1127

2. **Chakrabarty, A. M.** 1976. Plasmids in *Pseudomonas*. Annu. Rev. Genet. **10**:7–30.

3. **Clayton, D. A., R. W. Davis, and J. Vinograd.** 1970. Homology and structural relationships between the dimeric and monomeric circular forms of mitochondrial DNA from human leukemic leukocytes. J. Mol. Biol. 47:137–153.

4. **Clewell, D. B., and D. R. Helinski.** 1969. Supercoiled circular DNA-protein complex in *Escherichia coli*: purification and induced conversion to an open circular form. Proc. Natl. Acad. Sci. U.S.A. **62**:1159–1166.

5. **Clowes, R. C.** 1972. Molecular structure of bacterial plasmids. Bacteriol. Rev. **36**:361–405.

6. **Crosa, J. H., D. J. Brenner, and S. Falkow.** 1973. Use of a single-strand specific nuclease for analysis of bacterial and plasmid deoxyribonucleic acid homo- and heteroduplexes. J. Bacteriol. **115**:904–911.

7. **Crosa, J. H., L. K. Luttropp, F. Heffron, and S. Falkow.** 1975. Two replication initiation sites on R-plasmid DNA. Mol. Gen. Genet. **140**:39–50.

8. **Crosa, J. H., J. Olarte, L. J. Mata, L. K. Luttropp, and M. E. Penoranda.** 1977. Characterization of a R-plasmid associated with ampicillin resistance in *Shigella dysenteriae* type 1 isolated from epidemics. Antimicrob. Agents Chemother. **11**:553–558.

9. **Currier, T. C., and E. W. Nester.** 1976. Isolation of covalently closed circular DNA of high molecular weight from bacteria. Anal. Biochem. **76**:431–441.

10. **Eckhardt, T.** 1978. A rapid method for the identification of plasmid deoxyribonucleic acid in bacteria. Plasmid **1**:584–588.

11. **Finkelstein, M., and R. H. Rownd.** 1978. A rapid method for extracting DNA from agarose gels. Plasmid **1**:557–562.

12. **Grindley, N.D. F., G. O. Humphreys, and E. S. Anderson.** 1973. Molecular studies of R-factor compatibility groups. J. Bacteriol. **115**:387–398.

13. **Guerry, P., D. J. LeBlanc, and F. Falkow.** 1973. General method for the isolation of plasmid deoxyribonucleic acid. J. Bacteriol. **116**:1064–1066.

14. **Hansen, J. B., and R. H. Olsen.** 1978. Isolation of large bacterial plasmids and characterization of the P2 incompatibility group plasmids PMG1 and PMG5. J. Bacteriol. **135**:227–238.

15. **Helinski, D. R.** 1973. Plasmid determined resistance to antibiotics: molecular properties of R-factor. Annu. Rev. Microbiol. **27**:437–470.

16. **Hirt, B.** 1967. Selective extraction of polyoma DNA from infected mouse cell cultures. J. Mol. Biol. **26**:365–369.

17. **Humphreys, G. O., G. A. Willshaw, and E. S. Anderson.** 1975. A simple method for the preparation of large quantities of pure plasmid DNA. Biochim. Biophys. Acta **383**:457–463.

18. **Kleinschmidt, A. K.** 1978. Monolayer techniques in electron microscopy. Methods Enzymol. **12B**: 361–377.

19. **Kupersztoch, Y. M., and D. R. Helinski.** 1973. A catenated DNA molecule as an intermediate in the replication of the resistance transfer factor R6K in *Escherichia coli*. Biochem. Biophys. Res. Commun. **54**:1451–1459.

20. **Kupersztoch-Portnoy, Y. M., M. A. Lovett, and D. R. Helinski.** 1974. Strand and site specificity of the relaxation complex of the antibiotic resistance plasmid R6K. Biochemistry **13**:5484–5490.

21. **Lederberg, J.** 1952. Cell genetics and hereditary symbiosis. Physiol. Rev. **32**:403–430.

22. **Manis, J. J., and H. J. Whitfield.** 1977. Physical characterization of a plasmid cointegrate containing an *F' his gnd* element and the *Salmonella typhimurium* LT2 cryptic plasmid. J. Bacteriol. **129**:1601–1606.

23. **Myers, J. A., D. Sanchez, L. P. Elwell, and S. Falkow.** 1976. A simple agarose gel electrophoretic method for the identification and characterization of plasmid deoxyribonucleic acid. J. Bacteriol. **127**:1529–1537.

24. **Palchaudhuri, S.** 1977. Molecular characterization of hydrocarbon degradative plasmids in *Pseudomonas putida*. Biochem. Biophys. Res. Commun. **77**:518–525.

25. **Radloff, R., W. Bauer, and J. Vinograd.** 1967. A dye buoyant density method for the detection and isolation of closed circular duplex DNA: the closed circular DNA in HeLa cells. Proc. Natl. Acad. Sci. U.S.A. **57**:1514–1522.

26. **Sharp, P. A., M. T. Hsu, E. Ohtsubo, and N. Davidson.** 1972. Electron microscope heteroduplex studies of sequence relations among plasmids of *Escherichia coli*. I. Structure of R-prime factors. J. Mol. Biol. **71**:471–497.

27. **Skurray, R.A., H. Nagaishi, and A. J. Clark.** 1976. Molecular cloning of DNA from F sex factor of *Escherichia coli* K-12 Proc. Natl. Acad. Sci. U.S.A. **73**:64–68.

28. **Southern, E. M.** 1975. Detection of specific sequences among DNA fragments separated by gel electrophoresis. J. Mol. Biol. **98**:503–517.

29. **Thompson, R., S. G. Hughes, and P. Broda.** 1974. Plasmid identification using specific endonucleases. Mol. Gen. Genet. **133**:141–149.

30. **Watson, B., T. C. Currier, M. P. Gordon, M.-D. Chilton, and E. W. Nester.** 1975. Plasmid required for virulence of *Agrobacterium tumefaciens*. J. Bacteriol. **123**:255–264.

Section IV

METABOLISM

Introduction to Metabolism

W. A. WOOD

The metabolic activities of bacterial cells constitute the organized and regulated biosynthetic and biodegradative enzymatic sequences which are responsible for maintenance, growth, motion, and reproduction. These functions proceed at the expense of energy sources and other nutrients of extracellular and intracellular origin. The metabolic systems that support these functions are studied through a variety of biological, physical, chemical, and enzymatic methods. The biological methods prominently include nutrition and genetics, as described in other sections of this manual.

This section of the manual deals with the physical, chemical, and enzymatic methods used in the study of bacterial metabolism. Included in the following chapters are physical and instrumental techniques, fractionation and analysis of chemical components, preparation and assay of enzymes and their pathways, and measurement of permeability and transport processes. In each of these areas, the state of the art has been so highly developed as to justify the publication of books which are dedicated to that methodology and its applications. Here, however, the authors have restricted the scope to describe only reliable and easy methods which are widely applicable to basic studies of bacterial metabolism by a neophyte user.

Chapter 16

Physical Methods

W. A. WOOD

This chapter includes material on basic physical analytical methods and separation procedures, i.e., photometry, ion electrodes, chromatography, radioactivity, gel electrophoresis, and manometry. Their use in bacteriology has, in many cases, been major and almost characteristic of the field. The approach in this manual is to give a general description along with strengths and weaknesses of a method, specific techniques and preparations, and examples of application to investigations with bacteria.

In this chapter, especially, it has been necessary to confine the scope to the simple procedures which are appropriate for laboratory classes or are of value in initiating a research program; in-depth treatment of each subject can be found in books specifically dedicated to one technique. Some areas have been omitted or are mentioned only briefly; among these are thin-layer chromatography, analysis of macromolecules by sedimentation procedures, and several types of electrophoresis. On the other hand, some information is presented which is seldom visible to bacteriologists.

16.1. PHOTOMETRY

A wide variety of photometric methods have found important uses in bacteriological research and applications, including absorption photometry, fluorimetry, and nephelometry, as well as flame, emission, nuclear, and electron spin resonance methods (1, 2). These have facilitated identification of compounds, determination of structure, estimation of concentrations, and measurement of reaction rates.

16.1.1. Absorption Photometry

Ultraviolet-visible spectrophotometers measure the ability of solutes to absorb light at specified wavelengths. A plot of the light absorbed versus the wavelength (**absorption spectrum**) aids in the identification of compounds and gives information about the structure of the chromophore. The light absorbed at fixed wavelength can yield concentration information, and plots of light absorbed versus time can yield reaction rates.

The monochromator part of the instrument selects and transmits a narrow range of wavelengths from a radiant source through a sample to a photometer, which quantitates the energy received and expresses it as the ratio of the transmitted light (I_T) to the incident light (I_0). When I_0 is set at 100, the quantity measured directly is percent **transmittance** (%T). %T can be related to molar concentration (C) of an absorbing solute through the Beer-Lambert law

$$-\log \%T = \epsilon \cdot l \cdot C = A$$

where ϵ is the molar extinction coefficient of the absorbing species at a specified wavelength and l is the length of the light path in centimeters. Since both ϵ and l are constant, C is proportional to the negative logarithm of %T or directly proportional to **absorbance** (A). For this reason, newer spectrophotometers reading linearly in A are more convenient and more accurate for samples of moderate to high absorbance.

Factors affecting performance.

For laboratory spectrophotometers, quality is determined by: purity of the light presented to the sample, intensity and area of the beam (numerical aperture), accuracy of wavelength calibration, accuracy of photometer calibration, linearity of photometric response, noise, and short- and long-term drift. For some instruments, manufacturers go to great length and expense to achieve quality. However, for most biological experiments it is preferable to know the limitations of an ordinary instrument rather than to feel compelled to acquire a high-quality unit.

Effect of spectral bandwidth.

Spectral purity of light from a monochromator, or **natural bandwidth**, is the width in nanometers at one-half height of the emission energy peak at a specified nominal wavelength (Fig. 1A). This bandwidth is specified for the instrument and is not readily determined in the laboratory unless a very narrow band-pass (ca. 1 nm) interference filter is available. Bandwidth is reasonably constant across the wavelength range in monochromators where a grating is the dispersing element, but varies with wavelength for prism monochromators. Bandwidth varies with slit width in both cases. The accuracy of a measured absorbance, assuming no stray light (see below), depends on the ratio of the **spectral bandwidth** (a property of the instrument) to the natural bandwidth (the width in nanometers

at half-peak height) of the peak of the sample in the light beam (a property of the chromophore in solution) (Fig. 1B). Thus, in an inexpensive spectrophotometer with a spectral bandwidth at 340 nm of 8 nm, a peak of 80-nm natural bandwidth (giving a ratio of 0.1) allows absorbance measurements with 99.5% accuracy (4). Similarly, with an 8-nm band pass, 99% accuracy is achieved with a chromophore of 50-nm natural band pass. The most commonly measured material at 340 nm, reduced nicotinamide adenine dinucleotide (NADH), has a natural band pass of 58 nm at that wavelength. Therefore, it is readily apparent that NADH can be measured at better than 99% accuracy in this inexpensive spectrophotometer (4).

Effect of stray light.

Another consequence of wide spectral bandwidth, or low spectral purity, is an increased content of stray light (i.e., energies at wavelengths removed from the nominal value and not part of the spectral band-pass wavelengths). Since spectral band pass is related to both slit width and wavelength for prism monochromators, so also is stray light content. Stray light derives from incomplete removal of unwanted wavelengths due to faulty design or dirty optics. The stray light content of the exit beam diminishes the linear range of response to chromophore concentration, i.e., the range where the Beer-Lambert law applies. The operator should determine the useful linear range in the wavelength region where determinations are made.

Since stray light content increases as wavelength decreases, a conservative and convenient approach to determine the stray light general performance of the instrument is to use common solvents as sharp-cutoff filters (5). Figure 2A shows absorbance versus wavelength plots for a series of solvents. The sharp change in measured absorbance by a common laboratory spectrophotometer compared to the nominal value (dashed line) indicates that stray light energy predominates over that of the selected wavelength. The result is an inability to make measurements of the absorbance of solutes in the absorbance region where stray light content becomes appreciable. The effect of percent stray light on adherence to the Beer-Lambert law is shown in Fig. 2B (4). Percent stray light is the percent transmittance measured at a wavelength where a sharp-cutoff filter should be opaque.

Newer spectrophotometers in wide use can make accurate absorbance measurements in the 0.2- to 0.3-nm (2.0- to 3.0-Å) region, but this ability can be lost where optical components become dirty, the sample is fluorescent, or there is a light leak.

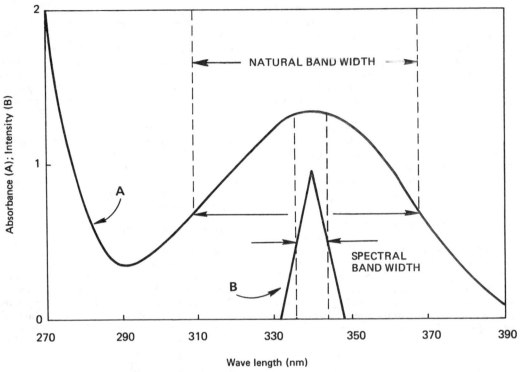

FIG. 1. *(A) Natural bandwidth diagram of NADH; the natural bandwidth is 58 nm. (B) Spectral bandwidth diagram (schematic) of spectrophotometer exit beam; the spectral bandwidth is 8 nm.*

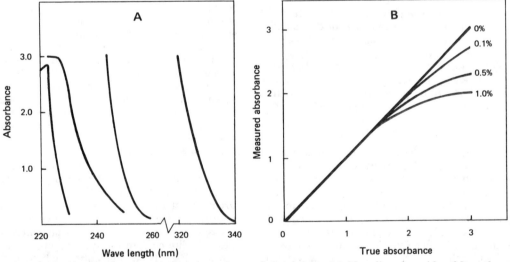

FIG. 2. *(A) Absorbance measurements on solvents (left to right: 0.1 M sodium bromide, chloroethanol, dichlorethane, acetone) for determination of stray light behavior of a monochromator. (B) Effect of various percentages of stray light on Beer-Lambert behavior of a chromophore.*

Photometric accuracy and linearity.

For certain spectrophotometers, adjustment of photometric calibration is not available and linearity of response is assumed. In at least one spectrophotometer, adjustment of calibration using supplied absorbance standards is a basic feature. In either case both absorbance accuracy and linearity of absorbance measurement should be established and an error table should be developed if necessary. For those instruments where absorbance calibration is possible, a series of graded neutral density filter standards of known absorbance at specified wavelengths can be used to adjust the photometer response to give the same absorbance value at one or a few absorbance values. With linearity between these points assumed, the total absorbance range is calibrated. The National Bureau of Standards and some instrument companies supply glass filter standards or information on preparation of liquid absorbance standards for this purpose (3, 7).

Linearity can be ascertained at fixed wavelength by various versions of the following procedure using either a standard of known absorbancy of about $0.4\ A$ (A_T) or a neutral density filter or solution in that absorbance range (A_N) as follows: (i) install the filter in a cuvette carrier so that either the air (or water) blank or the filter can be placed in the light beam, and null the photometer on the blank; (ii) place the filter in the beam and read the absorbance (A_{O_1}); (iii) return to the air blank and close the slit so that the blank has an arbitrary absorbance reading (A_X) of, for example, 0.1; (iv) return to the filter and read the absorbance (A_{O_2}). The true absorbance of this reading is A_T (or A_N) $+ A_X$. The observed absorbance is $A_{O_1} + A_X$. This process can then be repeated cyclically by closing the slit with the blank in the beam to give the absorbance reading of A_Y and then reading the filter (A_{O_2}). The result is a set of true and observed absorbance values increasing from A_T at intervals across the absorbance range. A lower value for A_T increases the number of points in the determination of linearity. Figure 3A shows

a plot of observed versus the true absorbance values for a particular instrument. When the increment is constant, the photometric response is linear. Figure 3B shows a plot of the error ($A_O - A_T$) in observed absorbance (A_O) versus true absorbance. This plot is especially useful in showing departure from linearity and the magnitude of the error.

Effect of slit width.

For prism monochromators working at fixed slit setting, there is a very marked nonlinear dependency of band pass on wavelength. Operating manuals from spectrophotometer manufacturers and books on spectrophotometry contain graphs of this function, and an example for one instrument is shown in Fig. 4. Thus, to make measurements at constant band pass with prism optics, the slit setting must be made according to the graph. To be rigorous, determination of solute concentration should be at the slit width setting used for determination of its extinction coefficient. Conversely, when reporting an extinction coefficient, the band pass should be stated. For prism instruments this involves both the slit width and the graph of band pass versus wavelength for that instrument.

Similarly, spectra should be acquired at constant band pass, not constant slit width. The absorbance values at fixed wavelength or the absorbance spectra determined at two different slit widths seldom are identical. The error is magnified when the peaks are sharp and may be negligible when peaks are broad. Thus, it is prudent to use the minimum slit width.

Slit width also is a function of sensitivity of the photometer, spectral characteristics of the

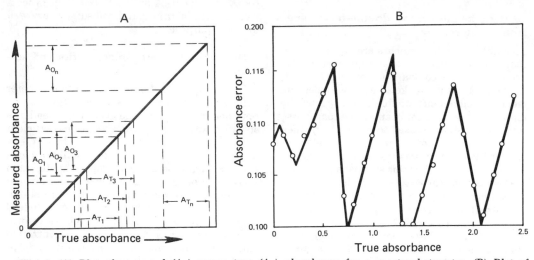

FIG. 3. *(A) Plot of measured (A_O) versus true (A_T) absorbance for a spectrophotometer. (B) Plot of absorbance error ($\Delta A_O/\Delta A_T$) versus true absorbance.*

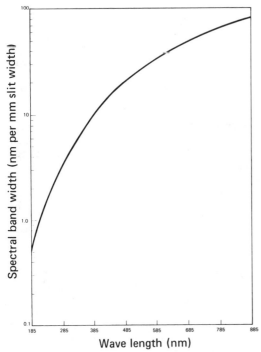

Wave length (nm)

FIG. 4. *Effect of wavelength on spectral band-width for a prism monochromator exemplified by data for a Gilford model 2400 spectrophotometer.*

photodetector, and the source. More importantly, intensity of the lamp, cleanliness of optical surfaces, and critical focusing of the source greatly affect slit width. As a rule of thumb, the average spectrophotometer should be capable of nulling by opening the slit to 2 mm at 210 nm or below. If this is not possible, one or more of the above factors may need attention.

Single-beam versus double-beam spectrophometers.

Double-beam instruments are constructed so that there are two light beams giving a simultaneous or parallel comparison between I_0 and I_T; single-beam units make these two measurements serially. Thus, double-beam spectrophotometers splilt the light beam so that part passes through the solution used to set the instrument at $I_0 = \%T = 100$ or $A_0 = 0$, while the second part of the beam passes through the sample to measure $\%T$ or A_1. For this approach to be valid, the parallel optical systems and their associated electronics must perform identically. Considerable attention to design is necessary to achieve identity, and this has been accomplished in many available instruments. In single-beam units, the same end is achieved by first setting $\%T = 100$ or $A = 0$ with the blank and then reading $\%T$ or A when the sample is introduced in the light beam.

There is a widespread belief that double-beam instruments are superior to single-beam units, the implication often being that single-beam units are less than adequate. Only in narrowly restricted situations is this the case. When rapid acquisition of spectra is required, or when there is rapid drift or instability in the light source of a single-beam unit, a double-beam unit is superior. Double-beam units are highly convenient for the direct acquisition of difference spectra. Conversely, there is no reason to use a double-beam unit for individual determinations of absorbance, as in colorimetric determinations and other routine spectrophotometric measurements.

Preparation of spectra.

Spectra are most conveniently obtained in a double-beam unit with an associated recorder. One cuvette contains a reagent (buffer) blank. The second contains the sample in the same reagent. Increased resolution is gained by keeping the slits as narrow as possible and by using a slow scan rate. For critical work, ascertain that the concentration of chromophore is within the linear response region, that fluorescence is not a problem at high concentrations (an indication can be obtained with a fluorimeter), and that the unit is in wavelength calibration.

For determination of spectra with a single-beam unit, manually set the wavelength at acceptable increments. Zero the instrument at each wavelength setting with the reagent blank and then record the absorbance of the sample. Construct a plot of wavelength (nanometers) on the X axis versus A on the Y axis. This process will be more laborious, but will suffice very well for most purposes.

Determination of concentration of solutes.

In a properly calibrated spectrophotometer with standard 1-cm cells, concentration can be obtained directly from the absorbance read at a specified wavelength if the extinction coefficient (ϵ) at that wavelength is known. Thus, $A/\epsilon =$ moles per liter or millimoles per milliliter. However, a high-quality spectrophotometer and knowledge of the extinction coefficient are not required if pure material is available to standardize even a filter colorimeter. A standard regression line is prepared which relates absorbance to concentration for a set of concentration standards which may be measured directly or have been run through the procedure for color development. Plots show the linear region and the variability of individual determinations. Standard statistical considerations apply, i.e., replication, least-square fits of the data to straight lines, determination of standard deviation from the mean, and standard error. Although linearity is highly desired, it is also possible to use a standard curve if the line is reproducible by taking values directly from the graph.

Determination of concentration of particles (6).

Photometric determination of the concentration of particles in suspension depends on attenuation of the light beam due to light scattering. A **nephelometer**, which is designed for this purpose, utilizes a highly sensitive photometer with the photodetector at 90° to about 160° from the incident beam. The photometric response does not follow the Beer-Lambert law; however, at very low particle concentrations, which can be measured in a nephelometer but not well in a transmission photometer, the response is nearly linear. Spectrophotometers and colorimeters can also be used at higher particle concentrations, but the method is based on attenuation of the incident beam. With both instruments a standard curve relating instrument response to bacterial cell concentrations is prepared (see 11.4).

Determination of reaction rates.

When either product or reactant of a reaction is a chromophore, the rate of chromophore production or utilization can be determined at fixed wavelength by reading the absorbance at time intervals or by following the absorbance change versus time with a strip chart recorder. A plot of absorbance versus time permits determination of velocity from the slope of the line. Specialized spectrophotometers and attachments to standard spectrophotometers have been developed for this purpose. Some of these, especially designed for determination of enzyme-catalyzed rates, present absorbance values to a chart recorder and have the capability to follow several reactions simultaneously.

16.1.2. Fluorimetry

Fluorescence is the emission of light by a fluorophore that absorbs light at a lower wavelength. Since the measurement involves the appearance of radiant energy compared to the very low level of light in the blank, the signal-to-noise ratio is very high, and hence fluorimetric determinations are inherently more sensitive than absorption methods. For instance, the fluorescent determination of NADH is roughly 100 times more sensitive than the spectrophotometric method. Fluorescence intensity depends on the concentration of fluorophore and, unlike absorption, also depends upon the intensity of the excitation beam. Since different wavelengths are involved in excitation and fluorescent emission, the instrument must be designed to prevent excitation energy from reaching the photodetector. This is usually accomplished by placing the detector at 90° to the excitation beam. Stray light arising from scattering in both the solution and from surfaces can be a major problem, but this can be minimized with sharp-cutoff filters which pass fluorescent energy but block excitation energy.

Since fluorescent intensity is dependent upon excitation intensity, lamp emission characteristics and stability are of major importance. Some instruments monitor beam intensity through a second photodetector and display the ratio of fluorescent to excitation intensities.

At low fluorescent emission intensities, photodetector output voltage is nearly linear with intensity; hence, concentration is proportional to the equivalent to transmittance, not absorbance, as in absorption photometry. Concentration of a fluorophore can be determined within limits by measuring relative fluorescence. The process basically involves preparation of a solvent blank and a set of standards, construction of a standard curve, and comparison with unknowns. Since sensitivity is greater, it is more likely that measurements can be affected by artifacts than in absorption photometry.

Instrument variation.

Care should be taken to correct for or prevent fluctuations in excitation intensity due to variations in lamp intensity. It may be necessary to introduce frequently the solvent blank and standards.

Absorbance of solutions.

High absorbance of solutions produces two kinds of errors: (i) reabsorption of fluorescent energy to give an apparently lower fluorescent yield and (ii) nonuniform intensity across the face (emitting surface) for a right-angle instrument. This results from variations due to high absorbance at the excitation wavelength. For this reason absorbance should not exceed $0.2\,A$, and independent absorbance measurements on standards and unknowns should be made at the excitation and emission wavelengths.

Fluorescence of impurities.

The limiting sensitivity is determined by background fluorescence. This is minimized through the use of pure reagents. In addition, errors can result from nonuniform contamination of samples and glassware by a variety of fluorescent materials, including rubber, grease, fingerprints, and extracts of filter paper or dialysis tubing. Turbidity contributes to error through scattering, which is measured as fluorescence. Thus, maintenance of scrupulously clean glassware for measurement and storage of solutions is necessary. Many of the above effects can be minimized by the use of blocking filters at the photomultiplier.

Deterioration and other artifacts.

Samples may deteriorate before or during measurement. Photochemical decomposition and adsorption to glass surfaces, especially for macromolecules, are quite common. The former can be detected as a loss of fluorescence with time, and the latter can be suspected if there is an apparent lag in fluorescence with increased concentration of sample or fluorophore.

Effect of temperature.

Since fluorescent phenomena are temperature dependent, all measurements should be made at constant temperature, preferably in a thermostated cell.

16.2. ION ELECTRODES

The Nernst equation

$$E = E_0 + \frac{RT}{nF} \log \frac{[ox]}{[red]}$$

can be used to measure the concentration (activity), or the ratio of concentrations, of certain molecules or ions. E is the measured potential, E_0 is the standard electrode potential, R is 8.315 J/degree, T is the absolute temperature, n is the number of electrons, F is 96,500 C, and [ox] and [red] are the concentrations of oxidized and reduced members of the half-cell pair, respectively.

This basic behavior has made possible the determination of a number of biologically important ions, e.g., hydrogen, oxygen, ammonia, chloride, and many other ions. Incorporation of these principles into a usable instrument involves: (i) availability of a measurement electrode capable of sensing the potential generated by the oxidant-reductant system, (ii) a standard half-cell of known potential, and (iii) an electrometer or potentiometer which measures the voltage between the two half-cells. Solution of the Nernst equation for the measuring half-cell results from the relationship:

$$E_{\text{(measuring half-cell)}} = E + E_0$$

The ability to measure a specific ion is dependent upon the construction of the measuring electrode. For instance, a glass electrode develops a potential proportional to hydrogen ion concentration, and several ion-selective electrodes develop potentials for specific ions because the membrane or solid material isolating the metallic electrode from the medium is ion selective. There are, in addition, many other strategies for determination of concentration of various ions based upon electrode behavior. The oxygen electrode, for instance, is a form of polarography because there must be a polarizing voltage to ionize the oxygen.

16.2.1. Hydrogen Ion Electrode

The glass electrode is the only practical means of measuring hydrogen ion activity. A thin membrane of special glass isolates a silver-silver chloride half-cell from the measured solution (11). The electrical potential of the glass electrode varies linearly with pH over a wide range, pH 1 to 11. In the region of pH 11 and above, there is a nonlinearity unless a special alkali-resistant glass is used. The reference is a standard calomel (Hg_2Cl_2) half-cell which is connected to the measured solution by a salt bridge of KCl. Both the glass electrode and the standard electrode are immersed in the sample, or they are assembled into a single, combination electrode. The latter can be constructed to enter long test tubes or to accommodate very small samples.

In addition to static determination of pH of samples, the pH meter or its automated version (the pH-stat) can be used to follow reactions with time in which a hydrogen ion is evolved or consumed. In this application alkali or acid is added to maintain constant pH; a plot of acid or alkali consumption versus time expresses the reaction progress. For instance, the oxidation of glucose and glucose-6-phosphate either enzymatically or chemically, the phosphorylation of substrates with adenosine triphosphate by kinases, and the hydrolysis of proteins by proteases are a few examples in which the reaction can be followed with the glass electrode.

Calibration of the pH meter.

The system must be calibrated by using buffers of known pH. A popular but expensive approach to calibration is to buy prepackaged liquid or solid buffers of specified pH, but this is by no means necessary or superior. Table 1 gives some standard buffers (10) which have dependable pH values. With linear electrode behavior, a single standard buffer should serve to calibrate across the useful range. However, electrode behavior is seldom ideal; hence, it is wise to use

TABLE 1. *pH standard buffers at 20°C (3)*

pH	Buffer
1.10	0.1 M HCl
2.16	0.01 M potassium tetroxalate
4.00	0.05 M potassium hydrogen phthalate
6.88	0.025 M KH_2PO_4 + 0.025 M Na_2HPO_4
9.22	0.01 M borax
10.02	0.025 M $NaHCO_3$ + 0.025 M Na_2CO_3

two standard buffers (for instance, pH 4 and pH 6.88) and to bracket the pH region of the unknown.

Problems.

The major problems in measurement of pH come from nonlinearity in the alkaline region, absorption of CO_2 in neutral and alkaline samples and in standards, and slow evaporation of standards. Most of the problems of unsatisfactory instrument behavior come from poor maintenance of electrodes. The glass electrode needs an equilibrium gel layer for accurate and constant pH measurement. Electrode manufacturers recommend that a new or cleaned electrode be soaked in water, or in buffer followed by water, for up to several days to establish the gel layer; they also recommend that the electrodes be maintained in distilled water between measurements. Electrodes must be cleaned after use by washing and wiping to remove adhering materials.

Unfortunately, incomplete cleaning plus storage in water can result in a buildup of contaminants and fouling of the glass surface. Lipid and protein contribute to the nonlinearity of electrode behavior and a slow response. In addition, bacterial growth may occur with both fouling of the glass surface and plugging of the asbestos fiber which functions as a salt bridge between the standard electrode and the solution being measured. The KCl level in the calomel electrode should be maintained, and the plug should be removed during measurement so that KCl drainage through the fiber will prevent impurities from lodging in the fiber.

Once the glass electrode is fouled, the temptation is to replace it with a new one. However, effective cleaning can be accomplished by soaking alternately in 0.1 N HCl and 0.1 N NaOH at 50°C for several cycles, followed by reestablishment of the gel layer by soaking. In difficult cases, immersion in 10% HF for a few seconds followed by 5 N HCl and then thorough soaking in distilled water is recommended.

Effect of salt.

High ionic strength of the solution being mea-

sured results in errors in pH measurement by as much as 0.5 pH unit. This problem is commonly encountered in preparing high-molarity ammonium sulfate solutions in the neutral and alkaline pH region. It is common practice to dilute such samples 1:20 before pH measurement. Dilution of solutions of completely dissociated salts has little effect on pH because the ratio of ions remains nearly constant. The magnitude of the salt effect for specific circumstances is ascertained by making up standards in the salt solutions involved.

16.2.2. Other Ion-Specific Electrodes

Ion-specific electrodes are available for electrometric determination of a large number of ions and gases (8). While most of these have applications in industrial processes, several determine biologically important ions with sufficient sensitivity and selectivity to be of use in bacteriological investigations. Ion-specific membrane probes fall into four categories: glass, solid or precipitate, gas-sensing, and liquid electrodes (Table 2). Electrode response is, broadly speaking, the result of an ion-exchange process, and potentials follow the Nernst equation or one of its expanded forms.

The **ammonia electrode** uses a gas-permeable membrane (14) through which ammonia diffuses to react with the filling solution:

$$NH_3 + H_2O = NH_4^+ + OH^-$$

The hydroxide level of the filling solution is then measured against a reference electrode. The electrode is specific for ammonia; ammonium ion is also measured after addition of NaOH to pH 11. Only volatile amines interfere, and the range for the electrode of the Orion Co. is 17 μg to 17 mg/ml with an accuracy of 2% or better. Nitrate and nitrite can also be measured after reduction, and Kjeldahl nitrogen can be measured after digestion and addition of NaOH (20.2.7). Urea N can also be measured in a special adaptation of the ammonia electrode in which a second membrane encloses a urease-containing solution. Alternatively, the samples may be pretreated with urease and measurement may be made with the normal ammonia electrode (14).

TABLE 2. *Some ion-selective electrodes with potential use in bacteriology (8)*

Glass	Solid	Gas sensing	Liquid membrane
H^+, monovalent cations	Halides, S^-, SO_4^{2-}	CO_2, NH_3, SO_2, NO_2, H_2S, HCN	Br^-, I^-, NO_3^-, CO_3^{2-}, SO_4^{2-}, K^+, Ca^{2+}, PO_4^{2-}, organics

Ion-specific electrodes have been adapted to measure activities of enzymes and concentrations of compounds which are substrates for certain enzymes (15). For instance, the Hg_2S/HgI crystal membrane electrode which is designed to monitor iodide ion can be used for the determination of both glucose oxidase and glucose according to the equations:

$$glucose + H_2O + O_2 \xrightarrow{\text{glucose oxidase}}$$

$$\rightarrow gluconic\ acid + H_2O_2$$

$$H_2O_2 + 2I^- + 2H^+ \xrightarrow{Mo(IV)} I_2 + 2H_2O$$

In addition, a flow-through iodide-sensing electrode has been developed so that up to 70 samples per hour can be analyzed with a commercial autoanalyzer. With excess glucose oxidase and limiting glucose in the sample, the potentiometric response measures glucose in the sample. With excess glucose and limiting glucose oxidase, the rate of potential change can be used to determine the amount of oxidase present.

Since there is no system for absolute calibration of probes as there is for pH using the hydrogen electrode, it has been the practice to use dilutions of stock solution standards made up of a completely dissociating salt (16).

With respect to sensitivity, measurements can be made over the range of 10^{-5} to 10^{-6} M to 1 M for the favored ion. Seldom is the sensor highly specific. In fact, interferences with other ions are quite common. For instance, a calcium electrode gives 10% response to equivalent amounts of Pb^{2+}, Zn^{2+}, Cu^{2+}, and Fe^{2+} in the 10^{-5} to 10^{-6} M range. Chloride electrodes respond similarly to iodide and bromide, i.e., 10% response at 10^{-5} to 10^{-6} M. On the other hand, the K^+ electrode is quite insensitive to Na^+, Li^+, Cs^+, and H^+. The specific details of sensitivity and selectivity depend on the particular electrode being used. Reference should be made to the manufacturer's literature for detailed information.

16.2.3. Oxygen Polarograph (12)

A platinum electrode at -0.5 to -0.8 V consumes oxygen according to the equation:

$$O_2 + 2H_2O + 4e = 4OH^-$$

The equation shows that the process occurs with the flow of electrons or current toward the electrode. This current flow through a resistor appears as a voltage across the resistor. The "potential" of such a half-cell, when used with a standard Hg_2Cl_2 or Ag-AgCl (silver wire in chloride-containing solution) half-cell, can be mea-sured if the two are connected by a salt bridge. Alternatively, the current may be measured with a galvanometer or voltage can be measured after amplification, and these do not require a reference electrode.

The platinum electrode may be stationary, rotated, or oscillated to minimize diffusion gradients. However, such electrodes tend to be poisoned in biological measurements. A popular solution to this problem is the Clark oxygen electrode (13) in which the platinum electrode is covered with a gas-permeable membrane. Whereas the platinum electrode measures the number of oxygen molecules present at its surface as governed mostly by collision theory, the Clark electrode response is proportional to the rate of diffusion across the membrane. In this arrangement a gradient across the membrane results from the zero oxygen concentration at the electrode surface where oxygen consumption occurs. Fortunately, the rate of diffusion is linearly dependent on oxygen concentration or partial pressure. The current output is also dependent on the platinum cathode area.

A plot of current versus polarizing voltage shows a flat or nearly constant response in the region of -0.5 to -0.8 V. Thus, there should be a polarizing voltage adjusted to the center of the constant current region. The behavior in response to polarizing voltage is the major distinction of the oxygen electrode.

Effects of temperature.

The sensor current is quite sensitive to temperature because the membrane material permeability has a high temperature coefficient (about 3 to 5% per degree). In addition, there is the well-known difference in oxygen solubility with temperature. For this reason, good temperature control and adequate equilibration are required.

Calibration and calculations.

Calibration is based upon the known oxygen solubility of a solution under defined conditions. Zero oxygen level or "bleed current" can be defined as the response in the presence of added sodium dithionite. Full response is set at 100% using an oxygen-saturated solution with the oxygen concentration calculated from the Bunsen coefficient, which is obtained from the *International Critical Tables* or *Handbook of Chemistry and Physics*. The oxygen content of solutions to be used as working standards can then be measured. For most work the actual oxygen concentration of standards is not needed; only a reproducible concentration to set the instrument is necessary, the measurements needed being differences between experimental samples. For example, if Ringer's solution contains 5 μl of O_2/ml when air saturated, 3 ml of solution giving an instrument response of 92% saturation contains 13.8 μl of O_2 (3 × 5.0 × 0.92). Rates of oxygen consumption or evolution are either calculated by taking measurements at two times or obtained from the slope of plots of percent full scale versus time involving several readings. For example, with an instrument response at 0 min and 5 min of 92% and 70%, the calculation is as follows: 92 − 70 = 22% consumed/5 min and 5.0 × 3 = 15 μl of O_2 × 0.22 = 3.3 μl/5 min = the reaction rate.

Unfortunately, it is unlikely that reaction mixtures under study have the same composition as the oxygen standard, and hence there will be some error. However, the error can be minimized by maintaining a low total solute concentration. When solutions of greater oxygen concentration are needed, increased partial pressures can be generated by saturation of standards and working solutions with oxygen or oxygen-N_2 mixtures.

The oxygen content of standards and reagents is influenced by temperature and altitude. Barometric pressure fluctuations seldom change oxygen concentration by more than ±3%. Correction of the Bunsen coefficient for barometric change involves the relationship:

$$\frac{\text{observed barometric pressure}}{760}$$

× solubility at 760 mmHg (ca. 101.3 kPa)

Operating information.

At the beginning of an experiment, the probe should be tested for (i) plateau setting of polarizing voltage, (ii) response time for a step change in polarizing voltage, (iii) noise, and (iv) drift. The lack of horizontal plateau indicates a dirty electrode; however, some slope can be tolerated in many instances. Response time for a step change in polarizing voltage should be 1 to 1.5 min. Noise may result from poor connections and grounding, a damaged membrane (folds, holes, KCl crystals, or drying), or from poor contact with the silver electrode, in which case cleaning with ammonia is recommended.

A properly functioning probe should have less than 2% drift per h. Drift may be caused by atmospheric changes or by contamination with organic material. Thus, proper care and cleaning of the probe are essential. In one report reproducible results were obtained when the probe and receptacle were washed with reagents used in the test.

Assembly of electrode membrane.

The demanding aspect of the Clark electrode is attaching the membrane. Directions and spare membranes are furnished by manufacturers of the probe. Care must be taken in assembly to ensure that the membrane is smoothly stretched over the probe, that there is a true seal at all points, and that no air bubbles are trapped on the electrode side of the membrane. To obtain good wetting in assembly, one manufacturer recommends addition of 3 to 4 drops of Kodak Photoflow per eyedropper bottle of KCl used in assembly.

16.3. CHROMATOGRAPHY

Chromatography and related separation techniques utilize many kinds of distribution processes which take place repeatedly in a small local environment (17, 21, 23). Thus, the operations of ion-exchange, solvent partition, adsorption-desorption, and affinity (association-dissociation) chromatography and the operationally related gel permeation techniques can be used in similar laboratory equipment to carry out separations of molecules based on widely different properties.

16.3.1. Ion-Exchange Chromatography

Sorption of solutes is based on ionic charge, although other types of bonding may have some influence on separation. Charged groups in a permeable matrix are able to exchange sorbed ions with those in solution

$$-(X) - SO_3^-Na^+ + K^+Cl^-$$
$$\rightarrow -(X) - SO_3^-K^+ + Na^+Cl^-$$

Matrix materials include cross-linked, substituted, or derivatized resins such as polystyrene or polymethacrylate as well as derivatized carbohydrate-containing natural polymers such as cellulose, dextran, or agarose.

Exchangers are available with a great variety of functional groups, support materials, "purity," porosity, capacity, size, shape, selectivity, and stability. It is therefore necessary to consult technical literature from manufacturers for specific information needed, for instance, Dow Chemical Co. for Dowex resins, Rohm and Haas Co. for Amberlites, Bio-Rad Laboratories for a wide variety of exchangers, Pharmacia Fine Chemicals for Sephadexes, and Reeve Angel Co. and Whatman Co. for cellulosic exchangers.

Resin exchangers.

Many synthetic resins are available with different degrees of cross-linking. Lower cross-linking results in higher permeability and water uptake and an ability to interact with larger molecules. Ion-exchange wet volume capacity and selectivity increase with higher cross-linking and also increase as the particle size decreases. Thus, fine-mesh exchangers have increased resolution per bed length simply because of the increased number of exchange cycles involved. However, increased packing of small particles decreases the flow rate, which in turn requires increased operating pressures.

Sorption, purification, and fractionation are major applications of exchangers. In sorption, a desired ion is selectively removed from extraneous material by ion exchange. In purification, the ion desired is selectively sorbed and eluted while other ions are not sorbed or are retained. Fractionation is the process of separating several similarly charged ions from one another by introduction of solvent-solute systems which progressively elute sorbed ions of increasing charge.

As a rule of thumb, resolution is related to bed length, whereas for a given resolution capacity is related to bed cross section. Since binding of soluble ions is stoichiometric and reversible, the amount of resin needed can be calculated from its published capacity. In general, strength of binding of inorganic ions is directly related to valency (number of charges) and inversely related to atomic number. For weak acids and bases, strength of binding is related to pK of the ionizing group if the pH of operation is in the dissociation range.

Exchanging counterions and deionization.

Since ion exchangers are charged with a counterion (for instance, for Dowex-1, this might be OH^-, Cl^-, formate, etc.), it is possible to exchange virtually completely an ion in the solvent with the charged counterion. For instance, chloride can be exchanged for OH^- with Dowex-1 and H^+ can be exchanged for Na^+ with Dowex-50. Thus, a major application of exchangers is the deionization of uncharged solutes. This can be accomplished by serial passages through cationic and anionic exchangers or through a mixed bed. The latter, however, is usually difficult to regenerate and is discarded. In deionization of carbohydrates, the noncharged sugars may behave as weak acids and be sorbed by a strong anionic exchanger; for this purpose, a weak basic exchanger should be used.

In exchanging ions there is a hierarchy of affinities of ions. For this reason it is difficult or inefficient to substitute directly a low-affinity ion for one of high affinity. For example, it is difficult to convert Dowex-1–Cl to Dowex-1–for-

mate simply by washing with formic acid or sodium formate. The more efficient route involves conversion of chloride to hydroxide form and from hydroxide to formate. The order of affinities can be obtained from the manufacturer's literature or from reference books.

Concentration.

As a consequence of the high affinity for ions plus the fact that fresh resin is encountered with movement of the sample through the column, it is possible to bind ions from a large volume of dilute solution. The ions sorbed can then be recovered in a small volume of eluant containing an ion of higher affinity or of higher concentration. In this way, any number of common and trace inorganic elements as well as radionuclides and trace organic compounds have been concentrated for analysis.

Purification and fractionation.

If two ions have different affinities for the exchanger at a given pH, separation is made by finding conditions which elute one ion but not the other. In such extreme cases there is a "frontal" elution of each ion. In cases where the affinity is similar, separation is still possible by finding conditions which selectively elute the ions. Development of the column with eluants then leads to a gradual separation. If the column length is sufficient, separation can be attained. However, diffusion of ions limits the ability of increased column length to effect resolution of components. Changing pH or ionic strength elutes sorbed ions of different pK values or of increasing affinities.

Gradient elution procedure.

Elution of sorbed molecules by continuous change of eluant composition is a useful way to maximize the separation potential of a chromatographic system. Rather than making step changes in concentration by application of different eluants, the change is continuous within given limits. Devices and procedures range from the very complicated, as in the nine-chamber elution for separation of amino acids on Dowex-50 (Na^+), to the relatively simple linear gradient involving mixing of two components. In recent years the simple linear gradient has been used most frequently. For this purpose two reservoirs (46) are connected so that the outlet of one is fed into the second reservoir (Fig. 5). Another outlet of the second reservoir feeds the column. A mixer in the second vessel keeps the eluant composition homogeneous. When both vessels are open to the atmosphere and each vessel has the same cross-sectional area, the eluant com-

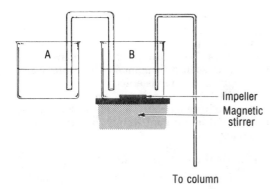

Impeller
Magnetic
stirrer

To column

FIG. 5. *Gradient maker which gives linear gradients. The surface areas of beakers A and B are equal.*

position will change linearly from the limit in the second (mixing) reservoir to the limit in the first. The steepness of the gradient is determined by the volume in the two reservoirs.

Suppose that it is desired to construct a gradient such that twenty-five 10-ml fractions can be collected in which the salt concentration increases linearly from 0.01 M potassium phosphate buffer, pH 7, to 0.01 M phosphate, pH 7, + 0.1 M KCl. The total volume needed for the gradient is 250 ml. Now arrange two 250-ml beakers level so that they can be connected with a loop of glass tubing (Fig. 5) or connect beakers modified to have outlets at the bottom, one for the first beaker and two for the second (mixing) beaker. Place the second beaker on a magnetic stirrer and connect the second outlet to the column. Add 125 ml of phosphate buffer + KCl to the first beaker and about 125 ml (should be equal weight) of phosphate buffer without KCl to the second. Independently run some 0.01 M phosphate buffer down the column to equilibrate the system with the starting eluant. Then start the eluant from the second beaker. The elution continues with stirring until the contents of both beakers have been transferred to the column. Independent determinations of chloride in the fractions will verify the linearity of the gradient. It is also necessary to prevent intermixing of the two reservoirs before elution is started. Therefore, pinch clamps at the inlet and outlet to the second beaker should be used.

When the ratio of cross section of beaker 1 to beaker 2 is greater than 1, the gradient will be convex, and when less than 1 it will be concave, as dictated by the equation

$$C_2 = C_1 - D(1 - v)^{A_1/A_2}$$

where C_2 is the concentration in the second vessel (lower limit) and C_1 is the concentration

in the first vessel (upper limit), D is the difference in concentrations, v is the fraction of initial volume withdrawn up to the point in question, and A_1 and A_2 are the cross-sectional areas of the vessel containing the higher concentration and the mixing vessel, respectively.

It should be noted that a common method of making gradients, which involves dropping eluant of higher concentration from a separatory funnel into a mixing reservoir of constant volume, produces convex gradients which are less useful.

More complex gradients can be constructed by placing more beakers in series with mixing in all but the farthest removed. The eluant in each beaker can be different to effect a compound gradient. A nine-chamber gradient device has been developed for this purpose (48).

Nucleic acids.

Nucleotides, nucleosides, and purine and pyrimidine bases are separated on anionic and cationic exchangers of low cross-linkage. The methods involve control of ionic strength, pH, and nonionic binding. With Dowex-50W-X4 (Dow Chemical Co.) or AG 50W-X4 (Bio-Rad Laboratories), samples in 0.1 to 1 ml of elution buffer, and a flow rate of 1 ml/min, 5'-nucleotides are separated on a 40-cm column with 0.1 M ammonium formate adjusted to pH 3.2 with formic acid. For 2'- and 3'-nucleotides, use a column length of 85 cm and elution buffer of 0.25 M ammonium formate adjusted to pH 4.1 with formic acid (33). Another system utilizing a salt gradient and ethanol to control nonionic binding separates ribo- and deoxyribooligonucleotides by base composition, chain length, and 3'- or 5'-terminal phosphate group (31). One system utilizes AG 1-X2 (−400 mesh, Bio-Rad Laboratories) in a 60 by 0.3 cm column. Elution is at 12 ml/h with 400 ml of 20% ethanol containing a linear gradient of 0 to 1.0 M NH₄Cl, pH 10. A second system with Dowex-1-X4 (400 mesh) in a 100 by 0.2 cm column involves a program of sample addition and elution as follows: 0.4 ml of 1 M NaOH, 1 ml of 20% ethanol, 1 ml of water and sample, and 1 ml of 20% ethanol followed by 200 ml of ethanol containing a linear gradient of 0 to 0.5 M NH₄Cl (pH 10); the flow rate is 6 to 10 ml/h.

For cyclic nucleotides in a homogenized, deproteinized, and neutralized sample, separation involves a 0.5 by 4 cm column of AG 1-X8–formate (200 to 400 mesh) equilibrated with 0.1 N formic acid. After sample addition, the elution involves 10 ml each of water, 2 M formic acid, and 4 M formic acid followed by 10 ml each of

8 M formic acid containing 20 mM, 100 mM, and 1 M sodium formate (42).

Citric acid cycle intermediates (44; see also 16.3.4 and 18.5.2).

The extract is adjusted to pH 6, and 0.5 ml is added to a 0.9 by 14 cm column of AG 1-X10–formate. Elution requires 150 ml of a gradient formed by adding 3 M formic acid to a 250-ml closed mixing flask filled with water. This is followed by 2 N ammonium formate; the flow rate is 2 ml/min. β-Hydroxybutyrate, acetoacetate, succinate, malate, pyruvate, citrate-isocitrate, α-ketoglutarate, and fumarate elute as separate peaks in that order.

Cellulose, dextran, and agarose exchangers (46).

For separation of macromolecules, modified natural polymers have been prepared which have capacities and other properties generally more suitable than resins. With dextrans and agarose, cross-linking and particle size distribution can be controlled, thereby controlling shrinkage and swelling, and providing high capacity for macromolecules at high and uniform flow rates.

Since proteins are amphoteric and have a large number of charges, often in clusters, it is possible to treat them as cations or anions by choosing the pH for ion-exchange separations one or more units above or below the isoionic point or the pI value. Cycles of binding and elution make use of a pH change which can cause disappearance of ionic groups, i.e., carboxyl or amino groups on the protein or on the exchanger. Alternatively, elution is effected by displacement with another ion.

The diethylaminoethyl derivative of cellulose, dextran, or agarose is a weak anion exchanger (pK of about 9.5) (Table 3). Hence, in the neutral pH region the amino group is protonated and binds proteins and peptides if the pH is near or above their pI values.

TABLE 3. *Derivatives of cellulose and dextran used in ion-exchange chromatography*

Derivative	Symbol	pK
Sulfopropyl	SP	Strong acid
Phosphocellulose	P	1.5, 6
Carboxymethyl	CM	4
Mixed amine	ECTEOLA	7.5
Benzoylated diethyl-aminoethyl	BD	9
Diethylaminoethyl	DEAE	9.5
Triethylaminoethyl	TEAE	9.5
Diethyl-2 hydroxypropyl aminoethyl	QAE	Strong base

The carboxymethyl derivative of the same supports is a cationic exchanger with a pK of about 3.5. Separations are generally carried out above pH 4. When the pH of the separation is below the isoelectric point of the protein and above pH 4, sorption is efficient.

ECTEOLA-cellulose and a similar derivative of dextran have a pK of 7.5. This exchanger has been useful in chromatography of nucleic acids and nucleoproteins.

Practical considerations.

Successful use of exchangers rests on the judicious choice of a number of parameters including exchange type, buffer type, pH of operation, ionic strength, and type of matrix as well as its particle size, shape, and porosity, column dimensions, and elution program. In addition, attention must be paid to exchanger packing, sample addition, maintaining low sample viscosity, and column design since these also determine the degree of resolution. Treatment of these aspects in adequate detail is beyond the space available here, and hence only general guides for each will be given. Reference 33 and the publications by Whatman, Inc., for exchanger celluloses and by Pharmacia, Inc., for the corresponding Sephadexes are recommended for the details.

In general the uniform spherical dextran beads (Sephadex) give faster flow rates although the cellulose derivatives now available also perform well. Smaller bead or particle sizes give slower flow rates and increased resolution. Increased porosity increases the capacity for a given molecule and is particularly important for work with macromolecules. The choice of exchanger is based on the ionic forms to be separated as described above. The pH chosen should be at least 1 pH unit removed from the isoelectric point or the pK of the dissociating group. When possible, cationic buffers [alkyl amines, ammonia, barbital, imidazole, and tris(hydroxymethyl)aminomethane (Tris)] should be used with anionic exchangers; anionic buffers (acetate, barbital, citrate, glycine, and phosphate) are used with cationic exchangers. The recommended pH range for diethylaminoethyl-Sephadex is 2 to 9 and for carboxymethyl-Sephadex is 6 to 10. Ionic strength influences the binding and therefore the capacity for ions. A relatively high ionic strength of ca. 0.1 M is recommended, but the concentration should be somewhat lower than is needed to elute the ions desired.

Column height determines the resolution achieved. When the zone of absorbed ions is 1 to 2 cm, a 20-cm bed height is recommended as a starting point. At fixed height, the column capacity is linearly related to cross-sectional area.

Elution is effected either by changing the pH to cause the ionic charge to diminish or disappear or by increasing the ionic strength to the point where the buffer ion displaces the ion of interest. These can be accomplished by either stepwise or gradient change. For anion exchangers the pH is decreased or the ionic strength is increased; for cation exchangers the elution is effected by increasing the pH or the ionic strength.

Either the sample total ionic content or column size must be determined first, because available capacity is the property of a specified volume of exchanger for each exchanger type. The ionic composition should be that of the starting buffer. When necessary, the ionic composition is changed to that desired by gel filtration on Sephadex G-25, dialysis, or dilution. Sample volume is of minor importance for all elution programs except when using only the starting buffer for development. Sample must be added without disturbing the bed and without becoming distributed in the oncoming developing buffer.

As with other types of chromatography, it is essential to use glassware designed for ease of packing, addition of eluant and sample, application of pressure if needed, regulation of flow rate, prevention of running dry, and low holdup volume above and below the exchanger bed. It is also necessary to establish criteria for achieving the objective in mind and to analyze the effluent to determine performance of the column.

16.3.2. Adsorption Chromatography

Adsorption involves van der Waals interactions as well as electrostatic, hydrophobic, and steric factors. Adsorption of a single component is typically described by Langmuir's adsorption isotherm, a rectangular hyperbola in a plot of solute concentration versus amount adsorbed. The same curve describes the saturation behavior of substrate for an enzyme and the "law of diminishing returns" among others. Surface sorption and elution of nonpolar materials have been successful with silica gel, kieselguhr, alumina, Florisil, and charcoal. Polar macromolecules such as proteins and nucleic acids have been purified by either batch treatment or column chromatography using hydroxyapatite, calcium phosphate gel, or alumina Cγ.

Types of adsorbent.

Silicic acid has been used for purification of simple and complex lipids, fatty acids, sterols, phenols, and hydrocarbons. Elution is effected by changing the solvent composition, usually from partially nonpolar toward polar character by use of a concentration gradient. Neutral alumina (pH 6.9 to 7.1) separates steroids, alkaloids, hydrocarbons, esters, organic acids and bases, and other organics in nonaqueous solution. Basic alumina (pH 10 to 10.5) has similar properties in adsorbing compounds from organic solvents. However, with aqueous media it acts as a strong cation exchanger sorbing basic substances such as amino acids and amines. Acid-washed alumina (pH 3.5 to 4.5) acts as an anion exchanger in aqueous medium.

Adsorbents for purification of macromolecules in aqueous solution are alumina Cγ, calcium phosphate gel, and hydroxyapatite. These are kept as concentrated suspensions which do not lend themselves well to column chromatography, although column methods for hydroxyapatite have been described (45).

Separation of proteins.

Batch separations are made by proper manipulation of adsorption and elution phases. The conditions for their use must be established by pilot tests with each batch of adsorbent and usually with each preparation of protein (enzyme or antibody) to be separated. Figure 6 represents the way that adsorption and elution cycles are utilized; these kinds of data are derived from pilot tests. In Fig. 6A are depicted three kinds of adsorption behavior influenced by the nature of the material being purified, the contaminants, and other conditions, i.e., pH and ionic strength. The protein of interest may be adsorbed either better than (left), the same as (center), or poorer than (right) the contaminating proteins. If better, an amount of adsorbent is added to remove most of the desired material from solution while leaving most of the other proteins behind (positive adsorption). If poorer, adsorbent is added and removed until the desired protein is slightly (i.e., 10%) adsorbed. In this way contaminating proteins are removed first (negative adsorption), after which the desired and partially purified material may now be adsorbed, or it can be subjected to another purification procedure. If both contaminant and desired proteins are adsorbed at equal rates, adsorption can provide little purification. Nevertheless, the protein being purified may be adsorbed and subjected to a useful elution cycle.

Similarly, Fig. 6B shows three types of behavior upon elution. The desired material may elute before (left), with (center), or after (right) the protein impurities. If the target material elutes preferentially, the adsorbent should be treated with only enough eluant or eluant cycles to

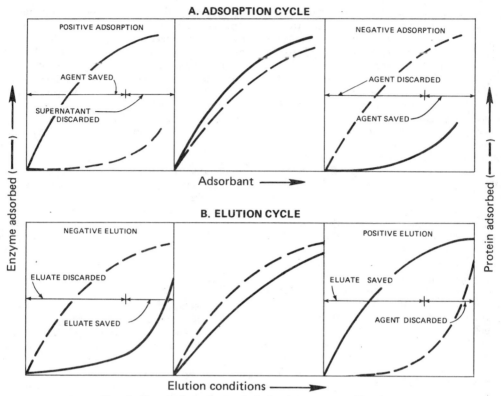

FIG. 6. *Use of adsorption and elution in enzyme purification.*

remove most of the adsorbed material, leaving behind the impurities. If the other proteins are eluted preferentially, the adsorbent should be treated with eluant with increasing ionic strength with or without pH change until significant amounts of desired material appear in the eluate. Then elution conditions are changed just enough to elute the desired material. It is obvious that if there is no differential behavior between the desired material and impurity on either adsorption or elution cycles this procedure should be discarded or new conditions should be found.

There is a great deal of empiricism in this kind of work, but construction of a diagram from small pilot runs showing the behavior of desired material versus impurities greatly aids in designing the total procedure. In general, increasing ionic strength and changing the pH toward the isoelectric point of a protein enhance elution. Further, substrates, effectors, and other ligands for enzymes and proteins often act as specific eluting agents.

Separation of lipids. See part 17.3.3.

Separation of phospholipids by thin-layer chromatography. See part 17.3.4.

16.3.3. Liquid-Liquid Partition Chromatography

When a two-phase solvent system is set up, a large number of solutes, partially soluble in each, reach an equilibrium distribution between the two phases which reflects the solubility in each phase. This has led to useful solvent extraction procedures, both batch and continuous. The Craig and Podbilniak countercurrent extractors are examples of these techniques at their best. However, the ability to separate similar materials is limited by the number of distribution cycles one can make. Fortunately, when one phase of a two-phase system is immobilized as a coating on a small particle, the distribution process can occur repeatedly in a column or paper chromatographic arrangement. Thus, paper and cellulose chromatography separate by partition between an aqueous phase coating the cellulose fiber and a nonpolar phase which flows through the interstitial spaces. Silicic acid or Celite is another support which has been widely used in partition chromatography, both in column and in thin layers. See part 18.5.2 for separation by silicic acid partition columns and part 17.3.4 for separation by thin-layer solvent partition. Gas chromatography (16.3.4) and some forms of high-pressure liquid chromatography (16.3.5) also separate by solute partition.

16.3.4. Gas-Liquid Partition Chromatography

A highly versatile tool for separation, identification, and quantitation of biologically important compounds is based on the partition of volatile solutes between a flowing carrier gas phase and a liquid phase coated on inert particles. In principle, any compound which is volatile, or can be made volatile at a temperature up to 375°C, can be analyzed by this method. For this reason, gas-liquid chromatography (GLC) has found wide use in bacteriology, particularly in analysis of cell constituents, as an aid to classification, and as a means of measuring metabolic products. The method has high sensitivity and in many cases accuracy and repeatability. The resolution, especially for capillary columns, is exceptional. However, relatively sophisticated equipment and some experience are required to achieve the potential of the method.

This type of partition chromatography has been subjected to considerable theoretical analysis which allows predictions of conditions needed to make separations and quantitate components. There are a large number of variables to be manipulated, including kinds and size of supports, kinds and composition of liquid phases, column dimensions and arrangement, choice of carrier gas and its flow rate, column temperature and its programming, and detector sensitivity and selectivity. While these have afforded flexibility and adaptability to studying problems in bacteriology, adequate treatment of these parameters is beyond the scope of this manual. There is, however, a general reference by Mitruka (26) which treats in detail the use of GLC in bacteriology. The following sections are intended only to give a general view of the method and present two protocols which are applicable to classification and analysis of products of metabolism.

Principle.

A gas containing a volatile solute is forced through a long tube packed with particles of inert material coated with a polar or nonpolar material which is liquefied at operating temperatures. The solute distributes between the gas and liquid phases. Although equilibrium between the phases is not reached, partition is related to relative solubilities. A band of solute moves through the column with a velocity reflecting its solubility in the liquid phase; solutes of lower solubility go at faster rates and vice

versa. The bands exit from the column distributed as roughly symmetrical peaks (depending on conditions) in order of their partition coefficients.

Instrument.

Although there is great variability in sophistication of instruments available, satisfactory performance is obtained in a unit which provides a heated inlet, a temperature-controlled oven for one or two columns with provision for controlled increase in temperature, a heated detector, carrier gas supply with flow regulator, and strip chart recorder. Cooling provision for the oven is also convenient in reducing the time between runs.

Among several types of detector, two have proved very useful in microbiological investigations. The thermal conductivity detector, which is especially useful in the analysis of gases such as those produced by bacteria, consists of an electrically heated filament whose resistance varies essentially linearly with temperature. Heat transfer to both carrier gas and to gas peaks imposed on the carrier are detected through a Wheatstone bridge. The sensitivity is moderate, and linearity covers a 10^4-fold concentration range.

The flame ionization detector consists of a hydrogen-oxygen flame into which the effluent gas is injected and organic samples are burned. Ions and electrons so formed pass through a charged electrode gap, and this results in an increase in current across the gap. The current is detected by an electrometer. This detector is insensitive to fixed gases and water. It is highly sensitive to mass flow rate (1 to 10 pg/s); the amount detectable is obtained by multiplying the sensitivity by the peak width.

Several other detectors are available, some specific for one atom and others responding to one class of molecules but not others. The electron capture detector, for instance, is extremely sensitive in detecting structures which capture slow electrons such as alkyl halides, conjugated carboyls, nitrates, etc., but does not respond to hydrocarbons, alcohols, ketones, etc., which do not capture electrons (25).

Sample preparation and derivatization.

Analysis of biological fluids, cell extracts, and other complex materials by GLC may fail as a result of the presence of highly polar, highly reactive components or the number of interfering substances. Further, such impure samples tend to prematurely foul the liquid phase. For this reason preliminary separations are usually made to remove organisms, culture medium components, proteins, gums, salts, etc. In some instances distilled or extracted samples may be applied directly to the columns; however, GLC analysis is not compatible with aqueous samples. In other situations drying, hydrolysis, or quantitative chemical processing to form volatile derivatives is required. These procedures often introduce additional errors in quantitation.

For samples of insufficient volatility in the temperature operating range, derivatization is necessary. This includes conversion to trimethylsilyl ethers of sugars, methyl glycosides (obtained by methanolysis) of polysaccharides and sugar-containing heteropolymers, methyl esters of fatty acids (except short-chain-length acids), and methyl esters plus N-acyl forms of amino acids and peptides, among many others. Specific procedures should be sought in the general references and elsewhere when undertaking an analysis requiring derivatization.

Factors affecting separation.

Column. Separation time increases as the square root of the packing length and elution time linearly with length, whereas capacity is proportional to bed cross section. For packed columns, length is limited by the maximum operating pressure and by oven capacity. Capillary columns with a layer of coating on the inside wall are less restricted and can be much longer. For most purposes 2 to 3 m by 2.5 mm (inner diameter) columns of glass or stainless steel, U-shaped or coiled, are satisfactory.

Carrier gas. Hydrogen and helium are used for rapid analysis at high flow rates when diffusion of sample components is rapid. N_2 is used to approach phase equilibrium by using lower flow rates. The higher density minimizes diffusion of solutes in the gas phase.

Temperature. Resolution is enhanced by lower and speed of analysis by higher temperatures. The temperature chosen must be a compromise between these factors.

Liquid phase and support material.

Although there are no clear-cut rules in liquid phase selection, choice is somewhat guided by solubility behavior, that is, polar solutes in polar solvents and vice versa. Thus, polar solutes elute more rapidly from a nonpolar liquid phase, and for a homologous series elution is in the order of the boiling points. Unless the investigator has special needs which require finding a satisfactory packing, it is more efficient to find a workable system in the literature.

Analysis of bacteria for carbohydrates.

The procedures in use are related to general methods for carbohydrates (49), glycosphingo-

lipids (43), and glycoproteins (47). First, sugars linked by glycosidic bonds are released by methanolysis to yield O-methylglycosides, and any free sugars are also converted to the methylglycosides in the same procedure. Methanolysis of glycosidic bonds proceeds with inversion of configuration; however, only one anomer is formed. With free sugars more than one anomer results as a consequence of the existence of a mixture of anomers in solution.

Incubate harvested and washed cells (i.e., 100 mg, dry weight) in stoppered test tubes with 2 μl of either 4 N methanolic HCl (dry HCl gas in methanol) at 75°C for 4 h or 0.5 N methanolic HCl for 24 h at 95°C. Resuspend the dried hydrolysate in methanol, and dry several times to remove the HCl. Dissolve the residue in 1 to 2 ml of dry pyridine, transfer to a small conical tube, and add 0.2 ml of hexamethyldisilazone and 0.1 ml trimethylchlorosilane. Shake the mixture for 30 s, and let it stand at room temperature overnight. Centrifuge, transfer the supernatant to a clean tube, and dry at 55°C under a stream of N_2. Extract the residue with 10 ml of hexane, and evaporate the extract to 0.25 ml.

Analyze on a gas chromatograph set up as follows: injector at 245°C, detector at 250°C, N_2 carrier gas, either a nonpolar column of 3% OV-1 or Gas-Chrom Q, 60 to 80 mesh, column 1.8 m by 2.5 mm with oven temperature at 120°C for 5 min followed by an increase to 200°C at 5°C per min. Alternatively, use a polar column of 15% Carbowax (20 M), 60 to 80 mesh, on acid-washed Chromosorb W at 120°C.

Alternate procedures (34, 35, 37, 38, 41, 51) have been successfully applied. Also see part 17.2.

Acid intermediates or products of metabolism (also see parts 16.3.1 and 18.5.2).

Acids of the citric acid cycle and other metabolic pathways including lactic, citric, isocitric, malic, succinic, *cis*-aconitic, fumaric, pyruvic, oxalacetic, and glyoxylic acids have been quantitated as methyl, ethyl, propyl, and trimethyl silyl esters. Reference 29 describes special problems encountered with some of these, i.e., unsaturated and keto acids. Although several methods have been described in detail, these have only made use of authentic samples of the acids. Extraction of acids from biological material, however, has proved difficult. The following procedure based on that of Alcock (29) appears to be applicable to quantitation of acids from bacteria and reaction mixtures.

Adjust the acidity of a 25-ml sample to pH 2 with sulfuric acid, add a known quantity of margoric acid (heptadecanoic acid) as internal standard, and dry. Add 5 ml of methanol-boron-trifluoride reagent (Applied Science Laboratories, State College, Pa.), and transfer the mixture quantitatively to a 50-ml centrifuge tube. Centrifuge and wash the residue twice with a small amount of methanol-boron trifluoride reagent, combine supernatants, and let stand overnight. Dilute with an equal volume of water, and extract three times with 2 ml of chloroform. Remove the chloroform layers, combine, and evaporate excess chloroform with dry nitrogen.

Chromatography can be done on 5% diethylene glycol adipate on purified (29) Chromosorb W (30 to 60 mesh; 244 by 0.64 cm, inner diameter) with a helium flow rate of 45 ml/min. Set the column temperature at 88°C for 7 min; then increase to 165°C at the rate of 7.5°C per min. Set the temperature of the detector and injector at 230°C.

With this procedure, difficulties will be encountered with oxalacetic, pyruvic, and α-ketoglutaric acids. However, success has been achieved in making the trimethylsilyloxime derivatives (40).

16.3.5. High-Pressure Liquid Chromatography

Resolution is increased in solid-liquid partition chromatography by decreasing the particle size. This increases the number of separation cycles that can be performed and also increases the equilibrium of solutes with the stationary phase. These gains are accompanied by greatly diminished flow rates which can be restored only by much higher pressures.

High-pressure liquid chromatography (HPLC) is a new and rapidly advancing field with great analytical and preparative capabilities whose components, including the solid-phase packing, are designed for operating pressures in the region of 7 to 550 kg/cm^2 (100 to 3,500 lb/in^2). Separations can be based on adsorption, partition including reversed-phase and ion pair techniques, and ion-exchange principles (see 24, 28).

Equipment.

A simple, relatively inexpensive system can be assembled from a high-pressure liquid pump and pressure gauge, sample addition valve, column, and fraction collector or stream absorbance monitor. Further, a simple gradient maker can feed the pump inlet. Stainless-steel tubing, fittings, and column are used. For some solid-phase materials, packing of columns can be performed in the laboratory. However, it is also advantageous in many instances to buy prepacked columns.

HPLC media.

Two types of HPLC media are available: spherical, porous layer (or coated), glass beads and irregular-shaped porous microparticulates. Porous layer glass beads (37 to 43 μm in diameter) have a thin coating (ca. 1 μm) bonded to the glass. The coatings perform partition, ion exchange, and in some cases adsorption separations. Because of the ease of packing, these spheres are recommended for general or less demanding uses and for packing columns in the laboratory. The porous microparticulates (5 to 20 μm, 10 μm most commonly used) are closely sized silica gel or alumina particles which are used for adsorption separations; silica gel also is a base for other coatings bonded through the silica hydroxyl groups. With much greater surface than beads (ca. 100 m^2/g), the capacity and resolution are much higher in a given column. Coatings include a cyano-amino polar phase, a C$_{18}$ reversed phase, and strong anion and cation exchangers. For glass beads, in addition to the above coatings, there are polyamide and weak anion exchange layers.

Choice of media and mobile phase.

A relatively large number of methods have appeared in the literature, and those applicable to the separations needed should be sought before an attempt is made to manipulate the many variables involved. When an appropriate method is not available, thin-layer chromatography is recommended as an aid to finding a suitable solvent. As a rough guide, water-insoluble solutes can be separated by adsorption with elution in an alcohol-alkane (5:95) or chloroform-heptane (10:90) mixture, or on cyano-amino polar phase and eluted with isopropanol-heptane (2:98). Water-soluble, ionizing solutes, if acidic in solution, are separated on a strong anion exchanger; if basic, they are separated on a strong cation exchanger. Elution may be satisfactory in 0.1 M KH$_2$PO$_4$, pH 3.0, or 0.2 M NH$_4$H$_2$PO$_4$, pH 3.5, respectively. For partition coatings, if the solute elutes with the front, make the eluant more polar; do the reverse if the solute elutes after 30 min. With ion-exchange packings, if the sample does not stick, decrease the ionic strength; do the reverse if the sample elutes late.

Amino acids are separated on porous microparticulate strong cation exchangers; carbohydrates, on cyano-amino polar phase coating; nucleotides and cyclic nucleotides, on a strong anion exchanger; oligonucleotides, on an aliphatic amine exchanger; and water-soluble vitamins, on a strong cation exchanger. These are a few examples among many.

Separation of tryptic peptides (39).

Analytical peptide mapping of tryptic digests of proteins can be carried out in a standard HPLC arrangement with a linear gradient maker. The column is a prepacked C_{18} μBondapak reversed-phase column (10 μm, 4 mm by 30 cm; Waters Associates, Milford, Mass.).

Prepare a tryptic digest of protein (dialyzed against distilled water and lyophilized) at 1 mg/ml in 0.2 N N-ethylmorpholine acetate (pH 8.1) at 37°C with a trypsin to protein ratio of 1:100 for 16 h. Terminate the incubation with acetic acid, and add a sample of the digest directly to the HPLC column. Elute the peptides at room temperature with a linear gradient from 0.1% orthophosphoric acid to acetonitrile at a flow rate of 2 ml/min and a pressure of 500 to 1,000 lb/in². The phosphoric acid is filtered through a 0.5-μm membrane filter (Millipore Corp., Bedford, Mass.); acetonitrile is used as such. Peptides can be monitored at 210 mm as fractions or via a flow detector.

16.3.6. Affinity Chromatography

Affinity chromatography makes use of highly specific binding equilibria found in biological systems to effect separations which usually cannot be accomplished by other means (19, 20, 22). The principle simply involves immobilization of one component of a reversible binding system on a solid support and mixing to separate the other component(s) of the equilibrium from contaminants. Thus, the basis of separation is the specificity of the binding system rather than differences in physical and chemical properties. Affinity chromatography techniques have been highly developed so that they make use of the exceptionally high specificity of antibody-antigen, hormone-binding protein, and enzyme-ligand specificities to effect separation. The major factors involved are the nature and chemistry of the matrix, the chemistry of ligand attachment, the requirement for a spacer arm, and conditions for adsorption and elution of the mobile binding member of the system. General sources of information are found in references 19 and 20 and in manufacturers' literature.

Many types of affinity materials are commercially available, including the matrix or support, either underivatized, with spacer arm terminating in a variety of functional groups, or with the latter bonded to certain widely used ligands such as adenosine 5'-monophosphate. Dextrans such as Sepharose 4B from Pharmacia Fine Chemicals and agaroses such as Bio-Gel A from Bio-Rad Laboratories have the necessary stabilities, flow rates, and pore size to be useful matrices. Chemistries have been developed which afford a wide variety of applications. For Sepharose 4B, coupling to ligands with amino groups is accomplished by activation of the matrix with cyanogen bromide (32). The beads can be prepared and stored or can be purchased in the activated form.

While small ligands can attach to activated Sepharose 4B directly, large ligands such as proteins are often hindered sterically, and reaction is slow and incomplete. When proteins do react, it is likely that several covalent bonds will be formed with the matrix. Spacer arms can be added to CNBr-activated Sepharose, and these can terminate in amino, carboxyl, thiol, and other functional groups. An agarose matrix can be derivatized (36) or purchased with an alkylamino arm which is terminated with succinyl-N-succinimide ester. This group reacts with alkyl and aryl amines.

With CNBr-activated Sepharose the most common arms are produced by reaction with 1,6-diaminohexane and 6-aminohexanoic acid. Ligand bonding then involves coupling using a water-soluble carbodiimide to form amides. For this reaction, 1-ethyl-3-(3-dimethylaminopropyl) carbodiimide hydrochloride or 1-cyclohexyl-3-(2-morpholinoethyl) carbodiimide metho-p-toluene sulfonate give good yields in aqueous and nonaqueous media at pH 4.5 to 6.0.

Use of cyanogen bromide-activated Sepharose.

Activated Sepharose can be purchased as a freeze-dried powder (1 g, ca. 3.5 ml of swollen gel), or it may be prepared by the method of Axen et al. (32). The dry material is washed and reswollen on a sintered-glass filter (G-3) with several wash cycles of 200 ml of 0.1 M HCl, the supernatant being removed by suction. It is then used immediately.

Procedure for bonding the ligand or spacer arm.

Dissolve the protein- or amino group-containing ligand in $NaHCO_3$ (0.1 M, pH 8.3) containing 0.5 M NaCl (for proteins only). Use 5 to 10 mg of protein per ml of gel, and remember that amino buffers cannot be used. Mix the gel and ligand slowly for 2 h at room temperature or overnight at 4°C; do not use a magnetic stirrer. Wash the excess protein away with coupling buffer, and remove the excess coupling groups by treatment with 1 M ethanolamine, pH 9, for 2 h. Other amines, including Tris buffer, can be used. Wash the final product four to five times with high- and low-pH buffers such as 0.1 M acetate, pH 4, and 0.1 M borate, pH 9, containing 1 M NaCl. Changes in pH are usually necessary to remove adsorbed protein. Store suspensions in a refrigerator (without freezing), and add a bacteriostat if long storage is planned. Of course, the above procedure should be modified to meet the stability requirements of any particular protein. As a guide it should be noted that CNBr-Sepharose is stable (not very reactive) at low pH and reactive at high pH.

With aminohexyl-Sepharose or 6-aminohexanoic acid–Sepharose, coupling is by carbodiimide methods. The dried material is weighed out and swollen as above in 0.5 M NaCl (1 g = 4 ml of gel). The ligand to be coupled, protein or small molecule, is dissolved in water or water-solvent, and the pH is adjusted to 4.5. 1-Ethyl-3-(3-dimethylaminopropyl) carbodiimide·HCl in water is added to give 10 mg of swollen gel per ml. Mix the gel and carbodiimide overnight at room temperature; do not use a magnetic stirrer. Remove excess reagents by washing with the same water-solvent mixture to be used in the ensuing ligand coupling step. This conjugate can be stored at 4 to 8°C until used. The ligand to be coupled is dissolved in water or a water-solvent mixture for which the pH of the water phase has been adjusted to 4.5; no buffer is used. The pH, which decreases mostly in the early part of the reaction, is maintained at 4.5 by the addition of NaOH. The liganded gel is poured into a column and thoroughly washed with coupling solution; blocking of excess groups is unnecessary. The gel is then washed in distilled water and finally in the buffer to be used in affinity chromatography. The material can be stored at neutral pH below 8°C, without freezing.

Derivatization of Agarose-Affi Gel 10.

The affinity material Agarose-Affi Gel 10 is available from Bio-Rad Laboratories as a wet slurry (1 g of dry weight/25 ml). The spacer arm and succinimide ester group are attached through a succinylated aminoaklyl arm to a Bio-Gel A matrix. The gel is washed with three bed volumes of water on a filter. The gel is mixed with buffer, i.e., 0.1 M acetate, morpholinepropanesulfonic acid (MOPS), N-tris(hydroxymethyl)methyl-2-aminoethanesulfonic acid (TES), or HCO_3^-, pH 5 to 8.5. Alternatively, ligand in nonaqueous solvent may be added directly to the moist gel pellet or to the gel washed and suspended in solvent.

Use of liganded affinity agents (20).

The amount of material needed is calculated from the furnished or determined capacity, making some assumptions as to the experimental conditions to be used, the binding affinity of the material to be adsorbed, etc. Usually, excess agent is used and the adsorbed material forms a band at the top of the bed; thus, small columns are used. Columns are packed as for gel filtration to achieve homogeneity and free flow.

For adsorption, the starting buffer should be chosen to maximize binding and mininize protein-protein interactions. Thus, the pH and presence of specific ions or factors should be considered, and the buffer should have a high ionic strength, e.g., 0.5 M NaCl. The sample is added in the starting buffer, and if necessary the buffer ion in the sample is exchanged either by dialysis or by membrane or gel filtration.

Elution may involve one or a combination of pH shift, change in ionic strength, or displacement by a solute related or identical to the immobilized ligand. For the first two methods, the specificity characteristic of affinity chromatography lies only in the adsorption phase. In elution by specific displacement of immobilized ligand, specificity is exhibited in both adsorption and elution phases.

Other aspects.

In instances where a highly reactive support material is to be coupled to macromolecules, which are highly sensitive to excessive multipoint coupling, it may be necessary to reduce the number of coupling groups in the matrix. With CNBr-Sepharose and some other materials, this is accomplished by controlled hydrolysis of the reactive group; in others, by controlled pretreatment with a small ligand. Even without such sensitivity, it is likely that some distortion of macromolecular structure occurs. For enzymes this is displayed by a lower specific activity and other changes in catalytic properties.

After completion of the procedure for binding a ligand to a matrix, any remaining reactive sites on the matrix should be destroyed by adding an excess of small molecule containing the same reactive group; i.e., for CNBr-activated Sepharose, add ethanolamine.

16.3.7. Gel Filtration (18, 27)

The availability of spherical beads of relatively homogeneous pore size distribution makes possible a number of kinds of molecular separations which are based to a great degree on molecular size. When a sample is layered on a gel filtration column, molecules much larger than the pore size of the beads proceed through the column by displacing the solvent without penetrating the beads and exit as a peak after the void or interstitial volume, V_0, has passed through. Small molecules, which freely equilibrate with the internal space of the beads, V_i, theoretically proceed down the column by eluant displacement as though beads were not present and exit after passage of the total column volume. Molecules intermediate to these extremes "partition" between interstitial and internal spaces largely on the basis of molecular size. Since a wide variety of pore sizes, and hence fractionation ranges, of gel filtration media are available, it is possible to separate molecules over a molecular weight range of 1×10^2 to 5×10^6. While many molecules behave "ideally" as described above, many others are retarded more than expected. This is due to adsorption, ion exchange with a small number of carboxyl groups, or an interaction between aromatic rings and the matrix. Thus, the total column volume, V_T, is the sum of the excluded volume, V_0, the internal volume, V_i, and the matrix or bead material volume, V_g.

$$V_T = V_0 + V_i + V_g$$

Filtration materials.

At least four kinds of gel beads are available, i.e., cross-linked dextrans and derivatives, polyacrylamides, agaroses and cross-linked agaroses, and styrene-divinyl-benzene copolymer. These are available in several bead sizes and porosities; from only two manufacturers 180 kinds are available. It is therefore essential to consult manufacturers' literature for details because listing of these with their properties is beyond the scope of this presentation. Further, such literature is an excellent source of detailed basic information.

Multicomponent analysis.

By proper choice of bead porosity and carefully worked out conditions, it is possible to separate closely related molecules or conversely to separate classes of widely different molecules. Analysis of mixtures places the greatest demand on separation, but under these conditions rapid flow rate is not as important. Since quantitative yields are usually obtained on small columns, separation of components into nonoverlapping bands followed by quantitative analysis by nonspecific methods is possible. In this way a series of simple sugars such as raffinose, maltose, and glucose can be separated and also desalted on Sephadex G-15 (28). Usually, such separations are based on molecular weight. However, such separations have also been made because particular molecules are retarded for other reasons, such as adsorption of a polar solute from a nonpolar solvent.

Determination of equilibrium constant.

The amount of a ligand reversibly or irreversibly bound to an unknown amount of macro-

molecule can be used to calculate the equilibrium constant or stoichiometry of binding, respectively (41) (Fig. 7). This can be determined by incubation of the macromolecule and ligand, followed by gel filtration on Sephadex G-50 which has been equilibrated with and developed in the same concentration of ligand as used in the binding experiment. During column development, the concentration of ligand in the protein band is higher than the base-line level used in the elution buffer, the excess being due to binding to the protein which is in the eluate at that time. In the region of the total volume or "salt" volume, there will be a trough. This region contains the original incubation mixture, which is depleted by the amount of ligand bound to the protein that has been removed. The area of the trough equals the area of the peak. Conversion of these areas to amount of ligand by integration of the area allows calculation of the equilibrium constant or stoichiometry of binding if the binding is very tight. Chromatography on a coarse or fine Sephadex G-50 (40 cm by 2 cm) column is effective. The conditions used in Fig. 7 (Hummel and Dreyer [41]) are a column of coarse Sephadex G-25 (0.4 by 10 cm), a sample size of 0.25 ml, and an elution rate of 0.3 ml/min. V_0 and $V_i + V_0$ should be determined with blue dextran and ferricyanide, respectively, to obtain the volume that separates the two peaks. Sample size can be increased so that the first peak occupies the region between the two peaks. However, because peak broadening occurs, no more than 66% of the maximum sample volume should be used. For instance, on a 5 by 50.8 cm column, KCl and albumin are separated in a 30.0-ml sample with fivefold dilution and a large elution volume between the peaks. The same degree of separation could also be obtained with a 300-ml sample with a 1.5-fold dilution; eluting agent can be water or buffer. It should be pointed out that the same procedure is used to remove small molecules after chemical treatment of proteins

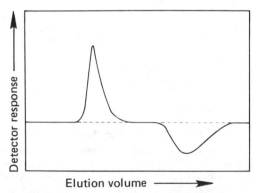

FIG. 7. *Elution diagram for determination of equilibrium binding of a ligand to a protein.*

(i.e., soluble fluorescence reagents) and to exchange buffers. For the latter, a "desalting" column is used, but it is equilibrated and developed with the new buffer.

Fractionation of proteins and determination of molecular weight.

For gel filtration on a preparative scale such as in the purification of enzymes and antibodies, it is often desirable to make a compromise between resolution and flow rate. Thus, some resolution is sacrificed by using a larger gel particle size in order to accommodate the volume involved. The porosity is chosen (often empirically) to resolve as many components as possible and to place the desired component intermediate in the elution diagram between V_0 and $V_0 + V_i$. Successive gel filtrations or a gel filtration step in combination with other separation procedures has been used effectively. Molecular weight of soluble proteins can be estimated by gel filtration on Sephadex G-75 and G-100 (30, 50). A regression line of the elution position V_x/V_0, where V_x is the volume to the peak of the unknown and V_0 is the void volume, versus the log of molecular weight is established with a group of enzymes and proteins of known molecular weight. The molecular weight of the unknown is taken from the graph. This method, while very useful, has largely been replaced by the more rapid and convenient disc gel electrophoresis method (see 16.5.1).

Concentration.

Macromolecules can be concentrated very simply without change in pH or ionic strength by addition of dry Sephadex G-25, coarse grade. After 10 or more minutes, the excluded volume is recovered by centrifugation in a basket centrifuge or with filter adapters for centrifuge tubes.

Choice of gel filtration medium.

It is necessary to choose a medium with a fractionation range to suit the application in mind. For total exclusion, use a gel with a molecular weight fractionation range well below that of the material, i.e., Sephadex G-25 (fractionation range, 100 to 5,000 molecular weight) for exclusion of most proteins or Sephadex G-10 (fractionation range, 0 to 1,500 molecular weight) for molecules above 3,000 molecular weight. For total equilibration with the gel interior, V_i, and elution with the "salt" volume, $V_0 + V_i$, select a gel with a fractionation range whose lower limit is above the molecular weight of the molecule of interest. For fractionation, the gel should have a molecular weight fractionation range such that the molecule(s) of interest will be partially retarded, i.e., for fractionation of serum proteins Sephadex G-200 (exclusion limit, 200,000) or for molecular-weight determination Bio-Gel A, 0.5 M (exclusion limit, 500,000).

Efficiency of resolution is a function primarily of flow rate which in turn is related to bead size and uniformity. Thus, very small bead size and slow flow rates give the highest resolution. For Bio-Gels P and A, flow rates of 2 to 10 ml/h per cm^2 of bead cross section give the highest resolution, and this is much slower than the maximum gravity flow rate for some bead sizes and porosities. With the most porous gels, for instance, Bio-Gels A 50 and 150 M (200 to 400 mesh), increased hydrostatic head is needed. With Bio-Gels P-2, P-4, and P-6, elevated pressures of several atmospheres may be used with 400-mesh beads capable of high-resolution separation of small molecules. The maximum recommended pressure should be ascertained from the manufacturer's literature. With some highly porous materials, collapse of the bead structure leads to very low flow rates.

Preparation of biological materials rarely utilizes chemical conditions which would threaten integrity of the gel filtration medium. However, the general stability characteristics are included here as a guide. Dextrans such as Sephadex are stable in salt solution, organic solvents, alkaline solutions, and weakly acidic solvents, but are hydrolyzed in strong acid. Dextrans are stable to 30 min at 121°C, wet or dry, at neutral pH.

Agaroses such as Sepharoses do not contain cross-links and hence should be maintained between pH 4 and 9 and between 2 and 30°C. They are stable to 1 M NaCl and 2 M urea. Cross-linked agarose, i.e., Sepharose CL, is stable between pH 3 and 9, but oxidizing conditions should be avoided. It is highly stable in 3 M potassium thiocyanate, a strong chaotropic agent. Sepharose CL can be transferred from water to a number of organic solvents with little change in pore size. It may be sterilized repeatedly at neutral pH at 110 to 120°C.

Polyacrylamides such as Bio-Gel P are reasonably stable in dilute organic acids, 8 M urea, 6 M guanidine·HCl, chaotropic agents, and detergents. Buffers with 0.5 M ionic strength are recommended for protein separations. Miscible solvents, i.e., alcohol up to 20%, do not materially change the pore size. Bio-Gel P is stable between pH 2 and 10 at room temperature and to autoclaving (121°C) at pH 6 to 8.

Column design and size.

Many gel filtration applications such as desalting do not require high resolution and hence can be carried out on homemade or disposable columns of ordinary design. However, success in critical applications requires, as it does for all other chromatographic separations, properly designed columns and associated equipment. Attention is called to the fact that "mixing reservoirs" above and below the gel bed must be avoided; otherwise, the gradient shape, or step change in eluant composition, as well as eluted peaks, is distorted by passing through the reservoirs. Commercial columns are available which have virtually no holdup volume below the gel bed. Others have fittings both above and below the gel bed to add eluant and collect effluent in the absence of reservoirs. Such devices are also essential for upward flow work. Sintered-glass supports are to be avoided, particularly for low cross-linked and high-porosity gels.

Column length affects resolution and column cross-sectional area is proportional to capacity. On the other hand, finer particle size increases resolution and requires shorter gel beds. Since sample size also dictates initial bandwidth, length and diameter become functions of the degree of separation needed for a given sample volume, longer for greater separation and wider for a narrower starting band. A 51-cm bed height will easily desalt a protein in a small volume, whereas very long columns (300 cm) are needed to partially separate a set of cyanogen bromide-produced peptides of an enzyme. An internal diameter of 2.5 cm may be a good starting point, although a somewhat narrower column can also be used. With very narrow columns there are wall effects caused by a greater proportion of laminar flow which broadens the bands. With excessive diameter, there will be unnecessary dilution and difficulty in maintaining horizontal zones.

Column packing.

The amount of dry gel needed is calculated from the swollen volume furnished in manufacturers' literature; for preswollen gels this step is unnecessary. The gel is then swollen in twice the expected bed volume of eluant. The length of time varies with volume regain. Highly cross-linked and lower-regain gels such as Sephadexes G-10, G-15, G-25, G-50, and LH-50 and Bio-Gel P require 3 to 4 h at room temperature, whereas Sephadexes G-75 to G-200 require 1 to 3 days. At 90 to 100°C these times are reduced to 1 to 5 h. Swelling in boiling water aids in removing bubbles, which at room temperature must be removed by vacuum deaeration. Preswollen gels are diluted enough to allow bubbles to escape; stirring should be avoided. The swollen gel is allowed to stand, and excess eluant is removed to give a thick slurry of a consistency which will still allow air bubbles to escape. Preswollen gels should be diluted if necessary to give a slurry of the same consistency.

The column is mounted and vertically aligned. The column is then filled with eluant, preferably from the bottom, until the dead space below the gel bed is occupied and all air bubbles are removed. With the outlet closed, pour the gel slurry smoothly down the column wall or down a rod all at one time. Add eluant to fill the column completely, connect an eluant reservoir, and remove all bubbles through an air vent in the column top piece. If the total gel suspension cannot be added at one time, use a column extension. Start the flow immediately. Bed formation from a thin suspension, addition of a slurry to a column containing eluate, or packing in stages is not recommended. Two or three column volumes of eluant are passed through at about the flow rate to be used. The column must not be allowed to go dry. This can be prevented by making a loop in the outlet tube which is higher than the gel surface.

Sample addition.

Before sample addition, a disk of filter paper or perforated plastic, or a sample application cup, is placed on the gel surface. The cup is a plastic cylinder with a nylon net bottom. Homogeneity of packing, as well as void volume (V_0) and fluid volume ($V_0 + V_i$), is then determined by passage of a mixture of blue dextran (2 mg/ml) and potassium ferricyanide or other colored small molecule through the column.

Sample addition is made to the gel surface after excess eluant has been removed. The sample is carefully layered on the bed, and the outlet is opened until the sample has moved into the bed. The column wall in the region of the surface is washed with a small amount of eluant. The column is then carefully filled with eluant and connected without bubbles in the column or line to a reservoir constructed to maintain a constant level of eluant (Mariotte flask). The Mariotte flask is a vessel with an outlet at the bottom to connect to the column via tubing and closed at the top with a stopper or plug through which passes an adjustable-height glass tube. The tube extends below the eluate surface and maintains a constant head.

It is also possible to layer the sample on the gel bed and beneath the eluant. The sample must be more dense than the eluant; if necessary, sucrose or salt is added. Layering is done by pipette or capillary tubing attached to a syringe. Pharmacia Inc. has developed flow adaptors for its columns. The adaptors consist of pluglike devices attached to threaded tubes which pass through the column ends and terminate in capillary tubing. Two of the adaptors with tight fits to the internal walls can be adjusted via the threaded rods to fit snugly at the bottom and top of the gel bed. With this arrangement, sample is added directly to the gel bed followed by eluant, and eluant is removed from the bed bottom.

Elution.

The flow rate can be regulated by hydrostatic pressure, i.e., by the difference between eluant surface and outlet, as long as the maximum pressure permitted for a particular gel is not exceeded. The effluent reservoir or Mariotte flask, any other devices, and the outlet tube end are then placed so that the difference in fluid level or fluid level in the adjustable vent tube and the column outlet end do not exceed the permissible difference in height (hydrostatic pressure). A pump can also be used under the same limitations. For higher cross-linked gels, the maximum gravity flow rates greatly exceed the flow rate for optimal separation or peak sharpness. With lower cross-linked gels, the flow rate may be limited by the permissible hydrostatic pressure. For gravity flow, eluant level and flow rate are maintained constant by the Mariotte flask.

Factors affecting peak sharpness.

Highest resolution is obtained when void and internal volumes of the beads are in equilibrium. In mixtures of similar-sized molecules, this is approached only at very slow flow rates. Thus, the flow rate established requires consideration of speed and resolution. In analytical work, resolution is primary; in preparative-scale work, resolution is usually secondary to speed of separation. For Sepharoses, values of 2 to 8 ml/cm^2 per min have been used. For desalting on Sephadex G-15, 12 ml/cm^2 per h has been used, and for Sephadex G-200, separations at 5 to 20 ml/cm^2 per h have been reported.

Elution patterns, i.e., peak shapes, are highly sensitive to viscosity of the sample. Thus, for maintenance of peak shape, samples containing sucrose or very high protein concentrations must be diluted so that the relative viscosity is not more than twice that of the eluant.

Storage.

If the column is to be inactive for more than 1 week at a time, it is necessary to prevent microbial growth by adding sodium azide (0.02%), Cloretone (0.05%), Merthiolate (0.005%), or chlorhexidine (0.002%). Chloroform, butanol, and toluene should not be used. Prior to returning the column to use, the above agents can be removed by passing eluant through the column.

16.4. RADIOACTIVITY (52–54)

Most bacteriological studies can utilize radioisotopes to advantage in the experimental program. The choice of radionuclide depends on half-life, the kind of decay process and its energy, and a number of other factors peculiar to the experiment. The half-life, short or long, determines the practicality of a given strategy as well as the necessity to correct for decay during the experiment. The type of decay and its energy determine the kind of counting instrument to be used as well as the resolution on photographic emulsions. Table 4 gives information on the isotopes most likely to be used in biological work. Unfortunately, there are no useful radioactive isotopes of nitrogen and oxygen.

Most measurements of radioactivity in biology are of β particles (electrons) and γ rays. Beta radiation, for instance, from tritium and ^{14}C, can be measured in a gas counting chamber attached

TABLE 4. *Physical properties of some radionuclides*

Radionuclide	Half-life	Type	Energy (MeV)	Specific Activity	
				Theoretical maximum (mCi/matom^a)	Common values for compounds (mCi/mmol)
Tritium (^{3}H, T)	12.26 yr	β	0.018	2.92×10^4	10^2–10^4
Carbon-14 (^{14}C)	5,730 yr	β	0.159	62.4	1–10^2
Sulfur-35 (^{35}S)	87.2 days	β	0.167	1.50×10^6	1–10^2
Chlorine-36 (^{36}Cl)	3×10^5 yr	β	0.714	1.2	10^3–10^1
Phosphorus-32 (^{32}P)	14.3 days	β	1.71	9.2×10^6	10–10^3
Iodine-131 (^{131}I)	8.06 days	β	0.81	1.62×10^7	10^2–10^4
		γ	0.08–0.72		
Iodine-125 (^{125}I)	60 days	γ	0.035	2.18×10^6	10^2–10^4

^{a} A milliatom is the atomic weight of the element in milligrams.

to an electrometer, by Geiger counter, by proportional methods (usually with barium carbonate), or by scintillation techniques. In recent years the liquid scintillation spectrometer has replaced all other methods and will be the only system described. Gamma counters are used primarily to measure ^{131}I and ^{125}I.

16.4.1. Safety

The use of radioisotopes for any purpose is strictly regulated by the Nuclear Regulatory Commission, and licensing and periodic inspection are usually required for the parent institution under which the research is conducted. In turn, the institution, through a radiation safety officer, ensures that safe practices and records are maintained, and it supplies working rules for the laboratory. These may be summarized as follows:

1. Work in a laboratory area that is appropriate for the kind of radioisotope compound involved. Often, this is a separate area with a hood if volatile radioactive materials, i.e., ^{3}H$_2$O, are involved. Control the amounts used and their disposition. Work in a setup which has specific waste and cleanup capabilities in a ready state.

2. Do not eat, drink, smoke, etc., in this area.

3. Do not pipette by mouth.

4. Wear a laboratory coat, gloves, and safety glasses.

5. Label all containers with specific radioisotope labels and give specific data.

6. Use disposable laboratory items when possible, and deposit them in designated solid-waste containers.

7. Monitor and decontaminate all laboratory glassware before returning it to general use.

8. Maintain an isotope inventory and record of isotope use.

9. Monitor the work area, as well as clothes and hands, before leaving the laboratory.

10. Report spills both to others in the laboratory and to the radiation safety officer.

A thin-window Geiger counter can monitor all isotopes except tritium, which requires using wipes and scintillation counting. Film badges are required for gamma emitters in quantities above 100 μCi and for persons using ^{32}P and other strong β emitters. ^{14}C, ^{3}H, and ^{35}S do not require film badges. However, for persons using millicurie levels of these isotopes, urinalysis should be made. This is often provided, along with a film badge service and waste disposal, by the radiation safety officer.

Protect counting equipment against contamination. This means a separate location and special attention to cleanliness of the *outside* of

counting vials, use of leak-free tops tightly sealed and maintenance of sample changer to prevent mishaps with vials. Distinguish between and separate high-level work and low-level work. Clean up. Do not leave contaminated glassware and equipment where others can handle it. Old radioactive materials are suspect as degraded until proven otherwise.

Tritium and ^{14}C in the quantities used in most experiments present no radiation danger. ^{3}H as in ^{3}H$_2$O is volatile, and ^{3}H$_2$ gas used in compound labeling is of very high specific activity; these require special precautions. ^{14}C, although a weak β emitter, has a very long half-life (5,000 years). Hence, its incorporation into biological material, i.e., deoxyribonucleic acid and bone, presents a long-term hazard. ^{35}S and ^{32}P are strong β emitters of relatively short half-lives (87.1 days and 14.3 days, respectively). High levels of these should be handled behind a radiation shield (lead bricks), and the radiation level at the worker location should be monitored.

16.4.2. Experimental Methods

Purity of isotopically labeled materials.

Isotopic materials supplied by commercial sources are of generally high quality. However, one should keep in mind that both radiochemical purity and chemical purity may be critical in the proposed experiment. At least, it should be known that impurities, if present, will have no effect. A compound which is radiochemically pure contains one kind of radioactive molecule, but determinations of radiochemical purity give no indication of nonradioactive contamination. Radiochemical purity may be determined chromatographically and by reverse isotope dilution (see below), that is, by addition of the pure nonradioactive compound (carrier) followed by isolation and determination of specific activity. If the compound is radiochemically pure, the new specific activity would be that expected by simple dilution with nonradioactive compound. If radioactive impurities were present, in reisolation with carrier these would be expected to be lost, and the specific activity would be lower than expected. If the impurity is not radioactive, the specific activity after reisolation would be higher than expected after dilution with carrier. If the impurity is known, its concentration can be determined by isotope dilution, i.e., by addition of pure nonradioactive contaminant, isolation, and determination of specific activity. Chemical purity is analyzed in the usual ways, i.e., spectrophotometry, optical rotation, gas-liquid chromatography, etc.

Substantial changes in the radiochemical and

chemical purities of a compound may occur during storage, due primarily to radiation damage from high specific activities. This process can be minimized by adherence to the specific conditions recommended by the supplier or more generally to one of the following: (i) storing at the lowest practical specific activity; (ii) subdivision into smaller lots; (iii) keeping dry if solid; (iv) storing in vacuo or under inert gas; (v) storing in pure benzene at 5 to 10°C; (vi) for water-soluble, benzene-insoluble compounds, adding 2 to 10% ethyl alcohol; and (vii) storing at the lowest possible temperature. Aqueous solutions of ^{3}H compounds should not be frozen. Storage of solids often does not require a vacuum or inert gas, and hence these can be stored as dry as possible in screw-cap vials. Volatile materials and liquids should be resealed in ampoules under vacuum or inert gas.

Isotope dilution techniques.

Radioactive compounds can be used to determine the amount of the same unlabeled compound in a mixture. It is necessary to know the specific activity (S.A.) and weight (wt) of the radioactive compound added and to isolate some of the pure compound for determination of its specific activity. The amount of unknown is derived from the equation:

wt of unknown (g)

$$= \text{wt of added radioactive compound (g)} \times \frac{\text{S.A. initial}}{\text{S.A. final}} - 1$$

Purity is attained when the specific activity after several reisolations (or recrystallizations) is constant.

A reverse approach can measure the total amount of a radioactive compound in a mixture. In this method, pure nonradioactive compound is added and the compound is reisolated. The initial specific activity of the compound and the amount of nonradioactive carrier must be known, and the final specific activity (after reisolation) is determined. The weight of isotopic material is calculated from:

wt of unknown (radioactive) (g)

$$= \text{total wt (carrier plus unknown) (g)} \times \frac{\text{S.A. final}}{\text{S.A. initial}}$$

A double-isotope dilution technique is valuable for isolation of very small quantities of, say, a sterol from a homogenate by a procedure involving many steps and partial recoveries. For instance, ^{14}C-sterol of known total activity can be added. Determination of total ^{14}C radioactivity of the pure isolated compound measures the recovery of the sterol in the procedure. In another procedure the sterol is derivatized either in the crude extract if practical or at the earliest possible purification stage. For instance, ^{3}H-acetic anhydride can be used to form the acetyl ester of the sterol. Then determination of ^{3}H in the isolated material gives the amount of sterol originally present. For this, only the specific activity of the acetic anhydride is needed; i.e.:

$$\text{sterol recovery} = \frac{\text{cpm recovered}}{\text{cpm added}}$$

amount of sterol (μmol)

$$= \frac{\mu\text{mol of }^3\text{H-compound}}{\text{recovery}}$$

$$= \frac{\dfrac{\text{cpm of }^3\text{H/counting efficiency}}{\text{S.A. of acetic anhydride }(\mu\text{Ci}/\mu\text{mol})}}{\text{recovery}}$$

$$1\ \mu\text{Ci} = 2.22 \times 10^6\ \text{dpm}$$

Calculation of the amount of radionuclide to use.

In general, the total amount of activity should be comfortably (twofold) above the minimum needed to make the final or lowest activity measurement. In this way radiation effects and laboratory contamination are avoided, and radiation safety is maximized. The easiest method, of course, is to use activities already reported for similar experiments, making adjustments for changes in volumes, instrument characteristics, etc. When starting from scratch, first decide on the minimum count rate which will be satisfactory (see part 16.4.4). Then calculate back to the start of the experiment to obtain the total amount (disintegrations per minute) which is required. To do this, it is necessary to bring into the calculations many of the following factors:

Half-life of nuclide
Duration of experiment
Initial sample size
Dilution factor
Counting efficiency
Molar specific activity
Size of sample counted
Biological half-life
Losses in isolation
Purity and stability of labeled compound
Metabolism of labeled compound

As an example, assume that it is desired to isolate on a column a tritium-labeled metabolite derived from the incubation of glucose in ^3H_2O. The material will be distributed in five 10-ml fractions from the column and will contain 1 atom of ^3H per molecule of metabolite. The yield from glucose added after the column is 5%, taking into account all factors of metabolite pools, side reactions, and losses in isolation. One-half of the total incubation mixture is used for this isolation.

In this experiment ^3H from water is incorporated into the material isolated. In the incorporation process, protons compete very effectively with $^3H^+$ both because there are many more and because the mass and the chemical characteristics of $^3H^+$ create an isotope effect. Let us call the isotope effect 4:1 against $^3H^+$.

The counting solution to be used functions well with a maximum of 0.5 ml of aqueous sample; we will use 0.2 ml. The counting efficiency for ^3H is 43%, and there is 20% quenching. Sometimes these two together are called counting efficiency. In the case of ^{14}C and ^3H, no calculations are needed to correct for decay of radioactivity.

From the standpoint of counting error (see below), we wish to count 5,000 events. Since we cannot tie up the scintillometer for more than 1 to 2 h, this count should be obtained in 10 min per sample. Thus, a 0.2-ml sample should contain radioactivity giving 500 cpm *for the lowest-activity fraction.* We can estimate that the total activity is distributed in the five fractions as 5, 20, 50, 20, and 5%; hence, the count rate in each 0.2-ml sample should be 500, 2,000, 5,000, 2,000, and 500 cpm, or 8,000 cpm for the five samples counted out of a total of 50 ml of eluate. To continue: 8,000 cpm $\times$ 50 (total sample) $\div$ 2 (total reaction mixture) $\div$ 0.43 (counting efficiency) $\div$ 0.8 (quench) $\times$ 4 (isotope effect) $\div$ 0.05 (yield) = 1.86×10^8 total dpm of ^3H_2O to be added to the reaction mixture. The ^3H_2O purchased contains 25 mCi/ml or 25 mCi $\times$ 2.22 $\times$ 10^9 (dpm/mCi) = 5.55×10^{10} dpm/ml in stock solution. $\dfrac{1.86 \times 10^8}{5.55 \times 10^{10}}$ = 0.00335 ml of 3.35 μl, or better 3.35 ml of 1:1,000 dilution of ^3H_2O stock solution.

Total activity calculations thus give information of quantities to be used. However, it is necessary to make specific activity calculations as well, i.e., disintegrations per minute per mole or disintegrations per minute per micromole to determine how many atoms of tritium have exchanged, or if other processes equilibrating with hydrogen atoms are involved. In the above example, it was calculated that 5.55×10^{10} dpm was present in 1 ml or 55.56 mmol of water or 111 meq of hydrogen per ml. The specific activ-

ity is $\dfrac{5.55 \times 10^{10}}{1.11 \times 10^2}$ or 5×10^8 dpm/meq of H.

In the 10-ml incubation the dilution of specific activity is $\dfrac{10 \text{ ml}}{0.00335 \text{ ml of } ^3H}$ = 2,985-fold, giving a specific activity of $5 \times 10^8/2,985 = 1.67 \times 10^5$ dpm/meq of H. If the metabolite isolated had the same disintegrations per minute per millimole, there was 1 eq of $^3H^+$ incorporated with the assumed isotope effect of 4. If the specific activity is higher or lower than this value, the magnitude of the isotope effect may be different, or there may be exchanges with hydrogen atoms not derived from ^3H_2O (lower specific activity), or more than one proton from ^3H_2O has been incorporated (higher specific activity). The isotope effect is significant with tritium, but is small enough with ^{14}C, ^{32}P, and ^{35}S to be ignored. It is often necessary as in this example to make an independent determination of the isotope effect by measuring the rate with H_2O or ^2H_2O and the rate with ^3H_2O.

Enzyme assays using isotopes.

A wide range of enzyme assays have been developed which depend on radioisotopes. In many cases there are no other convenient methods. Further, these have intrinsically high sensitivity which is demanded in many kinds of work. As detailed elsewhere in this manual (18.2.2), the amount of enzyme is proportional to the rate of substrate utilization or product appearance. Hence, determination of the concentration of either substrate or product with time is needed. For an assay to be valid, the initial rate must be constant for a short period and must also be proportional to the level of enzyme in the assay. In some instruments such as spectrophotometers and pH meters, rates can be followed continuously on a single assay mixture, often with the progress of the reaction being displayed as a strip chart recording. In other cases, however, such as with radioisotope enzyme assays, the rate must be established by measuring radioactivities on discrete aliquots of a reaction mixture at time intervals.

One of the major problems characteristic of such assays is the necessity of separating the radioactive substrate from the radioactive products. Usually, the labeled substrate is present in an amount required to saturate the enzyme and therefore to produce pseudo-zero-order kinetics where the rate of reaction is constant with time for an appreciable period. In sensitive and valid assays, the amount of product formed may be only 1% of the substrate present. It is therefore necessary to separate a small amount of radioactive product from a very large amount of radioactive substrate. Success of the method depends on finding an efficient separation that will give 0.1% or less cross-contamination. These procedures are often specialized and hence are outside the scope of this manual. The general types will be summarized. For details, it is recommended that the general reference by Oldham (54) be consulted.

The separation methods include: (i) precipitation of macromolecules synthesized or utilized, i.e., polynucleotides, polypeptides, and polysaccharides; (ii) release or uptake of a volatile radionuclide which can be separated by distillation or equivalent, i.e., ^3H_2O, $^{14}CO_2$, or volatile fatty acids; (iii) solvent extraction; (iv) ion exchange; (v) paper and thin-layer chromatography and electrophoresis; (vi) adsorption and elution; (vii) isolation of derivatives; (viii) conversion of unlabeled product to labeled derivative; (ix) dialysis and gel filtration; and (x) reverse isotope dilution.

The degree of cross-contamination of substrate into product in the isolation procedure limits the sensitivity of the method. For instance, assume that there is a 0.1% cross-contamination of substrate into product in the separation used. Further assume that the zero-time (i.e., with no product formed) count rate due to cross-contamination is 100 cpm. It is desirable to have the count rate due to product be at least equal to the rate at zero time or 100 cpm for 0.5% conversion of substrates. In this case the "signal-to-noise" ratio is 1:1, or the minimum sensitivity without very long count rates (see part 16.4.4). To attain these minimum count rates, 100,000 cpm in the substrate is the minimum to be used if the total sample is counted. The use of larger amounts of radioactivity will not improve the sensitivity because the zero-time value increases correspondingly. Of course, larger amounts of radioactivity in the product above the zero-time count rate decrease the error of the assay. However, a very substantial amount of product formation causes loss of the pseudo-zero-order condition needed for a valid assay.

From the example, it is clear that, while it is necessary to add enough substrate to saturate the enzyme, its radioactivity need not be high. On the other hand, if separation of substrate and product is complete, the sensitivity is proportional to the specific activity of the substrate and limited only by cost, practicality, and availability. Under these circumstances, investigators may prefer to use ^3H- rather than ^{14}C-labeled substrates because of availability of 100-fold higher specific activities at similar costs (see below for problems with ^3H substrates).

Dilution of radioactive substrate with unlabeled substrate affords a number of advantages in addition to facilitating saturation of the enzyme with substrate. It reduces (i) cost by using less isotope, (ii) effects of contaminants and radiation degradation products in radioactive substrates, (iii) the effect of an unwanted stereoisomer in radioactive substrate containing a mixture of isomers by adding carrier substrate of the desired stereo configuration, and (iv) radiation decomposition of the substrate. Such dilution also increases the accuracy of determinations.

Since enzyme activity is expressed in international units (IU), and 1 IU is 1 μmol of substrate used per min under standard conditions, it is necessary to convert count rates to micromoles per minute using the specific activity of the substrate (which will be converted to product presumably of the same specific activity). This value is subject to errors in measurement of both radioactivity and concentration. While the specific activity of the radionuclide as applied is usually known to ±5%, it is seldom used as such, but rather is diluted with large amounts of the nonradioactive form, with attendant errors of measurement and purity. It is wise, therefore, to determine carefully specific activity of the substrate as used in the assay. It may also be necessary to demonstrate radiochemical purity of the substrate or to purify it prior to use to remove decomposition products and other impurities.

Special problems with tritiated substrates.

Tritium shows isotope effects relative to hydrogen due to the large change in mass involved. The effect is most pronounced when making or breaking the bond between tritium or hydrogen and another atom is a rate-limiting step in the reaction. This effect may amount to virtual exclusion of reaction with the tritiated species, as has been observed in the reductive carboxylation of L-ribulose-5-phosphate by [^3H]NADH-phosphate and CO_2. The isotope effect may be displayed as a change in V_{max} of the enzyme when using a tritiated substrate or as a change in K_m. If the K_m is lower, the effect would not be observed when the enzyme is saturated with substrate. If the K_m is higher, the rate would decrease if the substrate concentration is no longer saturating.

When tritium is used as a label only to follow the conversion of substrate to product, it must be remembered that (unlike ^{14}C) ^3H often is released as $^3H^+$ into the medium in unrelated spontaneous reactions, giving substrate and products a lower specific activity and hence erroneously low enzyme activities. In some cases the side reaction is enzyme catalyzed, not necessarily accompanied by the complete reactions. In addition, it is characteristic of tritiated compounds to "leak" $^3H^+$ by chemical reaction. This varies with the location of ^3H in the molecule. For example, α-^3H-amino acids lose ^3H in many enzyme-catalyzed conversions, whereas ^3H in other positions, especially if remote from functional groups, is relatively stable. ^3H may also be lost in isolation of the product.

Some enzyme assays are based on release of ^3H from a specific position in the substrate. Errors result if ^3H is not exclusively located in the position assumed and/or in the correct stereo location, because it has been demonstrated many times that enzymes selectively or specifically labilize tritium in one of two or more bonding positions occupied by ^3H or H.

The low energy of ^3H radiation causes low counting efficiency and losses due to quenching and absorption by solid materials. These in turn lead to greater inaccuracies or even lower apparent enzyme velocities. Good efficiency and quench corrections made on a sample-by-sample basis often are required.

16.4.3. Liquid Scintillation Counting

Liquid scintillation counting has now supplemented Geiger, proportional, and gas counting methods for ^3H, ^{14}C, ^{35}S, and ^{32}P, even though the instruments are quite expensive, because of the elimination of self-absorption, more favorable counting efficiency, and greater convenience in sample preparation. In brief, the radiation event excites the emission of a light pulse from a fluorescent material which is then "seen" by a pair of photomultiplier tubes and registered as a radiation event or count. The sophistication of this instrument lies in the fact that both photomultiplier tubes must see the light flash, thereby eliminating a large number of spurious pulses. In addition, the photomultiplier produces voltage pulses, and the pulse height is proportional to the energy of the radiation event. By use of pulse height discriminator settings, the instrument counts only pulses appearing in the "window" corresponding to desired minimum and maximum radiation energies. Thus, the instrument can be set (i) to eliminate many low background pulses, (ii) to selectively count one isotope (one class of pulse heights), or (iii) to count all isotopes (all pulse heights), usually simultaneously. Current instruments accomplish this simultaneously via several parallel analyzer channels.

The radiation event is converted into a scintillation via solvent molecules which become activated and then transfer the excitation energy to a compound which fluoresces (fluor). The excited fluor either emits a photon of light at a characteristic wavelength or transfers the excitation energy to a secondary fluor which in turn emits a photon of light at a more advantageous wavelength with respect to the spectral sensitivity of the photomultiplier. 2,5-Diphenyloxazole (PPO) and 1,4-bis-2-(5-phenyloxazolyl)benzene (POPOP) and its dimethyl derivative are the most popular primary and secondary fluors. These are highly nonpolar substances which are dissolved in a nonpolar solvent such as toluene to make up the scintillation solution. The principal drawback in scintillation counting is the difficulty of introducing aqueous samples containing polar solutes into the nonpolar counting solution. Further, water and solvents of increasing polarity tend to diminish the counting efficiency; that is, they along with colored compounds quench the normal chain of events leading to light emission. Because of these problems, a variety of scintillation mixtures have been developed for polar and nonpolar materials in solution, for solid materials, and for other specific purposes (see below). In addition, high background counting rates can occur due to fluorescent light or sunlight-induced phosphorescence which may require hours or days to decay. Also chemiluminescence of samples in counting medium gives very large counts which decay slowly. Glass vials may contain ^{40}K which raises the background level. On the other hand, a number of materials quench the count rate; trichloroacetic acid, $HClO_4$, water, pyridine, and colored compounds are major problems.

Solid materials.

Small pieces of paper or removed sections of a thin-layer plate may be counted in scintillation medium. The main problems have to do with self-absorption by the particles, uniformity of orientation, uniformity of solubilization of radioactive material, and difficulties in using the external standard method for determining counting efficiency (see below). Since there is no direct way of determining self-absorption, only relative radioactivity of a series of samples counted under the same conditions can be obtained. Variability is greatly decreased if the material to be counted either completely dissolves or is completely insoluble in the scintillant. For instance, Hyamine hydroxide 10-X in methanol has been used to remove amino acids and sugars from paper, and then a toluene scintillant is added. When all of the material cannot be eluted, it may be necessary to combust the paper, trap the CO_2 in an organic base, or as sodium or barium carbonate, and count these materials. When papers containing insoluble radioactivity are counted, the most reproducible orientation is to place the paper in the bottom of the vial.

Glass filter disks and strips are now used to advantage because of higher counting efficiencies with tritium, i.e., 10% at best with ±5% reproducibility compared to 2 to 8% on cellulose paper. With stronger β emitters such as ^{14}C, reproducibility and efficiency are higher.

Polyacrylamide gels.

The main problem is to solubilize the gel rapidly. This is best accomplished by using a cross-linking reagent which is readily destroyed because linear polyacrylamide chains are soluble in scintillation mixtures. N,N'-Diallyltartardiamide has been used as a cross-linking agent because it is readily cleaved by periodic acid (56). A series of gels of 7% (wt/vol) acrylamide and 0.27% (wt/vol) cross-linking agent are treated with 0.5 ml of 2% periodic acid for 2 h. The scintillant is added, vigorously mixed on a Vortex mixer, and used as such, or a portion can be transferred to a scintillation vial. The scintillation cocktail is designed to maintain solubility (55).

Carbon dioxide.

Measurement of ^{14}C radioactivity as CO_2 is of major importance in metabolic studies and in the large number of applications which require combustion or digestion of carbonaceous material. For scintillation counting, the CO_2 is trapped in an organic base, for instance, Hyamine hydroxide 10-X, Primene 81-R, ethanolamine-ethylenediamine, or phenylethylamine, or in barium hydroxide or NaOH. With all of these, an apparatus or train containing the base is needed. The trapped CO_2 is then added to a scintillant.

For information on combustion methods, consult Jeffay (59).

Instrument operation.

Single isotope counting. The goal is to set the instrument for the highest efficiency and the lowest background. Two types of adjustment are provided for each channel: (i) high and low discriminator settings, and (ii) photomultiplier voltage and amplifier gain. Higher settings of the latter pair increase the pulse height. The high and low discriminator settings determine the pulse height range that will be accepted. The pulse height or energy spectrum of beta emission is determined by setting the upper and lower discriminator with a fixed differential and then counting a radioisotope at increasing positions of the discriminator band, i.e., 0 to 10, 10 to 20, 20 to 30, etc. A plot of the count versus discriminator range midpoint gives the spectrum (Fig. 8). Note that each isotope has a characteristic pulse height distribution. It is readily seen that discriminators in each channel may be set to include or exclude various parts of the isotope energy spectrum. Once this has been determined, the lower discriminator is set to exclude most of the background (i.e., set at 10) and the upper is set to just include the maximum pulse height of the isotope being measured. In this way much of the low-energy and high-energy background events are eliminated.

Determination of the pulse height spectrum at various photomultiplier voltage settings also gives a family of curves (Fig. 9) which are especially useful in separating the radiation of each isotope in dual-labeled samples. Note that lower voltage settings narrow the discriminator settings necessary to produce the window.

In all cases background count rate must be determined with a vial of scintillation solution with no radioisotope and subtracted from each sample count. It is assumed that quench of the background measured as above is nonexistent when high sample rates are measured. However, at very low count rates it may be necessary to determine the effect of sample quench by counting an identical nonradioactive solution.

Dual-label counting. Dual-label counting can be accomplished when the energies of the isotope pair are sufficiently different that there is resolution of the two pulse height spectra. This is practical for ^3H and ^{14}C, ^3H and ^{35}S, ^3H and ^{32}P, and ^{14}C and ^{32}P pairs. The more energetic isotope is counted exclusively in channel 1. However, the count for the weaker isotope in channel 2 will always have counts of the more energetic member of the pair. It is necessary to determine counting efficiency for each isotope separately, preferably by internal or external standard

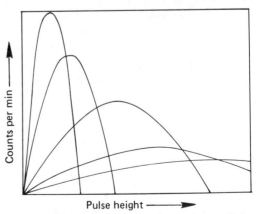

FIG. 8. *Pulse height spectra for various radioisotopes.*

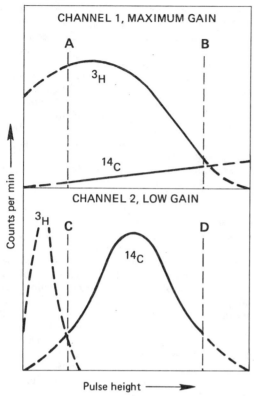

FIG. 9. *Pulse height spectrum as a function of photomultiplier voltage. A, B, C, and D represent the settings of four different discriminators.*

methods, and then to calculate the count for the second channel from simultaneous equations:

$$\text{dpm of }{}^{14}\text{C} = \frac{\text{net count in channel 1}}{\text{efficiency of }{}^{14}\text{C in channel 1}}$$

dpm of ^{3}H

$$= \frac{\begin{array}{c}\text{(net counts in channel 2)}\\ - \text{(dpm of }{}^{14}\text{C} \times \text{counting}\\ \text{efficiency of }{}^{14}\text{C in channel 2)}\end{array}}{\text{tritium efficiency in channel 2}}$$

The counting efficiency of channel 2 for the lower-energy isotope (^{3}H in this case) can be optimized first by adjustment of discriminator settings using a tritium-containing standard. Then the lower discriminator of channel 1 is set to exclude counts from ^{3}H and the upper is set to give maximum efficiency for ^{14}C.

The above procedure applies if the quench and efficiency of samples correspond to those of standards used. When quench is encountered, more of the higher-energy counts appear in the low-energy channel. The original relationship of channels can be reestablished by changing the discriminator settings if this is known from sim-

ilarly quenched standards. Alternatively, the gain on the photomultiplier or amplifier can be advanced to reestablish the pulse height pattern in the original window settings. Dual-label counting is fraught with difficulties and errors when low count rates are involved.

Quench correction and efficiency. Under ideal conditions not all radiation events are detected even in the best scintillation counters, although efficiencies of above 90% are obtained with intermediate and strong β emitters, i.e., ^{14}C, ^{32}P, etc. Instrument efficiency can be determined by several means described below using a nonpolar counting medium which has neither chemical nor physical quenchers. More commonly, however, a combined instrument efficiency and sample quench correction factor is produced by one of three methods, each of which has advantages and disadvantages.

Internal standard method. This original method is relatively simple and easy to understand. It is highly reliable and potentially very accurate for counting single isotopes. The correction includes both chemical and color quenching. It consists of first counting a sample (C) and then adding in a small volume containing a known number of disintegrations per minute (D_s) of the same isotope: i.e., for ^{14}C counting, [^{14}C]benzoic acid; for ^{3}H, tritiated toluene; etc. Then the sample plus standard (C_t) is recounted. From this, the overall efficiency is $(C_t - C)/D_s$, and the radioactivity of the sample is:

$$\text{sample dpm} = C \times \frac{D_s}{C_t - C}$$

The main disadvantages are that the vials must be removed from the counter to add the standard and then recounted. The sample alone cannot be recounted as an afterthought. Addition of the standard to the vial must be accurate. For large numbers of vials, this is now accomplished with improved accuracy with various kinds of automatic pipettes or syringes.

Channel ratio method. The channel ratio method is based on the observation that the spectrum of pulse heights in a given scintillant-isotope system shifts with the efficiency. In an instrument with two counting channels, the discriminators in each channel are set differently, and the count in each channel is determined simultaneously and expressed as a ratio. This method can also be used in a single-channel instrument; however, it is necessary to make two counts, each at a different discriminator setting. Using a set of standards prepared to have different efficiencies, a standard curve of efficiency versus channel ratio is constructed and used to correct the unknowns. In both channels the lower discriminator is set above the pulse height of background noise. One discriminator setting, either upper or lower, may be the same in both channels, with the other different. Alternatively, both upper and lower discriminator settings may be different for the two channels. Since quenching may be due to colorless chemicals such as water and CCl_4 and also to colored compounds, it may be difficult to prepare standards of the same quencher content as the unknowns. Thus, there will be some error if colored quenchers are present in the unknowns, especially at high concentrations. However, this method applies over a wide range of conditions (57). Dual-channel systems require less time unless low counts are involved. In that case, much more counting time is required to obtain a ratio of given precision than for the internal standard method. In this method, sample counting can be repeated.

External standard method. Newer instruments are equipped with a gamma source of known disintegrations per minute which upon call by the instrument program is positioned directly below the counting vial. Counts are made with and without the gamma source. A standard curve of external standard counts versus counting efficiency (produced by introducing quenchers) is constructed and used to correct counts in unknowns. The method is susceptible to variations in volume of scintillant, thickness of glass vial, and different scintillants. In addition, inhomogeneity of solutions or counting of paper strips produces errors. Long counts are not needed at low radioactivity levels as for the channel ratio method. It is possible to improve the external standard method described above by making a channel ratio approach with and without the gamma source. In this case the external standard is counted in two channels with discriminators set above the pulse height of the isotope being counted to obtain counting efficiency. This method requires the availability of four channels or two counting runs at different discriminator settings.

Counting solutions.

1. *Nonpolar applications* (minimum temperature, $-30°C$):

    2,5-Diphenyloxazole (PPO), 5 g/liter
    1,4-Bis-2-(5-phenyloxazolyl)benzene (POPOP), 0.3 g/liter
    Toluene
    Very high efficiency (90 to 95%). Can be used to count filter disks (Millipore Corp.) if dry. Disks must be translucent when in toluene; if white, drying is not complete.

2. *Homogeneous solutions or emulsions* (depending on amount of water; 4 to 14% water

produces a solution, 20 to 30% produces an emulsion) (61, 62):

    Toluene (667 ml) containing 0.4% PPO (2.67 g) and 0.1% POPOP (0.07 g)
    Triton X-100 (Rohm and Haas Co., Philadelphia, Pa.), 333 ml
    Counting efficiency for ^{14}C, 52 to 68%; for ^3H, 7 to 10%, but varies with batch of Triton X-100.

3. *Stable gel with high water content* (forms emulsions in about 35% water and a stable gel at low temperature [to $-5°C$], which resolubilizes when warmed) (61):

    Toluene (538 ml) containing 0.4% PPO (2.1 g) and 0.1% POPOP (0.05 g)
    Triton X-100, 462 ml
    Counting efficiency for ^{14}C = 43 to 55% at 43% water. Use with tritium not recommended unless a gel is needed. ^3H efficiency about 5%. Useful with tissue suspensions as well as powders such as barium carbonate and other white materials.

4. *Homogeneous solution with up to 33% water* (58):

    Triton X-100, 257 ml
    Ethylene glycol, 37 ml
    Ethanol, 106 ml
    Xylene, ca. 500 ml (to make 1 liter)
    PPO, 3 g (POPOP not used)
    Counting efficiencies unquenched are 87% for ^{14}C and 47% for ^3H. With maximum water, the value for ^{14}C is 82% and for ^3H is 35%. The relationship between efficiency and external standard ratio (see below) is linear.

5. *Solubilized polyacrylamides:*

    Triton X-114 (Rohm and Haas Co.), 250 ml
    Xylene, 750 ml
    PPO, 3 g (POPOP is not needed)
    Maximum counting efficiency is 93% for ^{14}C and 47% for ^3H. Up to 30% water may be used with a decrease in efficiency to 50 to 60% for ^{14}C and 10% for ^3H. A stable emulsion is formed at intermediate water concentrations.

6. *CO_2 counting* (60):

    Ethanolamine, 1 part
    Ethylene glycol monomethyl ether, 8 parts
    Toluene, 10 parts
    PPO, 5 g/liter

16.4.4. Statistics (52, 53)

Determinations of radioactivity from count and time measurements under ideal circumstances contain random errors which cause the measured count rate to depart from the true value. For this reason, one should have at hand statistical tools which aid in estimating error. The intelligent use of these methods requires that one be informed of the assumptions and terminology involved if not as to how the final useful expressions are derived. It is not within the scope of this manual to present the application of these tools with the appropriate background, because various tables and graphs are needed to perform the operations. Rather, it is recommended that the general references be consulted. Nevertheless, several important tools are listed and their equations are given below. These include: average of several counts, standard deviation estimates of single and multiple counts, percent standard error (% SE) of counts and count rates, and time required for a desired % SE of count rates of samples with and without background for equal and unequal times for counting background. Additional expressions and tables are available for determination of times required for counting background and for total count with a given standard error, limits of detection of radioactivity, and the correct way to eliminate a count value which varies widely from the mean.

There are a number of different kinds of error: i.e., probable error, where the true value is likely to exceed the error expressed 50% of the time, standard error, 68% of the time; and 99/100 error, 99% of the time. Since standard error has a coefficient of 1 in many statistical derivations and since this expression is most commonly used, only "standard" forms of equations will be given here. This means that, among members of a set of values, one value will not vary from the mean by ± standard deviation more than 32% of the time. Since count errors are independent of time, one can determine the number of counts required for a given % SE using the expressions below. For instance, for 1% SE, 10,000 counts are required; for 2.5% SE, 1,600; for 5% SE, 400; and for 10% SE, 100. From the corresponding expression below, the times required to determine both background and total count rates at 1% SE when both rates are equal (see calculations of the amount of radionuclide to use) are 804 min for background and 1,140 min for sample plus background; for 10% SE, 8 min and 11 min are required, respectively.

Statistical formulas.

1. *Average of several counts:*

$$\bar{X} = \frac{\Sigma \text{ individual counts } (X)}{\text{number of determinations } (n)}$$

2. *Standard deviation from several counting cycles:*

$$\text{SD} = \left(\frac{\Sigma (\text{deviations of } X \text{ from } \bar{X})^2}{n-1} \right)^{1/2}$$

3. *SE (%) in count rate:*

$$\text{SE of rate (\%)} = \frac{100}{\text{rate } (R) \times \text{time } (t)^{1/2}}$$

4. *Time required to give the desired SE (%):*

$$t = \frac{10^4}{R(\text{SE \%})^2}$$

5. *SE (%) of net sample count when the times of counting sample and background are equal:*

$$\text{SE (\%)} = \frac{100(X_{\text{total}} - X_{\text{background}})^{1/2}}{X_{\text{total}} - X_{\text{background}}}$$

6. *SE (%) of sample count without background when there are unequal counting times for background and sample:*

$$\text{SE (\%)} = \frac{200(R_t/t + R_b/t_b)^{1/2}}{R_t - R_b}$$

where R_t and R_b are the total rates and the background rate, respectively, and t_t and t_b are the times for R_t and R_b.

16.5. GEL ELECTROPHORESIS

16.5.1. Disc Gel Electrophoresis

Separation of macromolecules by disc gel electrophoresis in its many forms has become a very important tool in biological and medical research as well as in clinical diagnosis and monitoring of industrial processes. Since the inception of the discontinuous method by Ornstein (67) and Davis (64) in 1959, intensive investigation of all aspects has developed a large body of theory, detailed information on conditions, and many variants of the buffer and gel system (for general reference see 66). It is intended here to give only one basic system of wide applicability together with information on operation of the system to obtain good results. In addition, some of the major applications are described.

General method.

Electrophoresis is carried out either in cylindrical gel columns (in glass tubes) or in a gel slab about 10 cm by 10 cm by 1 to 4 mm thick. Three layers of gel are involved: separation gel at the bottom, spacer gel in the middle, and sample gel at the top. Each is composed of polyacrylamide of specified percent cross-linking. The gel is formed in a container (tube or flat plate) by mixing acrylamide (monomer), and N,N'-methylenebisacrylamide (BIS) with buffer, cross-linking reagent, and polymerizing agent. The polymerization is catalyzed either chemically with ammonium persulfate or photochemically by light which produces free radicals when oxygen is present.

Separation of proteins in the discontinuous system is the result of two major factors, differences in migration due to sieving effect of the cross-linked gel, resulting in size separation, and differences due to charge. In addition, unusually sharp bands occur due to concentration of the proteins into a very thin band or disk before entering the separation gel. Thus, molecular sieving is controlled by the degree of cross-linking which is in turn dictated by the concentration of cross-linking acrylamide. Concentrating proteins into a narrow band before they enter the separation gel is characteristic of the discontinuous system and results from an ingenious manipulation of discontinuities of pH, buffer ion composition, and gel pore size as the proteins pass through the three gel layers. Concentration is obtained by choosing a pair of buffer ions and pH such that the mobility times the degree of dissociation for one ion (Cl⁻) is greater than the same product for protein, which in turn is greater than the product for the second ion (glycine) (see p. 23 of reference 66). To accomplish this, a zone of low cross-linked stacking gel is layered

over the longer, smaller-pore-size separation gel. Above this, the sample in low cross-linked gel is layered over the stacking gel (Fig. 10). The separation gel is more alkaline, i.e., pH 8.9, whereas the stacking gel and sample are in a different buffer at pH 8.3. When a potential is applied, the proteins quickly traverse the spacer gel and form a very thin band at the top of the separation gel and sandwiched between Cl⁻ and glycinate. The proteins then slowly penetrate the separation gel. Migration rate in the separation gel is then a result of charge (dictated by pH) and molecular size (dictated by pore size or degree of cross-linking). Since the bands were originally very thin, resolution is exceptionally high. For instance, blood serum gives 15 to 20 bands. With an optimized discontinuous system, the separation of 62 serum components has been reported (81). Following electrophoresis, the gels may be analyzed in place by ultraviolet scanning or extruded and analyzed by ultraviolet scanning, staining, or sectioning. The sections can be soaked to elute components, or they are solubilized and radioactivity is determined. Gels or longitudinal sections may be dried for autoradiography.

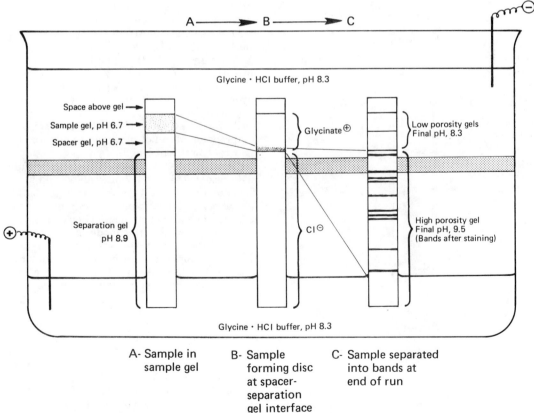

A- Sample in sample gel   B- Sample forming disc at spacer-separation gel interface   C- Sample separated into bands at end of run

FIG. 10. *Gel layers and ionic behavior in discontinuous polyacrylamide electrophoresis.*

Column method (66).

The column system requires very simple equipment, either commercial or home built. It involves a watertight holder for 1 to n glass tubes, 5 mm (inner diameter) by 65 to 70 mm long. The construction is such that buffer plus electrode (cathode) is above and in contact with the gel tube, and the same buffer plus anode is in contact with the bottom (Fig. 11). The edges of the tubes must be planar and the diameter must be uniform. However, these tubes can be made in the laboratory or glassblowing shop. The tubes must be scrupulously cleaned and may be dipped in a nonionic detergent (i.e., Kodak Photo-Flo) before drying.

The original sequence of filling tubes with acrylamide given by Ornstein (67) and Davis (64) started by polymerizing the separation gel in the bottom section. A reverse procedure has since been recommended (66) as producing a more uniform interface between the spacer and separation gels as follows. Transfer 0.2 ml of 40%

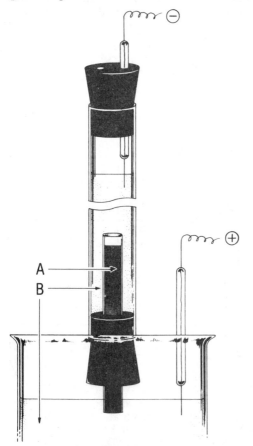

FIG. 11. *A simple arrangement for column electrophoresis. A, agar-solidified buffer; B, Buffer solution.*

sucrose to a rubber cap and press the gel tube firmly to the bottom of the cap; this prevents entrapment of air bubbles. Align the tube vertically; then prepare about 0.15 ml of spacer gel mixture (see below), take it up in a small syringe equipped with an 8-cm (0.6- to 0.8-mm inner diameter) needle, and layer it on the sucrose solution. Layer water very carefully over the spacer gel layer so that a sharp boundary is formed with no mixing. High performance results largely from a sharp uniform boundary between the spacer and separation gels. Polymerize the gel by irradiation with a fluorescent lamp for 20 to 30 min at a distance of 2 to 3 cm. Carefully and completely remove the water layer using a syringe and then a wick of lint-free material.

Next, prepare the separation gel. Rinse the space over the spacer gel quickly twice with separation gel mixture and fill the tube with separation gel mixture by syringe to give a convex meniscus, taking care not to entrap air bubbles. Then cover the surface tightly with plastic film. This must be accomplished within 5 min after completion of spacer gel preparation. After 30 to 40 min of polymerization time, carefully remove the cap at the bottom, invert the tube, carefully place the cap on the new bottom, remove the sucrose solution from the new top, and then rinse the top of the spacer gel with spacer gel mixture. Apply the sample in spacer gel mixture, and carefully cover with electrode buffer to the upper rim, taking care that the two layers do not mix. Then photolyze the sample layer as above. This procedure can be managed easily for many tubes if suitable carriers, etc., are prepared ahead of time, and the fluorescent light is of sufficient length.

Place the tubes in the apparatus, and carefully add precooled electrode buffer above and below the tube to at least 4 cm for the lower buffer. The upper buffer contains bromothymol blue to track the migrating boundary. Then connect the power supply so that the anode (positive terminal) is at the bottom, and adjust the current of the power supply to 1 mA per gel tube for 2 min and then to 2 to 5 mA per tube until the tracking dye disk has migrated to about 1 cm from the bottom of the tube.

The voltage drop across the gel produces heat which is proportional to current flow. Excessive heat makes uneven bands. A 40-min run starting at 20°C will increase the temperature to 30 to 40°C. The temperature rise can be diminished by lower current flow, i.e., 1 mA/gel with an increase in electrophoresis time (140 min); however, there will be some band broadening due to diffusion. Alternatively, electrophoresis can be run at 4°C and even with stirring of the lower electrode buffer.

Separation gels can be stored for several days, but after sample and spacer gels have been added, electrophoresis should be performed within 1 h. After electrophoresis, the electrode buffers may be removed and saved for future runs, but anode and cathode buffers must be kept separate. The buffers should be discarded after several runs or when the pH changes. The gels are then analyzed either before or after removal from the tubes. Since the conductivity of the gel column changes during electrophoresis, changes in voltage and current can be expected to occur during the run. It is recommended that a current-regulated power supply be used so that current can be maintained constant throughout the run. When n gels are run simultaneously, the current required is n times the value chosen for one tube and is equally divided among the tubes only if the cross section and composition are identical.

Slab method.

The same electrophoretic separations can be made in slabs of polyacrylamide of 100 mm by 100 mm by 1 to 3.5 mm. Special trays are used to form the slab with slots for each sample, and special equipment is needed for performing the electrophoresis. The apparatus provides a way to hold the slab vertically with the slots of the upper edge in contact with the upper buffer and the opposite edge in the lower buffer. Provision must be made to circulate coolant across the face of the slab on both sides. The main advantage of the slab system is the ability to run multiple samples under identical conditions so that comparisons among samples can be made easily. It is claimed that slab gels are easier to read in a densitometer, but a special apparatus is needed. Slabs are also easier to cut into strips and analyze for: (i) different components with different stains, (ii) radioactivity, and (iii) drying and storage. However, more acrylamide is needed, especially when only a few samples are run. As with tube gels, a wide variety of continuous and discontinuous buffer systems have been developed. The methods of preparing gels, running, and analyzing are similar to those for tube gels.

Preparation of gels.

Details for preparation of one of the more common discontinuous gel systems are presented here. Many others are given by Mauer (66). *CAUTION*: Acrylamide must not be inhaled or allowed to contact the skin.

Spacer and sample gel:
  1 part A (25.6 ml of H_3PO_4, 5.7 g of Tris/100 ml)
  2 parts B (10 g of acrylamide, 2.5 g of N,N'-methylenebisacrylamide [BIS]/100 ml)
  1 part C (4 mg of riboflavin/100 ml)
  4 parts D (40% sucrose)
Separation gel:
  1 part E (48 ml of 1 N HCl, 36.6 g of Tris, 0.23 ml of N,N,N',N'-tetramethylethylenediamine [TEMED]/100 ml)
  2 parts F (30 g of acrylamide, 0.8 g of BIS/100 ml)
  1 part water
  4 parts G (0.14 g of ammonium persulfate/100 ml)

Polymerization using persulfate yields gels of variable porosity and electrophoretic behavior. Thus, it is important to establish a standard set of polymerization conditions. Persulfate also may cause artifacts. These may be eliminated either by substitution of riboflavin in the gel mixture and photolytic polymerization or by use of reducing agents, i.e., thioglycolate (0.01 to 0.001 part per 1 part glycine in buffers), or by adding mercaptoethanol (5 mM) to the spacer gel, sample, and upper buffers. Pre-electrophoresis has also been used to remove persulfate and other impurities, but the discontinuous nature of the system is destroyed, resulting in a continuous electrophoretic system which may or may not have the required resolution.

A number of protein-dissociating agents can be incorporated into the gel formulations, i.e., 0.5% sodium dodecyl sulfate (see below), deoxycholate, Triton X-100, and urea.

Electrode buffers.

  0.6 g of Tris, 2.88 g of glycine, 2 ml of 0.001% filtered bromothymol blue (upper buffer only), pH 8.3

A large number of general and special systems have been reported which may be useful (see p. 44 of Mauer [66] and following).

The purity of acrylamide and Bis are important in obtaining good results, especially in ultraviolet scanning for proteins or nucleic acids (see below). High-purity reagents are available or can be prepared (see reference 66 for summary).

Sample.

As nearly as possible, the sample should have the same buffer composition, pH, and ionic strength as the spacer gel. High ionic strength must be avoided.

For a small number of bands, 5 to 50 μg of protein per gel is quite sufficient; a single band containing 1 μg of protein is normally visible after staining with Coomassie blue. Care must be taken that some of the sample is not lost to the electrode solution.

Analysis of gels.

Direct scanning. For both proteins and nucleic acids the possibility of direct scanning in a spectrophotometric device is often overlooked. This method is rapid, is amenable to repetitive scans at intervals during electrophoresis to obtain electrophoretic mobility, and is free from the damage that may result from expulsion of gels from tubes. The disadvantages are decreased sensitivity of two- to threefold (for proteins) compared to staining and a requirement for ultraviolet-transmitting tubes, pure reagents, and precise alignment in an optical system designed to work with a circular cross section. Inexpensive ultraviolet-transmitting tubes are easily made by cutting ordinary Vycor tubing (Corning Glass Works) of 5-mm inner diameter to the required length with a diamond saw and candling each tube for uniform optical characteristics. The problem of optical alignment of cylindrical gels is basically the same for both stained and unstained gels. Some scanners are designed to function optimally with such gels.

Analysis after removal of gel. Removed gels can be scanned with or without staining. Stained gels can be scanned or photographed. Unstained gels also can be sectioned and material can be eluted for analysis, or they can be solubilized for analysis, including determination of radioactivity. The gels are removed immediately and stained to avoid diffusion of the bands. The removal process is difficult and may require practice to prevent scratching the gel. One simple procedure involves slow insertion with rotary motion of a syringe needle about 8 cm long while discharging water or 50% glycerol between the wall and the gel. The needle is advanced until the gel is free or can be extruded with pressure from a medicine dropper bulb filled with water. The process can be performed over or in a tray of water.

Staining. Most analyses are done with prior staining. Amido black 10B, Procion brilliant blue RS, and Coomassie blue R250 are commonly used for proteins, with Coomassie blue being the most sensitive. Many other stains are available (see p. 73–79 in reference 66) for proteins, deoxyribonucleic acid, and ribonucleic acid. The literature contains many variants and improvements. For reproducible quantitation using a densitometer, Coomassie blue is said to be inferior to Procion brilliant blue (66).

To remove up to 0.1% sodium dodecyl sulfate, rinse the gels several times in 10% (wt/vol) trichloroacetic acid–33% (vol/vol) methanol in water before staining.

Protein stain (68). To 1 part of 0.2% (wt/vol) aqueous solution of Coomassie brilliant blue G250 add 1 part of 2 N H_2SO_4, and let stand for 3 h. Remove the precipitate by filtration through Whatman no. 1 paper, and add one-ninth volume of 10 N KOH. Then add 100% trichloroacetic acid to a final concentration of 12% (wt/vol).

The gel is placed in the stain solution for 5 to 8 h and then stored in water to give greater sensitivity; however, band intensity is lost by storage in acids. The method gives linear densitometer readings between 1 and 20 μg of protein (serum albumin).

Destaining. The above procedure gives clear backgrounds without destaining. However, many staining procedures require destaining. This can be accomplished either by electrophoresis or by leaching. In electrophoretic destaining, the stain must be stably bound to the protein; otherwise, it is possible that some bands will be lost. Also, an apparatus is needed; however, the process is rapid and gives a lower background. Leaching consists merely of allowing the gels to stand for some hours in a solvent for the dye, usually acetic acid.

Recording results and quantitation. As a minimum, sketches of stained gels can be recorded in a notebook. More quantitative measurements are obtained from: (i) photographs of stained gels followed by a densitometer scan of the negative; (ii) optical scanning of stained or unstained gels (280 nm) to give strip chart recordings of absorbance along the gel length; and (iii) sectioning, elution, and analysis of small (1 mm) cross sections or liquefaction into a stream and segmentation into fractions. With enzymes it is often possible to immerse the intact gel in a catalytic reaction mixture containing the substrate and color-producing system. After incubation for a few minutes, color develops in a disk where the diffusing reactants contact the enzyme. Alternatively, when reagents do not diffuse into the gel the reaction produces a ring around the gel where the band of enzyme is exposed.

Molecular weight of native proteins.

The method of Hedrick and Smith (74) is widely used to determine molecular weights of oligomeric proteins because sodium dodecyl sulfate-gel electrophoresis as described below would cause dissociation into random coiled monomers. The method for native proteins requires a plot of the log of R_m (the log of the ratio of migration of a protein to that of the dye front) for each protein and standard run at several gel concentrations, i.e., at different degrees of cross-linking (Fig. 12A). These plots are linear, and plots of their slopes versus molecular weight are also linear. Hence, the second plot is established with standards of known molecular weight. The slope of the unknown protein derived by plotting log R_m versus gel concentration can then be related to the molecular-weight plot.

Procedure. The gels are formed and run as described above for the general method except that the sample, diluted 1:1 in diluent (below), is carefully layered between the spacer gel and buffer. Following electrophoresis, the dye front is marked by insertion of a no. 32 copper wire and the gels are stained as above. R_m is then obtained for each gel, and a plot of the log of R_m versus gel concentration (at least four concentrations) is constructed. The slopes of these plots for each protein and standard are plotted against the molecular weight for each (Fig. 1). The molecular weight of unknowns is obtained by locating their log R_m values on the line.

Separation gels, pH 7.9:
    5 parts A (Tris·HCl, 0.48 M based on Cl⁻; 0.46 ml of *N,N,N′,N′*-tetramethylethylenediamine/100 ml, pH 7.9)
    10 parts B (24 g of acrylamide, 0.8 g of BIS/100 ml) gives 6% gel
    5 parts water

The gel concentration needed for this work varies between 3 and 15%. These are prepared by diluting a concentrated reagent B to obtain the desired concentration, i.e., the 6% level above. For example, a 3% gel contains 12 g of acrylamide and 0.4 g of BIS per 100 ml.

Spacer gel, pH 5.7:
    2 parts C (imidazole·HCl, 0.48 M in Cl⁻, pH 5.7)
    4 parts D (10 g of acrylamide, 2.5 g of BIS/100 ml)
    2 parts E (4 mg of riboflavin/100 ml)
    8 parts water
Diluent, pH 5.7:
    Imidazole·HCl, 0.06 M in Cl⁻ in 50% glycerol
Reservoir buffer:
    0.034 M asparagine, 5×10^{-5}% bromothymol blue neutralized to pH 7.3 with Tris.

Molecular weight of denatured monomeric proteins or subunits of oligomeric proteins.

An effective way to minimize charge effects of proteins so that migration is related only to size is to carry out the electrophoresis in sodium dodecyl sulfate (SDS). SDS disrupts tertiary and secondary structures. The procedure of Weber and Osborn (80) is widely used with purified proteins as well as membranes, ribosomes, etc., in which case SDS also releases and solubilizes

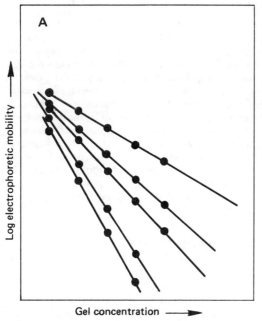
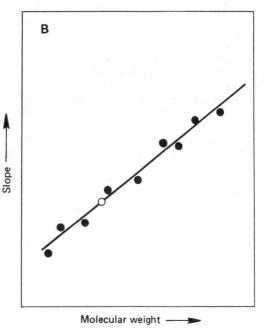

FIG. 12. *(A) Plot of relative mobility (R_m) versus degree of cross-linking (gel concentration) for several reference proteins. (B) Plot of the slopes of lines in (A) versus the molecular weight of the reference proteins. The open point represents the unknown protein.*

otherwise insoluble components. In this method only one gel concentration is needed, and a plot of relative mobility (distance of dye front/distance of protein band) versus the log of the molecular weight is linear over a fairly wide range of molecular weights. Thus, a plot constructed with proteins of known molecular weight is used to find the molecular weight of the unknown (Fig. 12B). For a given gel concentration, i.e., 10%, the plot is linear over a range of molecular weights, i.e., 10,000 to 100,000 (71). With other gel concentrations the range is either higher or lower.

Procedure. The separation gel containing SDS is poured, water layered, etc., as above for discontinuous gel electrophoresis. However, no spacer gel is used, and the sample is added directly to the gel surface.

Protein solutions (0.2 to 0.6 mg/ml) are incubated at 37°C for 2 h in 0.01 M sodium phosphate buffer, pH 7.0–1% SDS–1% β-mercaptoethanol. To obtain complete unfolding and dissociation, it may be necessary to boil the solution, add urea to 8 M, and/or reduce and carboxymethylate the thiol groups (70). In many cases the solution can be used directly, but if salts interfere, the sample is dialyzed for several hours at room temperature against 500 ml of 0.01 M sodium phosphate buffer, pH 7.0, containing 0.1% SDS and 0.1% β-mercaptoethanol.

For sample preparation, mix 3 μl of tracking dye (0.05% bromothymol blue), 1 drop of glycerol, 5 μl of mercaptoethanol, and 50 μl of reagent A (below) in a small tube. Add 10 to 50 μl of protein solution, mix, and apply to the surface of the gel. Reagent A diluted 1:1 is carefully layered over the sample to fill the tubes. The electrode compartments are filled with the buffer of reagent A diluted 1:1. Tubes 10 cm by 6 mm are used (80), and electrophoresis is performed with constant current of 8 mA per tube. Time required is 4 h. Gels are removed, stained, and measured as usual using a 32-gauge copper wire to mark the dye front. Usually 10 μg of protein is used, but this can be less if the stain is sensitive enough to detect the band. With 100 μg, the band has a sharp leading edge and a diffuse trailing edge. However, measurements using the leading edge give good results.

10% Separation gel:
15 parts A (0.78 g of $NaH_2PO_4 \cdot H_2O$, 3.86 g of $Na_2HPO_4 \cdot 7H_2O$, 0.2 g of SDS/100 ml), deaerated
13.5 parts B (22.2 g of acrylamide, 0.6 g of BIS/100 ml) filtered through Whatman no. 1 paper. Keep at 4°C in a dark bottle. For increased or decreased cross-linking, vary the amount of BIS accordingly. Deaerate after mixing A and B.

1.5 parts C (1.5 g of ammonium persulfate/100 ml), freshly prepared
0.045 parts TEMED

Molecular weight of nucleic acids.

The R_m (electrophoretic mobility) of low-molecular-weight ribonucleic acid (76) and ribosomal ribonucleic acid (75) is inversely related to sedimentation coefficient. Hence, the log molecular weight is inversely linear with mobility (78). The range of this proportionality depends on gel cross-linking.

Fractionation of low-molecular-weight oligodeoxynucleotides has been accomplished (72), but molecular weight measurement on high-molecular-weight deoxyribonucleic acids has not become routine.

Micromethod. Microgram (10^{-6} g) to picogram (10^{-12} g) amounts of protein can be separated on gel columns in capillary tubes of various sizes by the discontinuous method. The procedure is basically similar but requires apparatus adapted to small-scale filling, electrophoresing, extruding, and analyzing operations.

Preparative procedures. Conversion of the above method to a preparative procedure of similar attributes is not achieved simply by expanding dimensions of the gel column. For increased capacity, a large series of small gels can be run, but this is laborious. With large gel columns provision must be made for recovery of bands, dissipation of heat, support of a large gel column, etc. Commercial units in one way or another attempt to diminish the effect of these problems. In addition, retrieval of the desired component may require cutting out bands and elution from the gel by diffusion or electrophoresis. Alternatively, the bands are allowed to exit from the gel where they are swept away to a fraction collector by a continuous flow of buffer. Such units require that the gel be maintained in the column by hydrostatic equilibrium between upper and lower buffers, that the temperature rise be controlled as much as possible by a circulating system, and that volume of the elution chamber be small. A number of gel and buffer systems have been developed for preparative use (66).

16.5.2. Immunoelectrophoresis

See reference 65.

16.5.3. Isoelectric Focusing (63)

A special kind of electrophoresis involves migration of charged molecules in an electric field across a liquid column with a stable pH gradient. Amphoteric molecules such as proteins change charge as they traverse the pH gradient and come to a region where the pH is the same as the isoelectric point or pI of the protein. At this point of equal charge, the protein no longer migrates, and the band is maintained in an equilibrium between the forces of diffusion and electrophoretic movement. Thus, separation is based on different pI values. It is necessary to stabilize the column against convection by use of a sucrose density gradient, Sephadex, or polyacrylamide. The column is charged with a commercially produced material (Ampholine) which is a mixture of ampholytes of different pI values over a specified range. In an electric field the various ampholyte molecules arrange themselves according to pI to produce a stable pH gradient of high conductivity and buffering capacity. A mixture of proteins then introduced separates into zones at the pI values of the components of the mixture. Ampholines are available over wide and narrow pH ranges.

Isoelectric focusing is used in both preparative and analytical work. For analytical work procedures very similar to those of polyacrylamide disc gel electrophoresis are used, in which the polyacrylamide is an anticonvectant of low cross-linking. For preparative-scale separations, a commercial or homemade device can be used.

Analytical procedures.

Four polyacrylamide isoelectric focusing systems are described in reference 66. The system of Catsimpoolas (69) is given in the next section. Fill the gel tube and photopolymerize the polyacrylamide as in disc gel work. The protein sample can be added to the polymerizing system, in which case the water is diminished by the volume of the sample. Alternatively, add a layer of Ampholine (2.5% in 5% sucrose) and carry out electrophoresis for 30 min to set up the gradient. Then carefully layer the sample in 10% sucrose under the Ampholine at the top of the gel. For gel columns of 5 by 65 mm, 30 to 300 μg of protein should be applied. Perform the electrofocusing by running at a constant 150 V with the current decreasing from 5 to 0.5 mA.

Remove the gels and stain them by one of several procedures. The following is from Grässlin et al. (73). Stain the gel with gentle agitation for 30 to 60 min in 0.2% Coomassie brilliant blue G250 in ethanol-water-glacial acetic acid (45:50: 5, vol/vol). Destain for 2 h by treatment with three changes of ethanol-water-glacial acetic acid (40:55:5).

The pH gradient across the gel can be determined by making a longitudinal section of the gel column before staining. The slice is cut into sections, each section is soaked in a minimum of water, i.e., 0.5 ml or less, for 1 h, and the pH of the water is determined with a miniature or microcombination glass electrode.

For proteins of very high molecular weights, the molecular sieving effect can be diminished by use of 0.5% agarose-2% polyacrylamide (79). Methyl red (0.5 ppm) can be added to the lower electrode solution. It becomes yellow as it moves into the more alkaline region of the gel.

Gel for isoelectric focusing (69):
  4.5 parts water
  2 parts A (acrylamide, 25 g; BIS, 1 g/100 ml of water)
  0.5 part B (TEMED, 1.0 ml/100 ml of water)
  2.5 parts C (riboflavin, 4 mg/100 ml of water)
  0.5 part 40% Ampholine

The upper (anode) solution is 5% phosphoric acid and the lower (cathode) solution is 5% ethylenediamine.

16.5.4. Two-Dimensional Separation

It is well known that a single band on disc gel electrophoresis can contain more than one protein and that incomplete separation of complex mixtures can occur. The most effective approach is to base the separation on two strategies, i.e., size and pI. Thus, a sample may first be subjected to isoelectric focusing in polyacrylamide. The gel is then removed from the tube, or alternatively a narrow slice is cut from a slab gel. The gel strip is then laid along the top edge of a slab gel in the position where the normal slots would be and is held in place by adding molten agar. The second stage, SDS-gel electrophoresis, is carried out and the slab is stained. A particularly effective example of the method is described by O'Farrell (77), who separated 1,000 proteins from an *Escherichia coli* extract.

16.6. MANOMETRY

Volumetric measurement of gas exchanges has historically been a major tool in metabolic research. For many applications it still is the method of choice, and at times it is the only method available. However, the complexity of the system together with the advent of other methods such as ion-specific electrodes and gasliquid chromatography have greatly diminished the popularity of the manometric method until now it threatens to become a "lost art." Certainly, it is difficult to "teach" manometry in formal laboratories with the insufficient time available. Thus, the investigator becomes proficient with manometry only in a research setting. In view of the fact that manometric methods can measure oxygen, hydrogen, CO_2, methane, and hydrogen ion production or utilization (all essentially continuously), the method is directly applicable to a wide spectrum of uses with tissue homogenates, tissue cells, bacterial suspensions, and enzyme preparations. The spectrum is further broadened by indirect methods, such as assays for kinases based on hydrogen ion production.

The main source of information in this field and indeed an excellent reference on many aspects of metabolism is *Manometric Techniques* by Umbreit et al. (82).

16.7. LITERATURE CITED

Photometry

General references:
1. **Bauman, R. P.** 1962. Absorption spectrophotometry. John Wiley & Sons, Inc., New York.
2. **Brewer, J. M., A. J. Pesce, and R. B. Ashworth.** 1974. Experimental techniques in biochemistry. Prentice-Hall, Inc., Englewood Cliffs, N.J.
   Chapter 7 treats the theoretical aspects of light adsorption and fluorescence emission.

Specific articles:
3. **Burke, R. W., and R. Mavrodineonu.** 1977. Natl. Bur. Stand. Spec. Publ. 446.
4. **Lucas, D. H., and R. E. Blank.** 1977. Am. Lab. **9**(11):77–89.
5. **Slavin, W.** 1963. Anal. Chem. **35**:561–566.
6. **Vanous, R. D.** 1978. Am. Lab. **10**(7):67–79.
7. **West, M. A., and D. R. Kemp.** 1977. Am. Lab. **9**(3):37–49.

Ion Electrodes

General references:
8. **Buck, R. P.** 1978. Ion selective electrodes. Anal. Chem. **50**:17R–29R.
9. **Hitchman, M. L.** 1978. Measurement of dissolved oxygen. Chem. Anal. Ser. Monogr. Anal. Chem. **49**:1–255.
10. **Long, C.** 1961. Biochemists handbook, p. 21. Van Nostrand Co., Princeton, N.J.
11. **Westcott, C. C.** 1978. Biomedical pH measurements. Am. Lab. **10**(3):131–145.
12. **Westcott, C. C.** 1978. Selection and care of pH electrodes. Am. Lab. **10**(8):71–73.

Specific articles:
13. **Estabrook, R. W.** 1967. Methods Enzymol. **10**:41–47.
14. **LeBlanc, P. J., and J. F. Sliwinski.** 1973. Am. Lab. **5**(7):51–54.
15. **Rechnitz, G. A.** 1974. Am. Lab. **6**(2):13–21.
16. **Toth, K., and E. Pungor.** 1976. Am. Lab. **8**(6):9–18.

Chromatography

General references:
17. **Brewer, J. M., A. J. Pesce, and R. B. Ashworth.** 1974. Experimental techniques in biochemistry. Prentice-Hall, Inc., Englewood Cliffs, N.J.
18. **Determann, H.** 1968. Gel chromatography. Springer Verlag, New York.
19. **Dunlap, R. B. (ed.).** 1974. Immobilized biochemical and affinity chromatography. Plenum Publishing Corp., New York.
20. **Gelotte, B., and J. Porath.** 1967. *In* E. Heftman (ed.), Chromatography, 2nd ed., p. 343–369. Reinhold Publishing Corp., New York.
21. **Heftman, E.** 1975. Chromatography. Van Nostrand-Reinhold, New York.
22. **Jakoby, W. B., and M. Welchek (ed.).** 1974. Methods Enzymol. **34**:3–171.
23. **Khym, J. X.** 1974. Ion exchange procedures in chemistry and biology: theory, equipment, technology. Prentice-Hall, Inc., Englewood Cliffs, N.J.
24. **J. J. Kirkland (ed.).** 1971. Modern practice of liquid chromatography. Wiley Interscience, New York.
25. **McNair, H. M., and E. J. Bonelli.** 1967. Basic chromatography. Varian Aerograph Co., Walnut Creek, Calif.
26. **Mitruka, B. M.** 1975. Gas chromatographic applications in microbiology and medicine. John Wiley & Sons, Inc., New York.
   This book gives much emphasis to methods applicable to the study of microorganisms.
27. **Pharmacia Fine Chemicals.** Sephadex. Gel filtration in theory and practice. Pharmacia Fine Chemicals AB, Uppsala, Sweden.
28. **Whatman Inc.** 1977. Liquid chromatography guide. Bulletin No. 123. Whatman Inc., Clifton, N.J.
   This is one of several commercial sources of information, particularly on chromatography media.

Specific articles:
29. **Alcock, N. W.** 1969. Methods Enzymol. **13**:397–415.
30. **Andrews, P.** 1964. Biochem. J. **91**:222–233.
31. **Asteriadis, G. T., M. A. Armbruster, and P. T. Gilham.** 1976. Anal. Biochem. **70**:64–74.
32. **Axen, R., J. Porath, and S. Ernback.** 1967. Nature (London) **214**:1302–1304.
33. **Blattner, F. R., and H. P. Erickson.** 1967. Anal. Biochem. **18**:220–227.
34. **Brooks, J. B., V. R. Dowell, D. S. Forsby, and A. Y. Armfield.** 1970. Can. J. Microbiol. **16**:1071–1078.
35. **Brooks, J. B., C. W. Moss, and V. R. Dowell.** 1969. J. Bacteriol. **100**:528–530.
36. **Cuatrecasas, P., and I. Parikh.** 1972. Biochemistry **11**:2291–2299.
37. **Farshtchi, D., and C. W. Moss.** 1969. Appl. Microbiol. **17**:262–267.
38. **Fuller, A. J.** 1938. Br. J. Exp. Pathol. **19**:130–139.
39. **Fullmer, C. S., and R. H. Wasserman.** 1979. J. Biol. Chem. **254**:7208–7212.
40. **Horii, Z., M. Makita, and Y. Tamura.** 1965. Chem. Ind. (London) **34**:1494.
41. **Hummel, J. P., and W. J. Dreyer.** 1962. Biochim. Biophys. Acta **63**:530–532.
42. **Krishnan, N., and G. Krishna.** 1976. Anal. Biochem. **70**:18–31.
43. **Laine, R. A., W. J. Esselman, and C. C. Sweeley.** 1972. Methods Enzymol. **28**:159–178.

44. **LaNoue, K., W. J. Nicklas, and J. R. Williamson.** 1970. J. Biol. Chem. **245:**102–111.
45. **Levin, O.** 1962. Methods Enzymol. **5:**27–33.
46. **Peterson, E. A., and H. A. Sober.** 1962. Methods Enzymol. **5:**3–27.
47. **Reinhold, V. N.** 1972. Methods Enzymol. **25:**244–249.
48. **Sober, H. A., F. J. Gutter, M. M. Wykoff, and E. A. Peterson.** 1956. J. Am. Chem. Soc. **78:** 756–763.
49. **Sweeley, C. C., W. W. Wells, and R. Bentley.** 1966. Methods Enzymol. **8:**95–108.
50. **Whitaker, J. R.** 1963. Anal. Chem. **35:**1950–1953.
51. **Yamakawa, T., and N. Ueta.** 1964. Jpn. J. Exp. Med. **34:**361–374.

Radioactivity

General references:
52. **Brewer, J. M., A. J. Pesce, and R. B. Ainsworth.** 1974. Experimental techniques in biochemistry. Prentice-Hall, Inc., Englewood Cliffs, N.J.
    Chapter 8 gives a good theoretical treatment of measurement of radioactivity.
53. **Horrocks, D. D., and C.-T. Peng.** 1971. Organic scintillators and liquid scintillation counting. Academic Press Inc., New York.
54. **Oldham, K. G.** 1976. Radiochemical methods in enzyme assay. Amersham Searle Co., Arlington Heights, Ill.

Specific articles:
55. **Anderson, L. E., and W. O. McClure.** 1973. Anal. Biochem. **51:**173–179.
56. **Anker, H. S.** 1970. FEBS Lett. **7:**293.
57. **Bush, E. T.** 1963. Anal. Chem. **35:**1024–1029.
58. **Fricke, U.** 1975. Anal. Biochem. **63:**555–558.
59. **Jeffay, H.** 1962. Oxidation techniques for preparation of liquid scintillation samples. Packard Technical Bulletin 10, p. 1–8. Packard Instrument Co., Inc., Rockville, Md.
60. **Jeffay, H., and J. Alvarez.** 1961. Anal. Chem. **33:**612–615.
61. **Patterson, M. S., and R. C. Green.** 1965. Anal. Chem. **37:**854–857.
62. **Turner, J. C.** 1967. Sample preparation for liquid scintillation counting. The Radiochemical Centre, Amersham, England.

Gel Electrophoresis

General references:
63. **Catsimpoolas, N.** 1975. Isoelectric focusing: fun-damental aspects. Sep. Sci. **10**(1):55–76.
64. **Davis, B. J.** 1964. Disc electrophoresis. II. Method and application to human serum proteins. Ann. N.Y. Acad. Sci. **121:**404–427.
65. **Kochwa, S.** 1980. Electrophoretic and immuno-electrophoretic characterization of immuno-globulins, p. 121–134. *In* N. R. Rose and H. Friedman (ed.), Manual of clinical immunology, 2nd ed. American Society for Microbiology, Washington, D.C.
66. **Mauer, H. R.** 1971. Disc electrophoresis and related techniques of polyacrylamide gel electrophoresis. W. de Gruyter, New York.
67. **Ornstein, L.** 1964. Disc electrophoresis. I. Background and theory. Ann. N.Y. Acad. Sci. **121:** 321–403.

Specific articles:
68. **Blakesly, R. W., and J. A. Boezi.** 1977. Anal. Biochem. **82:**580–582.
69. **Catsimpoolas, N.** 1968. Anal. Biochem. **26:**480–482.
70. **Crestfield, A. M., S. Moore, and W. H. Stein.** 1963. J. Biol. Chem. **238:**622–627.
71. **Dunker, A. K., and R. R. Rueckert.** 1969. J. Biol. Chem. **244:**5074–5080.
72. **Elson, E., and T. M. Jovin.** 1969. Anal. Biochem. **27:**193–204.
73. **Grässlin, D., H. Weicker, and D. Z. Barwich.** 1970. Z. Klin. Chem. Klin. Biochem. **8:**288–291.
74. **Hedrick, J. L., and A. J. Smith.** 1968. Arch. Biochem. Biophys. **126:**155–164.
75. **Loening, U. E.** 1969. Biochem. J. **113:**131–138.
76. **McPhie, P., J. Hounsell, and W. B. Gratzer.** 1966. Biochemistry **5:**988–993.
77. **O'Farrell, P.** 1975. J. Biol. Chem. **250:**4007–4021.
78. **Peacock, A. C., and C. W. Dingman.** 1968. Biochemistry **7:**668–674.
79. **Riley, R. F., and M. K. Coleman.** 1978. J. Lab. Clin. Med. **72:**714–720.
80. **Weber, K., and M. Osborn.** 1969. J. Biol. Chem. **244:**4406–4412.
81. **Wright, G. L.** 1968. Canalco Newsletter, no. 10, p. 6. Canalco, Inc., Rockville, Md.

Manometry

General reference:
82. **Umbreit, W. W., R. H. Burris, and J. F. Stauffer.** 1957. Manometric techniques. Burgess Publishing Co., Minneapolis.

Chapter 17

Chemical Composition

R. S. HANSON AND J. A. PHILLIPS

The methods selected for this chapter are commonly used in research, but also are suited to advanced undergraduate and graduate teaching laboratories. Many other methods and more detailed descriptions of those selected are presented in publications listed at the end of the chapter. Techniques which require costly instrumentation or significant training are not described. Specialized methods (which vary with nature of the samples, equipment, and expertise) are treated by providing references to other sources of information rather than by describing experimental details.

17.1. MAIN CHEMICAL COMPONENTS: FRACTIONATION AND RADIOACTIVITY DETERMINATION

The incorporation of radioactively labeled compounds into bacterial cells is employed as a means of estimating the rates and amounts of synthesis as well as the distribution of the main small-molecule and macromolecule fractions of the cells (2). After the cells have been incubated with the labeled substrate and the various molecular components have been fractionated, the radioactivity in each fraction can be determined.

Macromolecules are precipitated by cold dilute solutions of trichloroacetic acid. The (denatured) macromolecules are separated from the small molecules by centrifugation. The supernatant fraction of small molecules contains the pooled metabolites of sugars, sugar derivatives, organic acids, amino acids, nucleotides, and coenzymes; oligonucleotides and basic proteins with fewer than 20 nucleotide or amino acid residues are also contained in this fraction (2). The precipitate fraction of macromolecules is sequentially extracted with organic solvents for lipids, alkali for ribonucleic acids (RNAs), and hot trichloroacetic acid for deoxyribonucleic acids (DNAs); the precipitate remaining after these extractions contains the cell proteins. A flow diagram of this fractionation procedure is shown in Fig. 1.

No single procedure gives complete separation of all of the chemical components and total recovery of a given type of macromolecule in a single fraction. The procedures described by Kennell (1) are presented below. They are modifications of the methods of Roberts et al. (2).

17.1.1. Reagents and Materials

Saline solution: 0.85% (wt/vol) NaCl in distilled water
Trichloroacetic acid solutions (*CAUTION*: Trichloroacetic acid causes skin irritation, penetrates skin, and is dissolved in lipids. Avoid contact by wearing protective gloves, appropriate laboratory clothing, and eye goggles.): 50, 20, 10, and 5% (wt/vol) trichloroacetic acid in distilled water
Ethanol (*CAUTION*: Ethanol is flammable. Avoid using near open flames and electrical appliances that cause sparks.): 70% (vol/vol) in distilled water
Diethyl ether (*CAUTION*: Ether is very flammable and may explode when vapors contact flames or electrical sparks. Keep the minimum amount necessary in a laboratory. Warm all solutions using

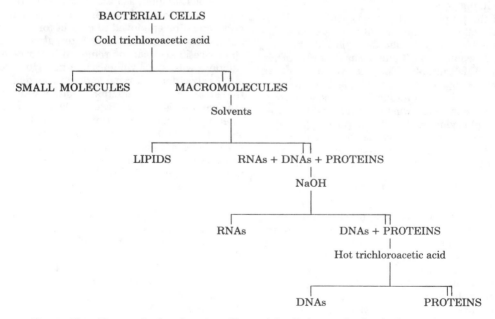

FIG. 1. *Flow diagram for fractionation of bacterial cells into main chemical components.*

sealed containers in water baths rather than incubators. Avoid inhalation of fumes by working in a proper hood.)

Sodium hydroxide (0.5 N): Dissolve 2.0 g in 100 ml of distilled water.

Bovine serum albumin: Dissolve 0.1 g in 10 ml of distilled water. Store in a refrigerator.

Glassware: 20-ml Pyrex glass beakers, 10-ml Erlenmeyer flasks and 25-ml volumetric flasks, capped 15- and 30-ml centrifuge tubes

Filters: Glass-fiber or synthetic membrane filters, 0.45-μm pore size and 25-mm diameter. The filters must be acid and solvent resistant. Pyrex filter holders with fritted-glass filter supports.

Vacuum filtering flask: 125-ml flask and test tubes cut to fit below the filter holder inside the flask. The test tubes are used to collect filtrates.

Heating lamp: 250-W infrared heat lamp, and a suitable reflector for sample drying

Ice-water bath

Water baths: Adjustable to 80°C

Refrigerated centrifuge: Capable of achieving 10,000 × g with a centrifuge head that holds 30-ml centrifuge tubes

17.1.2. Labeling of Cells

(*CAUTION*: The use of radioactive isotopes is strictly regulated. All personnel involved in their use must be familiar with the regulations [part 16.4.1]. All precautions should be observed.)

The volume of culture to be used will vary with the organism and growth conditions. For example, 30 ml of a mid-exponential-phase culture of *Escherichia coli* is adequate to provide the 10 mg of cells required.

If the purpose of the experiment is to determine cell composition, a uniformly labeled substrate such as [^{14}C]glucose (5 μCi/30 ml of a culture containing 1% glucose as the only carbon source) is suitable. The incorporation of an amino acid into protein can be measured by adding 2 to 5 μCi of a labeled amino acid to 30 ml of a culture growing on a mineral salts medium plus glucose. Ideally, an auxotrophic mutant that is unable to synthesize the amino acid should be used when incorporation into protein is measured. In this case, 10 μg of the unlabeled amino acid per ml and 2 to 5 μCi of the labeled amino acid should be added to the medium. It is important to use a non-metabolizable amino acid as a labeled protein precursor. Leucine, lysine, phenylalanine, and tyrosine are not significantly metabolized by most bacteria. When labeled thymidine is used as a precursor of DNA, and when uracil or uridine is used as a precursor of RNA, auxotrophs should also be employed in order to obtain incorporation of a larger fraction of the label added to the medium.

The cultures should be harvested during exponential growth when cell composition studies are performed. The cell composition changes during the lag and stationary phases of growth.

17.1.3. Small Molecules

1. Cool the culture rapidly by pouring it over ice. Centrifuge the radioactively labeled cells (approximately 10 mg, dry weight) from the growth medium (5,000 × g for 10 min at 5°C), and wash the sedimented cells twice by centrifugation with ice-cold (5°C) 0.85% saline (one-fifth of the culture volume).

2. Suspend the cells with 10 ml of distilled water in the centrifuge tube. Place the suspension in an ice bath until cold, and add 10 ml of ice-cold 20% trichloroacetic acid. Cap the centrifuge tube, and mix the contents by inversion. Place the tube in the ice bath for 5 min to allow complete precipitation.

3. Centrifuge (5,000 to 10,000 × g for 15 min) at 0 to 4°C. Carefully decant the supernatant fraction, without disturbing the precipitate, into a 25-ml volumetric flask.

4. Add 4 ml of ice-cold 10% trichloroacetic acid to the precipitate, suspend it, and centrifuge again. Decant the supernatant fraction into that obtained after the first centrifugation. Adjust the volume of the combined supernatant fractions to 25 ml with distilled water.

5. Determine the radioactivity of a sample of the combined supernatant fraction containing the small molecules (metabolite pools) directly in a solution used for determining radioactivity in aqueous samples, as described in the preceding chapter (16.4.2). Internal or external standards should be employed to correct for quenching by trichloroacetic acid. Alternatively, trichloroacetic acid can be removed by three extractions with equal volumes of ether. However, ether extraction will remove organic acids from the small-molecule fraction. Samples of the ether-extracted portion can be counted separately in a counting solution used for aqueous solvents or can be dried in vials with the assistance of a heat lamp; after drying, the residue can be counted in a toluene-based scintillation cocktail (see 16.4.3).

17.1.4. Lipids

1. Proceed to separate lipids from the precipitated fraction of macromolecules in the centrifuge tube (see above). Suspend the sediment in 10 ml of ice-cold 10% trichloroacetic acid with the aid of a Vortex mixer. (*CAUTION*: Cap the tube to avoid aerosols of trichloroacetic acid and radioisotopes.)

2. Filter 1 ml of the homogenous suspension, and wash the filter twice with 2 ml of ice-cold

10% trichloroacetic acid. Wash the filter with 3 ml of ice-cold 70% ethanol to remove residual trichloroacetic acid, which may cause solubilization of some protein in the next step. Discard the filtrate.

3. Place the filter holder (with the filter and suction flask) and a test tube for collecting the filtrate in an incubator at 45°C, and allow the assembly to warm to temperature. Then remove the filter flask from the incubator, quickly add 10 ml of 70% ethanol (warmed in a water bath to 45°C), and allow the solvent to pass slowly through the filter. Collect the filtrate in a test tube. The filtration should take approximately 10 min to effect the extraction.

4. Wash the filter in a hood with 2- to 5-ml volumes of ethanol-diethyl ether (1:1, vol/vol) warmed to 45°C in a water bath by allowing the solvents to pass slowly through the filter (10 min). Collect the filtrate in a test tube and combine it with a 45°C ethanol wash. Retain the filter containing the residual precipitated fraction of macromolecules for further fractionation.

5. Adjust the volume of the combined filtrates to 25 ml, and dry a sample in a scintillation vial under a stream of filtered air at 40°C. Air filtration can be accomplished by passing compressed air through glass wool.

6. Determine the radioactivity of the dried lipid sample by use of a toluene-based scintillation cocktail (see 16.4.3).

17.1.5. Ribonucleic Acids

1. Proceed to separate RNA from the precipitated fraction of macromolecules, with the lipids now removed. Remove the filter containing the precipitate from the filter holder (step 4 above), and place upside down in a 20-ml beaker. Allow the filter to air dry in a hood (about 10 min).

2. Add 2 ml of 0.5 N NaOH (warmed to 37°C) to the beaker, and incubate with occasional shaking at 37°C for precisely 40 min. Make sure the filter is completely covered with the NaOH solution. Longer incubation causes hydrolysis of some proteins (1, 3).

3. Transfer the beaker to an ice bath, and cool rapidly. Add 0.6 ml of ice-cold 50% trichloroacetic acid, mix, and leave the beaker in the ice bath for about 30 min.

4. Transfer the liquid in the beaker to a new filter held in a filter holder; apply a vacuum, and collect the acid-soluble products in a test tube. Rinse the beaker and original filter with 3 ml of ice-cold 5% trichloroacetic acid, transfer the liquid to the same filter, and collect the filtrate in the same test tube. Adjust the volume of the combined filtrates to 15 ml in a volumetric flask with distilled water. The filtrate contains about 95% of the ribonucleotides in RNA.

5. Determine the radioactivity of the hydrolyzed RNA fraction in the same way as for the small-molecule fraction (see above and 16.4.3).

17.1.6. Deoxyribonucleic Acid

1. After the removal of lipids and RNA from the macromolecular fraction, the DNA is separated from the precipitated fraction of macromolecules by treatment with hot trichloroacetic acid. Use a duplicate sample from the suspended sediment in cold trichloroacetic acid, as described in the first steps of the lipid fractionation (17.1.4). Repeat steps 1 through 4 for removal of lipids. After washing the filter with ethanol and ethanol-ether to remove lipids, place the filter in a beaker, add 2 ml of 0.5 N NaOH, mix, and incubate for 90 min at 37°C. The incubation time with NaOH is increased in order to give total hydrolysis of RNA. DNA is present in smaller amounts than RNA, and traces of unhydrolyzed RNA can significantly increase estimates of radioactivity in DNA. The solubilization of some proteins will not affect the estimations of DNA.

2. Cool the sample on ice, add 0.6 ml of cold 50% (wt/vol) trichloroacetic acid, and keep it on ice for 30 min; then pour the liquid in the beaker onto a new filter that has been presoaked in cold 5% trichloroacetic acid. Apply a vacuum to facilitate filtration. Wash the original filter twice with 3 ml of ice-cold 5% trichloroacetic acid, and discard the filtrate.

3. Place the two filters in a 10-ml Erlenmeyer flask, add 3 ml of 5% trichloroacetic acid, and heat for 30 min in a water bath at 80°C. Make sure the filters are entirely immersed in the liquid.

4. Add 0.1 ml of a solution (1.0 mg/ml) of bovine serum albumin, mix rapidly, cool the flask for 30 min in an ice bath, and filter the entire contents of the Erlenmeyer flask through a new filter. The bovine serum albumin ensures precipitation of minute amounts of macromoles which may otherwise remain in suspension. Use a slight vacuum to increase the filtration rate. Collect the filtrate in a test tube placed inside the vacuum flask. Wash the Erlenmeyer flask and filters twice with 3 ml of cold 5% trichloroacetic acid. Collect the filtrates in the same test tube. Adjust the volume of the combined filtrates to 15 ml with water in a volumetric flask. Discard the filters.

5. Determine the radioactivity of the hydrolyzed DNA fraction in the same way as for the small-molecule fraction (see above and 16.4.3).

17.1.7. Proteins

1. Proceed to separate proteins from the precipitated fraction of macromolecules. Use a du-

plicate sample from the suspended sediment in cold trichloroacetic acid, as described in the first step of the lipid fractionation. Transfer 1 ml into a 15 by 100 mm test tube. Dilute with 1 ml of water to give a trichloroacetic acid concentration of 5% (wt/vol).

2. Heat the diluted sample for 30 min at 80°C to hydrolyze both RNA and DNA.

3. Cool the hydrolyzed sample on ice for 30 min to precipitate proteins. Collect the precipitate on a filter, and discard the filtrate. Wash the test tube, and filter twice with 2 ml of ice-cold 10% trichloroacetic acid and once with 3 ml of ice-cold 70% ethanol, to remove residual trichloroacetic acid that would otherwise solubilize some protein in the next step.

4. Wash the filter twice with 5 ml of 70% ethanol (45°C) and twice with 5.0 ml of ethanol-diethyl ether at 45°C.

5. Air dry the filter, place it in a scintillation vial, and dry it under a heat lamp.

6. Determine the radioactivity of the dried protein by use of a toluene-based scintillation cocktail (see 16.4.3).

17.1.8. Applications

The techniques described above are usually applied to the fractionation of bacterial cells labeled with a medium component that is nonspecifically incorporated into all of the cell molecules. When bacterial cells are incubated in the presence of a specific precursor of a macromolecule in order to measure its rate of synthesis, it is important to determine that the macromolecule of interest contains all of the radioactivity. This can be determined by fractionating the cells and estimating the radioactivity of each fraction by the techniques described above. If all of the radioactivity occurs in the fraction of interest, subsequent experiments can be simplified. For example, if radioactive thymidine is used to measure the amount of DNA synthesis and preliminary experiments indicate that it is not found in other fractions, the cells in their culture medium can be cooled and treated with ice-cold 10% trichloroacetic acid. The precipitate containing the DNA can then be collected on filters, washed with cold ethanol, dried, and counted. This procedure is suitable for measuring the amount of DNA synthesis. The same technique can be applied to measure the amount of protein or RNA synthesis if the precursor appears only in the one fraction.

Other factors must be considered in calculating the rates of synthesis. A primary concern is the dilution of exogenously added substrates by unlabeled precursors in the cell pools. In kinetic experiments it is necessary to determine the rate

of saturation of the pool by radioactive precursors.

While the methods described above provide a means of obtaining an estimate of the distribution of carbon from a ^{14}C-labeled compound in bacteria, several cell components are not accounted for. Products of hydrolysis of teichoic acids and polyphosphates will appear in the RNA fraction, and peptidoglycan will be precipitated with protein.

Poly-β-hydroxybutyrate, if present, can be extracted with chloroform-methanol (2:1, vol/vol) at 60°C after the ether-ethanol washes. If this solvent is used, care should be taken to use filters that are stable in methanol and chloroform. (*CAUTION*: the extraction should be performed in a hood. See 17.3.1.)

Polysaccharides, if present, are partially extracted by cold trichloroacetic acid and can significantly affect estimates of the small molecules. When present in large amounts, polysaccharides present a variety of troublesome problems if nonspecific precursors of macromolecules such as radioactive glucose are used. When these problems occur, modifications of these procedures must be employed (3).

Some bacterial cells leak part of their pools of small molecules when chilled during centrifugation and washing. Some cells have high activities of nucleases or proteases that may be activated during harvesting and washing.

Significant losses of macromolecules occur when smaller amounts of material than indicated are used in the first steps. It is possible to add unlabeled bacterial cells as a carrier to facilitate the precipitation of macromolecules. The precipitation of DNA and protein can also be facilitated by the addition of serum albumin to the solutions.

Each organism potentially presents a unique problem in fractionation. Although the procedures described above have been widely used, the data obtained should be interpreted with proper regard for the problems of obtaining total separation of the small molecules and macromolecules (3).

Sequential methods for treating a single sample have been described (2, 3). However, they are apt to give greater cross-contamination of fractions than the procedure described here.

17.2. CARBOHYDRATES

Bacteria contain a large number of different carbohydrates: free sugars, sugar derivatives, simple polysaccharides (polymers of a single sugar or sugar derivative, like glycogen), complex polysaccharides (polymers of more than one different sugar, amino sugars, uronic acids, etc.), and macromolecules that contain both carbohydrate and noncarbohydrate components (lipopolysaccharides, peptidoglycans, teichoic acids, lipoteichoic acids, teichuronic acids, nucleic acids, glycoproteins, etc.).

The colorimetric methods based on the Molisch test for carbohydrates (4, 7, 8) are useful for estimating the amount of simple sugars and their polymers. Specific assays for pentoses in nucleic acids (ribose and deoxyribose) are described in the part below that deals with nucleic acids (17.5.1 and 17.5.2). More complete descriptions of analytical methods for the analysis of carbohydrates have been presented by Herbert et al. (5) and Colowick and Kaplan (4).

17.2.1. Total Carbohydrates by Anthrone Reaction

The most convenient assays for total carbohydrates involve the heating of media, cells, or isolated carbohydrates with sulfuric acid to hydrolyze the polysaccharides and dehydrate the monosaccharides to form furfurals from pentoses and hydroxymethyl furfurals from hexoses. The solutions of furfurals and hydroxymethyl furfurals are then treated with a reagent (an aromatic amine or phenol) to produce a colored compound, which is measured in a colorimeter or spectrophotometer. Many variations of this **Molisch test** have been described (4, 7, 8). Pentoses, hexoses, and heptoses, and their derivatives (except the amino sugars), yield colored products in these reactions. Trioses and tetroses do not yield chromogens.

None of the methods accurately measures all of the carbohydrate in bacteria because different sugars give different color intensities at any selected wavelength (5). The anthrone and the phenol methods (described below) are presented because of their simplicity, their relative insensitivity to interference by other cellular components, and their similar response to the most common hexoses in microbial cells, thus providing a reliable index of "total carbohydrates."

Either the anthrone or the phenol procedure is suitable for the estimation of total carbohydrate in bacterial cells that contain 10% or more hexose polymers. The pentoses in nucleic acid interfere when the hexose content of the cells is low.

Reagents and equipment.

Saline solution: 0.85% (wt/vol) NaCl in distilled water
Stock glucose solution: Dissolve 100 mg of glucose in 100 ml of 0.15% (wt/vol) benzoic acid, which is added as a preservative. Store at 5°C. This solution is stable for several months. Dilute 1:10 in distilled water just before use to give a solution containing

100 μg/ml. Prepare standards (10 to 100 μg/ml) from the diluted solution.
Sulfuric acid solution (75%, vol/vol): Add 750 ml of reagent-grade concentrated sulfuric acid to 250 ml of distilled water. (*CAUTION*: Add the acid to the water slowly. Use a Pyrex container. Do not mix by adding the water to the acid. Note other precautions described in part 17.2.2.)
Anthrone reagent: Add 200 mg of anthrone to 5 ml of absolute ethanol, and make up to 100 ml with 75% H_2SO_4. (*CAUTION*: See 17.2.2.) Stir until dissolved. Prepare fresh daily. Store in a refrigerator.
Thick-walled Pyrex boiling tubes: 6 inches by 1 inch diameter (15 by 2.5 cm)
Spectrophotometer or colorimeter
Glass cuvettes

Procedure.

Wash samples of bacterial cells free from medium constituents by centrifugation at 5,000 × g for 10 min. Resuspend in saline solution, and centrifuge again. Resuspend the sedimented cells in distilled water, and pipette samples (0.1 to 1.0 ml) of the cell suspension into the boiling tubes. Adjust the volume of all samples to 1.0 ml with distilled water. Prepare a blank of distilled water (1.0 ml) and standards containing 10 to 100 μg of glucose in 1.0 ml. Chill the tubes and the anthrone reagent in an ice-water bath until cold. Add 5.0 ml of cold anthrone reagent, mix rapidly by swirling the tubes in ice water, and continue the mixing in the ice-water bath for 5 min. Then transfer the tubes to a bath of boiling water for precisely 10 min. Return the tubes to the ice water. Measure the color formed in each tube in a spectrophotometer or colorimeter at a wavelength of 625 nm (16.1.1). Determine the concentration of glucose in the samples from a standard curve prepared by plotting the absorbances of the standards versus the concentration of glucose.

17.2.2. Total Carbohydrates by Phenol Reaction

Reagents and equipment.

Phenol reagent: Dissolve 5 g of reagent-grade phenol in 100 ml of distilled water. (*CAUTION*: Phenol causes severe skin and eye burns. Wear protective goggles, gloves, and clothing when using this reagent. Do not pipette by mouth.)
Reagent-grade concentrated sulfuric acid (*CAUTION*: See below.)
Glucose standard (see anthrone procedure, above)
Thick-walled Pyrex boiling tubes: 6 inches by 1 inch diameter
Spectrophotometer or colorimeter
Glass cuvettes

Procedure.

Pipette the samples to be analyzed into thick-walled Pyrex test tubes. Adjust the volume of all samples to 1.0 ml with distilled water. Prepare a reagent blank with 1.0 ml of distilled water and a set of glucose standards from 10 to 100 μg in 1.0 ml. Add 1 ml of the phenol reagent; mix rapidly and thoroughly. Add 5 ml of concentrated sulfuric acid, mix rapidly, and let stand for 10 min. (*CAUTION*: The addition of sulfuric acid to aqueous solutions causes heating and occasionally boiling. Severe burns occur when this solution contacts skin or eyes. Use protective goggles and gloves, and do not pipette by mouth. Be aware of first-aid measures employed when acids contact skin and eyes.) Place the tubes in a water bath at 25°C for 15 min. Read the absorbance of each tube at 488 nm against the blank prepared without glucose (16.1.1). Determine the concentration of glucose in the samples from a standard curve prepared by plotting the absorbances of the standards versus the concentration of glucose.

17.2.3. D-Glucose

A large number of monosaccharides are liberated by the hydrolysis of simple and complex polysaccharides in bacterial cells. Because the polymers exist as mixtures, except in the homopolymers like glycogen, the estimation of individual sugars presents difficult analytical problems that can sometimes be resolved only by resorting to gas-liquid chromatographic procedures (16). However, some sugars and sugar derivatives can be estimated by specific colorimetric reactions or enzymatic methods (4), with which the two most common sugars are determined as follows.

The enzyme glucose oxidase catalyzes the following reaction:

$$\beta\text{-D-glucose} + O_2 \rightarrow \text{gluconic acid} + H_2O_2$$

The hydrogen peroxide, in the presence of peroxidase and a reduced chromogen, causes the production of a brown- or blue-colored oxidized chromogen that can be measured spectrophotometrically. Different chromogenic compounds are used in kits. The enzyme mixture required for the assay can be purchased as a kit from several commercial sources.

The glucose oxidase test is specific, sensitive, and easily performed. Only glucose and 2-deoxyglucose are known to produce colored products in the assay.

Reagents.

Glucose stock solution: Dissolve 300 mg of D-glucose in 100 ml of 0.15% (wt/vol) benzoic acid. Store in a refrigerator. Dilute 1:10 with distilled water, and adjust to pH 7.0 before use.

Commercial glucose oxidase-peroxidase preparation: Prepare the dry enzyme mixture according to the manufacturer's instructions.

Chromogen: Available as part of the commercial kits. Prepare according to the manufacturer's instructions.

Procedure.

Dilute samples with distilled water so that they contain 0.05 to 0.3 mg of glucose per ml. Adjust the pH between 6.8 and 7.2 with 0.1 N KOH or 0.1 N HCl. A pH indicator paper is adequate. Prepare a blank and a set of standards containing 0.05 to 0.3 mg of glucose in 1.0 ml by dilution of the standard with distilled water. Add the enzyme and chromogen, and incubate the mixture according to the manufacturer's instructions. Add 1 drop of 4 N HCl with a Pasteur pipette, and read the absorbance at 400 nm in glass cuvettes (16.1.1). Determine the concentration of glucose in the samples from a standard curve prepared by plotting the absorbances of the standards versus the concentration of glucose.

17.2.4. Galactose

Galactose oxidase assay kits are also available from commercial sources. The kits contain galactose oxidase, chromogen, and standards. Since galactose oxidase oxidizes the C_6-hydroxymethyl group of galactose to produce galacturonic acid, the enzyme will also oxidize some galactosides.

Applications.

The oxidase tests for glucose and galactose are suitable for estimating the amount of each in growth media, in neutralized hydrolysates of polysaccharides, and in enzyme reaction mixtures, e.g., the measurement of glucose as a product of the enzymatic hydrolysis of cellulose and a variety of other applications. Glucose oxidase oxidizes only β-D-glucose, but in the presence of mutarotase (present in some commercial kits) it estimates both α- and β-D-glucose. It is wise to add a standard to each unknown solution to check for interference by salt and other inhibitors of oxidases when performing either assay.

17.2.5. Glycogen

It is difficult to quantify any simple polysaccharide except glycogen by unsophisticated methods. The method for the isolation of glycogen is presented because this polymer functions as a reserve material in many bacteria and can account for a significant fraction of the biomass in cultures. The methods for isolation of other commonly occurring complex polysaccharides (lipopolysaccharides and peptidoglycan) are presented in separate sections. Glycogen, like other polysaccharides, is resistant to hydrolysis by alkali, but is readily soluble in water and insoluble in ethanol. The quantitative isolation of the polymer takes advantage of these properties. After isolation, the amount of glucose can be estimated in acid hydrolysates by the anthrone or phenol reactions or by the glucose oxidase assay, after hydrolysis for 6 h in 4 N HCl at 100°C in a sealed ampoule.

Reagents and equipment.

Absolute ethanol

Potassium hydroxide (30% wt/vol) in distilled water (*CAUTION*: This reagent is caustic. Do not mouth pipette. Avoid contact of the reagent with skin and eyes.)

Refrigerated centrifuge: Capable of achieving 10,000 × g

30-ml glass centrifuge tubes

Procedure.

Harvest bacterial cells by centrifugation at 5,000 to 10,000 × g for 15 min, resuspend the cells in an equal volume of 0.85% (wt/vol) NaCl, and centrifuge again. Wash the cells by centrifugation once more, and lyophilize them (12.2.1). Treat 100 mg of dry cells in a Pyrex or Kimax screw-capped test tube with a Teflon-lined cap with 1 ml of 30% KOH at 100°C in a steamer or boiling-water bath for 3 h. If lyophilization is not practical, suspend 500 mg of packed wet cells in 0.6 ml of 50% (wt/vol) KOH and heat at 100°C to solubilize the glycogen. Cool without opening the tube. When cool, add 3 ml of distilled water and 8 ml of ethanol to precipitate the glycogen. Centrifuge at 10,000 × g for 15 min, and wash the precipitate with 8 ml of 60% (vol/vol in distilled water) ice-cold ethanol by centrifugation. Dry the precipitate in a vacuum desiccator. The washed precipitate can be analyzed for glucose by the anthrone or phenol method (17.2.1 and 17.2.2) or hydrolyzed to yield glucose, which can be measured by the glucose oxidase procedure (17.2.3). Mechanical disruption of gram-positive cells is occasionally necessary to obtain complete extraction of glycogen (9).

17.2.6. Hexosamines in Purified Peptidoglycans (5, 8–12)

Hexosamines occur in lipopolysaccharides, peptidoglycans, and teichoic acids of bacteria. The most common hexosamines are glucosamine, galactosamine, and muramic acid. All *N*-acetylated hexosamines react with *p*-dimethylaminobenzaldehyde to produce a red-colored product. The test is very sensitive and specific for hexosamines but does not distinguish among glucosamine, muramic acid, and galactosamine. Muramic acid can be estimated separately by other procedures (17.2.8).

The following procedure modified by Ghuysen et al. (11) is more precise and convenient than the Morgan-Elson procedure as originally described. It is applicable to the estimation of total hexosamines in purified peptidoglycans. However, the estimation of hexosamines in complex polymers or bacterial cells requires the separation of hexosamines from neutral sugars by ion-exchange chromatography, as described in the succeeding method.

Reagents, equipment, and supplies.

Morgan-Elson reagent: Dissolve 16 g of *p*-dimethylaminobenzaldehyde in sufficient acetic acid to give a volume of 95 ml. Add 5 ml of concentrated HCl. Dilute 1:5 with acetic acid to prepare the color reagent. (*CAUTION*: Avoid inhalation of the fumes of hydrochloric and acetic acids by working in a fume hood. Wear protective gloves, clothing, and goggles during preparation and use of this reagent.)

3 N HCl

3 N NaOH

Saturated solution of sodium bicarbonate: Add approximately 15 g of $NaHCO_3$ to 100 ml of distilled water, and stir for 20 min at room temperature.

5% (vol/vol) acetic anhydride in distilled water: Prepare just before use, and keep on ice until used. (*CAUTION*: Use in a fume hood to avoid contact of liquid or fumes with skin and eyes. Do not pipette by mouth.)

5% $K_2B_4O_7$ in distilled water

Glucosamine standard: Dissolve 17.92 mg of glucosamine in 10 ml of 3 N HCl. Store the stock solution in a refrigerator. Prepare fresh at weekly intervals. This standard contains 0.2 µmol in 20 µl. Dilute with 3 N HCl to obtain standards containing 0.01 to 0.2 µmol per 20 µl.

Ampoules: 1-ml lyophilization ampoules or 1-ml screw-cap test tubes with Teflon-lined caps

Pyrex test tubes: 3 ml

Spectrophotometer or colorimeter

Glass cuvettes: 1 ml

Procedure.

Lyophilize a sample or several samples containing between 0.01 and 0.1 µmol of hexosamine. Add 20 µl of 3 N HCl to each sample. Prepare standards of glucosamine (0.01 to 0.2 µmol per 20 µl of 3 N HCl). Heat the samples and standards at 95°C for 4 h in a sealed ampoule or tube. Cool and neutralize the hydrolysates with 20 µl of 3 M NaOH. Transfer a 30-µl sample of each hydrolysate and standard to separate 3-ml test tubes. Add 10 μl of a saturated solution of NaHCO$_3$ and 10 μl of freshly prepared 5% acetic anhydride. Allow the tubes to stand at room temperature for 10 min, during which time N-acetylation of the hexosamines occurs; then immerse them in boiling water for precisely 3 min to destroy the unreacted acetic anhydride. Add 50 μl of 5% K$_2$B$_4$O$_7$, mix, and heat in a boiling-water bath for 7 min. Cool, and add 0.7 ml of the color reagent (a 1:5 dilution of Morgan-Elson reagent). Mix again, and incubate at 37°C for 20 min. Read the absorbance of each tube at 585 nm in 1-ml glass cuvettes (16.1.1).

Determine the amount of total hexosamine in each sample from a standard curve prepared by plotting the absorbance at 585 nm versus the amount of hexosamine in each standard (16.1.1). Glucosamine, galactosamine, and muramic acid have the same molar extinction coefficients.

Applications.

As mentioned above, this procedure is not suited for analyses of hexosamines in complex mixtures of sugars and sugar derivatives. Hexosamine derivatives such as N-acetyl-muramic acid substituted with peptides produced by the action of lysozyme on cell walls yield the same colored products.

The nearly exact neutralization of the HCl after the hydrolysis of the samples is critical to the reaction. Therefore, it is important to prepare the HCl and NaOH carefully and to be sure that the ampoules or tubes are tightly sealed during hydrolysis.

Glucosamine, galactosamine, and muramic acid can be estimated separately by a procedure described by Stewart-Tull (15). Other assays for muramic acid are given in part 17.2.8.

17.2.7. Hexosamines in Complex Solutions

Hexosamines in complex solutions and those produced by the hydrolysis of mixed polymers can be separated from neutral sugars that interfere with the Morgan-Elson reaction by ion-exchange chromatography. The purified hexosamines can then be quantified by the modified Morgan-Elson reaction, as described above.

Reagents, equipment, and supplies.

4 N HCl
1 N NaOH
Lyophilization ampoules or Pyrex screw-cap test tubes with Teflon-lined caps.
Dowex-50 (H$^+$) columns: Add Dowex-50 resin (1 g for each sample to be analyzed) to 2 N NaOH (20 ml/g of resin). Mix for 5 to 10 min, and filter onto a Büchner funnel. Wash the resin with distilled water (50 ml/g), and slowly stir the resin with 3 N HCl (20 ml/g). Remove the HCl by filtration on a Büchner funnel, and wash the resin on the funnel with 2 to 5

liters of distilled water to remove the acid. Remove the water by applying suction, and suspend the resin in distilled water (1 g/5 ml). Transfer 5 ml of the suspended resin to a Pasteur pipette or 3-ml disposable syringe and 18-gauge needle, plugged at the bottom with glass wool. Attach a rubber tubing and clamp at the bottom to control the effluent flow. Do not allow the liquid surface to pass below the top of the resin bed.
Vacuum desiccator containing NaOH pellets
Sintered-glass funnel, fine porosity
Büchner funnel

Procedure.

Mix the sample to be assayed for hexosamines with 4 N HCl in a small lyophilization ampoule. The amount of material and volume of acid will vary with the material used. Samples should contain 50 to 100 mg (dry weight) of bacterial cells and lesser amounts of purified material in 3 ml of 4 N HCl. Each sample should contain 150 to 500 g of hexosamine in 3 ml. Hydrolyze the preparation for 6 h at 100°C in a vial which has been flushed with nitrogen gas and sealed.

Hydrolysis times will ideally be optimized by using several subsamples treated for different times. The amount of hexosamine will increase to a maximum and then decrease with time of hydrolysis as a result of destruction of hexosamines. Corrections for losses can be made by adding a known amount of hexosamine to one sample (17).

Dry the hydrolysate in vacuo over NaOH to remove HCl. Drying can be accomplished in a vacuum desiccator containing NaOH pellets. Add water, and redry. Dissolve the hydrolysate in water, and filter through a sintered-glass funnel to remove insoluble material. Add the filtered hydrolysate to a column made from a disposable Pasteur pipette containing 1 g of Dowex-50 resin. Wash the column with 15 ml of water, and discard the eluate. Elute the hexosamines with 10 ml of 2 N HCl, and dry the eluate in a vacuum desiccator containing NaOH pellets. Add 2 ml of water, and redry.

Drying of the hydrolysates and column eluates causes considerable destruction of muramic acid due to concentration of HCl. The losses of this compound are variable; thus, this procedure tends to underestimate total hexosamines.

Dissolve the dried hexosamine fraction in 0.1 ml of water, and proceed with the estimation of hexosamines as described above (17.2.6). Because hydrolysis has been completed, the treatment with 3 N HCl and subsequent neutralization can be eliminated.

Applications.

This procedure for the purification of hexosamines is necessary when complex polymers other than peptidoglycan (e.g., lipopolysaccharides) are analyzed. It can be applied to whole bacterial cells with the knowledge that muramic acid recoveries will be variable. It is also applicable to the estimation of hexosamines in culture media.

17.2.8. Muramic Acid

Muramic acid occurs only as a component of the peptidoglycan of the cell walls of bacteria and blue-green algae. It exists in all procaryotes except those organisms belonging to the proposed kingdom Archaebacteria (9, 11, 14). The amount of muramic acid in a bacterial cell suspension is a measure of the presence of and the amount of peptidoglycan. Assays for this compound have been used to estimate the bacterial biomass in soils and aquatic environments (10, 14).

The most convenient and specific assays for muramic acid involve treatment of acid hydrolysates of cell walls with alkali to remove D-lactate from the 3 position of muramic acid. The D-lactate is then estimated by an enzymatic method described by Tipper (17), which utilizes lactate dehydrogenase to determine lactic acid, or by the colorimetric method of Hadzija (12) given below.

Reagents, equipment, and glassware.

Copper sulfate solution: Dissolve 4 g of $CuSO_4$ in 100 ml of distilled water.
p-Hydroxydiphenyl reagent: Dissolve 1.5 g of p-hydroxydiphenyl in 95% (vol/vol) ethanol.
Muramic acid stock solution: Dissolve muramic acid (Sigma Chemical Co., St. Louis, Mo.) in water to give 0.25 mg/ml. Muramic acid contains 30% by weight of O-ether–linked lactic acid.
Screw-cap test tubes: 20 ml, with Teflon-lined screw caps
Spectrophotometer or colorimeter and glass cuvettes or colorimeter tubes
All glassware must be carefully cleaned and rinsed with sulfuric acid. After cleaning, contamination by dust and fingerprints must be prevented.

Procedure.

Neutralize samples of hydrolyzed peptidoglycan containing 5 to 50 μg of muramic acid in screw-cap test tubes with Teflon-lined caps by adding an equal volume of 4 N NaOH. Prepare standards containing 1 to 10 μg of lactic acid from the muramic acid stock solution; also prepare a reagent blank. Adjust the volume of the samples, standards, and the reagent blank to 1.0 ml with distilled water. Add an additional 0.5 ml of 1.0 N NaOH, and incubate the solution for 30 min at 37°C. Add 10 ml of sulfuric acid, seal the tubes, and heat them in boiling water for 10 min. After cooling, add 0.1 ml of the copper sulfate

reagent and 0.2 ml of the p-hydroxydiphenyl reagent, cap, and mix rapidly on a Vortex mixer. Incubate the tubes at 30°C for 30 min in a water bath, and read the absorbance at 560 nm. Determine the concentration of muramic acid in the samples from a standard curve prepared by plotting the absorbance of each standard versus the concentration of muramic acid in the standard.

Applications.

The method described is a simple, rapid, inexpensive procedure for the detection of muramic acid in circumstances where metabolic lactic acid is not present. Other components of peptidoglycan do not interfere nor do uronic acids and neutral sugars at concentrations up to 250 μg per sample.

17.2.9. 2-Keto-3-Deoxyoctanoic Acid

The 2-keto-3-deoxyoctanoic acid (KDO) in lipopolysaccharides is determined by the method of Weisbach and Hurwitz (18) as modified by Osborn (13). When oxidized with periodate, KDO yields formyl pyruvic acid which reacts with thiobarbituric acid to yield a chromogen with an absorption maximum at 550 nm. Deoxy sugars other than 2-keto-3-deoxy sugar acids produce different products which form chromogens in this assay that absorb light maximally outside the 545- to 550-nm spectral region.

Reagents.

Periodate reagent: Dissolve HIO_4 in 0.125 N H_2SO_4 to give a concentration of 0.025 N.
Dilute sulfuric acid: 0.02 N H_2SO_4 (CAUTION: See 17.2.2.)
Sodium arsenite reagent: 0.2 N sodium arsenite dissolved in 0.5 N HCl (CAUTION: Sodium arsenite is poisonous. Do not breathe dust. Wash hands well after use. Do not pipette this reagent by mouth.)
Thiobarbituric acid reagent: 0.3% thiobarbituric acid in distilled water; adjust to pH 2.

Procedure.

Dissolve lyophilized polysaccharide containing 0.01 to 0.25 μmol of KDO in 0.1 ml of 0.02 N H_2SO_4 and heat at 100°C for 20 min to release KDO from the polymer.

Adjust samples of the hydrolysate to 0.2 ml with 0.02 N H_2SO_4. Add 0.25 ml of the periodate reagent, mix, and let stand for 20 min at room temperature. Add 0.5 ml of 0.2 N sodium arsenite solution. Incubate for 2 min at room temperature, and add 2 ml of the thiobarbituric acid reagent. Mix, and heat at 100°C for 20 min. Cool, and read the absorbance at 548 nm.

One micromole of KDO gives an absorbance of 19.0 in a 1-cm cuvette.

Applications.

The procedure described above is suitable for the determination of KDO in purified lipopolysaccharide or cell wall preparations.

17.2.10. Heptoses

The product formed when a heptose is heated with sulfuric acid reacts with cysteine hydrochloride to produce a chromogen with an intense orange color. This chromogen changes to a pink compound which absorbs at 505 nm (13).

Reagents.

Surfuric acid reagent: Add 6 volumes of concentrated H_2SO_4 to 1 volume of water. (*CAUTION*: See 17.2.2.)
Cysteine hydrochloride reagent: 0.3 g of cysteine hydrochloride in 10 ml of distilled water.

Procedure.

Adjust sample volumes to 0.5 ml. Place in an ice bath. Add 2.25 ml of the sulfuric acid reagent to samples. Mix the tubes in the ice bath for 3 min. Transfer to a 20°C water bath for 3 min; then heat in boiling water for 10 min. Cool, add 0.05 ml of cysteine-HCl reagent, and mix well. Read absorbance at 505 nm and 545 nm 2 h after adding cysteine hydrochloride. Use a blank containing the same reagents described above, but lacking the cysteine hydrochloride reagent, to zero the instrument at each wavelength.

The absorbance at 505 nm is subtracted from the absorbance at 545 nm. One micromole of L-glycero-D-mannoheptose in 1 ml gives a value of 1.07 in a cuvette with a 1-cm light path.

Applications.

The procedure described by Dische (7) can be used to detect deoxyhexoses and deoxypentoses. The modification of Osborn (13), described here, is a rather specific method for the determination of heptoses in purified preparations of glycoproteins and lipopolysaccharides.

17.3. LIPIDS

Lipids are a chemically diverse group of biological molecules which are insoluble in water but soluble in organic solvents, and which commonly contain long-chain hydrocarbon groups in their molecules. This definition includes long-chain hydrocarbons, alcohols, aldehydes, fatty acids, and their derivatives including glycerides, phospholipids, glycolipids, and sulfolipids. Sterols, fatty acid esters of sterols, vitamins A, D, E, and K, carotenoids, and other polyisoprenoids often are considered in this class because they, like lipids, are associated with membranes and are extractable by lipid solvents. A survey of the types and composition of microbial lipids has been compiled by O'Leary (21).

The cytoplasmic membrane of all living cells contains lipids. Bacterial membranes are much less complex than those of eucaryotic cells. However, bacteria are known to contain a wide variety of lipids in addition to phospholipids. These include sphingolipids, neutral lipids, glycolipids, and a variety of unusual lipids associated with organisms of specific genera such as *Corynebacterium*, *Nocardia*, and *Mycobacterium* (21). Procaryotes were believed not to contain sterols. However, bacteria have recently been shown to contain squalene and a variety of sterols (22, 28). Gram-negative bacteria contain unique lipid constituents in lipopolysaccharide complexes as part of their outer membrane (21). Poly-β-hydroxybutyric acid is accumulated in large amounts in the cytoplasm of several bacteria and serves as a carbon and energy reserve.

Because of the complexity of bacterial lipids, a description of procedures that would provide a total analysis of each class is impractical here. For variations of the procedures described below and for detailed techniques required for a more complete analysis of bacterial lipids, the reader should consult one of the specialized articles on the subject (19, 20).

17.3.1. Crude Total Lipids

The procedure for extraction of total crude lipids described by Bligh and Dyer (23) is less time-consuming and requires less sample manipulation than other methods. This procedure has been widely used with bacteria. The bacterial cells are mixed with chloroform and methanol in amounts that yield a monophasic solution when combined with water in the tissue (24). Upon dilution with water or chloroform, a biphasic solution results. The lipids are retained in the chloroform phase, and nonlipid materials are retained in the methanol-water phase. The chloroform layer is isolated, and the lipids can be purified and separated into classes as described below. The recovery of total crude lipids after separation from the nonlipid contaminants should exceed 95%.

Hara and Radin (25) have described an alternative to this procedure which does not involve the use of chloroform and which has the advantage that centrifugation can be used instead of the filtration steps in the Bligh and Dyer procedure. A hexane-isopropanol mixture is used to extract lipids, and the nonlipid contaminants are removed by washing the extract with aqueous sodium sulfate. The lipid extract can be fractionated on silica gel columns by use of hexane, isopropanol, and water mixtures.

Reagents and materials.

Absolute methanol and chloroform: Use analytical-grade, freshly distilled reagents. (*CAUTION*: Chloroform is hepatotoxic as well as anesthetic. Contact with liquid or vapor should be avoided by working in a fume hood. Methanol is very flammable, and is toxic when swallowed.)

Filter paper: Whatman no. 1

Porcelain Büchner funnel: 43 mm

Suction flask: glass, 125 ml

Procedure.

Mix a thick cell paste prepared from 5 g of wet packed cells, or 1 g of lyophilized cells in 4 ml of distilled water, in a glass-stoppered tube with 5 ml of chloroform and 10 ml of methanol. Shake the mixture for 5 min, and leave it at room temperature with intermittent shaking for 3 to 4 h. Add another 5 ml of chloroform and 5 ml of distilled water, and shake the mixture at room temperature for another 30 min.

Rapidly filter the extract through Whatman no. 1 filter paper with slight suction. Transfer the filtrate to a glass graduated cylinder to allow separation of the layers.

Completely remove the methanol-water layer by aspiration.

Reextract the residue of cell material on the filter paper by shaking with 5 ml of chloroform. Filter the mixture through Whatman no. 1 filter paper, and rinse the filter with 12.5 ml of chloroform. Add the combined filtrates to the chloroform layer from the first extraction.

Add the chloroform extracts to a tared glass or porcelain container, and evaporate them to dryness under a stream of nitrogen in a water bath at 35 to 40°C to give a dry crude total lipid preparation. Work in a fume hood.

Measure the amount of crude total lipid with an analytical balance.

17.3.2. Purified Total Lipids (34)

Reagents and materials.

Solvents: Mix redistilled chloroform (200 parts) and methanol (100 parts) with 75 parts by volume of water. Shake vigorously in a separatory funnel, and allow the mixture to separate into an upper phase (solvent I) and a lower phase (solvent II). Collect the phases separately and store them at room temperature. (*CAUTION*: See 17.3.1.)

Sephadex G-25, fine bead form (Pharmacia Fine Chemicals, Inc., Piscataway, N.J.)

Chromatography tube: 1 cm by 15 cm with sintered-glass bottom and Teflon stopcock

Procedure (see 16.3.7 for principles of gel filtration).

Soak 25 g of Sephadex G-25 beads overnight in 100 ml of solvent I. Allow the beads to settle, decant the solvent, and rinse the beads four times with solvent I.

Prepare a thick slurry of beads (50-ml total volume) in solvent I. De-gas the slurry by placing it in a vacuum flask, and draw a vacuum with an aspirator for 10 to 15 min. Swirl occasionally. Pour the slurry of beads into a column containing a few milliliters of solvent I. Continue adding resin slurry until a resin bed 10 cm high is obtained. Apply pressure of 1 to 2 lb/in^2 from a nitrogen tank to settle the resin bed, and place a glass-fiber filter disk on top of the resin bed. Rinse the column with 10 ml of solvent I and 10 ml of solvent II.

Dissolve 150 to 200 mg of the dry crude total lipids (see preceding procedure) in 2 to 5 ml of solvent II. Filter the dissolved lipid fraction through a medium-porosity sintered-glass funnel directly onto the Sephadex column. Rinse the funnel with 2 ml of solvent II. Apply pressure of 2 to 3 lb/in^2 with a nitrogen tank to achieve a flow rate of 0.5 to 1.0 ml/min. Add solvent II, and elute the column until 25 ml of solvent is collected.

Concentrate the eluate under reduced pressure at room temperature until a cloudy aqueous emulsion forms. Lyophilize this emulsion to dryness. Dissolve the dried lipid in chloroform-methanol (2:1, vol/vol), transfer it to tared glass-stoppered vials, and evaporate the solvents under a stream of nitrogen at 35 to 40°C. Complete the drying in an evacuated vacuum desiccator containing paraffin shavings. After drying, weigh the vial containing the dried lipid extract.

Recovery of lipids can be checked by using known amounts of purified total lipids in the extraction and purification steps.

17.3.3. Classes of Lipids

Lipids can be bound to a solid adsorbent in a chromatography tube; the adsorbent materials most often used are silicic acid, alumina, and magnesium oxide (16.3.2). The lipids are bound by polar, ionic, and Van der Waal's forces. Separation is effected by using solvents of increasing polarities, which separates lipids into classes determined largely by the number and types of polar groups in the molecules of each class (19, 20).

The method described here employs acid-treated Florisil (a coarse grade of silicic acid). The solvents used to elute the adsorbed lipids are chloroform, a chloroform-acetone mixture, and methanol. These have been selected because of the relative ease and reproducibility of the technique and because all of the phospholipids are eluted in a single fraction.

Reagents and materials.

Acid-treated Florisil: Mix 100 g of Florisil (60 to 100 mesh, Fisher Scientific Co.) with 350 ml of concentrated HCl in a 1-liter Erlenmeyer flask, and heat on a steam bath in a hood for 3 h. (*CAUTION*: Avoid breathing HCl fumes. Handle all strong acids with precautions to avoid exposure of skin and eyes to liquid or vapor.) Decant the supernatant carefully, and wash the settled resin with 50 ml of concentrated HCl. Add another 350-ml quantity of concentrated HCl, and heat the slurry for 12 to 16 h (or overnight) in a round-bottom flask fitted with a condenser at 100°C. Be very careful to avoid exposure to fumes by working in a fume hood. Decant and discard the HCl, and wash the settled resin three times with distilled water by decantation and finally on a Büchner funnel until the wash water is neutral. Dry the residue, first by applying a vacuum to the funnel and then by heating of the Florisil at 120°C overnight in a glass dish. Repeat the acid treatment and water wash. Apply a vacuum to the resin in the funnel to remove the water, and add to the resin in succession: 150 ml of methanol, 150 ml of methanol-chloroform (1:1, vol/vol), 150 ml of chloroform, and 150 ml of ether. Dry the resin overnight at 120°C in a glass dish. (*CAUTION*: Ether can explode when in contact with flames or hot electrical wires. Remove as much ether from the resin as possible, and complete the drying in a fume hood.)

Freshly distilled chloroform (*CAUTION*: See 17.3.1.)

Acetone and methanol (*CAUTION*: See 17.3.1.)

Glass chromatographic tube approximately 1 by 30 cm, fitted with a solvent reservoir (separatory funnel with a Teflon stopcock) coupled to the top of the column by a 24/49 or standard taper greaseless glass joint. The column should be fitted with a fritted-glass disk or a glass-wool plug and greaseless Teflon stopcock at the bottom.

Procedure.

Make a slurry of 12 g of acid-treated Florisil in redistilled chloroform. Place the slurry in a vacuum flask or desiccator, and apply a vacuum for 10 min to de-gas it. Pour the de-gassed slurry into the chromatography tube. Allow the chloroform to drain through the column until the meniscus is about 1 mm above the resin.

Dissolve 150 to 200 mg of the purified total lipids (see preceding procedure) in 3 ml of chloroform. Add the solution to the column, and drain the column until the chloroform just enters the resin layer. Add an additional 5 ml of chloroform, and drain the column again until all the lipid just enters the resin bed. Carefully add 50 ml of chloroform to the column and reservoir without disturbing the resin bed surface. Allow the solvent to flow through the column into a glass collection vessel at a rate not exceeding 1 ml/min. When the chloroform just enters the resin bed, change collection vessels and add 60 ml of chloroform-acetone to the column and

reservoir. Collect the effluent at a rate not exceeding 1 ml/min in the second collection flask. When this solvent mixture just enters the resin bed, add 60 ml of acetone and collect in the same vessel; then add 60 ml of methanol to the column and collect the eluate in the third vessel.

The first collection vessel will contain neutral lipids (hydrocarbons, carotenoids, chlorophyll, sterols and sterol esters, glycerides, waxes, long-chain alcohols, and aldehydes) and fatty acids. The second reservoir will contain glycolipids, sulfolipids, and occasionally some phosphatides. The third reservoir will contain phospholipids. These can be detected and quantified by assaying for organic phosphate (17.3.5).

17.3.4. Fractionation of Phospholipids

There are many different solvent systems and choices of developing chambers for thin-layer chromatography plates (19, 20, 31).

Reagents and materials.

Silica gel glass plates: 20 by 20 cm, 250 μm thick. Pre-prepared plates are available from several manufacturers, or they can be prepared by coating silica gel onto glass plates (20). Develop the plates to the top in a chromatography chamber with redistilled acetone. Air dry in a fume hood, and heat for 30 min at 110°C. Cool in a desiccator, and use the same day.

Glass chromatography tanks: Thin-layer chromatography tanks or sandwich developing tanks can be purchased from several sources. Add freshly mixed solvent (60 ml) to multiplate tanks 2 h before developing the chromatograms. Line the tanks with sheets of filter paper saturated with solvent, and seal the tank to effect equilibration.

Solvents (*CAUTION*: See 17.3.1.)

    Solvent I: Chloroform–methanol–water, 65:25:4 (vol/vol), prepared from distilled solvents

    Solvent II: Chloroform–methanol–7 N ammonium hydroxide, 60:35:5 (vol /vol)

Nitrogen gas: Liquid in a tank

Micropipettes: 50 μl, graduated at 5- or 10-μl intervals

Procedure.

Dissolve the dry phospholipid fraction from the Florisil column in chloroform-methanol (1:1, vol/vol; about 100 mg/ml). Record the volume of the solution. Apply a measured amount of each solution as a spot, about 10 mm in diameter, 2.5 cm from the bottom and 2.5 cm from the right side of a separate chromatography plate. Use a graduated 50-μl disposable pipette. Dry between applications with a stream of nitrogen gas. Apply up to 0.1 ml of a solution to a plate. Dry the spot on the plate with a stream of nitrogen, and place in a developing tank containing solvent I. Seal the tank, and allow the chromatogram to develop until the solvent front reaches the top of the plate (1 to 2 h). Remove the plate, and air dry it. Place the plate, rotated 90° clockwise, relative to the first solvent flow direction in another tank containing solvent II, and develop in the second direction. Air dry the developed plate.

Stain the developed chromatogram with iodine vapors as described below to visualize the phospholipid spots. Wet the chromatogram slightly with a water mist, and scrape each spot with a razor blade. Transfer all the silica gel from each spot to a separate Pasteur pipette plugged with glass wool in the constricted end. Tap the Pasteur pipette until the silica gel forms a layer on the glass wool. Rinse the Pasteur pipette and silica gel with 2 ml of chloroform-methanol (1:1, vol/vol) followed by 2 ml of methanol to elute the phospholipids. Collect the eluates in glass-stoppered tubes. Distribute each phospholipid into three equal portions in two glass-stoppered 15-ml test tubes and a lyophile ampoule for further analysis. Evaporate the solvents at 37°C under a stream of nitrogen. One tube containing each of the separated phospholipids is to be assayed for total organic phosphorus in order to determine the amount of each phospholipid present. Other samples of each phospholipid can be subjected to acid and alkaline hydrolysis followed by identification of the products in order to establish their probable structures.

Iodine stain for detection of lipids on chromatograms (19, 20).

Add crystals of iodine to a chromatographic tank, cover the tank, and warm it in a water bath at 60°C to vaporize the iodine. (*CAUTION*: Iodine vapors are a skin irritant. They can cause burns and are highly toxic. Avoid contact with skin; avoid breathing vapors. Work in a hood.) Place the chromatograms in the tank on glass supports to hold them 2.5 cm off bottom. Cover tank for 1 min, remove the chromatogram, and view it in daylight and with an ultraviolet lamp (365 nm). (*CAUTION*: Ultraviolet light can cause severe retinal burns. Wear protective eyeglasses.) The lipids will appear as brown spots in daylight and very dark spots under ultraviolet light. Mark the spots with a pencil before the color fades.

17.3.5. Phosphate in Phospholipids

The amount of phosphate in each phospholipid fraction obtained by thin-layer chromatography can be estimated by the procedure of Bartlett, as described by Kates (20).

Reagents and materials.

Clear Pyrex 13 by 100 mm test tubes: Clean in hot 1 N nitric acid for 1 h, and rinse with distilled water.

All glassware must be carefully cleaned to remove traces of detergent and other sources of phosphorous contamination. (*CAUTION*: Wear rubber gloves, protective clothing, and protective goggles when working with hot nitric acid. Work in a fume hood.)

Acid-washed glass beads and marbles: Clean with the test tubes above.

Perchloric acid (72%, vol/vol) in distilled water (*CAUTION*: Perchloric acid liquid and vapors cause burns. Wear protective gloves, clothing, and goggles. This reagent may cause a fire or explode when heated to 150°C in the presence of oxidizable compounds.)

Ammonium molybdate solution (5%, wt/vol) in distilled water

Amidol reagent: Add 0.5 g of 2,4-diaminophenol dihydrochloride to 50 ml of 20% (wt/vol) sodium bisulfite solution. Filter.

Phosphorus standard (10 μg of phosphorus/ml): Dissolve 1.097 g of KH_2PO_3 in 250 ml of distilled water. Dilute 1:10 with distilled water to give 10 μg/ml.

Procedure.

Place a measured volume of lipid sample in test tubes, and evaporate solvent at 40°C under a stream of filtered air or nitrogen. Dry the phosphate standards containing 1 to 10 μg of phosphorus per ml (0.1 to 1.0 ml of standard) in similar fashion. Add 0.4 ml of 72% perchloric acid and an acid-washed glass bead. Heat at 100°C until the solution is clear and colorless. Work in a fume hood that has been approved for use with perchlorate. Add 4.2 ml of distilled water, 0.2 ml of ammonium molybdate solution, and 0.2 ml of the amidol reagent. Mix, and cover the tubes with acid-washed glass marbles. Heat in a boiling-water bath for 7 min. Cool rapidly, and read the absorbance at 830 nm after 15 min (see 16.1.1). A blank without phosphate is used to zero the spectrophotometer. Determine the concentration of phosphate in each sample from a standard curve prepared by plotting the concentration of phosphate in each standard versus the absorbance of the standard. The absorbance is linear up to 10 μg of phosphate per ml. The total phospholipid phosphate in each fraction can be converted into micromoles of each phospholipid by assuming that each phospholipid contains one phosphate group.

17.3.6. Identification of Phospholipids (19, 20, 31)

Phospholipids can be deacylated by mild alkaline hydrolysis to yield water-soluble glycerol phosphate esters and fatty acids, which are soluble in organic solvents. The glycerol phosphate esters can be identified by specific staining reactions and their R_f values as determined by paper chromatography (21). The fatty acids can best be identified by gas-liquid chromatography of methyl esters (see 16.3.4).

Further confirmation of the identity of the phospholipids can be obtained by analysis of the free bases liberated by acid hydrolysis. The free bases are identified by R_f values and color reactions after chromatography on paper (19, 31).

Reagents and materials.

Methanolic sodium hydroxide (0.2 N): Dissolve 0.4 g of NaOH in 50 ml of methanol. Prepare fresh. (*CAUTION*: See 17.3.1.)

Methanolic ammonium hydroxide (1.5 N): Dilute 10 ml of concentrated ammonium hydroxide to 100 ml with methanol (*CAUTION*: Ammonium hydroxide [fumes and liquid] is caustic. Wear protective gloves, clothing, and goggles. Work in a fume hood.)

Sodium hydroxide (1.25 N)

Chloroform (*CAUTION*: See 17.3.1.)

Methanol (*CAUTION*: See 17.3.1.)

Cation-exchange resin: Bio-Rad AG 50W × 8, Dowex-50 (H+), Amberlite IR-100 (H+), or a similar resin

Hydrochloric acid (1 N)

Water-saturated phenol: Place equal volumes of redistilled liquid phenol and water in a well-capped bottle, shake well, and remove the phenol layer. Store in a brown bottle at 4°C. Prepare fresh for each use from redistilled phenol stored at −20°C. (*CAUTION*: Phenol causes severe burns. Wear protective gloves, clothing, and goggles when using this reagent. Distill in a fume hood.)

Methanol-water (10:9, vol/vol)

Ethanol: Reagent grade, absolute

Acetic acid: Reagent grade, absolute

Hexane: Spectroscopic grade (Merck & Co., Inc., Rahway, N.J.)

Glass-stoppered 15-ml test tubes

Whatman no. 1 chromatography paper: The size of the paper depends on the developing tank available.

Tank for descending chromatography

Lyophile ampoules: 10 ml

Lyophilizer or flash evaporator

Phospholipid standards:

  Phosphatidyl choline, distearoyl ester
  Phosphatidyl serine from bovine brain
  Phosphatidyl ethanolamine from *E. coli*
  Phosphatidyl-*N,N*-dimethylethanolamine, dipalmitoyl ester
  Diphosphatidyl glycerol (cardiolipin) from bovine heart

Dragendorf spray reagent: Dissolve 1.7 g of bismuth nitrate in 100 ml of 20% acetic acid. Dissolve 10 g of potassium iodide in 25 ml of water. Just before use, mix 5 ml of the potassium iodide solution, 20 ml of the bismuth nitrate solution, and 70 ml of water to use as a spray reagent.

Ninhydrin reagent: Dissolve 0.25 g of reagent-grade ninhydrin in 100 ml of acetone-lutidene (2,6-dimethyl pyridine) (9:1, vol/vol). Prepare fresh for each use. (*CAUTION*: Some individuals are allergic to ninhydrin.)

Procedures.

Acid hydrolysis of phospholipids and identification of free bases. Add 2 ml of 1 N HCl to the lyophile ampoule containing the purified individual phospholipids from the thin-layer chromatographic plates (17.3.4). Seal the ampoule, and heat to 100°C for 4 h. Cool, open the ampoule, transfer contents to a 10-ml screw-cap tube, and add 2 ml of redistilled hexane. Mix on a Vortex mixer, and let stand to allow phases to separate. Remove the aqueous phase, and lyophilize to dryness or dry in a flash evaporator.

Dissolve the residue in 0.1 ml of distilled water. Spot three samples of each aqueous phase 10 mm apart on Whatman no. 1 chromatographic paper, and develop the chromatogram by descending chromatography using the solvent system phenol-ethanol-acetic acid (50:5:6, vol/vol). Known phospholipids treated identically, or choline, ethanolamine, serine, and *N,N*-dimethylethanolamine, can be used as standards for chromatography and identification of the free bases (19).

After the chromatogram is developed, dry overnight in a fume hood and cut it into strips. One strip containing a standard or an unknown is sprayed with ninhydrin reagent, and another strip with the same unknown or standard is sprayed with the Dragendorf reagent. Serine and ethanolamine will give blue spots with R_f values of approximately 0.27 and 0.51, respectively, after treatment with ninhydrin. Choline and *N,N*-dimethylethanolamine do not react with the ninhydrin spray. Choline gives an orange spot and dimethylethanolamine gives a weak orange spot with the Dragendorf spray reagent. The R_f values for choline and *N,N*-dimethylethanolamine are approximately 0.93 and 0.86.

Alkaline hydrolysis of phospholipids and identification of glycerol phosphate esters (19, 20). Add 0.2 ml of chloroform, 0.3 ml of methanol, and 0.5 ml of methanolic sodium hydroxide to 1 to 5 mg of phospholipid prepared by chromatography on Florisil (17.3.3). Mix in a glass-stoppered test tube on a Vortex mixer, and leave at room temperature for 15 min.

Add 1.9 ml of chloroform–methanol–1.25 N NaOH (1:4:4.5, vol/vol). Transfer to a conical centrifuge tube, mix on a Vortex mixer, and centrifuge for 1 min at 600 × g. Remove the upper methanol-water phase with a Pasteur pipette. Add cation-exchange resin to the methanol-water phase with vigorous mixing until the pH, measured with indicator paper, is slightly acidic (pH 5 to 6). Centrifuge and remove the liquid from the resin with a Pasteur pipette. Add 1 or 2 drops of methanolic ammonium hydroxide to make the supernatant slightly alkaline (pH 7.5 to 9.0). Check the pH with indicator paper. Concentrate to dryness under a stream of nitrogen at 30°C, and dissolve the residue in 0.1 ml of methanol-water (10:9, vol/vol).

The dissolved glycerol phosphate esters can be identified by co-chromatography with products of known phospholipids treated in an identical fashion (19). Three 25-μl aliquots of each sample should be spotted 10 cm apart on Whatman no. 1 paper and developed by descending chromatography using a solvent system of phenol-ethanol-acetic acid (50:5:6). After the chromatogram is developed, it is dried overnight in a fume hood and cut into strips. One of the strips containing the glycerol-phosphate ester sample is sprayed with the Dragendorf spray reagent, another is sprayed with the ninhydrin reagent, and another is stained with salicylsulfonic acid-ferric chloride spray reagent of Vorbeck and Marinetti (33) (see below). Phosphatidyl ethanolamine ($R_f = 0.66$) and phosphatidyl serine ($R_f = 0.28$) give mauve-colored spots when sprayed with the ninhydrin spray and left for a few hours at room temperature. Phosphatidyl choline ($R_f = 0.86$) and phosphatidyl dimethylethanolamine ($R_f = 0.80$) give orange spots and weak orange spots, respectively, with the Dragendorf reagent and no color with the ninhydrin spray reagent. All react with the salicylsulfonic acid-ferric chloride reagent described below.

Detection of glycerol phosphate esters on chromatograms by use of the salicylsulfonic acid-ferric chloride reagent (19). Dissolve 1.5 g of FeCl$_3\cdot$6H$_2$O in 30 ml of 0.3 N HCl. Dilute with 90 ml of acetone. Dip paper strips from chromatograms in this reagent, dry in a fume hood, and dip in a solution prepared by dissolving 1.5 g of salicylsulfonic acid in 100 ml of acetone. Glycerol phosphate esters appear as white spots on a mauve-colored background.

17.3.7. Fatty Acids

One of the most convenient methods for the identification of fatty acids in bacterial cells is by gas-liquid chromatography of their methyl esters prepared from phospholipids, total lipids, or other lipid fractions. Methanolysis of lipids is accomplished by heating the lipids in sulfuric acid and methanol. The methyl esters can be extracted into hexane, and quantitative measurements of each fatty acid can be made with a gas chromatograph equipped with the appropriate columns (16.3.4).

Several procedures are available for the preparation of methyl esters of fatty acids. One method selected for its simplicity and convenience is presented below. The references listed present alternative methods (26, 32).

Reagents and materials.

Gas chromatograph: Equipped with a flame ionization detector and suitable recorder. Several choices for

columns and temperature programs are available. Investigators should consult one of the references before selecting a gas-liquid chromatography procedure suitable for the instrument available and the material to be analyzed (20, 26, 29, 30, 32).

n-Hexane: Spectroscopic grade (Merck & Co., Inc., Rahway, N.J.)

Sulfuric acid-methanol (5%): Add 5 ml of reagent-grade concentrated sulfuric acid to 95 ml of water-free methanol. Prepare fresh before each use. (*CAUTION*: See 17.2.1 and 17.3.1.)

Standards: Methyl esters of straight- and branch-chain C$_{13}$ to C$_{20}$ fatty acids (Larodan Lipids, Malmo, Sweden; or Applied Science Laboratories, State College, Pa.)

Glass-stoppered 15-ml Pyrex test tubes

Procedure.

Transfer 0.5 to 1 mg of the purified total lipids (17.4.2) to glass-stoppered vials, and evaporate the solvent under a stream of nitrogen at 30 to 37°C. Add 4 ml of 5% (wt/vol) sulfuric acid in methanol to the dried lipid, and heat for 2 h at 70°C. After cooling, add 4 ml of *n*-hexane and shake the tube for 10 min. Remove and save the hexane phase. Reextract the methanol phase with 4 ml of hexane. The pooled hexane phases are used for gas chromatographic analysis (see 16.3.4). The fatty acid methyl esters can be identified by comparing retention times with known fatty acids. Quantitation is achieved by comparing peak areas with an internal standard added to each sample (26).

17.3.8. Poly-β-Hydroxybutyrate (27)

Poly-β-hydroxybutyrate, although a lipid, is not extracted from bacterial cells by the usual solvents used for lipid extraction. In the procedure described by Law and Slepecky (27), poly-β-hydroxybutyrate is extracted into hot chloroform from cell material that is obtained by first treating bacterial cells with hypochlorite to remove interfering substances. The extract is hydrolyzed to yield β-hydroxybutyrate, which undergoes dehydration to crotonic acid in concentrated sulfuric acid. Crotonic acid absorbs ultraviolet light; the absorption maximum of the double bond is 235 nm.

Reagents and materials.

Sodium hypochlorite solution: A commercial bleach solution such as Clorox or other bleach

Cuvettes: 1 cm, made of quartz or other material transparent to ultraviolet light.

Spectrophotometer: Capable of ultraviolet absorbance measurements

Polypropylene or glass centrifuge tubes: 12 by 100 mm, or 30 ml. The plastic centrifuge tubes should be washed with hot chloroform and ethanol to remove plasticizers. (*CAUTION*: See 17.3.1.)

Centrifuge: A refrigerated centrifuge capable of achieving 10,000 × g

Pyrex glass boiling tubes

Volumetric flask: 10 ml

Boiling-water bath

Reagent-grade sulfuric acid: The quality of the sulfuric acid is important. Poor grades contain materials that absorb ultraviolet light and cause high absorbance in the blanks used to zero the spectrophotometer. (*CAUTION*: See 17.2.2.)

Standard: Dissolve 3-hydroxybutyric acid (0.5 mg/ml) in ethanol. Dilute 1:10 in ethanol to obtain a solution of 50 µg/ml.

Procedure.

Centrifuge bacterial cells from a volume of culture containing 10 to 20 mg of dry weight for 15 min at 10,000 × g (4°C). Determine the dry weight of cells from an equal volume of culture (25.2.2). Add a volume of sodium hypochlorite equal to the original culture volume, and transfer the suspension to a centrifuge tube. Incubate the suspension at 37°C for 1 h. Centrifuge the suspension at 10,000 × g for 15 min to sediment the lipid granules. The temperature during centrifugation is not important. Suspend the sediment in water, and centrifuge at 10,000 × g for 15 min. Discard the supernatant fraction, and wash the lipid granules adhering to the walls of the centrifuge tube with 5 ml of acetone followed by 5 ml of ethanol. The washings which remove water are accomplished by suspending the granules in 5 ml of acetone with a Vortex mixer followed by centrifugation at 10,000 × g for 15 min. Suspend the sediment in 5 ml of ethanol using a Vortex mixer, and centrifuge again at 10,000 × g for 15 min.

Dissolve the washed granules with boiling chloroform by heating with 3 ml of chloroform in a boiling-water bath for 2 min. Cool the solution, and centrifuge it at 10,000 × g for 15 min at room temperature. Save the supernatant fraction in a 10-ml volumetric flask, and reextract the sediment twice with 3 ml of chloroform at 100°C for 2 min. If the solution is not clear, filter while hot through Whatman no. 1 filter paper (previously washed with hot chloroform). The chloroform extracts are pooled and made to 10 ml with chloroform. Work in a fume hood while extracting with chloroform. (*CAUTION*: See part 17.3.1 for precautions involved in the use of chloroform.)

Add samples of the chloroform extract containing 5 to 50 µg of the polymer and 5 to 10 standards containing 5 to 50 µg of 3-hydroxybutyric acid in ethanol to glass boiling tubes. Immerse the tubes in a boiling-water bath until all the solvents are evaporated. Add 10 ml of sulfuric acid, cap the tubes with a glass marble, and

heat them in a boiling-water bath for 10 min. Cool the solutions, and mix them thoroughly.

Read the absorbance of the solutions at 235 nm against a sulfuric acid blank (see 16.1.1). Calculate the amount of β-hydroxybutyrate from the standard curve. The amount of poly-β-hydroxybutyrate per gram (dry weight) of bacterial cells can be determined from the dry weight measurements of cells in the culture.

Applications.

The technique described above is suitable for measuring the amount of poly-β-hydroxybutyrate in freeze-dried cells or cell pastes obtained by centrifugation. Other β-hydroxy acids and some sugars interfere because they are converted to products that absorb at 235 nm. Sugars are removed by hypochlorite treatment followed by extraction of poly-β-hydroxybutyrate into chloroform. Other substances in cells occasionally produce products that absorb ultraviolet light. The absorption spectrum of the product from bacterial cells should be compared with that of the standard to ensure that the absorbance is due to crotonic acid. (See 16.1.1 for determining absorption spectra.)

17.4. CELL WALL POLYMERS

There are striking chemical and structural differences in the cell walls of gram-postive and gram-negative bacteria. Peptidoglycans may make up 40 to 90% of the weight of isolated cell walls of gram-positive bacterial cells (15 to 20% of the cell dry weight), usually less than 10% of the gram-negative bacterial cell walls, and as little as 1.2% of the cell wall of some pseudomonads (35–38, 44, 45, 58). This polymer controls the shape of procaryotes except mycoplasmas, halobacteria, methanogenic bacteria, and other species of procaryotes belonging to the proposed kingdom Archaebacteria (49). The peptidoglycan of most bacteria has a backbone containing alternate residues of N-acetylglucosamine and N-acetyl muramic acid linked by β-1-4-glycosidic bonds. Minor exceptions to this backbone structure are found in a few species of *Streptomyces* (38) in which N-glycoly-muramic acid exists, and a more modified peptidoglycan containing a muramyl lactam is found in the cortex of bacterial spores (55).

A tetrapeptide is attached to the carboxyl group of the 3-O-D-lactic acid linked to many of the N-acetyl muramic acid residues. The tetrapeptide most frequently contains amino acids in the order L-alanine, D-glutamic acid, a diamino acid, and a terminal D-alanine. The diamino acids found in the tetrapeptide include *meso*-

and LL-diaminopimelic acid, L-lysine, L-orni-thine, and L-diaminobutyric acid. Descriptions of these and several variations of the tetrapeptide sequences have been reviewed by Ghuysen (40), Rogers et al. (45), and Cummins (38).

Many of the tetrapeptide chains are cross-linked to one another. The extent of cross-linking and the nature of the linkages vary from organism to organism. The linkage may be a direct peptide bond from a dibasic amino acid on one tetrapeptide to a D-alanine on another tetrapeptide, or it may involve one or several amino acids as bridges between one tetrapeptide and another (36, 38, 45, 48).

The unique occurrence of N-acetyl muramic acid in this polymer provides a means of determining the presence of peptidoglycan in bacterial cells and can be used to estimate microbial biomass in soil and water samples. Diaminopimelic acid is found in macromolecules only as a component of peptidoglycan, and positive assays for this compound indicate the presence of peptidoglycan. However, not all peptidoglycan polymers contain this diamino acid (38).

Cell walls of gram-positive bacteria often also contain teichoic acids, teichuronic acids, proteins, and lipoteichoic acids. Teichoic acids are complex polymers of polyols (ribitol and glycerol) linked together in a backbone by phosphodiester bonds. Several variations in the backbone structure have been described (38, 47) These polymers, which make up 50% of the cell wall of some gram-positive bacteria, are good antigens, and their serological reactivity has been used to define antigenic groups of several genera (47). Cell wall teichoic acids are covalently linked to peptidoglycan. Teichoic acids of the glycerol phosphate polymer type also occur as lipoteichoic acids which contain covalently linked glycolipid. They are found in the plasma membrane after breakage of cells. The most common antigenic determinants are sugars or amino sugars attached through the hydroxyl groups of the polyols of the backbone. Alanine is often found attached to the polyols as well.

Teichuronic acids are acidic polysaccharides which contain uronic acids but no phosphate. Polymers of glucuronic acid with N-acetyl glucosamine and aminomannuronic acid and glucose occur in *Bacillus licheniformis* and *Micrococcus lysodeikticus*, respectively. Other uronic acid polymers have been observed as cell wall components of other gram-positive bacteria (38).

Cell wall polysaccharides and lipoteichoic acids of gram-positive bacteria, particularly the streptococci, are responsible for antigenic differences which have enabled the separation of these bacteria into serological groups (38). Cell wall polysaccharides of other gram-positive bacteria form a heterogeneous group of macromolecules that usually contain neutral sugars and occasionally amino sugars. These polysaccharides are of value in distinguishing between strains and species, especially when unusual or uncommon sugars are present.

Proteins occur widely as surface components of gram-positive bacteria. The surface antigens of streptococci include numerous acid-soluble proteins (38).

The cell walls of gram-negative bacteria are more complex than those of gram-positive cells. They have a comparatively low peptidoglycan content, and in enteric bacteria, the peptidoglycan is covalently bonded to lipoproteins which probably form a bond between the peptidoglycan and the outer membranous layer of the cell wall. The outer membranous layer is rich in lipopolysaccharide.

Lipopolysaccharides are extremely complex molecules with molecular weights higher than 10,000 (35, 36, 39, 41, 44, 54, 57). The lipopolysaccharide from *Salmonella* spp. is made up of three different parts: the lipid moiety, a core polysaccharide, and the O-polysaccharide. The lipid moiety, termed lipid A, contains a phosphorylated glucosamine disaccharide heavily esterified through hydroxyl and amino groups with fatty acids that are 10 to 22 carbon atoms in length. A major fatty acid component of the lipid A fraction is β-hydroxymyristic acid, which is unique to the lipid A portion of the lipopolysaccharide molecule.

The core polysaccharide serves as a link between lipid A and the O-polysaccharide. It is composed of four basal sugars, 2-keto-3-deoxyoctanoic acid, phosphate, and ethanolamine. 2-Keto-3-deoxyoctanoic acid is unique to this part of the lipopolysaccharide molecule.

In *Salmonella typhimurium* the sugars of the core polysaccharide are glucose, galactose, heptose, and glucosamine (35, 57). The core polysaccharide is covalently bonded through carbon-2 of 2-keto-3-deoxyoctanoic acid to carbon-2 of a glucosamine residue in lipid A. The core polysaccharide is also covalently linked to the variable O-polysaccharide chains. Several lipopolysaccharide subunits may be bonded together by pyrophosphate bonds in the lipid A portion of the molecule. The lipid A portion is embedded in the outer membrane with the O-polysaccharides extending outward (35, 36). The O-polysaccharides are the major antigenic determinants (the O antigen) of the surface of gram-negative bacteria. They often serve as receptors of bacteriophage attachment.

The fine structure and composition of the

outer membrane of gram-negative bacteria suggest that it is a bilayer containing phospholipids and proteins with variable amounts of lipopolysaccharide. The outer membrane encloses the periplasmic space, an enzyme-containing compartment bounded by the plasma membrane on the inside (35, 36).

Several gram-negative cells possess proteinaceous outer coats or thick carbohydrate layers outside the lipopolysaccharide zone extending from the outer membrane (35, 36).

It is impractical to present here all the methods employed for isolating and characterizing the structural components of cell walls. The methods employed will depend on the organism, the facilities available, and the objective of the investigation. For some investigations, intact highly purified polymers are not required; for others, macromolecular integrity and purity are required. Those persons who wish to thoroughly characterize one of these polymers from a bacterium should first consult the literature describing the different isolation techniques and the advantages and limitations of each (35, 36, 57, 58). The simplest methods for the isolation and preliminary characterization of peptidoglycans and lipopolysaccharides from the more ordinary laboratory strains of bacteria are presented here.

17.4.1. Peptidoglycan

Reagents and equipment.

Saline solution: 0.9% (wt/vol) NaCl in distilled water
Trypsin crystalline enzyme preparation
Tris buffer: Dissolve 1.21 g of tris(hydroxymethyl)aminomethane in 90 ml of distilled water. Adjust the pH to 7 with 5 N HCl, and add additional water to 100 ml.
Deoxyribonuclease: Dissolve 5 mg of the dry powdered enzyme preparation in 1.0 of Tris buffer.
Ribonuclease: Dissolve 5 mg of the dry powdered enzyme preparation in 1.0 ml of Tris buffer.
Refrigerated centrifuge: Capable of achieving 22,000 × g
Heavy-walled glass or Pyrex 30-ml centrifuge tubes, centrifuge bottles (100 to 250 ml), and rotors to accommodate each. (*CAUTION:* Thin-walled Pyrex glass tubes will not withstand 22,000 × g.)
Sonifier, French pressure cell, or colloid mill

Procedures.

Cool the bacterial cells in their growth medium to 4°C with ice, and harvest them by centrifugation in 100- to 250-ml centrifuge bottles (5,000 × g for 15 min). Resuspend the cells in 20 ml of cold saline solution, and centrifuge again. Use the packed cells immediately, or lyophilize them and store in a freezer to prevent autolysis. Suspend the cells homogeneously in 20 ml of cold water. If lyophilized, suspend the

dry cells in stages by first wetting the cells with cold water to form a paste and subsequently adding liquid slowly while stirring the suspension. Mix the cells mechanically or with a hand operated homogenizer to obtain complete dispersal of the cells. Sucking and blowing through the orifice of a cotton-plugged pipette is adequate if mechanical homogenizers are unavailable. Keep the suspension cold to prevent autolysis of cell walls during all manipulations.

Disintegrate the cells by ultrasonic vibrations using a Sonifier, mechanical grinding device, or French pressure cell (5.1).

As soon as possible after disruption of the cells, heat the suspension to 75°C. Keep the suspension at this temperature for 15 min to inactivate autolytic enzymes. Centrifuge at a low speed (1,000 × g for 10 min) to remove unbroken cells. Discard the sediment, and centrifuge the supernatant fraction in 30-ml centrifuge tubes at 22,000 × g for 15 min. A small opaque layer of unbroken cells under a translucent layer of cell walls should be obtained. Remove the upper layer carefully by gently washing the sediment with a stream of ice-cold 0.1 M Tris buffer, pH 7, delivered from a Pasteur pipette. When the upper layer is totally suspended in 15 ml of buffer, centrifuge again at 1,000 × g for 10 min, discard the sediment, and centrifuge the supernatant fraction at 22,000 × g for 15 min. Resuspend the upper layer in the Tris buffer. Repeat these steps until a homogeneous cell wall fraction, free from intact cells, is obtained.

Suspend the cell wall fraction in 100 ml of Tris buffer that has been saturated with chloroform. Add deoxyribonuclease (0.5 ml of a 1-mg/ml solution) and ribonuclease (1.0 ml of a 5-mg/ml solution), and incubate the suspension at 37°C for 30 min. Add 100 μg of crystalline trypsin, and continue the incubation for 6 h at 37°C. Sediment the cell walls by centrifugation, and wash them as described above. Continue the sedimentation and washing until a homogeneous layer is obtained. Trypsin is added to remove denatured proteins. This enzyme causes autolysis of the peptidoglycan of some bacteria. If this occurs, centrifugation will not yield a precipitate, and the trypsin treatment must be omitted. Purification of peptidoglycan from most gram-negative cells involves the use of much larger amounts of cell material (20 to 30 g) because of the small amount of peptidoglycan per unit weight of cells and the greater susceptibility of the peptidoglycan of these bacteria to autolysis. A procedure for preparation of 150 mg of peptidoglycan from *Escherichia coli* has been described by Weidel et al. (56).

The amounts of hexosamine and muramic

acid can be estimated after acid hydrolysis of the purified cell walls (see 17.2.7 and 17.2.8). The amino acid content can be determined with an amino acid analyzer or by paper chromatography (58). The presence of many amino acids indicates contamination by membrane, cell wall, or cytoplasmic proteins.

Good examples of approaches for a more complete characterization of cell walls using a variety of chemical assays to determine reducing sugars, unsubstituted amino groups of amino acids, etc., have been provided by Kottell et al. (51) and Ghuysen et al. (48).

17.4.2. Lipopolysaccharides (42)

Reagents, materials, and equipment.

Saline solution: 0.9% (wt/vol) NaCl in distilled water

Phenol solution: Add 10 ml of distilled water to 90 ml of warm (60°C) redistilled phenol. (*CAUTION:* Phenol causes severe burns when it contacts skin or eyes. Wear protective gloves, clothing, and goggles when using this reagent. The fumes are equally dangerous. Use hot phenol in a fume hood.)

Centrifuges, rotors, and tubes: A centrifuge capable of achieving 10,000 × g and an ultracentrifuge capable of achieving 100,000 × g, centrifuge bottles (100 to 250 ml) and a rotor for them to be used for harvesting cells from the medium, 30-ml centrifuge tubes and rotors to hold them for both centrifuges

Lyophilizer or vacuum desiccator and a vacuum source

Procedures.

Centrifuge the bacteria from approximately 3 liters of growth medium at 10,000 × g for 15 min. Suspend the sedimented cells in 300 ml of ice-cold saline solution, and centrifuge again. Repeat the resuspension and centrifugation to remove medium constituents. Lyophilize the bacteria in the final sediment, or store the packed cells in a freezer until used.

Thoroughly suspend 20 g (dry weight) of bacteria in 350 ml of water at 65 to 68°C. Heat 350 ml of phenol solution to 65 to 68°C, and add it to the bacterial cell suspension. Stir vigorously, and keep the mixture at 65°C for 15 min; then cool it to 10°C in an ice bath. Centrifuge the emulsion at 10,000 × g for 30 min. Three layers result: a water layer, a phenol layer, and a sediment. Sometimes a layer of denatured protein is formed at the phenol-water interphase. Remove the upper aqueous layer and save it. Remove and discard the material at the interface. Add another 350 ml of water to the phenol layer, and sediment and mix thoroughly while the temperature is held at 65 to 68°C for 15 min. Centrifuge the mixture again, and combine the upper water layer with that obtained from the first centrifugation step. Dialyze the aqueous suspension for

3 to 4 days against distilled water to remove the phenol and low-molecular-weight contaminants. Concentrate the dialyzed solution at 37°C under a vacuum to 5 ml. A rotary evaporator is useful for this purpose if one is available. Centrifuge the concentrate at 3,000 × g for 10 min to remove insoluble material. Discard the sediment. Half the suspended organic material in the supernatant fraction should be lipopolysaccharide. The major contaminant is RNA. Centrifuge the supernatant fraction at 100,000 × g for 1 h. Thoroughly resuspend the sediment in saline solution and centrifuge again at 100,000 × g for 1 h. Repeat the centrifugation and resuspension of the sediment five times to wash it free from contaminating material. Finally, dissolve the sediment in distilled water, and lyophilize the material until dry. If a lyophilizer is unavailable, the suspension can be dried in a vacuum desiccator. A yield of 100 to 500 mg of lipopolysaccharide should be obtained from 20 g (dry weight) of species of the *Enterobacteriaceae*. This lipopolysaccharide is a potent endotoxin (35, 39) and contains firmly bound lipid A.

17.4.3. Lipid A and Polysaccharide from Lipolysaccharides

The lipid A moiety can be obtained as a water-insoluble product after mild acid hydrolysis of purified lipopolysaccharide. The polysaccharide portion of the polymer is soluble in water.

Reagents, materials, and equipment.

Acetic acid solution (1%, vol/vol): Add 1 ml of acetic acid to 99 ml of distilled water. (*CAUTION:* Do not mouth pipette acetic acid. Avoid contact with liquid or vapors by working in a fume hood. Wear goggles and protective laboratory clothing.)

Lyophilized lipopolysaccharide preparation

Lyophilization ampoules (15 ml)

Tank of compressed nitrogen gas

Vacuum desiccator or lyophilizer

Centrifuge capable of achieving 5,000 × g, 15-ml centrifuge tubes, and an appropriate rotor. Temperature control is not important.

Procedure.

Suspend 0.1 g of the dried lipopolysaccharide, obtained as described above, in 10 ml of a 1% acetic acid solution that has been gassed with nitrogen for 10 min. Seal the suspension in an ampoule, and heat it in a steamer for 4 h. Cool and open the vial, transfer the contents to a centrifuge tube, and centrifuge at 5,000 × g for 10 min. Carefully remove the supernatant from the insoluble lipid A. Save the supernatant fraction which contains the polysaccharide portion of the lipopolysaccharide.

Resuspend the sediment in distilled water, and centrifuge again. Repeat the resuspension and centrifugation five times. Discard the supernatant fractions. Dry the sediment in a vacuum desiccator or by lyophilization.

Dialyze the polysaccharide remaining in solution after acid hydrolysis against 200 volumes of water and dry by lyophilization or in a vacuum desiccator. Between 40 and 100 mg of lipid-free polysaccharide and 30 to 150 mg of lipid A should be obtained from 100 to 500 mg of lipopolysaccharide.

Assays for 2-keto-3-deoxyoctanoate and heptose can be employed to estimate the amount of lipopolysaccharide or lipid-free polysaccharide (17.2.9). Hexosamine assays (17.2.6) can be used to estimate the amount of lipid A in purified preparations.

17.5. NUCLEIC ACIDS

The most widely used method for the estimation and identification of deoxyribose and ribose nucleic acids (DNA and RNA) in cell extracts involves extraction and hydrolysis with a hot solution of trichloroacetic (or perchloric) acid followed by specific colorimetric assays for pentose components of nucleic acids. A more convenient means of measuring the amounts of DNA and RNA in pure solutions takes advantage of their abilities to absorb ultraviolet light at 260 nm. Other procedures less applicable to the purposes of this manual are described in other reference books (59, 63).

The estimation of nucleic acids in bacterial cultures by the techniques described below requires that the cells are present in sufficient numbers in a noninterfering medium or that they can be harvested from the growth medium, washed free from interfering medium components, and resuspended at an appropriate concentration without lysis of the cells or hydrolysis of the nucleic acids by nucleases. Total extraction of nucleic acids from the cells is essential for accurate determination of their nucleic acid content.

17.5.1. DNA by Diphenylamine Reaction

The reaction of DNA with diphenylamine in a mixture of acetic and sulfuric acids is the most widely used of the colorimetric procedures for the determination and identification of this polymer in complex mixtures. The chemistry of the reaction and the nature of the chromophore formed remain unknown. The Burton method (61), described below, has the advantages that it is sensitive and that few substances in bacterial cells interfere with accurate determinations of DNA. Another method is described in part 22.4.3.

Reagents, materials, and equipment.

Perchloric acid: 0.5, 1.0, and 2.5 N solutions (*CAUTION:* Perchloric acid may cause fire or explosion when hot [150°C] and in contact with oxidizable compounds. Avoid contact with skin, eyes, and clothing. Work in an approved fume hood when handling this reagent.)

Diphenylamine reagent: Dissolve 1.5 g of purified diphenylamine, recrystallized from boiling hexane (white crystals, mp 52.9°C), in 100 ml of redistilled acetic acid, and add 1.5 ml of reagent-grade concentrated sulfuric acid. On the day it is used, add 0.1 ml of a 16-mg/ml aqueous solution of acetaldehyde (1 ml of acetaldehyde in 50 ml of water) to each 20 ml of the diphenylamine reagent. (*CAUTION:* Acetaldehyde is very volatile, and the vapors and liquid are toxic. Work in a fume hood.)

Saline-citrate solution (SSC): 0.15 N NaCl plus 0.015 N trisodium citrate, pH 7.0

Standards: Dissolve crystalline DNA (0.4 mg/ml) from one of several commercial sources in 5×10^{-3} M NaOH. Prepare standards each 2 weeks by mixing measured volumes of the stock DNA solution with an equal volume of 1 N perchloric acid, and heat at 70°C for 15 min. Prepare dilutions in 0.5 N perchloric acid.

Most DNA preparations contain at least 15% water, and it is difficult to obtain or prepare DNA totally free from water to use in the preparation of standards. One should be aware of the inaccuracies resulting from the use of DNA as a standard. As an alternative, a DNA standard can be hydrolyzed, and its phosphate content per milligram of DNA can be determined (62). The results can then be reported as micrograms (or milligrams) of DNA-phosphorus rather than micrograms of DNA.

Deoxyribose prepared as an aqueous solution (10^{-3} M) is also a convenient standard. Solutions are stable for 3 months in a refrigerator. Dilutions made in 0.5 N perchloric acid can be used to construct a standard curve for the diphenylamine assay. The results are reported as deoxyribose equivalents of DNA if this standard is used.

Spectrophotometer or colorimeter and glass cuvettes

Centrifuge: Capable of achieving a gravitational force of 10,000 × *g*. The centrifuge should either be refrigerated or placed in a cold room.

Centrifuge tubes: 15 or 30 ml

Water baths at 70°C and an incubator or water bath at 30°C

Test tubes: Pyrex glass, 15 by 125 mm

Procedure.

A suspension of 5 to 10 mg (dry weight) of bacterial cells per ml (about 5×10^{10} to 1×10^{11} cells per ml for bacteria the size of *Escherichia coli*) is required.

Harvest the cells from the medium if the culture does not contain sufficient cells or if the medium contains interfering compounds. Complex media contain nucleic acid components and sugars that must be eliminated by harvesting the cells from the media and washing them by centrifugation. Harvesting of most bacteria can be accomplished by centrifugation at 5,000 × g for 15 min at 4 to 5°C. Suspend the sedimented cells in one-tenth of the culture volume of SSC. Centrifuge the cells at 5,000 × g for 15 min at 5°C, and resuspend them in ice-cold SSC. If the cells are grown in mineral salts medium and the cell concentration is sufficient, samples of the culture can be used directly for extraction of DNA.

Acidify each sample of the bacterial culture or suspension of bacteria in SSC with sufficient 2.5 N perchloric acid to give a final concentration of 0.25 N perchlorate. Cool the acidified sample on ice for 30 min, and centrifuge at 10,000 × g for 10 min. Using a glass rod, suspend the precipitate containing the nucleic acids in 0.5 ml of 0.5 N perchloric acid. Add an additional 3.5 ml of 0.5 N perchloric acid, and heat the suspension at 70°C for 15 min. It is important to stir or mix the suspension occasionally during this step. Centrifuge the heated suspension at room temperature for 10 min at 5,000 × g, and carefully decant the supernatant fraction into a 10-ml graduated test tube. Reextract the precipitate in an identical fashion. Combine the two extracts, and mix. Record the volume of the combined extracts.

For the diphenylamine reaction, mix samples of various volumes from the combined extracts with sufficient 0.5 N perchloric acid to bring the final volume to 2 ml. It is usually wise to use a 10-fold range of sample sizes from each extract (0.2 ml, 1.0 ml, and 2.0 ml) to ensure that the amount of color developed in one sample falls on the standard curve. After the volumes of each sample have been adjusted to 2.0 ml, add 4.0 ml of the diphenylamine reagent to each tube, and mix thoroughly. Adjust the volume of the liquid in tubes containing dilutions of the standard (5 to 100 µg/ml) to 2.0 ml with 0.5 N perchloric acid, and mix them with 4.0 ml of the diphenylamine reagent. Prepare a blank for the spectrophotometer by adding 4 ml of the diphenylamine reagent to 2.0 ml of 0.5 N perchloric acid. Incubate the tubes at 30°C for 16 to 20 h (or overnight). If cloudy, centrifuge the contents of the tubes at 10,000 × g for 15 min to remove unhydrolyzed protein. Determine the absorbances of the supernatant fractions at 600 nm against the blank (16.1.1). Estimate the concentration of DNA in each tube from the absorbance values, using a standard curve prepared by plotting the absorbance of standard solutions against the DNA, deoxyribose, or phosphate concentrations of the standards.

It is not important to control the temperature carefully. If all tubes, including the standards, are incubated at the same temperature, the re-

sults should not be affected. A standard curve should be prepared with each set of determinations.

17.5.2. RNA by Orcinol Reaction

The reaction of aldopentoses with acidified orcinol to produce a green chromogen is a classical reaction of sugar chemistry. This reaction detects ribose moieties present in RNA. Deoxyribose gives 20% of the color of ribose. It is suitable for the measurement of RNA or ribonucleotides in solution as well as in tissue or cell extracts (62).

Reagents and equipment.

Ferric chloride reagent: Dissolve 100 mg of $FeCl_3 \cdot 6H_2O$ in 100 ml of reagent-grade concentrated HCl. (*CAUTION*: See 17.3.3.)

Ethanolic orcinol solution: Dissolve 6.0 g of orcinol in 100 ml of ethanol. Store in a dark bottle. Commercial orcinol should usually be crystallized from benzene. If the crystals are yellow, recrystallization is necessary.

Orcinol reagent: On the day the reagent is to be used, mix equal volumes of the ferric chloride reagent with the ethanolic orcinol solution. The complete reagent solution should be bright yellow when prepared.

Perchloric acid HCl, 0.1 N (*CAUTION*: See 17.5.1.)

Standard solutions for RNA analysis: Heat a solution containing 200 µg of RNA per ml with 0.25 N perchloric acid at 70°C for 30 min. Add an equal volume of 0.1 N HCl to give a final RNA concentration of 100 µg/ml.

Adenosine 5'-monophosphate (AMP) is a preferred standard because its concentration can be accurately measured. Prepare a solution containing 100 µg/ml by heating 200 µg/ml in 0.25 N perchloric acid. Dilute the solution with an equal volume of 0.1 N HCl.

Spectrophotometer or colorimeter and glass cuvettes or tubes

Water bath: Set at 90°C

Test tubes: Pyrex glass, 18 by 125 mm

Clear glass marbles: Approximately 200 mm in diameter

Procedure.

Prepare hydrolysates from suspensions of bacterial cells by treatment with perchloric acid exactly as described for the extraction of DNA (17.5.1). Dilute the cell hydrolysates with 0.1 N HCl so that they contain 10 to 100 µg of RNA per ml. (Cell hydrolysates from 0.8 to 1 mg [dry weight] of cells contain approximately 100 µg of RNA.) Mix the diluted perchlorate hydrolysates of the cells or standards with 2 volumes of the orcinol reagent, cover the tubes with marbles, and heat them at 90°C for 30 min. The volumes used can vary depending on the spectrophotometer or colorimeter available. For most spectrophotometers and colorimeters, mix 1.0 ml of the extracts or standards with 2 ml of the orcinol reagent. Cool the tubes under running tap water, and determine the absorbance of each tube at 665 nm. Determine the concentration of RNA or nucleotides from a standard curve prepared with known amounts of AMP or RNA (see 16.1.1). The standard curve should be repeated each time a sample or series of samples is assayed. The assay can reliably measure 15 to 100 μg of RNA in an extract of bacterial cells.

When AMP is used as a standard, the results should be reported as purine riboside equivalents of RNA.

17.5.3. DNA and RNA by Other Methods

Pure DNA or RNA solutions, free from contaminating protein and carbohydrate, can be measured by ultraviolet spectroscopy (see 22.4.1). A solution containing 1 mg of pure nucleic acid per ml has an absorbance at 260 nm on the order of 22 to 23.

A rapid and sensitive alternative to the colorimetric analysis of DNA in cells and tissues has been described by Zahn (63). This procedure is based on the assay of thymine in hydrolysates of cell samples by gas chromatography. This procedure has the disadvantage that access to an appropriate gas chromatograph is required. If DNA assays are routinely performed, the use of a gas chromatograph for this purpose can be justified because of the savings in time.

Applications.

The procedure described above extracts approximately 95% of the nucleic acid from bacterial cells and endospores. Alternative extraction procedures can be used if incomplete extraction is indicated by the presence of DNA in the soluble fraction after a third extraction of the precipitate from the second hot perchloric acid treatment. Cells can also be extracted conveniently with 5% (wt/vol in water) trichloroacetic acid by heating at 90°C for 15 min. The extract can be diluted to the desired volume with 5% trichloroacetate and assayed by the procedure described above.

If an estimate of the concentration of purified DNA in solution is to be made, the sample of the solution can be treated once in 0.5 N HClO$_4$ at 70°C for 15 min. Portions of the treated DNA solution can be assayed by the diphenylamine reaction described above. The precipitation of the nucleic acid with cold perchloric acid causes losses when small amounts of DNA or other macromolecules are present. Therefore, this step should be omitted when pure DNA solutions are assayed. If it is necessary to precipitate the DNA

from dilute solutions to remove interfering compounds like nucleotides, the addition of 100 μg of a protein, like bovine serum albumin, to each tube facilitates complete precipitation of the DNA.

The colorimetric and ultraviolet absorption assays are applicable to any class of RNA. All RNA classes have the same extinction coefficient when the colorimetric assay is used. Proteins, polysaccharides, and many other ultraviolet-absorbing compounds (including phenol, which is often used in the preparation of nucleic acids) compromise the accuracy of the ultraviolet absorption assay. Deviations from a normal spectrum indicate contamination with protein, carbohydrate, or other compounds.

The orcinol procedure for measuring RNA is subject to interference by hexoses, which produce a brown color. Thus, the ultraviolet absorbance assay is preferred for the measurement of RNA in sucrose gradients or solutions containing significant levels of hexoses or hexose polymers.

The ultraviolet absorbance assays are nondestructive, a matter of importance when small amounts of nucleic acids are available, and they are more convenient to perform.

It is possible to calculate the DNA content per cell from the DNA content of a suspension if the number of cells per unit volume is known. The nucleoid content per cell can be determined, and these data can be used to calculate the genome size of an organism (59).

17.6 NITROGEN COMPOUNDS

The choice of the most suitable method for the analysis of nitrogen in a compound depends on the form of nitrogen analyzed, the concentration of nitrogen, and the presence of interfering compounds. For the analysis of nitrite nitrogen, the most widely accepted method involves modification of the diazotization and coupling reactions. For nitrate nitrogen, ammonia nitrogen, and organic nitrogen, many other methods have been used. A comparison of the sensitivity and uses of some of these analyses is given in Table 1. For purposes of this manual, the simplest yet widely applicable method for each nitrogen compound will be described. For other methods, references are given in Table 1.

Because of the diversity of substances that may be analyzed for nitrogen (such as milk, fertilizer, soil, and plant and animal materials), many of the methods in Table 1 must be modified, or the sample must be pretreated to fit the method. Many variations of the procedures listed or described here have been published, and the reader should refer to the literature for

TABLE 1. *Comparison of analytical techniques available for assays of nitrogen compounds*

Nitrogen class	Analytical methods available	Minimum detectable levels of nitrogen	Interference	Major advantages	Major disadvantages	References
NO_2^-	Diazotization	1 µg/liter	Cl⁻ Colored samples Particulates Ions of Sb Bi Fe(III) Pb Hg Ag Cu(III) Auric Chloroplatinate Metavanadate	Simple.	Cl⁻ and Fe³⁺ interfere.	64, 68, 73, 80, 92
	Reduction to NH_4^+ by hot alkali or salicylic acid followed by analysis of NH_3[a] Others[b]	>2 mg/liter	NO_3^-, NH_4^+	Good when [NO_2] > 2,000 µg/liter.	Hot alkali, requires distillation, gives NO_2^- + NO_3 + NH_4^+.	61, 62, 68, 70
NO_3^-	Reduction to NO_2^- followed by analysis for NO_2^-[a]					97, 104
	Cd-Cu reduction	Up to 95% reduction	[S²⁻] > 2 mg/liter can activate columns	Rapid reduction once column is prepared.	Preparation of column.	73
	Cd-Hg reduction	Same as above	Same as above	Same as above	Same as above Use of Hg	68
	Zn reduction	≥95% reduction[a]		Does not require column preparation.	Longer reduction time.	81 92
	Absorbance at 220 nm	40 µg of N/liter	Many organic compounds NO_2^- Surfactants Chromate⁶⁻ Colored samples	Simple.	Requires ultraviolet spectrophotometer. Interference by many organic compounds.	68

TABLE 1—*Continued*

Nitrogen class	Analytical methods available	Minimum detectable levels of nitrogen	Interference	Major advantages	Major disadvantages	References
	Brucine	100 μg of N/liter	Colored samples, Particulates, Strong oxidants, Strong reductants, NO_2^-, High organics, Cl^-, Fe ions, Mn^{4+}		Brucine is highly toxic alkaloid—must be weighed out in a hood. Low sensitivity. Interference from oxidants and reductants.	68
	Chromotropic acid	100 μg/liter	Colored samples, Particulates, Ba, Pb, Sr, Iodide, iodate, Selenite, selenate, Chromate	Color is stable for 24 h.	Low sensitivity.	68
	Nitrophenol-disulfonic	20 μg/liter	Organics, Cl^-	Simple and direct.	Requires drying. Organic compounds interfere. Preparation of reagents requires boiling H_2SO_4.	64, 65, 93, 100
	Reduction to NH_4^+ by hot alkali or salicylic acid followed by analysis of NH_4^+	>2 mg/liter	NO_2^-	Good when NO_3^--N > 2 mg/liter.	Use of hot alkali. Requires distillation. Gives NO_3^- + NO_2^- + NH_4^+	64, 65, 68, 84
	Others[b]		NO_2^-, NH_4^+			70, 75, 76, 84, 103, 104
NH_4^+	Oxidation to NO_2^- followed by analysis of NO_2^{-a}		Amino acids, Atmospheric NH_3	No distillation required.	Some amino acids yield NO_2^-.	73
	Indophenol blue (also known as phenate)	10 μg/liter	Particulates, Colored substances, Excess acid or alkali, Small-molecular-weight nitrogen compounds, Cl^-, Phenol	Simple. Preferred colorimetric method following Kjeldahl design.	May require distillation prior to reaction. Use of phenol. Reagents unstable.	64, 68, 99

TABLE 1—*Continued*

Nitrogen class	Analytical methods available	Minimum detectable levels of nitrogen	Interference	Major advantages	Major disadvantages	References
	Nessler reagent	20 µg/liter	Particulates Colored substances Ions of Mg Fe Ca S Mn Cl Small-molecular-weight organic compounds	Satisfactory for pure water samples.	May require distillation prior to reaction. Use of Hg.	64, 68
	Acidimetric	2 mg/liter	Cl^-	After the distillation step the procedure is simple. Good when $[NH_4^+]$ is high. Preferred method following Kjeldahl digestion.	Requires distillation.	64, 68, 70
	Pyridine-pyrazole	50 µg/liter	Ions of FE Zn Ag Cu Cyanate	Specific for NH_4^+	Use of pyridine.	73, 88
Total organic N	Kjeldahl digestion followed by analysis for NH_4^+ or by oxidation to NO_2^- and analysis for NO_2^{-a}		NH_4^+ (must determine separately and subtract or distill off)	Quantitative.	Use of Hg (however, see references). Requires distillation. H_2SO_4 fumes.	64, 65, 66, 68, 70, 71, 73, 79, 98
Protein nitrogen	Lowry	30 mg/liter	Amino acid or peptide buffers Mercaptans $(NH_4)_2SO_4$	Very sensitive.	Results vary depending on amino acid composition; does not follow Beer's law.	67, 82, 91
	Biuret	1 g/liter	Amino acid or peptide buffers	Rapid.	Not highly accurate or sensitive. Results vary depending on amino acid composition.	67, 70, 89

TABLE 1—*Continued*

Nitrogen class	Analytical methods available	Minimum detectable levels of nitrogen	Interference	Major advantages	Major disadvantages	References
	Ultraviolet absorbance	10 mg/liter	Nucleic acids Phenolics	Simple.	Requires ultraviolet spectrophotometer. Nucleic acid interference.	86, 87, 89, 105
	Dye binding	30 mg/liter	Detergents	Response not as dependent on protein content as Lowry. Rapid.	Use of phosphoric acid.	78, 95, 101
	Fluorometric assay	5 μg/liter	Amino acid or peptide buffers	Rapid, high sensitivity, small sample required, good for purified proteins.	Results vary, depending on amino acid composition; use of dioxane.	77, 94
Total N	Kjeldahl digestion followed by reduction of NO_3^- and NO_2^- to NH_4^+ and analysis for NH_4^+[a]			Assays total N.	Time-consuming.	64, 73, 79, 98
	Separate determinations of nitrogenous compounds, then addition[a]			Can use reagents prepared for the separate determinations.		

[a] For further information, see analysis listed under appropriate nitrogen class.
[b] Techniques referenced here include new techniques and/or some techniques that may not be easily applicable to most samples.

procedures directly applicable to a particular kind of sample.

17.6.1. Nitrite (also see 16.2.2)

The analysis of nitrite nitrogen is based on the reaction of NO_2^- with sulfanilamide in an acid solution to form a diazo compound. This compound reacts with N-(1-naphthyl)-ethylenediamine to form a colored dye whose absorbance is measured at 543 nm in a spectrophotometer (73, 80). Sulfanilic acid and α-naphthylamine can be substituted for the two analytical chemicals; however, α-naphthylamine has been identified as a carcinogen and should be used only under carefully controlled conditions.

This reaction can detect as little as 1 μg of NO_2^- per liter when 10-cm light-path cuvettes are used. The absorbance of the colored compound follows Beer's law (16.1.1) up to 180 μg of NO_2^- per liter (68). Ions which interfere by causing precipitation are listed in Table 1. The presence of any of these may indicate the use of an alternate method of NO_2^- detection, such as reduction and subsequent analysis as NH_4^+. As with any colorimetric procedure, colored or particulate substances in the sample may interfere; this is particularly important with 10-cm cuvettes. Interference by colored substances may be lessened by dilution prior to the reaction if the NO_2^- concentration is high enough to permit this. Alternatively, treatment with activated charcoal often is effective for removal of interfering compounds. Particulate substances (such as bacterial cells) can be removed by centrifugation or filtration through a small-pore (0.45-μm) filter.

Sampling.

Samples should be assayed immediately, or frozen at $-20°C$ until such time as the assay may be run. Samples for NO_2^- assays should not be preserved with acid. Neutralization of all samples to pH 7.0 is suggested before analysis.

Reagents.

NO_2^--free water: Most sources of distilled or demineralized water are free from NO_2^-. If not, NO_2^--free water may be prepared as follows: add 1 ml of concentrated H_2SO_4 (*CAUTION*: caustic; avoid spilling; do not mouth pipette) and 0.2 ml of $MnSO_4$ (36.4 g of $MnSO_4 \cdot H_2O$ per 100 ml of distilled water) to each liter, and make pink with 1 to 3 ml of $KMnO_4$ (400 mg of $KMnO_4$ per liter).

Sulfanilamide reagent: Add 5 g of sulfanilamide to 300 ml of distilled water containing 50 ml of concentrated HCl. (*CAUTION*: Concentrated HCl is caustic. Add concentrated HCl slowly to the water. Avoid breathing fumes.) Adjust the volume to 500

ml with distilled water. The solution is stable at room temperature for many months.

N-(1-naphthyl)-ethylenediamine dihydrochloride solution: Dissolve 500 mg of N-(1-naphthyl)-ethylenediamine dihydrochloride in 500 ml of water. Store in the dark. A brown coloration appears after about 1 month, and the reagent should be discarded.

Standard NO_2^- solutions (68): Prepare a NO_2^- stock solution by dissolving 0.1411 g of $NaNO_2$ in 500 ml of distilled water. Use a new bottle of $NaNO_2$, as NO_2^- is readily oxidized in the presence of water. Keep the stock bottle tightly stoppered. One milliliter contains 50 μg of NO_2^--N. Dilute the stock to prepare standards for the standard curve.

Procedure.

Add 1 ml of the sulfanilamide solution to 50 ml of sample. Allow it to react for a minimum of 2 min, but no longer than 8 min. Undesirable side reactions and decomposition are significant after 10 min (73). Add 1 ml of N-(1-naphthyl)-ethylenediamine solution, and mix immediately. Allow color development for at least 10 min; then measure the absorbance at 543 nm (16.1.1).

Prepare a set of standards from the 50-μg/liter NO_2^- stock solution to give 0 to 25 μg of NO_2^--N per liter. Plot $[NO_2^-]$ versus absorbance at 543 nm (16.1.1.). Using the standard curve, determine the concentration of NO_2^--N in the sample by reading the concentration from the curve or by determining the extinction coefficient (ϵ) (the slope of the line) using linear regression analysis. This extinction coefficient can be used in successive determinations with the same reagents. A new standard curve should be prepared when new reagents are prepared.

17.6.2. Nitrate (also see 16.2.2)

The method of analysis of NO_3^- outlined here was chosen because of its relative simplicity, and the reagents prepared for the NO_2^- determinations are used in the assay. The basis for this method is the reduction of NO_3^- to NO_2^-, followed by the analysis of NO_2^-. Two procedures have been commonly used for the reduction of NO_3^- to NO_2^-, reduction by zinc (81, 92) and reduction by cadmium (68, 86); both are described here.

The reduction of NO_3^- to NO_2^- using Zn is a simple procedure involving the reaction of the sample with Zn dust for 30 min, followed by filtration. The disadvantage of this procedure is the manipulation time required for each sample.

The reduction of NO_2^- using Cd amalgamated with Cu is performed by passing the sample treated with NH_4Cl (to prevent deactivation of the column) through a previously prepared column of Cd-Cu^{2+} filings. This procedure requires little effort at the time of analysis, but requires prior preparation of the reduction column.

After either of the reduction procedures described here, the analysis for NO_2^- is performed as outlined above in the method for nitrite. Since the analysis is for *total* NO_2^-, part of the sample must be analyzed for NO_2^- before reduction, and this amount must be subtracted from the total NO_2^- to determine NO_3^-.

Sampling.

Samples should be treated as described above for nitrite.

Zinc reduction procedure.

Reagents. Use of NO_2^--free distilled water (17.6.1) is recommended in the preparation of all reagents.

Buffer: Mix 50 ml of 0.2 M KCl with 34 ml of 0.2 M HCl, and dilute to 400 ml. Adjust the pH to 1.5 with HCl.

NH_4Cl (2%, wt/vol)

Zinc powder

Procedure. Add 1.0 ml of 2% NH_4Cl and 2 ml of buffer to 25 ml of each sample in 50-ml stoppered flasks, and bring to 20 to 25°C. Add 0.2 g of zinc powder, and let stand for 30 min with occasional shaking. Filter out the zinc particles using a low-porosity filter paper.

Cadmium reduction procedure.

Reagents.

Copper sulfate reagent: 2% (wt/vol) $CuSO_4 \cdot 5H_2O$

Concentrated NH_4Cl: Add 175 g of NH_4Cl to 500 ml of distilled water

Dilute NH_4Cl: Dilute 1 volume of concentrated NH_4Cl with 40 volumes of NO_2^--free water to give 4.38 g of NH_4Cl in 500 ml of water.

Preparation of columns. Columns should be 1 by 30 cm with an appropriate opening (e.g., thistle tube or funnel) at the top so that the total sample can be placed on the column at once. The bottom of the column should have a burette-type stopcock or be attached to rubber tubing with a pinch clamp to allow the flow to be stopped before the column runs dry. The bottom of the column should be packed with glass wool or fine Cu "wool" turnings. A small amount of glass wool or copper filings is often added to the top of the column to prevent disturbance of the amalgamated cadmium-copper filings when adding the sample.

Cadmium filings can be prepared by filing sticks of reagent-grade cadmium metal with a coarse metal hand file (about second cut). Collect the fraction that passes a 2-mm opening sieve, but is retained on a 0.5-mm opening sieve. Stir approximately 50 g of filings with 250 ml of 2% $CuSO_4 \cdot 5H_2O$ until the blue color disappears from solution and semicolloidal Cu particles begin to enter the liquid above the sediment. Allow the suspension to settle for 10 min. Fill the

column with dilute NH_4Cl solution or with the supernatant fraction from the Cd-Cu amalgamation; then slowly pour in the Cd-Cu filings and let them settle until a column height of about 30 cm is achieved. Wash the column three times with 100 ml of the dilute NH_4Cl solution. The flow rate should be adjusted to approximately 10 ml/min.

Procedure. Add 2 ml of concentrated NH_4Cl reagent to 100 ml of the sample. Add 5 ml of this solution to the column, and allow the liquid to drain to the top of the column. This small addition ensures that when the bulk of the sample is added to the column no error results from mixing with the previously run sample.

Add the remainder of the sample NH_4Cl solution. Collect and discard the first 25 to 30 ml of the eluting liquid. Then collect 50 ml for the analysis. Discard the remainder of the eluting liquid. A column can be reused up to 100 times and usually need not be rinsed between samples. However, if the column will not be used for a few hours, rinse it with dilute ammonium chloride.

The 50 ml of eluate collected for analysis is assayed for NO_2^- (17.6.1) as soon as possible after reduction.

17.6.3. Ammonium by Indophenol Blue Reaction (100)

Many methods are available to determine the concentration of NH_4^+ in samples (see Table 1). An easy determination of NH_4^+ is based on ammonia, hypochlorite, and phenol reacting to form indophenol blue. The reaction is catalyzed by Mn^{2+}. The color is measured in a colorimeter or spectrophotometer at 630 nm (64, 68, 99). The major disadvantages of this method are the use of phenol and the instability of the reagents.

Another easy determination of NH_4^+ is Nessler's method, in which the NH_4^+ reacts with mercury and iodide in an alkaline solution to yield a color complex which is measured in a colorimeter or spectrophotometer (64, 68). The major disadvantage of this method is the use of mercury compounds. Both Nessler's method and the indophenol method are given here.

An alternative and almost as simple method, which can use the reagents prepared for NO_2^- determination, involves the oxidation of NH_4^+ to NO_2^- (73) with subsequent analysis of NO_2^-. However, the oxidation requires several reagents, as well as 3.5 h for the oxidation; therefore, the indophenol blue method or Nessler's method is recommended.

If a technique very specific for NH_4^+ is required, the pyridine-pyrazole method (73, 88) may be more satisfactory than the other techniques. Also see part 16.2.2.

Samples.

Samples should be treated as described in part 17.6.1. Samples high in interfering compounds (see Table 1) may require distillation before the analysis is performed (68).

Reagents.

Ammonium-free water: Add 0.1 ml of concentrated H_2SO_4 (*CAUTION:* Caustic. Do not mouth pipette. Avoid spills.) to each liter of distilled water, and redistill. Prepare fresh for each sample group, as NH_4^+ fumes from the laboratory will rapidly contaminate the water. Alternatively, an ion-exchange column may be used. Resins should be selected for maximum binding of NH_4^+ and other interfering organic compounds.

0.003 M $MnSO_4$: Dissolve 50 mg of $MnSO_4 \cdot H_2O$ in 100 ml of distilled NH_4^+-free water.

Hypochlorite reagent: Add 10 ml of commercial bleach (5% sodium hypochlorite solution) to 40 ml of NH_4^+-free distilled water. Adjust the pH to 6.5 to 7.0. Prepare fresh weekly.

Phenate reagent: Dissolve 2.5 g of NaOH and 10 g of phenol in 100 ml of NH_4^+-free distilled water. Prepare fresh weekly. (*CAUTION:* Phenol is rapidly absorbed through the skin and causes severe burns. Gloves, apron, and goggles should be worn while weighing out the solid. Avoid breathing the vapors by working in a fume hood. Do not mouth pipette the liquid.)

Standard NH_4^+ solution: Prepare a stock NH_4^+ solution by dissolving 381.9 mg of anhydrous NH_4Cl (dried at 100°C) in NH_4^+-free water, and make to 1 liter. One milliliter contains 100 μg of N and 122 μg of NH_3. To prepare standards, dilute the stock solution to obtain a range of 0.1 to 5 μg of NH_3-N per 10 ml.

Procedure.

Place 10 ml of each sample and 0.05 ml (1 drop) of the $MnSO_4$ solution in a small beaker or flask. With constant and vigorous stirring, preferably with a magnetic stirrer, add 0.5 ml of the hypochlorite reagent. Immediately add 0.6 ml of phenate reagent dropwise.

The reaction is complete in 10 min, and the color is stable for 24 h. Measure the color in a colorimeter or spectrophotometer at 630 nm. Prepare a blank and a standard curve each day, as the color intensity may change with the age of reagents.

17.6.4. Ammonium by Nessler Reaction

Samples.

Samples should be treated as described in part 17.6.1. Samples high in interfering compounds (see Table 1) may require distillation before the analysis is performed (68).

Reagents.

Ammonium-free water: Prepare as described in part 17.6.3.

Reagents for removing large amounts of interfering cations: Prepare $ZnSO_4$ solution by dissolving 100 g of $ZnSO_4 \cdot 7H_2O$ in ammonium-free water to a final volume of 1 liter. Prepare 6 N NaOH by dissolving 240 g of NaOH in ammonium-free water to a final volume of 1 liter.

Reagent for removing trace amounts of interfering cations: Dissolve 50 g of Na_2EDTA (ethylenediaminetetraacetic acid, disodium salt) in 60 ml of ammonium-free water containing 10 g of NaOH. (Heat may be required to dissolve the chemicals.) Adjust the volume of the cooled solution to 100 ml with ammonium-free water.

Nessler's reagent: Dissolve 100 g of HgI_2 and 70 g of KI in a minimal amount of water (100 to 200 ml). Add this slowly and with stirring to 500 ml of 8 N NaOH at 20 to 25°C. Adjust the volume of this mixture to 1 liter. Store in a rubber-stoppered glass vessel in the dark. The reagent is stable for up to 1 year. (*CAUTION:* This reagent is toxic. Do not mouth pipette. Avoid spills. Dispose of mercury solutions properly.) Check the reagent monthly by comparing a standard curve of NH_4^+-N using the reagent with the standard curve prepared when the reagent was new. When the curves deviate significantly, discard the reagent.

Standard NH_4^+ solution: Prepare a stock NH_4^+ solution as outlined in part 17.6.3. To prepare the standards, dilute the stock solution to obtain a range of 20 to 500 μg/50 ml.

Procedure.

To remove interfering ions of Ca, Mg, S, and Fe, two methods are employed. For large quantities of interfering ions, add 1 ml of the $ZnSO_4$ reagent to 100 ml of sample. Adjust the pH to 10.5 with 0.4 to 0.5 ml of 6 N NaOH by using a pH meter. Precipitation of the interfering ions occurs within 15 min. The precipitate will usually sediment other suspended matter as well. Remove the precipitate by filtration (use NH_4^+-free filter paper) or by centrifugation. If experience with the samples indicates that only trace amounts of the interfering ions are present, add 0.05 ml (1 drop) of the EDTA solution, with thorough mixing, to 50 ml of sample. Samples pretreated with $ZnSO_4$ and alkali are often also treated with EDTA. Samples with no interfering cations may be treated with Nessler's reagent without pretreatment.

The Nessler reaction is performed by adding 1 ml of Nessler's reagent to 50 ml of sample. If EDTA has been added to the sample, 2 ml of Nessler's reagent should be used. Mix the sample-reagent solution well, and let it stand for at least 10 min at room temperature. (For low NH_4^+, 30 min is preferred.) Keep the temperature and time of reaction constant for all samples, standards, and blanks. Read the absorbance of the resulting colored compound at a single wavelength between 400 and 425 nm for samples and standards with low NH_4^+ (20 to 250

$\mu g/50$ ml) and at a wavelength between 450 and 500 nm for those with high NH_4^+ (250 to 500 $\mu g/$ 50 ml). This reaction can be used with proportionally smaller volumes of sample, standards, and reagents as required by the investigator. The sensitivity of the method can be increased to 5 $\mu g/50$ ml if a 5-cm-light path cuvette is available. Calculate the amount of ammonia in each sample from a standard curve prepared by plotting the absorbance of each standard versus the concentration of ammonia in the standard solution (see 16.1.1).

17.6.5. Protein by Folin Reaction

The most widely used colorimetric method for estimation of proteins is that of Lowry et al. (90), using the Folin reaction. The blue color is due to the reaction of protein with copper ion in alkaline solution (the biuret reaction) and the reduction of the phosphomolybdate-phosphotungstic acid in the Folin reagent by the aromatic amino acids in the treated protein.

The following procedure is useful for proteins that are already in solution or that are soluble in dilute alkali.

Reagents.

Reagent A: Dissolve 20 g of Na_2CO_3 in 1 liter of 0.1 N NaOH.

Reagent B: Dissolve 0.5 g of $CuSO_4 \cdot 5H_2O$ in 100 ml of a 1% (wt/vol) aqueous solution of sodium tartrate.

Reagent C: Just before use, mix 50 ml of reagent A and 1 ml of reagent B. Discard after 1 day.

Reagent D: Diluted Folin reagent. Concentrated Folin-Ciocalteu reagent can be purchased commercially from several sources (e.g., Fischer Scientific Co., Pittsburgh, Pa.) and is diluted according to the manufacturer's instructions. (CAUTION: This reagent contains a strong volatile acid and should not be pipetted by mouth. Protective gloves, clothing, and goggles should be worn, particularly if a Vortex mixer is used for mixing.)

Standard protein solution: Any one of several commercially available proteins or commercial standards are suitable. Crystalline bovine serum albumin or prepared solutions of bovine serum albumin that can be purchased from biological supply houses produce suitable linear standard curves. The stock standard solution should contain 0.5 mg of protein per ml in distilled water. Different proteins give different standard curves. The protein standard used should be specified when the results are reported.

Procedure.

Add up to 1.0 ml of a sample containing 50 to 250 μg of protein to a clean 10-ml test tube. Adjust the volumes of each sample to 1.0 ml with distilled water. Add 5.0 ml of reagent C, and mix well. Allow this mixture to stand for at least 10 min at room temperature. Add 0.5 ml of

reagent D, and mix immediately. The reactivity of this reagent to freshly added protein lasts only seconds after dilution. After 30 min or longer at room temperature, measure the absorbance at 500 nm in a spectrophotometer or colorimeter. Greater sensitivity can be achieved by measuring the absorbance at 750 nm if an appropriate spectrophotometer is available.

If 1-ml cuvettes and a suitable instrument are available, all volumes can be reduced by a factor of five, and reagents and sample can be conserved. The samples and standards can be adjusted to 0.2 ml, and 1.0 ml of reagent C and 0.1 ml of reagent D are then added at the time intervals indicated above.

When protein is to be measured in bacterial cell suspensions or when proteins are insoluble in dilute alkali, it is necessary to heat the suspension at 90°C in 1 N NaOH for 10 min to obtain complete solubilization. The standard should be treated in the same way because this treatment reduces the intensity of the color. When this procedure is applied to dissolve or extract proteins, a reagent containing 20 g of Na_2CO_3 in 1 liter of water (no alkali) is substituted for reagent A. The assay procedure is otherwise identical.

Ammonium sulfate in amounts greater than 0.15% reduces color development. Investigators should be aware of this when this assay is used for proteins purified by fractionation procedures based on differential solubilities in ammonium sulfate solutions. High concentrations (1 to 5 M) of many other salts, some sugars, glycine, Tris buffer, some chelating agents, and compounds with sulfhydryl groups also interfere with the assay. Aromatic amino acids produce color with the phenol reagent, and glycine interferes with color development.

A modification of the Lowry assay that permits its use with samples containing sucrose or ethylenediaminetetraacetic acid (EDTA) and with membrane and lipoprotein preparations without prior solubilization of the proteins has been described by Markwell et al. (91). The procedure employs a detergent (sodium dodecyl sulfate) in the alkaline carbonate reagent to improve solubilization of samples containing lipoidal material and an increase in the amount of the copper tartrate reagent to overcome interference by sucrose and EDTA. This procedure has several applications, including assays of proteins in fractions from sucrose gradients.

17.6.6. Protein by Biuret Reaction (68, 82)

Peptide bonds of proteins form blue-colored chelates with copper ions in alkaline solutions. This reaction is much less sensitive than the Lowry procedure, but has the advantage that it is less time-consuming and can be used in the presence of ammonium salts and other materials that interfere with the Lowry procedure. There is also less variability between proteins in the amount of color formed per unit weight of protein.

The modification of the biuret procedure (83) described below involves the use of tartrate to form a copper complex that is soluble in NaOH. The protein displaces the copper ion from the complex to form a copper-protein complex with a different color and absorption intensity.

Reagents.

Dissolve 1.5 g of cupric sulfate ($CuSO_4 \cdot 5H_2O$) and 6 g of sodium potassium tartrate $\cdot 4H_2O$ in 500 ml of water. Add, with stirring, 300 ml of a 10% (wt/vol) solution of sodium hydroxide in water. Adjust the volume of the mixture to 1 liter with distilled water. Discard if a reddish or black precipitate is evident.

Procedure.

Add 4.0 ml of the above biuret reagent to a neutralized sample (1.0 ml) containing 1 to 10 mg of protein. Mix thoroughly, and incubate at room temperature for 30 min. Time and temperature for all samples, standards, and blanks should be the same. Measure the absorbance at 500 nm. A blank is prepared using 1.0 ml of water or aqueous solution used to dissolve proteins and 4.0 ml of biuret reagent. A standard curve can be prepared using any completely soluble protein such as crystalline bovine serum albumin. The response is not linear with protein concentration because of competition between the tartrate and protein for copper ions.

Applications.

A modification of this procedure described by Stickland (102) is suitable for the analysis of total protein in bacterial cells. In this procedure the cellular protein is dissolved in 1.0 N NaOH, and then $CuSO_4$ without tartrate is added. The insoluble cellular material and insoluble $Cu(OH)_2$ are removed by centrifugation, leaving the colored Cu-protein complex in solution.

17.6.7. Protein by Coomassie Blue Reaction

Proteins bind several dyes, causing a shift in the absorption spectrum of the dye. In the procedure described below, Coomassie brilliant blue is used as the dye. This dye exists in two forms: the red anionic form is converted to a blue form when the dye binds to amino groups of proteins. This procedure is often more sensitive and less subject to interference by many compounds that restrict the use of other assays (78). The dye assay can be performed more rapidly than those described above. The sensitivity of the dye assay is comparable to that of the Lowry procedure. The response to different proteins is less variable than the color reactions in the other assays described above.

Reagent (commercially prepared reagent kits are also available).

Dissolve 100 mg of Coomassie brilliant blue-G250 in 50 ml of 95% ethanol. Add 100 ml of 85% (wt/vol) phosphoric acid. (*CAUTION:* This is a strong acid. Do not pipette by mouth. Wear protective goggles, gloves, and laboratory clothing.) Adjust the volume to 1 liter with distilled water.

Procedure.

Add a protein solution containing 10 to 100 μg of protein in 0.1 ml to a test tube. Add 5 ml of the dye reagent, and thoroughly mix the contents. Measure the absorbance at 595 nm after 2 min and before 1 h against a blank containing 0.1 ml of the buffer or salt solution and 5 ml of the dye reagent.

The protein content of the sample is determined from a standard curve obtained by plotting the absorbance of standard solutions containing 5 to 100 μg of protein per ml in the same buffer used to dissolve the unknown samples versus the concentration of protein in each standard solution (see 16.1.1).

Applications.

This procedure is applicable to the assay of many soluble proteins. Detergents interfere with color development; thus, glassware should be rinsed well. The assay is linear from 25 to 75 μg of protein per sample. Other limitations of the assay are described by Pierce and Suelter (95).

A microassay procedure (78) and another variation that has high reproducibility and can detect 1.0 μg of protein have been described (101).

17.6.8. Protein by Ultraviolet Absorption

The absorption maximum of proteins is 280 nm and is due to the presence of the aromatic amino acids: tyrosine and tryptophan. These two amino acids are present in nearly all proteins, and their proportions relative to other amino acids usually vary over a narrow range. The absorption of ultraviolet light is a suitable means of estimating proteins in solution if the protein solution does not contain more than 20% by weight of other ultraviolet light-absorbing compounds, such as nucleic acids or phenols, and if the solution is not turbid.

Correction for the absorbance of nucleic acids

in protein can be made by the method of Warburg and Christian (105) or the method of Kalckar (87).

Equipment.

Ultraviolet spectrophotometer: Calibrate with an absorbance standard (16.1.1).

Cuvettes: Made from quartz or silica, which do not prevent transmittance of ultraviolet light.

Procedure.

Absorbance measurements are made at 280 nm and 260 nm with the protein dissolved in a suitable buffer. A solvent blank containing the buffer is used to zero the instrument at each wavelength. If the ratio of absorbance at 280 nm/absorbance at 260 nm is not greater than 1.70, except with solutions known to contain pure protein, the following equation (87) should be used to calculate the concentration of protein, or an appropriate table (89) should be consulted. The equation is used to subtract the contribution of nucleic acids to the ultraviolet absorbance of the solution.

protein concentration (mg/ml)

$$= 1.45A_{280} - 0.74A_{260}$$

Applications.

When preservation of a protein sample is important because of its limited availability, this nondestructive assay procedure is the method of choice, provided that the protein is sufficiently pure to permit its application. Different proteins can yield different extinction coefficients, and a standard prepared from an identical protein (if available) should be used if precise measurements are important.

17.6.9. Total Nitrogen

The analysis of a sample for total nitrogen may be accomplished by adding the results of analyses for the separate nitrogenous components (NO_2^-, NO_3^-, NH_4^+, and organic nitrogen). However, if one is interested only in the total nitrogen, performing all these analyses can be time-consuming. The most commonly used analysis for total nitrogen is the Kjeldahl digestion of organic nitrogen compounds to yield NH_4^+, which is then assayed by one of the techniques for NH_4^+ analysis. NO_3^- and NO_2^- are included in this analysis by reduction to NH_4^+ with hot alkali (68) or salicylic acid and zinc (65). An alternative method, which allows use of reagents already prepared for the NO_2^- analysis, is based on the oxidation of the NH_4^+ from the digestion to nitrite and subsequent analysis of nitrite.

The procedure used to digest the organic material depends on the state of the sample (solid or liquid). Because the procedures for the Kjeldahl digestions are variable, complex, and easily available in various texts, the reader is referred to the literature that treats the specific procedure. Kjeldahl digestion techniques developed for liquid samples include those for seawater (73), lake water (64, 68), and wastewater (64, 68). Techniques for solids include those for soil (64–66, 71), sediment (68), plant materials (64, 70), fertilizers (70), animal materials (70), and microbes (68).

17.6.10. Molecular Nitrogen

The need to detect molecular nitrogen in a biological system is usually limited to nitrogen fixation studies in which nitrogen (dinitrogen) is reduced by an enzyme system (called nitrogenase) to ammonia, which can then be assimilated by the cell. It would be possible to quantitate this reaction by the disappearance of nitrogen, detected by atomic absorption, or the appearance of ammonia using analyses described (17.6.3). However, it is easier to measure nitrogen fixation by the ability of nitrogenase to reduce acetylene (C_2H_2) to ethylene (C_2H_4). Acetylene and ethylene can be detected and quantified by use of a gas chromatograph. Acetylene reduction is usually tested in systems that have sufficient nitrogenase activity to allow a short time (hours) of exposure to the acetylene, which precludes interference due to nonbiological reduction of acetylene. However, this interference can be compensated for with proper controls in longer-term assays. Although no system of rapid acetylene reduction other than by nitrogenase enzymes has been discovered, it is important to confirm the requirement of nitrogenase enzymes by ammonia suppression of the acetylene reduction system (96).

Since N_2 fixation can be accomplished by both aerobic and anaerobic organisms, the experimental set-up will vary (96). It is also important to consider the proper incubation conditions, including light and temperature requirements. For qualitative determination of nitrogenase in systematic bacteriology, and for further discussion, see part 20.2.7.

Equipment and reagents.

Gas chromatograph with a flame ionization detector, with appropriate column to detect acetylene and ethylene

Plastic syringes and appropriate needles

Teflon syringe valve which can be opened to take a gas sample and closed to retain the sample in the syringe

High-purity commercial acetylene, in an inert gas (1 to 4%, vol/vol). (*CAUTION:* The gas tank should be attached to a bench. The regulator should be the proper one for that tank.)

Procedure.

Flask or tube cultures of suspected organisms or enrichments are prepared in nitrogen-free broth. For best results, the aerobic cultures should be just barely turbid. The flask or tube should have 5 to 10 volumes of gas space above the culture. The aerobic vessel should be sealed before acetylene is added. Acetylene is then added to a final pC_2H_2 of 0.05 to 0.10 atm, which is equivalent to the saturation of nitrogenase enzymes produced by 0.8 atm of N_2 (85). The remaining gas mix in the gas phase is dependent on the sample type (aerobic versus anaerobic). To standardize the acetylene injected and to detect nonbiological acetylene reduction, similar vessels with uninoculated media or formaldehyde-killed cells should be treated with acetylene and tested immediately and after the incubation period for acetylene and ethylene.

After a few hours of incubation at appropriate light and temperature conditions (3 h are usually sufficient with pure cultures), 50-μl samples of the gas phase should be taken and assayed for ethylene by use of a gas chromatograph. The retention times for acetylene and ethylene are determined by the gas chromatograph set-up as specified by the manufacturer or supplier of the packing material, or can be determined by using acetylene and ethylene standards. The amount of ethylene produced is determined from the peak height (including consideration of range and attenuation settings on the chromatograph) compared to the standard. The amount of ethylene produced should be equal to acetylene reduced, to be sure that the ethylene produced is not a natural by-product of metabolism for the sample organisms. To determine the moles of N_2 fixed, the moles of C_2H_2 reduced must be divided by 3 to account for the electron requirement for total fixation of N_2 to ammonia.

Applications.

The procedure outlined can be used for pure or enrichment cultures of nitrogen-fixing organisms. A modification of this procedure for enzyme preparations has been described (96).

17.6.11. Specific Nitrogenous Compounds

Nitrogenous compounds such as peptides, nucleic acids, pesticides, and pollutants can be specifically assayed for both qualitative and quantitative determination by use of mass spectrometry. The advantages of this technique are its specificity, ability to detect small amounts, use of small quantities of sample (micrograms, microliters), and ability to assay gases, liquids, and solids. The major disadvantage is the equipment requirement and the difficulty of analysis of the data obtained. The theory and use of mass spectroscopy have been well described (69, 72, 74).

17.7. ORGANIC ACIDS AND ALCOHOLS

A number of short-chain organic acids and alcohols are formed in bacterial cells as intermediates or end products of the citric acid, glycolytic, and other metabolic pathways. These compounds may be accumulated in the cytoplasm or excreted into the medium. In bacterial cell fractions, these compounds occur in the small molecule, supernatant fluid fraction after precipitation of the macromolecules with cold trichloroacetic acid (Fig. 1 in part 17.1). The small molecule fraction may include acetate, acetoacetate, β-hydroxybutyrate, citrate-isocitrate, α-ketoglutarate, formate, lactate, succinate, pyruvate, malate, fumarate, ethanol, and many others.

These small molecules may be analyzed by ion-exchange chromatography (16.3.1), absorption chromatography (18.5.2), or especially gasliquid partition chromatography (16.3.4). Lactate may be also analyzed by enzymatic or colorimetric methods (see 17.2.8).

17.8. ELEMENTS

The amounts of different elements in bacterial cells, and the biological roles of trace elements, are described in references 107 and 109; references 107 and 111 describe sample preparation and storage. Metal analysis in samples of interest to bacteriologists can be accomplished by colorimetric methods, atomic absorption, photometry, spectrophotometry, flame emission photometry, nuclear activation analysis, the use of ion-specific electrodes, and other means; these and other methods for determining metals are described in references 106 to 109.

A rapid, accurate, and convenient titrimetric procedure that requires no specialized equipment can be applied to the analysis of some elements such as calcium. Interfering elements and compounds present in small amounts can sometimes be removed by treatment of the samples. These procedures are described in references 109 and 110.

Flame emission photometry (107, 108) is ideally suited to the analysis of alkali and alkaline earth metals. Atomic absorption spectroscopy can be used to detect parts per million of some

40 elements in a variety of samples (107). The principles of the techniques are described in references 107 and 108. Atomic absorption analysis (106, 108) is an accurate and relatively convenient method for the analysis of several metals. These methods require experienced personnel and a specialized spectrophotometer. The inaccessibility of proper advice or the instrument may limit the use of the technique. Several elements and compounds interfere with atomic absorption analysis of an element, but proper treatment of samples can reduce or eliminate interferences.

Nuclear activation analysis (108) involves the bombardment of a sample in order to make one or more elements radioactive. The radioactive species are identified and measured quantitatively. This method of elemental analysis has the advantages of sensitivity, and many elements can be assayed in a single small sample. The use of this technique requires access to a facility that has a high-energy neutron source (a nuclear reactor or accelerator), sophisticated isotope detection equipment, a suitable computer, and expert personnel to operate the facility. When available, this method is relatively inexpensive in comparison with other methods if time and other costs are taken into account and if several elements are determined simultaneously in a single sample. This is particularly true when the availability of sample material for analysis is limited.

X-ray emission spectroscopy (108) is a nondestructive method of analysis applicable to the estimation and identification of some elements that are present in sufficient abundance in biological material. Electron probe X-ray microanalysis can be applied to the estimation and location of some elements in situ. The resolution of the technique is suitable for use in individual bacterial cells or spores (112). This technique also requires sophisticated equipment and expertise. Thus, its application is generally limited to research laboratories.

Methods for colorimetric and turbidimetric analysis of metals and inorganic compounds, including automated analysis, have been described by Snell and Snell (109) and in publications from the U.S. Environmental Protection Agency (110).

Commercially available selective ion electrodes also provide a very convenient and inexpensive means of measuring many elements. Their application is limited to a few types of samples and elements, however (see 16.2.2).

17.9. LITERATURE CITED

Fractionation and Radioactivity

1. **Kennell, D.** 1967. Methods Enzymol. **12A:**686–
692.
2. **Roberts, R. B., D. B. Cowie, E. T. Bolton, P. H. Abelson, and R. J. Britten.** 1955. Biosynthesis in *Escherichia coli.* Carnegie Inst. Washington Publ. 607.
3. **Sutherland, I. W., and J. F. Wilkinson.** 1971. *In* J. R. Norris and D. W. Ribbons (ed.), Methods in microbiology, vol. 5B, p. 346–383. Academic Press, Inc., New York.

Carbohydrates

General references:
4. **Colowick, J., and N. Kaplan (ed.).** 1966. Methods Enzymol. **8:**1–759.
5. **Herbert, D., P. J. Phipps, and R. E. Strange.** 1971. *In* J. R. Norris and D. W. Ribbons (ed.), Methods in microbiology, vol. 5B, p. 209–344. Academic Press, Inc., New York.
6. **Work, E.** 1971. *In* J. R. Norris and D. W. Ribbons (ed.), Methods in microbiology, vol. 5A, p. 361–418. Academic Press, Inc., New York.

Specific articles
7. **Dische, Z.** 1953. J. Biol. Chem. **204:**983–997.
8. **Dische, Z.** 1962. Methods Carbohydr. Chem. **1:** 477–514.
9. **Elson, L. A., and W. T. Morgan.** 1933. Biochem. J. **27:**1824–1828.
10. **Fox, G. E., L. J. Magrum, W. E. Balch, R. S. Wolfe, and C. R. Woese.** 1977. Proc. Natl. Acad. Sci. U.S.A. **74:**4537.
11. **Ghuysen, J.-M., D. J. Tipper, and J. L. Strominger.** 1966. Methods Enzymol. **12A:** 695–699.
12. **Hadzija, O.** 1974. Anal. Biochem. **60:**512–517.
13. **Osborn, M. J.** 1963. Proc. Natl. Acad. Sci. U.S.A. **50:**499–506.
14. **Rondle, C. J. M., and W. T. J. Morgan.** 1955. Biochem. J. **61:**586–590.
15. **Stewart-Tull, D. E. S.** 1968. Biochem. J. **109:** 13–18.
16. **Sweeley, C. C., W. W. Wells, and R. Bentley.** 1966. Methods Enzymol. **8:**95–107.
17. **Tipper, D. J.** 1968. Biochemistry **7:**1441–1449.
18. **Weisbach, A., and J. Hurwitz.** 1959. J. Biol. Chem. **234:**705–712.

Lipids

General references:
19. **Dittmer, J. C., and M. A. Wells.** 1969. Methods Enzymol. **14:**482–530.
20. **Kates, M.** 1972. *In* T. S. Work and E. Work (ed.), Laboratory techniques in biochemistry and molecular biology, vol. 3, p. 269–610. American Elsevier Publishing Co., Inc., New York.
21. **O'Leary, W.** 1974. *In* A. I. Laskin and H. A. Lechevalier (ed.), Handbook of microbiology, vol. 2, p. 275–327. Chemical Rubber Co., Cleveland, Ohio.

Specific articles:
22. **Bird, C. W., J. M. Lynch, S. J. Pirt, W. W. Reid, C. J. Brooks, and B. S. Middleditch.** 1971. Science **230:**473–474.
23. **Bligh, E. G., and W. J. Dyer.** 1951. Can. J.

Biochem. Physiol. **37**:911-917.
24. **Folch, J.** 1957. J. Biol. Chem. **266**:497-509.
25. **Hara, A., and N. S. Radin.** 1978. Anal. Biochem. **90**:420-426.
26. **Johnson, A. R., and R. B. Stocks.** 1971. *In* A. R. Johnson and J. B. Davenport (ed.), Biochemistry and methodology of lipids, p. 195-218. Wiley Interscience, New York.
27. **Law, J. H., and R. A. Slepecky.** 1961. J. Bacteriol. **82**:33-36.
28. **Patt, T. E., and R. S. Hanson.** 1978. J. Bacteriol. **134**:634-636.
29. **Patton, J. C., E. J. McMurchie, B. K. May, and W. H. Elliot.** 1978. J. Bacteriol. **136**:754-759.
30. **Rilfors, L., A. Wieslander, and S. Stahl.** 1978. J. Bacteriol. **135**:1043-1052.
31. **Skipski, V. P., and M. Barklay.** 1969. Methods Enzymol. **14**:541-548.
32. **Stein, R. A., V. Slawson, and J. F. Mead.** 1976. *In* G. V. Marinetti (ed.), Lipid chromatographic analysis, 2nd ed., p. 857-896. Marcel Dekker, Inc., New York.
33. **Vorbeck, M. L., and G. V. Marinetti.** 1965. Biochemistry **4**:296-305.
34. **Wuthier, R. E.** 1966. J. Lipid Res. **7**:558-561.

Cell Wall Polymers

General references:
35. **Braun, V.** 1978. Symp. Soc. Gen. Microbiol. **28**:111-138.
36. **Braun, V., and K. Hantke.** 1974. Annu. Rev. Biochem. **43**:89-121.
37. **Costerton, J. W., J. M. Ingram, and K. J. Cheng.** 1974. Bacteriol. Rev. **38**:87-110.
38. **Cummins, C. S.** 1974. *In* A. I. Laskin and H. A. Lechevalier (ed.), Handbook of microbiology, vol. 2, Microbial composition, p. 251-284. Chemical Rubber Co., Cleveland, Ohio.
39. **Elin, R. J., and S. M. Wolff.** 1974. *In* A. I. Laskin and H. A. Lechevalier (ed.), Handbook of microbiology, vol. 2, Microbial composition, p. 674-731. Chemical Rubber Co., Cleveland, Ohio.
40. **Ghuysen, J.-M.** 1968. Bacteriol. Rev. **32**:425-464.
41. **Leive, L. (ed.).** 1973. Bacterial membranes and walls. Marcel Dekker, Inc., New York.
42. **Luderitz, O., A. M. Staub, and O. Westphal.** 1966. Bacteriol. Rev. **30**:192.
43. **Milner, K. C., J. A. Rudbach, and E. Ribi.** 1971. *In* G. Weinbaum, S. Kadis, and S. Ajl (ed.), Microbial toxins, vol. 4, p. 1. Academic Press, Inc., New York.
44. **Reaveley, D. A., and R. E. Burge.** 1972. Adv. Microb. Physiol. **7**:1-81.
45. **Rogers, H. J., J. B. Ward, and I. D. J. Burdett.** 1978. Symp. Soc. Gen. Microbiol. **28**:139-176.
46. **Schleifer, K. H., and O. Kandler.** 1972. Bacteriol. Rev. **36**:407-477.

Specific articles:
47. **Davidson, A. L., and J. Badiley.** 1964. Nature (London) **202**:874.

48. **Ghuysen, J. M., D. Tipper, and J. L. Strominger.** 1965. Biochemistry **4**:2245-2256.
49. **Kandler, O., and H. Konig.** 1978. Arch. Microbiol. **118**:141-152.
50. **Knox, K. W.** 1966. Biochem. J. **100**:73-78.
51. **Kottel, R. H., K. Bacon, D. Clutter, and D. White.** 1975. J. Bacteriol. **124**:550-557.
52. **Osborn, M. J.** 1963. Proc. Natl. Acad. Sci. U.S.A. **50**:499-514.
53. **Strominger, J. L., and D. J. Tipper.** 1975. *In* E. F. Osserman, R. E. Canfield, and S. Beychak (ed.), Symposium on Lysozyme.
54. **Taylor, A., K. W. Knox, and E. Work.** 1966. Biochem. J. **99**:53-61.
55. **Warth, A. W., and J. L. Strominger.** 1969. Proc. Natl. Acad. Sci. U.S.A. **64**:528-535.
56. **Weidel, W., H. Frank, and H. H. Martin.** 1960. J. Gen. Microbiol. **22**:158-166.
57. **Westphal, O., and K. Jan.** 1965. Methods Carbohydr. Chem. **5**:83-91.
58. **Work, E.** 1971. *In* J. R. Norris and D. W. Ribbons (ed.), Methods in microbiology, vol. 5A, p. 361-418. Academic Press, Inc., New York.

Nucleic Acids

General references:
59. **DeLey, J.** 1971. *In* J. R. Norris and D. W. Ribbons (ed.), Methods in microbiology, vol. 5A, p. 301-311. Academic Press, Inc., New York.
60. **Parish, J. H.** 1972. Principles and practice of experiments with nucleic acids. Longman Group, Ltd., London.

Specific articles:
61. **Burton, K.** 1957. Biochem. J. **62**:315-323.
62. **Griswold, B. C., F. L. Humoller, and A. R. McIntyre.** 1951. Anal. Chem. **23**:192-194.
63. **Zahn, R. Z.** 1970. FEBS Lett. **6**:141.

Nitrogen Components

General references:
64. **Allen, S. E., H. M. Grimshaw, J. A. Parkinson and C. Quarmby.** 1974. *In* S. E. Allen (ed.), Chemical analysis of ecological materials, p. 184-206. Blackwell Scientific Publications, London.
65. **Black, C. A. (ed.).** 1965. Methods of soil analysis, part 2, Chemical and microbiological properties. American Society of Agronomy, Inc., Madison, Wis.
66. **Bremner, J. M., and D. R. Keeney.** 1965. Anal. Chim. Acta **32**:485-495.
67. **Cooper, T. G.** 1977. The tools of biochemistry. John Wiley & Sons, Inc., New York.
68. **Franson, M. A. (ed.).** 1976. Standard methods for the examination of water and wastewater, 14th ed. American Public Health Association, Washington, D.C.
69. **Gudinowicz, B. J., M. J. Gudinowicz, and H. F. Martin.** 1976. Fundamentals of integrated GC-MS. Part II: Mass spectrometry. Marcel Dekker, Inc., New York.
70. **Horowitz, W. (ed.).** 1975. Official methods of analyses of the Association of Official Analytical Chemists, 12th ed. Association of Official Ana-

lytical Chemists, Washington, D.C.

71. **Jackson, M. L.** 1958. Soil chemical analysis, Prentice-Hall, Inc., Englewood Cliffs, N.J.

72. **Safe, S., and O. Hutzinger.** 1973. Mass spectrometry of pesticides and pollutants. CRC Press, Cleveland, Ohio.

73. **Strickland, J. D. H., and T. R. Parsons.** 1960. A practical handbook of seawater analysis. Fisheries Research Board of Canada, Ottawa.

74. **Waller, G. R. (ed.).** 1972. Biochemical applications of mass spectrometry. John Wiley & Sons, Inc., New York.

Specific articles:

75. **Armstrong, F. A. J.** 1963. Anal. Chem. **35**:1292.

76. **Baca, P., and H. Freiser.** Anal. Chem. **49**:2249-50.

77. **Bohler, P., S. Stein, W. Dairman, and S. Udenfriend.** 1973. Arch. Biochem. Biophys. **155**:213-220.

78. **Bradford, M. M.** 1976. Anal. Biochem. **72**:248-254.

79. **Bremmer, J. M.** 1960. J. Agric. Sci. **55**:11-33.

80. **Canney, P. J., D. E. Armstrong, and J. H. Wiersma.** 1974. Determination of nitrite and nitrate ions in natural waters using aromatics or diamines as reagents. Technical Report, University of Wisconsin Water Resources Center, Madison.

81. **Chow, T. J., and M. S. Johnstone.** 1962. Anal. Chim. Acta **27**:441-446.

82. **Coakley, W. T., and C. J. James.** 1978. Anal. Biochem. **85**:90-97.

83. **Gornall, A. G., C. S. Bardawill, and M. M. David.** 1949. J. Biol. Chem. **177**:751-756.

84. **Guiraud, G., J. C. Fardeau, G. Llimous, and M. A. Barral.** 1977. Ann. Agron. **28**:329-333.

85. **Hardy, R. W. F., and A. H. Gibson (ed.).** 1977. A treatise on dinitrogen fixation. Section IV. Agronomy and ecology. John Wiley & Sons, Inc., New York.

86. **Kalb, V. F., Jr., and R. W. Bernlohr.** 1977. Anal. Biochem. **82**:362-371.

87. **Kalckar, H. M.** 1947. J. Biol. Chem. **167**:429-475.

88. **Kruse, J. M., and M. G. Mellon.** 1953. Anal. Chem. **25**:1188-1192.

89. **Layne, E.** 1957. Methods Enzymol. **3**:447-454.

90. **Lowry, O. H., N. J. Rosebrough, A. L. Farr, and R. J. Randall.** 1951. J. Biol. Chem. **193**:265-275.

91. **Markwell, M. A., S. M. Haas, L. L. Bieber, and N. E. Tolbert.** 1978. Anal. Biochem. **87**:207-210.

92. **Matsunaga, K., and M. Nishimura.** 1969. Anal.

Chim. Acta **45**:350-353.

93. **Mubarek, A., R. A. Howald, and R. Woodriff.** 1977. Anal. Chem. **49**:857-860.

94. **Nakamura, H., and J. J. Pisano.** 1976. Arch. Biochem. Biophys. **172**:102-105.

95. **Pierce, J., and C. H. Suelter.** 1977. Anal. Biochem. **81**:478-480.

96. **Postgate, J. R. (ed.).** 1971. The chemistry and biochemistry of nitrogen fixation. Plenum Press, New York.

97. **Raganowicz, E., and A. Niewiadomy.** 1976. Pol. Arch. Hydrobiol. **23**:1-4.

98. **Rexroad, P. R., and R. D. Cathey.** 1976. J. Assoc. Off. Anal. Chem. **59**:1213-1217.

99. **Russel, J. A.** 1944. J. Biol. Chem. **156**:457-461.

100. **Snell, F. D., and C. T. Snell.** 1945. Colorimetric methods of analysis. D. Van Nostrand Co., Inc., New York.

101. **Sedmak, J. J., and S. E. Grossberg.** 1977. Anal. Biochem. **79**:544-552.

102. **Stickland, H. L.** 1951. J. Gen. Microbiol. **5**:698-703.

103. **Tan, Y. L.** 1977. Anal. Chim. Acta **91**:373-374.

104. **Terado, K., H. Honnami, and T. Kiba.** 1977. Bull. Chem. Soc. Jpn. **50**:132-7.

105. **Warburg, O., and W. Christian.** 1942. Biochem. Z. **310**:384-421.

Elements

General references:

106. **Christian, G. D., and F. J. Feldman.** 1970. Atomic absorption spectroscopy. Applications in agriculture, biology and medicine. Wiley Interscience, New York.

107. **Herrman, R., and C. T. J. Alkemade.** 1963. Chemical analysis by flame photometry, 2nd ed. Translated by P. T. Gilbert, Jr. Interscience Publishers, Inc., New York.

108. **Morrison, G. H. (ed.).** 1956. Trace analysis. Interscience Publishers, Inc., New York.

109. **Snell, D. S., and C. T. Snell.** 1963. Colorimetric methods of analysis, vol. 2, 3rd ed. D. Van Nostrand Co., Inc., Princeton, N.J.

110. **U.S. Environmental Protection Agency.** 1974. Methods for chemical analysis of water and wastes. Office of Technology Transfer, Washington, D.C.

111. **Weinberg, E. D. (ed.).** 1977. Microorganisms and minerals. Marcel Dekker, Inc., New York.

Specific articles:

112. **Stewart, M., A. P. Somlyo, A. V. Somlyo, H. Shuman, J. A. Lindsay, and W. G. Murrell.** 1980. J. Bacteriol. **143**:481-491.

Chapter 18

Enzymatic Activity

W. J. DOBROGOSZ

The first enzyme was crystallized by Sumner in 1926, and knowledge concerning these indispensable biological catalysts has grown at an astounding rate during the intervening years. Advances in biochemical technology have been steadily incorporated into enzymological research with the result that today one can detect and measure hundreds of enzymes in both eucaryotic and procaryotic cells. In some cases, as little as a few molecules per cell can be detected, and reaction rates in the order of microseconds can be analyzed. Perhaps more than any other facet of modern biochemical research, work with individual enzymes and enzyme systems has contributed to the sophisticated understanding one now can have of the fundamental bioenergetic and biosynthetic processes that underlie all living organisms.

Throughout these years, the study of bacterial systems has played a key role in essentially all phases of enzymological research, so much so that a thorough understanding of theoretical and practical enzymology must be included in the modern bacteriologist's repertoire. This is true whether one's interests are in basic studies or in any of the medical, agricultural, industrial, or other applied fields.

Acknowledging this, one must also recognize that such expertise cannot be derived from a single exposure to the subject matter. This chapter will be no exception in this regard. Its purpose is twofold: first, to provide the reader with a general, holistic perspective on the practical study of bacterial enzyme systems; and second, to focus on some specific laboratory techniques that can be used to analyze the function, formation, and regulation of bacterial enzyme systems.

The various types of cell preparation that are commonly used in enzyme studies are described in part 18.1, which is followed by a description of how enzyme activities are measured generally and also specifically for the six main classes of enzymes (part 18.2). Various control processes altering activity may be imposed on an enzyme; these are considered in general, and examples of the two major types of controls (allosteric and covalent) are described in part 18.3. Whether or not certain enzymes exist in a particular bacterium depends, of course, on its genetic capabilities. Even when particular genes are present, however, they may or may not be transcribed and translated into corresponding enzyme molecules. Factors involved in the regulation of gene expression and thus of enzyme synthesis must be taken into account and are considered in part 18.4. Finally, inasmuch as enzyme reactions rarely exist in bacterial cells as isolated entities but rather are parts of a delicately interwoven and interdependent series of metabolic pathways and cycles, consideration is given to some general and specific methods of pathway analysis (part 18.5).

There are many methodologies used today in studying bacterial metabolism that "ought" to be included in this chapter. It would be helpful, for example, to include a section on methods used to identify and quantify the various respiratory enzymes and other components through whose interconnections and loops electron flow occurs and chemiosmotic gradients are generated. A section dealing with general techniques now available for analyzing specific catabolic and anabolic processes would also be most helpful to many readers. Unfortunately, these and numerous other procedures cannot be accommodated within the limits of this manual.

The following common abbreviations are used throughout this chapter: cAMP, cyclic adenosine 3',5'-monophosphate; AMP, ADP, and ATP, adenosine mono-, di-, and triphosphate, respectively; F6P, fructose-6-phosphate; FDP, fructose-1,6-diphosphate; G6P, glucose-6-phosphate; GDP and GTP, guanosine di- and triphosphate, respectively; EDTA, ethylenediaminetetraacetic acid; NAD and NADH, oxidized and reduced nicotinamide adenine dinucleotide, respectively; NADPH, reduced nicotinamide adenine dinucleotide phosphate; Tris, tris(hydroxymethyl)aminomethane.

18.1. PREPARATION OF CELLS

The ultimate purpose of research on enzyme systems in bacteria is to obtain as much infor-

mation as possible concerning the various reactions that occur and to understand how they are interwoven into the fabric of the life process. Ideally, one would want to conduct such studies on growing cells that are structurally intact and metabolically undisturbed, and in many instances, such optimum conditions can be attained. More often than not, however, it is necessary to use a less ideal source of cell material.

Each researcher must judge which of the three general types of preparation is best suited to meet the experimental objective. The countless variety of experiments being conducted on hundreds of different species of bacteria preclude making any general recommendation as to a "method of choice." Only a general rule of thumb applies: use disintegrated cell preparations when permeabilized cells are impossible, and use permeabilized cells when intact resting or growing cells are infeasible.

18.1.1. Intact Cells

Growing cells.

Actively growing bacteria carry out all the major life processes, including replication, in a highly organized and regulated fashion and, in many ways, thereby constitute the ideal system for enzyme analyses. During growth, the entire metabolic "gestalt" is in operational flux as nutrients and ions are drawn into the cells and end products are dispelled. Whenever possible, the use of such intact growing cells is recommended.

Resting cells.

Growth can be halted by removing the cells (by centrifugation or filtration) from their source of nutrients and suspending them in a "neutral," osmotically appropriate environment such as buffered physiological saline or a mineral salts basal medium devoid of carbon and nitrogen sources (25.1). Cells thus treated are termed resting cells. They contain the entire enzymatic machinery of their growing counterparts but are in a relatively inactive or resting state. With such cells, one can focus on and analyze only those reactions, or groups of reactions, that are of particular interest. One can analyze, for example, the flow of electrons from substrates such as D-lactate or succinate through the respiratory system under conditions in which the complexities and metabolic demands of growth are minimum. Resting cells of some species are able to synthesize proteins and have been used for such studies. Resting cells are also used routinely in various transport studies involving translocations of substrates and ions. For this purpose, washed cells are usually suspended in an osmotically appropriate environment containing chloramphenicol (ca. 50 to 100 $\mu g/ml$) to inhibit protein synthesis. In some experiments, such as those concerned with transport analyses, it is often advisable that the cells contain abundant endogenous energy reserves (19.2).

Starved resting cells.

In other cases, it is imperative that the cells be thoroughly starved of endogenous energy reserves. An excellent example of the use of starved resting cells is in measuring the intracellular ATP generated by chemiosmotic membrane potentials; the background level of ATP synthesis in unstarved cells tends to obscure the ATP synthesized via the chemiosmotic gradient (49). An excellent starvation procedure for *Escherichia coli* was devised by Koch (25). This procedure allows for rapid utilization of metabolizable reserves by exploiting a cyclic phosphorylation/dephosphorylation of α-methylglucoside, which in this species is vectorially phosphorylated by the phosphoenolpyruvate:phosphotransferase system. The cells are harvested by centrifugation at $4°C$ and washed with basal medium or 120 mM Tris-hydrochloride (pH 8.0). They are then suspended at a density of about 5 mg of cell dry weight per ml in basal medium containing 20 mM α-methylglucoside and 40 mM sodium azide and are incubated at $37°C$ for 45 to 120 min, depending on the strain used. Cells are again centrifuged, washed, and resuspended in an appropriate diluent (25.1).

A more general means of reducing endogenous metabolism, at least for aerobic or facultative bacteria, consists merely in vigorously aerating a heavy suspension of cells (suspended in basal medium or buffer) for 2 to 4 h, or longer if necessary. The reduction of reserves in the cells can be monitored by measuring the decrease to a minimum of the endogenous $Q(O_2)$ value relative to the $Q(O_2)$ value obtained for oxidation of some appropriate substrate such as glucose.

18.1.2. Permeabilization

The use of intact cells has limitations; chief among these is the fact that, whether growing or resting, they are impermeable to most of the substrates, cofactors, and metabolites that are generally used in enzyme measurements. It is widely agreed that it is best to study enzymatic reactions under conditions in which the protein-protein (or protein-membrane) interactions of the enzyme, both with itself (homologous interactions) and with other proteins (heterologous interactions), mimic the in situ state (42). Although the intact cell provides this condition, it is of little value if the necessary substrates cannot gain entry through the cell wall-membrane structure.

Treatment with solvents.

A number of techniques have been used, with varying degrees of success, to make whole cells more permeable and thereby prepare them for enzyme analyses. Solvent (toluene or benzene) treatment has been successfully used in allowing β-galactosides to enter *E. coli* and undergo hydrolysis by β-galactosidase (4). Solvents have also been used to assay the phosphoenolpyruvate:phosphotransferase system in *E. coli* (17). Various combinations of toluene dissolved in ethanol, with or without accompanying freezing and thawing, have also been employed (6).

Treatment with chelating agents.

Treatment of enterobacteria with a chelating agent (Tris-EDTA) has been used to make them permeable and thereby sensitive to inhibition by actinomycin D (26). It has been suggested (42) that permeabilization be extended and used as much as possible to study complex systems such as membrane-associated and interconvertible enzyme systems and, furthermore, that enzyme activities in such cells be compared with those in intact cells whenever possible.

18.1.3. Disintegrated Cell Preparations

Whole cells, whether rendered permeable or not, are inappropriate or too complex for use in many enzyme studies. When this is so, one turns to use of disintegrated cell preparations. Such material as a source of enzymes was first described by Buchner in 1897 and is acclaimed as "one of the basic archetypal experiments of modern biochemistry ranking with that of Lavoisier in chemistry" (22).

Methods for the disintegration of bacterial cells are also presented, in conjunction with the isolation of cell structures, in Chapter 5 of this manual (5.1).

Osmotic disruption.

A wide range of disruptive intensities can be applied to bacterial cells. Perhaps the gentlest of all mechanical methods of cell disruption involves the application of hydrostatic pressure (osmotic pressure) against the cell membrane. This method has limited use with most bacteria unless their cell walls are first weakened by enzymatic attack or other means.

Osmotic shock has proven valuable for study of certain enzymes in gram-negative bacteria which apparently exist outside the cytoplasmic membrane but within the matrix of the double membrane envelope of these organisms (32, 33),

so-called "periplasmic" enzymes. These are usually hydrolases such as alkaline phosphatase,
ribonuclease I, and cyclic phosphodiesterase.
The method generally used to extract these enzymes from *E. coli* cells has been described by
Hughes et al. (22) and involves the following
steps: (i) wash cells thoroughly with fresh medium or appropriate buffers; (ii) suspend the
washed pellet in 80 parts of 0.5 M sucrose; (iii)
centrifuge, and decant the supernatant solution;
(iv) rapidly disperse the remaining pellet by
vigorous shaking in 80 parts of cold 5×10^{-4} M
$MgCl_2$; (v) centrifuge and then examine the supernatant fraction for the presence of periplasmic enzymes.

A lysozyme-EDTA procedure also is an excellent means for rapid, gentle lysis of gram-
negative cells and can be carried out in the cold
or at room temperature. Spheroplasts are produced when cells are thus treated and maintained in an osmotically protective environment
(otherwise, total disruption occurs). This
method can be scaled up to 30-liter batches of
cells. It can be gentle enough to isolate the total
population of polyribosomes. A wide variation
in buffer, EDTA, and lysozyme concentrations
can be used in optimizing breakage for each
particular species of organism. The following
procedure is that originally described by Repaske (37) for disruption of *E. coli*. Suspend
washed cells in a 3-ml volume containing 100 M
Tris-hydrochloride buffer (pH 8.0), 0.8 mg of
EDTA (pH 7.5), 50 µg of lysozyme, and water;
start lysis by addition of lysozyme. Lysis can be
conducted at room temperature and is generally
completed within 5 min, or it can be carried out
in the cold.

Gentle osmotic disruption also is used in the
preparation of protoplasts and membrane vesicles from gram-positive bacteria. These preparations have proven extremely useful for study
of transport across the bacterial membrane. Procedures for their preparation are discussed and
described in the succeeding chapter (19.3).

Disintegration.

Reviews of the various physical and chemical
techniques that can be used to disintegrate bacterial cells are presented by Coakley et al. (6)
and by Hughes et al. (22); readers are referred
to these sources, particularly the latter, for further information on the wide variety of procedures that can be used for this purpose. Of the
many options available, only the French press
and similar units, vibration mills, ultrasonic disintegrators, some hand-grinding techniques, and
the Hughes press have sufficient shear force
intensities for effective disruption of bacteria

(and yeasts). Most of the other techniques have
specialized uses or are effective only with organisms (such as protozoans or animal tissues)
which are easier to disrupt than are bacteria. Of
the chemical procedures available, the most useful and effective involve lytic agents (such as
lysozyme or lysozyme combined with chelating
agents such as EDTA). The following is a brief
description of the more effective physical techniques currently in use for bacterial enzymology.

French press. The French press is a very
efficient device for disruption of many species of
microorganisms and is available in various
forms. The simplest unit is one that is placed in
a laboratory hydraulic press capable of delivering 10 to 20 tons of total pressure. It consists of
a chamber in which the cell suspension is placed
and then subjected to pressure with a piston.
Gradual extrusion of the cells from the pressure
chamber through a tiny orifice results in effective shear rupture. The pressure can be maintained with one arm handling the hydraulic
pump and the other operating the needle valve
that controls the extrusion rate. The unit is also
available with an automated hydraulic press
(American Instrument Co., Silver Spring, Md.)
or as a refined modification (Sorvall-Ribi Fractionator, Ivan Sorvall, Inc., Norwalk, Conn.).

Precool and fill the French press with the cell
suspension. Preset the piston to provide a chamber volume of 5, 10, 20, or 50 ml. Mount the unit
(piston end down) on the tripod stand provided
by the manufacturer. Add a slight excess of
suspension, and have the needle valve open so
that, when the cap is applied over the chamber,
the excess fluid flows out through the orifice.
This will ensure displacement of air from the
chamber. (*CAUTION:* If this precaution is not
taken, the trapped air will squirt out explosively
with the last of the suspension, a potentially
dangerous situation.) Close the needle valve, and
place the assembled unit (piston end up) on the
hydraulic press. Apply 1×10^4 to 3×10^4 lb/in^2
in the pressure cell, and maintain the pressure
while letting the suspension slowly seep out of
the orifice through some attached tubing and
into an appropriate chilled receptacle.

Most bacteria, even some considered difficult
to break (e.g., mycobacteria), are susceptible to
disruption by this method. Gram-positive cocci
tend to be among those resistant to breakage
and often require a second passage through the
press.

In addition to being a convenient and effective
means for cell breakage, a minimal amount of
enzyme inactivation occurs during this disruption relative to other procedures. In this connection, however, it is recommended that relatively

thick cell suspensions be used: a range of 0.5 g of cell wet weight per ml or greater is suggested (22). Thick suspensions tend to produce better breakage and less inactivation of enzymes than do more dilute suspensions.

Vibration mills. The Mickle (Mickle Ltd., Mill Works, Gomshall, Surrey, England) and the Nossal (McDonald Engineering Co., Bay Village, Ohio) instruments are the two most commonly used vibration mills. An inexpensive vibration mill for small volumes of cells can be obtained with use of a dental amalgamator (e.g., model LP-60, Crescent Co., Lyons, Ill.). Small glass beads (Superbrite glass beads, grade 110, Minnesota Mining & Manufacturing Co., Minneapolis, Minn.; or Ballotini Beads, English Glass Co., Ltd., Leicester, England) provide the shear surface. They are added to a cell suspension, and the mix is placed in the glass or metal capsule provided by the manufacturer. An empirically determined, but usually high, ratio (e.g., 1:1, vol/vol) of beads to cells is used to obtain optimum disruption. Volumes are generally limited to 25 ml or less, and lengthy shaking treatments are required with the Mickle mill. Shorter periods are required for the Nossal mill, which functions with a higher vibrational energy. Heat is rapidly generated during shaking and must be dissipated by some type of cooling system; a CO_2 cooling system is used in the Nossal unit.

Ultrasonication. Cell disintegration by ultrasound is caused by cavitational forces producing shock waves, chemical attack by free radicals, or cavitational microstreaming. Whatever the mechanism of disintegration, the ultrasonic technique continues to be an effective means for rupturing bacterial cells. A variety of sonicators are commercially available.

Although relatively thick suspensions of cells can be used, breakage efficiency is decreased considerably in viscous solutions. As large molecules such as deoxyribonucleic acid are released from broken cells, viscosity of the suspension increases and a decrease in the rate of breakage follows. For this reason, more dilute cell suspensions will undergo more rapid disruption by ultrasonication. Sonication time is generally determined empirically, keeping in mind the need to dissipate generated heat and the sensitivity of the enzyme(s) to inactivation.

Cool the cell suspension by placing the sonication cup in ice or an ice-salt mixture, or by circulating coolant through a jacketed cup. Place the tip of the sonicator probe a few millimeters beneath the liquid surface. Then set the generator at the recommended power setting and tune to the resonant frequency (a sharp, crisp, sizzling sound indicates the most effective frequency for cell disruption). After the required exposure, detune the instrument and turn it off. Some form of acoustic shielding around the transducer and probe is recommended to baffle the intense noise emitted by the instrument.

Sonication is perhaps the quickest and easiest of the disruption techniques to use. This, combined with its effectiveness in breaking a wide variety of bacteria, accounts for its extensive use in bacteriology laboratories.

Alumina grinding. Alumina grinding, which involves solid-shear rupture of cells, is one of the more inexpensive means available for breakage of bacterial cells. Only a mortar and pestle are needed plus a supply of aluminum oxide (e.g., Polishing Alumina, grade A-301, Fisher Scientific Co.). Wash the alumina and dry it at 100°C. Add a frozen (or chilled) cell pellet to the precooled mortar at a ratio of 1:2 parts by weight of precooled alumina to cell paste. Grind vigorously for 2 to 5 min, during which the powdery texture of the mixture will change to a tacky, claylike consistency which is indicative of cell breakage. Add appropriate buffer, resuspend the contents of the mortar, and centrifuge to remove alumina and unbroken cells. This grinding technique can be used on relatively large volumes of cells and can be carried out rapidly. Most bacteria, however, are more effectively disrupted by other methods.

Hughes press. A much more effective technique for solid shear disruption of bacteria is that described by Hughes et al. (22); it is based on solid shear resulting from the action of ice crystals within the cells or of abrasive powder outside the cells. The Hughes press consists of a device that expresses a frozen suspension or paste of cells (with or without inclusion of an abrasive, e.g., an equal volume of Pyrex glass powder) through a small orifice into a receiving chamber. The expression is carried out at high pressure (10,000 to 80,000 lb/in^2) using a hydraulic press. A number of modifications of the original Hughes press are available.

Chill the press in a freezer or with dry ice to −25 to −30°C. Place the sample, as a slurry or a frozen pellet, into the chamber. Mount the piston over the sample, and apply pressure until the sample is forced through the orifice.

Freeze pressing is one of the most efficient of the disruption techniques. It is effective over the entire range of cell materials from bacteriophages to whole animal tissues. Bacterial cells that normally are resistant to rupture (such as gram-positive cocci) tend to be susceptible to this technique.

In general, the more commonly used physical methods can be ranked according to their disintegrative force in the following sequence: Hughes press > vibration mills > French press > ultrasonication > alumina grinding. A general pattern of cell resistance to physical disintegration is the following: yeasts > fungal mycelia > gram-positive cocci > gram-positive rods > gram-negative rods > halobacteria > mycoplasmas.

The choice of a cell disruption technique is determined largely on an empirical basis, varying with the kind of cell employed, the volume of extract required, the sensitivity of the enzyme(s) to inactivation or alteration during disruption, and the subsequent storage. Trial and error is often the only means for making these assessments.

18.2. MEASUREMENT OF ACTIVITY

18.2.1. General Considerations

An enzyme assay should, if possible, be accurate, sensitive, continuous, convenient, and specific for the reaction under study. More often than not, however, less ideal conditions are encountered. Consequently, precautions must be taken, and numerous control and standardization tests must be conducted before an assay is judged suitable for use. Among these general considerations are the following.

Variable assay conditions.

Most assays are developed under specific conditions using one particular species of microorganism or some other biological source. It is incorrect to assume that such an assay can be used per se to analyze for that enzymatic activity in another species of organism. Requirements for ions and cofactors can vary considerably. The pH and temperature optima, K_m values, susceptibility to inhibitory substances, and equilibrium of the reaction can vary appreciably from species to species. Bacterial succinic dehydrogenase is a classical example of this type of variation; assay conditions for this enzyme vary so widely that some authors are reluctant to prescribe any formula for the reaction components (35).

Inhibition of activity.

Inability to detect enzyme activity in a crude extract does not necessarily mean that the organism from which the extract was prepared is devoid of that enzyme. Activity present in the cells can be partially or totally lost during preparation of the extract. In its intracellular environment an enzyme may be protected from inactivation by oxidants and other substances, or it may enjoy protection owing to numerous pro-

tein-protein interactions which are destroyed or dissipated during cell disruption. In this connection, it should be noted that the protein content of most cells is in excess of 100 mg of protein per ml (28). Yet, most in vitro analyses employ dilute aqueous solutions of enzymes because these are the conditions which are most experimentally manageable. This need to consider the effect of protein concentration on enzyme activity is particularly important when one is dealing with regulatory enzymes (28).

If susceptibility to oxygen is suspected of causing inactivation, antioxidants such as reduced glutathione, β-mercaptoethanol, or dithiothreitol can be added to the cells before disruption. Or the breakage and subsequent manipulations and assays can be conducted in anaerobic chambers. This is an expensive and inconvenient recourse, but one used successfully in some laboratories. Chelating agents such as EDTA are often included in cell extracts to minimize inactivation caused by divalent ions. Inasmuch as many enzymes in vitro are susceptible to heat inactivation, extracts are generally prepared and held at 0 to 4°C. It should be noted, however, that some enzymes are sensitive to low temperature and undergo rapid inactivation at 0 to 4°C.

When an enzyme is thought to be present in a crude extract but the activity cannot be detected, a number of possibilities need to be recognized. First, there is the possibility that the presumption of its presence is incorrect. Second, the enzyme may be present in the extract but will not be expressed unless the reaction mixture is modified to include components which are missing or nullify those which are inhibitory. Suggestions along these lines have already been described earlier, and additional suggestions are included in most reference sources on this subject. In addition, however, it is recommended that a positive control be carried out if possible, namely, that a known bacterial source of the enzyme be added directly to the extract being tested (50). If the added control enzyme shows expected activity under these experimental conditions, one has reasonable grounds (but not full assurance) to assume that the necessary reaction components are present and that no general inhibitors are present in the extract material. Even when this control is conducted and the added activity is demonstrated, the absence of the original activity can be attributed to numerous unappreciated factors. Among these is the possibility that the enzyme protein is in fact present in the extract, but in an inactive state. Covalent modification reactions are known to exist by which active enzymes are rendered inactive by covalent alterations of the protein moiety (see 18.3.4).

Competing side reactions and use of coupled enzyme assays.

Enzyme activity can be obscured by competing side reactions. The concentrations of substrates or products can be so drastically altered by such competing reactions that accurate measure of activity is impossible. The activity of NADH oxidase or adenosine triphosphatase in crude extracts, for example, can be so high as to interfere with assays designed to measure production or utilization of NADH or ATP, respectively. Appropriate controls need to be exercised to take such side reactions into account or means must be sought to either inhibit or circumvent these undesirable side reactions.

On the other hand, competing reactions have been exploited in development of coupled enzyme assays. These involve the coupling of a primary enzyme (the enzyme whose activity is being measured) system to one or two auxiliary enzymes added to the assay reaction mixture:

$$A \xrightarrow[\text{enzyme}]{\text{primary}} B \xrightarrow[\text{enzyme}]{\text{auxiliary}} C$$

$$A \xrightarrow[\text{enzyme}]{\text{primary}} B \xrightarrow[\text{enzyme}_1]{\text{auxiliary}} C$$

$$\xrightarrow[\text{enzyme}_2]{\text{auxiliary}} D$$

The following general assay for kinases described by Wood (50) demonstrates the use of the coupled assay procedure and also some techniques for dealing with competing side reactions. It is a general assay for kinases but is applicable to a wide variety of substrates, including gluconic and 2-ketogluconic acids, pentoses, and polyols, as well as hexoses.

The procedure is based on a coupling of the ADP formed by the reaction of ATP-kinase (the primary enzyme) to pyruvate kinase and lactate dehydrogenase (the two auxiliary enzymes) as follows:

$$ROH + ATP \xrightarrow{\text{kinase}} ROP + ADP$$

$$ADP + \text{phosphoenolpyruvate}$$

$$\xrightarrow[\text{kinase}]{\text{pyruvate}} ATP + \text{pyruvate}$$

$$\text{pyruvate} + NADH$$

$$\xrightarrow[\text{dehydrogenase}]{\text{lactate}} \text{lactate} + NAD$$

The velocity of the primary enzyme is established as the rate-limiting step and is obtained from a continuous measurement of the absorbance change at 340 nm with time. An extinction

coefficient of $6.22 + 10^6$ cm^2/mol and the reaction volume are used to determine this velocity as micromoles per minute, with the specific activity then calculated as micromoles of substrate phosphorylated per minute per milligram of protein.

The resynthesis of ATP in this system is of distinct advantage in that low levels may be used, thereby diminishing side reactions involving ATP and at the same time diminishing possible allosteric inhibition of the kinase by ATP. The concentration of ATP remains essentially constant. This method can be conducted conveniently using a recording spectrophotometer, and it yields data that are more reliable than those obtained with the single-point velocity methods.

Prepare the following reagents for this assay: 0.7 M Tris-hydrochloride buffer (pH 7.5), 0.1 M $MgCl_2$, 0.05 M Na_2ATP, 0.05 M trisodium or tricyclohexylammonium phosphoenolpyruvate, 0.15 M sodium glutathione (reduced), 0.005 M Na_2NADH, 0.15 M lactic dehydrogenase, 0.15 M pyruvate kinase, and 0.15 M substrate. For use in a microcuvette (1.25 by 1.25 by 2.5 cm, with volume = 0.5 ml; Pyrocell Manufacturing Co., Westwood, N.J.), add the following volumes using a micropipette or gas chromatography syringe: Tris buffer, 0.05 ml; $MgCl_2$, ATP, phosphoenolpyruvate, glutathione, and NADH, 0.01 ml each; lactic dehydrogenase, 0.5 μl; and water, kinase fraction, and substrate to a final volume of 0.15 ml. With this reaction volume, a correctly positioned microaperture (Pyrocell Manufacturing Co.) is needed to confine the light beam to the fluid volume. For routine measurements of a large number of samples, combine the reagents to reduce the number of additions required. Initiate the reaction by addition of the substrate. With each determination, include controls (i) for the combined NADH oxidase and adenosine triphosphatase rates by conducting the reaction in the absence of substrate and (ii) for the NADH-linked substrate reduction (if applicable) by conducting the reaction in the absence of ATP.

A useful variant of this procedure greatly diminishes the rate of NADH oxidation by substituting NADPH for NADH. Lactic dehydrogenase utilizes NADPH at approximately 10% of the rate with NADH, and since NADPH oxidase activity is generally very low in bacterial extracts, a compensatory increase (ca. 100-fold) in lactic dehydrogenase is all that is needed to circumvent a bothersome high rate of NADH oxidation.

When using a coupled assay involving only one auxiliary enzyme, it is generally satisfactory in standardizing the assay to add successively larger amounts of the auxiliary enzyme until the linear initial velocity of the primary reaction can be determined accurately. A procedure developed and described by McClure (29) is available for determining the time required to produce this linear reaction velocity and also for determining the amount of each auxiliary enzyme to be added to the reaction mixture.

General comments on assay procedures (with particular emphasis on those described below).

It is recommended by the Commission on Enzymes of the International Union of Biochemistry that "units of enzyme activity" always should be defined and referred to as micromoles of substrate utilized (or product formed) per minute. In practice, however, units of activity are often expressed on a different basis in the original publications that appear in the various journals. For the latter reason, different expressions of activity are used (as described by the various authors) in the procedures described below. However, I recommend the eventual adoption of usage of the international unit whenever possible.

Also, the reader should be cautioned again that the compilation of enzyme assay procedures cannot be considered as dogma. Each procedure was devised (in most cases) for an enzyme in a given bacterium (E. coli in these instances), and it cannot be stressed too strongly that conditions necessary to obtain maximum activity in one species are not necessarily optimum for the same enzyme in a different species. To apply these methods to all situations without appreciation of the subtleties involved can lead to erroneous conclusions.

18.2.2. General Standardization Procedures

Whenever an enzyme is to be analyzed and its relative concentration (specific activity) in the cells is to be determined, certain preliminary standardization steps must be performed. First of all, it is necessary that saturating concentrations of substrates be used in the assay systems. The calculation of units of enzyme activity present in a particular preparation is then generally possible, but only when the assay meets the additional criterion of linear proportionality with respect to both the reaction time and the amount of enzyme protein present. These time-protein constraints are established empirically in each case. The following series of experiments illustrate the types of manipulations generally used for this purpose. They also demonstrate

quite clearly the differences between enzyme activity on the one hand and enzyme specific activity on the other. This difference is self-evident to the experienced worker, but may be a source of considerable confusion to the beginning student.

In the following illustrations, the β-galactosidase assay is used with toluene-treated E. coli cells as the enzyme source. The specific assay procedure and the toluene treatment are described elsewhere in this chapter (see 18.2.8).

Standardization of the β-galactosidase assay with E. coli cells.

Use any convenient wild-type strain of E. coli, and grow two 100-ml cultures (A and B) with vigorous shaking overnight (37°C). Induce β-galactosidase synthesis in both cultures by including 2.5 mM isopropyl-β-D-thiogalactopyranoside in the culture medium as an inducer. Have culture B synthesize less of the enzyme by also including 20 mM glucose in the medium. A convenient basal medium contains, per liter: KH_2PO_4, 2 g; K_2HPO_4, 7 g; $(NH_4)_2SO_4$, 1 g; $MgCl_2 \cdot 6H_2O$, 0.1 g; and vitamin-free casein hydrolysate, 2.5 g. Adjust to pH 7.2. Under these conditions, β-galactosidase synthesis will occur at a relatively nonrepressed rate. Culture A yields more than enough cells for the following experiments. Culture B will produce cells containing considerably less β-galactosidase as a result of catabolite repression by the glucose.

In addition to these two sources of enzyme, there is needed some means for accurately determining the biomass of the cells being used, e.g., a standard curve relating absorbancy at 420 nm (determined with a colorimeter) of cells suspended in 0.05 M sodium phosphate buffer (pH 7.2) to the actual dry weight of organisms present. Cultures are always diluted appropriately to ensure that absorbancy is linearly related to dry weight (see 25.2.2). In the following experiments, one can use absorbancy units per se as a relative measure of cell mass, provided that the measurements are made at those cell concentrations in which absorbancy and mass are relatively linear.

Harvest cultures A and B by centrifugation, wash once with 0.05 M sodium phosphate buffer (pH 7.2), and resuspend in the same buffer. Samples are removed and diluted appropriately in order to determine the cell mass (e.g., as absorbancy at 420 nm [A_{420}] or micrograms of cell dry weight per milliliter) present per unit volume. The cells are then treated with toluene as described later (see 18.2.8), after which they are ready to be used in the following standardization procedures.

Enzyme activity versus time.

In duplicate, prepare 10 to 20 enzyme reaction mixtures complete with a suitable mass of cells (a few preliminary trials with different dilutions of cells will generally be needed). Start the reaction by addition of o-nitrophenyl-β-D-galactopyranoside, and then stop the reaction by addition of the Na_2CO_3 at 0, 1, 2, 4, 6, 8, 10, 15, 20, 25, and 30 min. Read the A_{420}, and determine how long the reaction remains linear with respect to time by plotting the A_{420} (ordinate) against time of incubation (abscissa). Choose an experimentally convenient time (e.g., 15 min) within the linear portion of the curve thus obtained for use in subsequent assays.

Enzyme activity versus biomass.

Prepare another series of reaction mixtures in which the amount of cells added to the tubes is incrementally varied from zero to relatively high levels. At the higher cell levels any turbidity (A_{420}) caused by the cells themselves will have to be controlled by using appropriate cell blanks; i.e., the A_{420} of reaction mixtures containing cells minus o-nitrophenyl-β-D-galactopyranoside is subtracted from the A_{420} of the same system but which also contained the o-nitrophenyl-β-D-galactopyranoside. All of the reactions are terminated at the predetermined time described above (e.g., 15 min). The units of activity present in each case are then calculated and plotted as a function of the cell mass present in each reaction system, as shown in Fig. 1. Provided that this assay is always conducted within the constraints of the linear relationships thus established, valid measurements of the enzyme can be made.

These same standardization experiments should be performed with culture B cells. When the final units of enzyme activity per unit of cell mass (milligrams or micrograms of cell dry weight or A_{420}) are calculated and compared with those obtained from culture A cells, a determination of the "relative amount" (specific activity) of enzyme present in each culture will have been made. Culture B cells, being catabolite repressed, will have lower activity of β-galactosidase per unit of cell mass than will culture A cells.

18.2.3. General Assay Procedures

It is not feasible to describe in detail here the many ways in which many enzymes can be assayed. Suffice it to say that a reaction can be monitored without disturbance in a **continuous assay**, or by a **discontinuous assay** in which samples are periodically removed for measuring changes in substrate utilization or product for-

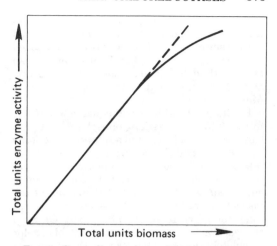

FIG. 1. *Generalized relationship between enzyme activity and biomass.*

mation. For example, fumarase can be measured directly in continuous assay by using a spectrophotometric process or indirectly by coupling it with another enzyme system (see glutamine synthetase assay using a coupled assay method, part 18.2.11). More often, however, discontinuous sampling assays must be used. These assays can be made in a variety of ways including use of spectrophotometric, manometric, polarimetric, chromatographic, and electrode methodologies, as well as with a wide assortment of chemical and radiotopic procedures (see Chapter 16).

Some of the various procedures used are exemplified in the following assays. The eight examples utilize enzymes from each of the six classes: oxidoreductases, transferases, hydrolyases, lyases, isomerases, and ligases. Each class is involved in metabolic regulatory processes. Methods used for studying the control of enzymatic activity and the regulation of enzyme synthesis will be described later in this chapter. All of the following enzymes are found in commonly used, wild-type strains of *E. coli* such as Crooks strain (ATCC 8739), strain W (ATCC 9637), or strain K-12 (ATCC 14948).

18.2.4. Oxidoreductases (Nitrate Reductase)

Nitrate reductase (cytochrome) (EC 1.9.6.1, ferrocytochrome:nitrate oxidoreductases) catalyzes the following oxidoreductase reaction:

ferrocytochrome + nitrate
$$\rightarrow \text{ferricytochrome + nitrite}$$

Nitrate reductase is a membrane-bound respiratory enzyme closely associated with formate dehydrogenase and cytochrome b. Induction of activity occurs only under anaerobic growth conditions in which NO_3^- can function as an electron acceptor for anaerobic respiration. Enzymatic activity in disrupted cells can be measured when a suitable reducing agent (such as reduced methyl viologen) is supplied. The following is an assay method described by Lowe and Evans (27) and summarized in the *Worthington Enzyme Manual* (9).

For the control blank add equal volumes (0.1 ml) of 0.15 M potassium phosphate (pH 7.0), 0.02% methyl viologen, and a solution containing 23 mM sodium dithionite freshly prepared in 48 mM sodium bicarbonate and glass distilled water. To this blank and to a test system containing equal volumes (0.1 ml) of 0.10 M sodium nitrate instead of the water, add the enzyme source (after temperature equilibration to 30°C). Stop the reaction by vigorous aeration (the blue color completely removed). Determine the amount of nitrite formed colorimetrically as follows: quickly add 0.5 ml of 58 mM sulfanilamide (prepared in 3 N HCl) and 0.5 ml of 0.39 mM N-(1-naphthyl)ethylenediamine hydrochloride to 0.4 ml of the assay mixture. Add 1.5 ml of water, and incubate at room temperature for 10 min. Read at 540 nm versus a blank. Determine micromoles of NO_2^- produced from a previously prepared standard curve.

18.2.5. Oxidoreductases (Succinic Dehydrogenase)

Succinic dehydrogenase [EC 1.3.99.1, succinate:(acceptor) oxidoreductase] catalyzes the oxidation of succinate via a number of artificial electron acceptors by the following oxidoreductase reaction:

succinate + acceptor →

  fumarate + reduced acceptor

Succinic dehydrogenase is a membrane-bound flavoprotein component of the tricarboxylic cycle. It functions only under aerobic conditions, and its synthesis seems to be partially under cAMP control, at least in *E. coli* cultures. Enzyme activity in disrupted cells or in purified membrane preparations can be measured by a continuous spectrophotometric assay (1) involving the reduction of cytochrome *c* as follows:

succinate + phenazine methosulfate (PMS)

  → fumarate + $PMSH_2$

$PMSH_2$ + cytochrome *c* (oxidized) →

  PMS + cytochrome *c* (reduced)

The rate of cytochrome reduction is measured in a single-beam spectrophotometer at 550 nm. A chart recorder is used to obtain the absorbance versus time data.

Prepare an "assay mix" containing 50 ml of 0.1 M potassium phosphate buffer (pH 7.6), 2.5 ml of 1 M succinate, 10 ml of 0.01 M EDTA (pH 7.6), 0.75 ml of KCN (5.8 mg/ml), and 42.5 ml of distilled water. Place 2.5 ml of this "assay mix" into a test tube containing 5 to 100 μl of enzyme sample, and incubate at 23°C for 15 min. Transfer 1.95 ml of this solution to a cuvette, add 15 μl of cytochrome *c* (Sigma type III, horse heart, 250 mg/2.4 ml), and mix. Set the spectrophotometer to zero by adjusting the slit width. Start the reaction by adding 25 μl of phenazine methosulfate (12 mg/ml in water; prepare fresh and protect from light by storing in a dark-brown bottle). Run the assay in a room as dark as possible. The reaction rate is taken from the slope of the absorbance tracing (change in absorbance [ΔA] per minute). A unit of succinic dehydrogenase is expressed as micromoles of cytochrome *c* per minute (ΔA ÷ extinction coefficient [$\epsilon_{550} = 29.9$]) as determined under these conditions. 2,6-Dichlorophenol indophenol reduction rather than cytochrome *c* reduction can be used in this assay (1).

18.2.6. Transferases (Phosphofructokinase)

Phosphofructokinase (EC 2.7.1.11, ATP:D-fructose-6-phosphate 1-phosphotransferase) can be assayed by the following procedure developed for yeast phosphofructokinase (43). It is a continuous spectrophotometric assay employing a coupled system involving the following transferase reactions:

F6P + GTP → FDP + GDP

$$FDP \xrightarrow{\text{aldolase}} \text{glyceraldehyde-3-P}$$

  + dihydroxyacetone-P

glyceraldehyde-3-P → dihydroxyacetone-P

2 dihydroxyacetone-P + 2 NADH →

  2 glycerol-3P + 2 NAD

$\overline{}$

F6P + GTP + 2 NADH →

  GDP + 2 glycerol-3-P + 2 NAD

GTP is used as the substrate instead of ATP to avoid allosteric inhibition by ATP. The composition of the reaction mixture is as follows (final concentrations): F6P, 1 mM; GTP, 1 mM; $MgCl_2$, 5 mM; potassium phosphate (pH 6.5), 25 mM; ethanethiol, 5 mM; NADH, 0.15 mM; aldolase, 0.1 U/ml; glycerolphosphate dehydrogenase, 1 U (per 2 ml); triosephosphate isomerase, 1.5 U/ml. Oxidation of the added NADH is followed in a 1-cm light-path cuvette containing a final volume of 2.0 ml. The reaction is started with the addition of phosphofructokinase (containing no more than ca. 10 mU of enzyme) and is followed by measuring the decrease in optical density at 340 nm. A blank without GTP must be included in each experiment, particularly with crude preparations, to measure the NADH oxidase rate, which must be subtracted.

For quantitative studies on allosteric regulation by ATP, ADP, and AMP (energy charge factors) in crude extracts which contain glucose phosphate isomerase, it is important to avoid marked changes in the concentration of F6P by isomerization during the assay. This can be prevented by the use of a 0.2 M neutralized solution of G6P with some 20 U of G6P isomerase per ml. Within 1 h at room temperature, this solution will be approximately 0.05 M F6P–0.15 M G6P. An international unit is the amount of enzyme that phosphorylates 1 μmol of F6P per min in the above condition at about 25°C (ΔA_{340} $\div$ 2 $\div$ 6.22 $\times$ 10^3 = micromoles per milliliter $\times$ 2 = micromoles per cuvette).

ATP is a substrate as well as an allosteric inhibitor, and its effect is dependent on F6P concentrations. AMP can reverse ATP inhibitions; citrate, particularly in the presence of ATP, also inhibits, whereas NH_4^+ can activate the enzyme.

18.2.7. Transferases (Ornithine Transcarbamylase)

Ornithine transcarbamylase (EC 2.1.3.3, carbamoylphosphate:L-ornithine carbamoyltransferase) catalyzes the sixth step in the synthesis of L-arginine in *E. coli* and is subject to endproduct repression by this amino acid. The enzyme carries out the following transferase reaction:

ornithine + carbamyl phosphate →

citrulline + inorganic phosphate

It is assayed by measuring appearance of a derivatized form of the reaction product as described by Neidhardt and Boyd (31). As in the case of β-galactosidase, toluene-treated cells (or crude extracts) can be used as enzyme source.

To 2 ml of an appropriate dilution of toluene-treated cells, add 0.3 ml of 62 mM carbamyl phosphate and 0.3 ml of 50 mM $MgCl_2$; to start the reaction, add 0.3 ml of 0.1 M L-ornithine. Incubate at 37°C for 20 min, and terminate the reaction by chilling on ice and adding 2.0 ml of 0.25 N HCl. Mix well, and centrifuge to remove the cells. Decant the supernatant fluid into test tubes. Determine the amount of citrulline formed by removing appropriate samples (0 to 2.0 ml) of this supernatant fluid and treating

them as follows. Dilute samples to 2.0 ml with distilled water (water alone for blank control), and then add 1.0 ml of a sulfuric-phosphoric acid mixture (1:3) (use a Pro-pipette or another safe pipetting device) and 0.13 ml of 3% 2,3-butadienemonoxime. Mix the contents well, cap with marbles, and heat in a covered boiling-water bath for 10 min. Cool for 10 min in a covered cold-water bath (light causes a side reaction to occur). Read absorbancy at 490 nm, and calculate moles of citrulline produced by reference to a standard curve prepared in an identical fashion with L-citrulline using a concentration range of 0 to 0.25 μmol. One unit of the enzyme is the amount that produces 1 μmol of citrulline per min under these conditions.

18.2.8. Hydrolases (β-Galactosidase)

β-Galactosidase (EC 3.2.1.23, β-D-galactoside galactohydrolase) catalyzes the following hydrolase reaction:

ONPG + H_2O → galactose + o-nitrophenol

Instead of lactose, the natural substrate of this bacterial enzyme, an artificial substrate, ONPG (o-nitrophenyl-β-D-galactopyranoside), can be used; ONPG is colorless but upon hydrolysis yields o-nitrophenol, which is yellow in alkaline solution. The reaction can be carried out using either cell-free extracts or toluene-treated cells and is stimulated by Na^+ ions and thiol reagents. The procedure described below is a modification of that described by Pardee et al. (34).

The reaction mixture consists of 4.10 ml of 0.05 M sodium phosphate buffer (pH 7.5) and 0.20 ml of 0.032 M reduced glutathione. Allow this mixture to equilibrate to temperature (30°C). Add the enzyme source within a 0.20-ml volume, and follow by adding 0.50 ml of 0.01 M ONPG (also preincubated) to start the reaction. After an appropriate incubation period (generally 15 min but dependent on the level of enzyme present in the cells), stop the reaction and intensify the color by adding 1 ml of 1 M Na_2CO_3. Shake, and determine the absorbancy at 420 nm. A unit of enzyme activity is defined as the amount that produces 1 μmol of o-nitrophenol per h under these assay conditions. Prepare a standard curve relating absorbancy at 420 nm to different o-nitrophenol concentrations under conditions identical to those employed in the assay except for omitting the ONPG.

The toluene-treated cells used for this assay can be prepared as follows. Culture samples of 1.0, 0.5, or 0.20 ml are added to 4.0, 4.5, or 4.8 ml of 0.05 M potassium phosphate buffer (pH 7.5) to obtain 1:5, 1:10, or 1:25 dilutions, respectively. The absorbancy of these dilutions can be used to determine the mass (dry weight) of cells present in each sample with reference to a previously prepared standard curve. These culture dilutions can then be held in an ice bath for up to 6 h prior to assay for β-galactosidase. Just prior to assay, they are removed from the ice bath and treated with 1 drop of toluene and 1 drop of 0.1% sodium deoxycholate, stoppered, shaken at 37°C for 15 min, and then put back in the ice bath. Appropriate samples (generally 0.20 ml) are then added to the assay system described above for β-galactosidase activity determinations.

18.2.9. Lyases (Threonine Deaminase)

Threonine deaminase (EC 4.2.1.16, L-threonine hydro-lyase [deaminating]) catalyzes the following lyase reaction:

$$\text{L-threonine} \rightarrow \alpha\text{-ketobutyric acid} + NH_4^+$$

In many bacteria two such enzymes exist—one biodegradative and the other biosynthetic in function. The biodegradative enzyme is distinct from the biosynthetic threonine deaminase; the former is an inducible enzyme that is synthesized in the absence of glucose and oxygen but in a medium containing L-threonine plus some other amino acids of the branched-chain family, although it is formed best in a more complex amino acid mixture. It is activated by AMP. The biosynthetic enzyme is subject to end-product repression, and its activity is inhibited by L-isoleucine. Both enzymes are assayed by a determination of α-ketobutyrate formed as its 2,4-dinitrophenylhydrazone. The biodegradative enzyme can be assayed in cell extracts or in toluene-treated cells by the following procedure (20).

Prepare a reaction mixture (1 ml, final volume) to contain 0.20 ml of 0.5 M potassium phosphate buffer (pH 7.4), 0.10 ml of 0.1 M AMP (neutralized), and 0.20 ml of 0.5 M L-threonine. Initiate the reaction at 37°C by addition of the enzyme preparation, and terminate by addition of 1 ml of 1 N HCl. Then add 0.2 ml of 0.1% 2,4-dinitrophenylhydrazine, and incubate the mixture at 37°C. After about 10 min, add 2 ml of 2 N NaOH, and determine the absorbancy of the hydrazone colorimetrically at 416 nm. A standard curve relating absorbancy to concentration of authentic α-ketobutyrate (sodium salt) is constructed under identical conditions. One unit of enzyme is defined as the amount producing 1 μmol of α-ketobutyrate per min under these conditions.

The biosynthetic enzyme can be assayed with a lysed-cell procedure or with a cell extract (18). The following procedure is used for assay with extracts. Prepare a reaction mixture (1.0 ml, final volume) containing 0.10 ml of 1 M Tris-hydrochloride buffer (pH 8.0), 0.10 ml of 1 M NH$_4$Cl, 0.1 ml of 1 mM pyridoxal phosphate, 0.20 ml of 0.2 M L-threonine, 0.10 ml of extract, and water to 1.0 ml. Use a blank lacking the L-threonine with each set of assays. Incubate the tubes at 37°C for a predetermined period (ca. 10 to 20 min), and stop the reaction by addition of 0.10 ml of 50% trichloroacetic acid. Measure production of α-ketobutyrate as described above.

18.2.10. Isomerases (L-Arabinose Isomerase)

L-Arabinose isomerase (EC 5.3.1.4, L-arabinose ketol-isomerase) catalyzes the following isomerase reaction:

$$\text{L-arabinose} \rightarrow \text{L-ribulose}$$

This enzyme can be assayed in a continuous system (48) described above for assay of kinases when coupled to the following reactions:

$$\text{L-ribulose} + \text{ATP} \xrightarrow{\text{L-ribulokinase}} \text{L-ribulose-5-P} + \text{ADP}$$

$$\text{phosphoenolpyruvate} + \text{ADP} \xrightarrow[\text{kinase}]{\text{pyruvate}} \text{pyruvate} + \text{ATP}$$

$$\text{pyruvate} + \text{NADH} \xrightarrow[\text{dehydrogenase}]{\text{lactate}} \text{lactate} + \text{NAD}$$

The isomerase can also be assayed by a simple spectrophotometric technique based on determination of the ketopentose product formed as measured by the cysteine-carbazole test of Dische and Borenfreund with L-arabinose as the substrate (41). This latter assay can be carried out with crude cell extracts or toluene-treated cells, as in the case of the β-galactosidase assay.

The reaction mixture (1.0 ml) consists of 10 μmol of L-arabinose, 43 μmol of Tris buffer, pH 7.5, and suitable dilutions of the enzyme preparation (extract or toluene-treated cells). A control mixture without enzyme is included in each set of assays. After 10 min of incubation at room temperature (or 30°C), 0.20 ml of a 1.5% solution of cysteine hydrochloride is added followed by 6 ml of a mixture of 190 ml of water and 450 ml of concentrated H$_2$SO$_4$ and then immediately afterward by 0.2 ml of a 0.12% of an alcoholic solution of carbazole. The mixture is shaken and left at room temperature for 20 min; it is then read in an appropriate colorimeter or spectrophotometer at 540 nm. A unit of activity is defined as the amount required to produce 1 μmol of L-ribulose per min under the conditions of the assay. Commercially available L-ribulose o-nitrophenylhydrazone is used as such to construct a standard curve (range of 0 to 0.25 μmol) relating absorbancy at 540 nm to concentration of L-ribulose under these conditions.

18.2.11. Ligases (Glutamine Synthetase)

Glutamine synthetase (EC 6.3.1.2, L-glutamate:ammonia ligase [ADP-forming]) constitutes the first step in a highly branched pathway leading to the biosynthesis of a large number of essential end products. In *E. coli* and *Klebsiella aerogenes* it is a highly regulated enzyme subject to (i) repression/derepression, (ii) complex feedback inhibition and other allosteric modulations, and (iii) covalent modification. The enzyme catalyzes the following ligase reaction:

$$\text{glutamate} + \text{NH}_3 + \text{ATP} \xrightarrow{\text{Mg}^{2+}}$$

$$\text{glutamine} + \text{ADP} + \text{inorganic phosphate}$$

Its activity can be measured several different ways (40), including a coupled assay for kinases described above involving NADH$^+$ oxidation:

$$\text{glutamate} + \text{NADH}^+ + \text{NH}_3^+$$
$$+ \text{phosphoenolpyruvate} \longrightarrow$$

pyruvate kinase
lactate dehydrogenase
glutamine synthetase

$$\text{glutamine} + \text{NAD} + \text{lactate} +$$

inorganic phosphate

Although ATP is added and ADP is a product of the glutamine synthetase reaction, these do not appear in the net equation.

In this assay, a continuous recording of catalytic activity is achieved by coupling the production of ADP by glutamine synthetase to oxidation of NADH$^+$ by addition of phosphoenolpyruvate, pyruvate kinase, and lactate dehydrogenase in excess. Although a convenient, sensitive, direct assay for various kinetic measurements, it is not suitable for feedback inhibition types of analyses because the various effectors could also influence the activities of the coupling enzymes.

A convenient assay of considerable value to studies dealing with the covalent regulation (via adenylylation [deadenylylation]) of glutamine synthetase activity is the transferase assay involving the reaction:

$$\text{glutamine} + \text{hydroxylamine} \xrightarrow{\text{ADP, arsenate}}$$

$$\gamma\text{-glutamylhydroxamate} + \text{NH}_4^+$$

This activity is used to measure the total amount of glutamine synthetase present, since both the adenylylated and deadenylylated forms of the enzyme are active in this assay. In addition, the assay can be conducted using whole cells made permeable to the reactants by treatment with hexadecyltrimethylammonium bromide. Assay of this enzyme is described in detail by Stadtman et al. (45) for use with *E. coli* cultures and by Bender et al. (3) for use with *Klebsiella aerogenes*. The procedure described below, taken from the latter source (3), is adapted from Shapiro and Stadtman (40).

Prepare a fresh concentrated assay mixture daily by mixing the following volumes of stock solutions (final concentrations of each in the reaction mixture are given in parentheses): 7.53 ml of water; 2.25 ml of 1 M imidazole HCl, pH 7.15 (135 mM); 0.37 ml of 0.80 M hydroxylamine-hydrochloride (18 mM); 0.045 ml of 0.1 M MnCl$_2$ (0.27 mM); 1.5 ml of 0.28 M potassium arsenate, pH 7.15 (25 mM); 0.15 ml of 40 mM sodium ADP, pH 7.0 (0.36 mM); and 1.5 ml of hexadecyltrimethylammonium bromide (1 mg/ml) (90 μg/ml). The hexadecyltrimethylammonium bromide can be replaced with water when crude extract or purified enzyme is used.

Adjust this concentrated mixture to pH 7.55 at room temperature with 2 M KOH, and cool to 4°C if it is not to be used immediately. Add sample and water to 0.40 ml of this concentrated assay mixture to give a volume of 0.45 ml. Equilibrate for 5 min at 37°C, and initiate the reaction by addition of 0.050 ml of 0.20 M L-glutamine. After appropriate periods of incubation, terminate the reaction by adding 1.0 ml of "stop mix" containing 55 g of FeCl$_3$·6H$_2$O, 20 g of trichloroacetic acid, and 21 ml of concentrated HCl per liter. Centrifuge the samples to remove any precipitate, and measure absorbancy at 540 nm. Under these conditions, 1 μmol of glutamyl hydroxamate gives 0.532 U of absorbancy at 540 nm. One unit of enzyme activity is defined as the amount of enzyme producing 1 μmol of glutamyl hydroxamate per min, and specific activity is defined as units of enzyme per milligram of cell protein. Prepare a blank with each assay in which the arsenate and ADP solutions are replaced with water. This allows correction for γ-glutamyl hydroxamate produced by any glutaminase which may be present. When measured as just described, both the adenylylated and the unadenylylated forms of the enzyme are active. For assay of unadenylylated enzyme only, 60 mM MgCl$_2$ at a pH of 7.15 is added to the reaction mixture.

Using conditions similar to those described here, Stadtman and co-workers (40, 45) were able to calculate the $\bar{n}$ value (state of adenylation) for this enzyme in *E. coli*. This is given by the expression:

$$\bar{n} = 12 - 12 \frac{\text{activity in 0.4 M MnCl}_2}{\text{activity in 0.4 M MnCl}_2, \text{pH 7.2}}{\text{activity in 0.4 M MnCl}_2, \text{pH 7.38}}$$

18.3. CONTROL OF ACTIVITY

18.3.1. General Considerations

When an accurate and sensitive assay is available and has been standardized, one can proceed to determine whether that enzyme is subject to one or more of several metabolic control processes. Two major types of controls, allosteric and covalent, have been demonstrated in bacteria:

Allosteric controls of enzyme activities
  Simple end-product inhibition
  Cumulative end-product inhibition
  Sequential end-product inhibition
  Precursor stimulation
  Precursor inhibition
  Product stimulation
  Product inhibition
  Energy charge
  Catabolite inhibition
Covalent controls of enzyme activities
  Phosphorylation
  ADP-ribosylation
  Adenylylation
  Uridylylation
  Sulfhydryl conversion

The subtypes of allosteric controls are named on the basis of where the effector is positioned in a metabolic pathway (see references 2, 16, 21, 40, 44, and 46 for further discussion). At the present state of biochemical technology, it is difficult, if not impossible, to analyze some of these control processes in vivo. Consequently, it is often difficult to determine whether they exist per se in bacterial cells or only in the researcher's imagination. No judgment is made here in this regard. The subtypes or categories of controls listed above should thus be viewed with caution. In vitro experiments are generally used to study allosteric control. Suspected control compounds, called effectors, are added to an in vitro assay reaction mixture, and their influence on the overall reaction is examined. In the case of allosteric interactions, the effects are essentially immediate, reversible, and directly dependent on the effector concentration.

Some allosterically modifiable enzymes tend to exhibit sigmoidal rather than rectangular hyperbolic curves when their initial reaction velocities are plotted against increasing substrate con-

centrations, probably because the substrate functions in an effector capacity as well as in a substrate capacity. This, however, is not a prerequisite for obtaining a sigmoid response. The influence of positive (accelerative) and negative (inhibitory) effectors on allosteric controls is illustrated in Fig. 2. These effects have been described in detail elsewhere (2, 16). In comparison to a control situation where no effector is present, greater concentrations of substrate are required to achieve a given reaction velocity in the presence of a negative allosteric effector. Conversely, less substrate is required when positive effectors are present. All of these interactions are thought to involve physical binding of the various effectors to specific recognition sites on the enzyme. This physical interaction is believed to yield a net conformational change in the protein molecule which is reflected in altered enzymatic function with respect to substrate binding properties and reaction velocities.

The control of enzymatic activity via covalent modification (21), on the other hand, involves a covalent change rather than an allosteric (reversible) binding of the regulatory ligand to the enzyme. This covalent change is probably always catalyzed by another enzyme, as depicted in Fig. 3. An active (interconvertible) enzyme is rendered inactive by another enzyme (a converter enzyme, or "inactivating" enzyme) which covalently modifies the first enzyme. Another specific converter enzyme (an activating enzyme) catalyzes the reactivation of the original enzyme activity by removal of the original covalent moiety.

Experiments exemplifying both types of control processes (control by allosteric and by covalent binding of effectors) are described below.

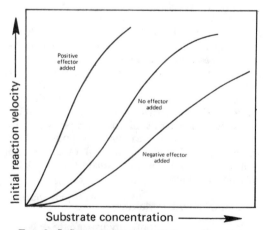

FIG. 2. *Influence of positive and negative effectors on allosteric control of enzymatic activity.*

18.3.2. Allosteric Controls (Simple End-Product Inhibition)

Threonine deaminase (biosynthetic in *E. coli*) can be assayed using extracts or lysed cells as described earlier. The ability of L-isoleucine to allosterically inhibit this enzyme can be demonstrated by performing the following experiment (H. E. Umbarger, personal communication).

Prepare tubes containing the components shown in Table 1. Incubate at 37°C for up to 20 min (depending on activity of the extract), and stop the reaction by adding 0.1 ml of 50% (wt/vol) trichloroacetic acid. Dilute a fraction (again, depending on the activity of the enzyme) of the above reaction mixture to 1.0 ml. To this, add 3.0 ml of 0.025% 2,4-dinitrophenylhydrazine in 0.5 M HCl, and allow the mixture to stand at room temperature for 15 min. Add 1 ml of 40% KOH, shake the contents, and then read in a Klett-Summerson colorimeter using the 54 filter or in any standard spectrophotometer or colorimeter at 540 nm. Use relative absorbance values or calculate micromoles of α-ketobutyrate formed in the reaction mixture from a previously prepared standard curve. The same basic experiment can be conducted with various concentrations of substrate and effector. The data are plotted as shown in Fig. 2.

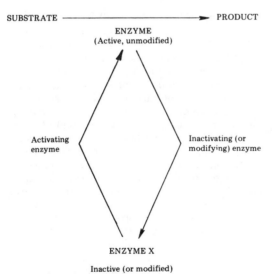

SUBSTRATE ⟶ PRODUCT

ENZYME
(Active, unmodified)

Activating enzyme

Inactivating (or modifying) enzyme

ENZYME X

Inactive (or modified)
via covalent linkage

FIG. 3. *Covalent control of enzymatic activity. In its simplest form this process involves an "inactivating" or modifying converter enzyme which catalyzes the covalent (E + X → EX) change in a substrate enzyme, and an "activating" converter enzyme which catalyzes the reverse process (EX → E + X), thereby restoring the original enzyme activity.*

18.3.3. Allosteric Controls (Energy Charge)

Atkinson (2) has proposed that various key bioenergetic and biosynthetic enzymes are allosterically regulated in response to the cell's energy charge. The energy charge is defined as the mole fraction of ATP plus half the mole fraction of ADP (since 2 mol of ADP can be converted to 1 mol of ATP via adenylate kinase) per total moles of adenylates:

energy charge

$$= (ATP + \tfrac{1}{2} ADP)/(ATP + ADP + AMP)$$

The basis for this hypothesis is predicated on the fact that the adenylates constitute a common end-product pool that interconnects and thereby regulate various key energy-regenerating (R) and energy-utilizing (U) reactions as depicted in Fig. 4. ATP is an end product of R reactions, while ADP and AMP are end products of U reactions, the former representing a high energy charge situation and the latter representing one that has a low charge. Enzymes such as phosphofructokinase, pyruvate kinase and citrate synthetase are thought to be affected by the energy charge.

To measure the response of an enzyme to variations in the energy charge, an enzyme reaction mixture is prepared with a constant total adenylate pool (ATP + ADP + AMP) at the desired values of energy charge. The size of this pool, for different kinds of cells, is generally in the range of 2 to 5 mM. A three-component mixture at a predetermined value of energy charge can be produced by making up an AMP-ATP mixture with the mole fraction of ATP equal to the desired energy charge, adding adenylate kinase, and then incubating long enough to ensure equilibrium. The concentration of Mg^{2+} should be varied as well as the concentration of the substrate(s). The use of several substrate concentrations is necessary because the

TABLE 1. *Preparation of tubes for demonstrating inhibition of threonine deaminase by L-isoleucine*

Component	Amt (ml)			
	Tube 1	Tube 2	Tube 3	Tube 4
Tris-hydrochloride (pH 8), 1 M	0.1	0.1	0.1	0.1
NH₄Cl, 1 M	0.1	0.1	0.1	0.1
Pyridoxal phosphate, 1 M	0.1	0.1	0.1	0.1
L-Threonine, 0.2 M ..	0.2	0.2	0.2	0.0
Isoleucine	0.0	0.1ᵃ	0.1ᵃ	0.0
Extract	0.1	0.1	0.1	0.1
Water	0.4	0.3	0.3	0.6

ᵃ The concentration of isoleucine for tube 2 is 10 mM and that for tube 3 is 1 mM.

most common response of enzymes to variations in the energy charge is believed to be a change in the affinity for one or more substrates.

The initial velocity of the reaction at each substrate concentration can be plotted as a function of the energy charge. Repeating this plot for each substrate level tested should produce a family of curves that describe the effect of energy charge on the enzyme under study.

It is suggested that phosphofructokinase be tested in this way with the assay procedure described earlier. Another enzyme known not to be sensitive to energy charge regulation can be used as a comparative control. These are not simple, quick experiments to be undertaken with only a casual interest. They can, however, be informative and useful in obtaining practice in assaying enzymes and ascertaining some of their control parameters.

18.3.4. Covalent Controls (Adenylylation)

Glutamine synthetase is of central importance in the nitrogen metabolism of *E. coli* and other bacteria. This importance is reflected in the fact that both its synthesis and its catalytic activity are under a variety of control processes. Its synthesis appears to be under repression/derepression control; its activity is controlled by a number of end products in a complex way by a series of effectors. Its activity is also controlled by a covalent modification process involving a complex series of activating and inactivating converter proteins and various effector signals as shown in Fig. 5.

Inactivation of this enzyme via adenylylation can be demonstrated in vitro as described by Mecke et al. (30) and others. It can also be observed in vivo under conditions of ammonia "shock" as follows.

Grow a wild-type strain of *E. coli* in a mineral salts medium (as described above) with 0.4% glucose as carbon source and 0.2% L-glutamine as nitrogen source. Grow the culture to the mid-exponential phase, remove a sample for analysis, and split the culture into two portions. Leave one portion untreated, but add 15 mM (final concentration) $(NH_4)_2SO_4$ to the other; continue incubation for 5 to 10 min, and then remove samples for permeabilization with hexadecyltrimethylammonium bromide and analyze for glutamine synthetase activity using the whole-cell glutamyl transferase assay as described earlier (18.2.11). In each case, the sample being removed (10 ml) is immediately placed in a flask containing 90 μg of hexadecyltrimethylammonium bromide per ml (final concentration), shaken at 30°C for 1 to 3 min, and then prepared for assay as described (18.2.11). Determine en-

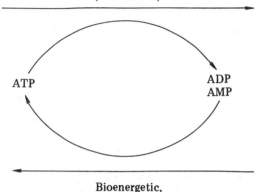

Biosynthetic,
energy-utilizing reactions
(U reactions)

ATP

ADP
AMP

Bioenergetic,
energy-regenerating reactions
(R reactions)

FIG. 4. *Allosteric control of enzymatic activity by energy charge.*

zyme activity in the presence and absence of 60 mM $MgCl_2$ as indicated in order to determine the extent of covalent modification caused by the ammonium "shock." In both *E. coli* and *K. aerogenes*, this "shock" is known to promote adenylylation of glutamine synthetase (3, 40, 45).

18.4. REGULATION OF SYNTHESIS

18.4.1. General Considerations

The oracle at Delphi admonished us all to "Know Thyself." A similar sense of awareness should prevail before we search out a bacterial species or strain with which to conduct a particular metabolic study. "Know Your Bacterium" would seem an apt paraphrase in this regard.

In this age of specialization, it is not uncommon for many bacteriologists to spend years or even entire careers working with a single species of organism. In many instances this is justifiable. On the other hand, occasions can arise in which one ought to take advantage of the seemingly endless diversity existing in the microbial world and choose another organism better suited to meet the needs of a specific research goal. Rather than clinging to those few familiar species, one ought occasionally to explore other cell types.

One would not, for example, want to continue working with a lactic acid bacterium when undertaking a project concerned with pathways for amino acid biosynthesis. These organisms have very limited abilities for amino acid biosynthesis in this regard, and their amino acids must be supplied in the culture fluid. An organism such as *E. coli*, which synthesizes all its amino acids

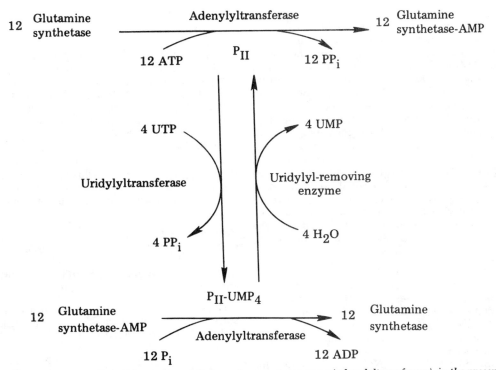

FIG. 5. *Covalent control of glutamine synthetase. A converter enzyme (adenylyltransferase), in the presence of 12 ATPs, catalyzes the adenylylation of each of the 12 subunits of glutamine synthetase. This covalent modification causes inactivation of the Mg^{2+}-dependent glutamine synthetase activity. The adenylyltransferase is controlled by the metabolite-directed interconversion of a P_{II} protein. The P_{II} protein activates adenylyltransferase activity. When the P_{II} protein is covalently modified by another converter enzyme (uridylyltransferase) to its uridylylated form [$P_{II} - UMP_4$] it activates the adenylyltransferase-catalyzed deadenylylation of glutamine synthetase, thereby restoring the latter to its normal biosynthetic activity (44).*

from glucose and NH_4^+, would obviously be a more suitable agent for this purpose. On the other hand, if one wanted to study regulation of bioenergetic processes in an organism largely unaffected by oxygen metabolism, a lactobacillus and not *E. coli* might be preferable. In each case, the suitability of the organism for use in a particular study is determined largely by its genetic potential. In some cases this potential can be altered to meet one's specific research needs. This is done through the production and screening of suitable mutant strains. Selection of such mutants is an invaluable procedure which has become another staple in the bacteriologist's methods repertoire, as described elsewhere in this manual (Chapter 13).

Even when an organism has the genetic potential for production of a particular enzyme, synthesis (transcription and translation) of that enzyme is not ensured. Many enzymes and enzyme systems are synthesized or not depending on the presence or absence of certain regulatory signals or "triggers" which are made endogenously or provided exogenously in the culture medium. Some substances, called inducers, stimulate transcription, and the process is termed **induction; de-induction** occurs when these inducers are not available. Conversely, some substances, called repressors, prevent transcription, and the process is termed **repression; de-repression** occurs when these repressors are not available. Among the bacteria various types of repression systems have been described: simple feedback or end-product repression; multivalent repression, which occurs with certain enzymes involved in synthesis of the branched-chain amino acids; and coordinate repression, wherein all of the enzymes in a biosynthetic pathway are coordinately repressed when the end product (tryptophan or histidine, for example) is present in high concentration. The following experiments demonstrate some of the induction and repression types of regulation in the syntheses of bacterial enzymes.

18.4.2. Inductions (Nitrate Reductase)

Using an appropriate wild-type *E. coli* strain (e.g., ML 30 or K-12) and the nitrate reductase assay described earlier (18.2.4), one can demonstrate the induction of nitrate reductase by adding NO_3^- to the medium, if the culture is also made anaerobic. The accumulation and subsequent utilization of NO_2^- in the culture medium can also be measured as an indicator of regulation over activity of the nitrate (and nitrite) reductase system.

With a fresh overnight culture as inoculum, inoculate a series of flasks containing a basal medium such as the following (per liter): KH_2PO_4, 2 g; K_2HPO_4, 7 g; $(NH_4)_2SO_4$, 1 g; $MgSO_4 \cdot 6H_2O$, 0.1 g (pH 7.2); add presterilized 0.02 M glucose and 0.25% (final concentrations) vitamin-free casein hydrolysate just prior to inoculation. Let the cultures grow with vigorous shaking to approximately 50 μg of cell dry weight per ml of culture or longer (visible turbidity), and then treat as follows.

Culture A: Add 20 mM $NaNO_3$; keep aerobic.
Culture B: Add 20 mM $NaNO_3$, and make anaerobic by transfer to a tightly sealable vessel and sparging with N_2 or a mixture of 95% N_2–5% CO_2.
Culture C: Do not add $NaNO_3$, but make anaerobic as above.
Culture D: Add 20 mM $NaNO_3$; keep aerobic for a few hours, and then make anaerobic (as in B). If desired, switch the culture back to an aerobic state again a few hours later.

A similar experiment can be designed and performed to study regulation of biodegradative threonine deaminase. This enzyme, like nitrate reductase, is synthesized only under anaerobic conditions but is subject to cAMP/catabolite repression control as well.

Samples of the culture in either case (centrifuge if the turbidity gets too high and causes interference in subsequent measurements) are removed and assayed for NO_2^- accumulation according to the following procedure. Add samples of each culture (0.2 ml) to 1 ml of reagent A, which contains 0.80% sulfanilic acid in 5 N acetic acid. Dilute the sample to 10 ml with 0.05 M phosphate buffer (pH 7.5), and add 1.0 ml of reagent B (6 ml of dimethyl-α-naphthylamine in 1 liter of 5 N acetic acid). After 30 min of color development, read the samples at 540 nm; calculate the concentration of NO_2^- present in each sample from a previously prepared standard curve.

One can also remove culture samples, centrifuge the cells, and then measure the cellular nitrate reductase activity as described earlier (18.2.4). The enzyme is synthesized only under anaerobic conditions, and it requires the presence of NO_3^- for induction to occur (removal of the NO_3^- results in de-induction).

18.4.3. End-Product Repressions (Ornithine Transcarbamylase)

This important regulatory process can be readily demonstrated in most *E. coli* strains by comparing cultures grown in a glucose-mineral salts basal medium with and without added L-arginine (31). Ornithine transcarbamylase catalyzes the sixth step in the sequence of reactions culminating in synthesis of the end product, L-arginine. In the presence of L-arginine, ornithine transcarbamylase synthesis is repressed. To demonstrate this, grow a wild-type strain overnight in any suitable mineral salts medium (e.g., as described elsewhere in this chapter) containing:

Culture 1: 0.02 M glucose
Culture 2: 0.02 M glucose plus 100 μg of L-arginine per ml
Culture 3: 0.02 M glucose plus 100 μg of another amino acid (e.g., histidine) per ml

Harvest, wash with 0.1 M Tris-hydrochloride buffer (pH 7.8), and assay for ornithine transcarbamylase as described earlier.

18.4.4. Multivalent Repressions (L-Threonine Deaminase)

In a similar series of experiments, one can demonstrate the sensitivity of L-threonine deaminase (the biosynthetic enzyme) to multivalent repression by the end products (L-isoleucine, L-leucine, and L-valine). All three amino acids must be present to cause this repression (H. E. Umbarger, personal communication).

Culture 1: 200 ml of minimal medium plus 0.02 M glucose
Culture 2: 200 ml of minimal medium plus 0.02 M glucose and 0.8 M L-valine, 0.4 mM L-leucine, and 0.4 mM L-isoleucine
Culture 3: 200 ml of minimal medium plus 0.02 M glucose and 0.8 mM L-valine, 0.4 mM L-leucine, and 0.15 mM L-isoleucine

Using the threonine deaminase assay described earlier, measure the specific activities of the enzyme in cells grown overnight under these three conditions. Culture 2 should show a repressed level of enzyme relative to culture 1. Culture 3, owing to eventual decrease in intracellular L-isoleucine concentration (it is present initially at a lower level in culture 3 and its uptake tends to be inhibited by the other two amino acids as well), will undergo a derepression relative to culture 2.

For an excellent, thorough, and detailed review of overall methodologies used in studying enzyme and pathway regulation, the reader is referred to an article by Clarke (5).

18.4.5. Inductions Plus Repressions (*lac* Operon)

The *lac* operon in *E. coli* involves both repression (negative) and induction (positive) genetic controls as depicted in Fig. 6. The binding of the *lac* repressor (*i* gene product) to the operator site prevents ribonucleic acid polymerase transcription (negative control) of the *lac* genes. The inducing agent (any of certain β-galactosides) modifies the repressor protein sufficiently to prevent its binding. This prerequisite alone, however, is not sufficient to initiate the transcription process. A positive regulatory element, the cAMP receptor protein (the product of the *crp* locus) upon modification by cAMP binding, must be available and bound to the promoter locus in order for ribonucleic acid polymerase to initiate its catalytic function in transcription. Thus, in the presence of inducers and cAMP, the *lac* operon is turned on and transcription commences. The absence of either the inducer (de-induction) or decreased cAMP production via inhibition of adenylate cyclase activity (catabolite repression) prevents this transcription.

Adenylate cyclase is a membrane-associated enzyme which catalyzes the conversion of ATP to cAMP and pyrophosphate. This enzyme is generally active during growth of the cells in various media, and cAMP is thus available for their use. However, when certain substrates such as glucose or mannitol are present, its activity is inhibited and cAMP production is curtailed, resulting in the regulatory phenomenon known as catabolite repression.

The major features of these regulatory phenomena can be demonstrated experimentally in the following series of experiments. In each case, β-galactosidase synthesis is measured as described above (using toluene-treated cells) to monitor control over expression of the *lac* operon. *E. coli* K-12, *E. coli* ML 30, or other appropriate wild-type strains can be used. The cells will grow readily in a mineral salts basal medium containing 0.25 to 0.50% vitamin-free casein hydrolysate (acid hydrolyzed) as the sole carbon and energy source (18.4.2), and under these conditions yield relatively nonrepressed rates of β-galactosidase synthesis when supplied with 1 mM isopropyl-β-D-thiogalactopyranoside as the gratuitous inducer.

Grow an overnight aerobic culture (15 to 17 h,

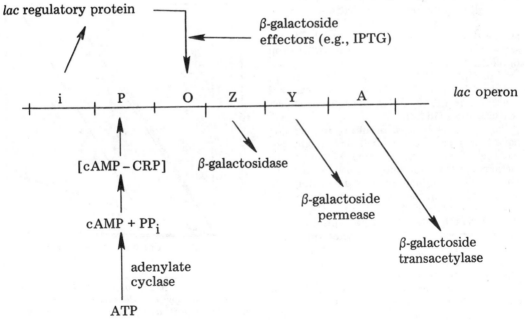

FIG. 6. *Positive and negative regulation of the lac operon in E. coli. Components involved in transcription of the lac operon include: the Z, Y, and A genes containing the code for β-galactosidase, β-galactoside permease, and β-galactoside transacetylase, respectively; the regulatory gene (i) that codes for production of the lac regulatory protein; a promoter (P) and operator (O) site; and also a cya gene and a crp gene which contain the code for synthesis of adenylate cyclase, and the cAMP binding protein (CRP), respectively. IPTG, Isopropyl-β-D-thiogalactopyranoside.*

TABLE 2. *Preparation of cultures for demonstration of cAMP regulation of the lac operon in E. coli*

Ingredient	Vol (ml) added to flask:					
	1	2	3	4	5	6
Basal medium (1.25× strength)	16	16	16	16	16	16
Casein hydrolysate (10%)	0.5	0.5	0.5	0.5	0.5	0.5
Distilled water	3.5	2.5	2.1	1.6	2.1	1.6
Glucose (1 M)	—	—	0.4	0.4	0.4[a]	0.4
cAMP (20 mM)[b]	—	—	—	0.5	—	0.5[c]
IPTG (20 mM)	—	1.0	1.0	1.0	1.0	1.0
Inoculum[d] (approximately 1 mg of cell dry wt/ml)	1.0	1.0	1.0	1.0	1.0	1.0

[a] In this culture, add the glucose at approximately 150 min after IPTG (isopropyl-β-D-thiogalactopyranoside) addition.
[b] cAMP is solubilized by neutralization with NaOH.
[c] In this culture, add the cAMP about 150 min after IPTG addition.
[d] Add the cells to the basal medium, water, and casein hydrolysate medium, and let grow for approximately 60 min. At the end of that time, add IPTG and other compounds as indicated; this is recorded as zero time for the experiment.

37°C) under these conditions, harvest by centrifugation, suspend in sterile 0.05 M potassium phosphate buffer (pH 7.5) (or physiological saline), and inoculate into six 125-ml flasks containing the ingredients listed in Table 2.

At 0 min and at 20- to 30-min intervals thereafter, remove 1.0-, 0.50-, or 0.20-ml samples, and add to 4.0, 4.5, or 4.8 ml of 0.05 M sodium phosphate buffer (pH 7.5), respectively, to obtain 1:5, 1:10, and 1:25 dilutions of the cultures as needed. Start with a 1:5 dilution, and switch to higher dilutions when previous dilution exhibits an A_{420} of approximately 0.50. The A_{420} for each culture dilution is recorded and used to calculate the dry weight of cells based on a standard curve relating A_{420} to dry weight. These same suspensions are then treated with 1 drop of toluene and 1 drop of 0.1% sodium deoxycholate, put into a 37°C water bath, shaken for 15 min, and then placed in an ice bath until samples are withdrawn (0.20 ml) for subsequent β-galactosidase assay (18.2.8).

Plot the growth of each culture (on semilog paper) as micrograms of cell dry weight per milliliter of culture versus time. Plot on arithmetic paper the total units of enzyme per milliliter of culture (ordinate) versus the micrograms of cell dry weight per milliliter of culture (abscissa). The slope of this plot is the P value or the differential rate of enzyme synthesis. Calculate the specific activity of each determination (units of enzyme per microgram of cell dry weight), and plot as a function of incubation time on arithmetic graph paper.

A comparison between cultures 1 and 2 will demonstrate the need for a β-galactoside inducer; comparison between cultures 2 and 3 will demonstrate glucose repression; and culture 4 will show reversal of this repression by exogenously added cAMP. Cultures 5 and 6 will show this repression and depression in a different time frame and somewhat more dramatically. Data

like those shown in Fig. 7 are generally obtained from such experiments.

These experiments can be varied in many ways through use of other substrates, introduction of anaerobic-aerobic shock transitions, and use of various mutant strains (e.g., *cya* and *crp* mutants, etc.). They can be applied to study of other enzyme systems such as those transcribed by the L-arabinose (*ara*) operon or the biodegradative threonine deaminase which is formed only under anaerobic conditions.

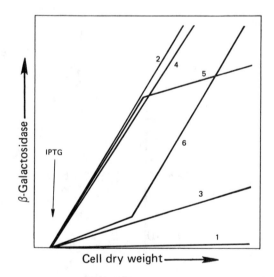

FIG. 7. *Cyclic AMP regulation of the lac operon in E. coli. All cultures induced with IPTG (isopropyl-β-D-thiogalactopyranoside) except no. 1. Culture 2 = casein hydrolysate as sole carbon and energy source. Culture 3 = same as culture 2 but glucose also added. Culture 4 = same as culture 3 but cAMP also included in growth medium. Culture 5 = same as culture 3 but the glucose added 1 to 2 h after start of induction period. Culture 6 = same as culture 4 but cAMP added 1 to 2 h after start of induction period.*

18.5. ANALYSIS OF PATHWAYS

Enzymes do not exist as independent entities carrying out their catalytic functions in random array, but rather as a series of interlinked, sequential functions. A bacterial cell can be viewed, in fact, as one great multienzyme system functioning with intricate harmony. While the wholeness of the system ought not to be forgotten, we are usually forced for practical reasons to deal with smaller clusters of reactions known as pathways. A **metabolic pathway** is generally defined as a series of coordinated reactions carrying out a particular, well-defined function of either biosynthetic or bioenergetic importance, e.g., the linkage of electron transport carriers, the glycolytic pathway, or the branched-chain amino acid biosynthetic pathway. It has been the study of such pathways, more than the analyses of individual enzymes, which has contributed richly to our present understanding of bacterial metabolism.

The methodologies currently available for studying metabolic pathways have been extensively analyzed and evaluated by Dagley and Chapman (8). In the remainder of this chapter, the general methods described by these authors are outlined, and a general "radiosotope inventory" is detailed. These can be used to provide an overview perspective.

18.5.1. General Methods

Identification of chemical intermediates.

During the catabolism of some substrates by either growing or resting bacterial cells, chemical intermediates accumulate in the medium where they can be readily identified by various chemical, chromatographic, or radioisotopic techniques. One or more of these intermediates will accumulate if its rate of synthesis exceeds its rate of dissimilation by subsequent reactions in that pathway. This accumulation is often followed by utilization when the original substrate becomes depleted, as shown in Fig. 8. *E. coli* cultures, for example, growing in a rich medium with glucose as the major source of carbon and energy will accumulate and subsequently utilize pyruvate in this manner. Considerable information concerning the pathways involved in degradation of aromatic substances has been obtained by these types of analyses.

Use of metabolic inhibitors.

Using the same principle, one can make use of inhibitors in pathway analyses. When an inhibitor of some step in the pathway is added to the culture, one or more of the metabolites posi-

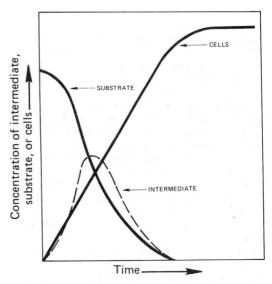

FIG. 8. *Transient appearance of an intermediate from metabolism of a substrate during growth of a bacterial culture.*

tioned prior to the inhibited step may accumulate and thus be easily identified. This information can be used to help establish the reaction sequence. For example, much of our knowledge of the chemical nature of microbial cell walls derives from observations that various nucleotides involved in their biosynthesis accumulate when *Staphylococcus aureus*, *Streptococcus faecalis*, or *E. coli* cells are treated with antibiotics such as penicillin, bacitracin, or cycloserine. Obviously, the cells must be permeable to such inhibitors/metabolic poisons if they are to be effective for this purpose, and caution always must be exercised when interpreting the significance of data derived from inhibitor studies, primarily because all of the sites of inhibition may not be known.

Use of substrate analogs.

The enzymes in a particular pathway may differ in their substrate specificities. The first three enzymes, for example, may be able to carry out their respective reactions with the analog, but the fourth may be unable to do so. If this is the case, the partially metabolized analog would accumulate, and its identification and assignment in the metabolic sequence would be made easier.

Simultaneous adaptation.

The technique of simultaneous adaptation is based on the following logic. If an inducible dissimilatory pathway (e.g., A → B → C → D

→ tricarboxylic acid cycle) is induced by growth substrate A, the cells will simultaneously develop the capacity to dissimilate metabolites B, C, and D as well as A, but not other metabolites that are not members of this pathway. If this pathway, for example, involves an oxidation of all these components, one would expect to see oxidation (measured with a Warburg [16.6] or electrode oxygraphic system [16.2.3]) of substrates A, B, C, and D but not of substrates E, F, and G unless one or more of these latter substances were utilized constitutively by the cells. This latter possibility could be determined by testing for their oxidation with cells grown on a substrate other than A. Another flaw in this technique is the possibility that, even though substrate B, C, or D may be a component of the pathway, the substrates may be unable to penetrate the cells and thus not be oxidized for this reason. The use of cell extracts or permeabilized cells would be needed to evaluate this possibility.

Use of cell extracts.

Most of the analyses just described above can be carried out using cell extracts in place of whole cells. Accumulation of intermediates and/or cofactors can occur spontaneously or can be promoted by use of metabolic poisons or substrate analogs. Permeability barriers are avoided, thus simplifying the analysis in this respect. However, the enzymes of the pathway under study are no longer in their natural environment and thus may display distorted activities or become rapidly inactivated.

Use of radioisotopes.

The use of radioisotopes, particularly those compounds containing ^{14}C, ^{32}P, ^{3}H, or ^{35}S, has proven an invaluable tool in pathway analyses. The various types of procedures available for this purpose have been described and evaluated by Dagley and Chapman (8), Quayle (35), Wang (47), and Eaton (14), and will be considered here only in a general manner.

First, it should be pointed out that their use is not without difficulties and pitfalls. Perhaps foremost among these is the assurance that pure substances are being used. "Before a radiotracer is used a careful examination of its purity is essential, for study with impure starting material may be worse than no study at all" (8). (This admonition is now posted and underscored in my laboratory; considerable time was wasted attempting to track down a second transport system for glucose-6-phosphate only to find that it was an artifact resulting from a low level of free [^{14}C]glucose contaminant present in all but

freshly prepared samples of the [^{14}C]glucose-6-phosphate that were used as the transport substrate.)

Even when pure materials are used, it is no simple matter to trace the fate of one labeled molecular species among the myriad of interconnected and often reversible flows of metabolic reactions. The more readily metabolites enter active metabolic pools, the more difficult it is to trace their specific functions. Because actively dividing bacteria (in contrast to resting cells) are endowed primarily for synthesis, turnover and redistribution of metabolites tend to be minimal under these conditions. For this reason use of actively dividing bacteria is recommended whenever possible in radioisotopic analyses of pathways. The use of appropriate mutant strains having a defect which would minimize or eliminate distribution and/or turnover of metabolites might be helpful in this respect.

A study of the amino sugar biosynthetic pathway illustrates the benefits to be derived from use of appropriate mutant strains coupled with radiotracer methods. N-Acetylglucosamine is transported via the phosphoenolpyruvate:phosphotransferase system and thereafter undergoes (i) dissimilation via entry into the glycolytic pathway via deamination of glucosamine-6-phosphate to NH_4^+ plus fructose-6-phosphate, and (ii) assimilation into amino sugar macromolecular moieties according to the pathway shown in Fig. 9.

Mutant strains of E. coli K-12 lacking N-acetylglucosamine-6-phosphate deacetylase or glucosamine-6-phosphate deaminase activity are available (48). Growth of the deacetylaseless mutant in a medium containing a suitable source of carbon and energy (e.g., gluconate) in the presence of N-acetyl-[1-^{14}C]glucosamine results in N-acetylglucosamine-6-phosphate accumulation and no ^{14}C assimilation. The deaminaseless mutant, on the other hand, will assimilate the label specifically into the amino sugar biosynthetic pathway. Some experiments along these lines have been conducted (12, 13), but full advantage of this specific pathway labeling technique has not yet been taken.

Rapid sampling techniques have been used successfully in a number of cases. They were invaluable in the study of the $^{14}CO_2$ fixation pathway in unicellular green algae and in the discovery of the glyoxalate cycle in *Pseudomonas* and other organisms. This procedure involves addition of radioisotopic materials to cells for very brief (seconds or minutes) periods, followed by direct examination of labeled products.

Isotope competition techniques can also be used in pathway analyses. These techniques are

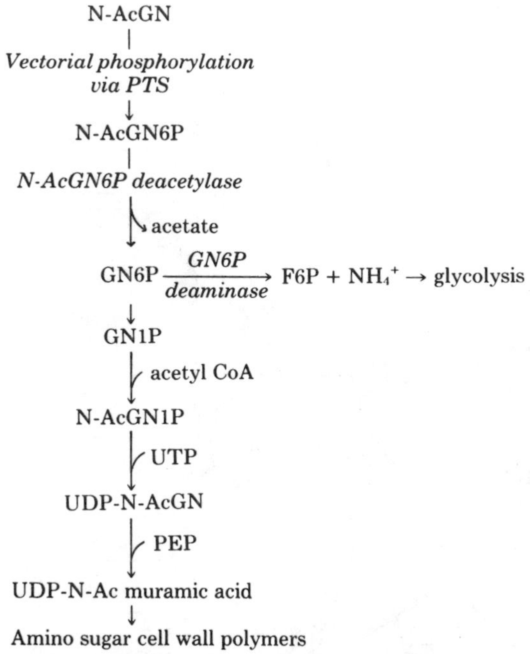

N-AcGN
|
Vectorial phosphorylation
via PTS
↓
N-AcGN6P
|
N-AcGN6P deacetylase
↘ acetate

GN6P $\xrightarrow[\text{deaminase}]{GN6P}$ F6P + NH$_4^+$ → glycolysis
↓
GN1P
↙ acetyl CoA
N-AcGN1P
↙ UTP
UDP-N-AcGN
↙ PEP

UDP-N-Ac muramic acid
↓
Amino sugar cell wall polymers

FIG. 9. *Amino sugar biosynthetic pathway. N-AcGN, N-acetylglucosamine; PTS, phosphoenolpyruvate:phosphotransferase system; N-AcGN6P, N-acetylglucosamine-6-phosphate; GN6P, glucosamine-6-phosphate; F6P, fructose-6-phosphate; GN1P, glucosamine-1-phosphate; CoA, coenzyme A; N-AcGN1P, N-acetylglucosamine-1-phosphate; UTP, uridine triphosphate; UDP-AcGN, uridine diphosphate N-acetylglucosamine; PEP, phosphoenolpyruvate; UDP-N-Ac muramic acid, uridine diphosphate N-acetylmuramic acid.*

based on the logic that, if all the components being channelled into a pathway are labeled, the addition of a nonlabeled intermediate to that pathway would cause a proportionate decrease in the isotope specific activity in all metabolites from the point of that intermediate forward in the pathway. Roberts et al. (38) used this technique extensively and skillfully in their classic studies on the biosynthesis of the various families of amino acids in *E. coli*.

Radiorespirometry, a technique developed by Wang (47), refers to a class of methods by which the respiratory activity of a biological system can be measured using radiotracer methods. Specifically, it involves a study of the rate and extent of conversion of a ^{14}C-labeled carbon atom of a given substrate (e.g., [1-^{14}C]glucose, [3,4-^{14}C]glucose, or DL-[1-^{14}C]glutamate) to respiratory CO_2. This technique has been used to analyze catabolic respiratory pathways in yeasts and a variety of bacteria, including among others *Brevibacterium, Arthrobacter, Xanthomonas,*

Corynebacterium, Neisseria, and *Escherichia*. It has been extensively used to evaluate the flow of carbon through concurrent pathways such as the Embden-Meyerhof-Parnas pathway, the Entner-Doudoroff pathway, the pentose phosphate pathway, the glucuronic acid pathway, and the tricarboxylic acid cycle.

Like other research methodologies, radiorespirometry has certain disadvantages and limitations. Generally, one has to construct the necessary apparatus and pieces of equipment since these are not commercially available. This, however, can be done with minimal cost, as demonstrated by the six-vessel unit constructed in my laboratory (Fig. 10). A glassblower can readily convert 250-ml Erlenmeyer flasks into radiorespirometer vessels and simple chromatographic columns into CO_2-trapping units. Gas flowmeters, an incubator-shaker, a "monkey bar" rig for mounting the flowmeters and CO_2 traps, and counting equipment such as a liquid scintillation counter complete the list of required items. Limitation stems from the fact that some of the compounds specifically labeled with ^{14}C that one would want to use in certain studies are not commercially available.

Use of induced auxotrophs

The principles employed in the use of auxotrophic mutants for pathway analyses, particularly biosynthetic pathways, are well known, and details for selection and use of such mutant cultures are described elsewhere in this manual (13.6.2). Given a biosynthetic pathway sequence such as

$$ A \xrightarrow{E_1} B \xrightarrow{E_2} C \xrightarrow{E_3} D \xrightarrow{E_4} E $$

in which E is the end product required for growth of the organism, the following situation obtains. Any mutation affecting enzyme E_1, E_2, E_3, or E_4 will render the organism incapable of growth unless a metabolite distal to that lesion is supplied. That metabolite, of course, must be capable of being transported into the cells. Metabolites proximal to the lesion will in some circumstances accumulate in high concentrations if sufficient amounts of growth substrates are available. Identification of the metabolites that accumulate and those that stimulate growth of the mutant cells, along with appropriate enzyme analyses, can yield considerable information as to the sequence of reactions in that pathway. Such mutant methodologies have been used with great success in study of the branched-chain and aromatic amino acid biosynthetic pathways among others.

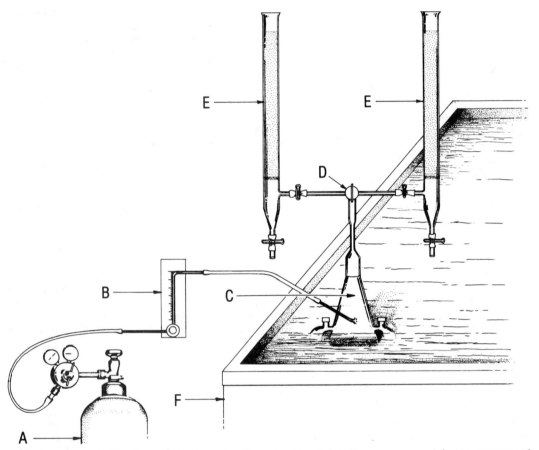

FIG. 10. *Construction of a radiorespirometry apparatus. Gas (e.g., nitrogen or oxygen) from a compressed tank (A) is passed through a gas flowmeter (B) to regulate gas passage through the radiorespirometer flask (C) at a constant rate. The flask (C) can be constructed from a 125- to 250-ml Erlenmeyer flask. A gas import tube is built into the flask so that the atmosphere is controlled and the constant flow of gas carries the metabolically produced $^{14}CO_2$ up from the flask into a three-way stopcock (D) which directs the gas into one of the two CO_2 traps (E). After an appropriate time interval, the stopcock is turned and the $^{14}CO_2$ is collected in the alternate trap. The flask (C) is modified also to contain side arms for use in substrate additions or removal of samples through serum stoppers mounted on the side arms. At least six such vessels can be placed in a controlled water bath shaker (F) and run simultaneously.*

18.5.2. Isotopic Carbon Inventory

General considerations.

Imagine the following hypothetical situation: you isolate a culture of a new bacterium about which nothing is known except that it will grow on glucose in an otherwise complex, undefined medium. You are challenged to describe the general type of metabolic pathways utilized by this organism. How would you set about undertaking this task? What types of bioenergetic pathways are involved? How much of the glucose carbon, if any, is assimilated for biosynthetic purposes? Into what macromolecular components does the carbon flow? Answers to

questions such as these (and many more, of course) would be needed to meet the challenge successfully. A "radioisotopic inventory" experiment would provide a sound foundation for this undertaking. The culture is grown for a period of time sufficient to dissimilate 30 to 70% of the glucose added, which is labeled with [^{14}C]glucose (either uniformly labeled or specifically labeled in the C-1, C-3,4, or C-6 positions). At the end of that time, a ^{14}C inventory is conducted, with the data thus derived serving as an excellent starting point in assessing the metabolic pathways involved.

This type of experiment has been successfully carried out with wild-type *E. coli* ML 30 cultures as follows. A culture of *E. coli* in log growth is transferred to a radiorespirometer vessel (125-ml Erlenmeyer flasks are convenient) constructed in such a way as to allow continuous gassing (with N_2, N_2-CO_2, air, O_2, etc.) and collecting of the $^{14}CO_2$ produced (Fig. 10). Construction of side arms in the flasks (sealed with gas-tight closures such as serum stoppers) is convenient for the addition of substrate. If two gas traps are used per culture vessel, connected via a three-way stopcock, the $^{14}CO_2$ can be collected at intervals during the experiment and used in kinetic analyses essentially as described by Wang (47) for pathway analysis via radiorespirometry.

Harvest an actively growing *E. coli* culture; wash and resuspend it in a fresh basal salts medium (18.4.2) with NH_4^+, NH_4^+ plus casein hydrolysate, or whatever N source is desired. The cell mass of 50 to 100 μg of cell dry weight per ml of culture at the start of the experiment is sufficient for rapidly growing cells like *E. coli*; a higher cell mass would be used for slower-growing organisms. Place a solution of [^{14}C]glucose (uniformly labeled at a specific activity in the range of 1×10^5 cpm/μmol of glucose or higher) in a side arm, and tip into the culture vessel (final concentration, 10 mM) to start the experiment. Turn on the gassing system, shake the culture (the entire unit can be mounted in and over a laboratory water-bath shaker with temperature control), and begin gas collection in the collection traps. For $^{14}CO_2$ collection use a 10-ml solution of monoethanolamine-absolute ethyl alcohol (1:2). At 20- to 30-min intervals, quantitatively transfer the trapping solutions from the columns (which can be constructed from nothing more than sintered glass-based chromatographic columns modified with addition of a gas entry port as indicated in Fig. 10), and dilute them to 20 ml with absolute ethyl alcohol. Samples are then removed and the radioactivity is determined (with appropriate quench corrections made) in a liquid scintillation counter (16.4.3). In this way an accounting of the total $^{14}CO_2$ resulting from the metabolism of the labeled glucose is obtained.

After 2 to 3 h of growth, H_2SO_4 is tipped into or injected into the culture to stop further reaction and release dissolved $^{14}CO_2$. The gassing is continued for 15 min, and units are then dismantled. The culture is then treated in the following manner.

Determination of unused substrate.

The concentration of glucose at the start and at the end of the growth period is determined by any standard procedure. The difference is the amount of glucose utilized by the culture and,

because the radio-specific activity of the glucose is known, one can calculate the percentage of the total counts per minute in the culture attributable to unused glucose. The remainder of the radioactivity must therefore be accountable as CO_2, nongaseous end products, and carbon assimilated into the cells.

Determination of assimilated ^{14}C.

Various dilutions of the above culture are filtered onto membrane filters (0.45-μm pore size) and washed, and the total radioactivity is determined and recorded as that carbon assimilated into cellular material.

A sample of the culture can also be removed (via a syringe) prior to termination of growth by the H_2SO_4 addition. This sample is centrifuged, washed with fresh medium, and recentrifuged. The cell pellet thus obtained can now be fractionated to determine the distribution of ^{14}C among the various cellular components by the procedure described by Roberts et al. (38). This procedure is useful for many studies, but modifications of the procedure are available and may be more suitable than the original technique in certain cases (see 17.1). The pellet is first suspended in 5 ml of 5% trichloroacetic acid and kept in an ice bath for 30 min. A sample of the total suspension is removed, and the total radioactivity in the cells is determined. The suspension is then centrifuged, and the supernatant fraction is removed.

The radioactivity in this supernatant fluid is determined as an indication of the **cold trichloroacetic acid-soluble fraction**, which represents low molecular materials. The pellet is resuspended in 5 ml of 75% ethanol, heated at 45°C for 30 min, and then centrifuged. The supernatant fluid is decanted, and a sample of it is removed to determine the radioactivity present in the **ethanol-soluble fraction**. The pellet is resuspended in 5 ml of 7.5% ethanol-diethyl ether (1:1), heated at 45°C for 30 min, and centrifuged. The supernatant fluid is removed, and the radioactivity of a sample is determined as the **alcohol-ether-soluble fraction**. These two solvent-soluble fractions contain the total cellular lipids and some alcohol-soluble proteins. The pellet is resuspended in 5% trichloroacetic acid, placed in a boiling-water bath for 30 min, and recentrifuged. The supernatant fluid is decanted, and a sample of this **trichloroacetic acid-soluble fraction** (the nucleic acids) is removed for determination of radioactivity. The **hot trichloroacetic acid-insoluble pellet fraction** (proteins and cell wall) is resuspended in 0.01 N NaOH, and a sample is counted. By using this procedure, one obtains an estimate not only of the percentage of the total radioactive substrate that was assimilated into the cells but also an indication of those particular macromolecular components into which this assimilation occurred.

Determination of nongaseous end products (also see parts 16.3.1 and 16.3.4).

Prior to termination of growth in the culture vessel, remove a 5-ml (or more) sample and add to 5 ml of a mixture of cold carrier compounds containing sufficient H_2SO_4 for a final pH of 1.5 to 1.8. This stops further reactions and converts any metabolic acids present to their acid form. The cold carrier mixture (for *E. coli* studies) contains the following components dissolved in basal growth medium: ethanol, acetic acid, pyruvic acid, formic acid, fumaric acid, lactic acid, succinic acid, and citric acid. Because of the small amounts of end products present in only 5 to 10 ml of culture fluid, the use of a cold carrier mixture is essential for obtaining quantitative recoveries of the radioactive end products produced by the cells. All carrier substances were present at 0.1 M concentrations except fumaric acid, which was at 0.05 M. The cold carrier-culture mixture is centrifuged to remove the cells, and the supernatant fluid is frozen until analyzed. Other cold carrier mixtures would be prepared for analysis of other organisms. Substances included would depend, of course, on the types of end products expected to accumulate in the culture fluid.

The nongaseous end products in 1 ml of this sample can be determined by silicic acid chromatography as follows. The 1-ml sample is mixed with 1.8 g of dry silicic acid (chromatography grade), and the mixture is quantitatively transferred to the top of prepared columns with 5 ml of benzene. The column is prepared by mixing 10 g of silicic acid and 6.5 ml of 0.5 N H_2SO_4 to a dry powder. Benzene is added to form a slurry which is poured into the column and rinsed with benzene. The column is packed under a pressure of 2 lb/m^2, making sure that the packing material does not run dry or separate from the column wall. (*CAUTION:* Benzene is classified as a weak carcinogen.)

The columns are developed by elution with 50 ml of chloroform, followed by 600 ml of a solvent gradient consisting of 300 ml of chloroform in the mixing chamber and 300 ml of 5% *t*-butanol in chloroform in the addition chamber of the gradient apparatus. After gradient elution is completed, an additional 50 ml of 5% *t*-butanol in chloroform is added, followed by 100 ml of 10% *t*-butanol in chloroform. All solvents are passed through the columns at a rate of 2 ml/min with the use of a solution metering pump.

All solvents are pretreated with 20 ml of 0.5 N H_2SO_4 (per liter of solvent) in a separatory funnel and then "dried" by filtering through two layers of dry filter paper (the aqueous layer is not allowed to go through the filter).

Fractions of 10 ml each can be collected with the use of a siphon assembly attachment to an automatic fraction collector. Appropriate samples are taken from each fraction and placed in scintillation vials; the radioactivity is determined in an appropriate counting mixture or fluid. Precautions are taken to correct for all quenching (chloroform quenches) that may occur (16.4.3). Known samples of ^{14}C-labeled ethanol, acetate, pyruvate, formate, lactate, and succinate are treated in the above manner to determine their elution sequence and recovery percentage. In all cases essentially 100% recovery is obtained if all corrections and controls are properly observed.

A typical pattern of nongaseous end products formed during anaerobic growth of *E. coli* ML 30 on [^{14}C]glucose determined as described above is shown in Fig. 11.

Inventory calculations and assessments.

By calculating the total (and subtotal) radioactivity associated with each of the steps described above, a number of inventory relationships can be calculated. From these relationships one can derive a general overview and assessment as to some of the pathways that may or may not be utilized by the organism under the growth conditions used in the study.

At the end of the growth period, all the original radioactivity (R added) must be accounted for as the sum of unused substrate (Su), gaseous end products (G), assimilated carbon (A), and nongaseous end products (E). If not, a technical error of some sort has occurred.

The total radioactivity present in the culture at the end of the growth period represents all but the gaseous carbon (R − G, or Su + A + E). If the cells are removed by centrifugation at the end of the experiment, the total radioactivity present in the supernatant fraction should represent only unused substrate (Su) plus nongaseous end products (E). If one calculates the counts present in this fraction as unused substrate (Su), the remainder must be accountable as nongaseous end products (E). This value (E) must be accounted for by the total of all substances recovered from the silicic acid column. If not, the presence of an end product not detectable under these conditions would be indicated.

Once the general inventory relationships such as these are satisfied and the nature of the carbon flow that occurred under these condi-

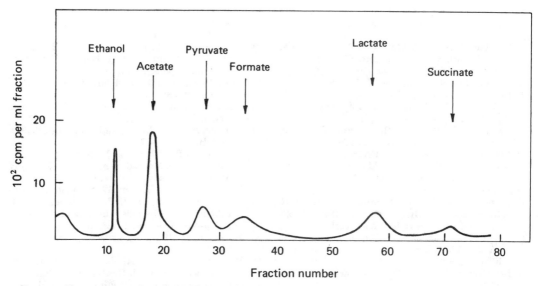

FIG. 11. *Nongaseous end products of anaerobic glucose metabolism by E. coli. End products resulting from the anaerobic degradation of [U-^{14}C]glucose are separated by silicic acid chromatography and determined by radioisotopic counting procedures.*

TABLE 3. *Key enzymes in main metabolic pathways*

Pathway	Key enzymes	Reference
Embden-Meyer-hof-Parnas (gly-colytic pathway)	Kinases, fructose-1,6-di-phosphate aldolase	
Hexose monophos-phate and pen-tose phosphate pathways	Glucose-6-phosphate de-hydrogenase, transaldo-lase, transketolase	50
Phosphoketolase pathway	Phosphoketolase	
Entner-Doudoroff pathway	2-Keto-3-deoxy-6-phos-phogluconate aldolase	
Direct oxidation pathways	Glucose and gluconate dehydrogenases	
Tricarboxylic (Krebs) acid cy-cle and glyoxal-ate cycle	Citrate synthase, citrate oxalacetate-lyase, acon-itate hydratase, isocit-rate dehydrogenase, ox-alsuccinate decarboxyl-ase, α-ketoglutarate de-hydrogenase, lipoyl re-ductase transsuccinyl-lase, dehydrolipsoyl de-hydrogenase, succinyl-coenzyme A-synthe-tase, succinyl-coenzyme A-hydrolase, fumarate hydrotase, malate de-hydrogenase, isocitrate lyase, malate synthase, the pyruvate dehydro-genase system	36
CO$_2$ metabolic pathway	Various carboxylases, de-carboxylases, etc.	39

ance of radioactivity in the hot trichloroacetic acid-insoluble cell fraction would indicate oper-ation of amino acid biosynthetic pathways and/or operation of cell wall biosynthetic reactions. The types of end products, gaseous and non-gaseous, formed and their ratios to each other can give clues to the types of bioenergetic path-ways utilized under the specific conditions of growth employed in the experiment. Performing these analyses under varied growth conditions (aerobic versus anaerobic atmospheres, rich me-dium versus poor medium, with and without various inhibitors, etc.) can lead to additional clues in this regard.

18.5.3. Key-Enzyme Occurrence

Another step, and indeed a necessary one, in establishing existence of a metabolic pathway is to demonstrate the occurrence of at least those enzymes that distinguish one pathway from an-other, i.e., to determine the presence of key enzymes in a pathway. For many classical path-ways of carbohydrate dissimilation, techniques are available for assays of the key enzymes in-volved. An excellent general description of such assays can be found in *Methods in Microbiology*. Specific detailed information needed to conduct such assays can be found in *Methods in Enzy-mology*. Key enzymes in main metabolic path-ways are listed and appropriate references are cited in Table 3.

18.6. LITERATURE CITED

1. **Arrigoni, D., and T. P. Singer.** 1962. Nature (London) **183:**1256–1258.

tions is assessed, some general statements con-cerning the bioenergetic and biosynthetic path-ways become possible. For example, the appear-

2. **Atkinson, D. E.** 1977. Cellular energy metabolism and its regulation. Academic Press, Inc., New York.
3. **Bender, R. A., K. A. Janssen, A. D. Resnick, M. Blumberg, F. Foor, and B. Magasanik.** 1977. J. Bacteriol. **129:**1001–1009.
4. **Buttin, G.** 1963. J. Mol. Biol. **7:**164–182.
5. **Clarke, P. H.** 1971 *In* J. R. Norris and D. W. Ribbons (ed.), Methods in microbiology, vol. 6A, p. 269–326. Academic Press, Inc., New York.
6. **Coakley, W. T., A. J. Bater, and D. Lloyd.** 1977. Adv. Microb. Physiol. **16:**279–341.
7. **Cota-Robles, E. H., and S. Stein.** *In* A. I. Laskin and H. A. Lechevalier (ed.), Handbook of microbiology, vol. 2, p. 833–843. CRC Press, Cleveland, Ohio.
8. **Dagley, S., and P. J. Chapman.** 1971. *In* J. R. Norris and D. W. Ribbons (ed.), vol. 6A, p. 269–326. Academic Press, Inc., New York.
9. **Decker, L. A. (ed.).** 1977. Worthington enzyme manual. Worthington Biochemicals Corp., Freehold, N.J.
10. **Dobrogosz, W. J.** 1966. J. Bacteriol. **91:**2263–2269.
11. **Dobrogosz, W. J.** 1968. J. Bacteriol. **95:**578–584.
12. **Dobrogosz, W. J.** 1968. J. Bacteriol. **95:**585–591.
13. **Dobrogosz, W. J.** 1969. J. Bacteriol. **97:**1083–1092.
14. **Eaton, N. R.** 1972. *In* J. R. Norris and D. W. Ribbons (ed.), Methods in microbiology, vol. 6B, p. 231–246. Academic Press, Inc., New York.
15. **Eisenberg, R. C., and W. J. Dobrogosz.** 1967. J. Bacteriol. **93:**941–949.
16. **Ferdinand, W.** 1976. The enzyme molecule. John Wiley & Sons, Inc., New York.
17. **Gachelin, G.** 1969. Biochem. Biophys. Res. Commun. **34:**382–387.
18. **Hatfield, G. W., and H. E. Umbarger.** 1971. Methods Enzymol. **17B:**561–566.
19. **Hess, B.** 1963. *In* H. H. Bergmeyer (ed.), Methods of enzymatic analysis, p. 43–55. Academic Press, Inc., New York.
20. **Hirata, M., M. Tokushige, A. Tnagaki, and O. Hayaiski.** 1965. J. Biol. Chem. **240:**1711–1717.
21. **Holzer, H., and W. Duntze.** 1971. Annu. Rev. Biochem. **40:**345–374.
22. **Hughes, D. E., J. W. T. Wimpenny, and D. Lloyd.** 1971. *In* J. R. Norris and D. W. Ribbons (ed.), Methods in microbiology, vol. 5B, p. 1–54. Academic Press, Inc., New York.
23. **Kaback, H. R.** 1971. Methods Enzymol. **22:**99–120.
24. **Kingdom, H. S., and E. R. Stadtman.** 1967. J. Bacteriol. **94:**949–957.
25. **Koch, A. L.** 1971. J. Mol. Biol. **59:**447–459.
26. **Leive, L.** 1968. J. Biol. Chem. **243:**2373–2380.
27. **Lowe, R. H., and H. J. Evans.** 1964. Biochim. Biophys. Acta **85:**377–389.
28. **Masters, C. J.** 1977. Curr. Top. Cell. Regul. **12:**75–105.
29. **McClure, W. R.** 1969. Biochemistry **8:**2782–2786.
30. **Mecke, D., K. Wulff, and H. Holzer.** 1966. Biochim. Biophys. Acta **128:**559–567.
31. **Neidhardt, F. C., and R. F. Boyd.** 1965. Cell biology: a laboratory text. Burgess Publishing Co., Minneapolis, Minn.
32. **Neu, H. C., and L. A. Heppel.** 1964. J. Biol. Chem. **239:**3893–3900.
33. **Neu, H. C., and L. A. Heppel.** 1965. J. Biol. Chem. **240:**3685–3692.
34. **Pardee, A. B., F. Jacob, and J. Monod.** 1959. J. Mol. Biol. **1:**165–178.
35. **Quayle, J. R.** 1972. *In* J. R. Norris and D. W. Ribbons (ed.), Methods in microbiology, vol. 6B, p. 157–183. Academic Press, Inc., New York.
36. **Reeves, H. C., R. Rabin, W. S. Wegener, and S. J. Ajl.** 1971. *In* J. R. Norris and D. W. Ribbons (ed.), Methods in microbiology, vol. 6A, p. 425–462. Academic Press, Inc., New York.
37. **Repaske, R.** 1956. Biochim. Biophys. Acta **22:**189–191.
38. **Roberts, R. B., P. H. Abelson, D. B. Cowie, E. T. Bolton, and R. J. Britten.** 1957. Studies of biosynthesis in *Escherichia coli*. Carnegie Inst. Washington Publ. 607.
39. **Scruton, M. C.** 1971. *In* J. R. Norris and D. W. Ribbons (ed.), Methods in microbiology, vol. 6A, p. 479–541. Academic Press, Inc., New York.
40. **Shapiro, B. M., and E. R. Stadtman.** 1970. Methods Enzymol. **17:**910–922.
41. **Smyrniotis, P. Z.** 1962. Methods Enzymol. **5:**344–347.
42. **Sols, A., R. E. Reeves, and C. Gancedo.** 1973. *In* E. H. Fischer, E. G. Krebs, H. Neurath, and E. R. Stadtman (ed.), Metabolic interconversion of enzymes, p. 393–399. Springer-Verlag, Berlin.
43. **Sols, A., and M. L. Salas.** 1966. Methods Enzymol. **9:**436–442.
44. **Stadtman, E. R., P. B. Chock, and S. P. Adler.** 1976. *In* S. Shaltiel (ed.), Metabolic interconversion of enzymes, p. 142–149. Springer-Verlag, Berlin.
45. **Stadtman, E. R., P. Z. Smyrniotis, J. N. Davis, and M. E. Wittenberger.** 1979. Anal. Biochem. **95:**275–285.
46. **Umbarger, H. E.** 1974. *In* A. I. Laskin and H. A. Lechevalier (ed.), Handbook of microbiology, vol. 4, p. 35–42. CRC Press, Cleveland, Ohio.
47. **Wang, C. H.** 1972. *In* J. R. Norris and D. W. Ribbons (ed.), Methods in microbiology, vol. 6B, p. 185–230. Academic Press, Inc., New York.
48. **White, R. J.** 1968. Biochem. J. **106:**847–858.
49. **Wilson, D. M., T. F. Alderte, P. C. Maloney, and T. H. Wilson.** 1976. J. Bacteriol. **126:**327–337.
50. **Wood, W. A.** 1971. *In* J. R. Norris and D. W. Ribbons (ed.), Methods in microbiology, vol. 6A, p. 411–424. Academic Press, Inc., New York.
51. **Yamanka, K., and W. A. Wood.** 1966. Methods Enzymol. **9:**596–602.

Chapter 19

Permeability and Transport

R. E. MARQUIS

Although the molecular details of solute passage across the permeability barriers of bacterial cells are not really known, still it is worthwhile to distinguish two general types of processes for entry and exit:

Permeation denotes passive movement of a solute into (uptake) or out of (efflux) a cell by diffusion, which may be facilitated by chemical interaction between the solute and some cell structure, e.g., a membrane. A cell may be permeable or impermeable to a particular solute. **Permeability** is a property of the cell and not of the solute. The following equation applies to permeation of a solute through a membrane of a cell:

$$dS/dt = PA(\Delta C)$$

where dS/dt is the change in amount of internal solute per unit of time, P is the permeability coefficient (which has units of distance divided by time), A is the area of the membrane, and ΔC is the difference in solute concentration between the interior and the exterior of the cell. If a plot is made of the internal solute concentration (C_{in}) as a function of time, the resulting curve is parabolic with a plateau at $C_{in} = C_{out}$, that is, when the internal concentration equals the external concentration or, more exactly,

when the internal activity of the solute equals the external activity. For such a curve, $C_{in} = $ (maximum C_{in}) $(1 - e^{-kt})$, where k is a constant and t is the time.

Transport denotes active movement of a solute into or out of a cell by a process which is coupled to metabolism and may involve energized carriers. Transport is a property of either the cell or the solute. The two most thoroughly studied classes of transport systems in bacteria are phosphotransferase systems and so-called permease systems (15, 34). The phosphotransferase system occurs primarily in facultatively anaerobic bacteria for transport of sugars (39), while essentially all bacteria have permease systems for transport of amino acids, inorganic ions, nucleic acid precursors, and sugars.

A curve relating solute uptake to time for a transport process does not differ qualitatively from one for a permeation process. However, transportive uptake does not stop when $C_{in} = C_{out}$ but continues until a substantial concentration gradient has developed, and the process becomes one of **concentrative transport**. Maximum uptake is then determined by the rates of the inward and outward processes.

As indicated, transport may result in movement of solutes into or out of cells. For many solutes, a transport process of so-called **exchange diffusion** occurs in which there is a one-for-one exchange of internal and external solutes across the membrane. This exchange is generally assessed by use of radioactively labeled solutes; details of the technique are presented by Maloney et al. (3).

Concentrative solute uptake does not require a membrane and transport catalysts. Ion-exchange resins, for example, can take up and concentrate ions. Most of the organic components of a cell are not in aqueous solution but are in solid phases in the cell wall, membranes, ribosomes, the nuclear body, etc., and so may take up solutes in the same manner as a resin. This aspect of solute transport and permeation has been reviewed by Damadian (12).

19.1. PERMEABILITY

19.1.1. Assay Procedure

The most commonly used technique for assessing solute permeation into bacterial cells or cell structures is the space or thick-suspension technique described by Conway and Downey (11) and modified by Mitchell and Moyle (32), by MacDonald and Gerhardt (26), and by others. Large masses of cells are commonly used for the assay—generally 3 g or more (wet weight) per assay tube. Obtaining this quantity of cells can be a problem, but not a great one when common laboratory bacteria are used. Techniques which require fewer cells and are based on use of membrane filters or a minicentrifuge are described in the part on transport (19.2). In any case, careful attention should be paid to details of growth and harvest since the permeability of bacteria may change greatly during the culture cycle in batch cultures or as a result of washing procedures used to rid the cells of medium constituents. There is often advantage in using cells obtained from continuous cultures. Changes in growth medium, growth temperature, pH, etc., can also be expected to alter permeability.

The space technique is designed specifically to give an estimate of the fractional space (volume) of cells that can be penetrated by the solute being tested. The technique is based simply on determination of the degree of dilution of a solution containing the solute due to mixing the solution with a pellet of centrifuged cells. The following information is needed: the initial volume and concentration of the test solution, the volume of the cell pellet, and the final concentration of the solute in the extracellular phase of the mixture of pellet and solution. Cell pellets contain not only cells but also interstitial (intercellular) space. The fractional volume of

the interstitial space is generally estimated by use of polymeric solutes of high molecular weight which can penetrate the interstitium but not the cell.

The stepwise procedure for the space technique is as follows:

1. Washing of cells.

After harvest, it is generally necessary to wash the cells before they can be used for permeability assays. Usually, a single wash with a large volume of fluid is sufficient to nearly rid the cells of medium constituents but not to deplete them excessively of internal solutes. Often two washings are done, but there is little justification for a third wash. Many bacteria, especially gram-positive ones, are hardy and can be washed with water or dilute buffers without great harm, especially if the wash fluids are chilled. However, other bacteria, particularly gram-negative types, are much more sensitive and can be damaged in major ways by washing with water. Moreover, with gram-negative bacteria and some gram-positive ones, it is often best to use warm wash fluids instead of chilled ones. Commonly used wash solutions for gram-negative bacteria contain magnesium ion (usually 50 mM), a buffer [e.g., 50 mM phosphate buffer or tris(hydroxymethyl)aminomethane (Tris) buffer], and an osmotic stabilizer (e.g., 0.4 M NaCl or 0.5 M sucrose). Maloney et al. (3) recommend using growth medium 63 for washing *Escherichia coli* cells, but with deletion of the carbon source. With any previously untested bacterium, it is worthwhile to try a variety of wash solutions to find the best one. The criterion for efficacy is that the cells should lose medium constituents but retain internal pools of potassium, amino acids, or inorganic phosphate. No wash solution is perfect, and generally some compromise must be accepted. When dense cell suspensions are used for permeability assays, the composition of the wash (and suspension) fluid is not as critically important as when dilute suspensions are used for transport assays. Also see 10.4.2 and 25.1.

2. Centrifugation of cells.

Once the cells have been washed, they must be centrifuged to obtain a workably tight pellet. The centrifugation schedule varies depending on the particular organism. A packing curve of the type shown in Fig. 1 is generally prepared by plotting pellet weight against centrifugation time, using a series of cell suspensions to obtain the curve. The centrifugation time chosen for assays should be well out into the plateau region so that minor variations in centrifugation time

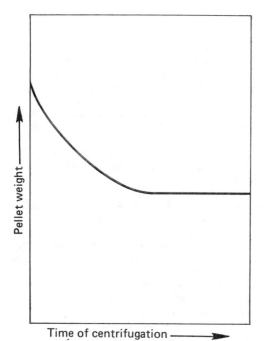

FIG. 1. *Typical packing curve for a cell pellet.*

do not affect the pellet weight. The final centrifugation before solute addition is generally done with a tared centrifuge tube so that the pellet weight can be obtained by difference. Tubes with nominal 50-ml capacity are best for pellets of about 10 g (wet weight) and 15-ml tubes are best for pellets of about 3 g. Weighing to within 0.01 g is sufficiently accurate for routine assays. Prior to weighing, it is best to decant the supernatant fluid carefully, to invert the tubes for maximum and uniform draining, and to wipe excess moisture from the inside of each tube with absorbent tissue. However, attempts to blot surface water from the pellet are best avoided because cells may be removed with the water.

3. Measurement of pellet weight and volume.

Pellet wet weight is usually an adequate indicator of pellet volume and may be used instead of volume in calculations, since the wet densities of bacterial vegetative cells are generally about 1.02 to 1.04 g/ml. However, bacterial spores have significantly higher wet densities, as high as 1.22 g/ml for *Bacillus stearothermophilus* spores, and pellet weight is not a good substitute for pellet volume. Certain inclusion granules (e.g., poly-β-hydroxybutyrate) of vegetative cells also have relatively high wet densities and may significantly increase the difference between pellet weight and volume.

Pellet volumes can be estimated readily by use of a standard pycnometer or a volumetric flask. For example, suppose one has a cell pellet that weighs about 3 g and wishes to know its volume. The pellet can be mixed with water and transferred to a 50-ml, tared pycnometer vessel or volumetric flask, and the mixture is made up to volume with water. Suppose that the weight of the suspension determined with an analytical balance is found to be 50.110 g (weight of mixture plus pycnometer minus weight of pycnometer) and the pycnometer has a calibrated volume of exactly 50.000 ml at 25°C, the temperature of weighing. Suppose also that the pellet has been found with the analytical balance to weigh exactly 3.000 g. Then the weight of water added to the pellet must be 50.110 − 3.000, or 47.110 g. At 25°C, this weight of water would occupy a volume of 47.249 ml, based on the water density at that temperature obtained from tables in a standard chemical handbook. Therefore, the pellet volume must be 50.000 − 47.249, or 2.752 ml. The density of the pellet is then 3.000 g/ 2.752 ml, or 1.090 g/ml. Once density is known, and if standard centrifugation conditions are used, it is possible readily to calculate pellet volume from pellet weight.

Pellet volume can also be measured directly in a tared centrifuge tube closed with a tared microscopy cover glass (10). The open ends of glass tubes need to be ground flat, but polycarbonate tubes are manufactured with flat ends. The total volume in the tube is obtained by filling it completely with distilled water and closing it with the cover glass, taking care to exclude air bubbles and to wipe excess water from the external surfaces. The weight of water in the tube is measured and converted to volume with the density of water at the ambient temperature.

Pellet volumes also have been estimated directly by use of calibrated centrifuge tubes (14) which yield cytocrit data equivalent to the hematocrit data obtained routinely in hematology; this cytocrit method can give useful volume estimates, but it has difficulties in practice.

4. Mixing and incubation of cells with solute.

The cell pellet to be used for a permeability assay is brought to the desired experimental temperature and mixed with a solution containing the test solute by use of glass or Teflon-coated stirring rods, which do not absorb fluid. The volume of solution used should be about equal to the pellet volume. Basically, one wants to have sufficient dilution of the solution so that concentration differences are readily determined. After mixing of pellet and solution, the suspension should be incubated for a set time at the experimental temperature to allow for diffusional equilibration, generally 15 min to 1 h. However, it is best to determine that equilibrium has in fact been reached during the incubation period by use of multiple pellets and incubation times.

5. Recentrifugation.

After each mixture of pellet and test solution has been incubated, it should be centrifuged sufficiently to obtain a cell-free supernatant fluid for assay. The supernatant fluid may require an additional, clarifying centrifugation.

6. Determination of solute concentrations.

The concentrations of solute must be determined both in the initial solution that was mixed with the cell pellet and in the final solution that was obtained after the recentrifugation. These measurements generally can be made on small samples by means of gravimetric or refractometric techniques. However, these techniques are not specific, and other substances purposely added to or leached from the cells may interfere. If so, a specific chemical or radioactive determination must be used for the test solute.

19.1.2. Determination of Interstitial Space

It is necessary to know the interstitial volume of the cell pellet in order to be able to calculate the volume of the cells themselves available to the test solute. Consequently, one pellet in each set should be mixed with a solution of a large polymeric preparation in which all of the component molecules are too large to penetrate through cell walls and move only into the interstitium. A dextran preparation with an average molecular weight of 2,000,000 (Dextran T2000, Pharmacia Fine Chemicals, Inc., Piscataway, N.J.) best serves this purpose. Dextrans are polydisperse, but even the smallest molecules in the distribution of this large a dextran preparation do not penetrate even relatively porous bacterial cell walls (41).

Dextran concentrations generally can be determined gravimetrically, by means of refractometry, or chemically by the phenol-sulfuric acid method (13). However, these assays are not specific for dextrans, and other substances purposely added to or leached from the cells may interfere. It would be convenient to use radioactively labeled dextrans, but the largest ones commercially available (e.g., from New England Nuclear Corp.) have average molecular weights of only 70,000. The smallest molecules in this dextran preparation are sufficiently large to be excluded by the walls of some bacteria but not others. Therefore, they may or may not be useful, and before using them for a particular bacterium, it is necessary to determine the dextran exclusion threshold. Of course, it would be possible to label larger dextran molecules with [14C]cyanide by conventional chemical techniques to obtain carboxyl-dextran. Blue dextrans (Pharmacia) have been used for colorimetric assays, but bacteria can become stained after contact with blue dextran because of transfer of the dye.

A number of other polymeric molecules have been used instead of dextrans. Proteins (such as serum albumin) have an advantage in being essentially monodisperse and specifically assayable. However, they have a disadvantage in that they may bind to cells, and this binding would lead to anomalously high uptake values. Inulin also has been used, but it is capable of penetrating the cell walls of many bacteria, especially gram-positive ones (41).

Yet another problem in determining interstitial volumes arises from the swelling-shrinking responses of cells to many test solutes, especially to salts. It may be necessary to assess interstitial volume in the presence of such a test solute in addition to the dextran. If the additional solute is of small molecular size (say, NaCl or KCl), it is possible to dialyze the supernatant fluid with retention of the large dextran molecules within the dialysis sac and loss of the small solutes. Then all of the dextran can be recovered from the dialysis sac for assay.

19.1.3. Expression of Results

The most commonly used permeability index, at least in bacteriology, is the space value or S value, which can be calculated by use of the formula

$$S = \left(\frac{V_s}{V_p}\right)\left(\frac{C_0}{C_f} - 1\right)$$

where V_s is the volume of the solution added to the pellet, V_p is the volume of the pellet (commonly, the wet weight of the pellet, W_p, is used in place of V_p), C_0 is the initial concentration of solute, and C_f is the final concentration. **S values** indicate the fraction of the cell pellet that is penetrated by the solute.

However, the desired information is the fraction of the actual cell volume rather than that of the pellet volume; this fraction is called the **R value**. It is calculated by use of the formula

$$R = (S_{sol} - S_{dex})/(1 - S_{dex})$$

where S_{sol} is the space value for the solute under

study, and S_{dex} is the dextran or interstitial space. A zero value for R indicates that the cells are impermeable to the solute. Values from zero to one indicate various degrees of penetration. Values greater than one indicate that the cells have concentrated the solute.

When pellet weight is used as a measure of pellet volume, a superscript w is often used with the symbols S and R. Also, it is often desirable to express results with units of cell dry weight or cell water as the base, e.g., the sucrose-permeable volume per gram of cell dry weight or per milliliter of cell water.

19.1.4. Determination of Cell Volume

Useful cytological information can be gained from space values. For example, the technique offers by far the most accurate means available for determining the average cell size of a bacterium. The pellet volume minus the interstitial volume is equal to the volume of the cells. If the pellet is then resuspended in water to some known volume (say, 50 ml) in a volumetric flask, the number of cells can be assessed by use of a Petroff-Hausser counter, and the average volume of a cell can be calculated. The technique also can be used to determine the volume of the protoplast in a cell (19.3.2).

19.1.5. Determination of Cell Water Content (also see part 25.2)

The resuspended cells can also be used for a dry weight determination from which one can estimate the dry weight per cell. The amount of cell water is equal to the difference between the wet weight of the cell and the dry weight. Since the weight or volume of the interstitial water is known, it is possible to estimate cell water from information about the amount of pellet water. Sometimes it is convenient to estimate the amount of cell water directly from the R value for solutes such as ^2H_2O and ^{14}C-labeled urea or glycerol, which readily permeate, but are not concentrated by, resting cells. Tritiated water is used sometimes also to assess the amount of cell water, but one should be aware that tritium may exchange rapidly with hydrogen in various cell polymers and monomers, especially with ionizable components.

19.1.6. General Considerations

The major advantages of the space technique include its simplicity and reliability. However, it is not very useful for kinetic measurements because of the relatively long periods of time required to mix the cells and test solutions and then to separate them. Thus, the technique is good for equilibrium but not rate determina-

tions. Because of the very thick suspensions used, there are generally no problems due to excessive leaching of physiologically important solutes (such as potassium ion) from cells. However, the thickness of the suspensions makes it difficult to control pH or to provide adequate aeration for aerobic organisms. Thus, the technique is most useful for resting cells and much less useful for actively metabolizing ones. As mentioned previously, there is often a problem in obtaining the large masses of cells required. However, the accuracy of the technique depends on having large quantities, especially since solute uptake is estimated from determinations of the change in concentration of the suspending medium.

19.2. TRANSPORT

19.2.1. Assay Procedure

1. Preparation of cells.

For studies of the rates of solute transport, it is necessary to have some means rapidly to separate cells from their suspending medium. This separation is generally done by filtration through membrane filters and, therefore, it is necessary to use rather dilute suspensions, generally with less than about 200 μg (dry weight) per ml if 1-ml samples are to be filtered through filters 25 mm in diameter.

As with permeation assays, it is necessary for transport assays to pay close attention to controlling the conditions of cell growth and harvest. Wash and suspension media tend to be more critical in preparing adequate cell suspensions for transport assays because cells in dilute suspensions are more likely to be harmed by unfavorable environments than are cells in dense suspensions, in which the environment contains high levels of materials leached from the cells.

For transport studies, it may be necessary to pretreat the cells. For example, they may be starved to deplete endogenous pools of amino acids. This starvation may be of particular advantage when radioactively labeled solutes are added exogenously, as it reduces changes in specific activity resulting from mixing with endogenous unlabeled solute. However, starvation may also deplete cells of adenosine triphosphate or phosphoenolpyruvate or may diminish the proton motive force across the membrane. Therefore, some fuel for the transport process may have to be added along with the test solute.

2. Mixing and incubation of cells and solute.

The cell suspension and the test solution are brought to the experimental temperature separately and then mixed together at zero time. Rapid mixing is generally required because of the rapidity of most transport reactions, and the incubation of the reaction mixture is generally of short duration (e.g., 10 min).

3. Sampling and separation.

At short intervals (e.g., 1 min), samples of the reaction mixture are withdrawn. The cells and the test solution are rapidly separated, generally by filtration through a membrane filter with 0.45-μm-diameter pores, which should be pre-wetted with wash solution or test solution. Lower porosity filters may be used, but they offer more resistance to fluid flow and slow down the filtration process. The 0.45-μm filters are adequate for filtration of most microbial cells. Generally, filters of cellulose acetate and cellulose nitrate are used, especially if scintillation counting is planned. Vacuum is developed within a filter flask by use of line vacuum in the laboratory or, if that is not adequate, with a vacuum pump. The common clip-on funnel attachment for the filter unit may be replaced with a custom-made heavy brass ring to hold down the filter disk. The ring can then be rapidly placed and removed to allow for repeated use of a single filter unit. Also, it is possible to purchase or manufacture manifold filter units for very rapid sampling.

There may be advantage to mixing ingredients in a syringe with a Swinney filter attachment, from which samples of cell-free medium can be expressed. The obvious disadvantage of this method is that one must depend on determination of differences in solute concentration in the suspending medium when estimating uptake. Moreover, the sample taking results in progressively increased cell concentration in the remaining mixture.

In some instances, it is possible to use centrifugation for separating the cells, even when rate data are required. Microcentrifuges, which accommodate small tubes of 250- to 1,500-μl capacity, can attain speeds of 12,000 $\times$ g in only a few seconds. Levin and Freese (25) described the use of a so-called minifuge technique to assay solute uptake by *Bacillus subtilis*.

Also, it is often possible to stop transport reactions by adding samples directly to ice or iced buffer. Centrifugation is then carried out in the cold. For this procedure to be effective, chilling must stop influx but not induce efflux. Satisfactory results are more likely to be obtained with gram-positive than with gram-negative bacteria.

In some cases, it is not necessary to separate the cells from the test solution. For example, ion-specific electrodes allow for continuous monitoring of pH, partial pressure of oxygen (pO_2), potassium, or other ions (16.2). Moreover, in studies of salt uptake, it is often necessary only to follow changes in conductivity of the suspending medium with a conventional conductivity cell.

4. Washing of cells.

The cells that are retained on the filter must generally be washed to remove adherent and interstitial medium. For gram-negative bacteria especially, the wash liquid should be osmotically buffered because the transport systems are osmotically sensitive. Concentrated sugar solutions are often used for washing (e.g., 0.5 M sucrose solution). Magnesium ions (usually 50 mM) may also be added as the chloride or sulfate salt. An ideal wash solution should remove test solute from the cell pad and filter primarily in the first wash, but not in subsequent washes. Some compromise must generally be tolerated (see 19.1.1 for additional information).

It is possible sometimes to avoid the problems of washing cells by centrifuging them through a layer of liquid silicones, as described by Hurwitz et al. (17). The cells pass through the silicone to form a pellet that does not contain interstitial water. Some adherent water may accompany the cells, and the cell wall water will remain undisturbed.

5. Removal and assay of cells for solute content.

The cells may be eluted from the filter, resuspended to a known volume, and assayed for content of the test solute by a chemical (Chapter 17) or radiometric method (16.4).

When radioactive solutes are used, it is more common to dry the filters and then place them directly into scintillation fluid for counting. It is possible also to estimate the amount of solute incorporated into polymeric materials by treating the cells on the filter with cold trichloroacetic acid solution (17.1) and then washing them with water before drying. There is advantage in using filters which dissolve in the scintillation fluid or, alternatively, which dissolve in a solvent that is miscible with the scintillation fluid. Commercial brochures or articles on scintillation counting should be consulted for advice.

Membrane filters may bind test solutes, especially ionized ones, and it may be necessary to correct for this binding. An estimate of the extent of binding can be obtained by passing some of the test solution without cells through a filter, washing in the same way that cells will be washed, and then assaying the filter for test solute.

19.2.2. Expression of Results

Transport uptake or release is generally expressed in terms of base units of cell weight, cell volume, or cell water. Transport results may also be expressed on a per cell basis, and such values are useful in comparing transport capacities of various cells. The most commonly used base for expressing the results is the cell water.

It is of value also to calculate concentration ratios, i.e., the concentration of the test solute within the cell divided by its concentration in the suspending medium, so that assessment can be made of whether or not concentrative transport has occurred.

When using radioactively labeled solutes, it is highly desirable to convert data from counts per minute to molar units.

Transport processes are generally considered to be a subclass of enzyme-catalyzed reactions, and the principles and techniques of enzyme kinetics can be applied to transport studies. For example, Lineweaver-Burk double-reciprocal plots of the inverse of transport rate versus the inverse of solute concentration commonly yield straight lines from which one can calculate values for K_m (Michaelis constant) and V_{max} (maximum velocity). For a specific example, see the paper by Ames (1). Alternatively, one can prepare Eadie-Hofstee plots of the velocity of transport versus the velocity divided by substrate concentration. The complications of kinetics that arise in the study of enzyme kinetics arise also in the study of transport.

It is not uncommon for cells to have more than one transport system for a particular solute. For example, many bacteria have a high-affinity transport system that is highly specific for an individual aromatic amino acid as well as a low-affinity system that is much less specific. Methods for assessing these multiple systems are described by Ames (1).

In addition, many solutes can enter cells passively by diffusion (permeation) as well as actively by transport. The diffusion component generally does not show saturation at high substrate concentrations, and its extent can be estimated from plots such as that shown in Fig. 2. With intact bacteria, the diffusion component may include diffusion of solute into the cell wall water (29), which can account for some 50% of the cell volume in organisms such as *Micrococcus lysodeikticus* (*M. luteus*).

19.2.3. General Considerations

A major advantage to the use of dilute suspensions of cells is the ease with which kinetic data can be obtained. Also, with dilute suspen-

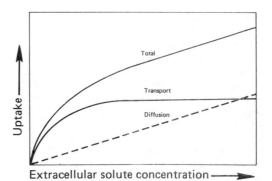

FIG. 2. *Graphic estimation of the diffusion component in the total uptake of a solute by bacterial cells. The transport uptake is the total minus the passive uptake. The uptake parameter may be either the rate or the extent of uptake.*

sions, it is possible more readily to control pO_2, pH, ionic strength, etc., and to maintain nearly constant solute concentration in the suspending medium.

The disadvantages include problems associated with solute leaching from cells during initial suspension or washing, the general need to assess solute uptake or loss from changes in intracellular rather than extracellular concentration, and the inherent vagaries in the use of dilute suspensions in which the cells may be highly sensitive to changes in environmental conditions.

19.3. SPECIAL METHODS

19.3.1. Proton Motive Force

The proton motive force is considered to be a major means for energy conversion and transfer in biological systems (31). It is defined as

$$\Delta p = \Delta \psi - 2.3 \frac{RT}{F} (\Delta pH)$$

where Δp is the proton motive force, $\Delta \psi$ is the membrane potential, R is the gas constant (8.318 volt coulombs/°K equivalent), T is the Kelvin temperature, F is the faraday (96,500 C/equivalent), and ΔpH is the difference in pH between the interior and the exterior of the cell. At about 25°C, the value of 2.3 RT/F (which is often given the symbol z) is approximately 59 mV. Thus, in order to estimate Δp, it is necessary separately to estimate $\Delta \psi$ and ΔpH.

Techniques for estimating $\Delta \psi$ generally involve measurements of the distribution of some ion to which the cell membrane is freely permeable. For example, Harold and Altendorf (16) suggested that chloride distributions could be used as a basis for calculating $\Delta\psi$, which would be equal to $2.3\,RT/F[\log(\text{Cl}_{out})/(\text{Cl}_{in})]$, where (Cl_{out}) and (Cl_{in}) refer to chloride concentrations outside and inside of the cell. These concentrations can be assessed by means of the space technique or variations of it. Actually, chloride activities should be used instead of concentrations, but it is impossible to determine the chloride activity coefficient within the cell. It is probably only slightly below 1.0 so that activity and concentration are not greatly different. Chloride activities can be assessed with the chloride electrode (Orion Research Inc., Cambridge, Mass.), and chloride concentrations can be measured chemically by the method of Schales and Schales (40).

Harold and Altendorf (16) say that a useful indicator for $\Delta\psi$ must "diffuse rapidly across the membrane . . . be fully dissociated by physiological pH, metabolically innocuous and not subject to translocation by a biological transport system." These criteria can be met by K^+ in cells that have been treated with 1 to 10 μM valinomycin to render the membrane permeable to K^+. Since the membrane potential in bacterial cells is generally about 180 mV with the interior negative, the potassium concentration in the cytoplasm of valinomycin-treated cells is some 20 times the concentration in the suspending medium. Potassium concentrations can be determined by means of atomic absorption spectrophotometry, by flame photometry, or, less accurately, by use of an ion-specific electrode (16.2.2). The uptake of K^+ is determined by the space technique, or a variation of it, and the concentration is calculated in terms of moles of K^+ taken up per unit of cell water.

$\Delta\psi$ can be assessed also by determining the distribution of permeant organic cations or by use of fluorescent dyes. Details are given in the article by Maloney et al. (3). The use of thiocyanate as a permeant anion has become popular, especially since the ^{14}C-labeled compound is available. An example of its use is described in the paper by Sorgato et al. (42).

For determination of ΔpH, a weak acid or base to which the cell is permeable is used. The compound chosen must dissociate at physiological pH values so that its distribution between the cell and its environment will depend on the pH difference between the two phases. The membrane of the cell should then be permeable to the undissociated form but impermeable to the ionized form. Compounds that are commonly used for ΔpH estimates include acetate, dimethyl sulfoxide, methylamine, and salicylate. Considering the Henderson-Hasselbach equation for a weak acid, let A stand for the anionic

form, let HA stand for the undissociated form, and use the subscripts i and o to refer to concentrations inside and outside of the cell. Therefore

$$pH_i = pK_i + \log[A_i]/[HA_i]$$
$$pH_o = pK_o + \log[A_o]/[HA_o]$$

It is generally assumed that $pK_i = pK_o$ and, since the membrane is permeable to the undissociated form, $[HA_i] = [HA_o]$. Therefore

$$pH_i - pH_o = \log[A_i]/[HA_o] - \log[A_o]/[HA_o]$$
$$= \log[A_i]/[HA_i] + \log[HA_o]/[A_o]$$
$$= \log\{[A_i]/[HA_i] \times [HA_o]/[A_o]\}$$
$$\Delta pH = \log[A_i]/[A_o]$$

The ΔpH value can then be assessed from estimates of the distribution of, say, acetate between cells and suspending medium by the methods described above for permeability assays. From the external acetate concentration and pH, one can calculate the external concentration of HA, which is equal to the internal concentration. The internal concentration of A is equal to the total internal acetate concentration minus the HA concentration. From the ratio $[A_i]/[HA_i]$ and a knowledge of pK_i, an estimate of the internal pH can be derived with the Henderson-Hasselbach equation.

Again, the space technique can be used to estimate the uptake of the pH-probing compounds. It is possible also to use techniques that do not require large quantities of cells if one uses some nonpermeant compound to estimate noncytoplasmic water. For example, on page 29 of the article by Maloney et al. (3) an example is presented of an experiment in which [^{3}H]sorbitol is used to estimate the noncytoplasmic water in a suspension of Streptococcus lactis cells and [^{14}C]dimethyl sulfoxide is used as a pH probe. One should realize that ΔpH and $\Delta\psi$ determinations are only estimates and that there are problems associated with specific and nonspecific binding of charged species to biopolymers. However, one is often most interested in changes in ΔpH or $\Delta\psi$ rather than in their absolute magnitudes.

A flow dialysis method that does not require separation of the cells and the suspending medium has been developed and described in detail by Ramos et al. (36) and Ramos and Kaback (35). It is particularly useful for work with membrane vesicles.

Recently, methods have been developed (33) for estimation of the intracellular pH of bacterial cells by use of nuclear magnetic resonance techniques. pH_i has also been assessed by using fluorescent dyes such as pyramine (20). If one knows pH_i, pH_o is readily determined with a glass electrode, and ΔpH can be calculated.

19.3.2. Osmotically Sensitive Cells

When a solute is taken up by a cell, the process results in a lowering of the water activity in the interior and a consequent influx of water, with resultant swelling of the cell. Biological membranes are highly permeable to water, and the swelling reaction is an extremely rapid one. For example, Matts and Knowles (30) used a stopped-flow spectrophotometer to show that the osmotic movement of water from E. coli cells transferred into a 0.4 M MgCl₂ solution is complete within 50 ms at 37°C. Since water movement is so rapid, the net rate of swelling that accompanies solute uptake can be related directly to transport. However, there are some cautions that must be considered, especially for bacterial cells with cell walls that are sufficiently elastic to resist swelling. (Yes, it is their actual elasticity rather than supposed rigidity which resists swelling. The physical properties of bacterial cell walls have been reviewed by Rogers [38].)

The swelling and shrinking of an ideal osmometer can be predicted by use of the van't Hoff-Boyle equation

$$V - b = a/\Pi$$

where V is the total volume of the cell, b is the so-called osmotically dead space (which generally is approximately equal to the calculated volume of the dry matter of the cell), a is a constant (the number of osmoles in the cell), and Π is the osmolality of the suspending medium. For an ideal osmometer, a plot of V against $1/\Pi$ yields a straight line with intercept b and slope a, while the actual behavior of bacterial cells yields a curved line (Fig. 3) Clearly, bacteria are not truly ideal osmometers.

Osmotic responsiveness is markedly reduced in media of high osmolality, and one can obtain an experimentally determined value for b only by extrapolation. It is not known just why the cell behaves anomalously in concentrated media. The behavior could have to do with bound water and gelling of the cytoplasm or with decreased water permeability.

When placed in dilute media, bacterial cells are resistant to osmotic swelling primarily because they have an elastic cell wall. When placed in concentrated media, bacterial cells become plasmolyzed; that is, the protoplast shrinks so that the membrane is pulled away from the wall and plasmolysis vacuoles are formed between the wall and the protoplast membrane. These vacuoles can often be seen microscopically, es-

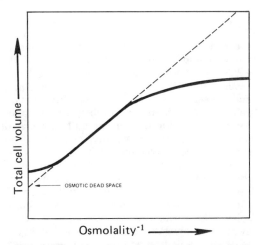

FIG. 3. *Osmotic volume changes for bacterial cells. The dashed line indicates the ideal behavior predicted by the van't Hoff-Boyle equation. The solid curve indicates the actual behavior.*

pecially in gram-negative bacteria. However, even if they cannot be seen, their presence can be detected by permeability determinations. Thus, one can assess the total cell volume in terms of the dextran-impermeable space. The protoplast volume is essentially equal to the raffinose-impermeable volume under conditions in which raffinose does not pass through the protoplast membrane. The aqueous volume in the cell wall and the plasmolysis vacuoles together is then equal to the dextran-impermeable volume minus the raffinose-impermeable volume. A reasonable estimate of the aqueous wall volume alone can be made in terms of the raffinose-permeable space of turgid nonplasmolyzed cells.

If the plasmolyzed cells are now returned to more dilute media, water flows into the protoplast across the water-permeable membrane, the protoplast swells to occupy the volume previously occupied by the plasmolysis vacuoles, and the optical density of the suspension increases. There is a direct relationship between the protoplast volume and the inverse of optical density (27). Initially, when the protoplast swells, there is little resistance to swelling, and the process is nearly ideal; i.e., it follows the van't Hoff-Boyle relationship. However, when the protoplast abuts against the cell wall, the elasticity of the wall resists further swelling, and wide deviation from ideality occurs. Now water uptake involves an increase not only in volume of the protoplast but also in that of the whole cell, with stretching of the wall. A break in the curve for V versus $1/\Pi$ occurs at what is called the point of incipient plasmolysis. Further reductions in medium osmolality result in the buildup of turgor pressure. It is considered that the medium osmolality at the point of incipient plasmolysis is approximately equal to the internal osmolality of the cell, as discussed primarily for plant cells by Stadelmann (6). It is possible also to obtain another estimate of internal osmolality from the value of the constant a (number of ideal molar equivalents within the cell) and a knowledge of protoplast volume. Generally, estimates from the two sources agree reasonably well.

The elasticity of the cell wall does not affect the swelling of isolated protoplasts, which have been used often for transport studies, and is thought to contribute little to the swelling of spheroplasts, which retain wall remnants. However, if protoplasts are made to swell slowly so that their membranes do not undergo brittle fracture, the membrane can become taut and the magnitude of swelling is not predictable from the van't Hoff-Boyle equation (27).

As mentioned, the inverse of optical density can be used as an indicator of cell volume and, therefore, of solute uptake, although of course one should always be aware that it is actually water movement that is being monitored. It is important also to know that optical density depends on the difference in refractive index between the cell and the suspending medium and that corrections may have to be made for changes in medium refractive index due to increased or decreased solute concentration. Ideally, one would like to have suspending media of varying solute concentration but constant refractive index. In many studies of solute uptake, the external concentration does not change greatly, and so no corrections are needed. However, in studies of osmotic volume changes, concentrations generally do vary widely. If salt solutions are used, the changes in refractive index due to changes in solute concentration are smaller than when sugar solutions are used. However, in either case, turbidity values may have to be corrected for changes in refractive index. It is useful to prepare matching solutions of polymeric dextrans that have high refractive increments but low osmolalities. Thus, for example, it would be possible to prepare a concentrated sucrose solution of high refractive index and high osmolality and a solution of dextran T2000 of the same high refractive index but nearly nil osmolality. The two solutions could be mixed in various proportions to give solutions of various osmolalities but all with the same refractive index. The refractive increment (a) for carbohydrates is approximately 0.00143. Therefore, the refractive index of a carbohydrate solution can be calculated by means of the formula $n = n_s + \alpha C$, where n is the refractive index, n_s is the refractive index of water (1.33252 at 25°C), and C is the carbohydrate concentration in grams per 100 ml of solution. Of course, it is possible also to determine the refractive indices of the solution with a refractometer.

The use of plasmolyzed cells for transport studies has been described in detail by Maloney et al. (3). The use of protoplasts has been described by Abrams (7).

19.3.3. Nonmetabolized Analogs for Separation of Transport from Metabolism

For many studies, there is advantage in dissociating transport of the test solute from its subsequent metabolism, especially if one wants to demonstrate concentrative uptake. With phosphotransferase systems, complete dissociation is not possible. There are three major ways of making a separation: by the use of non-metabolizable analogs that can be transported, by use of mutants blocked in metabolism but not transport, and by the use of membrane vesicles that lack cytoplasmic enzymes.

A list of a few commonly used substrate analogs, with a pertinent literature citation for each, is presented in Table 1.

19.3.4. Metabolism-Blocked Mutants

The selection of mutants for histidine transport studies in *Salmonella typhimurium* is described in detail by Ames (9). The zy^+ mutants of *E. coli*, which lack β-galactosidase but retain the transport system for β-galactosides, have been used extensively and can be obtained from genetic stock culture collections (13.10).

TABLE 1. *Examples of nonmetabolized analogs used for transport studies of bacteria*

Metabolized substrate	Analogs	Reference
Lactose (β-galacto-side)	Thiomethylgalactoside Thiodigalactoside Thiophenylgalactoside Isopropylthiogalacto-side	21
Glucose	2-Deoxyglucose α-Methylglucoside	39 44
Glycine Alanine	α-Aminoisobutyrate	29, 43
Phenylalanine	p-Fluorophenylalanine	28
K^+	Rb^+	5, 37

19.3.5. Membrane Vesicles

The use of membrane vesicles has become increasingly popular since they were introduced by Kaback and Stadtman (19). Their use in transport studies has recently been reviewed by Konings (23).

Generally, the first step in preparing vesicles is to convert the test bacteria into osmotically sensitive forms, which are then lysed under conditions that allow for membrane resealing. Spheroplasts of gram-negative bacteria such as *E. coli* can be prepared by the lysozyme-ethylenediaminetetraacetate method (18), which leaves the outer membrane of the cell wall associated with the spheroplast. Protoplasts of gram-positive bacteria can often be completely freed from cell wall structures by treating the cells with a muralytic enzyme. Lysozyme has been used for *Bacillus megaterium, Bacillus subtilis, Streptococcus faecalis,* and *M. lysodeikticus*; lysostaphin has been used for *Staphylococcus aureus* (23).

Kobayashi et al. (22) isolated vesicles from the ATCC 9790 strain of *S. faecalis*, and their procedure is illustrative. They first suspended cells in an osmotic stabilizing solution of 500 mM glycylglycine plus 2 mM $MgSO_4$ at pH 7.2, added 0.5 mg of lysozyme per ml, and incubated the mixture at 37°C for 1 h (the protoplasts can also be stabilized with sucrose instead of glycylglycine). The protoplast suspension was centrifuged, resuspended in a hypotonic solution of 50 mM Tris-maleate, pH 7.4, 250 mM sucrose, and 5 mM $MgSO_4$, and then treated with deoxyribonuclease I, which requires magnesium ion for activity. The suspension was chilled, homogenized with a tissue homogenizer, centrifuged at 5,000 × g for 10 min to remove debris, and then centrifuged at 27,000 × g for 10 min to pellet the membranes. They were resuspended in the buffer solution of Tris-maleate plus sucrose plus $MgSO_4$ and then passed through a French press (10,000 lb/in^2). The resulting suspension was twice centrifuged, once at 27,000 × g to remove large particles and once at 48,000 × g for 90 min to pellet the vesicles.

There are many possible variants of this procedure, and some preliminary experimentation is needed to determine optimum procedures for a particular organism. A shortened, "one-step" procedure for use with gram-positive bacteria, notably *B. subtilis*, was described by Konings et al. (24).

Vesicle preparations tend to be heterogeneous in regard to size, and the preparation of Kobayashi et al. (22) contained both right-side-out and wrong-side-out vesicles, as indicated by their appearance in electron micrographs of freeze-fracture preparations. With *E. coli*, it is possible to prepare populations of vesicles all of which are right side out. For example, Kaback (18) described a procedure involving osmotic lysis without passage through a French pressure cell. *E. coli* vesicles prepared with use of the French pressure cell may contain mainly wrong-side-out forms which accumulate Ca^{2+} instead of extruding it and which cannot take up solutes such as proline (8).

The membrane continuity of vesicles is best tested by assessing the extent of swelling or shrinking in response to changes in osmolality of the suspending medium. The procedures outlined in a previous section (19.3.2), including determinations of changes in light scattering, can be used. The functionality of the vesicles is tested by determining their abilities to transport solutes such as proline.

19.4. LITERATURE CITED

19.4.1. General References

1. **Ames, G. F.-L.** 1974. Two methods for the assay of amino acid transport. Methods Enzymol. **32:** 843–849.
   This article gives detailed information on the assay of histidine uptake by S. typhimurium. Methods are described for use with growing cells or starved, nongrowing cells; the methods can be applied to the study of many other bacterial transport systems.

2. **Lamanna, C., M. F. Mallette, and L. Zimmerman.** 1973. Basic bacteriology, 4th ed. The Williams & Wilkins Co., Baltimore.
   This standard textbook contains sections in chapter 6 on osmosis and permeability with attention to methodology.

3. **Maloney, P. C., E. R. Kashket, and T. H. Wilson.** 1975. Methods for studying transport in bacteria. Methods Membrane Biol. **5:**1–49.
   This excellent article reviews basic techniques for studying bacterial transport systems, presents sample data, and discusses many important technical details. It focuses on lactose transport, but the techniques described are generally useful ones.

4. **Rosen, B. P.** 1978. Bacterial transport. Marcel Dekker, Inc., New York.
   This entire book on transport contains an opening chapter on methodology by M. Futai, and many of the subsequent chapters have descriptions of assay techniques.

5. **Silver, S., and P. Pattacharyya.** 1974. Cations, antibiotics, and membranes. Methods Enzymol. **32:**881–893.
   This article contains valuable descriptions of techniques for assaying cation uptake and includes a table of useful isotopes.

6. **Stadelmann, E. J.** 1966. Evaluation of turgidity, plasmolysis, and deplasmolysis of plant cells. Methods Cell Physiol. **2:**144–216.
   Plant cells are the subject of this article, but much

of the material presented can be applied to bacterial cells, which are usually walled cells also. Especially helpful are the discussions of plasmolysis and turgor.

19.4.2. Specific References

7. **Abrams, A.** 1960. J. Biol. Chem. **235**:1281–1285.
8. **Altendorf, K. H., and L. A. Staehelin.** 1974. J. Bacteriol. **117**:888–899.
9. **Ames, G. F.-L.** 1974. Methods Enzymol. **32**:849–856.
10. **Arnold, W. N., and J. S. Lacy.** 1977. J. Bacteriol. **131**:564–571.
11. **Conway, E. J., and M. Downey.** 1950. Biochem. J. **47**:347–355.
12. **Damadian, R. V.** 1973. Crit. Rev. Microbiol. **2**:377–422.
13. **Dubois, M., K. A. Gilles, J. K. Hamilton, P. A. Rebers, and F. Smith.** 1956. Anal. Chem. **28**:350–356.
14. **Epstein, W., and S. G. Schultz.** 1965. J. Gen. Physiol. **49**:221–234.
15. **Hamilton, W. A.** 1977. *In* B. A. Haddock and W. A. Hamilton (ed.), Microbial energetics, p. 185–216. Cambridge University Press, New York.
16. **Harold, F. M., and K. Altendorf.** 1974. Curr. Top. Membr. Transp. **5**:1–50.
17. **Hurwitz, C., C. B. Braun, and R. A. Peabody.** 1965. J. Bacteriol. **90**:1692–1695.
18. **Kaback, H. R.** 1971. Methods Enzymol. **22**:99–120.
19. **Kaback, H. R., and E. R. Stadtman.** 1966. Proc. Natl. Acad. Sci. U.S.A. **55**:920–927.
20. **Kanos, K., and J. H. Fendler.** 1978. Biochim. Biophys. Acta **509**:289–299.
21. **Kepes, A.** 1970. Curr. Top. Membr. Transp. **1**:101–134.
22. **Kobayashi, H., J. Van Brunt, and F. M. Harold.** 1978. J. Biol. Chem. **253**:2085–2092.
23. **Konings, W. N.** 1977. Adv. Microb. Physiol. **15**:175–251.
24. **Konings, W. N., A. Bisschop, M. Veenhuis, and C. A. Vermeulen.** 1973. J. Bacteriol. **116**:1456–1465.
25. **Levin, B. C., and E. Freese.** 1978. J. Environ. Sci. Health, Part C, Environ. Health Sci. **13**:1–31.
26. **MacDonald, R. E., and P. Gerhardt.** 1958. Can. J. Microbiol. **4**:109–124.
27. **Marquis, R. E.** 1967. Arch. Biochem. Biophys. **118**:323–331.
28. **Marquis, R. E.** 1970. Handb. Exp. Pharmacol. **XX/2**:166–192.
29. **Marquis, R. E., and P. Gerhardt.** 1964. J. Biol. Chem. **239**:3361–3371.
30. **Matts, T. C., and C. J. Knowles.** 1971. Biochim. Biophys. Acta **249**:583–587.
31. **Mitchell, P.** 1966. Biol. Rev. **41**:445–502.
32. **Mitchell, P., and J. Moyle.** 1959. J. Gen. Microbiol. **20**:434–441.
33. **Ogawa, S., R. G. Shulman, P. Glynn, T. Yamane, and G. Navon.** 1978. Biochim. Biophys. Acta **502**:45–50.
34. **Postma, P. W., and S. Roseman.** 1976. Biochim. Biophys. Acta **457**:213–257.
35. **Ramos, S., and H. R. Kaback.** 1977. Biochemistry **16**:848–854.
36. **Ramos, S., S. Schuldiner, and H. R. Kaback.** 1976. Proc. Natl. Acad. Sci. U.S.A. **73**:1892–1896.
37. **Rhoads, D. B., A. Woo, and W. Epstein.** 1977. Biochim. Biophys, Acta **469**:45–51.
38. **Rogers, H. J.** 1979. Adv. Microb. Physiol. **19**:1–62.
39. **Romano, A. H., S. J. Eberhard, S. L. Dingle, and T. D. McDowell.** 1970. J. Bacteriol. **104**:808–813.
40. **Schales, O., and S. S. Schales.** 1941. J. Biol. Chem. **140**:879–884.
41. **Scherrer, R., and P. Gerhardt.** 1971. J. Bacteriol. **107**:718–735.
42. **Sorgato, M. C., S. J. Ferguson, D. B. Kell, and P. John.** 1978. Biochem. J. **174**:237–256.
43. **Thompson, J., and R. A. MacLeod.** 1973. J. Biol. Chem. **248**:7107–7111.
44. **Winkler, H. H.** 1971. J. Bacteriol. **106**:362–368.

Section V

SYSTEMATICS

Introduction to Systematics

NOEL R. KRIEG

Systematics in bacteriology is concerned with three interrelated topics: (i) **classification**, the arranging of bacteria with similar phenotypic or genetic characteristics into groups; (ii) **nomenclature**, the naming of bacteria according to an international code of principles, rules, and recommendations (3) which allow precise communication between bacteriologists concerning the various kinds of bacteria; and (iii) **identification**, the comparing of unknown organisms with bacteria that have already been classified, for the purpose of determining the identities or names of the unknown organisms. Systematics requires first learning as much as possible about the characteristics of bacteria and then deciding how the bacteria should be arranged or grouped. Only after bacteria have been characterized and classified can reliable identification schemes be constructed for determining the identity of unknown organisms. In this way, systematics seeks to bring order from chaos and to bring the many kinds of bacteria into a coherent, understandable, and useful form.

The basic unit used in systematics is the **species**. In bacteriology, a species consists of a type strain together with all other strains that are deemed to be sufficiently similar to the type strain as to be included with it in the species. The **type strain** is a strain that has been designated as the permanent example of what is meant by the species, although it may not necessarily be the most typical or representative strain in the species. All other strains being considered for inclusion in a species must be compared to the type strain of that species. Cultures of type strains can be purchased from various collections such as the American Type Culture Collection (Rockville, Md.) or the National Collection of Type Cultures (London, England).

In the above definition of a bacterial species, it is the phrase "sufficiently similar" that causes most of the difficulties in the classification of bacteria, because what is deemed sufficiently similar by one person may not necessarily be considered so by another person. Yet, the more that is known about a group of bacteria, the more likely it is that various investigators can agree on a suitable classification scheme. The

historic way to characterize bacteria is to describe qualitatively as many phenotypic properties as possible, such as morphology, structure, cultivation, nutrition, biochemistry, metabolism, pathogenicity, antigenic properties, and ecology. Routine and special tests for many of these characteristics are described in Chapter 20. Numerical taxonomy methods, described in Chapter 21, are useful for quantifying similarities based on phenotypic characteristics and can achieve the objectivity that is sometimes lacking when bacteria are grouped according to intuitive judgments. Phenotypic similarities do not necessarily indicate phylogenetic relationships or relatedness (relationships based on the ancestry of the organisms); however, in recent years nucleic acid homology methods have been developed which provide a basis for grouping bacteria according to relatedness. These methods enable bacteria to be compared with respect to the nucleotide sequences of their deoxyribonucleic acid or ribonucleic acid and are described in Chapter 22.

Although no "official" or internationally accepted classification scheme for bacteria exists, the most popular and widely used classification scheme is that given in *Bergey's Manual of Determinative Bacteriology* (1). Recently, a short version of *Bergey's Manual* has been published which is specifically designed for use in identification (2). *Bergey's Manual* owes its usefulness to several factors: (i) although many classification schemes are available for one or another group of bacteria, *Bergey's Manual* is the only one that covers all of the known genera and species of bacteria; (ii) it is based on the work of a large number of contributors who have extensive familiarity with the specific groups of bacteria; (iii) there is provision for change and improvement in subsequent editions as new information becomes available; and (iv) it includes tables and keys useful for the identification of bacteria.

The methods to be described in the subsequent chapters of this section have been selected on the basis of their interest and usefulness for persons who have had little experience with bacterial systematics but who wish to cope with this area of bacteriology. Emphasis is given to the practical performance of the various methods rather than to the principles and theoretical considerations involved. However, general references and specific articles are provided at the end of each chapter for those who wish to pursue any topic further.

LITERATURE CITED

1. **Buchanan, R. E., and N. E. Gibbons (ed.).** 1974. Bergey's manual of determinative bacteriology, 8th ed. The Williams & Wilkins Co., Baltimore.
2. **Holt, J. G. (ed.).** 1977. The shorter Bergey's manual of determinative bacteriology. The Williams & Wilkins Co., Baltimore.
3. **LaPage, S. P., P. H. A. Sneath, E. F. Lessel, V. B. D. Skerman, H. P. R. Seeliger, and W. A. Clark (ed.).** 1975. International code of nomenclature of bacteria. American Society for Microbiology, Washington, D.C.

Chapter 20

General Characterization

ROBERT M. SMIBERT AND NOEL R. KRIEG

Some preliminary precautions apply universally in the application of methods useful for characterization of bacteria. The most important is that one must use only pure cultures throughout. This may seem obvious, but the pitfalls involved in obtaining a pure culture are not always apparent. In many cases the single picking of a colony from a plate and transferring it to another medium may not ensure purity. For example, colonies of a slime producer may contain contaminants trapped in the slime or, in the case of actinomycetes or of *Bacillus* species, contaminants may be enmeshed in the chains. Difficulties may also occur when isolation is done on selective media. Here, contaminants may exist in a bacteriostatic condition and cause contamination of the colonies that are picked and subcultured. Purification is best achieved on nonselective media. Even here, colonies should not be picked too soon because of the possibility of the occurrence of slow-growing contaminants whose presence has not yet been revealed. See Chapter 8 for pure culture methods.

During a long-term study of a group of strains, stock cultures may be subject to genetic variation. Thus, the cultures of an organism that are used near the end of a study may not contain quite the same organism as that used earlier. If cultures are properly preserved at the beginning of the study, one can always have a source of the organism with its original characteristics. See Chapter 12 for various preservation methods.

In identifying bacteria, certain general characteristics are of primary importance for determining to which major group the new isolate most likely belongs. The investigator should determine whether the organism is phototrophic, chemoautotrophic, or chemoheterotrophic and whether it is aerobic, anaerobic, microaerophilic, or facultative. Similarly, certain morphological features should be determined: the Gram reaction and the cell shape (e.g., rod, coccus, vibrioid, helical, or other). Special morphological features, if they occur, are valuable (e.g., endospores, exospores, true branching, acid fastness, sheaths, cysts, stalks or other appendages, fruiting bodies, gliding motility, or budding division). The arrangement of the cells may also be helpful: e.g., cocci occurring in chains or in irregular clusters. Three physiological tests are of primary importance: the oxidase test, the catalase test, and the determination of whether sugars are oxidized or fermented. Special physiological characteristics, if they occur, may narrow the field of possibilities considerably (e.g., nitrogen-fixing ability, degradation of cellulose, bioluminescence, requirement for seawater for growth, production of pigments, or a heme requirement).

It is desirable to use an orderly approach to identification, based on common sense and on the use of only those tests that are pertinent. Avoid a "shotgun" approach, where all sorts of tests are performed in the desperate hope that perhaps some of them may be helpful. The table of contents and the major keys given in the 8th edition of *Bergey's Manual of Determinative Bacteriology* (2) will be of great assistance, as will the general descriptions of the various orders, families, and genera of bacteria. After the initial possibilities have become more restricted, consult the detailed keys and tables from various sources (many of which are listed in the general references [1–13] in 20.5.1) for the identification of genera and species. In the final steps of identification, compare the new strain with either the type strain of the suspected species or with some other named strain whose inclusion in the species has been firmly established.

In many cases, especially with pathogenic bacteria, one may be able to make a rapid presumptive identification of a genus or species by the use of antisera. However, such tests often are not absolutely specific, and additional physiological tests are usually required for confirmation of the identification. For example, agglutination of gram-negative enteric rods by polyvalent *Salmonella* serum is presumptive evidence for a *Salmonella* strain, but this must always be confirmed by a number of biochemical tests.

In performing characterization tests, the organisms used for inoculation of test media should be from fresh transfers and in good physiological condition. Old stock cultures, which may contain mostly dead or dying cells, are not satisfactory. Incubate the inoculated test media under conditions that are optimum for the organisms or for the characteristic being tested (e.g., optimum temperature, pH, gaseous conditions, and ionic conditions). Furthermore, apply characterization methods not only to the organism to be tested but also to strains known to be positive and negative for the characteristic, as a check on the reliability of the reagents and media.

At the end of this chapter is listed a selection of general references (20.5.1) that provide identification schemes for various groups of bacteria, further information on methodology, or additional characterization methods.

20.1. ROUTINE TESTS

20.1.1. Acid-Fast Stain

See 3.3.6 for procedure. A positive reaction indicates *Mycobacterium* or *Nocardia*.

20.1.2. Acidification of Carbohydrates

Method 1. Incorporate a pH indicator into the medium during its preparation. A list of the most commonly used indicators and their properties is given in Table 1.

Method 2. In cases where an indicator may be toxic to the bacteria or may be reduced during their growth, add drops of indicator (0.02 to 0.04% alcoholic solution) to the culture after growth is completed.

Method 3. For the same reasons given above and also to provide greater precision, measure the pH of the culture after it has grown, using a pH meter. Use a long, thin, combination pH electrode which can be inserted directly into a culture tube. Affix a flat, rubber washer with a diameter larger than that of the culture tube to the upper portion of the electrode to prevent the electrode tip from striking the bottom of the culture tube when the electrode is inserted. When determining the pH of a number of cul-

tures of pathogenic bacteria, dip the electrode into a beaker of distilled water between cultures. This water should later be autoclaved. When finished with the electrode, rinse it with a suitable disinfectant (e.g., 3% hydrogen peroxide or 70% ethanol).

In the case of anaerobes cultured in prereduced PY broth (20.3.28) containing carbohydrates, the oxygen-free carbon dioxide used to purge the tube during their inoculation will usually lower the pH of the medium to 6.2 to 6.4. Consequently, pH values for PY carbohydrate broth cultures are usually interpreted as acidification when they are 5.5 to 6.0 (weak acid) or below 5.5 (strong acid). For PY broth cultures containing arabinose, ribose, or xylose, a pH of 5.7 or below usually indicates acidification, because the sterile media purged with carbon dioxide may have a pH of 5.9 after 1 to 2 days of incubation (7). In any case, the pH of cultures in PY broth lacking any carbohydrate should be determined as a control, because acids may be formed from peptones in certain cases.

20.1.3. Esculin Hydrolysis

Grow cultures in broth or agar media supplemented with 0.01% esculin and 0.05% ferric citrate. Positive test: medium becomes brownish black.

Examples: positive, *Streptococcus faecalis*; negative, *Streptococcus mitis*.

20.1.4. Ammonia from Arginine

Inoculate arginine broth (see 20.3.2 and 20.3.3). Also inoculate a control lacking arginine. After incubation for 2 to 3 days, test samples of the culture in a spot plate with Nessler's solution (20.4.10). Positive test: yellow or orange color as compared with control.

Examples: positive, *Streptococcus faecalis*; negative, *Streptococcus salivarius*.

20.1.5. Arginine Dihydrolase

Method 1. This method is widely used to distinguish among members of the family *Enterobacteriaceae*. Inoculate Møller's broth base (20.3.23) supplemented with 1% L-arginine monohydrochloride (or 2% of the DL form). Also inoculate a control lacking arginine. After inoculation, overlay the broth with 10 mm of sterile mineral (paraffin) oil. Examine daily for 4 days. Positive test: violet or reddish-violet color. Weak reactions are bluish gray.

Examples: positive, *Enterobacter cloacae*; negative, *Proteus vulgaris*.

Method 2. This method is suitable for a wide variety of facultative bacteria. Inoculate Thornley's semisolid medium containing arginine (see 20.3.32) and also a control lacking arginine. After stab inoculation, seal the medium with a layer of sterile melted petrolatum and incubate. Positive test: a change from yellow-orange to red within 7 days.

Method 3 (88). This method is suitable for aerobic and facultative bacteria. Make a dense suspension of the bacteria in 0.033 M phosphate buffer, pH 6.8 (see 20.4.16). Purge 4 ml of the suspension by bubbling nitrogen through the suspension for several minutes, and add 1 ml of 0.001 M L-arginine monohydrochloride. After purging again, stopper the tubes, incubate for 2 h, and heat at 100°C for 15 min. After removing the cells by centrifugation, determine the arginine in the supernatant by the method of Rosenberg et al. (77) as follows: mix 1 ml of the sample with 1 ml of 3 N NaOH, 2 ml of developing solution (see 20.4.4), and 6 ml of water; read the tubes at 30 min against a blank prepared without arginine, using a colorimeter equipped with a green filter (540 nm); compare the readings to those obtained with an uninoculated control containing arginine. Positive test: disappearance of some or all of the arginine.

20.1.6. Aromatic Ring Cleavage (88)

Grow the culture in chemically defined medium containing 0.1% sodium p-hydroxybenzoate as the carbon source (and also grow it on a yeast extract agar to determine whether the enzymes are constitutive or inducible). Scrape growth from the agar and suspend it in 2 ml of 0.02 M Tris buffer (2-amino-2-hydroxymethyl-1,3-propanediol), pH 8.0. Shake the tubes with 0.5 ml of toluene and add 20 μmol (3.5 mg) of sodium protocatechuate. A yellow color within a few minutes indicates *meta* cleavage. If no color appears, shake the tubes for 1 h at 30°C. Add 1.0 g of $(NH_4)_2SO_4$, 1 drop of 1.0% sodium nitroprusside (nitroferricyanide), and 0.5 ml of ammonia solution (specific gravity, 0.880, or 28 to 30%). A purple color indicates *ortho* cleavage.

Examples: *meta* cleavage, *Pseudomonas acidovorans*: *ortho* cleavage, *Pseudomonas fluorescens*; negative, *Escherichia coli*.

20.1.7. Arylsulfatase

Aseptically add a sufficient amount of a filter-sterilized 0.08 M solution of tripotassium phenolphthalein disulfate (ICN Biochemicals, Cleveland, Ohio) to a sterile liquid medium to give a final concentration of 0.001 M. Media containing methionine as the sole source of sulfur are best for synthesis of arylsulfatase; sulfur sources such as sulfate, sulfite, thiosulfate, or cysteine may repress synthesis (14, 60). Inoculate the medium, incubate for 7 days, and add 1 N NaOH or 1 M Na_2CO_3 drop by drop. Positive test: faint-pink to light-red color. An uninoculated control similarly treated should remain colorless.

Examples: positive, *Proteus rettgeri*; negative, *Chromobacterium violaceum*.

20.1.8. Bacitracin Test

Use sterile commercially available differentiation (not sensitivity) disks (Difco Laboratories, Detroit, Mich.; BBL Microbiology Systems, Cockeysville, Md.) or sterile paper disks impregnated with 0.04 U of bacitracin (Sigma Chemical Co., St. Louis, Mo.). Place a disk on an inoculated blood agar plate (see 20.3.7) and incubate for 24 h. Positive test: zone of growth inhibition around disk.

Examples: positive, *Streptococcus pyogenes*; negative, other beta-hemolytic streptococci.

20.1.9. Bile Solubility

Centrifuge a 24-h culture grown in 10 ml of Todd-Hewitt broth (see 20.3.33), and discard the supernatant into a flask of disinfectant. Suspend the cells in 0.5 ml of 0.067 M phosphate buffer, pH 7.0 (20.4.16). Add 0.5 ml of 10% sodium deoxycholate and incubate at 37°C for 15 to 30 min. Positive test: the suspension becomes clear.

Examples: positive, *Streptococcus pneumoniae*; negative, other alpha-hemolytic streptococci.

20.1.10. Bile Tolerance

Method 1. Streak a plate of bile agar (20.3.6). Compare the growth on the plates with that occurring in the absence of the oxgall.

Examples: positive (10 and 40% bile), *Streptococcus faecalis*; positive (10 but not 40% bile), *Streptococcus salivarius*; negative (neither 10 nor 40% bile), *Streptococcus dysgalactiae*.

Method 2. Use this method for anaerobes (7). Inoculate PY broth (20.3.28) containing 2% oxgall and 1% glucose. Inoculate tubes with a Pasteur pipette while flushing them with oxygen-free carbon dioxide. Stopper the tubes and incubate. Observe the growth response and compare with that occurring in the absence of the oxgall.

Examples: growth, *Bacteroides oralis*; no growth, *Bacteroides melaninogenicus*.

20.1.11. Casein Hydrolysis

Combine sterile (autoclaved) skim milk at 50°C with an equal volume of double-strength nutrient agar (see 20.3.25) or other carbohydrate-free agar medium at 50 to 55°C. Incubate streaked plates up to 14 days, and look for clear zones surrounding the growth. Confirm by flooding the plates with 10% HCl. *Note:* Acid production from the lactose in the milk may inhibit the casein hydrolysis and necessitate prior dialysis of the skim milk.

Examples: positive, *Bacillus polymyxa*; negative, *Bacillus macerans*.

20.1.12. Catalase Test

Method 1. Inoculate a nutrient agar slant (see 20.3.25) or other medium lacking blood. After incubation, trickle 1 ml of 3% hydrogen peroxide down the slant. Examine immediately and after 5 min for the evolution of bubbles, which indicates a positive test. Alternatively, add a few drops of 3% peroxide to colonies on a plate or to the heavy growth that may occur at or near the surface of semisolid media. For broth cultures, add 0.5 ml of 3% hydrogen peroxide to 0.5 ml of culture and observe for continuous bubbling. *Note:* Media containing blood must not be used for catalase tests, because blood contains catalase activity unless the blood has been heated (Method 3). Also note that some bacteria (e.g., certain lactic acid organisms) make a nonheme "pseudocatalase" in media containing low levels of glucose or no glucose (94). Pseudocatalase can be prevented by incorporating 1% glucose into the medium. When anaerobes are tested for catalase activity, it is important to expose the cultures to air for 30 min before adding peroxide (7).

Examples: positive, *Staphylococcus epidermidis*; negative, *Streptococcus lactis*.

Method 2. This method eliminates problems of penetration that might occur when peroxide is added to colonies on a plate or to growth on a slant. Scrape the growth from a slant or plate with a nonmetallic instrument, and suspend it in a drop of 3% hydrogen peroxide on a slide. Examine immediately and at 5 min for bubbles, either macroscopically or with a low-power microscope.

Method 3. Use this method for certain bacteria that can make catalase only if grown on a heme-containing medium, e.g., certain lactic acid bacteria (94). To sterile blood agar base (20.3.31) containing 1% glucose to inhibit pseudocatalase formation, add 5% (vol/vol) of a 1:1 mixture of defibrinated blood (20.3.7) and sterile water. Heat the medium at 100°C for 15 min to inactivate blood catalase, cool to 45 to 50°C, and dispense into plates. Test growth on the plates directly with 3% hydrogen peroxide or use Method 2.

20.1.13. Cellulolytic Activity

Method 1 (45). This method of preparing cellulose gives the best form of native cellulose for testing cellulolytic activity; for other methods, see references 12 and 13. Incorporate finely divided cellulose into appropriate carbohydrate-free agar media. To prepare the cellulose, wet-grind 3% (wt/vol) Whatman no. 1 filter paper in a pebble mill as follows: place 30 g of the paper (torn into small pieces) into a porcelain jar (ca. 4-liter capacity) with 1 liter of water; add enough flint pebbles (porcelain balls are not as satisfactory) so that the liquid just covers them, roll the jars for 24 h at 74 rpm, or until a very fine state of suspension has been achieved and the suspension has become viscous; stop before the viscosity decreases and copper-reducing substances appear. (To test for the latter, remove 1 ml of the suspension and test as described in 20.1.30 with 1 ml of Benedict's reagent [20.4.1].) Positive test: clear zones occur around the colonies on the cellulose agar. Long periods of incubation may be required.

Examples: positive, *Cellulomonas* species; negative, *Escherichia coli*.

Method 2. This method is less sensitive than Method 1. Place a strip of Whatman no. 1 filter paper in tubes of carbohydrate-free broth before autoclaving. A portion of the strip should extend above the level of the broth. Positive test: partial or complete disintegration of the paper strip during growth of the culture.

20.1.14. Citrate Utilization

Method 1. Prepare a dilute suspension of the organisms in sterile water or saline. Make a single streak up a slant of Simmons' citrate agar (20.3.30). Incubate for up to 7 days. Positive test: a blue color indicates the utilization of citrate as a sole carbon source.

Examples: positive, *Enterobacter aerogenes*; negative, *Escherichia coli*.

Method 2. From a dilute suspension, streak the entire surface of a slant of Christensen citrate agar (20.3.11). Incubate for up to 7 days. Positive test: a red or magenta color. *Note:* A positive reaction indicates that citrate is used, but not necessarily as a sole carbon source; i.e., an organism could give a positive reaction on Christensen agar and a negative reaction on Simmons' citrate agar.

20.1.15. Coagulase

Mix one loopful of growth from an agar slant, 0.1 ml of broth culture, or a single colony from an agar plate with 0.5 ml of undiluted rabbit plasma or plasma diluted 1:4 with saline. Incubate at 37°C and examine at 4 and 24 h. Positive test: solid clot or a loose clot suspended in the plasma. Granular or ropy formations are inconclusive.

Examples: positive, *Staphylococcus aureus*; negative, *Staphylococcus epidermidis*.

20.1.16. Coccoid Bodies

Examine cultures that are held static (not shaken during growth) for up to 4 weeks by phase-contrast microscopy. Coccoid bodies may occur as early as 2 or 3 days. Positive test: a predominance of round refractile forms which have a cell diameter greater than that of the original cells and which lack a thickened cell wall. Many of the forms have a discrete, dark peripheral region of cytoplasm in an otherwise empty-appearing cell. Coccoid bodies occur in certain spirilla, vibrios, and campylobacters.

Examples: positive, *Aquaspirillum itersonii*; negative, *Aquaspirillum serpens*.

20.1.17. Colonies

Measure colony diameter in millimeters, describe pigmentation, and describe the form, elevation, and margin as indicated in Fig. 1. Low-power microscopy may be necessary for observation of the margin. Also indicate whether the colonies are smooth (shiny glistening surface), rough (dull, bumpy, granular, or matte surface), or mucoid (slimy or gummy appearance). Re-

cord the opacity of the colonies (transparent, translucent, or opaque) and their texture when tested with a needle: butyrous (butter-like texture), viscous (gummy), or dry (brittle or powdery). Describe the colonies from both young and old cultures. The medium, age of the culture, gaseous conditions, exposure to illumination, and other cultural conditions may affect the colony characteristics.

20.1.18. Cysts and Microcysts

Examine cultures daily by phase-contrast microscopy. Initially rod-shaped cells become spherical with thickened walls in older cultures. The spherical forms may be optically dense or may be refractile. They do not have the heat resistance of endospores (20.1.52), except for those formed by certain species of *Nocardia*, but they are extremely resistant to desiccation. In *Azotobacter* the forms are termed "cysts." In the order *Myxobacteriales* they are termed "microcysts"; here, the term "cyst" refers to the sporangium, if any, which contains the microcysts. In *Nocardia* the forms are termed "microcysts" or "chlamydospores."

See also 3.3.8 for staining methods for cysts.

20.1.19. Flagella

Use flagellar staining with light microscopy (see 3.3.10) or electron microscopy (see 4.1.1 and 4.1.4). Agitate the bacteria as little as possible

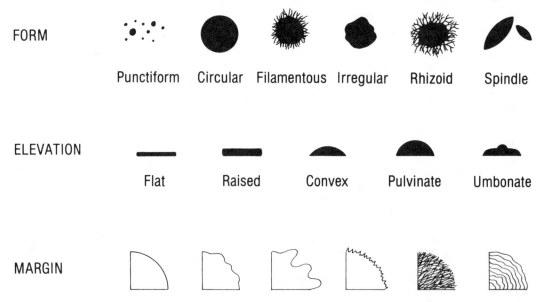

FIG. 1. *Diagram illustrating the various forms, elevations, and margins of bacterial colonies. (Adapted by permission of the McGraw-Hill Book Co. from reference 74.)*

during their preparation, because flagella may be easily broken off the cells. Flagellar stains with light microscopy may occasionally be misleading when applied to polar-flagellated bacteria because a fascicle of several flagella may sometimes seem to be a single flagellum (97). Cultural conditions may be important; e.g., some bacteria, such as *Vibrio parahaemolyticus*, form a single polar flagellum when grown in liquid media but also form lateral flagella of shorter wavelength when grown on solid media (82). Some bacteria, such as *Listeria monocytogenes*, may not form flagella at certain temperatures, yet do form them when grown at a lower temperature (50). In some instances glucose-containing media may inhibit the formation of flagella (15).

20.1.20. Fluorescent Pigment

Make a single streak across a plate of medium B (see 20.3.5). Examine plates with the covers removed at 24, 48, and 72 h under ultraviolet light (below 260 nm). A short-wave lamp used for examination of mineral specimens is suitable. Plates should not be reincubated after examination because the ultraviolet illumination may be bactericidal. Positive test: a fluorescent zone in the agar surrounding the growth. Fluorescent pseudomonads produce a yellow-green pigment; certain members of *Azotobacter*, *Azomonas*, and *Beijerinckia* may form green or white fluorescent pigments.

Examples: positive, *Pseudomonas fluorescens*; negative, *Escherichia coli*.

20.1.21. Formate-Fumarate Requirement

Use this test to distinguish certain anaerobes that can oxidize formate and concomitantly reduce fumarate to succinate. Prepare formate-fumarate (F-F) stock solution (20.4.6), and add 1 drop of the solution per ml of prereduced PY broth (20.3.28) when inoculating the medium under oxygen-free carbon dioxide. Compare the growth response with that occurring in medium lacking the F-F. Positive test: growth is greatly stimulated by the F-F. In some cases, growth may not occur at all unless the supplement is added.

Examples: positive, *Bacteroides corrodens*; negative, *Bacteroides oralis*.

20.1.22. β-Galactosidase (56)

Culture the organism for 18 h at 37°C on a triple sugar iron agar slant (20.3.34). Remove a loopful of growth from the slant, and suspend the cells in 0.25 ml of saline (0.85% NaCl) to make a dense suspension. Add a drop of toluene, agitate the suspension, and incubate for 5 min at 37°C. Add 0.25 ml of ONPG test reagent (20.4.13). Incubate in a water bath at 37°C, and examine at intervals up to 24 h. A control lacking the cell suspension should also be prepared and incubated. Positive test: a yellow color develops.

Examples: positive, *Escherichia coli*; negative, *Proteus vulgaris*.

20.1.23. Gas Production from Sugars

Method 1. Before autoclaving the sugar broth, place a small vial (ca. 10 by 75 mm) in an inverted position in the tube. After autoclaving, the vial will become completely filled with medium. For thermolabile sugars add filter-sterilized stock solutions aseptically after the base medium has been autoclaved and cooled; allow tubes to stand for a day or so to allow diffusion of the sugar into the inverted vials. Inoculate the medium and incubate. Positive test: accumulation of gas within the vial. *Note:* If media are stored in a refrigerator before use and then are inoculated and incubated, dissolved gases in the medium may be liberated and accumulate in the vial, giving a false-positive reaction. Sterile controls should be used to detect this occurrence. Also, if the organism forms only small amounts of carbon dioxide, detection may be difficult because of the great solubility of carbon dioxide and because of rapid diffusion into the air. Methods 2 and 3 are designed to eliminate this difficulty.

Method 2. Same as Method 1, but after inoculation add ca. 1 ml of sterile mineral (paraffin) oil to prevent diffusion of carbon dioxide into the air.

Alternatively, use sugar-containing agar media cooled to 45°C after autoclaving (or, for thermolabile sugars, add a filter-sterilized stock solution of the sugar to the base agar medium which has been melted and cooled to 45°C). Inoculate the tubes and allow them to solidify. Overlay with ca. 6 mm of sterile nonnutrient 2% agar (i.e., agar prepared only with water) and then a thin layer of sterile mineral oil. Incubate the cultures. Positive test: cracks in the medium and upward movement of the agar overlay.

Method 3 (7). Use this method only for anaerobes. Cool melted prereduced PY agar (20.3.27) containing 1% glucose to 45°C, and inoculate it with 2 to 3 drops of culture. Cover tubes with sterile aluminum foil (do not stopper), allow to solidify, and incubate. Positive test: formation of gas bubbles in the medium or breaks in the agar.

20.1.24. Gelatin Hydrolysis

Method 1. Use broth supplemented with 12% gelatin.

(A) Incubate at 20 to 22°C for up to 6 weeks. Positive test: liquefaction of at least a portion of the medium. Evaporation should be prevented during prolonged incubation.

Examples: positive, *Proteus vulgaris*; negative, *Escherichia coli*.

(B) Incubate at the optimum growth temperature for the organism if greater than 22°C; then chill in a refrigerator along with an uninoculated control tube. If the test culture fails to solidify, the organism has hydrolyzed the gelatin. If the test culture does solidify, remove the culture and the control tube, and incubate in a tilted position at room temperature or in a 30°C incubator for ca. 30 min. If the culture liquefies before the control tube, a weakly positive reaction is indicated. However, if the test culture takes as long as the control to liquefy, this indicates a negative reaction.

(C) As for Method 1B, but use 4% gelatin. This is a more sensitive method than 1B.

Method 2. This method is more sensitive than Method 1A or 1B. Streak plates of agar medium supplemented with 0.4% gelatin. Incubate at the optimum temperature for the organism. Flood plates with gelatin-precipitating reagent (15% $HgCl_2$ in 20% [vol/vol] concentrated HCl). Positive test: clear zones around colonies.

Method 3. This method is sensitive and provides a more readily observable reaction. Culture the organism in a tube of broth containing a charcoal-gelatin disk (20.4.3). Positive test: the charcoal granules become dispersed in the medium.

20.1.25. Gram Stain

See 3.3.5 for procedure.

20.1.26. Hemolysis

Method 1. Streak a plate of blood agar (20.3.7). Incubate at 37°C for 24 to 48 h. *Beta (β) hemolysis:* a clear, colorless zone surrounds the colonies. The erythrocytes in the zone are completely lysed. *Alpha (α) hemolysis:* an indistinct zone of partial destruction of the erythrocytes surrounds the colonies, often accompanied by a greenish to brownish discoloration of the medium. *Alpha prime (α′) hemolysis:* a small halo of intact or partially lysed erythrocytes occurs adjacent to the colony, with a zone of complete hemolysis extending farther out into the medium; the colony can be confused with a beta-hemolytic colony when examined only macroscopically. *Note:* Some bacteria, such as certain staphylococci, produce beta hemolysis only if the inoculated plates are first incubated for 48 h at 37°C and then chilled in a refrigerator for 1 h ("hot-cold hemolysis").

Method 2. Streak a plate of blood agar base (without added blood). Pour an overlay of 10 ml of blood agar (at 45 to 50°C) onto the streaked plate. Hemolysis (alpha or beta) may be more pronounced than in Method 1 because the colonies are submerged, offering some protection for hemolytic enzymes (hemolysins) that may be oxygen labile.

20.1.27. Hippurate Hydrolysis

Method 1. Prepare a broth medium supplemented with 1.0% sodium hippurate. Mark the level in each tube after autoclaving. Inoculate the medium and incubate along with a sterile control tube for 4 days. If evaporation occurs, restore the volumes in the tubes to their original levels with distilled water at the end of the incubation period. Place 1.0 ml of the sterile control broth in a small tube, and add 0.10-ml increments of ferric chloride solution (12 g of $FeCl_3 \cdot 6H_2O$ in 100 ml of 2% HCl). Agitate the tube after each addition. At first, a precipitate of iron hippurate will form, but upon further addition of ferric chloride this will dissolve, leaving a clear solution. Determine the smallest total amount of ferric chloride that will allow the initial precipitate to dissolve. Centrifuge the broth culture, add 1.0 ml of the supernatant to another small tube, and then add an amount of ferric chloride solution equivalent to that determined previously for the control. Positive test: formation of a heavy permanent precipitate of iron benzoate.

Examples: positive, *Streptococcus agalactiae*; negative, *Streptococcus pyogenes*.

Method 2 (46). Use this method for a more rapid test. Prepare a 1% solution of sodium hippurate in distilled water, and dispense 0.4-ml portions into tubes. Store the corked tubes in a freezer at −20°C until needed. To test an organism, thaw a tube and mix a large loopful of the growth from the surface of a solid medium into the hippurate solution to form a very cloudy suspension. Incubate the tube in a heating block or water bath at 37°C for 2 h. After incubation, add ca. 0.2 ml of ninhydrin reagent (3.5 g of ninhydrin in 100 ml of a 1:1 [vol/vol] mixture of acetone and butanol). Do not shake the tube (35). Continue incubation at 37°C for 10 min. Positive test: formation of a deep-purple color, due to the glycine which is formed from hippurate hydrolysis. Negative test: no color reaction or only a faint tinge of purple color. Controls using known positive and negative organisms should be included.

20.1.28. Hydrogen Sulfide Production

Method 1. Use a deep tube of an agar medium supplemented with 0.025% ferric ammonium citrate or 0.015% ferrous sulfate. The medium should contain a source of organic sulfur (usually peptone) and also sodium thiosulfate (0.008 to 0.03%), since some bacteria may form H_2S from one source but not the other. Sulfide production usually occurs best under anaerobic or semianaerobic conditions; therefore, the medium should be inoculated by stabbing the agar. Positive test: blackening of the medium during growth, due to formation of iron sulfide. (See also Triple Sugar Iron Agar Reactions, part 20.1.54.)

Examples: positive, *Proteus vulgaris*; negative, *Escherichia coli*.

Method 2. This method is more sensitive than Method 1. Cut filter paper into strips 5 by 20 mm, and soak them in a solution of 5% lead acetate. Sterilize the strips in cotton-stoppered test tubes and dry in a drying oven. Inoculate a tube of liquid medium containing sources of sulfur (see Method 1). Before incubating, suspend a lead acetate strip in the tube so that the end of the strip is approximately 1.3 cm above the level of the medium; the top of the strip can be held by the cotton plug or other tube closure. The strip must not be allowed to come into contact with the medium. Prepare uninoculated controls in a similar manner. Positive test: blackening of the lower portion of the strip after incubation for several days.

20.1.29. Indole Production

Method 1. Inoculate carbohydrate-free and nitrite-free liquid medium containing a source of tryptophan (e.g., 1% tryptone broth or media supplemented with 0.1% L-tryptophan). Test cultures of different ages by the following method: add 1 ml of xylene (or water-saturated toluene), agitate the culture vigorously to extract the indole, and then allow the xylene to form a layer at the top of the broth. With the tube held in a slanted position, trickle 0.5 ml of Ehrlich's reagent (20.4.5) down the wall of the tube to form a layer between the xylene and the broth. Allow to stand for 5 min. Positive test: formation of a red ring just below the xylene layer. A tube of uninoculated medium should also be tested as a control. *Note:* If the medium in which the organisms are growing contains a carbohydrate, synthesis of the enzyme tryptophanase, responsible for indole formation, may be repressed (21). Also, nitrite at levels of 0.75 mg/ml or greater may block detection of indole (87); therefore, the presence of nitrate in the medium may lead to a false-negative test if the organism can reduce the nitrate to nitrite.

Examples: positive, *Escherichia coli*; negative, *Enterobacter aerogenes*.

Method 2. Method 2 is not as sensitive as Method 1; however, Method 2 eliminates the necessity of using flammable and toxic solvents. Use Kovacs' reagent (20.4.7) rather than Ehrlich's reagent, and omit extraction with xylene. Add 0.5 ml of the reagent to the broth culture and shake the tube gently. Positive test: red color.

Method 3. Cut filter paper into strips, and soak in a warm, saturated solution of oxalic acid. Crystals of the acid will form on the strips upon cooling. Dry the strips thoroughly (sterilization by heat seems unnecessary), and insert aseptically into the inoculated culture tube before incubating the culture. The strip should not come into contact with the medium but rather should be bent so that it presses against the wall of the tube and remains near the mouth when the cotton stopper or screw cap is replaced. Positive test: the paper strip becomes pink or red during growth of the culture.

20.1.30. 2-Ketogluconate from Glucose Oxidation

Grow the culture in a medium such as Haynes' broth (20.3.15) containing 4% potassium gluconate (sterilized by filtration and added aseptically). After various periods of incubation, remove 1-ml samples and add 1 ml of Benedict's reagent (20.4.1). Heat the mixture in a boiling-water bath for 10 min, and cool rapidly. Positive test: yellowish-brown precipitate of cuprous oxide.

Examples: positive, *Pseudomonas aeruginosa*; negative, *Escherichia coli*.

20.1.31. 3-Ketolactose from Lactose Oxidation (19)

Grow the organism initially on slants of the following medium: 1% yeast extract, 2% glucose, 2% $CaCO_3$, and 1.5 to 2.0% agar. With a loop needle, scrape growth from the slant and deposit it as a small heap, approximately 5 mm in diameter, on a plate of lactose agar (1% lactose, 0.1% yeast extract, 2% agar). Four to six different strains may be tested on the same plate. After incubation for 1 to 2 days, flood the plate with a shallow layer of Benedict's reagent (20.4.1). Positive test: formation of a yellow ring of cuprous oxide around the growth, reaching a maximum in 2 h (approximately 2 to 3 cm).

Examples: positive, *Agrobacterium tumefaciens* and *A. radiobacter*; all other bacteria so far studied are negative.

20.1.32. Lecithinase

Streak a plate of egg yolk agar (20.3.14) to obtain well-isolated colonies. Positive test: cloudy (opaque) zone within the medium around the colony. See also Lipase (20.1.33).

Examples: positive, *Clostridium perfringens*; negative, *Clostridium butyricum.*

20.1.33. Lipase

Method 1 (7). Streak a plate of egg yolk agar (20.3.14). After incubation, remove the petri dish cover, and examine the growth closely under oblique illumination. Positive test: oily, iridescent sheen or pearly layer over colony and on surface of surrounding agar.

Examples: positive, *Clostridium sporogenes*; negative, *Clostridium butyricum.*

Method 2. Use this method for palmitic, stearic, and oleic acid esters (83). Use an agar medium supplemented with 0.01% $CaCl_2 \cdot H_2O$. Sterilize Tween 40 (palmitic acid ester), Tween 60 (stearic acid ester), or Tween 80 (oleic acid ester) by autoclaving for 20 min at 121°C. Add sterile Tween to the molten agar medium at 45 to 50°C to give a final concentration of 1% (vol/vol). Shake the medium until the Tween is completely dissolved. Dispense into petri dishes, allow to solidify, and streak the plates. Positive test: an opaque halo occurs around the colonies. Microscopically, the halo consists of crystals of calcium soaps.

Controls: positive (Tween 80), *Pseudomonas aeruginosa*; negative, *Pseudomonas putida.*

Method 3. Use this method for fats of various types (12). Prepare fat of the desired type stained with Nile blue sulfate (20.4.11). Add 1 ml of melted fat emulsion to 20 ml of an agar growth medium which has been melted and cooled to 45 to 50°C. Mix well and dispense into petri dishes. Allow to solidify, and streak the plates. Positive test: formation of a blue halo around the colonies.

20.1.34. Lysine Decarboxylase

See Method 1 for arginine dihydrolase (20.1.5). Substitute 1% L-lysine dihydrochloride (or 2% of the DL form) for the arginine.

Examples: positive, *Serratia marcescens*; negative, *Proteus vulgaris.*

20.1.35. Malonate Utilization

Inoculate a tube of malonate broth (20.3.21) from a young agar slant or broth culture. Incubate at 37°C for 48 h. Positive test: change from green to deep blue.

Examples: positive, *Klebsiella pneumoniae*; negative, *Escherichia coli.*

20.1.36. Meat Digestion by Anaerobes (7)

Inoculate prereduced chopped meat broth (20.3.10) while purging the tube with oxygen-free carbon dioxide. Incubate for up to 21 days.

Positive test: meat particles disintegrate, leaving a fluffy powder at bottom of tube.

Examples: positive, *Clostridium sporogenes*; negative, *Clostridium butyricum.*

20.1.37. Methyl Red Test

Inoculate a tube of MRVP broth (20.3.24). Incubate at 30°C for 5 days (in many cases incubation at 37°C for 48 h may also be satisfactory). Add 5 to 6 drops of methyl red solution (dissolve 0.1 g of methyl red in 300 ml of 95% ethanol, and make up to 500 ml with distilled water). Positive test: bright-red color, indicating a pH of 4.2 or less. Negative test: yellow or orange. Weakly positive test: red-orange.

Examples: positive, *Escherichia coli*; negative, *Enterobacter aerogenes.*

20.1.38. Milk Reactions

Method 1. Inoculate a tube of litmus milk (20.3.20). Reactions of cultures are as follows. Acid reaction: pink color. Alkaline reaction: blue color. Reduction of litmus: milk appears white due to discoloration of the litmus; usually occurs in bottom portion of tube. Acid curd: a firm pink clot which does not retract and is soluble in alkali. Rennet curd: a soft curd which retracts and expresses a clear grayish fluid (whey) and which is insoluble in alkali. Peptonization (proteolysis): the milk becomes translucent as a result of hydrolysis of the casein, which often begins at the top of the medium; also, the medium frequently becomes alkaline.

Method 2 (7). Use this method for anaerobes. Inoculate a tube of prereduced milk medium (20.3.22) while the tube is being purged with oxygen-free carbon dioxide. Incubate for up to 3 weeks. Check for curd and/or peptonization.

20.1.39. Motility

Prepare wet mounts from broth cultures which are young (e.g., still in the exponential phase) and which are grown at two temperatures (e.g., 37 and 20°C). In some instances glucose-containing media may inhibit the formation of flagella (15). For methods of light microscopic observation, see 3.3.4. Examine wet mounts immediately after they are prepared. In the case of anaerobes, the central portion of the mount is best for examination. For aerobes, areas near air bubbles or near the edge of the cover slip are best. Not all the cells in the mount need be motile. For a bacterial strain to be considered motile, at least one cell should be seen to alter its location in relation to at least two other cells (12). *Note:* Motility must be distinguished from Brownian motion (random "jiggling about" of cells) and from the passive motion of cells being carried about by currents under the cover slip.

Anaerobes may also be tested by drawing up some of the broth culture into a capillary tube and immediately sealing both ends of the capillary in a narrow oxygen flame (3). Examine the capillary with a high-power dry objective.

Some bacteria, such as the members of the orders *Myxobacteriales* and *Cytophagales*, lack flagella but exhibit "gliding motility" when in contact with a solid surface. For methods of detection and observation of gliding motility, see 3.3.4.

20.1.40. Nitrate Reduction and Denitrification

See 20.4.12 for preparation of reagents and caution about their carcinogenic properties.

Method 1 (66). This method is quantitative for nitrite. Inoculate a suitable broth medium supplemented with 0.1% KNO_3 and 0.17% agar (the latter to give a semisolid consistency and promote semianaerobic conditions). Gently mix the inoculum with the medium to distribute it throughout the tube. Examine the cultures periodically during incubation for gas bubble formation, which is presumptive evidence for denitrification. Control cultures grown without nitrate should not exhibit gas and should not give a positive test for nitrite by the procedure given below. To detect the ability of the organisms to reduce nitrate to nitrate, remove 0.1-ml samples from cultures of various ages, and add 2 ml of test reagent (20.4.12). Add water to make the final volume up to 4 ml, and allow the color to develop for 15 min. Estimate the intensity of the purple-red color visually on a scale from 0 to 5. The intensity of the color can also be estimated by using a spectrophotometer at 540 nm. Accumulation of nitrite with increasing age of the culture indicates the reduction of nitrate to nitrite, whereas an initial formation of nitrite followed by a decrease in its level indicates a further ability to reduce the nitrite. If no red color develops, add up to 5 mg of zinc powder per ml of the mixture. If the nitrate in the medium has not been reduced, the zinc will reduce it chemically to nitrite and a red color will appear. If no color appears, the organism has reduced all of the nitrite, and denitrification has probably occurred. In this case an earlier examination of the culture might give a positive reaction for nitrite. *Note:* Zinc powder may become inactive with long storage, and its ability to reduce nitrate to nitrate should be checked periodically with uninoculated medium.

Examples: denitrification,   *Pseudomonas*

aeruginosa; nitrate to nitrate, *Escherichia coli*; negative, *Bacillus sphaericus*.

Method 2. This method is qualitative for nitrite. Grow the organisms as described for Method 1. Examine the cultures periodically for gas bubbles, which are presumptive evidence for denitrification. To cultures of different ages, add 1 ml of solution A and 1 ml of solution B (20.4.12). A pink or red color indicates the presence of nitrite. If no red color occurs, test the culture with zinc powder as described in Method 1 to see if nitrate is present. If neither nitrite nor nitrate can be demonstrated, denitrification has probably occurred. The controls described for Method 1 also apply to Method 2.

An alternative way to demonstrate gas production is to grow the organism in nitrate broth (no agar) containing a small, inverted vial. This allows the visible accumulation of gas during denitrification.

Method 3 (66). This method allows the detection of nitrous oxide formed during denitrification and provides stronger evidence for denitrification than Method 1 or 2. Inoculate tubes or flasks of semisolid nitrate medium. Seal the vessels with serum bottle stoppers. Inject acetylene gas, generated as described in 20.2.7, through the stopper to give a final concentration of 1% (vol/vol). The acetylene will prevent N_2O formed during denitrification from being further reduced to N_2 (18). After the culture has grown, remove gas samples with a syringe, and determine the presence of N_2O by gas chromatography, using a Porapak Q (80 to 100 mesh) column (274 cm by 4.8 mm) at 35°C with helium carrier gas and a thermal conductivity detector.

20.1.41. Nucleic Acid Hydrolysis

Method 1. Use this method to test for both deoxyribonuclease (DNase) and ribonuclease (RNase). Dissolve deoxyribonucleic acid (DNA) or ribonucleic acid (RNA) (Sigma Chemical Co., St. Louis, Mo.) in distilled water at a concentration sufficient to give 0.2% when added to an agar growth medium. DNA is readily soluble in water. To dissolve RNA in water, slowly add 1 N NaOH, but do not allow the pH to become greater than 5.0. Add the nucleic acid solution to the agar medium just prior to autoclaving (121°C for 15 min), and pour plates when the medium has cooled to 50°C. Make a single streak of the organism across the solidified medium. After incubation, flood the plate with 1 N HCl. Positive test: a clear zone around the growth indicates DNase or RNase activity.

Examples: positive (DNase), *Serratia marcescens*; negative (DNase), *Enterobacter aerogenes*.

Method 2. Use this method to test for only DNase. Add 100 mg of toluidine blue O (Allied Chemical Corp., New York, N. Y.) per liter of DNA-containing test medium (see Method 1) prior to autoclaving (79). Alternatively, add methyl green solution (20.4.9) to give a final concentration of 0.005% (85). For the toluidine blue medium, make a single streak of the test organism across the plate; for the methyl green medium, streak the plate to obtain isolated colonies. Positive test: with the toluidine blue plates, a bright rose-pink zone surrounds the growth; with the methyl green plates, the colonies are surrounded by a colorless zone on an otherwise green plate. These methods provide a more readily observable reaction for DNase than Method 1. *Note:* The two dye methods should not be used with bacteria grown under anaerobic conditions because the indicator sytem may be nullified (75).

Method 3 (32). Use this method as a more rapid test for DNase although it may not detect very weak DNase activity (32). Prepare a test medium by combining 31.5 mg of DNA (Difco Laboratories), 5.0 ml of 0.1 M $CaCl_2$, and 5.0 ml of 0.1 M $MgCl_2$ in 100 ml of Tris buffer (2-amino-2-hydroxymethyl-1,3-propanediol), pH 7.4. Make a heavy (milky) suspension of the organism in 0.5 ml of this medium, and incubate for 2 h at 37°C. Add 1.6 ml of methyl green solution (20.4.9) to 100 ml of distilled water; then add 2 drops of this solution to the bacterial suspension. Mix the contents of the tube, and incubate for an additional 2 h at 37°C. Positive reaction: colorless tube. Negative test: blue-green tube.

20.1.42. O/129

Place one or two crystals of 2,4-diamino-6,7-diisopropylpteridine phosphate (Sigma Chemical Co., St. Louis, Mo.) on an inoculated agar plate. Alternatively, soak sterile paper disks (6 to 8 mm in diameter) in a solution of the reagent (0.1 g in 100 ml of acetone), dry the disks, and apply to an inoculated agar plate. Positive test: zone of growth inhibition around the crystals or disks.

Examples: positive, *Vibrio parahaemolyticus*; negative, *Escherichia coli*.

20.1.43. Optochin

Place a sterile, commercially available (Difco Laboratories, BBL Microbiology Systems) optochin differentiation disk or a sterile paper disk impregnated with 0.02 ml of a 0.025% solution of optochin (ethylhydrocupreine hydrochloride; Sigma Chemical Co., St. Louis, Mo.), on an inoculated agar plate. Positive test: zone of growth inhibition around disk.

Examples: positive, *Streptococcus pneumoniae*; negative, other alpha-hemolytic streptococci.

20.1.44. Ornithine Decarboxylase

See Method 1 for arginine dihydrolase (20.1.5). Substitute 1% L-ornithine dihydrochloride (or 2% of the DL form) for the arginine.

Examples: positive, *Enterobacter aerogenes*; negative, *Proteus vulgaris*.

20.1.45. Oxidase

Method 1. Moisten a piece of filter paper with a few drops of a 1% solution of tetramethyl-*p*-phenylenediamine dihydrochloride (Sigma Chemical Co., St. Louis, Mo.), prepared on the same day as it is to be used. With a platinum loop (ordinary Nichrome wire loops may give false-positive reactions), remove growth from the surface of an agar medium and smear it on the moistened paper. Positive test: development of a violet or purple color in 10 s.

Examples: positive, *Pseudomonas aeruginosa*; negative, *Escherichia coli*.

Method 2. This method is less sensitive than Method 1, but may be more convenient to use. To colonies on an agar plate add a few drops of a 1:1 (vol/vol) mixture of 1% α-naphthol in 95% ethanol and freshly prepared 1% aqueous dimethyl-*p*-phenylenediamine oxalate (Difco Laboratories). Alternatively, add 0.2 ml of the α-naphthol and 0.3 ml of the dimethyl-*p*-phenylenediamine to a broth culture, and mix vigorously. Positive test: colonies or broth culture become purplish blue within 10 to 30 s.

20.1.46. Oxidation/Fermentation

A test carbohydrate (usually glucose) should be prepared as a stock solution, sterilized by filtration, and added aseptically to an autoclaved semisolid basal medium at 45 to 50°C to give a final concentration of 0.5 to 1.0%. The most widely used basal medium is that of Hugh and Leifson (44) (20.3.17). If the organism cannot grow in this medium because of complex nutritional requirements, the further addition of 0.1% yeast extract or 2% blood serum may permit growth. In cases in which acid production may be masked by the production of alkaline substances from the peptone, a medium such as that devised by Board and Holding (20) (20.3.8) may be suitable. In cases in which the bromothymol blue indicator present in the medium is inhibitory to the organism being tested, substitution of another indicator such as phenol red or bromocresol purple may be necessary. For marine bacteria, the modified O-F (MOF) medium of Leifson (55) (20.3.19) is useful.

To perform the test, place the tubes of the semisolid carbohydrate medium in a boiling-water bath for a few minutes to remove most of the dissolved oxygen; cool the tubes quickly, and inoculate them by stabbing with a straight needle. After inoculation, overlay the medium in only one of the tubes with 10 mm of sterile melted petrolatum; this tube is the "anaerobic" tube. No overlay is used for the second tube (the "aerobic" tube). Prepare the following controls to detect nonspecific reactions: (i) inoculate similar media lacking the carbohydrate, and (ii) incubate sterile media containing the carbohydrate.

Interpret results as follows. (i) If only the aerobic tube is acidified, the organism catabolizes the carbohydrate by oxidation (an "O" reaction). Growth should be evident in the aerobic tube. (ii) If both the aerobic and the anaerobic tubes are acidified, the organism is capable of fermentation (an "F" reaction). Growth should be evident in both tubes. (iii) If neither tube becomes acidified, the organism is unable to catabolize the carbohydrate. If no growth is evident, the medium may be lacking some nutrient required for the organism being tested.

Examples: F reaction with glucose, *Staphylococcus epidermidis*; O reaction with glucose, *Micrococcus varians*; negative with glucose, *Pseudomonas lemoignei*.

20.1.47. Phenylalanine Deaminase

Inoculate heavily a phenylalanine agar slant (20.3.26). After incubation, allow 4 to 5 drops of 10% ferric chloride solution to run down the slant. Positive test: green color, indicating formation of phenylpyruvic acid.

Examples: positive, *Proteus vulgaris*; negative, *Escherichia coli*.

20.1.48. Phosphatase

Method 1. Use this method to test for both acid and alkaline phosphatases. To an agar growth medium such as nutrient agar (20.3.25), melted and cooled to 45 to 50°C, aseptically add a sufficient amount of a filter-sterilized 1% solution of the sodium salt of phenolphthalein diphosphate (Sigma Chemical Co., St. Louis, Mo.) to give a final concentration of 0.01%. Incubate streaked plates for 2 to 5 days. Place 1 drop of ammonia solution (specific gravity 0.880, or 28 to 30%) in the lid of the inverted plate, and replace the culture over it to allow the ammonia fumes to reach the colonies. Positive test: colonies become red, as a result of the presence of free phenolphthalein.

Examples: positive, *Staphylococcus aureus*; negative, *Micrococcus varians*.

Method 2 (22). Use this method to test for only acid phosphatase. Prepare a very dense suspension of cells in saline (0.85% NaCl). Add 0.3 ml of the suspension to 0.3 ml of a substrate solution consisting of 0.1 M citrate buffer, pH 4.8, containing 0.01 M disodium *p*-nitrophenyl phosphate (Sigma Chemical Co.). Incubate the mixture for up to 6 h at 37°C. Then add 0.3 ml of 0.04 M glycine buffer which has been adjusted to pH 10.5 with NaOH. Also use a control lacking the bacteria. Positive test: a yellow color indicates acid phosphatase activity.

Method 3. Use this method to test for only alkaline phosphatase. This method is similar to Method 2, but the *p*-nitrophenyl phosphate is dissolved in the glycine buffer instead of the citrate buffer. No further addition of glycine buffer is necessary to develop the yellow color.

20.1.49 Poly-β-Hydroxybutyrate Presence (53, 90)

Grow a broth culture of the organism under conditions conducive to the formation of poly-β-hydroxybutyrate (PHB). For aerobic bacteria these conditions usually are nitrogen limitation of exponential growth in the presence of excess carbon and energy source, and also oxygen limitation.

Staining procedures may be used as presumptive evidence for PHB inclusions (see 3.3.11); however, chemical analysis is required for a definitive test.

Chemical analysis. Place 1 ml of the culture into a conical glass centrifuge tube, add 9 ml of alkaline hypochlorite reagent (see 20.4.19), and incubate at room temperature for 24 h (90). Alternatively, add 8 ml of commercial 5% hypochlorite (Clorox bleach) to 2 ml of culture and incubate (72). The hypochlorite will digest most cell components but will not digest the PHB granules. Centrifuge the mixture to collect the granules, and discard the clear supernatant fluid. Wash the sediment by suspending it in 10 ml of distilled water, centrifuging, and discarding the supernatant fluid. Wash the sediment once more with water, twice with 10-ml portions of acetone, and finally twice with 10-ml portions of diethyl ether. Dry the sediment, and add 2 ml of concentrated sulfuric acid. Place the tube in a boiling-water bath for 10 min. Cool to room temperature. Using an ultraviolet spectrophotometer and "far ultraviolet" quartz cuvettes, determine the absorption spectrum of the sample at 5-nm increments from 215 to 255 nm against a blank of plain concentrated sulfuric acid. (The sample may need to be diluted 1:10 or more with sulfuric acid if its absorbance at 235 nm is greater than 1.0.) Positive test: occurrence of an absorption peak at 235 nm, due to crotonic acid which is formed from PHB by the action of the sulfuric acid.

Examples: positive, *Pseudomonas lemoignei*; negative, *Pseudomonas aeruginosa*.

20.1.50. Poly-β-Hydroxybutyrate Hydrolysis (88)

Hydrolysis of PHB is very useful for distinguishing among the members of the genus *Pseudomonas*; however, the PHB granule suspension required for the test cannot be purchased commercially and is tedious to prepare. Details of its preparation are given in 20.4.15.

Prepare a plate of mineral agar, such as that described in 20.2.5, lacking carbon sources. To another portion of the medium that has been melted and cooled to 45 to 50°C, add a sufficient amount of the sterile concentrated PHB granule suspension to give a final concentration of 0.25% (wt/vol) in the medium. Pour the granule-agar suspension over the surface of the plates of solidified medium to form a thin overlay.

After the overlay solidifies, inoculate a washed suspension of the organism to be tested onto a small area of the plate. (Several strains can be tested on a single plate.) Incubate the plate for several days. Positive test: growth of the organism at the site of inoculation and clearing of the surrounding medium due to depolymerization of the PHB granules.

Examples: positive, *Pseudomonas lemoignei*; negative, *Pseudomonas acidovorans*.

20.1.51. Pyocyanine and Other Phenazine Pigments

Inoculate a slant of medium A (20.3.1). Examine the slant at 24, 48, and 72 h. Positive test: pigments that diffuse into the agar and stain it. In some cases the pigments may be only slightly water-soluble and may precipitate as crystals in the medium.

Examples of pyocyanine: blue color which stains the medium; produced by *Pseudomonas aeruginosa*. Example of phenazine-1-carboxylic acid: orange-yellow, may crystallize in the colonies and surrounding medium; produced by *Pseudomonas aureofaciens*. Example of oxychlororaphin: stains the medium pale yellow; produced by *Pseudomonas chlororaphis*. Example of chlororaphin: only very sparingly soluble, accumulates as isolated green crystals in the growth at the butt of the slant; produced by *P. chlororaphis*.

20.1.52. Spores (Endospores)

For staining of spores, see 3.3.7. The media used for growth of cultures should contain ade-

quate levels of Ca^{2+} and Mn^{2+}. Media with and without carbohydrates should be inoculated, and cultures of different ages should be examined. *Note:* Inclusion granules (e.g., PHB) may be mistaken for endospores.

If sporelike structures are observed, a heat test should be performed for confirmation. Inoculate a tube of broth with growth from solid or liquid media in which sporulation is suspected to occur. Take care not to touch the wall of the tube with the inoculum. Place the tube in a water bath at 80°C along with an uninoculated tube of broth containing a thermometer. Make sure the level of the water in the bath is higher than the level of the broth. Incubate for 10 min, beginning when the thermometer reaches 80°C. Cool the tube, and incubate at the optimum growth temperature to see if growth occurs. *Note:* Some endospores, such as occur with certain strains of *Clostridium botulinum*, may be killed at 80°C, necessitating the use of lower temperatures for testing heat resistance (e.g., 75 or 70°C). Also, apply the "popping" test (3.3.7). Also, certain species of *Nocardia* form "chlamydospores" or "microcysts," which can survive 80°C for several hours, and yet are not true endospores.

20.1.53. Starch Hydrolysis

Prepare an agar medium on which the organism can grow well, and add 0.2% soluble starch before boiling. The medium should preferably contain no glucose, as this may diminish starch hydrolysis (93). Sterilize the medium at 115°C for 10 min. Make a single streak of the organism across the center of a plate of the starch agar. After incubation, flood the plate with iodine solution (the iodine normally used for the Gram stain is suitable). Positive test: colorless area around the growth.

Examples: positive, *Bacillus subtilis*; negative, *Escherichia coli*.

20.1.54. Triple Sugar Iron Agar Reactions

Inoculate a slant of triple sugar iron agar (20.3.34) by using a straight needle. First, stab the butt down to the bottom, withdraw the needle, and then streak the surface of the slant. Use a cotton stopper or loosely fitting closure to permit access of air. Read results after incubation at 37°C for 18 to 24 h. Three kinds of data may be obtained from the reactions.

1. *Sugar fermentations*

   Acid butt, alkaline slant (yellow butt, red slant): glucose has been fermented but not sucrose or lactose.

   Acid butt, acid slant (yellow butt, yellow slant): lactose and/or sucrose has been fermented.

Alkaline butt, alkaline slant (red butt, red slant): neither glucose, lactose, nor sucrose has been fermented.

2. *Gas Production*
Indicated by bubbles in the butt. With large amounts of gas, the agar may be broken or pushed upward.

3. *Hydrogen sulfide production*
Indicated by a blackening of the butt, due to reaction of H_2S with the ferrous ammonium sulfate to form black ferrous sulfide.

20.1.55. Urease

Method 1. Inoculate the surface of a Christensen urea agar slant (20.3.12) and incubate. Positive test: formation of a red-violet color.
Examples: positive, *Proteus vulgaris*; negative, *Escherichia coli*.

Method 2 (47). This method is useful for organisms that cannot grow on Christensen urea agar. Prepare the following solution: BES buffer [*N,N*-bis-(2-hydroxyethyl)-2-aminoethanesulfonic acid; ICN Pharmaceuticals, Cleveland, Ohio], 0.1065%; urea, 2.0%; and phenol red, 0.001%; pH 7.0. Sterilize by filtration, and aseptically dispense 2.0-ml volumes into small tubes. Centrifuge a broth culture of the organism, and suspend the cells in sterile distilled water or saline to a dense concentration. Add 0.5 ml of the cell suspension to a tube of the urea medium. A control medium lacking urea should be inoculated similarly. Incubate the tubes for 24 h. Positive test: red-violet color.

20.1.56. Voges-Proskauer Reaction

Inoculate two tubes of MRVP broth (20.3.24). Incubate one tube at 37°C and the other at 25°C. After 48 h, remove 1 ml of culture to another tube, add 0.6 ml of 5% (wt/vol) α-naphthol dissolved in absolute ethanol, and mix thoroughly. Then add 0.2 ml of 40% aqueous KOH. Mix well, and incubate the tube in a slanted position to increase the surface area of the medium (the reaction is dependent upon oxygen). Examine after 15 and 60 min. Positive test: strong red color that begins at the surface of the medium.
Examples: positive, *Enterobacter aerogenes*; negative, *Escherichia coli*.

20.1.57. X and V Factor Requirements

Inoculate a plate of soybean-casein digest agar (20.3.31): moisten a sterile cotton swab in the suspension, squeeze out excess fluid by pressing and rolling the swab against the wall of the tube, and roll the swab across the entire surface of the plate (do not "scour" the surface of the agar). Use sterile paper strips impregnated with the X factor (heme) or the V factor (NAD [nicotinamide adenine dinucleotide], DPN [diphosphopyridine nucleotide]). The strips are obtainable commercially (Difco Laboratories, Detroit, Mich.; BBL Microbiology Systems, Cockeysville, Md.). With a flamed forceps place an X strip on the surface of the seeded agar, flame the forceps, and then place a V strip on the agar approximately 2 cm from the X strip. Incubate the plates for 1 to 2 days, and interpret the results as follows. A requirement for either X or V factor, but not both, is indicated by growth only around the appropriate strip. A requirement for both factors is indicated by occurrence of growth only between the two strips. If growth occurs over the entire plate, neither factor is required.
Examples: positive for both X and V, *Haemophilus haemolyticus*; positive for V only, *Haemophilus parainfluenzae*; negative for both X and V, *Escherichia coli*.

20.2. SPECIAL TESTS

20.2.1. Metabolites by Gas Chromatography
(7; also, see part 16.3.4)

Acids and alcohols.

The following procedures were designed for identification of anaerobes, but they can also be used for aerobic and facultative organisms. Acidify 1 ml of culture medium containing glucose or other fermentable carbohydrate in a 12 by 75 mm tube by adding 0.2 ml of 50% (vol/vol) H_2SO_4; samples thus acidified can be stored for future analysis. Add 0.4 g of NaCl and 1 ml of diethyl ether. Stopper the tube with a cork stopper. Mix the contents of the tube by gently inverting the tube about 20 times to extract the free fatty acids into the ether phase. Centrifuge the mixture for a few minutes to break the emulsion and remove the ether layer (upper) from the aqueous layer. It is advisable not to remove too much of the ether layer in order to avoid contaminating the ether with the water. Place the ether layer into a new 12 by 75 mm tube, and add anhydrous $CaCl_2$ (4 to 20 mesh) to about one-fourth the volume of the ether, to remove traces of dissolved water. Inject 14 μl into the gas chromatograph in order to detect short-chain volatile fatty acids and alcohols. Uninoculated medium should be acidified, extracted, and examined as a control.

Pyruvic, lactic, fumaric, and succinic acids are nonvolatile and must first be converted to their methyl esters for detection by gas chromatography. Place 1 ml of the culture in a 12 by 75 mm tube, and add 2 ml of methanol and 0.4 ml of 50% H_2SO_4 (2 ml of BF_3 methanol [boron trifluoride methanol, obtainable from any chromatographic supply house] may be added in place of plain methanol). Heat the mixture at 60°C for 30 min. Add 1 ml of water and 0.5 ml of chloroform to the tube, stopper the tube tightly, and gently invert the tube about 20 times. Remove the chloroform layer (bottom), and inject 14 μl into the gas chromatograph.

A gas chromatograph equipped with a thermal conductivity (TC) detector is recommended. A flame ionization (FI) detector may be used; however, it will not detect formic acid. Use a column 6 ft (1.8 m) long and 0.25 in. (6 mm) in diameter, of stainless steel, aluminum, or glass. Pack the column with Supelco SP-1000, 100 to 200 mesh (Supelco Inc., Bellefonte, Pa.), or with 5% FFAP on Chromosorb-G. Both volatile and methylated fatty acids are detected with the same column. Set the column oven temperature at 135 to 145°C. The carrier gas is helium at a flow rate of 120 cm^3/min. If the instrument is equipped with separate injector and detector oven heaters, set these 20 to 30°C higher than the column temperature. Use a 1-mV recorder with a chart speed of 0.5 in. (1.3 cm)/min. Prepare standards for volatile acids, alcohols, and nonvolatile acids (20.4.18), and extract them in the same manner as the culture samples. Such standards may also be purchased from chromatography supply houses.

The presence or absence, as well as the approximate amounts, of the various metabolic end products constitute patterns that are very helpful in identifying bacteria. More quantitative information can be obtained by use of careful technique and by use of standard curves based on known amounts of each acid or alcohol. The order of elution of alcohols and volatile fatty acids from the gas chromatograph column is as follows: ethanol, n-propanol, isobutanol, n-butanol, isopentanol, n-pentanol, acetic acid, formic acid, propionic acid, isobutyric acid, n-butyric acid, isovaleric acid, n-valeric acid, isocaproic acid, n-caproic acid, and heptanoic acid. For methyl esters of the nonvolatile acids, the order of elution is as follows: pyruvic acid, lactic acid, oxalacetic acid, oxalic acid, methylmalonic acid, malonic acid, fumaric acid, and succinic acid.

Hydrogen and methane.

Analyze cultures grown in a medium containing glucose or other fermentable carbohydrates for the production of hydrogen and methane. Stopper the cultures with a rubber stopper and incubate. After the cultures have grown, remove gas samples from the culture tubes by pushing the needle of a syringe through the rubber stopper and into the gas phase of the culture. Use a 1- or 5-ml disposable plastic syringe with a 1.5-in. (3.8-cm) needle and start the needle into the culture tube at a point where the rubber stopper and glass wall meet; push the needle through the stopper at an angle that will permit the tip of the needle to enter the gas phase of the culture. Withdraw 0.7 to 1.0 ml of the gas phase, and inject it into the gas chromatograph.

The gas chromatograph must be equipped with a thermal conductivity (TC) detector. Use either nitrogen or argon as the carrier gas, with a flow rate of 25 to 30 cm^3/min. Use an aluminum or stainless-steel column with a length of 6 ft and a diameter of 0.25 in., packed with silica gel (grade 12, 80 to 100 mesh; Supelco Inc., Bellefonte, Pa.). Set the column oven temperature to 25 to 30°C; if the instrument is equipped with separate injector and detector ovens, set these at the same temperature as the column oven or a few degrees higher. Standards for gas analysis can be purchased from any chromatography supply house; natural gas can be used to determine the elution point of methane. For mixtures of hydrogen and methane, hydrogen will be eluted from the column before methane.

20.2.2. Lactic Acid Optical Rotation (7, 24)

To 10 ml of the culture, and also to 10 ml of uninoculated medium, add 0.2 ml of 50% (vol/vol) H_2SO_4. Use 1 ml of the acidified culture to determine the concentration of lactic acid (milliequivalents per 100 ml) by gas chromatography (see 20.2.1). This concentration must be at least 1 meq/100 ml in order to determine optical rotation by the method below.

See 20.4.8 for preparation of perchloric acid solution, buffer, lactic dehydrogenase solution, and nicotinamide adenine dinucleotide (NAD) solution.

Prepare the L-(+)-lactic acid working standard as follows. Add 0.0296 g of L-(+)-lactic acid (calcium salt) to 5.0 ml of acidified uninoculated medium. Add 5.0 ml of distilled water. Add 1.0 ml of this solution to 1.0 ml of perchloric acid solution. Centrifuge and save the supernatant, which is the working standard (=1.0 μmol of L-(+)-lactic acid per 0.1 ml).

To 1.0 ml of acidified culture, add 1.0 ml of distilled water and 2.0 ml of perchloric acid solution. Centrifuge and save the supernatant, which is the sample to be tested for optical rotation.

To 1.0 ml of acidified uninoculated medium, add 1.0 ml of distilled water and 2.0 ml of perchloric acid solution. Centrifuge and save the supernatant, which is the blank stock solution.

Use seven glass tubes, 12 by 75 mm. Add 0.1

ml of the blank stock solution to tube 1. Add 0.02, 0.04, 0.06, 0.08, and 0.10 ml of the working lactic acid standard to tubes 2 through 6, respectively. Make up the volume in each tube to 0.10 ml by adding 0.08, 0.06, 0.04, 0.02, and 0.00 ml of blank stock solution. To tube 7, add 0.10 ml of the sample to be tested. Add 3.0 ml of buffer to each of the seven tubes, followed by 0.03 ml of lactic dehydrogenase solution. Then add 0.1 ml of NAD solution, cover each tube with Parafilm, and invert gently several times to mix (do not shake the tubes). Incubate at 30°C for 60 min, being careful to start timing from the time the NAD was added to the first tube.

Read the absorbance of the contents of each tube. Use a wavelength of 340 nm if possible. If the instrument is not capable of reading at this wavelength, use a wavelength as near to 340 nm as possible; for example, if a Spectronic 20 spectrophotometer (Bausch & Lomb, Inc., Rochester, N.Y.) is used, set the wavelength to 366 nm. Use tube 1 as the blank to set the instrument to 0.00 absorbance. Read the absorbance values, being careful to use the same sequence as that used for addition of the NAD. Use the same cuvette for all readings; pour out the contents after making a reading, rinse out the cuvette, and fill it with next solution to be read.

Plot micromoles of L-(+)-lactic acid (0.0, 0.2, 0.4, 0.6, 0.8, and 1.0 μmol, respectively, for tubes 1 through 6) versus absorbance on linear graph paper. Draw the line of best fit. By comparison with this standard curve, determine the micromoles of L-(+)-lactic acid in the test sample (tube 7). Multiply this value by 4 to obtain milliequivalents of L-(+)-lactic acid per 100 ml of culture. Calculate the percent L-(+)-lactic acid by the following formula:

$$\frac{\text{meq of L-(+)-lactic acid (from enzyme reactions) per 100 ml}}{\text{meq of total lactic acid (from gas chromatography) per 100 ml}} \times 100$$

20.2.3. Commercial Multitest Systems

A number of convenient and rapid systems are commercially available for the identification of various groups of bacteria, especially the family *Enterobacteriaceae*. Complete information concerning the methodology can be obtained from the manufacturers. Charts, tables, coding systems, or characterization profiles designed to be used with the systems are also available from the manufacturers. A brief description of some of these systems is given below, together with references which the reader may consult to estimate the accuracy and reliability of the systems.

The API 20E system for *Enterobacteriaceae* and nonfermenters (Analytab Products, Plainview, NY 11803) consists of a plastic strip of 20 microtubes containing various dehydrated media, with which 23 biochemical tests can be performed within 18 to 24 h. The media are reconstituted by adding a dilute suspension of the organism to be tested to the microtubes. Drying of the suspension is prevented by the addition of water to the plastic chamber containing the tubes. Sterile mineral oil is added to certain microtubes to provide a seal against oxygen. Evaluations of reliability may be found in references 16, 31, 40, 68–70, 76, 78, and 81.

The API 20A system for anaerobes (Analytab Products) is similar to the API 20E system. A colony is suspended in an anaerobic medium which is provided, and the suspension is used to reconstitute various dehydrated media in microtubes under aerobic conditions. The strip of microtubes is then placed in an anaerobe jar or anaerobic chamber and incubated for 24 to 48 h at 35 to 37°C (62).

The Entero-Set 20 system for *Enterobacteriaceae* (Fisher Scientific Co., Diagnostics Div., Orangeburg, NY 10962) is similar to the API system but with microtubes designed to allow escape of air while filling with bacterial suspension. Oil is added to certain microtubes to seal against oxygen.

The Minitek system for *Enterobacteriaceae*, nonfermenters, *Neisseria*, and anaerobes (BBL Microbiology Systems, Cockeysville, MD 21030) consists of a plastic plate containing 12 wells. A dispensing apparatus deposits into the wells various disks which have been impregnated with substrates. A regimen of disks selected by the manufacturer, or by the user, is employed. A suspension of the test organism prepared in an appropriate Minitek broth is added to each well by means of an automatic pipette that has a sterile, disposable tip. Some wells are overlaid with oil. The incubation time is 18 to 24 h for *Enterobacteriaceae*, 24 to 48 h for nonfermenters, and 4 h for *Neisseria*. A plastic humidor prevents evaporation during incubation. For anaerobes, a cell suspension prepared in an anaerobic broth is dispensed into the disk-containing wells, and the plates are subsequently incubated for 48 h in an anaerobe jar or anaerobic chamber (17, 40, 41, 49, 89).

The PathoTec Rapid I-D system for *Enterobacteriaceae* and other groups (General Diagnostics Div., Warner-Lambert Co., Morris Plains, NJ 07950) consists of 10 test strips impregnated with reagents for such tests as phenylalanine deaminase, lysine decarboxylase, malonate utilization, etc. The strips are placed in tubes containing a suspension of the test organism or, in some cases, are rubbed with cells from colonies. The reactions are read in ca. 4 h, except for the oxidase test, which is read in 30 s (40, 68, 69).

The Enterotube system for *Enterobacteriaceae* (Roche Diagnostics, Nutley, NJ 07110) consists of a tube containing eight compartments, each with a different agar medium. The compartments are traversed by a channel that contains a thin, metal rod. The protruding end of the rod is inoculated from a colony and then drawn through each compartment, thereby inoculating the various media. Some of the media have been overlaid with wax to seal the media from air. Results are read after incubation for 24 h (39, 63, 68, 69, 92).

The Oxi/Ferm system for nonfermentative gram-negative rods (Roche Diagnostics) is similar to the Enterotube system but is designed for nonfermentative organisms (31, 48, 71, 80).

The Corning r/b Enteric Differential system for *Enterobacteriaceae* (Corning Medical Microbiology Products, Roslyn, NY 11576) consists of two conventional-sized tubes of agar media with a constriction in the middle to separate each tube into two compartments. The system is capable of testing nine different characteristics; two additional tubed media provide six more tests if necessary. The tubes are inoculated by stabbing to the bottom (through the constriction) and streaking the slanted surface after withdrawal of the needle. Results are read in 18 to 24 h (59, 78, 86).

In the Corning N/F system for oxidative, gram-negative bacteria (Corning Medical Microbiology Products), inoculation of two tubed media (GNF tube and 42P tube) allows determination of five initial screening reactions in 18 to 24 h. Further tests are performed by use of a circular plastic dish (Uni-N/F-Tek plate) which has wells filled with various prepared agar media, allowing 12 additional reactions in 24 to 48 h. The wells of agar are inoculated with a drop of bacterial suspension.

The Micro-ID system for *Enterobacteriaceae* (General Diagnostics Div., Warner-Lambert Co.) consists of a card with 15 test chambers containing filter paper disks impregnated with reagents and substrates. A suspension of bacteria is used to inoculate the chambers. Oil overlays to seal against oxygen are not needed. The results are read after incubation at 37°C for 4 h (16).

20.2.4. Multipoint Inoculators

The widespread commercial availability of sterile, plastic microtiter plates (plastic dishes with 96 miniature wells, each having a volume of ca. 0.2 ml) has led to development of miniaturized procedures for determination of acid production from carbohydrates, gas production, and other biochemical reactions.

One can make a multiple inoculator by simply inserting the pointed ends of 27-mm stainless-steel pins into wax-filled wells or a microtiter plate (36). One can also make an inoculator by inserting pins or brads into a wood, plastic, or metal template so that they are in the exact pattern of the wells of a microtiter plate (37). An inoculator made entirely of metal has the advantage of being autoclavable, but one can effectively "sterilize" the other inoculators by dipping the pinheads into alcohol and then flaming for 1 to 2 s with the pinheads pointing upwards (36). Charge an inoculator by lowering the device onto a master microtiter plate where each well contains a different strain of bacteria; in this way each pinhead becomes charged with ca. 0.006 ml of culture. Then lower the inoculator onto a microtiter plate with its wells filled with carbohydrate broth, thereby causing each pinhead to inoculate its particular strain of bacteria into a separate well of broth. Return the inoculator to the master plate for recharging, and use it to inoculate a second microtiter plate that has its wells filled with a different carbohydrate broth, etc. To determine gas production from carbohydrates, cover each inoculated well with a sterile overlay of a sealant such as 1 part of Amojell (American Oil Co.) and 1 part of mineral oil (38). Cover the inoculated plates with their lids and incubate them. Acid production is determined by a change in the color of the pH indicator in the medium. Gas production is indicated by the accumulation of bubbles under the overlay.

Filling the wells of a microtiter plate with sterile media can be greatly facilitated if one uses a multichannel automatic pipette (Cooke Engineering Co., Alexandria, Va.) or similar apparatus. One can construct a homemade device such as that described by Wilkins et al. (96) as follows. Remove the tips and hubs from eight 18-gauge hypodermic needles and braze the needles into holes in a 13.5-cm-long piece of 7-mm stainless-steel tubing so that the needle centers are the same distance apart as the microtiter wells (Fig. 2). Close off one end of the tubing and braze a Luer-Lok connecter onto the other end. After sterilizing the assembly, aseptically connect the Luer-Lok to a sterile automatic pipetting syringe (Cornwall type; Becton, Dickinson & Co., Rutherford, N.J.) equipped with a sterile filling hose (Fig. 2). With the end of the filling hose immersed in a flask of sterile medium, depress the plunger of the adjustable

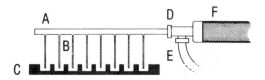

FIG. 2. *Diagram of apparatus for filling eight micro-titer plate wells simultaneously. (A) Stainless-steel tubing; (B) 18-gauge hypodermic needles brazed to tubing; (C) microtiter plate with wells; (D) Luer-Lok connector brazed to end of tubing; (E) filling hose attached to Cornwall syringe; (F) barrel and plunger of Cornwall syringe. By depressing the plunger, the eight wells can be filled simultaneously with sterile medium.*

syringe several times to fill the syringe. Then place the eight-needle manifold into a row of eight wells on a microtiter plate, and depress the plunger of the syringe to deliver 0.2 ml of medium simultaneously to each of the eight wells. Continue to fill rows of wells in this manner until all of the wells on the plate have been filled. Rinse the assembly by flushing it with sterile water (or boiling water, if an agar medium has been dispensed), and then flush it several times with the next medium to be dispensed.

Biochemical tests other than acid and gas production can be performed with microtiter plates. For example, perform the indole test by adding 2 drops of Kovacs' reagent to the well cultures, or for the Voges-Proskauer test add one loop of α-naphthol followed by one loop of KOH (37). Some tests may not be suitable for microtiter plates; e.g., in testing for urease, the ammonia produced by a urease-positive organism may diffuse to adjacent wells to give false-positive reactions.

Pinhead inoculators may be used to inoculate the surface of diagnostic agar media contained in conventional petri dishes (36, 57). Also, inoculators with the pointed ends of brads or pins projecting from the block may be used to pierce the agar surface (65). However, when one deposits various strains on a single petri dish, problems of overgrowth, spreading, or mutual interference may occur. These problems can be eliminated if one uses a petri dish divided into compartments (Dynos Plastics Ltd., Surbiton, Surrey, England) (57), mini-ice cube trays (36), or microtiter plates (37, 96). Tiny slants, such as triple sugar iron agar slants, may be prepared in microtiter plates and inoculated by a piercing inoculator (37).

For multipoint inoculation of anaerobes, dispense sterile semisolid media into microtiter plate wells, and incubate the plates with their lids in position overnight aerobically at 37°C to check for contamination and also to dry the plates slightly (95). Place the plates in an anaer-

obic glove box, and allow them to reduce for 2 days prior to inoculation. Inoculate the plates with a piercing inoculator within the glove box. Although each plate contains 96 wells, inoculate only 48 of them (in a checkerboard pattern) and leave 48 wells uninoculated to serve as controls to detect nonbiological pH changes or possible migration of motile bacteria from one well to another. After inoculation, cover each plate with a sheet of sterile plastic tape (nontoxic plate sealer, Cooke Engineering Co.) by laying the tape with adhesive side up on a rubber mat and pressing the inverted plate onto it. Roll the plates with the tape roller to assure a firm seal on each well. To allow for the escape of gases during growth, punch a pinhole into the tape over each well by means of the tape perforator. Incubate the plate within the anaerobic glove box for 3 days. Determine acid production by means of a narrow pH electrode (such as an S-30070-10 electrode, Sargent-Welch Co., Skokie, Ill.) inserted into the medium of each well. Do not rinse the electrode between wells, because each insertion pushes the previous agar medium upward away from the electrode tip (95).

20.2.5. Velveteen for Multiple Inoculation

The multiple inoculation method using velveteen is designed to determine the nutritional characteristics of a large number of strains. The procedure given below, in which the ability of 267 strains of pseudomonads to use 146 different organic compounds as carbon and energy sources are determined (88), exemplifies the method.

Prepare a mineral agar base with the following composition: 40 ml of Na_2HPO_4-KH_2PO_4 buffer (1.0 M, pH 6.8), 20 ml of Hutner's vitamin-free mineral base (20.3.18), 1.0 g of $(NH_4)_2SO_4$, 20 g of agar, and 940 ml of distilled water. Supplement the medium with 0.1 to 0.2% of the carbon source to be tested, and dispense the medium into petri dishes. Dry the sterile plates by incubating them at 37°C for 2 to 3 days, and store them at room temperature in plastic bags until used (up to 4 or 5 weeks of storage). Prepare a master plate by "patching" (depositing inoculum in a small area) up to 23 different strains on a plate of mineral agar base containing 0.5% yeast extract. After growth occurs, prepare submaster plates by the velveteen replication method originally devised by the Lederbergs (54). Also see 13.6.2. In this method, a piece of sterilized velveteen cloth is affixed to a piston having a diameter slightly smaller than that of the petri dish. By gentle pressing of the velveteen onto the master plate and then onto a series of sterile plates, the plates become inoculated with the various strains in exactly the same pattern as on the master plate. Nine submaster plates can be prepared from a single master plate. After growth occurs on the submaster plates, use each plate to print test plates for nine different carbon sources, followed by a terminal yeast extract agar plate to verify that transfer of all the strains has occurred. Read the plates at 48 and 96 h, and score the results as follows: 0, growth no greater than on a control plate lacking any carbon source; +, good growth; ±, scanty growth but significantly greater than on the control plate; and +M, late growth of a few colonies in the area of the patch, presumed to arise from mutants in the inoculum. When strains of markedly different nutritional character are being tested together on a plate, cross-feeding (diffusion of nutrients produced by one strain across to adjacent strains, resulting in the latter strains giving a false-positive growth response) or interference between strains may occur. Consequently, follow up each initial random screening by a second screening, using a different arrangement or pattern of strains on the master plate.

20.2.6. Auxanographic Nutritional Method

Seed a molten basal agar medium (45 to 50°C) lacking carbon sources with a washed suspension of the organism to be tested. After the plates have solidified, dip the edges of sterile filter paper disks into sterile solutions of various carbon sources, allowing the fluid to be absorbed by capillary action. Such disks may be applied immediately to the surface of the seeded agar plates or they may be dried and stored in a sterile container until needed. Place each disk near the periphery of the seeded plate, using three to four disks per plate, or, alternatively, place the disks at suitable distances on larger sheets of seeded agar (e.g., in a Pyrex baking dish). After incubation, determine whether a carbon source has been used by looking for a halo of turbid growth around the disk. In some cases growth may occur only as a line between two disks, indicating that the organism requires both carbon sources in order to grow. If growth occurs over the entire plate, the inoculum was probably too dense; use a more dilute inoculum to seed the agar.

20.2.7. Nitrogenase Activity and Nitrogen Fixation (also see 17.6.10)

The development of the acetylene reduction method has made the testing of bacterial cultures for nitrogenase activity relatively easy in comparison with the use of $^{15}N_2$ incorporation methods, although the latter still remain the most definitive. The principle of the acetylene reduction method is that the nitrogenase complex of enzymes is capable of reducing acetylene

to ethylene. The organism to be tested is cultured under conditions conducive to nitrogenase formation. Acetylene gas is then added to the culture vessel, and the formation of ethylene is determined after a period of incubation. Controls using media containing ammonium ions, which repress nitrogenase synthesis (except in the case of *Rhizobium*), should exhibit little or no ethylene production in comparison. Comprehensive reviews of the acetylene reduction method and its application are available (4, 5, 11).

It is important to culture the organisms under conditions that are favorable for nitrogenase synthesis. This involves the use of media free from fixed nitrogen but containing a suitable carbon and energy source and also a source of molybdenum. Also, an appropriate gaseous atmosphere should be supplied. For anaerobes such as *Clostridium pasteurianum*, anaerobic conditions are essential. Anaerobic conditions are also the most favorable for nitrogenase synthesis by facultative anaerobes such as *Klebsiella*. Many nitrogen fixers require oxygen for growth and energy production, but the oxygen tolerance for growth under nitrogen-fixing conditions varies considerably among such organisms. For example, *Azospirillum* grows best under nitrogen-fixing conditions at ca. 1% oxygen and does not grow at all under an air atmosphere (21% oxygen), whereas *Azotobacter* can tolerate an air atmosphere. Even aerobic nitrogen fixers, however, fix nitrogen best when grown at oxygen levels lower than that of air.

Aerobic nitrogen fixers.

Grow the organisms in a mineral salts medium (liquid or solid) in cotton-stoppered tubes or serum bottles. An example of such a medium is the Burk medium for *Azotobacter* (20.3.9). It is sometimes desirable to add a low level of a source of fixed nitrogen (e.g., 0.0001 to 0.005% yeast extract) in order to initiate growth. At the end of the incubation period, replace the cotton stopper with a sterile serum bottle stopper. Generate the acetylene gas (caution: flammable!) in a fume hood by the following procedure (11): add a lump (ca. 1 g) of calcium carbide to a tube half filled with 15 ml of water; stopper the tube immediately with a one-hole stopper fitted to a piece of rubber tubing, and immerse the distal end of the tubing in a beaker of water. After air has been flushed out of the tubing by the acetylene but while acetylene is still being formed, use a syringe equipped with a 26-gauge hypodermic needle to withdraw acetylene from the tubing. Inject the acetylene into the culture vessel through the serum bottle stopper to give 10% acetylene (vol/vol). After various periods of incubation, remove 1-ml gas samples from the culture vessel with a syringe and test for the presence of ethylene by gas chromatography (see below).

Facultative and anaerobic bacteria.

Bubble oxygen-free filtered nitrogen gas through sterile culture media in tubes, and inoculate the media while the tubes are being flushed with nitrogen. Seal the tubes with sterile serum bottle stoppers. Strictly anaerobic conditions may not be required for facultative bacteria such as *Klebsiella* because the organisms themselves may use up small residual amounts of oxygen and create highly anaerobic conditions. For examples of media, see 20.3.35 for *Klebsiella* medium, 20.3.13 for *Clostridium pasteurianum* medium, and 20.3.16 for *Bacillus* medium.

Alternatively, use 40-ml Pankhurst tubes (Astell Laboratory Service Co., Ltd., London, S.E. 6, England) (23). The Pankhurst tube (Fig. 3) consists of two test tubes, one larger (Fig. 3A) and one smaller (Fig. 3C), with a horizontal connecting tube (Fig. 3B) filled with nonabsorbent cotton as a gas filter. Dispense 5- to 10-ml volumes of culture medium into the larger tube, and temporarily stopper the tube with cotton. Sterilize the medium by autoclaving. Inoculate the cooled medium, and then aseptically seal both the larger and the smaller tubes with sterile

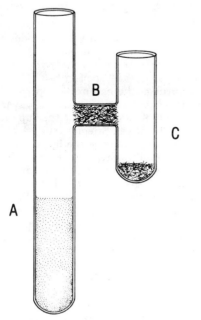

FIG. 3. *Pankhurst tube. (A) Larger test tube (1.9 by 15.7 cm) containing culture; (B) connecting tube (0.8 by 1.8 cm) with nonabsorbent cotton as gas filter; (C) smaller test tube (1.9 by 6.0 cm) with absorbent cotton to hold alkaline pyrogallol.*

serum bottle stoppers. Flush the assembly with nitrogen gas (this step may not be required; see reference 11) by inserting hypodermic needles into both stoppers, one to serve as an inlet for the flushing gas and the other as an exit. After flushing, remove the needles and inject 0.75 ml of a saturated solution of pyrogallol into the smaller tube, followed by 0.75 ml of a solution containing 10% (wt/vol) NaOH and 15% (wt/vol) K_2CO_3. These additions are soaked up by a little absorbent cotton previously placed in the bottom of the small tube as shown in Fig. 3C. The reagents react with residual oxygen and use it up, creating anaerobic conditions. If the Pankhurst tube has not been flushed with nitrogen, 2 to 3 h is required to establish anaerobic conditions, and ca. 12 ml of nitrogen should be added at this time to replace the oxygen that has been removed (11).

After the culture has grown, inject acetylene through the serum bottle stopper to give a final concentration of 10% (vol/vol). Incubate the culture for various periods, and use a syringe to withdraw 1-ml gas samples for gas chromatography to determine the presence of ethylene.

Bacteria requiring microaerobic conditions for nitrogen fixation.

Semisolid nitrogen-free media have been successfully used for demonstrating acetylene reduction by nitrogen-fixing spirilla; e.g., see part 20.3.4 for a medium for *Azospirillum*. Semisolid media can be incubated in an air atmosphere because they provide a stratified oxygen gradient from the surface of the medium downward. This allows microaerophilic bacteria to initiate growth at whatever oxygen level is most suitable. Growth begins as a thin band or pellicle several millimeters below the surface of the medium; as cell numbers increase, the pellicle becomes denser and migrates closer to the surface. The tube should not be agitated during growth or during subsequent addition of acetylene, as any disturbance of the pellicle may allow air to reach the organisms and inactivate the nitrogenase.

After growth has occurred, replace the cotton stopper of the tube or vial with a serum bottle stopper, and inject acetylene to give a final concentration of 10% (vol/vol). After incubation for various periods, remove 1-ml gas samples for gas chromatography (see below).

Gas chromatography (also see 16.3.4).

A gas chromatograph equipped with a hydrogen flame detector should be used. The column is usually ca. 5 ft (1.5 m) in length with a 2-mm inside diameter, filled with Porapak R or N packing (Waters Associates Inc., Framingham, Mass.) Keep the column oven temperature at

50°C, although any constant temperature between room temperature and 80°C will be satisfactory. Use nitrogen as the carrier gas, with a flow rate of ca. 50 cm^3/min. Calibrate the chromatograph with dilutions of commercial ethylene (99% purity) in nitrogen.

Use plastic disposable syringes for injection of gas samples and ethylene standards. Because of retention of some ethylene on the plastic surfaces, discard a syringe that has previously contained ethylene and replace it with a fresh syringe. Rubber serum bottle stoppers may also become contaminated with ethylene and should not be used again.

Hypodermic needles may punch out a hole in a serum bottle stopper or in the rubber septum of the gas chromatograph injector port. To prevent this, bend the tip of the needle as shown in Fig. 4.

When chromatographing gas samples containing both acetylene and ethylene, the acetylene will take longer to pass through the Porapak column than the ethylene. Do not inject new samples into the gas chromatograph until the acetylene from the previous sample has passed from the column.

20.2.8. Serology

Slide agglutination tests.

Slides with ceramic rings of ca. 14-mm diameter are preferred. However, glass plates marked off into 2.5-cm squares with a wax glass-marking pencil will suffice. Rehydrate the lyophilized antiserum according to the manufacturer's direction. Place a drop of the antiserum into one of the rings, and in another ring place a drop of saline (0.85% NaCl) or normal rabbit serum. Mix

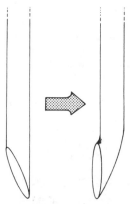

FIG. 4. *Diagram showing how the end of a hypodermic needle can be bent with a needle-nose pliers so that the needle will not punch out a hole in a serum bottle stopper or in the rubber septum of a gas chromatograph injector port.*

a loopful of the growth from the surface of an agar medium into each drop, so that a homogeneous suspension of cells without lumps is obtained. The suspension should be visibly turbid, but not extremely dense. Rock the slide in a circular fashion for 1 min to further mix the cells and liquid, being careful not to spill over the boundary of the ring. Examine the slide by eye; for best visibility hold the slide near a bright light and view against a dark background. Positive test: the cells mixed with the antiserum show clumping (i.e., the suspension appears to be granular or, with strong reactions, even to "curdle"). Clumps are best seen by tilting the slide slightly so that a fluid drains down toward the lower boundary of the ring. The saline or normal serum control should not show clumping. If it does, this usually indicates a rough (R) strain instead of a smooth (S) strain. *Note*: The bacteria-antiserum mixture must not be allowed to dry, as no indication of agglutination can be obtained if drying occurs. If drying is a problem, repeat the test using a larger drop of antiserum.

CAUTION: After completion, discard the slide into a container of disinfectant (e.g., 1% Lysol). Wash hands and bench surface with disinfectant. This is important because during the mixing of cells with antiserum small droplets may be scattered about on the slide and bench surface. The antiserum will not kill the bacteria.

Precipitin test for identification of streptococci.

Culture the organism in 40 ml of Todd-Hewitt broth (20.3.33) contained in 50-ml plastic (polycarbonate or polypropylene) screw-capped centrifuge tubes. After incubation at 37°C for 18 to 24 h, centrifuge the culture at top speed in a clinical-type centrifuge with a "swinging-bucket" type of rotor. Decant the supernatant fluid carefully into an Erlenmeyer flask containing disinfectant, and wipe the lip of the centrifuge tube with a towel moistened with disinfectant. Be sure to decant all of the supernatant fluid without suspending the sediment (cells) at the bottom of the tube. Add 0.5 ml of saline (0.85% NaCl) to the sediment, and suspend the organisms by gently agitating the tube from side to side. Replace the screw cap tightly, and autoclave the tube for 15 min at 121°C. Centrifuge again at top speed to sediment the autoclaved cells. Using care not to disturb the packed cells, withdraw most of the supernatant fluid with a Pasteur pipette into a small tube. The fluid must be perfectly clear; if it is cloudy, it must be recentrifuged. The supernatant fluid contains the soluble cell wall antigens which have been extracted from the streptococcal cell walls.

Rehydrate the antiserum according to the directions supplied by the manufacturer. Dip a capillary tube (0.7 to 1.0 mm inside diameter by 75 to 90 mm) in a tilted position into the antiserum, so that 2 to 3 cm of serum rises into the capillary. Place the forefinger over the opposite end of the capillary, remove the capillary from the serum bottle, and wipe off excess serum from the outside of the capillary. Dip the end of the capillary tube into the antigen extract, and remove the forefinger to allow a rise of 2 to 3 cm of antigen. There must be no air space between the antiserum and the extract in the capillary as the extract rises. Replace the forefinger, and withdraw the capillary. Tilt the capillary to a nearly horizontal position, and with finger removed, allow the column of fluid to move to the center part of the capillary. Leave an air space of at least 1 cm at each end. Plunge the end of the capillary into a small lump of modeling clay. Wipe the surface of the capillary free from fingerprints, and incubate for 10 to 15 min. Examine by holding the capillary near a strong light and viewing against a dark background with a hand lens. Positive test: development of a milky haze near the center of the capillary at the junction of the antiserum and extract. With further incubation this haze will become a precipitate that settles to the bottom of the column of fluid.

Quellung reaction.

Spread a loopful of broth culture on a clean slide and allow it to air dry without heating. (If a similar smear stained with crystal violet shows more than 15 to 25 organisms per oil immersion field, dilute the culture, as too high a concentration of cells may obscure the results.) Rehydrate the antiserum, and place a large loopful of it on a cover slip. Mix a loopful of 1% aqueous methylene blue into the antiserum, and then place the cover slip, fluid side down, onto the dried smear on the slide. Examine by oil immersion. Positive test: the capsules around the blue-stained cells will appear distinctly outlined. A comparison should always be made with a control slide prepared using normal rabbit serum instead of antiserum. If results of the test are negative, reexamine after 1 h; prevent the preparation from drying by incubating it in a moist chamber or by sealing the edges with melted Vaspar (a mixture of equal volumes of petrolatum and mineral oil).

Fluorescent-antibody test (direct method).

Rehydrate the fluorescent antiserum (termed "conjugate") and also fluorescent normal serum. Prepare a series of twofold dilutions of the flu-

orescent antiserum (e.g., 1:5, 1:10, 1:20, 1:40, and 1:80) in phosphate-buffered saline (PBS; see 20.4.17). Using the staining procedure given below with a known culture of the organism, and using a suitable fluorescence microscope system (part 1.4), determine the degree of fluorescence of the bacterial cells as obtained with each antiserum dilution. Rate the fluorescence on a scale of 0, 1+, 2+, 3+, and 4+ (maximum). The working dilution of the antiserum is one-half of the highest dilution that gives 4+ fluorescence. For example, if a 1:20 dilution of the serum is the highest dilution that gives 4+ fluorescence, the working dilution would be 1:10. An equivalent dilution of the fluorescent normal serum should give no fluorescence.

Centrifuge a young broth culture of the organism to be tested, and decant the supernatant fluid. Suspend the cells in PBS and recentrifuge. Suspend the cells to a dense concentration in a small amount of PBS. With a diamond or tungsten carbide pencil, circumscribe a circle about the diameter of a quarter on a glass slide. Clean the slide thoroughly so that it is free from grease, and smear a loopful of bacterial suspension within the circle. Allow the smear to air dry without heat. Place the slide in a jar of 95% ethanol for 1 min, remove, and allow to air dry.

Add several drops of the working dilution of fluorescent antiserum to the smear. Distribute the antiserum over the entire smear with an applicator stick, but do not touch the smear itself with the stick. Place the slide in a petri dish in which a moistened disk of filter paper has been affixed to the lid; this serves as a humid chamber to prevent evaporation of the antiserum. Incubate at 37°C for 15 to 20 min.

Drain off the excess antiserum by tilting the slide onto a piece of absorbent paper. Place the slide in a jar of PBS for 10 min, occasionally moving the slide up and down. Repeat this two more times, using a fresh jar of PBS each time. After the third washing in PBS, dip the slide only once into a jar of distilled water to remove the PBS and then immediately blot the slide (without rubbing) with bibulous paper. Allow the slide to dry. Place a small drop of buffered glycerol mounting fluid (20.4.2) on the smear, and apply a cover slip. Examine the smear under the high-power dry lens or oil immersion lens, using a fluorescence microscope (see 1.4). When using oil immersion, be sure to use only nonfluorescing oil (see 1.5).

At the same time the test organism is being stained, prepare two controls by a similar procedure. Make one control with the working dilution of the fluorescent antiserum and a known culture of the organism. Make the second control with the organism being tested, but use a dilution of fluorescent normal serum. The first control should give 4+ fluorescence, and the second should give no fluorescence.

Note: When examining smears under the fluorescence microscope, be sure to move the field often, because the fluorescence fades quickly (in ca. 15 s) when looking at any particular field. Fluorescence is always brightest when a fresh field is moved into position.

Note: The working dilution of the fluorescent antiserum should be prepared fresh daily. The undiluted stock antiserum is stable when stored in a refrigerator below 8°C.

20.3. RECIPES FOR MEDIA

20.3.1 Medium A (51)

Peptone	20.0 g
Glycerol	10.0 ml
K$_2$SO$_4$ (anhydrous)	10.0 g
MgCl$_2$ (anhydrous)	1.4 g
Agar (dried)	13.6 g
Distilled water	1,000 ml

Combine ingredients and adjust pH to 7.2. Boil to dissolve agar, and sterilize by autoclaving at 121°C for 15 min. Cool in a slanted position.

20.3.2. Arginine Broth of Niven et al. (67)

Yeast extract	5.0 g
Tryptone	5.0 g
K$_2$HPO$_4$	2.0 g
Glucose (dextrose)	0.5 g
L-Arginine·HCl	3.0 g
Distilled water	1,000 ml

(*Note:* Although the original formula designated the D form of arginine, the L form is satisfactory [3].)

Combine the ingredients and dissolve with heating. Adjust pH to 7.0. Boil, filter, and sterilize by autoclaving.

20.3.3. Arginine Broth of Evans and Niven (33)

Tryptone	10.0 g
Yeast extract	5.0 g
NaCl	5.0 g
L-Arginine·HCl	3.0 g
Distilled water	1,000 ml

Combine ingredients and dissolve with heating. Adjust to pH 7.0. Boil, filter, and sterilize by autoclaving.

20.3.4. *Azospirillum* Semisolid Nitrogen-Free Malate Medium (30)

KH$_2$PO$_4$	0.4 g
K$_2$HPO$_4$	0.1 g
MgSO$_4$·7H$_2$O	0.2 g
NaCl	0.1 g
CaCl$_2$·2H$_2$O	0.026 g
FeCl$_3$·6H$_2$O	0.017 g
Na$_2$MoO$_4$·2H$_2$O	0.002 g
L-Malic acid	3.58 g
(or sodium malate)	(5.00 g)
Bromothymol blue, 0.5% alcoholic solution	5.0 ml
Agar	1.75 g
Distilled water	1,000 ml

Combine gradients and adjust pH to 6.8 with KOH. Sterilize by autoclaving. *Note:* Some strains of *Azospirillum* may require addition of 0.005% yeast extract as a source of vitamins.

20.3.5. Medium B (51)

Pancreatic digest of casein USP	10.0 g
Peptic digest of animal tissue USP	10.0 g
K$_2$HPO$_4$	1.5 g
MgSO$_4$·7H$_2$O	1.5 g
Agar (dried)	14.0 g
Distilled water	1,000 ml

Combine ingredients and adjust pH to 7.2. Boil to dissolve agar and sterilize by autoclaving at 121°C for 15 min. Cool in a slanted position.

20.3.6. Bile Agar (3)

Beef extract	10.0 g
Peptone	10.0 g
NaCl	5.0 g
Distilled water	1,000 ml

Combine ingredients and dissolve with heating. Adjust pH to 8.0 to 8.4 with 10 N NaOH. Boil for 10 min. Filter, adjust pH to 7.2 to 7.4, and add 10.0 g of dehydrated oxgall. Readjust pH if necessary. Add 15.0 g of agar, boil, and sterilize by autoclaving. Cool to 55°C, and aseptically add 50 ml of sterile blood serum. Mix and dispense into plates. *Note:* 1% dehydrated oxgall = 10% bile (vol/vol); to make "40% bile agar," use 4% dehydrated oxgall.

20.3.7. Blood Agar

To sterile blood agar base (see 20.3.31) which has been melted and cooled to 45 to 50°C, add 5% (vol/vol) sterile defibrinated blood, mix well, and dispense into plates. Avoid getting plates with bubbles or froth on the surface. Defibrinated sheep or rabbit blood is preferred. Horse blood may give incorrect hemolytic reactions. Human blood bank blood contains citrate and glucose; the citrate may be inhibitory to some bacteria, and the glucose may cause false greening rather than true alpha-hemolysis. Antibiotics or chemotherapeutic agents should not be present in blood used for making blood agar.

Prepare defibrinated blood by drawing whole blood aseptically with a syringe equipped with an 18-gauge needle (to prevent hemolysis caused by mechanical damage) and dispensing the blood immediately into a sterile flask containing a layer of sterile glass beads (ca. 3 mm). Shake the flask from side to side for 10 min. The fibrin formed during clotting will be deposited on the beads. Decant the supernatant containing blood cells and serum into a sterile container, and store in a refrigerator.

20.3.8. Board and Holding Medium (20)

NH$_4$H$_2$PO$_4$	0.5 g
K$_2$HPO$_4$	0.5 g
Yeast extract	0.5 g
Bromothymol blue	0.03 g
Mineral solution (see 20.3.18)	20 ml
Agar	5.0 g
Distilled water	880 ml

Dissolve the first three ingredients in the distilled water; add the indicator and the mineral solution. Adjust pH to 7.2. Add the agar, and boil to dissolve. Disperse 9.0-ml volumes into tubes, and sterilize by holding momentarily at 22 lb/in^2. Cool to 45 to 50°C. A precipitate may form during autoclaving but should redissolve as the solution cools. To each tube, add 1.0 ml of a 5% solution of carbohydrate (usually glucose) which has previously been sterilized by filtration. Mix, and allow the medium to cool.

20.3.9. Burk Medium, Modified (73)

Solution A:

MgSO$_4$·7H$_2$O	0.2 g
CaSO$_4$·2H$_2$O	0.1 g
FeSO$_4$·7H$_2$O	0.005 g
Na$_2$MoO$_4$·2H$_2$O	0.00024 g
Glucose	10.0 g
Distilled water	500 ml

Solution B:

Potassium phosphate buffer, 0.005 M, pH 7.1	500 ml

Autoclave solutions A and B separately, and combine equal volumes when cool.

20.3.10. Chopped meat (CM) Broth, Pre-reduced (7)

Ground beef, fat-free: use lean beef (cutter-canner grade is suitable) or horsemeat	500 g
Distilled water	1,000 ml
NaOH, 1 N	25 ml

Remove fat and connective tissue before grinding meat. Mix meat, water, and NaOH; bring to a boil while stirring. Cool to room temperature, skim fat off surface, and filter—retaining both meat particles and filtrate. To the filtrate, add

sufficient distilled water to restore the original 1-liter volume. To this filtrate add:

Trypticase	30.0 g
Yeast extract	5.0 g
K$_2$HPO$_4$	5.0 g
Resazurin solution, 0.025%	4.0 mi

Boil in a flask equipped with a chimney (to prevent boiling over). For a 1-liter flask, use 750 ml of the broth; for a 750-ml flask, use 500 ml of broth. Boil until the resazurin changes from pink to colorless. Remove the flask from the heat, and replace the chimney with a two-hole stopper that has a gas cannula in one hole. Bubble oxygen-free CO$_2$ through the medium while cooling to room temperature in an ice bath. Add the following ingredients (per liter):

Cysteine	0.5 g
Hemin solution (see 20.3.28)	10 ml
Vitamin K$_1$ solution (see 20.3.28)	0.2 ml

Adjust pH to 7.2 with 8 N NaOH. Continue bubbling CO$_2$ until pH is lowered to 7.0. Bubble oxygen-free nitrogen through the medium while dispensing into 18 by 142 mm tubes (Bellco Glass, Inc., Vineland, N.J.) that are being flushed with oxygen-free nitrogen. The tubes should already contain the meat particles prepared above so as to give 1 part of meat particles per 4 to 5 parts of added broth. Stopper tubes with black rubber stoppers (size no. 1) as the gas cannula is being removed from the tubes. Autoclave the tubes in a press (to prevent the stoppers from being blown off) for 30 min. Cool. Discard any tubes that develop a light pink color on standing (indicating oxidation of the medium).

20.3.11. Christensen Citrate Agar (26)

Sodium citrate	3.0 g
Glucose	0.2 g
Yeast extract	0.5 g
Cysteine hydrochloride	0.1 g
Ferric ammonium citrate	0.4 g
KH$_2$PO$_4$	1.0 g
NaCl	5.0 g
Sodium thiosulfate	0.08 g
Phenol red	0.012 g
Agar	15.0 g
Distilled water	1,000 ml

Combine ingredients and adjust pH to 6.7. Sterilize by autoclaving. Cool in slanted position for a 2.5-cm butt and 3.8-cm slant.

20.3.12. Christensen Urea Agar (25)

Peptone	1.0 g
Glucose	1.0 g
NaCl	5.0 g
KH$_2$PO$_4$	2.0 g

Phenol red	0.012 g
Agar	20.0 g
Distilled water	1,000 ml
Yeast extract (optional)	(0.1 g)

Combine ingredients and adjust pH to 6.8 to 6.9. Boil to dissolve agar, and sterilize by autoclaving. Cool to 50°C. Aseptically add sufficient 20% solution of urea (sterilized by filtration) to give a final concentration of 2% urea. Mix, and aseptically dispense 2- to 3-ml volumes into sterile small tubes. Cool in a slanted position to give a 1.3-cm butt and a 2.5-cm slant.

20.3.13. *Clostridium pasteurianum* Nitrogen-Free Medium (28)

$MgSO_4 \cdot 7H_2O$	0.0493 g
$FeCl_3 \cdot 6H_2O$	0.0541 g
$MnSO_4 \cdot H_2O$	0.0034 g
$Na_2MoO_4 \cdot 2H_2O$	0.0048 g
$ZnSO_4 \cdot 7H_2O$	0.00058 g
$CuSO_4 \cdot 5H_2O$	0.00050 g
$CoCl_2 \cdot 6H_2O$	0.00048 g
K_2HPO_4	1.132 g
Biotin	Trace
Sucrose	20.0 g
$CaCO_3$	3.0 g
Distilled water	1,000 ml

(*Note:* $CaCO_3$ can be increased to 30 g for increased buffering.)

Combine ingredients and autoclave. The medium must be made anaerobic before inoculating by bubbling oxygen-free nitrogen through it.

20.3.14. Egg Yolk Agar (7)

Peptone	20.0 g
Na_2HPO_4	2.5 g
NaCl	1.0 g
$MgSO_4$, 0.5% (wt/vol) solution	0.1 ml
Glucose	1.0 g
Agar	12.5 g
Distilled water	500 ml

Combine ingredients and adjust pH to 7.3 to 7.4. Boil to dissolve agar, and sterilize by autoclaving. Cool to 60°C in a water bath. Disinfect the surface of an egg with alcohol, and allow the egg to dry. Crack the egg, and separate the yolk from the white. Add the yolk aseptically to the melted agar, and mix to obtain a homogeneous suspension. Dispense into plates, and allow to solidify. For anaerobes, use plates within 4 h or store plates in an anaerobe jar until needed.

20.3.15. Haynes Broth (42)

Tryptone	1.5 g
Yeast extract	1.0 g
K_2HPO_4	1.0 g
Potassium gluconate	40.0 g

Combine ingredients, and adjust pH to 7.0. Sterilize by filtration.

20.3.16. Hino and Wilson Nitrogen-Free Medium for *Bacillus* (43)

Solution A:

Sucrose	20.0 g
$MgSO_4 \cdot 7H_2O$	0.5 g
NaCl	0.01 g
$FeSO_4 \cdot 7H_2O$	0.015 g
$Na_2MoO_4 \cdot 2H_2O$	0.005 g
$CaCO_3$	10.0 g
Distilled water	500 ml

Solution B:

p-Aminobenzoic acid	10 μg
Biotin	5 μg
K_2HPO_4-KH_2PO_4 buffer, 0.1 M, pH 7.7	500 ml

Autoclave solutions A and B separately. When cool, combine equal volumes aseptically.

20.3.17. Hugh and Leifson O-F Medium (44)

Peptone (pancreatic digest of casein)	2.0 g
NaCl	5.0 g
K_2HPO_4	0.3 g
Bromothymol blue	0.03 g
Agar	3.0 g
Distilled water	1,000 ml

Combine ingredients, and adjust pH to 7.1. Boil to dissolve agar and dispense 3- to 4-ml volumes into 13 by 100 mm tubes. Sterilize by autoclaving. Cool to 45 to 50°C. Add sufficient 10% carbohydrate solution (usually glucose) sterilized by filtration to each tube to give a final concentration of 1.0%. Mix, and allow tubes to cool.

20.3.18. Hutner Mineral Base (27)

Mineral solution:

Nitrilotriacetic acid	10.0 g
$MgSO_4$	14.45 g
$CaCl_2 \cdot 2H_2O$	3.335 g
$(NH_4)_6Mo_7O_{24} \cdot 4H_2O$	0.00925 g
$FeSO_4 \cdot 7H_2O$	0.099 g
Stock salts solution (see below)	50 ml
Distilled water	950 ml

Stock salts solution:

Ethylenediaminetetraacetic acid (EDTA)	2.5 g
$ZnSO_4 \cdot 7H_2O$	10.95 g
$FeSO_4 \cdot 7H_2O$	5.0 g
$MnSO_4 \cdot H_2O$	1.54 g
$CuSO_4 \cdot 5H_2O$	0.392 g
$Co(NO_3)_2 \cdot 6H_2O$	0.248 g
$Na_2B_4O_7 \cdot 10H_2O$	0.177 g
Distilled water	1,000 ml

Add a few drops of H_2SO_4 to the stock salts solution to decrease precipitation. To prepare the mineral solution, dissolve the nitrilotriacetic acid in the distilled water, and neutralize with approximately 7.3 g of KOH. Add the remaining ingredients, and adjust the pH to 6.8.

20.3.19. Leifson Modified O-F (MOF) Medium (55)

Casitone (Difco)	1.0 g
Yeast extract	0.1 g
$(NH_4)_2SO_4$	0.5 g
Tris buffer (2-amino-2-hydroxymethyl-1,3-propanediol)	0.5 g
Phenol red	0.01 g
Agar	3.0 g
Artificial seawater (see 20.3.29)	500 ml
Distilled water	500 ml

Dissolve ingredients in the distilled water, adjust the pH to 7.5, and sterilize by autoclaving. Autoclave the seawater separately. After cooling to 45 to 50°C, mix the two solutions aseptically. Add sufficient 10 or 20% solution of the desired carbohydrate (usually glucose) sterilized by filtration to give a final concentration of 0.5 or 1.0%. Dispense 2.5- to 3.0-ml volumes of the medium aseptically into 13 by 100 mm tubes, and allow to cool.

20.3.20. Litmus Milk (12)

Alcoholic litmus solution. Grind 50 g of litmus in a mortar with 150 ml of 40% ethanol. Transfer to a flask, and boil gently on a steam bath for 1 min. Decant the fluid and save it. Add another 150 ml of 40% ethanol to the residue, and boil again for 1 min. Decant and combine the two supernatants. Allow to settle overnight, and dilute up to 300 ml with 40% ethanol. Add 1 N HCl drop by drop to adjust the pH to 7.0.

Preparation of medium. Add sufficient litmus solution to skim milk to give a bluish-purple color (usually 40 ml per liter). Adjust the pH to 7.0 with 1 N NaOH. Sterilize by autoclaving at 115°C for 10 min. Overheating will result in caramelization.

20.3.21. Malonate Broth (34)

Yeast extract	1.0 g
$(NH_4)_2SO_4$	2.0 g
K_2HPO_4	0.6 g
KH_2PO_4	0.4 g
NaCl	2.0 g
Sodium malonate	3.0 g
Glucose	0.25 g
Bromothymol blue	0.025 g
Distilled water	1,000 ml

Combine ingredients, and adjust pH to 7.0. Sterilize by autoclaving.

20.3.22. Milk Medium, Prereduced (7)

Fresh skim milk	100 ml
Resazurin solution, 0.025%	0.4 ml
Hemin solution (see 20.3.28)	1.0 ml
Vitamin K_1 solution (see 20.3.28)	0.02 ml

Add milk and resazurin to a flask, and prepare as described for PY broth (20.3.28). Add the hemin and vitamin K_1 after boiling and cooling. The pH after bubbling with CO_2 should be 7.1. After dispensing into tubes being purged with nitrogen, stopper and autoclave in a press for 12 min at 121°C.

20.3.23. Møller Broth Base (61)

Peptide digest of animal tissue USP (see Note)	5.0 g
Beef extract	5.0 g
Bromocresol purple solution, 1.6%	0.625 ml
Cresol red solution, 0.2%	2.5 ml
Glucose	0.5 g
Pyridoxal	0.005 g
Distilled water	1,000 ml

(*Note:* The original formula designates Orthana special peptone; however, Thiotone [BBL; peptic digest of animal tissue] is satisfactory [10].)

Combine ingredients, and adjust pH to 6.0 or 6.5. Dispense into 13 by 100 mm tubes, and sterilize by autoclaving.

20.3.24. MRVP (Methyl Red–Voges-Proskauer) Broth

Polypeptone or buffered peptone	7.0 g
K_2HPO_4 (see Note)	5.0 g
Glucose	5.0 g
Distilled water	1,000 ml

(*Note:* For testing the Voges-Proskauer reaction of members of the genus *Bacillus*, replace the phosphate with 5.0 g of NaCl [2].)

Combine ingredients, and adjust pH so that after autoclaving and cooling the pH will be 6.9. Dispense in 5-ml volumes, and sterilize at 121°C for 10 min.

20.3.25. Nutrient Agar

Beef extract	3.0 g
Peptone	5.0 g
Agar	15.0 g
Distilled water	1,000 ml

Combine ingredients, and adjust pH to 6.8. Sterilize by autoclaving.

20.3.26. Phenylalanine Agar

Yeast extract	3.0 g
L-Phenylalanine	1.0 g
Na_2HPO_4	1.0 g
NaCl	5.0 g
Agar	12.0 g
Distilled water	1,000 ml

Combine ingredients and adjust pH to 7.3. Boil to dissolve agar, dispense into tubes, and sterilize by autoclaving. Cool tubes in a slanted position.

20.3.27. PY Agar, Prereduced (7)

Prepare PY broth (20.3.28), and dispense 10-ml volumes into tubes containing 0.2 g of agar while purging the tubes with oxygen-free nitrogen. Stopper the tubes while removing the gas cannula, and autoclave in a press for 15 min using fast exhaust. After autoclaving, mix by inverting the tubes in the press several times.

20.3.28. PY Broth, Prereduced (7)

Peptone	5.0 g
Trypticase	5.0 g
Yeast extract	10.0 g
Resazurin solution, 0.025%	4.0 ml
Salts solution (see below)	40.0 ml
Hemin solution (see below)	10.0 ml
Vitamin K_1 solution (see below)	0.2 ml
Cysteine·HCl	0.5 g
Distilled water	1,000 ml

Salts solution: $CaCl_2$ (anhydrous), 0.2 g; $MgSO_4$·$7H_2O$, 0.48 g; K_2HPO_4, 1.0 g; KH_2PO_4, 1.0 g; $NaHCO_3$, 10.0 g; NaCl, 2.0 g. Mix the $CaCl_2$ and $MgSO_4$ in 300 ml of distilled water until dissolved. Add 500 ml of distilled water and, while swirling, add remaining salts. After the salts are dissolved, add 200 ml of distilled water.

Hemin solution: Dissolve 0.05 g of hemin in 1 ml of 1 N NaOH. Dilute to 100 ml with distilled water. Autoclave at 121°C for 15 min.

Vitamin K_1 solution: Dissolve 0.15 ml of vitamin K_1 in 30 ml of 95% ethanol.

To prepare PY broth, place dry ingredients (except cysteine) into a flask equipped with a chimney to prevent boiling over. For a 1-liter flask, use ingredients sufficient to prepare 750 ml of medium; for a 750-ml flask, prepare 500 ml of medium. Add water, salts solution, and resazurin. Boil until the resazurin changes from pink to colorless. Remove the flask from the heat, and replace the chimney with a two-hole stopper that has a gas cannula in one hole. Bubble oxygen-free CO_2 through the medium while cooling to room temperature in an ice bath. Add hemin solution, vitamin K_1 solution, and cysteine. Adjust pH to 7.1 with 8 N NaOH. Continue bubbling CO_2 until pH is lowered to 6.9. Then bubble oxygen-free nitrogen through the medium while dispensing into 18 by 142 mm tubes (Bellco Glass, Inc.) that are being flushed with nitrogen. Stopper with black rubber stoppers (size no. 1) as the gas cannula is being removed from the tubes. Autoclave the tubes in a press (to keep the stoppers from being blown off) for 15 min, using fast exhaust. Cool. Discard any tubes that develop a light pink color (indicating oxidation) upon standing.

20.3.29. Seawater, Artificial

Formula 1 (74):

NaCl	27.5 g
$MgCl_2$	5.0 g
$MgSO_4$	2.0 g
$CaCl_2$	0.5 g
KCl	1.0 g
$FeSO_4$	0.001 g
Distilled water	1,000 ml

Formula 2 (100):

NH_4NO_3	0.002 g
H_3BO_3	0.027 g
$CaCl_2$	1.140 g
$FePO_4$	0.001 g
$MgCl_2$	5.143 g
KBr	0.100 g
KCl	0.690 g
$NaHCO_3$	0.200 g
NaCl	24.320 g
NaF	0.003 g
Na_2SiO_3	0.002 g
Na_2SO_4	4.060 g
$SrCl_2$	0.026 g
Distilled water	1,000 ml

20.3.30. Simmons citrate agar (84)

Sodium citrate	2.0 g
NaCl	5.0 g
$MgSO_4$	0.2 g
$NH_4H_2PO_4$	1.0 g
K_2HPO_4	1.0 g
Bromothymol blue	0.08 g
Agar	15.0 g
Distilled water	1,000 ml

Combine the ingredients, and adjust to pH 6.9. Boil to dissolve agar, dispense into tubes, and sterilize by autoclaving. Cool in slanted position.

20.3.31. Soybean-Casein Digest Agar USP (see Note)

Pancreatic digest of casein USP	15.0 g
Papaic digest of soy meal USP	5.0 g
NaCl	5.0 g
Agar	15.0 g
Distilled water	1,000 ml

Combine ingredients and adjust pH to 7.3. Boil to dissolve agar and sterilize by autoclaving. (*Note:* This medium is also known as tryptic-soy agar or Trypticase-soy agar; it can be used as a blood agar base.

20.3.32. Thornley Semisolid Arginine Medium (91)

Peptone	1.0 g
NaCl	5.0 g

K_2HPO_4	0.3 g
Phenol red	0.01 g
L-Arginine·HCl	10.0 g
Agar	3.0 g
Distilled water	1,000 ml

Combine ingredients, and adjust pH to 7.2. Boil to dissolve agar, and sterilize by autoclaving.

20.3.33. Todd-Hewitt Broth, modified (35)

Beef heart, infusion from	1,000 ml
Neopeptone	20.0 g

Combine, and adjust pH to 7.0 with 1 N NaOH. Then add the following:

NaCl	2.0 g
$NaHCO_3$	2.0 g
Na_2HPO_4	0.4 g
Glucose	2.0 g

Adjust pH to 7.8. Boil for 15 min, filter through paper, and dispense into tubes. Sterilize by autoclaving at 121°C for 10 min.

20.3.34. Triple Sugar Iron Agar

Pancreatic digest of casein USP (see Note)	10.0 g
Peptic digest of animal tissue USP (see Note)	10.0 g
Glucose	1.0 g
Lactose	10.0 g
Sucrose	10.0 g
Ferrous sulfate or ferrous ammonium sulfate	0.2 g
NaCl	5.0 g
Sodium thiosulfate	0.3 g
Phenol red	0.024 g
Agar	13.0 g
Distilled water	1,000 ml

(*Note*—The following combination of ingredients can substitute for the first two components listed: beef extract, 3.0 g; yeast extract, 3.0 g; and peptone, 20.0 g.

Combine ingredients, and adjust pH to 7.3. Boil to dissolve agar, and dispense into tubes. Sterilize by autoclaving at 121°C for 15 min. Cool in a slanted position to give a 2.5-cm butt and a 3.8-cm slant.

20.3.35. Yoch and Pengra Nitrogen-Free Medium for *Klebsiella* (99)

Solution A:

Na_2HPO_4	6.25 g
KH_2PO_4	0.75 g
Distilled water	500 ml

Solution B:

$MgSO_4·7H_2O$	6.25 g
$FeSO_4·7H_2O$	0.04 g
Na_2MoO_4 (anhydrous)	0.005 g
Sucrose	20.0 g

NaCl	8.5 g
Distilled water	500 ml

Autoclave solutions A and B separately, and aseptically combine equal volumes when cool.

20.4. REAGENT SOLUTIONS

20.4.1. Benedict Reagent

Sodium citrate	17.3 g
Na_2CO_3	10.0 g
$CuSO_4·5H_2O$	1.73 g

Dissolve the first two ingredients in 80 ml of water by heating. Filter. Dilute to 85 ml. Dissolve the CuSO₄ in 10 ml of water, and add to the carbonate-citrate solution with stirring. Dilute to 100 ml with distilled water.

20.4.2. Buffered Glycerol Mounting Fluid

Glycerol	90 ml
Phosphate-buffered saline (PBS) (20.4.17)	10 ml

Combine ingredients, and adjust to pH 8.0 with NaOH. The solution should be tightly stoppered to prevent absorption of carbon dioxide. Fresh solutions should be prepared monthly.

20.4.3. Charcoal-Gelatin Disks (52)

Mix 15 g of dehydrated nutrient gelatin (Difco Laboratories) with 100 ml of tap water. Heat to dissolve. Add 3 to 5 g of finely powdered charcoal, shake thoroughly, and pour into petri dishes to form a layer about 3 mm thick. Pour the mixture when quite cool so that it sets quickly before the charcoal can sediment. It is advisable to smear the bottom of the dish very thinly with petrolatum before pouring the gelatin. After the mixture has set, lift off the whole sheet from the dish and place it in 10% Formalin for 24 h. Then punch the formalinized sheet into disks (about 1 cm in diameter) or cut it into strips (about 5 to 8 mm by 20 mm). Wrap the pieces in gauze, and place in a basin under running tap water for 24 h. Then place the pieces into screw-capped bottles, and cover them with water. Sterilize by steaming for 30 min or by repeated heating in a water bath at 90 to 100°C for 20 min each time. Do not autoclave. After sterilization, decant the water, and add the pieces aseptically to sterile broths (one piece per tube). Incubate the broths to test for sterility. The prepared media are then ready for inoculation.

20.4.4. Developing Solution for Arginine (77)

α-Naphthol, 25% (wt/vol) solution in *n*-propanol	20 ml

Diacetyl, 1% (wt/vol) in *n*-propanol ... 2.5 ml

Combine the ingredients and dilute to 100 ml with *n*-propanol.

20.4.**5. Ehrlich Reagent**

p-Dimethylaminobenzaldehyde	1.0 g
Ethanol, 95%	95 ml
HCl, concentrated	20 ml

Dissolve the aldehyde in the ethanol and then add the acid. Protect from light, and store in a refrigerator.

20.4.**6. Formate-Fumarate (F-F)** (7)

Sodium formate	3.0 g
Fumaric acid	3.0 g
Distilled water	50 ml

Combine ingredients. Add 20 pellets of NaOH with stirring until the pellets are dissolved and the fumaric acid is in solution. Adjust pH to 7.0 with about 15 drops of 4 N NaOH. Sterilize by filtration.

20.4.**7. Kovacs Reagent**

p-Dimethylaminobenzaldehyde	3.0 g
Pentanol or butanol (amyl or butyl alcohol)	75 ml
HCl, concentrated	25 ml

Dissolve the aldehyde in the alcohol at 50 to 55°C. Cool, and add the acid. Protect from light, and store in a refrigerator.

20.4.**8. Lactic Acid, Reagents for Optical Rotation** (7)

Perchloric acid (PCA) solution:

Perchloric acid, 70%	5.8 ml

Make up to 100 ml with distilled water.

Buffer solution:

Glycine	3.75 g
Hydrazine sulfate	5.2 g

Add ingredients to 60 ml of distilled water. Adjust pH to 9.0 with 8 N NaOH, and make up to 100 ml with distilled water.

Lactic dehydrogenase (LDH) solution:

Lactic dehydrogenase, rabbit muscle (catalog no. L-2375; Sigma Chemical Co., St. Louis, Mo.)	0.8 ml
Distilled water	0.8 ml

Make up solution in an ice bath, and store in ice. Solution should be prepared the same day it is used.

NAD solution:

Nicotinamide adenine dinucleotide (NAD, catalog no. N-7004; Sigma Chemical Co.)	0.160 g
Distilled water	8.0 ml

Dissolve the NAD in the water. Store in a refrigerator. Solution should be freshly prepared.

20.4.**9. Methyl Green for Deoxyribonuclease (DNase) Test** (85)

Methyl green (catalog no. M295, Certified Biological Stain, C.I. no. 43590; Fisher Scientific Co., Pittsburgh, Pa.)	0.5 g
Distilled water	100 ml

Dissolve the dye in the water. Extract the solution with approximately equal volumes of chloroform until the chloroform is colorless (usually six to eight extractions). Store the resulting aqueous solution of methyl green at 4°C. The dye may be added to DNase medium before autoclaving, or it may be sterilized by filtration and added aseptically to the autoclaved medium at 45 to 50°C.

20.4.**10. Nessler Solution**

KI	5.0 g
Distilled water, ammonia-free	5.0 ml

Dissolve the KI in the water and add a saturated solution of $HgCl_2$ (about 2.0 g in 35 ml) until an excess is indicated by the formation of a precipitate. Then add 20 ml of 5 N NaOH, and dilute to 100 ml. Let any precipitate settle out, and draw off the clear supernatant.

20.4.**11. Nile Blue Sulfate Fat** (12)

Nile blue sulfate solution.

Prepare a saturated aqueous solution of Nile blue sulfate. Add 1 N NaOH drop by drop until precipitation is complete. Filter, and wash the precipitate with distilled water at pH 7.5. Dry, and store for use.

Staining of fat.

Prepare a saturated solution of the Nile-blue sulfate oxazine base, using the prepared dried precipitate prepared above. Mix 1 ml of the solution with 10 ml of the fat (tripropionine, tributyrin, tricaproine, tricaprylin, triolein, beef tallow, butterfat, coconut oil, corn oil, cottonseed oil, lard, linseed oil, or olive oil, as desired). If necessary, work in a heated water bath to liquefy the fat and maintain it in a liquid state throughout the subsequent washing procedure.

Washing procedure.

With fats that are liquid at room temperature, add 2 volumes of diethyl ether to the dye-fat mixture in a separatory funnel; separate the red ether-soluble layer from the water layer, and wash several times with water. (*CAUTION*: Ether is flammable! For all operations involving ether, use a fume hood and avoid any flames or sparks that could cause an explosion.) Finally, separate the ether-fat layer and evaporate off the ether. Separate the fat from the residual water and sterilize by autoclaving. Store in a refrigerator. For use, add 1 ml of the dyed fat to 10 ml of sterile, melted basal medum, mix well, and dispense into plates.

With fats that are solid at room temperature, wash the dyed fat with several changes of hot water, and then disperse the fat in a melted, neutral 0.5% agar solution, using 10 ml of fat per 90 ml of agar. Sterilize by autoclaving. For use, melt the fat emulsion and add 1 ml of fat emulsion to 20 ml of the sterile, melted basal medium. Mix well, and dispense into plates.

20.4.12. Nitrite Test Reagents

For Method 1.

Solution A (see Caution):
N-(1-Naphthyl)-ethylenediamine dihydrochloride (Sigma Chemical Co.) .... 0.02 g
HCl, 1.5 N solution .... 100 ml
Dissolve by gentle heating in a fume hood.

Solution B:
Sulfanilic acid .... 1.0 g
HCl, 1.5 N solution .... 100 ml
Dissolve by gentle heating.

Mix equal volumes of solutions A and B just before use to make the test reagent.

For Method 2.

Solution A (see Caution):
N,N-Dimethyl-1-naphthylamine dihydrochloride (Sigma Chemical Co.) .... 0.6 g
(or use α-naphthylamine) .... (0.5 g)
Acetic acid, 5 N solution .... 100 ml
Dissolve by gentle heating in a fume hood.

Solution B:
Sulfanilic acid .... 0.8 g
Acetic acid, 5 N solution .... 100 ml
Dissolve by gentle heating in a fume hood.

CAUTION: The Occupational Safety and Health Administration (OSHA) has declared α-naphthylamine to be a carcinogen. Although N-(1-naphthyl)-ethylenediamine and N,N-di-

methyl-1-naphthylamine have not been listed as carcinogenic, one would suspect on the basis of structural similarity that they might also be dangerous. All precautions pertinent to the use, handling, and subsequent disposal of carcinogens should be observed for these compounds.

20.4.13. ONPG Test Reagent

Solution A:
$NaH_2PO_4 \cdot H_2O$ .... 6.9 g
NaOH, 30% (wt/vol) solution ...... ca. 3 ml
Distilled water .... 45 ml
Dissolve the sodium phosphate in the distilled water. Add the NaOH to adjust the pH to 7.0. Make the volume to 50 ml with distilled water. Store at 4°C.

Test reagent:
o-Nitrophenyl-β-D-galactopyranoside (ONPG; Sigma Chemical Co.) .... 0.080 g
Distilled water .... 15 ml
Solution A (above) .... 5 ml
Dissolve the ONPG in the distilled water at 37°C. Add the sodium phosphate solution. Store at 4°C. Warm to 37°C before use.

20.4.14. pH Indicators for Culture Media

The pH indicators for use in culture media are listed in Table 1. Also see part 6.1.2.

20.4.15. Poly-β-Hydroxybutyrate Granule Suspension (29; also see part 17.3.8)

Bacillus megaterium strain KM (ATCC strain 13632) was originally used as the source of the granules, but strains of *Pseudomonas* may be used alternatively and may give higher yields of PHB (N. J. Palleroni, personal communication). Culture the bacterial strain under conditions conducive to PHB production. For *B. megaterium* these conditions are described by Macrae and Wilkinson (58); for pseudomonads, see Stanier et al. (88). When the level of PHB is sufficiently high, as judged by staining with Sudan black (see 3.3.11), harvest the cells by centrifugation and suspend them to a concentration of 15% (wt/vol) in ice-cold 0.02 M phosphate buffer, pH 7.2 (see 20.4.16). Disrupt the cells by sonication (5.1.2), and centrifuge at high speed. Suspend the pellet in the alkaline hypochlorite reagent of Williamson and Wilkinson (see 20.4.19) to a density of ca. 8 mg (dry weight) per ml or less (98). Alternatively, commercial 5% hypochlorite (Clorox bleach) can be used (N. J. Palleroni, personal communication). Incubate at room temperature for 2 days with stirring. Either allow the suspension to settle and carefully remove and discard the supernatant fluid, or col-

TABLE 1. *pH indicators for culture media*

Name	pK'	pH range and colors[a]	Concn usually employed in media (g/liter)[b]	Amt (ml) of 0.01 N NaOH needed to dissolve 0.1 g
Phenol red	7.8	6.9(Y)–8.5(R)	0.010–0.030	28.2
Bromothymol blue	7.1	6.1(Y)–7.7(B)	0.010–0.032	16.0
Bromocresol purple	6.2	5.4(Y)–7.0(P)	0.010–0.032	18.5
Chlorophenol red	6.0	5.1(Y)–6.7(R)	0.015	23.6
Bromocresol green	4.7	3.8(Y)–5.4(B)	0.020	14.3

[a] Symbols: Y, yellow; B, blue; P, purple; R, red.
[b] For addition to culture media, dissolve in alcohol or prepare an aqueous solution using 0.01 N NaOH.

lect the granules by centrifugation. Suspend the granules in water and dialyze against running tap water until free from chloride (use $AgNO_3$ on a small portion; there should be no precipitate of AgCl formed). Dialyze further against distilled water. In a series of centrifugations, wash the granules several times with acetone. Place the granules in a Soxhlet apparatus and continuously extract them with acetone-ether (2:1, vol/vol) for 3 days to remove non-PHB lipids. Further extract the granules with hot diethyl ether, pulverize them, and dry them under a vacuum. Determine the dry weight of the granules.

After drying, disperse the granules in 0.02 M phosphate buffer by means of a tissue grinder and sonic oscillator to form a stable, concentrated suspension. Sterilize the suspension by autoclaving, and store it in a refrigerator. Calculate how much of the concentrated suspension one needs to add to molten mineral agar to give a final concentration of 0.25% (wt/vol) in the overlay described in 20.1.50.

20.4.16. Potassium Phosphate Buffer

Prepare 0.2 M solutions of (i) K_2HPO_4 and (ii) KH_2PO_4. Mix the solutions in the ratios shown in Table 2 to obtain the appropriate pH value. The mixture (0.2 M phosphate buffer) can be further diluted to give the desired molarity. Also see part 6.1.2.

20.4.17. Phosphate-Buffered Saline (PBS)

10× Stock solution:
Na_2HPO_4, anhydrous, reagent grade	12.36 g
$NaH_2PO_4 \cdot H_2O$, reagent grade	1.80 g
NaCl, reagent grade	85.00 g

Dissolve ingredients in distilled water to a final volume of 1,000 ml.

Working solution (0.01 M phosphate, pH 7.6):
Stock solution	100 ml
Distilled water	900 ml

TABLE 2. *Number of parts of K_2HPO_4 solution and KH_2PO_4 solution required to obtain potassium phosphate buffer of various pH levels*

K_2HPO_4	KH_2PO_4	pH
49	51	6.8
55	45	6.9
61	39	7.0
67	33	7.1
72	28	7.2
77	23	7.3

20.4.18. Standard Solutions for Gas Chromatography (7)

Standard mixture of volatile fatty acids (ca. 1 meq/ 100 ml):
Formic acid	0.037 ml
Acetic acid, glacial	0.057 ml
Propionic acid	0.075 ml
Isobutyric acid	0.092 ml
Butyric acid	0.091 ml
Isovaleric acid	0.109 ml
Valeric acid	0.109 ml
Isocaproic acid	0.126 ml
Caproic acid	0.126 ml
Heptanoic acid	0.142 ml
Distilled water	100 ml

When using the standard mixture, acidify and extract 1 ml of the solution with ether for gas chromatography.

Standard mixture of alcohols. The following list shows the amount of each component to be added to 100 ml of distilled water to obtain the final concentration shown in parentheses.
Ethanol	0.1 ml	(1.7 mM)
Propanol	0.035 ml	(0.5 mM)
Isobutanol	0.005 ml	(0.05 mM)
Butanol	0.01 ml	(0.1 mM)
Isopentanol	0.005 ml	(0.05 mM)
Pentanol	0.005 ml	(0.05 mM)

Extract 1 ml of the mixture with ether for gas chromatography.

Standard mixture of nonvolatile acids (ca. 1 meq/ 100 ml):

Pyruvic acid	0.068 ml
Lactic acid, 85% syrup	0.84 ml
Oxalacetic acid	0.06 g
Oxalic acid	0.06 g
Methylmalonic acid	0.06 g
Malonic acid	0.06 g
Fumaric acid	0.06 g
Succinic acid	0.06 g
Distilled water	100 ml

Methylate 1 ml of the mixture and extract with chloroform.

20.4.19. Williamson and Wilkinson Hypochlorite Reagent (98)

Triturate 200 g of fresh bleaching powder (chlorinated lime) with a little distilled water, and make the volume up to 1 liter. With stirring, add 1 liter of 30% (wt/vol) Na_2CO_3. Allow the mixture to stand for 2 to 3 h with shaking at intervals. Filter through paper. Adjust the pH of the filtrate to 9.8 with concentrated HCl. Warm the solution to 37°C, and remove the precipitate by filtration. Store the clear filtrate in a stoppered bottle in a refrigerator. The solution is stable for several months.

20.5. LITERATURE CITED

20.5.1. General References

1. **Bodily, H., E. L. Updyke, and J. O. Mason (ed.).** 1970. Diagnostic procedures for bacterial, mycotic and parasitic infections, 5th ed. American Public Health Association, Inc., New York.
   Contains microscopic, cultural, and biochemical methods for the identification of pathogenic organisms.
2. **Buchanan, R. E., and N. E. Gibbons (ed.).** 1974. Bergey's manual of determinative bacteriology, 8th ed. The Williams & Wilkins Co., Baltimore.
   The most comprehensive and widely used taxonomic treatment of bacteria. Contains detailed descriptions of nearly all of the described species together with differentiation tables, keys, and illustrations.
3. **Cowan, S. T.** 1974. Cowan and Steel's manual for the identification of medical bacteria, 2nd ed. Cambridge University Press, London.
   Useful not only for medical bacteriologists but for generalists as well. Contains a wealth of information about characterization methods, media and the principles of classification, identification and nomenclature.
4. **Hardy, R. W. F., R. C. Burns, and R. D. Holsten.** 1973. Applications of the acetylene-ethylene assay for measurement of nitrogen fixation. Soil Biol. Biochem. **5**:47–81.
5. **Hardy, R. W. F., R. D. Holsten, E. K. Jackson,** and R. C. Burns. 1968. The acetylene-ethylene assay for N_2 fixation: laboratory and field evaluation. Plant Physiol. **43**:1185–1207.
   These two articles provide extensive practical information for performance of the acetylene reduction technique under both laboratory and field conditions. The chromatographic analysis of ethylene and the various factors affecting acetylene reduction are discussed in detail.
6. **Hedén, C., and T. Illéni (ed.).** 1975. New approaches to the identification of microorganisms. John Wiley & Sons, Inc., New York.
   Automation, computers, and miniaturized techniques relating to characterization and identification of bacteria.
7. **Holdeman, L. V., E. P. Cato, and W. E. C. Moore (ed.).** 1977. Anaerobe laboratory manual, 4th ed. Virginia Polytechnic Institute and State University, Blacksburg.
   Identification charts, characterization methods, and media for anaerobic bacteria; describes in detail the preparation and use of prereduced media and culture techniques.
8. **Holding, A. J., and J. G. Colee.** 1971. Routine biochemical tests, p. 2–32. *In* J. R. Norris and D. W. Ribbons (ed.), Methods in microbiology, vol. 6A. Academic Press, Inc., New York.
   Principles and procedures for a wide variety of routine biochemical characterization tests.
9. **Holt, J. G.** 1977. The shorter Bergey's manual of determinative bacteriology. The Williams & Wilkins Co., Baltimore.
   Contains all of the tables, generic descriptions, and illustrations from the complete Bergey's Manual, but lacks the detailed descriptions of species. Designed specifically for use in the identification of bacteria.
10. **Lenette, E. H., A. Balows, W. J. Hausler, Jr., and J. P. Truant (ed.).** 1980. Manual of clinical microbiology, 3rd ed. American Society for Microbiology, Washington, D.C.
    Contains microscopic, cultural, biochemical and serological methods for the identification of pathogenic microorganisms.
11. **Postgate, J. R.** 1972. The acetylene reduction test for nitrogen fixation, p. 343–356. *In* J. R. Norris and D. W. Ribbons (ed.), Methods in microbiology, vol. 6B. Academic Press, Inc., New York.
    A succinct review of the principles and procedures involved in the acetylene reduction test for nitrogenase activity.
12. **Skerman, V. B. D.** 1967. A guide to the identification of the genera of bacteria. The Williams & Wilkins Co., Baltimore.
    Based on the 7th edition of Bergeys Manual. The methods section is a goldmine of useful procedures and media for many kinds of bacteria.
13. **Skerman, V. B. D.** 1969. Abstracts of microbiological methods. Wiley-Interscience, New York.
    A wide variety of routine physiological and bio-

chemical characterization methods, abstracted from original articles.

20.5.2. Specific Articles

14. **Adachi, T., Y. Murooka, and T. Harada.** 1973. J. Bacteriol. **116:**19–24.
15. **Adler, J.** 1966. Science **153:**708–716.
16. **Aldridge, K. E., B. B. Gardner, S. J. Clark, and J. M. Matsen.** 1978. J. Clin. Microbiol. **7:** 507–513.
17. **Back, A. E., and T. R. Oberhofer.** 1978. J. Clin. Microbiol. **7:**312–313.
18. **Balderston, W. L., B. Sherr, and W. J. Payne.** 1976. Appl. Environ. Microbiol. **31:**504–508.
19. **Bernaerts, M. J., and J. De Ley.** 1963. Nature (London) **197:**406–407.
20. **Board, R. G., and A. J. Holding.** 1960. J. Appl. Bacteriol. **23:**xi.
21. **Botsford, J. L., and R. D. DeMoss.** 1971. J. Bacteriol. **105:**303–312.
22. **Bürger, H.** 1967. Zentralbl. Bakteriol. Parasitenkd. Infektionskr. Hyg. Abt. I Orig. **202:**97–109.
23. **Campbell, N. E. R., and H. J. Evans.** 1969. Can. J. Microbiol. **15:**1342–1343.
24. **Cato, E. P., and W. E. C. Moore.** 1965. Can. J. Microbiol. **11:**319–324.
25. **Christensen, W. B.** 1946. J. Bacteriol. **52:**461–466.
26. **Christensen, W. B.** 1949. Res. Bull. No. 1., Weld County Health Department, Greeley, Colo.
27. **Cohen-Bazire, G., W. R. Sistrom, and R. Y. Stanier.** 1957. J. Cell. Comp. Physiol. **49:**25–68.
28. **Daesch, G., and L. E. Mortensen.** 1972. J. Bacteriol. **110:**103–109.
29. **Delafield, F. P., M. Doudoroff, N. J. Palleroni, C. J. Lusty, and R. Contopoulos.** 1965. J. Bacteriol. **90:**1455–1466.
30. **Döbereiner, J., and J. M. Day.** 1976. *In* W. E. Newton and C. J. Nymans (ed.), Symposium on nitrogen fixation, p. 518–538. Washington State University Press, Pullman.
31. **Dowda, H.** 1977. J. Clin. Microbiol. **6:**605–609.
32. **Elder, B. L., I. Trujillo, and D. J. Blasevic.** 1977. J. Clin. Microbiol. **6:**312–313.
33. **Evans, J. B., and C. F. Niven.** 1950. J. Bacteriol. **59:**545–550.
34. **Ewing, W. H., B. R. Davis, and R. W. Reavis.** 1957. Public Health Lab. **15:**153.
35. **Finegold, S. M., W. J. Martin, and E. G. Scott.** 1978. Bailey and Scott's diagnostic microbiology, 5th ed. C. V. Mosby, St. Louis.
36. **Fung, D. Y. C., and P. A. Hartman.** 1972. Can. J. Microbiol. **18:**1623–1627.
37. **Fung, D. Y. C., and P. A. Hartman.** 1975. *In* C. G. Hedén and T. Illéni (ed.), New approaches to the identification of microorganisms, p. 385–392. John Wiley & Sons, Inc., New York.
38. **Fung, D. Y. C., and R. D. Miller.** 1970. Appl. Microbiol. **20:**527–528.
39. **Gallien, R.** 1975. *In* C. G. Hedén and T. Illéni (ed.), New approaches to the identification of microorganisms, p. 385–392. John Wiley & Sons, Inc., New York.
40. **Guthertz, L. S., and R. L. Okuluk.** 1978. Appl. Environ. Microbiol. **35:**109–112.
41. **Hansen, S. L., D. R. Hardesty, and B. M. Meyers.** 1974. Appl. Microbiol. **28:**798–801.
42. **Haynes, W. C.** 1951. J. Gen. Microbiol. **5:**939–950.
43. **Hino, S., and P. W. Wilson.** 1958. J. Bacteriol. **75:**403–408.
44. **Hugh, R., and E. Leifson.** 1953. J. Bacteriol. **66:**22–26.
45. **Hungate, R. E.** 1966. The rumen and its microbes. Academic Press Inc., New York.
46. **Hwang, M., and G. M. Ederer.** 1975. J. Clin. Microbiol. **1:**114–115.
47. **Hylemon, P. B., J. S. Wells, Jr., N. R. Krieg, and H. W. Jannasch.** 1973. Int. J. Syst. Bacteriol. **23:**340–380.
48. **Isenberg, H. D., and J. Sampson-Scherer.** 1977. J. Clin. Microbiol. **5:**336–340.
49. **Kiehn, T. E., K. Brennan, and P. D. Ellner.** 1974. Appl. Microbiol. **28:**668–671.
50. **Killinger, A. H.** 1974. *In* E. H. Lennette, E. H. Spaulding, and J. P. Truant (ed.), Manual of clinical microbiology, 2nd ed., p. 135–139. American Society for Microbiology, Washington, D.C.
51. **King, E. O., M. K. Ward, and D. E. Raney.** 1954. J. Lab. Clin. Med. **44:**301–307.
52. **Kohn, J.** 1953. J. Clin. Microbiol. **6:**249.
53. **Law, J. H., and R. A. Slepecky.** 1961. J. Bacteriol. **82:**33–36.
54. **Lederberg, J., and E. M. Lederberg.** 1962. J. Bacteriol. **63:**399–406.
55. **Leifson, E.** 1963. J. Bacteriol. **85:**1183–1184.
56. **Le Minor, L., and F. Ben Hamida.** 1962. Ann. Inst. Pasteur (Paris) **102:**267–277.
57. **Lovelace, T. E., and R. R. Colwell.** 1968. Appl. Microbiol. **16:**944–945.
58. **Macrae, R. M., and J. F. Wilkinson.** 1958. J. Gen. Microbiol. **19:**210–222.
59. **McIlroy, G. T., P. K. W. Yu, W. J. Martin, and J. A. Washington II.** 1972. Appl. Microbiol. **24:**358–362.
60. **Milazzo, F. H., and J. W. Fitzgerald.** 1967. Can. J. Microbiol. **13:**659–664.
61. **Møller, V.** 1955. Acta Pathol. Microbiol. Scand. **36:**158–172.
62. **Moore, H. B., V. L. Sutter, and S. M. Finegold.** 1975. J. Clin. Microbiol. **1:**15–24.
63. **Morton, H. E., and M. A. J. Monaco.** 1971. Am. J. Clin. Pathol. **56:**64–66.
64. **Moussa, R. S.** 1975. *In* C. G. Hedén and T. Illéni (ed.), New approaches to the identification of bacteria, p. 407–420. John Wiley & Sons, Inc., New York.
65. **Neal, J. L., Jr., K. C. Lu, W. B. Bollen, and J. M. Trappe.** 1966. Appl. Microbiol. **14:**695–696.
66. **Neyra, C. A., J. Döbereiner, R. LaLande, and R. Knowles.** 1977. Can. J. Microbiol. **23:**300–305.
67. **Niven, C. F., Jr., K. L. Smiley, and J. M. Sherman.** 1942. J. Bacteriol. **43:**651–660.
68. **Nord, C. E., A. S. Lindberg, and A. Dahlback.** 1974. Med. Microbiol. Immunol. **159:**211–220.
69. **Nord, C. E., T. Wadstrom, and A. Dahlback.** 1975. *In* C. G. Hedén and T. Illéni (ed.), New

approaches to the identification of microorganisms, p. 393–406. John Wiley & Sons, Inc., New York.

70. Nord, C. E., B. Wretlind, and A. Dahlback. 1977. Med. Microbiol. Immunol. 163:93–97.

71. Oberhofer, T. R., J. W. Rowen, G. F. Cunningham, and J. W. Higbee. 1977. J. Clin. Microbiol. 6:559–566.

72. Okon, Y., S. L. Albrecht, and R. H. Burris. 1976. J. Bacteriol. 128:592–597.

73. Page, W. J., and H. L. Sadoff. 1976. J. Bacteriol. 125:1080–1087.

74. Pelczar, M. J., Jr. (ed.). 1957. Manual of microbiological methods. McGraw-Hill Book Co., New York.

75. Porschen, R. K., and S. Sonntag. 1974. Appl. Microbiol. 27:1031–1033.

76. Robertson, E. A., G. C. Macks, and J. D. MacLowry. 1976. J. Clin. Microbiol. 3:421–424.

77. Rosenberg, H., A. H. Ennor, and V. F. Morrison. 1956. Biochem. J. 63:153–159.

78. Rutherford, I., V. Moody, T. L. Gavan, L. W. Ayers, and D. L. Taylor. 1977. J. Clin. Microbiol. 5:458–464.

79. Schreier, J. B. 1969. Am. J. Clin. Pathol. 51:711–716.

80. Shayegani, M., A. M. Lee, and D. M. McGlynn. 1978. J. Clin. Microbiol. 7:533–538.

81. Shayegani, M., P. S. Maupin, and D. M. McGlynn. 1978. J. Clin. Microbiol. 7:539–545.

82. Shinoda, S., and K. Okamoto. 1977. J. Bacteriol. 129:1266–1271.

83. Sierra, G. 1957. Antonie van Leeuwenhoek J. Microbiol. Serol. 23:15–22.

84. Simmons, J. S. 1926. J. Infect. Dis. 39:201–214.

85. Smith, P. B., G. A. Hancock, and D. L. Rhoden. 1969. Appl. Microbiol. 18:991–993.

86. Smith, P. B., K. M. Tomfohrde, D. L. Rhoden, and A. Balows. 1971. Appl. Microbiol. 22:928–929.

87. Smith, R. F., R. R. Rogers, and C. L. Bettge. 1972. Appl. Microbiol. 23:423–424.

88. Stanier, R. Y., N. J. Palleroni, and M. Doudoroff. 1966. J. Gen. Microbiol. 43:159–271.

89. Stargel, M. D., F. S. Thompson, S. E. Phillips, G. L. Lombard, and V. R. Dowell, Jr. 1976. J. Clin. Microbiol. 3:291–301.

90. Stockdale, H., D. W. Ribbons, and E. A. Dawes. 1968. J. Bacteriol. 95:1798–1803.

91. Thornley, M. J. 1960. J. Appl. Bacteriol. 23:37–52.

92. Tomfohrde, K. M., D. L. Rhoden, P. B. Smith, and A. Balows. 1973. Appl. Microbiol. 25:301–304.

93. Wheater, D. M. 1955. J. Gen. Microbiol. 12:123–132.

94. Whittenbury, R. 1964. J. Gen. Microbiol. 35:13–26.

95. Wilkins, T. D., and C. B. Walker. 1975. Appl. Microbiol. 30:825–830.

96. Wilkins, T. D., C. B. Walker, and W. E. C. Moore. 1975. Appl. Microbiol. 30:831–837.

97. Williams, M. A. 1960. Int. Bull. Bacteriol. Nomencl. Taxon. 10:193–196.

98. Williamson, D. H., and J. F. Wilkinson. 1958. J. Gen. Microbiol. 19:198–209.

99. Yoch, D. C., and R. M. Pengra. 1966. J. Bacteriol. 92:618–622.

100. Zobell, C. E. 1946. Marine microbiology. Chronica Botanica Co., Waltham, Mass.

Chapter 21

Numerical Taxonomy

RITA R. COLWELL AND BRIAN AUSTIN

The principles of numerical taxonomy are well established, and several hundred publications concerning the application of these techniques to bacterial classification and identification have been published since the feasibility of the application of computer methods to the numerical taxonomy of bacteria became evident in 1957 (12). The usefulness of numerical techniques for bacterial taxonomy is attested by the large number of taxa which have been defined or clarified according to the principles of numerical taxonomy. These principles, based on the concept of Adansonian taxonomy (6), state that maximum information content should be achieved: i.e., all possible tests should be studied for the strains, the tests should be weighted equally, and taxa should be defined on the basis of overall similarity according to the results of **phenetic analyses** (analysis based on the observed characters of the organisms rather than on their ancestry). General references are listed at the end of this chapter for those readers who wish to pursue the concepts and methods of numerical taxonomy further (1–5).

Many workers have been reluctant to accept the concept of equal weighting of tests. Although it could be argued that some tests are more important than others, in reality present knowledge is such that it is not possible, objectively, to assign a weighting value to each test. Numerical taxonomy has permitted homogeneous taxa to be defined and, from overall characteristics, *a posteriori* weighting of the characters enables a few tests to be chosen and included in identification tables. Such tables are **polythetic**: i.e., although the variety of characters used for identification of a given taxon is distinctive for the taxon, a particular character need not necessarily be exhibited by every member of the taxon.

A numerical taxonomy study involves five essential steps: (i) selection of strains, (ii) selection of tests, (iii) coding and arraying of test results in a format suitable for analysis by a computer, (iv) computer analysis of the relationships between strains and the clustering of related strains, and (v) presentation and interpretation of the results. Several coefficients can be employed for the computer analyses, and these have been evaluated for microbiological application (4, 5, 7).

21.1. STRAIN SELECTION

With the availability of appropriate computer programs and adequate resources, numerical methods permit the analysis of data collected for a large number of bacterial strains. Over 300 strains usually can be examined in a single set. Larger data sets may need to be divided, but an overall comparison can subsequently be achieved in comprehensive inter- and intragroup similarity analyses. One method for dealing with large volumes of data is to employ the hypothetical median organism for data reduction (11). Thus, it is possible to compare data accumulated for sets of 1,000 or more strains.

There are no rules controlling the range of diverseness among the strains to be examined by numerical taxonomy. Selection of strains may be a simple or tedious task, depending on whether the strains represent one bacterial species, several species from the same genus, or several genera. The set of strains may include cultures of historical, pathological, or environmental importance. The set should include strains which have been identified and bear a scientific name; whenever possible, these known strains should include authentic "type" strains (permanent bench mark strains for taxa) as well as additional strains whose identity has been established ("reference strains") for comparative purposes. Type and reference strains can be obtained from culture collections such as the American Type Culture Collection (Rockville, Md.) or the National Collection of Type Cultures (London, England). It is advantageous to include at least two strains of each species used for comparison, to exclude the possibility of accidental contamination, mislabeling, or loss of viability of the strains, thereby negating their comparative value. If intraspecific, intergeneric, or interfamily relationships are being investigated, type and reference cultures for all the subspecies, genera, and families involved should be included.

The prerequisite step before subjecting the strains to extensive taxonomic studies is to ensure the isolation and preservation of pure cultures. See Chapter 8 for pure-culture methods and Chapter 12 for preservation methods.

21.2. TEST SELECTION

The selection of general characterization tests is an integral part of a numerical taxonomy study. Select routine tests that represent a broad spectrum of the biological activities of the organism and include morphological, colonial, biochemical, nutritional, and physiological characters (part 20.1). Special tests may also be useful, viz., tests determining serological and ecological characteristics (part 20.2). Any test that provides admissible quantitative and qualitative information about a strain under study is suitable for inclusion in an array of tests for numerical taxonomy. Do not consider the relative importance of one test over another; such a priori weighting of tests is not consistent with the basic principles of numerical taxonomy. However, some tests routinely employed in conventional approaches to bacterial taxonomy have been found to be a source of error when employed in numerical taxonomy. For example, the oxidase, Voges-Proskauer, and gelatin-hydrolysis tests are not highly reproducible (13–15) and should be avoided.

All taxonomy studies should consider test error, and whenever appropriate, tests that are not highly reproducible or that are subject to errors of interpretation should be eliminated. If "bad" tests form only a small proportion of a larger number of tests in a numerical taxonomy analysis, the results can be reliable, i.e., statistically valid. The larger the number of characteristics included in an analysis, the better will be the approximation to total representation of the phenotype of the organism. Too few tests (e.g.,

less than ca. 40 tests) may yield results that are not statistically valid. An optimum number of tests for numerical taxonomy is considered to be in the range of 100 to 200. This number of tests can be managed effectively in most laboratories, especially with the development of rapid screening procedures, such as multipoint inoculating devices for test tubes or plates (see 20.2.4) and replica plating techniques (see 20.2.5).

The choice of tests to be included in an analysis ultimately rests with the individual investigator. Nevertheless, it is preferable to use tests designed to provide information about a single property, i.e., a unit character. A **unit character** is defined as a taxonomic character of two or more states which cannot be subdivided logically, except for changes in the method of coding (5, 16). In numerical taxonomy, phenotypic characters provide the basic source of information for classification and identification; genetic characters (Chapter 22) are not commonly used.

In constructing data sets, do not use inadmissible tests, such as tests for redundant characters or for characters positive or negative in all the strains of the set (i.e., cases where such characters provide no useful discriminatory information). Include type or reference strains, for which test results are known, in every set of strains as a means of verification of media and methods. Include uninoculated controls in testing, since these can serve to detect accidental contamination. Such controls improve the accuracy of the data.

Standardize the treatment of all strains in a test set whenever possible. For example, if one strain in a set of strains requires 0.5% (wt/vol) sodium chloride, media for all strains in the set should contain this concentration of salt, provided that it is not inhibitory for any of the other strains.

Also standardize the inoculation and incubation procedures. Use inoculation loops of known diameter or Pasteur pipettes delivering droplets of known volume for transferring cultures. Use cultures in the logarithmic phase of growth. Incubation times are governed by growth rates of the strains being examined. Allow an adequate time of incubation for slow-growing strains, since a negative result for a test may be wrongly concluded if insufficient incubation time has been allowed. Usually, examine and record results after incubation for 1, 2, 3, 7, 14, 21, and 28 days. Provide incubation temperatures that are optimum for the strains being examined.

21.3. DATA CODING

Maintain an up-to-date, permanent record of test procedures and results. Loose-leaf pages with hastily scribbled notes have a tendency to become lost or indecipherable. Tablets containing forms marked off into 80 columns are available from university bookstores or computer supply centers; such forms are suitable for transferring taxonomic data from laboratory records onto punched cards or tape. The standard IBM card contains 80 numbered columns, with the only restriction on use being that the data format must be acceptable to the computer and consistent within a given analysis.

Eventually, convert all results for a given set of strains to a numerical format. There are essentially three types of number codes in use with numerical taxonomy. The simplest is the binary code for unit characters where a positive response (plus) is coded as 1 and a negative response (minus), as 0. There usually is provision for noncomparable (NC) characters (e.g., if the strain is incapable of growing in the test medium for H_2S production); the code for this symbol varies among computers and computer users. Thus, the results of a test may be plus, minus, or NC, with the numerical form 1, 0, or 3, respectively.

Another coding method applies to quantitatively identified characters determined by the strength or extent of reaction (e.g., strong versus weak acid production from glucose or other carbohydrates and strong versus weak oxidase reactions). The arrangement of such data for subsequent conversion to a numerical format is exemplified in Table 1.

A third type of coding applies to qualitatively identified characters. Pigmentation, Gram reaction, and cell shape are good examples for this type of coding (3).

Data coded in numerical form are ready to be keypunched onto computer cards. Verify punched cards for accuracy, as even the most experienced typist can make mistakes. Now the data are ready for computation.

21.4. COMPUTER ANALYSIS

Numerical taxonomy requires many computations to be made of the data, which can be

costly if a large number of strains are to be examined. Thus, the demand on computer resources can be great. The total number of arithmetic computations required (N) for the data set under study can be calculated by the equation

$$N = \tfrac{1}{2}n(n - 1)$$

where n is the total number of strains in the set. For example, a comparison of data for 100 strains requires 4,950 arithmetic computations. The number of computations expands exponentially with a linear increase in the number of strains represented in a data set.

The computations allow a logical stepwise examination of the data, including a determination of the overall similarity between strains and a hierarchical ordering of the strains according to their similarity. Many cluster-analysis computer programs are available. Some sources of such programs are Peter Sneath, University of Leicester, Leicester, England; Rita R. Colwell, University of Maryland, College Park, Md.; James Rohlf, State University of New York, Stony Brook, N.Y.; and Theodore Crovello, University of Notre Dame, Notre Dame, Ind. After installation of any of the available programs into a computer, even the novice taxonomist has the potential to carry out a numerical taxonomy study after a few brief lessions.

In the early stages of a numerical taxonomy study, enter the data into a data file in the computer. These files can be accessioned (i.e., called up from a remote terminal and updated). Examine the data, and calculate the affinities between each and every strain—the latter being referred to as an operational taxonomic unit (OTU).

Employ a coefficient of similarity. A large assortment of such coefficients is available (1, 5, 7, 8), but two are the most widely used in numerical taxonomy of bacteria. The simple matching coefficient (S_{SM}) is defined (14) as

$$S_{SM} = \frac{a + c}{a + b + c}$$

and the Jaccard coefficient (S_J) is defined (10) as

$$S_J = \frac{a}{a + b}$$

where a and c are the number of positive and negative matches, respectively, and b is the number of dissimilar tests between OTUs. The S_{SM} coefficient considers a total match between strains by including all of the positive and negative matches, and thereby it provides an estimation of overall similarity. The S_J coefficient

TABLE 1. *Arrangement of data for quantitatively identified characters for subsequent conversion to a numerical format*

Strain no.	Characters		
	Slight susceptibility to penicillin	Moderate susceptibility to penicillin	Very susceptible to penicillin
14	−	NC[a]	NC
15	+	−	−
16	+	+	−
17	+	NC	+

[a] NC, Noncomparable.

does not include negative matches in the equation. Although the merits of both equations have been adequately discussed elsewhere (4, 5), it needs to be emphasized that the S_{SM} coefficient reflects the presence of a large number of negative attributes between strains, thereby giving an erroneous indication of homogeneity. This can be a problem in studies involving large communities of bacterial strains, such as occur in the natural environment (9). In practice, data usually are examined by use of both the S_{SM} and S_J coefficients.

The analyses carried out in numerical taxonomy studies generate similarity matrices containing information about relationships among the strains. Using the output of the similarity computations, cluster analysis can then be done to arrange the strains in a hierarchical order, according to overall similarities. For cluster analysis, indices of similarities between all possible pairs of OTUs within the data set are calculated. With the simple "single-linkage" method, the data matrix is scanned to find the pair of organisms with the highest similarity (S) value (10). The matrix is subsequently rescanned to find additional pairs of OTUs with the next highest S value. Such a pair may be organisms joining the emerging cluster, or if one of the pair has already been allocated to an existing cluster, the other strain can also be added to the cluster. The scanning process is repeated by the computer until all the strains have been transferred from the unsorted similarity matrix to a cluster. Thus, the strains are sorted into a hierarchical order, and similarity matrices and dendrograms (see below) can be prepared.

21.5. PRESENTATION AND INTERPRETATION OF RESULTS

Upon initial examination of the output from a numerical taxonomy study, a large triangle of numbers is obtained, usually to three decimal places of accuracy. A sorted similarity matrix is shown in Fig. 1A, with comparisons between strains indicated. Multiplication by 100 yields the percentage of similarity between strains. The triangular array of numbers accurately portrays the relationships among the strains, but is not necessarily helpful. A more readily assimilable form of data arrangement is a shaded diagram form of the similarity matrix (Fig. 1B). Here, the similarity values are divided into ranges (viz., 90 to 100%, 85 to 89%, etc.) and assigned a different shading according to the assigned range. This makes it easier to see relationships between strains. It is helpful if a greater depth of shading reflects higher S values. The example given in Fig. 1B shows relationships among gram-negative, oxidase-positive bacteria isolated from a shellfish hatchery. From the diagram it is easy to see the three identified clusters of strains corresponding to *Pseudomonas* spp., *Lucibacterium* spp., and *Vibrio parahaemolyticus*, respectively.

In addition to the sorted similarity matrices, dendrograms (Fig. 1C) can be generated by the computer. These branched diagrams are usually constructed with the highest S values linking a pair or group of organisms. A simplified dendrogram (Fig. 1D) shows the relationships between clusters of strains and is used for displaying an analysis prepared from a very large data set. Dendrograms prepared from results of phenetic data analysis do not necessarily indicate phylogenetic relationships between strains (i.e., relationships based on ancestry of the organisms) and should not be treated as such unless molecular genetic evidence is provided, as in polyphasic taxonomy studies (10).

Clusters are defined on the basis of overall similarity. From the sorted similarity matrices and dendrograms, clusters of similar strains can be accurately defined (Fig. 1B) and similarity(ies) between clusters can be determined. Clusters of strains can be further characterized by tabulating the frequency of occurrence of each character among members of the clusters. Thus, identification tables, based on selected discriminatory tests, can be formulated for all the clusters for subsequent identification of fresh isolates. A median organism for each cluster can be calculated. This hypothetical median organism (HMO) will possess characters present in at least 50% of the members of the group (11). Members of homogeneous clusters will demonstrate high S values when compared with the HMO. Strains bearing only marginal similarity to a few members of the cluster can be eliminated from the cluster or tested further to define more reliably their relationship to the cluster. Clusters can be identified and named if type or reference strains have been included in the study or if comparison is made of the features of the clusters with those characteristics employed in diagnostic keys and tables. Sometimes, clusters defined by numerical taxonomy cannot be equated with existing known taxa, reflecting the limits of conventional approaches to taxonomy as well as indicating the potential inadequacies of existing diagnostic procedures. It is also possible that such clusters may represent new, undescribed taxa.

After analysis of a set of strains has been accomplished, the data from the numerical taxonomy study can be added to the data from other numerical taxonomy studies, thereby permitting construction of taxonomic data banks for future reference.

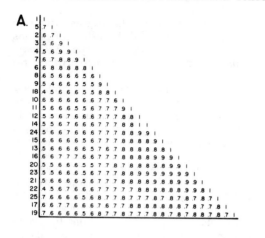

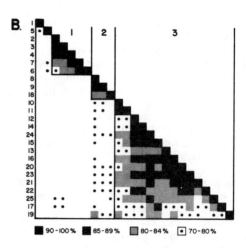

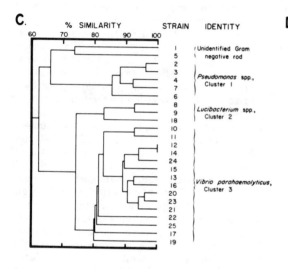

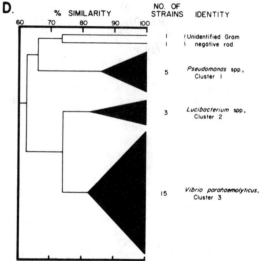

FIG. 1. *Presentation of output from a numerical taxonomy analysis of bacteria, in which a sorted similarity matrix has been calculated (A); the numbers have been rounded up to one decimal for convenience. From this information, a shaded diagram (B), dendrogram (C), and simplified dendrogram (D) can be prepared.*

21.6. LITERATURE CITED

21.6.1. General References

1. **Clifford, H. T., and W. Stephenson.** 1975. An introduction to numerical classification. Academic Press, Inc., New York.
2. **Colwell, R. R.** 1970. Numerical analysis in microbial identification and classification. Dev. Ind. Microbiol. **11**:154–160.
3. **Lockhart, W. R., and J. Liston (ed.).** 1970. Methods for numerical taxonomy. American Society for Microbiology, Bethesda, Md.
4. **Sneath, P. H. A., and R. R. Sokal.** 1973. Numerical taxonomy. W. H. Freeman & Co., San Francisco.
5. **Sokal, R. R., and P. H. A. Sneath.** 1963. Principles of numerical taxonomy. W. H. Freeman & Co., San Francisco.

The preceding references deal with the theoretical and practical aspects of numerical taxonomy and are of interest to the novice as well as to the experienced taxonomist. Sources of data, types of phenotypic characters, processing of data, computer methods, and interpretation of data are covered in detail.

21.6.2. Specific Articles

6. **Andanson, M.** 1763. Familles des plantes, vol. 1. Vincent, Paris.
7. **Austin, B., and R. R. Colwell.** 1977. Int. J. Syst. Bacteriol. **27**:204–210.
8. **Cheetham, A. H., and J. E. Hazel.** 1969. J. Paleontol. **43**:1130–1136.
9. **Colwell, R. R.** 1970. J. Bacteriol. **104**:410–433.
10. **Goodfellow, M., B. Austin, and D. Dawson.** 1976. *In* C. H. Dickinson and T. F. Preece (ed.), Microbiology of aerial plant surfaces, p. 277–292.

Academic Press, London.

11. **Liston, J., W. J. Wiebe, and R. R. Colwell.** 1963. J. Bacteriol. **85:**1061–1070.

12. **Sneath, P. H. A.** 1957. J. Gen. Microbiol. **17:**201–226.

13. **Sneath, P. H. A.** 1974. Int. J. Syst. Bacteriol. **24:** 508–523.

14. **Sneath, P. H. A., and V. G. Collins.** 1974. Antonie van Leeuwenhoek J. Microbiol. Serol. **40:** 481–527.

15. **Sneath, P. H. A., and R. Johnson.** 1972. J. Gen. Microbiol. **72:**377–392.

16. **Sokal, R. R., and C. D. Michener.** 1958. Univ. Kans. Sci. Bull. **38:**1409–1438.

Chapter 22

Genetic Characterization

JOHN L. JOHNSON

Elucidation of the structure and the physical properties of deoxyribonucleic acid (DNA) and of ribonucleic acids (RNAs) has enabled investigators to determine taxonomic relationships among bacteria by comparing their genomes. A crude comparison can be made with respect to the overall DNA **base composition**: of the total number of nucleotide pairs present in the DNA from a particular organism, that percentage represented by guanine plus cytosine (mole percent G+C) is a characteristic and constant feature of that organism. A large difference in the mole percent G+C values from DNAs of two organisms indicates the lack of a close genetic similarity; however, a similarity in the two mole percent G+C values does not necessarily mean that the organisms are similar with regard to the sequence of nucleotides in their DNAs. In such cases, a much more precise method than DNA base composition is required for comparing the genomes, viz., nucleic acid homology. Homology studies allow comparison of organisms with respect to the linear arrangements of the nucleo-

tides along (i) their entire DNA strands (**DNA homology**) or (ii) along those portions of the DNA stands that code for certain types of RNA (**RNA homology**).

. DNA homology values are average measurements of similarity which use the entire genome of each organism being compared. RNA homology values are specific for each type of RNA. Messenger RNA (mRNA) homology values are similar to those obtained by DNA homology (at least for bacteria) because a large portion of the genome is used for transcribing the mRNA molecules. In contrast, ribosomal RNA (rRNA) and transfer RNA (tRNA) are coded for by only a small fraction of the genome; therefore, in homology experiments using these two types of RNA, only those small fractions of the genome are being compared. In all groups of bacteria that have been systematically investigated, the arrangement of nucleotides in the rRNA and tRNA cistrons of the DNA appears to have evolved less rapidly than the bulk of the cistrons. Therefore, DNA homology methods are used to detect similarities between closely related organisms, whereas rRNA homology methods are used to detect similarities between more distantly related organisms. General references (1–6) dealing with the principles and specificity of homology methods and the use of homology methods as a guide to the genetic relatedness between organisms are given at the end of this chapter (22.8.1).

In theory and in practice, DNA and RNA homology methods are easy to do, once the materials for performing the methods have been prepared. The technical problems are associated with the isolation of DNA or RNA and with its radioactive labeling. These problems are usually unique for a given group of microorganisms, and, therefore, there are no universal solutions to them.

The methods described in this chapter are, for the most part, those that have been successfully employed in my laboratory, and familiarity has contributed to the selection of the specific methods. For isolation and characterization of plasmid DNA, see Chapter 15.

22.1. CELL DISRUPTION FOR NUCLEIC ACID ISOLATION

The first step in nucleic acid isolation is the disruption of the bacterial cells. Depending on the nature of the cell wall, bacterial cells can be disrupted (i) with a detergent alone, (ii) by a combination of detergents and hydrolytic enzymes, or (iii) by physical methods such as sonic oscillation, shearing release from a pressure cell, or shaking in the presence of glass beads.

Gram-negative bacteria are likely to be susceptible to lysis by a detergent alone, although there are many exceptions. In contrast, nearly all gram-positive bacteria must be digested with lysozyme (a hydrolytic enzyme that acts on the peptidoglycan of the cell wall) before they can be lysed by a detergent. Consequently, cell disruption of these two groups of bacteria will be considered separately.

22.1.1. Gram-Negative Bacteria

Centrifuge the cells from a culture that is in the late logarithmic or early stationary phase of growth. Resuspend the cells in a volume of saline-EDTA buffer (0.15 M NaCl, 0.01 M sodium ethylenediaminetetraacetate, pH 8.0) equal to about one-tenth to one-fortieth the volume of the original culture. Add sodium dodecyl sulfate (SDS) from a 20% (wt/vol) stock solution to give a final concentration of 1%. Swirl the flask in a 50 to 60°C water bath to increase the rate of lysis. Lysis will be signaled by a rapid increase in viscosity and a change in the cell suspension from turbid to opalescent.

For those bacteria that do not give significant lysis under these conditions, several alternative procedures can be tried. (i) Use a more dilute cell suspension. (ii) Make certain that the cells are in the log phase at the time of harvest, because rapidly growing cells are often more susceptible to lysis. (iii) Add lysozyme (Sigma Chemical Co., St. Louis, Mo.), pronase (Calbiochem-Behring, La Jolla, Calif.), or proteinase K (E M Laboratories, Inc., Elmsford, N.Y.) to the saline-EDTA suspension and incubate at 37°C before adding SDS. Remove samples periodically during incubation and test them for lysis by SDS. When using lysozyme, suspend the cells in a 1:5 dilution of the saline-EDTA buffer, as solutions of high ionic strength will inhibit the action of the enzyme. Remove samples periodically and test for lysis by SDS. Before adding SDS to the main cell suspension, increase the ionic strength of the buffer to restore the original concentration.

For gram-negative bacteria that will not lyse by the above detergent procedures, one must resort to one of the various physical methods. Rod-shaped bacteria of moderate length can usually be disrupted by passage through a French pressure cell (American Instrument Co., Silver Springs, Md.) at 12,000 to 16,000 lb/in²; shorter rods or cocci may require 30,000 to 34,000 lb/in². For additional information, see 5.1.1 and 18.1.3. Ultrasonic oscillators are usually less effective in disrupting cells than a French pressure cell; for additional information on ultrasonic disintegration, see 5.1.2 and 18.1.3. For those organisms that cannot be disrupted by these methods, a Bronwill cell homogenizer (Bronwill Scientific Co., Rochester, N.Y., distributor for B. Braun, Germany) will usually be effective. Operate the homogenizer at 4,000 cycles per min, using equal volumes of cell suspension and glass beads (0.1-mm-diameter glass beads are available from Bronwill Scientific Co.; 0.074- to 0.110-mm-diameter beads are available from Cataphote Div., Ferro Corp., Jackson, Miss., Class IV, no. 1420, type C). For additional information, see part 5.1.3.

22.1.2. Gram-Positive Bacteria

The usual procedure for rendering gram-positive bacteria susceptible to lysis by SDS is to partially digest the cell walls with lysozyme, with the cells suspended in a 1:5 dilution of saline-EDTA buffer as described previously. The incubation period required will range from several minutes to several hours. Remove samples periodically and test for lysis by SDS. Before adding SDS to the main suspension, increase the ionic strength of the buffer to full strength. As with gram-negative bacteria, it may be important to use cells that are in the log phase of growth. Other procedures that have been used to render gram-positive bacteria more susceptible to lysis by SDS are adding penicillin to actively growing cultures or growing the organisms in the presence of high concentrations of glycine (60).

If one is not concerned about the fragment size of the DNA, physical disruption of the cells is quick and easy. In general, the French pressure cell is less effective for disruption of gram-positive bacteria than for gram-negative bacteria. However, in my experience, shaking with glass beads (i.e., with a Bronwill homogenizer, as previously described) has invariably been effective.

In general, fragmented DNA will work satisfactorily in all of the procedures outlined in this chapter, although high-molecular-weight DNA is preferred for immobilization on nitrocellulose filters (see 22.6.5) and for in vitro labeling (see 22.6.3).

In the case of thermal denaturation (T_m) determinations (see 22.5.1), all of the DNA preparations, including the reference DNA, should be prepared in a similar manner.

22.2. DNA ISOLATION

The most time-consuming part of a DNA homology study, and also the part that is most subject to technical problems, is the isolation of DNA. Two methods will be given for isolating DNA from 0.5- to 1.0-liter amounts of culture (more volume may be required for bacteria that do not grow to a high density).

22.2.1. Marmur Method (44)

The Marmur method employs chloroform to remove protein from the lysate and is probably the most widely used method. The following is an abbreviated outline of this method.

Reagents.

Saline-EDTA buffer: 0.15 M NaCl, 0.01 M sodium EDTA, pH 8.0.
SDS: 20% (wt/vol) solution.
Sodium perchlorate: 5 M solution.
Chloroform-isopentanol: 24:1 (vol/vol).
SSC: 0.15 M NaCl, 0.015 M trisodium citrate, pH 7.0. *Note:* Other concentrations of SSC are indicated in the text by a number such as 20× (20-fold concentration) or 0.1× (one-tenth the concentration).
Ribonuclease (RNase): Bovine pancreatic RNase (Sigma Chemical Co., St. Louis, Mo.), 1 mg/ml in 0.15 M NaCl, pH 5.0. Heat the solution at 80°C for 10 min to inactivate any traces of deoxyribonuclease (DNase). Store in a freezer.

Procedure.

1. Suspend the centrifuged cells in 50 to 100 ml of saline-EDTA buffer and disrupt them by one of the methods previously described (22.1).

2. Add sodium perchlorate to a final concentration of 1 M.

3. Add 0.5 volume of chloroform-isopentanol, and shake on a wrist-action shaker in a ground glass-stoppered flask for 30 min, using a shaking speed just sufficient to produce an emulsion. Very vigorous shaking is not desirable.

4. Centrifuge the emulsion at 17,000 × *g* for 10 min in a refrigerated centrifuge at 0 to 4°C.

5. Carefully decant and/or pipette the upper aqueous layer from each tube, being careful not to collect any of the white precipitate (protein) at the interface between the two phases. For pipetting, use an inverted 10-ml serological pipette, with the tip inserted into a Propipette (Fisher Scientific Co., Pittsburgh, Pa.). The aqueous layer contains the DNA and is therefore very viscous. It is helpful to move the pipette continually back and forth in the tube to avoid collecting any of the protein at the interface.

6. Extract the lysate again by repeating steps 3 through 5.

7. Place the aqueous phase in a beaker and slowly overlay with cold 95% ethanol (an amount equal to about 2 volumes of the aqueous phase). Collect the precipitated DNA with a glass stirring rod by gently stirring the two phases while spinning the rod. The DNA will adhere or "spool" onto the rod (Fig. 1). Remove excess ethanol by pressing the rod against the side of the beaker; then allow the rod to drain vertically (with the DNA end up) for a few minutes to remove most of the remaining ethanol. In an alternative procedure, the aqueous phase is overlaid with ice-cold ethanol in a flask. Swirl the flask to form a loose clot of DNA. Trap the clot by pouring the contents of the flask into a 50-ml polypropylene centrifuge tube having a number of small perforations around the side and a larger hole at the bottom (Fig. 2). Make sure the large hole is plugged with a small serum bottle stopper before pouring in the DNA. After pouring, rinse the precipitated DNA one or two times with lysing buffer-ethanol mixture (1 part and 2 parts), and then allow the tube to drain until nearly all of the ethanol is gone.

8. Dissolve the spooled DNA in 10 to 20 ml of 0.1× SSC. Allow the glass rod to stand in the SSC until the DNA loosens and can be slipped off the rod. If the clot method is used, remove the stopper from the large hole at the bottom of the perforated tube, and place the tube in a small beaker (or into another tube just slightly larger than the polypropylene tube), and add 10 ml of 0.1× SSC. The DNA clot will start to

FIG. 1. *Spooling DNA onto a glass rod during precipitation by ethanol. Photo by D. Arbour.*

dissolve and will slip through the bottom hole as the polypropylene tube is withdrawn. Rinse the perforated tube with a second 10-ml volume of buffer, and then pool it with the first. The clot of DNA will dissolve faster than the "spooled" DNA. After the DNA is completely dissolved, adjust the SSC concentration to 1× by adding a suitable volume of 20× SSC.

9. Add RNase to the DNA preparations (50 μg/ml), and incubate at 37°C for 30 min.

10. Add chloroform-isopentanol (5 to 10 ml) to the DNA solution, and shake as in step 3 on a wrist-action shaker for 15 min. Centrifuge the emulsion as in step 4, and draw off the supernatant as in step 5. Repeat the chloroform-isopentanol extraction until very little protein is observed at the interface between the two phases after centrifugation.

11. Precipitate the DNA with ethanol by "spooling," and dissolve it in 0.1× SSC. Repeat this step two or three times to remove ribonucleotides. Remember to adjust the SSC concentration to 1× before each ethanol precipitation.

12. Finally, dissolve the DNA in 0.1× SSC and store in a freezer at −20°C, or place 2 to 3 drops of chloroform in the tube and store in a refrigerator.

22.2.2. Hydroxylapatite Method

Britten et al. (13) were the first to isolate DNA by adsorbing it to hydroxylapatite. Since then, several modifications of the method have been described (32, 35, 43, 47). The following is a useful adaptation of this method.

Reagents.

Saline-EDTA buffer (see 22.2.1).
SDS, 20% (wt/vol) (see 22.2.1).
RNase (see 22.2.1).

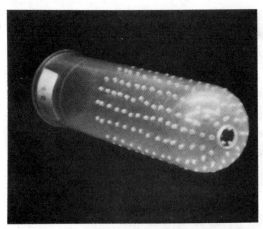

FIG. 2. *Perforated 50-ml polypropylene centrifuge tube for collection of precipitated DNA. The small perforations are made with an electric drill and small bits obtained from a hobby shop; an exact size is not important. The larger hole is 0.25 in. (6 mm) in diameter and is plugged with a serum bottle stopper prior to pouring in the DNA. Photo by N. Krieg.*

Liquefied phenol, chromatography grade, water saturated.

Hydroxylapatite, DNA grade Bio-Gel HTP (Bio-Rad Laboratories, Richmond, Calif.).

Phosphate buffer, 1.0 M, pH 6.8 (prepare by mixing equal volumes of 1 M Na_2HPO_4 and 1 M NaH_2PO_4). This is used for preparing the lower concentrations as indicated in the text.

Procedure.

1. Suspend the centrifuged cells in 25 ml of saline-EDTA buffer, and add 0.5 ml of RNase solution.

2. Add 1.2 ml of SDS. Swirl the flask, and warm in a 50 to 60°C water bath until lysis is complete. For those bacteria not disrupted by detergent alone, see part 22.1 for alternative methods.

3. Reduce the viscosity of the lysate by briefly subjecting it to sonic oscillation. Pronase b or proteinase K may be added at this time (50 µg/ml). For some organisms, proteinase digestion is not needed, but for others it is essential. If a proteinase is added, incubate it with the lysate at 50°C for 1 h.

4. Add 7 ml of water-saturated liquefied phenol, shake by hand to get the two phases well mixed, and then shake the flask on a wrist-action shaker for 20 min.

5. Centrifuge at 17,000 × g in a polypropylene centrifuge tube, using a refrigerated centrifuge at 0 to 4°C. Carefully draw off the upper (aqueous) layer (step 5 of 22.2.1), and return it to the flask.

6. Repeat steps 4 and 5.

7. After again returning the aqueous phase to the flask, add 2.0 ml of 1.0 M phosphate buffer. Then add 2 g (1 measuring teaspoon is convenient and sufficiently accurate) of dry hydroxylapatite. Suspend well, and gently shake on a rotary or reciprocal shaker for 1 h at a speed sufficient to keep the hydroxylapatite from settling out.

8. Transfer the suspension to a 50-ml polypropylene centrifuge tube, and centrifuge for 2 to 3 min at 5,000 × g at room temperature. Return the supernatant layer (lysate) to the flask (this can be used for a second DNA adsorption cycle), and use the sedimented hydroxylapatite (to which DNA is now adsorbed) in step 9.

9. To the sedimented hydroxylapatite add 8 ml of 0.10 M phosphate buffer. Suspend the hydroxylapatite with the aid of a Vortex mixer. Immediately add an additional 24 ml of the phosphate buffer (an automatic pipettor works well for these additions). The phosphate buffer should be added with some force, so that the hydroxylapatite will become evenly mixed for maximum dilution of nucleotides and phenol. Allow the hydroxylapatite to settle for 1 to 2 min, and then centrifuge for 2 to 3 min at 5,000 × g. Discard the supernatant.

10. Repeat step 9 six or seven times, or until the absorbance of the supernatant is less than 0.05 at 270 nm (the absorption maximum of phenol).

11. Suspend the hydroxylapatite in 5.0 ml of 0.5 M phosphate buffer to desorb the DNA. Centrifuge as before, but this time save the DNA-containing supernatant.

12. If most of the DNA has not been removed from the lysate in step 8, wash the hydroxylapatite from step 11 once with distilled water, and then add it to the lysate for a second adsorption cycle. Alternatively, fresh hydroxylapatite may be used. Add the DNA obtained from a second adsorption cycle to that from the first cycle.

13. Filter the DNA preparation through a glass fiber filter (Reeve Angel Inc., Clifton, N.J.; 2.4-cm diameter, type 934AH) used in a syringe-type filter holder (Gelman Instrument Co., Ann Arbor, Mich.) to remove any remaining hydroxylapatite particles.

14. Dialyze the DNA preparation against 0.02 M NaCl–10^{-3} M *N*-2-hydroxyethylpiperazine-*N'*-2-ethanesulfonic acid sodium salt (HEPES) buffer, pH 7.0. Cut 13- to 15-cm lengths of dialysis tubing (Fisher Scientific Co., Pittsburgh, Pa.), and wash out ultraviolet-absorbing materials. This is done by boiling the tubing in a 2 to 5% solution of sodium carbonate for a few minutes, thoroughly rinsing the lengths of tubing under running tap water, and then rinsing them under distilled water. The tubing can be stored

in a refrigerator for 2 to 3 days in distilled water. Cellulolytic organisms may hydrolyze the tubing if it is stored for a longer time. Tie a knot at one end of the tubing length, pour in the DNA solution, and either tie another knot at the other end or tie it off with a string. Wrap a piece of tape around the string for a convenient label. Dialyze in a 400- to 500-volume excess of the buffer for about 3 h, change the buffer, and continue dialyzing overnight. Store the DNA preparations in a freezer, or add a few drops of chloroform and store in a refrigerator.

Comments.

If the lysate is digested with pronase or proteinase K, a single phenol extraction will usually be sufficient.

An important feature of the hydroxylapatite procedure is that the RNA species are degraded to such an extent that they will not compete with DNA for adsorption sites on the hydroxylapatite. For some groups of organisms, RNase T1 may be required in addition to the pancreatic RNase to sufficiently degrade the RNA. Add 50 to 100 Sankyo units of RNase T1 (Calbiochem-Behring, La Jolla, Calif.) in step 1. For other groups of organisms, the RNase is inhibited when added in step 1. In such cases dialyzing the lysate against saline-EDTA buffer for several hours will remove the inhibitory effect.

The major contaminant of the DNA preparations, other than RNA, is polysaccharide. Many polysaccharides do not adsorb to hydroxylapatite and therefore are easily separated from DNA by this procedure. Others, however, can bind to hydroxylapatite and inhibit DNA adsorption.

A more dilute cell suspension of some organisms may be needed for this method to be efficient. The dilution appears to affect the completeness of lysis and/or RNA degradation. Scale up the volumes four or five times, and follow the above steps until the hydroxylapatite adsorption. Here, after adding the hydroxylapatite to part of the lysate, pour the hydroxylapatite suspension into a 60-ml sintered-glass filter. After the hydroxylapatite has partially settled, let all of the lysate pass through it. Then transfer the hydroxylapatite to a centrifuge tube, and complete the procedure as described above.

Many additional procedures and modifications of existing procedures have been published, and several are described in detail in reference 30.

22.3. RNA ISOLATION

The bulk of the nucleic acid within a bacterial cell is RNA (mostly rRNA and lesser amounts of tRNA and mRNA); therefore, it is easy to isolate RNA in large quantities. The major problem is to obtain RNA preparations free from RNase. This enzyme is rather ubiquitous and also is very heat stable. Glassware can best be rendered free from RNase by baking in an ashing oven or a dry-heat sterilizing oven. Aqueous buffers and solutions should be treated first by autoclaving and then by adding 0.2% diethyl pyrocarbonate (22).

Kirby (38) has described in detail procedures for isolating RNA. The following is a variation of his procedures that works well for the isolation of bacterial RNA that is essentially free from DNA.

Reagents.

Sodium naphthalene 1,5-disulfonate, 10% (wt/vol).

Phenol-cresol mixture: 550 ml of liquefied phenol, 70 ml of m-cresol, and 0.5 g of 8-hydroxyquinoline.

Diethyl pyrocarbonate (this reagent is very unstable in water; add it just prior to lysing the cells).

SSC (see 22.2.1). Also prepare SSC containing 1% SDS.

SDS, 20% (see 22.2.1).

Procedure.

1. Harvest the cells by centrifugation, and wash them once with distilled water. Suspend the cells in approximately 20 ml of cold distilled water. Measure the volume of the suspension with a graduated cylinder.

2. Add 0.05 ml of 10% naphthalene disulfonate for each 1.0 ml of the suspension; also add diethyl pyrocarbonate to a final concentration of 0.2%. Disrupt the cells immediately by passage through a French pressure cell at 10,000 to 12,000 lb/in^2 into a mixture containing 15 ml of phenol-cresol, 10 ml of 0.5% naphthalene disulfonate, and 0.02 ml of diethyl pyrocarbonate.

3. Shake the flask for 20 min on a wrist-action shaker. Place the mixture in a polypropylene centrifuge tube, and centrifuge at 17,000 × g for 10 min in a refrigerated centrifuge at 0 to 4°C. Carefully draw off and save the upper (aqueous) layer.

4. Add 20× SSC in a ratio of 1 part to 20 parts of the aqueous phase; also add SDS to a final concentration of 1%.

5. Add 15 ml of phenol-cresol, shake, and centrifuge at 17,000 × g in a refrigerated centrifuge. Save the aqueous layer.

6. Add 2 volumes of ice-cold 95% ethanol, mix, and allow to stand in a freezer for 30 to 60 min. Centrifuge at 4,000 × g in a refrigerated centrifuge for 10 min. Decant the supernatant, and allow the centrifuge bottle to drain well.

7. Dissolve the sediment (RNA) in 30 ml of SSC. Repeat the ethanol precipitation, centrifugation, and dissolution until the supernatant after centrifuging has a negligible absorbance at 270 nm. Store the RNA in SSC containing 1% SDS at −20°C or colder.

The total RNA, as isolated above, is usually used as competitor in competition experiments (22.7.3), although the components can be fractionated if desired. Preparations of radioactively labeled RNA are usually fractionated (22.7.1), and specific components are used in the hybridization experiments.

22.4. NUCLEIC ACID CONCENTRATION AND PURITY

22.4.1. Ultraviolet Spectrophotometry

The most common method for determining nucleic acid concentrations in nucleic acid solutions at room temperature is to measure the absorbance at 260 nm. The following formulas can be used provided that cuvettes with a 1-cm light path are employed (5, 59).

For native (double-stranded) DNA:

$$\text{mg per ml} = \frac{\text{absorbance at 260 nm}}{20}$$

For RNA or denatured (single-stranded) DNA:

$$\text{mg per ml} = \frac{\text{absorbance at 260 nm}}{23}$$

In practice, most stock DNA preparations will range from 0.5 to 2 mg of DNA per ml, and it is convenient to make a 1:20 dilution (0.1 ml of DNA solution added to 1.9 ml of buffer). This is within the absorbance range of most spectrophotometers, and the absorbance then reads directly as milligrams of DNA per milliliter in the stock preparation.

Ultraviolet spectrophotometry has also been used for detecting contaminating materials. (i) Proteins absorb light strongly at 280 nm. The absorbance at 260 nm of a nucleic acid solution divided by its absorbance at 280 nm (A_{260}/A_{280} ratio) has been used as an indicator of protein contamination. The extinction coefficient for protein, however, is very low compared to that for nucleic acids, and consequently the A_{260}/A_{280} ratio is not a very sensitive indicator. (ii) The absorption spectrum for nucleic acids has maxima at approximately 208 and 260 nm and a minimum at 234 nm. The peak at 208 nm is not specific for nucleic acids because many compounds will absorb light of that wavelength. If

contaminating material is present in a nucleic acid preparation and contributes to a large peak at 208 nm, the peak will overlap the 234-nm minimum for the nucleic acid. Therefore, the A_{234}/A_{260} ratio is a rather sensitive indicator of contaminating material in a DNA preparation. (iii) The absorption maximum of phenol is 270 nm; consequently, the A_{270}/A_{260} ratio is quite sensitive for detecting phenol contamination in a nucleic acid preparation.

22.4.2. Hyperchromic Shift

The major contaminant of DNA preparations is usually RNA, and the two nucleic acids cannot be differentiated directly by their ultraviolet absorbance. The hyperchromic shift that occurs when DNA is thermally denatured (see 22.5) can be used to estimate the fraction of the preparation that is DNA. A preparation of pure native DNA will increase in absorbance at 260 nm by approximately 40%. If the secondary structure of the contaminating RNA has been destroyed by RNase, the RNA will have no hyperchromic shift, but if the secondary structure is intact, the RNA will have a hyperchromic shift of approximately 30%. The hyperchromic shift for RNA occurs over a lower and wider temperature range than for DNA, so the profiles for RNA and DNA can be easily distinguished. Therefore, to estimate the percent DNA in a nucleic acid preparation, divide the hyperchromicity due to DNA by 0.4 and multiply by 100.

22.4.3. DNA by Diphenylamine Reaction

The diphenylamine reaction is specific for the deoxyribose in DNA. Although the Burton procedure (17.5.1) has often been used, the more sensitive modification of Giles and Meyers (27) is given below:

1. Place 1.0 ml of the DNA sample in a 16 by 125 mm screw-capped tube, and add 1.0 ml of 20% (wt/vol) perchloric acid.
2. Add 2.0 ml of glacial acetic acid containing 4% (wt/vol) diphenylamine.
3. Add 0.2 ml of 0.16% (wt/vol) solution of acetaldehyde.
4. Mix and incubate overnight at 30°C.
5. Read the absorbance at 595 and 700 nm, and calculate the difference. Compare this to values obtained with standards containing known amounts of a pure standard DNA (calf thymus or salmon sperm DNA; Sigma Chemical Co., St. Louis, Mo.) to determine the concentration of DNA in the sample. The diphenylamine procedure works well for purified DNA solutions in the range of 5 to 50 µg/ml.

The diphenylamine procedure can also be

used for quantifying the DNA bound to nitro-cellulose membranes in homology studies. It is particularly useful in rRNA homology experiments where one needs to know the amount of DNA present on the membrane at the end of the hybridization step. After the radioactivity of the membranes has been measured, remove the membranes from the scintillation fluid (use a toluene-based "cocktail" so that the membranes will not be dissolved and the radioactivity will not be eluted), and dry them in air. Then place the membranes in 2.0 ml of 10% perchloric acid (duplicate membranes can be placed together in the same tube), and heat in a boiling-water bath for 5 min. After cooling to room temperature, proceed with step 2 above.

22.4.4. RNA by Orcinol Reaction

Orcinol is used for determining RNA concentrations, although it reacts to a limited extent with DNA. For details of the procedure, see 17.5.2.

22.5. DNA BASE COMPOSITION

The mole percent guanine plus cytosine (mol% G+C) of DNA can be determined by several different methods: (i) hydrolysis and subsequent separation of the nucleotides or the purine and pyrimidine bases; (ii) buoyant density centrifugation; (iii) midpoint determination of the thermal melting (denaturation) profiles; (iv) bromination (58); (v) depurination (37); (vi) the ratio of absorbancies at 260 and 280 nm in low-ionic pH 3 buffer (25); and (vii) high-pressure liquid chromatography of nucleotides or free bases (14, 39). Contaminating RNA will interfere with all of the above methods except for the buoyant density and thermal melting (T_m) methods; consequently, these two methods have become the most popular.

22.5.1. Buoyant Density Method

The buoyant density method involves an analytical ultracentrifuge, the use of which is beyond the scope of this chapter. The procedure has been described by Mandel et al. (42).

22.5.2. Thermal Melting Method

The midpoint temperatures of the thermal melting profiles (T_m) were first correlated with the base composition of DNA preparations by Marmur and Doty (45). Mandel et al. (41) and De Ley (18) have reexamined the correlations and have also correlated the T_m and buoyant density values. Methods for estimating the T_m values from the denaturation profiles have been compared by Ferragut and Leclerc (23). The T_m values are greatly affected by the ionic strength of the buffer used, and this effect is the same on all DNA preparations. Consequently, one must either know the ionic strength of the buffer that is used or include a reference DNA preparation at the same ionic strength as used for the unknown samples. The latter is easy to do: one merely dialyzes all of the DNA preparations (including the reference DNA) together in a single batch of buffer.

Procedure.

The following procedure is reliable and may be adapted to other buffers or to buffer concentrations other than those given.

1. Prepare 2 to 4 liters of 0.5× SSC (22.2.1) at pH 7.0. Save some of the buffer for diluting the DNA preparations and for rinsing out cuvettes just prior to putting the DNA samples into them. Divide the rest into two equal parts so the buffer can be changed once during dialysis (see step 3).

2. Prepare 2- to 5-ml samples (depending upon the size of the cuvettes to be used) of DNA at 50 μg/ml, using 0.5× SSC to dilute the stock DNA preparations. Prepare a 10- to 20-ml volume of the reference DNA (e.g., the DNA of *Escherichia coli* b has a G+C content of 51 mol% and a melting point or T_m of 90.5°C in SSC), because it will have to be included in each instrument run.

3. Dialyze all of the preparations together overnight in the 0.5× SSC. Change the buffer once after the first few hours of dialysis. After dialysis, return the DNA preparations to screw-capped test tubes.

4. Determine the melting profile with an automatic recording spectrophotometer having a sample chamber that is heated by a circulating bath containing ethylene glycol or that has an electronically heated cuvette holder. Start at 60°C and increase the temperature linearly at 1.0°C per min over the entire range. A temperature sensor can be placed in the cuvette or in the cuvette chamber. In the case of the electronically heated cuvette holder, the sensor is located within the holder.

5. Determine the T_m values as in Fig. 3.

6. Calculate the moles percent G+C of the sample DNA by the following equation (26):

$$\text{mol\% G+C}_x = \text{mol\% G+C}_{ref} + 1.99(T_{m(x)} - T_{m(ref)})$$

where mol% G+C$_{ref}$ is the known moles percent G+C of the reference DNA and where $T_{m(x)}$ and $T_{m(ref)}$ are the T_m values determined for the sample DNA and the reference DNA, respectively, under the experimental conditions.

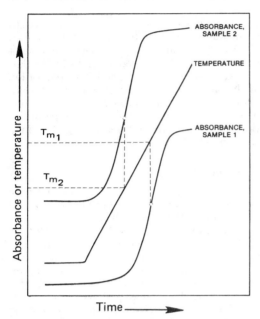

FIG. 3. *Hyperchromic shift tracings for two samples of DNA in an automatic recording spectrophotometer. The points at which any vertical line crosses the absorbance and temperature curves give values for these two parameters at a given time. The inflection points are estimated at one-half of the hyperchromic shift and the temperature at that time is the melting temperature (T_m).*

22.6. DNA HOMOLOGY

Similarities in the linear arrangement of nucleotides along DNA (or RNA) strands from different organisms are determined by homology and by thermal stability studies. Homology studies are used to measure the fraction of the genomes (or the fraction of a specific class of genes, such as rRNA genes) that can form duplexes under specific conditions of ionic strength and temperature. Thermal stability studies are used to estimate the degree of base mispairing in heterologous duplexes (hybrids).

There are two general methods for homology studies. The first, often referred to as the **membrane method**, involves immobilizing denatured (single-stranded) unlabeled DNA on a nitrocellulose membrane filter, treating the filter so that additional DNA will not adsorb to it, and then incubating the filter in the presence of radioactive denatured DNA fragments (or RNA, in the case of RNA homology). Duplexes of radioactive DNA (or DNA/RNA hybrids in the case of RNA homology) that form because of specific binding of the fragments to the DNA bound to the filter are stable to the subsequent washing of the filter, and the amount of duplex formation can be estimated by measuring the radioactivity of the filter.

The second homology method, often referred to as the **free-solution method**, involves the reassociation of the single-stranded nucleic acids in solution, as opposed to the membrane method where one component is fixed. Reassociation of DNA may be monitored optically by using ultraviolet spectrophotometry, or it may be measured by incubating a small amount of denatured DNA having a high specific radioactivity with a large excess of unlabeled, denatured DNA. In the latter instance, the ability of the labeled fragments to form duplexes with unlabeled DNA is assayed by adsorption to hydroxylapatite or by resistance to hydrolysis by an enzyme known as S1 nuclease.

The optical methods for assaying free solution reactions will not be discussed in this chapter. For these, refer to references 1, 6, 12, 19, 51, and 59.

22.6.1. Special Apparatus

Some specially constructed equipment for DNA homology methods is shown in Fig. 4 through 7. The reaction vials (Fig. 5) are constructed from 6-mm (outer diameter) Pyrex tubing cut into ⅞-in. (2.2-cm) lengths and sealed at one end with a flame. The vial stoppers (Fig. 5) are serum vial stoppers which have had the outside flange cut off. Solutions are added to the reaction vials by means of automatic micropipettes with disposable tips (Fig. 4E). The total volume of liquid in each vial is 110 μl. This volume just covers membranes added to the vials in membrane filter homology methods (Fig. 5); for free-solution homology methods, this volume allows removal of 100-μl samples in the hydroxylapatite or S1 nuclease procedures. When adding components to the reaction vials, the vials are placed in a Plexiglas holder (Fig. 4A) that has double rows of holes to help one keep track of the additions. After being filled, the vials are stoppered and placed in a Plexiglas holder (Fig. 5) for incubation in a circulatory water bath. This water bath must be large enough to contain the submerged holder and should be able to maintain temperatures within ±0.1°C. If the holder is not totally submerged, there is danger of distillation and collection of pure water drops on the vial sides and in the cap, which will greatly alter the ionic strengths and nucleic acid concentrations. For membrane filter homology methods, oblong-shaped membrane filters, 3 by 9 mm, are cut from larger membranes (up to 15 cm in diameter) by use of a punch (Fig. 4D: keysort punch, catalog no. 5302, McBee Systems, Roanoke, VA 23224). Although other membrane shapes can be used, the advantage of this shape is that it stands at an angle in the reaction vial (see Fig. 5). This shape allows for maximum contact with the solution in

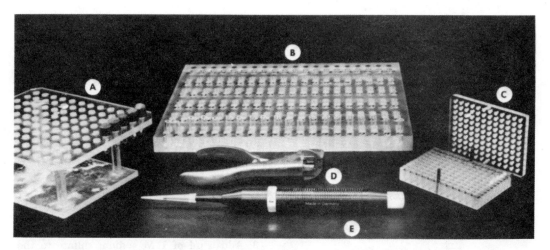

FIG. 4. *Equipment used in membrane homology methods. (A) Plexiglas holder for vials, used during incubation of vials in a water bath. Five reaction vials can be seen in the holder (see also Fig. 5). (B) Plexiglas holder for reaction vials during filling of the vials. (C) Plexiglas washing chamber for membrane filters. The top and bottom pieces are shown. Five filters are present in the right-hand side of the bottom piece (see also Fig. 5). (D) Punch for cutting oblong membrane filters. (E) Automatic micropipette with disposable tip, used for adding components to reaction vials. Photo by N. Krieg.*

FIG. 5. *Close-up of a portion of the reaction vial holder depicted in Fig. 4A. Membrane filters can be seen submerged in the reaction mixture within the vials. The vials are capped tightly with serum vial stoppers that have had the flanges cut off. Photo by N. Krieg.*

the vial, provides a high surface-to-volume ratio, and is easy to place in or remove from the vial. The Plexiglas washing chamber for membrane filter (Fig. 4C and 6) contains 120 compartments

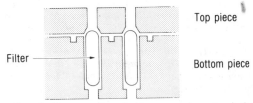

FIG. 6. *Cross-sectional diagram of part of the Plexiglas washing chamber for membrane filters.*

and is used to wash membrane filters free from unreacted radioactive fragments. The chamber consists of a top and bottom piece; the membrane filters are placed in the bottom piece, after which the top piece is bolted on. The assembly should be small enough to fit in a 400-ml beaker.

The filtration devices used for preparing DNA membranes (membranes to which denatured DNA is bound) are constructed mainly from Plexiglas and have a porous polyethylene filtration surface. The base of the apparatus is a short Plexiglas cylinder with an outside diameter the same as that of the membrane. The porous polyethylene is mounted on the inside and flush with the top. The base has a single exit tube, the flow rate of which can be controlled with an adjustable pinch clamp. The top part of the device consists of a Plexiglas cylinder with a neoprene (⅛ in. [3 mm] thick) gasket glued to the bottom rim. The top part is placed on top of the membrane and is held on by spring tension. All of the Plexiglas apparatus can be made by a local machine shop.

The thermal stability of membrane-bound duplexes can be determined with the aid of the device depicted in Fig. 7.

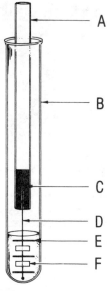

A

B

C

D

E

F

FIG. 7. *Diagram of apparatus for testing the thermostability of hybrid duplexes. A piece of 7-mm-diameter glass tubing (A) is sealed with a rubber plug (C), which is made from a rubber stopper by use of a cork borer. Membrane filters (E) and Teflon washers (F) are impaled on an insect pin (D) close to the head of the pin. The pointed end of the pin is inserted into the rubber plug. This assembly is placed into a 13 by 10 mm glass test tube (B) containing 1.2 ml of 0.5× SSC. The test tube is then placed in a heated water bath.*

22.6.2. Preparation of Labeled DNA: In Vivo Method

The DNA used for homology methods can be labeled by either in vivo or in vitro methods.

The in vivo labeling of DNA is dependent upon the medium required for culturing the organism and also upon the nucleic acid precursors which the organism will take up. The ^{3}H- or ^{14}C-labeled nucleic acid precursors most often taken up by bacteria are adenine, thymidine, and deoxyadenosine. Their uptake is not usually depressed by yeast extract up to 0.1% in the medium which, in some cases, may be needed to suppress biosynthetic pathways and allow incorporation of the precursors into DNA. The suitability of a given labeled precursor can be easily tested by adding it to a tube of medium inoculated with the organism in question. After growth, measure the radioactivity of portions of the culture, and also of the supernatant broth after the cells have been removed by centrifugation, by means of a liquid scintillation counter. The uptake of label is indicated by a decrease in counts in the supernatant compared with the total culture. Incorporation of $^{32}PO_4{}^{3-}$ or $^{33}PO_4{}^{3-}$ is most efficient in a peptone-free medium con-

taining no more than 1.5×10^{-4} M phosphate. When peptones and, in particular, yeast extract are required for growth, phosphate can be removed by precipitating it as the barium, calcium, or magnesium salt.

Procedure.

The following is a procedure described by Hayes and Gros (31).

1. Dissolve 100 g of dehydrated nutrient broth (Difco) and 50 g of Casamino Acids (Difco) in 900 ml of water.

2. Add 50 ml of 1 M barium acetate. (When other peptones or yeast extract is used, the exact amount of barium acetate required will have to be determined by titration.) Remove the precipitate by filtration.

3. Add 5 ml of 1 M sodium sulfate to the filtrate, let it stand for 15 min, and then filter again. Test for residual barium with rhodizonic acid (the red precipitate resulting from the barium salt of rhodizonic acid can be seen at about a 10 times lower concentration than can the $BaSO_4$ salt).

4. Adjust the volume to 1 liter, and use it as a 10× concentrate in preparing the labeling medium.

5. Isolate the DNA from the labeled cells by using a method described in 22.2. Dissolve the final DNA preparation in 0.1× SSC. For membrane homology experiments (22.6.5) use DNA concentrations in the range of 50 to 100 µg/ml (specific activities should be 10^3 cpm/µg or greater). For homology experiments using the hydroxylapatite or S1 procedures (22.6.4), use DNA concentrations of 5 µg or less per ml (specific activities should be in the range of 0.5×10^5 to 2.5×10^5 cpm/µg). When the labeled DNA is at a low concentration, include about 30 µg of sheared salmon sperm DNA per ml in the preparation to avoid adsorbance of labeled DNA to glass. Pass the DNA preparation twice through a French pressure cell at 16,000 lb/in^2.

22.6.3. Preparation of Labeled DNA: In Vitro Method

DNA can be iodinated with ^{125}I to greater than 10^7 cpm/µg. The procedure is not described here; details are given in references 16, 49, 52, and 56.

Another procedure to label DNA in vitro, which is becoming more commonly used, is the "nick translation" procedure (15, 37, 48, 50). The DNA is judiciously incubated with pancreatic DNase I to generate single-strand nicks. The DNase is then usually inactivated, and the nicked DNA is incubated with *Escherichia coli* polymerase I and the sodium salts of the four deoxyribonucleotide triphosphates (with one or more of them being radioactive, usually labeled with ^{3}H).

Reagents and components.

Tris-MgCl$_2$ buffer: 0.05 M tris(hydroxymethyl)-aminomethane–5 mM MgCl$_2$, pH 7.5. Also prepare the buffer as a 10× concentrate; be sure that the pH is correct after dilution.

Bovine pancreatic DNase I (electrophoretically purified; Sigma Chemical Co., St. Louis, Mo.). Reconstitute at 1 mg/ml in 70 mM K$_2$HPO$_4$–KH$_2$PO$_4$ buffer (pH 7.4) containing 1 mg of bovine serum albumin (fraction V, Sigma Chemical Co.) per ml and 50% glycerol. Store at −20°C. Use the same buffer when preparing 0.1- and 0.01-mg/ml dilutions.

E. coli DNA polymerase I. Reconstitute or dilute to 200 U/ml in the same buffer used for the DNase above. Store at −20°C. Preparations from Worthington Biochemical Corp. (Freehold, N.J.) and Miles Biochemicals (Elkhart, Ind.) have worked well in my laboratory.

Labeled thymidine 5′-triphosphate (TTP, tetrasodium salt) is shipped in 50% ethanol-water at concentrations of 0.5 to 1 mCi/μl.

DNase treatment for preparing nicked DNA.

Although high-molecular-weight DNA is preferable, DNA isolated by the hydroxylapatite method will also work. It is important that the DNA be very pure. The DNase treatment is as follows.

1. Dilute the stock DNA solution (it can be dissolved in 0.1× SSC) to 0.2 mg/ml with Tris-MgCl$_2$ buffer. If the stock DNA solution is near 0.2 mg/ml, add 10× concentrate of buffer.

2. Add 10 μl of 0.01-mg/ml DNase I to each 1 ml of the DNA solution, and incubate at 30°C for 1 to 2 min.

3. Inactivate the enzyme by heating the reaction vial in a 60°C water bath for 10 min. Store the nicked DNA at −20°C.

Procedure.

1. The radioactive nucleotide triphosphate is most stable when stored in 50% ethanol. Prevent any toxic effects of ethanol on the polymerase reaction by adding the labeled TTP to the reaction vial first and then evaporating it to dryness with a gentle stream of N$_2$ before adding the rest of the components.

2. Add the rest of the components listed in Table 1.

3. Incubate the reaction mixture (0.1-ml total volume) for 2 h at 30°C.

4. Stop the reaction by adding 10 μl of 10% SDS (Bio-Rad Laboratories, Richmond, Calif.) and heating in a 60°C water bath for 10 min.

TABLE 1. *Composition of polymerase I reaction mixture*

Component	Amt
[^{3}H]thymidine 5′-triphosphate (New England Nuclear, Boston, Mass.) *Evaporate to dryness before adding the rest of the components.*	10–15 μCi
Nicked DNA	50 μl
Tris buffer (2-amino-2-hydroxymethyl-1,3-propanediol), 0.7 M, pH 7.8	10 μl
MgCl$_2$, 0.14 M	5 μl
Dithioerythritol, 0.01 M (Sigma Chemical Co., St. Louis, Mo.)	5 μl
dXTP solution—contains the following deoxynucleoside triphosphates, each at 2 mM (all from Sigma Chemical Co.):	10 μl
2-Deoxyadenosine 5′-triphosphate	
2-Deoxycytidine 5′-triphosphate	
2-Deoxyguanosine 5′-triphosphate	
Thymidine 5′-triphosphate (Sigma Chemical Co.), 0.80 mM	10 μl
Polymerase I (Miles Biochemicals, Elkhart, Ind.), 200 U/ml	10 μl

5. Pass the reaction mixture through a 100- to 200-mesh Bio-Gel P100 polyacrylamide (Bio-Rad Laboratories) column, 1 by 7.5 cm, using 0.1× SSC buffer containing 0.1% SDS. Collect 0.25- to 0.28-ml fractions (14 drops = 0.27 ml with our fraction collector), and add 0.5 ml of 0.1× SSC containing 30 μg of sheared salmon sperm DNA per ml to each. The front peak containing labeled DNA and the retained peak of unreacted [^{3}H]TTP will elute in about 30 fractions.

6. Locate the peaks by spotting 5 μl from each fraction on strips of Whatman no. 1 filter paper (1.3 cm wide and marked at 2.5-cm intervals). Dry the strips, cut them at the marks, place them into scintillation vials, and measure the radioactivities.

7. Pool the front peak fractions and pass them twice through a French pressure cell at 16,000 lb/in^2. The labeled DNA should have a specific activity of about 2 × 10^5 cpm/μg and can be diluted out to about 2.6 μg/ml in 0.1× SSC containing 30 μg of salmon sperm DNA per ml.

8. Dilute the preparation to an activity of 6 to 8,000 cpm/10 μl, using the salmon sperm DNA-containing buffer employed above in step 5. Store at −20°C.

Comments.

The conditions for the successful in vitro labeling of DNA are somewhat arbitrary and may at times include some luck. The following are some suggestions for avoiding and solving some of the commonly encountered problems.

Before the actual labeling of a DNA preparation, test it using TTP of low activity. Dilute stock [³H]TTP to 20 μCi/ml with a 1:1 (vol/vol) mixture of ethanol-water, and use 10 μl of this (0.2 μCi) per reaction mixture (Table 1). Include a control vial containing DNA that you know will incorporate TTP. After incubating the reaction vials for 1 to 2 h, treat 25 μl of the sample with 5 to 10% trichloroacetic acid, collect the precipitate on a membrane filter, and measure the radioactivity in a scintillation counter and compare it with that for the control. Pass the rest of the sample through a P100 gel filtration column (see above), collect fractions 4 through 14 in Bio-vials (Beckman Instruments, Inc., Fullerton, Calif.), and measure the radioactivity in a scintillation counter using an appropriate cocktail for aqueous samples. The maximum front peak activity should be in fraction 5 or 6. From 30 to 50% of the added [³H]TTP should be incorporated into the trichloroacetic acid-precipitable DNA. High-quality labeled DNA will elute from the gel filtration column as a sharp front peak. Radioactivity in fractions 9 through 12 should be less than 5% of the peak fraction. The radioactivity in fractions 9 through 12 represents short trichloroacetic acid-precipitable DNA fragments which will not form duplexes.

Various batches and sources of DNase may have different amounts of activity; therefore, one should titrate the DNase by incubating 10-μl amounts of different concentrations (i.e., 0.1 and 0.01 mg/ml) with 1-ml amounts of a DNA preparation for various lengths of time and then testing the samples for maximum incorporation of label.

The DNA preparations appear to be the major variable affecting [³H]TTP incorporation. Endogenous DNase activity in the lysates may cause substantial nicking of the DNA. If the DNA preparations are free from nicks caused by endogenous DNase, little TTP will be incorporated unless they are first treated with DNase (see above). Other DNA preparations have enough nicks so that about the same amount of TTP is incorporated with or without the DNase treatment. In this case, the DNase treatment often generates too many nicks, which in turn cause the generation of short labeled DNA fragments (see above). Therefore, test each DNA preparation with and without the DNase treatment, and determine the amount of TTP uptake and the quality of the labeled material as outlined above.

The fine adjustment of TTP incorporation can be made by altering the concentration of unlabeled TTP. For example, if the incorporation of [³H]TTP into a non-DNase-treated DNA sample is only 15 to 20% but of good quality and the product from a DNase-treated part of the same sample is of poor quality, use the non-DNase-treated material, and lower the concentration of the unlabeled TTP to 0.4 mM (Table 1).

22.6.4. Free-Solution Method

Free-solution reassociation of denatured DNA fragments can be measured optically by use of a recording spectrophotometer; if radioactive DNA is included, it can be measured by adsorption to hydroxylapatite, or it can be measured by resistance to S1 nuclease. A general scheme for performing these procedures is given in Fig. 8. Both require high-specific-activity labeled DNA (a very small amount) to be incubated with a large excess of unlabeled DNA fragments. The specific activity must be in the range of 5×10^4 to 2.5×10^5 cpm/μg and the DNA should be at concentrations of less than 5 μg/ml.

Preliminary procedures.

Unlabeled DNA. For both the hydroxylapatite and S1 nuclease procedures, prepare unlabeled DNA as follows.

1. Dilute stock DNA (22.2.2) to about 0.9 to 1.0 mg/ml.

2. Sonicate each preparation for two 30-s time periods (e.g., we use a Biosonic III sonicator with a 0.375-in. [9.5-mm] probe and an energy setting of 60).

3. Denature the DNA, and hydrolyze any contaminating RNA by adding 5 N NaOH to a final concentration of 0.25 N. Heat for 15 min at 50°C, cool, and neutralize with an equivalent amount of HCl.

4. Dialyze the preparations overnight against 0.02 M NaCl–10^{-3} M HEPES buffer (pH 7.0), and adjust the concentrations to 0.6 mg/ml.

Labeled DNA. Both the hydroxylapatite and S1 nuclease procedures require a small amount of denatured labeled DNA with a high specific radioactivity to be incubated with a large excess of denatured unlabeled DNA fragments. The specific activity must be in the range of 0.5×10^5 to 2.5×10^5 cpm/μg, and the DNA is used at concentrations of 5 μg/ml or less.

Titration of S1 nuclease. Determine the nuclease activity of the stock S1 nuclease (Calbiochem-Behring, La Jolla, Calif.) in the following manner.

1. Place 50-μl amounts of a 0.6-mg/ml sheared, denatured bacterial DNA ("spiked" with [³H]DNA) preparation (dissolved in 0.02 M NaCl–1 mM HEPES buffer, pH 7.0) into a series of tubes (12 by 75 mm, polypropylene) containing 1-ml volumes of assay buffer (0.05 M sodium acetate–0.3 M NaCl–0.5 mM ZnCl₂, pH 4.6; a modification of Vogt's buffer A [57]).

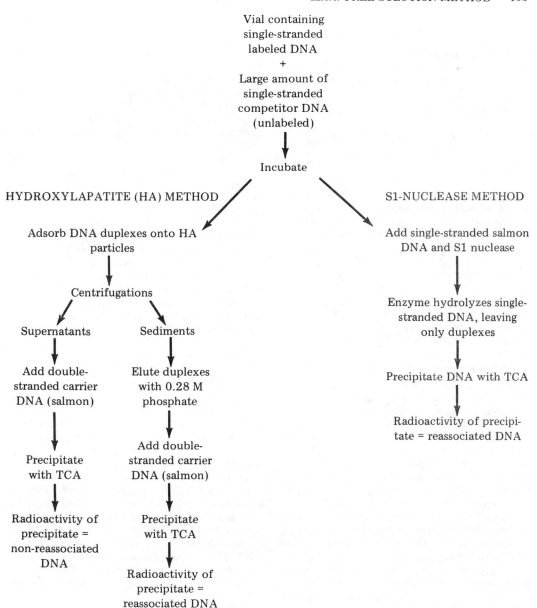

FIG. 8. *Scheme for performing the free-solution DNA homology method by use of hydroxylapatite and S1 nuclease procedures.*

2. Add 50-μl amounts of twofold dilutions of the S1 nuclease. Use the assay buffer for making the dilutions. Incubate the tubes for 1 h at 50°C.

3. Add 30 μg of sheared native salmon sperm DNA (50 μl of 0.6 mg/ml), precipitate with trichloroacetic acid (5% final concentration), collect the precipitate on nitrocellulose filters, and measure the radioactivity.

Perform the homology assays using a dilution

of the enzyme that is twice as concentrated as that required to effectively hydrolyze the 30 μg of titrating DNA. Dilute from the stock enzyme preparation to give only enough enzyme for each experiment. Titrate the stock S1 nuclease every 6 months and store it continuously at −20°C.

Reassociation mixture. When starting an experiment, heat as much of the labeled DNA as is needed in a boiling-water bath for 5 min and then cool it with ice. Allow the DNA preparation to warm to room temperature before dispensing the 10-μl amounts. Keep the temperature constant during dispensing to avoid the effect of temperature and sample size. Prepare the reaction mixtures in 6 by 22 mm vials as follows:

Labeled DNA	10 μl
0.88 M NaCl–10^{-3} M HEPES buffer (pH 7.0)	50 μl
Competitor DNA, 0.6 mg/ml (or native salmon DNA, 0.6 mg/ml)	50 μl
Total volume	110 μl

Place sheared native salmon sperm DNA (which has no homology with bacterial DNA) into two control vials for measuring the amount of self-renaturation of the labeled fragments. Use four vials for homologous reassociation and two vials for each of the heterologous DNA preparations. Incubate the vials at 25°C below the T_m of the reference DNA (as determined in SSC) for 20 to 24 h. After incubation, remove 100 μl from each reaction vial for determining the extent of reassociation by either the hydroxylapatite or the S1 nuclease procedure (see below).

Hydroxylapatite procedure.

The hydroxylapatite procedure separates double-stranded DNA from single-stranded DNA by selective adsorption to hydroxylapatite (Bio-Rad Laboratories, Richmond, Calif.). Double-stranded DNA will adsorb to hydroxylapatite in 0.14 M phosphate buffer whereas single-stranded DNA will not (11). There appears to be variability in the adsorptive capacity of different lots of hydroxylapatite, and therefore more than one lot may need to be tried.

1. Place 0.5 g of dry hydroxylapatite into a 13 by 100 mm test tube, add 3.0 ml of 0.14 M phosphate buffer (equimolar NaH$_2$PO$_4$ and Na$_2$HPO$_4$) containing 0.4% SDS, suspend the hydroxylapatite with a Vortex mixer, and warm the tube in a 60°C water bath.

2. Add 100 μl of the reaction mixture, mix on the Vortex mixer, and return to the water bath.

3. Centrifuge the tube at 5,000 rpm for 3 to 4 min, decant the buffer into an 18 by 150 mm collection tube, and add another 3 ml of 0.14 M phosphate buffer to the hydroxylapatite. Repeat the warming, mixing, and centrifugation, and decant into the same collection tube.

4. Add 3 ml of 0.28 M phosphate buffer to the hydroxylapatite, mix, and centrifuge. Decant the supernatant into a second collection tube. Add another 3 ml of buffer to the hydroxylapatite, mix, centrifuge, and decant the supernatant into the same collection tube.

5. To each of the two collection tubes add 30 μg of carrier DNA (50 μl of 0.6-mg/ml salmon

sperm DNA) and mix. Add trichloroacetic acid to a final concentration of 5%, and again mix well on a Vortex mixer. Cool in a refrigerator for 1 h.

6. Collect the precipitates on nitrocellulose filters, dry the filters, place them into scintillation vials, and measure the radioactivity.

7. A summation of the counts from the two scintillation vials represents the total number of counts per minute in the reaction vial. The counts per minute from the first scintillation vial represent the counts per minute of non-reassociated DNA, while the counts per minute from the second scintillation vial represent reassociated DNA.

Calculate homology values by dividing the amount of heterologous reassociation by the amount of homologous reassociation and multiplying by 100. A hypothetical example of the calculation is given in Table 2.

Brenner et al. (11) have described an effective hydroxylapatite procedure in which larger amounts of DNA are used.

S1 nuclease procedure.

Under carefully controlled conditions, S1 nuclease will have little effect on double-stranded DNA, but will hydrolyze single-stranded DNA (17). Therefore, the extent of duplex formation between radioactively labeled DNA fragments and an excess of unlabeled DNA fragments can be determined by measuring the amount of S1-resistant (i.e., trichloroacetic acid-precipitable) radioactivity.

1. Place a 100-μl sample from each reaction vial into a 12 by 75 mm polypropylene test tube containing 1 ml of 0.05 M sodium acetate–0.3 M NaCl–0.5 mM ZnCl$_2$ buffer (pH 4.6) and 50 μl of 0.5-mg/ml denatured sheared salmon sperm DNA. Mix the tube contents on a Vortex mixer, and then add 50 μl of S1 nuclease. Mix again, and incubate the tubes in a 50°C water bath for 1 h.

2. Remove the tubes from the water bath, add equal volumes of 10% trichloroacetic acid, mix well on a Vortex mixer, and cool in a refrigerator for 1 h.

3. Collect the precipitates on nitrocellulose filters, dry the filters, place them into scintillation vials, and count the radioactivity. Only the S1-resistant fragments (duplexes) are detected.

Calculate homology values by dividing the counts per minute of the heterologous S1-resistant DNA by the counts per minute of the homologous S1-resistant DNA and multiplying by 100. A hypothetical example of the calculation is given in Table 2.

TABLE 2. *Hypothetical examples of the calculations of percent DNA homology values in the free-solution method by the hydroxylapatite (HA) and by the S1 nuclease procedures*

HA procedure:

Reaction vial	cpm not adsorbed	cpm adsorbed	% Adsorbed	Net % adsorbed	% Homology
Labeled DNA fragments only	900	100	10	—	—
Homologous reassociation	100	900	90	80	100
Heterologous reassociation	400	600	60	50	62

S1 nuclease procedure:

Reaction vial	cpm S1 resistant	Net cpm S1 resistant	% Homology
Labeled DNA fragments only	100	—	—
Homologous reassociation	900	800	100
Heterologous reassociation	600	500	62

Thermal stability procedure.

The extent of base-pair mismatching can be estimated by comparing the thermal stability of heterologous duplexes with the stability of homologous duplexes. In this procedure the decrease of S1-resistant radioactively labeled DNA is measured as the reassociated DNA is heated stepwise in a circulating water bath to denaturing temperatures. Reassociate in the usual manner, except that the reaction mixtures must have a greater volume. Prepare reaction mixtures in 13 by 100 mm screw-cap test tubes as follows:

Labeled DNA	0.12 ml
Unlabeled DNA	0.60 ml
0.88 M NaCl–10^{-3} M HEPES buffer	0.60 ml
Total volume	1.32 ml

1. Incubate the tubes for 20 to 24 h at 25°C below the T_m of the labeled DNA (as determined in SSC).
2. Cool the tubes to room temperature, and add 1.32 ml of 50% (vol/vol) formamide in distilled water.
3. Place the tubes in a circulating water bath. Maintain the bath water level just below the cap to avoid concentration changes in the tubes due to evaporation and condensation on the tube walls.
4. Start the thermal stability profile at 45°C, and maintain the tubes at that temperature for 5 min. Then remove a 200-μl sample, place it in a tube containing 1.0 ml of acetate–NaCl–ZnCl$_2$ buffer, pH 4.6 (see above, Hydroxylapatite procedure), and return the reaction mixture tube to the water bath. Remove the sample with a standard pipette or a positive displacement micropipette (other micropipettes will be inaccurate as a result of the temperature variations).
5. Increase the water bath temperature by 5°C, and allow about 5 min for the bath to reach temperature. After 5 more min, remove the next

sample. Continue in this manner up to 90°C to give 10 samples for each reaction mixture.

Digest the samples with S1 nuclease in the usual manner (see above, S1 nuclease procedure). Plot the S1-resistant counts per minute versus the temperatures. The temperature at which 50% of duplexes become S1 sensitive (i.e., have dissociated) is called the $T_{m(i)}$ (irreversible separation of strands). The difference between the homologous and heterologous $T_{m(i)}$ values is the $\Delta T_{m(i)}$ value. Estimates for the amount of base mispairing for each degree Celsius $\Delta T_{m(i)}$ range from 1 to 2.2%.

Determination of C_0t values.

The initial reassociation of DNA in solution follows second-order reaction kinetics (1, 6, 12, 19, 59). The reassociation rate is a function of the genome size and has been used to estimate genome sizes. The usual equation for a second-order reaction (equation 1) is rearranged (equation 2) for plotting the so-called C_0t curves (12).

$$1/C - 1/C_0 = kt \qquad (1)$$

$$C/C_0 = 1/(1 + kC_0t) \qquad (2)$$

where C is the concentration of single-stranded DNA at time t in moles of nucleotides per liter, C_0 is the concentration of single-stranded DNA at zero time in moles of nucleotides per liter, t is the time in seconds, and k is the reassociation rate constant.

Calculate the moles of nucleotides per liter by dividing the milligrams of DNA per milliliter by 331 mg of DNA per mmol of nucleotide. The units millimoles per milliliter will be equal to moles per liter (M). Equation 2 is for a hyperbolic curve for which the general formula is:

$$y = 1/(1 + ax) \qquad (3)$$

When $y = \frac{1}{2}$ in equation 3, $x = 1/a = x_{1/2}$ and a

$= 1/x_{1/2}$. If the logs of the x values are plotted, when $y = \frac{1}{2}$, $\log a = \log 1/x_{1/2} = -\log x_{1/2}$. A general $\log C_0 t$ plot is shown in Fig. 9. There is an almost linear region on the curve that extends for about 2 logs. This is because any significant change in C/C_0 occurs between $1/(1 + 0.1)$ and $1/(1 + 10)$, i.e., when $kC_0 t$ is between 0.1 and 10. The rate constant $k = 1/C_0 t_{1/2}$. The units for the rate constant are: $k = 1 \text{ mol}^{-1} t^{-1}$.

Denatured DNA in free solution may be reassociated for $kC_0 t$ values well beyond 10, and it is not critical to know either the exact time of incubation or the exact concentrations of the DNA fragments. When determining $C_0 t_{1/2}$ values, however, one must know these parameters. This is because a series of points between $kC_0 t = 0.1$ and $kC_0 t = 10$ must be determined. In practice, for bacterial DNA preparations, these will be $C_0 t$ values from about 0.1 to 50. A particular $C_0 t$ value can be reached by altering the time of incubation, the concentration of the unlabeled DNA, or both. Listed in Table 3 are $C_0 t$ values for four DNA concentrations and eight reassociation times.

Set up reassociation mixtures in the usual manner (22.6.4, Preliminary procedures) with duplicate mixtures for each DNA concentration at each temperature. After incubation is completed for a given set of reaction vials, place them into a freezer until all have been incubated. Assay for the extent of duplex formation by either the hydroxylapatite or S1 nuclease procedure (22.6.4).

Plot the fractions of DNA fragments not bound to hydroxylapatite or the fractions which are S1 nuclease sensitive on the y axis and $C_0 t$ values on the x axis of four- to five-cycle semilog graph paper. The rate constant k can be calculated by using equation 2 when the duplexes are measured by the hydroxylapatite procedure. The results obtained by the S1 nuclease procedure, however, do not reflect second-order kinetics because the ends of duplexed fragments are hydrolyzed. The relationship between the two procedures is:

$$\frac{S}{C_0} = [1/(1 + kC_0 t)]^{0.45} \qquad (4)$$

where S is the concentration of S1 nuclease-sensitive DNA (54). Equation 4 can be rearranged as:

$$[S/C_0]^{2.222} = 1/(1 + kC_0 t) \qquad (5)$$

If one plots the S1 nuclease results according to equation 5, one will generate a $C_0 t$ curve comparable to that obtained by the hydroxylapatite procedure. Also important to keep in mind is that the reassociation rates are greatly affected by the ionic strengths of the reassocia-

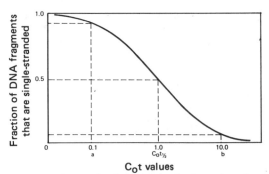

FIG. 9. *Generalized $C_0 t$ plot. At $C_0 t$ values less than or equal to a, $1/(1 + C_0 t)$ values are 0.9 or greater. At $C_0 t$ values equal to or greater than b, $1/(1 + C_0 t)$ values are less than 0.1.*

tion mixtures, and therefore comparisons should be made using a single ionic strength.

For additional information regarding reassociation kinetics and $C_0 t$ curves (including optical reassociation), see references 1, 12, 19, 51, and 59.

22.6.5. Membrane Methods

The membrane homology method differs from the free-solution method in that unlabeled DNA is immobilized on a nitrocellulose membrane filter in the former. A general scheme for performing the membrane method is given in Fig. 10.

Preliminary procedures.

The immobilization of DNA on nitrocellulose membrane filters and the preparation of competitor DNA can be done months prior to a homology experiment, whereas the preincubation of the membrane is a part of the experiment.

Preparation of competitor DNA.

1. Pass the DNA (2 mg/ml in 0.1× SCC or 0.02 M NaCl-10^{-3} M HEPES buffer, pH 7.0) through a French pressure cell two times (16,000 lb/in^2) or sonicate it (22.6.4, Preliminary procedures).

2. Denature the DNA by heating in a boiling-water bath for 5 min. Cool the DNA quickly by placing the tube containing the DNA in ice.

3. Dialyze the denatured DNA to equilibrium against cold 2.2× SSC.

4. Adjust the DNA concentration to 1.5 mg/ml using 2.2× SSC.

Alternatively, if the DNA preparations are heavily contaminated with RNA or if they have a high mole percent G+C, denature them with NaOH (22.6.4, Preliminary procedures).

Immobilization of DNA on membrane. The following adaption of the Gillespie and Spie-

TABLE 3. C_0t values for various DNA concentrations and reassociation times

Unlabeled DNA			Time		
$\mu g/110\ \mu l$	Equivalent $\mu g/ml$	Equivalent mol/liter	h	s	C_0t
1	9.1	0.27×10^{-4}	0.5	1.8×10^3	0.049
5	45	1.37×10^{-4}	0.5		0.247
20	182	5.49×10^{-4}	0.5		0.99
30	272	8.24×10^{-4}	0.5		1.48
1			1.0	3.6×10^3	0.097
5			1.0		0.49
20			1.0		1.98
30			1.0		2.97
1			2,0	7.2×10^3	0.194
5			2.0		0.986
20			2.0		3.95
30			2.0		5.93
1			4.0	1.44×10^4	0.388
5			4.0		1.97
20			4.0		7.90
30			4.0		11.9
20			8.0	2.88×10^4	15.8
30			8.0		23.7
20			16	5.76×10^4	31.6
30			16		47.5
20			20	7.2×10^4	39.5
30			20		59.3
20			24	8.64×10^4	47.4
30			24		71.2

gelman (28) procedure is for a 15-cm-diameter BA 85 nitrocellulose membrane filter (Schleicher & Schuell Co., Keene, N.H.) with an effective filtering surface of about 173 cm². The immobilization of 4.35 mg of DNA will result in 25 $\mu g/cm^2$. For smaller membranes, reduce the amount of DNA and the buffer volumes accordingly.

1. Dilute 4.35 mg of DNA to 50 $\mu g/ml$ with 0.1× SSC (approximately 90 ml).

2. Denature the DNA by heating the solution (in a 125-ml Erlenmeyer flask) in a boiling-water bath for 10 min, and then cool it quickly by pouring it into 800 ml of ice-cold 6× SSC (to give about 5 μg of DNA per ml).

3. Float the 15-cm nitrocellulose membrane filter on distilled water so that the pores will be filled with water (if the membrane is submersed immediately, air pockets will form and cause uneven filtration).

4. Place the wet membrane filter on the filtration device (22.6.1), and wash the membrane with 500 ml of cold 6× SSC, using a flow rate of approximately 30 ml/min. Then pass the denatured DNA through the filter and wash again with 500 ml of 6× SSC.

5. Let the membrane dry at room temperature and then overnight at 60°C.

6. Label the membrane on the edge with a pencil, cut the filter in half, separate the halves with half of a sheet of the paper that is used to separate the membranes in the shipping box, and place them into an envelope. Store at room temperature over $CaSO_4$ in a desiccator.

Handle the membranes by the outside edge, and avoid touching the surface on which the DNA is bound. When cutting small membranes out of the large one, avoid getting too close to the edge.

Preincubation of the DNA-containing membranes.

The Denhardt preincubation mixture (21) contains the following ingredients dissolved in 2× SSC:

Bovine serum albumin (Fraction V; Sigma
  Chemical Co., St. Louis, Mo.) 0.02%
Polyvinylpyrrolidone   (Calbiochem-Behr-
  ing, La Jolla, Calif.) 0.02%
Ficoll 400 (Pharmacia Fine Chemicals,
  Uppsala, Sweden) 0.02%

DIRECT-BINDING METHOD   COMPETITION METHOD

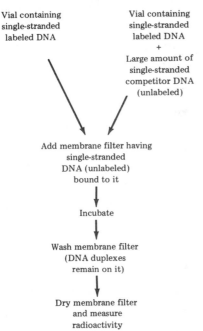

FIG. 10. *Scheme for performing the membrane DNA homology method by use of direct binding and competition procedures.*

1. Cut 3 by 9 mm membranes (about 5 μg of DNA per membrane) from a 15-cm membrane with a keysort punch (22.6.1).

2. Preincubate the membranes at the incubation temperature of the experiment for 0.5 to 2 h. Agitate the membranes several times during the preincubation period.

3. Remove the membranes, place on a paper towel to blot off excess liquid, and then place into the reaction vials.

Direct binding procedure.

The direct binding procedure measures the ability of a labeled DNA to form duplexes with DNA preparations immobilized on membrane filters. Larger amounts of labeled DNA (as compared to the free solution experiments) are needed to assure substantial reassociation with the immobilized DNA. Therefore, only low-specific-activity DNA preparations (2,000 to 10,000 cpm/μg) are required. Reaction mixtures contain the following:

Labeled DNA (50 to 100 μg/ml) . . . .   10 μl
2.2× SSC . . . .   100 μl

1. Preincubate the required number of DNA membrane filters. Keep each kind separate.

2. Make up the reaction mixtures and mix them on a Vortex mixer.

3. Place the membranes in vials, stopper, and incubate at 25°C below the T_m (as determined in SSC) of the labeled DNA. Incubate for 12 to 15 h.

4. Remove the membranes from the vials and place in the washing chamber (Fig. 4C and 5). Wash in 2× SSC at the incubation temperature. Use two 300-ml volumes of washing buffer, and wash for 5 min in each. Move the washing chamber back and forth in the beaker several times during the washing period to assure buffer flow through each compartment. Shake the washing chamber free from buffer. Remove the membrane filters, and dry on a paper towel under a heat lamp.

5. Skewer the membranes on insect pins (if duplicates, place both on the same pin as shown in Fig. 11).

6. Place into scintillation vials (Fig. 11) and count.

Because of variability of DNA membrane filters, direct experiments are not very quantitative. The calculations are similar to those used for the S1 procedure (Table 2) in that the counts bound by the heterologous DNA membrane are divided by the counts bound by the homologous DNA membrane. This ratio is then multiplied by 100 to give the percent homology.

Thermal stability procedure.

1. Follow the direct binding procedure through the washing (step 4).

2. Remove the membrane filters from the washing chamber, skewer on an insect pin near the head of the pin, insert the pin into the end of a rubber-stoppered tube (Fig. 7), and place into the first elution tube. When using more than one filter, separate them with Teflon disks.

3. Use 13 by 100 mm elution tubes containing 1.2 ml of 0.5× SSC. Place the tubes into a water bath, and do stepwise elution of the duplexes by increasing the bath temperature 5°C at 10-min intervals. At the end of each time period, transfer the filters to the next elution tube and increase the temperature by 5°C. Start the elution profile about 10°C below the incubation temperature used for the direct binding procedure. Use nine elution temperatures, and count the residual radioactivity on the filters (i.e., 10 scintillation vials per stability profile).

4. Empty the contents of each elution tube into a scintillation vial, and wash the tube with 0.5 ml of 0.5× SSC, adding the washings to the vial.

5. Add 10 ml of Triton X-100 counting solution (3 parts of toluene counting solution and 2 parts of Triton X-100) to each scintillation vial, mix, and count.

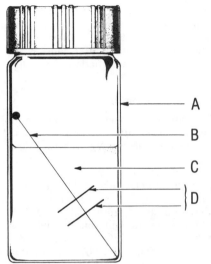

FIG. 11. *Apparatus for counting radioactivity on membrane filters. (B) Insect pin, (D) duplicate filters impaled on pin, (A) scintillation vial, (C) toluene-based cocktail.*

The integral elution profile curves are obtained by summing the radioactivities at each temperature and dividing by the total radioactivity (i.e., averaging all 10 samples). Although these curves will be reciprocal to those obtained by the free solution method, the $T_{m(i)}$ values are calculated in the same manner.

In obtaining the profile curves, place the 1.2 ml of $0.5\times$ SSC in all of the Wasserman tubes at the same time, using an automatic pipette. Keep one elution tube ahead in the water bath so that each is preheated. Have an extra tube with a thermometer in it, and record the exact tube temperature at the middle and the end of each temperature interval.

Competition procedure.

In this procedure, unlabeled denatured DNA fragments are used to compete with denatured labeled DNA for duplex formation with DNA bound to the membrane filter. The competitor DNA preparations contain 1.5 mg of DNA per ml in $2.2\times$ SSC and are used at two levels in the experiments, 75 or 150 μg per reaction vial. Volumes for the various reaction mixtures are given in Table 4. The manipulations are the same as for the direct binding procedure except that the same kind of DNA filter is used in all of the reaction vials, and the kind and amount of competitor DNA is varied. Duplicate each level of competitor DNA, and obtain six replicates of the vials containing no competitor DNA at the start of the procedure.

Calculate the percent homology values by dividing the depression in binding caused by the heterologous unlabeled DNA fragments by the depression resulting from the homologous unlabeled DNA fragments at each competitor concentration and then by multiplying the ratio by 100. A hypothetical example of the calculation is shown in Table 5.

22.7. RNA HOMOLOGY

rRNA homology procedures differ from DNA homology procedures because only a small fraction (0.1 to 0.3%) of the total DNA is complementary with the RNA. Because the RNA is complementary to only one of the DNA strands, there is no association between the RNA molecules. A major precaution with RNA procedures is that the buffers and glassware must be free from RNase. The procedures described here are for directly measuring binding of radioactive RNA to DNA immobilized on membrane filters or by indirectly measuring competition in the binding. Additional information on procedures and on interpretation of results can be found in references 9, 10, 11, 20, 26, 28, 29, and 36.

22.7.1. Preliminary Procedures

As with the membrane DNA homology method, there are several procedures that must be done before the analogous RNA method.

Preparation of RNase-free glassware and buffers (22).

Heat all glassware in an ashing oven or a dry-heat sterilizing oven. Autoclave all plasticware submerged in 0.1% detergent (e.g., Isoclean, Isolabs Co., Akron, Ohio). Autoclave all buffers and then treat by adding 0.2% diethylpyrocarbonate (Sigma Chemical Co., St. Louis, Mo.).

Preparation of labeled RNA.

The procedures for labeling rRNA are very similar to those used for DNA (see 22.6.2). Precursors used for in vivo RNA labeling are usually ^{14}C- or ^{3}H-labeled uracil or uridine, or $H_3^{32}PO_4$. During the last 0.5 h of incubation, the cultures are often incubated with a large excess (100-fold) of unlabeled precursor to decrease the specific activity of mRNA. Isolate the RNA by using the same procedure that is described for unlabeled RNA (22.3). RNA molecules are also amenable to in vitro labeling with ^{125}I (see 22.6.3).

Fractionation of RNAs.

Since there are several classes of RNA, the labeled RNA or RNAs that one wishes to use in hybridization experiments are usually separated from the rest. Although one may isolate the ribosomes, separate the subunits, and then isolate rRNA from them, total RNA is usually fractionated. Isolate total RNA, and separate the RNA components by polyacrylamide gel electrophoresis (8, 24, 40, 53) or by sucrose gradient centrifugation (46). Fractionate the 23S rRNA by centrifugation through 5 to 20% sucrose gradients prepared in 1× SSC. The fractionation of 16S rRNA may be complicated by cosedimentation of 23S degradation products.

Competitor RNA.

Dilute stock preparations of RNA to 1 mg/ml in 0.1× SSC–10^{-3} M HEPES buffer, pH 7.0. Heat the preparations for 5 min in a boiling-water bath, and after rapid cooling in an ice bath, store the preparations at −80°C.

Immobilization of DNA and preincubation of DNA-containing membranes.

DNA is immobilized on nitrocellulose membrane filters as described in part 22.6.5, Preliminary procedures. Use autoclaved and diethylpyrocarbonate-treated buffers, and if the DNA preparations are contaminated with RNase, treat them with diethylpyrocarbonate or proteolytic enzymes before denaturing. Preincubate 3 by 9 mm oblong or 10-mm round filters with the Denhardt preincubation mixture. Use RNase-free buffer.

Recently, investigators have immobilized DNA and RNA fragments or molecules by covalently attaching them to diazobenzyloxymethyl paper (7, 55). The immobilization of DNA in this manner may be very useful for

thermal stability experiments. The paper is available from two manufacturers (Enzobond, Enzo Biochemicals, Inc., New York, N.Y.; Schleicher & Schuell Co., Keene, N.H.). Both companies supply instructions for using the paper.

Use of formamide.

Formamide has been the solvent of choice for lowering the denaturation temperatures of DNA duplexes and RNA-DNA hybrids. This allows the procedure to be run at lower temperatures, which reduces the amount of thermal degradation of the nucleic acids. The T_m of DNA is lowered by 0.60°C per percent formamide in buffers containing from 0.035 to 0.88 M NaCl (34). Gillespie and Gillespie (29) have used 50% formamide in 6× SSC and an incubation temperature of 38°C for rRNA filter hybridization, whereas De Ley and De Smedt (20) have used 20% formamide in 2× SSC and an incubation temperature of 50°C.

For hybridization experiments containing 20% formamide, prepare the 5.5× SSC–10^{-3} M HEPES buffer (Table 6) with 55% formamide and the 3.5× SSC–10^{-3} M HEPES buffer (Table 5) with 35% formamide. Prepare the 3.5× SSC–10^{-3} M HEPES buffer with 87.5% formamide for experiments employing 50% formamide.

22.7.2. Membrane Method

Direct binding procedure

Depending upon the specific activity of an rRNA preparation, it may be necessary to use larger (10-mm-diameter) DNA-containing membranes because of the small fraction of the DNA that will be complementary to the RNA.

1. For direct binding using the smaller oblong membranes, 3 by 9 mm, follow the procedure described previously for the direct binding of DNA. Use 0.5 μg of labeled RNA per vial. Wash the membranes in the same manner as for the DNA procedure. To remove the unpaired segments of rRNA from the hybrids, place the washing chamber in a solution of RNase (pancreatic, 25 μg/ml of 2× SSC) and incubate at 37°C for 30 min. Wash the membranes once

TABLE 4. *Composition of the reaction mixtures used in the competition procedure of membrane method for DNA homology*

Component to be added	Amt of competitor DNA:		
	0	75 μg	150 μl
Labeled DNA (50 to 100 μg/ml)	10 μl	10 μl	10 μl
Competitor DNA in 2.2× SSC	—	50 μl	100 μl
2.2× SSC	100 μl	50 μl	—
Total volume	110 μl	110 μl	110 μl

TABLE 5. *Hypothetical example of the calculation of percent DNA homology values in the membrane method by the competition procedure*

Membranes from	cpm	cpm depression	% Homology
Direct binding, no competitor	1,000	—	—
Homologous competitor (75 μg)	120	880	100
Homologous competitor (150 μg)	90	910	100
Heterologous competitor (75 μg)	400	600	(600/880)100 = 68
Heterologous competitor (150 μg)	380	620	(620/910)100 = 68

more. After determining the radioactivity on the membrane, measure the amount of DNA on it (see 22.4.3). Calculate the results as counts per minute or micrograms (if specific activity is known) of RNA binding per microgram of DNA.

2. For direct binding using 10-mm filters, use a single-hole paper punch, making sure that it is free from lubricating oil. Incubate in 4-ml vials (Wheaton Scientific, Millville, N.J.) with a reaction volume of 0.35 ml. Use 1 to 2 μg of labeled rRNA per vial.

3. For determining the thermal stability of hybrids, use the procedure described for the thermal stability of membrane-bound DNA duplexes (22.6.5).

Competition procedure

Depending on the size of the membranes, use the reaction mixtures given in Table 6. For additional details, see reference 36.

Incubate the vials overnight at 65°C, and wash the membranes in the same manner as for the direct binding procedure. One may treat the membranes with RNase, although similar results are obtained whether or not they are treated. Again measure the amounts of DNA on the membranes. Calculate the results in a manner similar to that described for DNA homology experiments, except that the counts should be expressed as counts per minute per microgram of membrane-bound DNA.

22.8. LITERATURE CITED

22.8.1. General References

1. **Britten, R. J., D. E. Graham, and B. R. Neufeld.** 1974. Analysis of repeating DNA sequences by reassociation. Methods Enzymol. 24E:363–406.
   Although concerned mostly with repetitive eucaryotic DNA, this reference contains much practical information that is directly applicable to bacterial DNA.
2. **Kennell, D. E.** 1971. Principles and practices of nucleic acid hybridization. Prog. Nucleic Acid Res. Mol. Biol. 11:259–301.
   Reviews the hybridization of various RNA species by use of membrane methods.
3. **McCarthy, B. J., and R. B. Church.** 1970. The specificity of molecular hybridization reactions. Annu. Rev. Biochem. 39:131–150.
   Useful for providing an understanding of the principles of DNA and RNA hybridization reactions (homologies).
4. **Moore, R. L.** 1974. Nucleic acid reassociation as a guide to genetic relatedness among bacteria. Curr. Top. Microbiol. Immunol. 64:105–128.
   Reviews the use of nucleic acid similarities in bacterial taxonomy.
5. **Parish, J. H.** 1972. Principles and practice of experiments with nucleic acids. John Wiley & Sons, Inc., New York.
   Provides general and specific information for a wide range of experiments.
6. **Wetmur, J. G.** 1976. Hybridization and renaturation kinetics of nucleic acids. Annu. Rev. Biophys. Bioeng. 5:337–361.
   This review deals primarily with free-solution reassociation kinetics.

22.8.2. Specific Articles

7. **Alwine, J. C., P. J. Kemp, and G. R. Stark.** 1977. Proc. Natl. Acad. Sci. U.S.A. 74:5350–5354.
8. **Bishop, D. H. L., J. R. Claybrook, and S. Spiegelman.** 1967. J. Mol. Biol. 26:373–387.
9. **Bishop, J. O.** 1969. Nature (London) 224:600–603.
10. **Bishop, J. O.** 1970. Biochem. J. 116:223–228.
11. **Brenner, D. J., G. R. Fanning, A. V. Rake, and K. E. Johnson.** 1969. Anal. Biochem. 28:447–459.
12. **Britten, R. J., and D. E. Kohne.** 1968. Science 161:529–540.
13. **Britten, R. J., M. Pavich, and J. Smith.** 1969. Carnegie Inst. Washington Yearb. 68:400–402.
14. **Cashion, P., M. A. Holder-Franklin, J. McCully, and M. Franklin.** 1977. Anal. Biochem. 81:461–466.
15. **Chelm, B. K., and R. B. Hallick.** 1976. Biochem-

TABLE 6. *Composition of reaction mixtures used in competition procedure of membrane method for RNA homology*

Component to be added to vial	Amt of competitor RNA in experiments using 3 by 9 mm filters:			Amt of competitor RNA in experiments using 10-mm-diameter filters:		
	0	25 μg	50 μg	0	50 μg	100 μg
Labeled rRNA (25 μg/ml)	20 μl	20 μl	20 μl	50 μl	50 μl	50 μl
Competitor rRNA (1 mg/ml)	—	25 μl	50 μl	—	50 μl	100 μl
0.1× SSC–10^{-3} M HEPES buffer	50 μl	25 μl	—	100 μl	50 μl	—
5.5× SSC–10^{-3} M HEPES buffer	40 μl	40 μl	40 μl	—	—	—
3.5× SSC–10^{-3} M HEPES buffer	—	—	—	200 μl	200 μl	200 μl
Total volume per vial	110 μl	110 μl	110 μl	350 μl	350 μl	350 μl

istry **15**:593–599.

16. **Commerford, S. L.** 1971. Biochemistry **10**:1993–2000.

17. **Crosa, J. H., D. J. Brenner, and S. Falkow.** 1973. J. Bacteriol. **115**:904–911.

18. **De Ley, J.** 1970. J. Bacteriol. **101**:738–754.

19. **De Ley, J., H. Cattoir, and A. Reynaerts.** 1970. Eur. J. Biochem. **12**:133–142.

20. **De Ley, J., and J. De Smedt.** 1975. Antonie van Leeuwenhoek J. Microbiol. Serol. **41**:287–307.

21. **Denhardt, D. T.** 1966. Biochem. Biophys. Res. Commun. **5**:641–646.

22. **Ehrenberg, L., and I. Fedorcsak.** 1976. Prog. Nucleic Acid Res. Mol. Biol. **16**:189–262.

23. **Ferragut, C., and H. Leclerc.** 1976. Ann. Microbiol. (Paris) **127A**:223–235.

24. **Franklin, R. M.** 1966. Proc. Natl. Acad. Sci. U.S.A. **55**:1504–1511.

25. **Fredericq, E., A. Oth, and F. Fontaine.** 1961. J. Mol. Biol. **3**:11–17.

26. **Galau, G. A., R. J. Britten, and E. H. Davidson.** 1977. Proc. Natl. Acad. Sci. U.S.A. **74**: 1020–1023.

27. **Giles, K. W., and A. Meyers.** Nature (London) **206**:93.

28. **Gillespie, D., and S. Spiegelman.** 1965. J. Mol. Biol. **12**:829–842.

29. **Gillespie, S., and D. Gillespie.** 1971. Biochem. J. **125**:481–487.

30. **Grossman, L., and K. Moldave (ed.).** 1968. Methods Enzymol. **12B**:87–160.

31. **Hayes, D. H., and F. Gros.** 1968. Methods Enzymol. **12B**:771.

32. **Hontebeyrie, M., and F. Gasser.** 1977. Int. J. Syst. Bacteriol. **27**:9–14.

33. **Huang, P. C., and E. Rosenberg.** 1966. Anal. Biochem. **16**:107–113.

34. **Hutton, J. R.** 1977. Nucleic Acids Res. **4**:3537–3555.

35. **Johnson, J. L.** 1978. Int. J. Syst. Bacteriol. **28**: 245–256.

36. **Johnson, J. L., and B. S. Francis.** 1975. J. Gen. Microbiol. **88**:229–244.

37. **Kelly, R. B., N. R. Cozzarelli, M. P. Murray, P. Deutscher, I. R. Lehman, and A. Kornberg.** 1970. J. Biol. Chem. **245**:39–45.

38. **Kirby, K. S.** 1968. Methods Enzymol. **12B**:87–99.

39. **Ko, C. Y., J. L. Johnson, L. B., Barnett, H. M. McNair, and J. R. Vercellotti.** 1977. Anal. Biochem. **80**:183–192.

40. **Loening, U. E.** 1967. Biochem. J. **102**:251–257.

41. **Mandel, M., L. Igambi, J. Bergendahl, M. L. Dodson, Jr., and E. Scheltgen.** 1970. J. Bacteriol. **101**:333–338.

42. **Mandel, M., C. L. Schildkraut, and J. Marmur.** 1968. Methods Enzymol. **12B**:184–195.

43. **Markov, G. G., and I. G. Ivanov.** 1974. Anal. Biochem. **59**:555–563.

44. **Marmur, J.** 1961. J. Mol. Biol. **3**:208–218.

45. **Marmur, J., and P. Doty.** 1962. J. Mol. Biol. **5**: 109–118.

46. **McConkey, E. H.** 1967. Methods Enzymol. **12A**: 620–634.

47. **Meinke, W., D. V. Golstein, and M. R. Hall.** 1974. Anal. Biochem. **58**:82–88.

48. **Nonoyama, M., and J. S. Pagano.** 1973. Nature (London) **242**:44–47.

49. **Orosz, J. M., and J. G. Wetmur.** 1974. Biochemistry **13**:5467–5473.

50. **Rigby, P. W. J., M. Dieckmann, C. Rhodes, and P. Berg.** 1977. J. Mol. Biol. **113**:237–251.

51. **Seidler, R. J., and M. Mandel.** 1971. J. Bacteriol. **106**:608–614.

52. **Shaposhnikov, J. D., Y. P. Zerov, E. A. Ratovitski, S. D. Ivanov, and Y. V. Bobrov.** 1976. Anal. Biochem. **75**:234–240.

53. **Sogin, M. L., C. R. Woese, B. Pace, and N. R. Pace.** 1973. J. Mol. Evol. **2**:167–174.

54. **Smith, M. J., R. J. Britten, and E. H. Davidson.** 1975. Proc. Natl. Acad. Sci. U.S.A. **72**: 4805–4809.

55. **Stark, G. R., and J. G. Williams.** 1979. Nucleic Acids Res. **6**:195–203.

56. **Tereba, A., and B. J. McCarthy.** 1973. Biochemistry **12**:4675–4679.

57. **Vogt, V. M.** 1973. Eur. J. Biochem. **33**:192–200.

58. **Wang, S. Y., and J. M. Hashagen.** 1964. J. Mol. Biol. **8**:333–340.

59. **Wetmur, J. G., and N. Davidson.** 1968. J. Mol. Biol. **31**:349–370.

60. **Yamada, K., and K. Komagata.** 1970. J. Gen. Appl. Microbiol. **16**:215–224.

Section VI

LABORATORY SAFETY

Introduction to Laboratory Safety

G. BRIGGS PHILLIPS

Absolute control of bacteria by sterilization and practical control by containment and disinfection are important aspects of laboratory work. The methods of general bacteriology (morphology, growth, genetics, metabolism, and systematics) all are predicated on pure cultures, which in turn are dependent on prior sterilization and subsequent containment of bacteria in media, containers, and tools. Bacteriological methods also are predicated on the decontamination afterwards of residues that might infect other cultures or, more importantly, people. The dual purpose of laboratory safety procedures is to protect both the experiment and the experimenter, but human safety is the paramount consideration.

The generalist may consider that the bacteria usually handled are residents of the natural environment and therefore harmless to man. In fact, there is no distinction between nonpathogenicity and pathogenicity; rather, among bacteria there is a broad spectrum in degrees of virulence relative to human susceptibility. Yesterday's saprophyte becomes today's parasite and tomorrow's pathogen (as exemplified by the histories of *Escherichia coli, Proteus vulgaris, Pseudomonas aeruginosa, Serratia marcescens,* and *Bacillus cereus*). The lesson is precaution: handle *all* bacteria as potential hazards to human health.

In the first chapter of this section methods are described for sterilizing materials by use of moist heat, dry heat, gases, radiation, and filtration. The second chapter is composed of methods for physical containment and chemical disinfection, both in general and for specific laboratory operations.

Throughout this manual, a safety *CAUTION* is noted in context wherever a method may chemically or biologically imperil a user. Radioactivity safety is summarized in part 16.4.1.

475

Chapter 23

Sterilization

MICHAEL S. KORCZYNSKI

The objective of a sterilization process is to destroy or remove all living organisms in or on an item. Generally, in microbiological laboratories, steam is used for the sterilization of culture media, equipment, and glassware; dry heat is used for glassware and metallic equipment; gas is used for instruments; filtration is used for solutions; and radiation is used for limited purposes. In hospitals, pharmacies, and industries, sterilization has wider applications. Changes in health-care delivery services, the types of medical products requiring sterilization, and government regulations cause changes in sterilization procedures. Sterilization is an evolving technology, with constant improvements in equipment design and control. There are many general references available for sterilization methods (1–13).

Sterilization is a probability function. Bacterial cells usually die at an exponential (logarithmic) rate similar to a first-order chemical reaction (Fig. 1). Sterilization efficiency is expressed as the statistical probability of the process yielding a survivor (35). Terminal processes (moist heat, dry heat, ionizing radiation, and ethylene oxide) should have a probability of 10^{-6} or less of a survivor being present.

With the exception of filtration, each sterilization method can be characterized by the rate of the killing process. The **D-value** is used to characterize sterilization by heat and represents the time required at a given temperature to cause a decimal (90%) reduction in a microbial population; the temperature is often indicated as a subscript (e.g., D_{100}). The D-value can be derived graphically by plotting the logarithm of the number of survivors against time of exposure and is measured as the time on the abscissa corresponding to a decade of reduction in numbers on the ordinate (Fig. 1). The D-value can also be calculated mathematically (15). The **z-value** is the number of degrees Celsius or Fahrenheit required for a decade change in the D-value. The **F_0 value** is a standardized measure of the bacterial inactivation capability of moist heat and is calculated by converting the time required at a given temperature of heat exposure to the equivalent time at 121°C (250°F), using a z-value of 10°C (18°F).

The two methods frequently employed for determining the number of survivors involve plate counts of viable-cell numbers and estimations of most probable numbers by the fraction-negative method. Plate count methods are described elsewhere in this manual (11.2). The log of the plate counts is plotted against time at a given temperature, and a linear regression line is visually plotted (11.5.7) or calculated by least squares (11.5.8) to best fit a line through the data points. The D-value is then determined from this best-fit line. Death usually is logarithmic. However, deviations from logarithmic behavior also are observed; under these circumstances, calculating a D-value is not very meaningful. Explanations offered for nonlogarithmic bacterial death include heterogeneity of the ini-

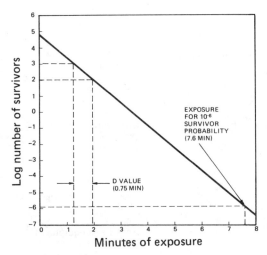

FIG. 1. *Exponential death plot, showing the graphic determination of a D-value and a survivor probability.*

tial population in terms of resistance properties, experimental variation (such as clumped or occluded cells), and adaptation during the heating process (24).

Alternatively, the fraction-negative method may be used to obtain the most probable number (MPN) of survivors at intervals of exposure, and thence a D-value (10, 13). The experiment should be designed so that all units are positive at the shortest exposure time and all units are negative at the longest exposure time. Intermediate exposure times should be planned to permit some positives to occur. A minimum of three temperatures should be used. The equation used to calculate bacterial survivors per unit is as follows:

$$B = \ln r/q = 2.303 \log_{10} (r/q)$$

where B is the probable number of survivors, r is the number of units tested per time interval, and q is the number of sterile units (23).

To establish sterilization cycles for industrial purposes using the conventional modes of sterilization, microbiological validation studies have to be conducted using microorganisms with known resistance to the process, in most cases bacterial spores. Spore suspensions should be evaluated periodically to assure that their resistance properties persist. Sterilization cycle studies should include the use of at least three replicate units for each specific exposure time and condition that is to be evaluated. The accuracy of the result is a function of the number of replicate samples used. The equipment should be thoroughly evaluated for performance to as-

sure consistent sterilization results. An excellent review exists on this topic (25).

The number and type of microorganisms associated with an item prior to sterilization (**bioburden**) must be determined. The required sterilization process time can be estimated by multiplying the D-value of the bioburden by their total estimated number in the sterilizer load (or on a unit basis when using radiation). Once the time required to inactivate the bioburden is determined, the probability of nonsterility for the most resistant microorganism can be calculated.

Biological indicators (BIs) contain known concentrations of microorganisms, usually bacterial spores, which exhibit a predictable death rate when exposed to a defined treatment. BIs may be prepared on various adventitious substrates (e.g., filter paper, thread, or porcelain) or may be directly inoculated on or in the product being sterilized. Table 1 shows the species of spores that are most frequently employed for monitoring sterilization processes. Table 2 lists some commercial suppliers of BIs.

A probability of nonsterility for a BI microorganism may be calculated in the same manner as for the bioburden. The probability of nonsterility for BIs should be known relative to that of the bioburden to establish a safety factor. Frequently, the D-value of the BI is substituted for the D-value of the bioburden in the probability equation. This permits estimation of the processing time required to sterilize in worst-case conditions.

A systems approach to producing sterile products involves the following:

1. Prepare, select, and compound the material in a manner that minimizes the microbial load.

2. Use a sterilization process that is compatible with the item and its packaging.

3. Select a package that will not be a barrier to the process and will permit the maintenance of sterility following the process.

4. Use process parameter measurements to verify the adequacy of the process.

TABLE 1. *Frequently used biological indicators*

Species of spores	Sterilization process	Approx D-value
Bacillus subtilis	Ethylene oxide (600 mg/liter at 50% relative humidity and 54°C)	3.0 min
Bacillus stearothermophilus	Moist heat (≥121°C)	1.5 min
Clostridium sporogenes	Moist heat (≤121°C)	0.2–0.8 min
Bacillus subtilis	Dry heat (170°C)	0.8 min
Bacillus pumilus	Ionizing radiation	0.17 Mrads

TABLE 2. *Commercial sources of biological indicators*

Type	Source (23.6)
Spore strip containing one species of bacterium	American Biological Control Co.
	American Sterilizer Co.
	BBL Microbiology Systems
	Castle Co. Div., Sybron Corp.
	Gilbraltar Biological Laboratories, Inc.
	North American Science Associates, Inc.
Spore strip containing two species of bacteria	American Sterilizer Co.
	Castle Co. Div., Sybron Corp.
	North American Science Associates, Inc.
	Propper Manufacturing Co., Inc.
Spore strip with graduated population	American Sterilizer Co.
Spore strip containing culture medium unit	3M Co.
Spore suspension in culture medium	BBL Microbiology Systems
	EM Laboratories, Inc.
Spore suspensions	American Sterilizer Co.
	Castle Co. Div., Sybron Corp.
	North American Science Associates, Inc.

TABLE 3. *Items sterilized by various processes*

Process	Item
Moist heat	Parenteral solutions
	Instruments
	Aseptic process equipment
	Culture media
	Rubber stoppers
Dry heat	Glassware
	Oils
	Instruments
	Needles
	Petrolatum
	Powders
	Aseptic processing equipment
Ethylene oxide	Anesthesia equipment
	Catheters
	Diagnostic equipment
	Implantable prosthetic devices
	Laboratory equipment
	Respiratory therapy equipment
	Surgical equipment
	Surgical supplies
	Telescopic instruments
	Plastic tubing
	Packaging materials
	Spices
Ionizing radiation	Powders
	Bandages
	Blood collection tubes
	Blood lancets
	Brushes
	Burn ointments
	Burn pads
	Centrifuge tubes
	Dialysis units
	Surgical sutures
	Laboratory animal bedding
	Surgical gowns
Filtration	Gases
	Liquids
	Low-viscosity ointments and oils

5. Store items appropriately after sterilization.

6. Deliver, open, and use the material without contamination.

Table 3 shows conventional sterilization methods and items usually sterilized by each. Factors influencing microbial death which must be considered in the selection of a sterilization process include the following:

1. The design and/or chemical composition of the item.

2. The biological and physical state of the microorganism prior to sterilization.

3. The inherent resistance of the microorganism to the sterilizing process and the resulting death rate of the microorganism.

4. The initial population of microorganisms associated with the item to be sterilized.

5. The intensity of the sterilizing agent.

6. The duration of exposure to the sterilizing environment.

23.1. MOIST HEAT

Steam sterilization is accomplished by using an autoclave with saturated steam under pressure for 15 min at a temperature of 121°C. Other time and temperature relationships can result in the same microbial lethality. The subject of equivalent lethality is treated in detail elsewhere (2, 9–11, 14, 34, 36).

Most laboratory autoclaves are gravity- or downward-displacement devices (Fig. 2) which depend on the difference in density between air and steam. Problems occur with moisture pen-

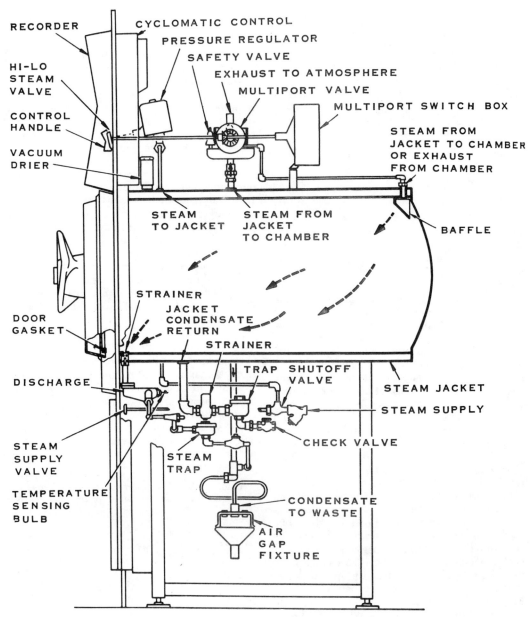

FIG. 2. *Schematic of a downward-displacement steam autoclave. Reprinted from Perkins (8), courtesy of American Sterilizer Co., Erie, Pa., and Charles C Thomas, Publisher, Springfield, Ill.*

etration, superheating, entrainment or removal of air, and heat and/or moisture damage. Correct preparation of materials and their proper loading is important. The manufacturer of the autoclave can provide directions for installing thermocouples in materials to be sterilized to determine loading patterns within the autoclave.

Automatic autoclaves, equipped with cycle timers and automatic temperature controls, are most suitable for laboratory use. Several other moist-heat systems are available which reduce trapped air by high vacuum, steam and vacuum pulsations, and hot-water cascade or immersion. (Vacudyne-Altair, American Sterilizer Co., Castle Co., FMC, and Rexham Co.)

Air causes the chamber to heat slower and results in lower temperatures. Table 4 shows the lower temperature of several air/steam mixtures

TABLE 4. *Influence of incomplete air discharge on autoclave temperatures (8)*

Gauge pressure (lb/in²)	Temp (°C)			
	Saturated steam, complete air discharge	Two-thirds air discharge	One-third air discharge	No air discharge
5	109	100	90	72
10	115	109	100	90
15	121	115	109	100
20	126	121	115	109
25	130	126	121	115
30	135	130	126	121

compared to the temperatures derived from pure steam. Laboratory autoclaves should be controlled by temperature readings (which are dependent on saturated steam pressures) and by pressure readings.

It is important to know the effects on steam penetrability of loading patterns, viscosity of solutions, size of containers, and densities. Also, the chemistry of the solution may influence microbial kill (18). Cool zones within loads of materials must be identified. Packages should be loosely arranged to facilitate steam penetration. Articles should be arranged to permit the downward displacement of the heavier air.

Table 5 illustrates the effect of the liquid volume and the numbers of containers within a load upon the time required for sterilization. The time is shorter for small volumes. Thus, sterilization cycle times must be adjusted for the sizes or numbers of containers.

Precautions must be taken when sterilizing solutions in glass containers. A controlled rate of cooling and slow release of pressure should be used to prevent glass shock when the cycle is completed. Some sterilizers use controlled water cool-down systems to avoid glass breakage. Care must be taken to assure that an autoclave has returned to ambient pressure before opening. Most modern autoclaves prevent this problem through automatic control. When sterilizing flasks or bottles, container closure fittings should be left loose to permit venting and prevent container breakage. Industrial processes frequently avoid this by using air-over-pressure systems. When the container is removed from the autoclave, a vacuum may result and cause an influx of air into the container, which can contaminate the solution unless the closure is made tight or other preventive measures are taken.

23.2. DRY HEAT

Materials that cannot be sterilized by moist heat are often sterilized by dry heat (e.g.,

greases, lubricants, mineral oil, waxes, and powders). Hot forced air and infrared energy have been used for dry-heat sterilization. There are a variety of commercial gas and electric dry-heat sterilizers, including batch ovens and continuous heating tunnels.

Dry heat is less efficient than moist heat and requires higher temperatures for longer durations. Table 6 presents examples of time and temperature relationships. The principal advantages of dry heat are that it is not corrosive for metals and instruments, does not affect glass surfaces, and is suitable for sterilizing powders and nonaqueous, nonvolatile viscous substances. Disadvantages include slow heat penetration and longer sterilization times. In dry heat, the higher temperatures may adversely affect some materials, and temperature stratification can occur unless the air is circulated (7).

Dry heat inactivates microorganisms by oxidation of intracellular constituents. Spores of bacteria are more resistant than vegetative cells. Examples of typical dry-heat D-values for spores are available (10).

23.3. GASES

Gases are used in sterilization because some medical materials and supplies cannot be sterilized by other means without damage or destruction. Gaseous processes are more difficult to control than other modes of sterilization because of the various process parameters. Factors such as gas stratification, temperature stratification, availability of moisture, chemically reactive bar-

TABLE 5. *Effect of liquid volume and number of containers on time required for liquid to reach 121°C in an autoclave (38)*

Liquid vol/container (liters)	No. of containers/load	Liquid temp at initiation of cycle (°C)	Time for liquid to reach 121°C (min)	Total time of cycle (min)
0.5	30	29	19	29
1.0	20	26	34	44
2.0	10	27	37	47
3.0	8	26	43	53
4.0	5	26	52	62
5.0	5	26	60	70
6.0	4	26	62	72

TABLE 6. *Frequently used times and temperatures for dry-heat sterilization (8, 22, 38)*

Temp	Time (h)
170°C (340°F)	1.0
160°C (320°F)	2.0
150°C (300°F)	2.5
140°C (285°F)	3.0

riers, physical diffusion barriers (such as packaging materials), devaporization, and polymerization must be considered when selecting a gaseous process (4, 16, 17).

23.3.1. Ethylene Oxide

Although a variety of other gases have been shown to possess germicidal properties (formaldehyde, propylene oxide, β-propiolactone, ozone, peracetic acid, and methyl bromide), ethylene oxide (EtO) has been widely employed in the sterilization of medical devices because of its wide compatibility with various materials (Table 3). Important properties of EtO are as follows:

Cyclic ether molecule
Ether-like odor
Colorless gas at ambient temperature
Boiling point, 10.8°C
Very reactive alkylating agent
Readily polymerizes
Highly flammable; flash point, −17°C
Explosive limits in air, 3 to 100%
Molecular weight, 44.05 g
Density, 0.87 g/cm^3 at 20°C
Vapor pressure, 21.2 lb/in^2 at 20°C

Mixtures of dichlorodifluoromethane and EtO are frequently employed to minimize flammability. Some typical commercially available gaseous EtO combinations are listed in Table 7. Although 100% EtO processes are used in the pharmaceutical industries, they are rarely used for laboratory purposes.

EtO is used as a sterilant in specially designed chambers or vessels. The essential elements of a sterilizer are shown in Fig. 3. Detailed instructions by the manufacturer of EtO sterilization equipment must be closely followed.

Subatmospheric pressure conditions are used when 100% EtO is employed, and conditions that exceed ambient pressures are used when EtO is in combination with an inert gas. *CAUTION*: EtO is explosive by itself, and its residues are mutagenic.

The use of EtO as a sterilization agent requires an understanding of the parameters affecting its activity. The diffusion of EtO, moisture, and heat into materials can be a limiting factor. Prehumidification overcomes diffusion barriers in packaging materials and permits better penetration of gas. Humidification prior to sterilization also assures hydration of spores.

EtO effectively kills all microbial vegetative cells and spores (28, 29). At concentrations of 220 to 880 mg of EtO per liter at temperatures of 5 to 37°C, the microbial death rate is logarithmic. The time required for sterilization is approximately halved by doubling the concentration of EtO, and the temperature coefficient of the reaction is 2.7 for each 10°C rise (37).

Concentrations for sterilization range from 250 to 1,200 mg of EtO per liter of chamber space. The relative humidity of the atmosphere should be maintained in the range of 35 to 60%, but the humidity is not as critical as the EtO concentration (27). The exact length of the sterilization process depends upon the temperature, the EtO concentration, the type of materials to be sterilized, and the degree of bioburden contamination. Generally, a treatment of 700 mg of EtO per liter for 5 to 8 h at 100°F (37.7°C), or for 3 to 4 h at 130°F (54.4°C), will sterilize clean medical items in a 20-ft^3 (0.56-m^3) sterilizer. Table 8 lists the ranges of typical EtO sterilization parameters.

23.3.2. Formaldehyde

Although gaseous formaldehyde sterilization is not widely used, this agent is a good space and surface sterilant. One of the first uses of formaldehyde was to fumigate sickrooms. It is effec-

TABLE 7. *Commercially available ethylene oxide sterilants*

Trade name	Composition
Oxyfume Sterilant-12^a	12% ethylene oxide, 88% dichlorodifluoromethane
Oxyfume Sterilant-20^a	20% ethylene oxide, 80% carbon dioxide
Oxyfume Sterilant-30^a	70% ethylene oxide, 30% carbon dioxide
Pennoxide^b	12% ethylene oxide, 88% dichlorodifluoromethane
Steroxide-12^c	12% ethylene oxide, 88% dichlorodifluoromethane
Cryoxicide^d	11% ethylene oxide, 79% trichloromonofluoromethane, 10% dichlorodifluoromethane
Benvicide^e	11% ethylene oxide, 54% trichloromonofluoromethane, 35% dichlorodifluoromethane
Carboxide^a	10% ethylene oxide, 90% carbon dioxide

^{a} Linde Div., Union Carbide Corp.
^{b} Pennsylvania Engineering Co.
^{c} Castle Div., Sybron Corp.
^{d} American Sterilizer Co.
^{e} Matheson Gas Products, Div. of Will Ross.

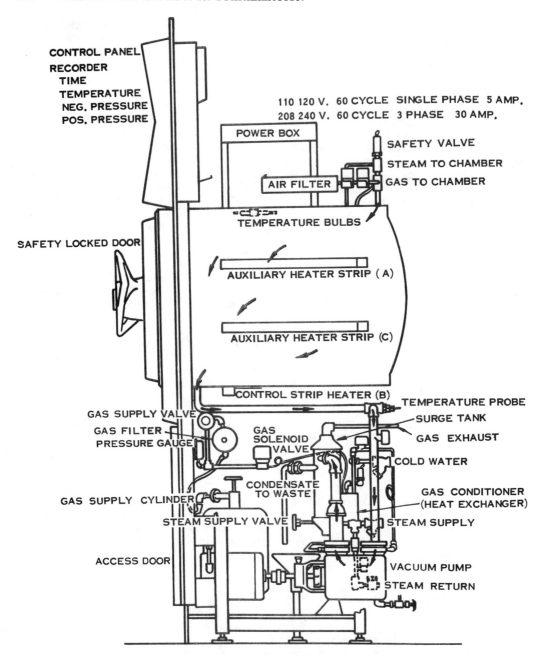

FIG. 3. *Schematic of essential elements of an ethylene oxide sterilizer. Reprinted from Perkins (8), courtesy of American Sterilizer Co., Erie, Pa., and Charles C Thomas, Publisher, Springfield, Ill.*

tive against bacteria, fungi, viruses, and rickettsiae, as well as insects and other animal life (38).

Formaldehyde gas for sterilization purposes can be generated by heating paraformaldehyde, which is a mixture of polyoxymethylene glycols containing 90 to 99% formaldehyde. The chem-

ical composition of paraformaldehyde is HO–$(CH_2O)_n$–H, where n represents 8 to 100 formaldehyde molecules. Paraformaldehyde depolymerizes to release gaseous formaldehyde when heated. Sterilization by formaldehyde gas generated from paraformaldehyde is more effective

TABLE 8. *Typical ranges of ethylene oxide (EtO) sterilization parameters*

Parameter	Range
Prehumidification	40 to 50% relative humidity
EtO concentration	250 to 1,200 mg/liter
Temperature	Ambient up to 125°F (52°C)
Time	2 to 16 h
Relative humidity	40 to 50%

than sterilization by vaporized formaldehyde solution.

Formaldehyde generated from paraformaldehyde can be used to sterilize stainless steel, rubber, plastic, and glass items wrapped in either cloth or paper (39). Formaldehyde reacts with primary amino and amido groups, secondary amide indole and imidazole groups, mercaptan radicals, and phenolic nuclei (40). Viruses may be inactivated by aldehyde-nucleic acid reactions (20). Neutralization procedures using ammonia gas to form hexamethylenetetramine should be used to minimize personnel exposure to the gas.

23.4. RADIATION

Radiation may be either nonionizing or ionizing. Nonionizing radiations include such forms as infrared, ultraviolet, ultrasonic, and radiofrequency. Ionizing radiations may be particulate (beta particles or electrons) or electromagnetic (X rays or gamma rays). Gamma rays generated from nuclides such as cobalt-60 or cesium-137 are examples of electromagnetic radiation. Electromagnetic radiation causes ionization through mechanisms that result in the ejection of an orbital electron with energy transferred from the incident gamma ray. These ejected electrons then behave similarly to the beta particles in ionizing reactions. Thus, both particulate and electromagnetic radiations are considered as ionizing radiations. Ionizing radiation represents a reproducible, consistent mode of sterilization.

The unit of radiation dosage is the rad, which is equivalent to an energy absorption of about 100 ergs/g. Sterilization doses are usually expressed in megarads (10^6 rads).

The efficacy of radiation is dependent on the absorbed dose, and the choice of radiation dose should be determined by the bioburden and the configuration and composition of the material to be sterilized. Solutions of ferrous sulfate, ferrous cupric sulfate, or ceric sulfate and films of dyed plastics have been used to quantitate the amount of absorbed radiation dose.

It is important to establish the radioresistance of the bioburden associated with the item to be sterilized and to evaluate the relative resistance of the bioburden to the resistance of an indicator

microorganism (such as spores of *Bacillus pumilus*. However, once satisfactory knowledge of the sterilization efficacy relative to the specific item being sterilized is known, dosimetry readings can be evaluated to assess sterilization assurance.

23.4.1. Electrons

Powerful electron accelerators are commercially available which focus an output of electrons into a narrow beam. This beam results in a highly uniform concentration of electron energy that can be used for sterilization. Most electron accelerators use a cathode tube to generate electrons, which are accelerated in a vacuum tube by electrostatic forces or microwaves to increase penetrability of the electrons. The use of electron accelerators requires careful control of electron energy, electron current, beam scan width, and exposure time.

A disadvantage of the system is the relatively low penetrating power of electrons compared to gamma rays. Because of charge and mass, electrons have shorter mean ranges than electromagnetic radiation of equivalent energy. Gamma rays of 1 million electron volt (meV) energy in water have a range approximately 80 times greater than that of electrons of 1 meV.

Electron accelerators have been used for industrial sterilization of low-density items, such as sutures and similar dressings. This process has advantages because products can be rapidly irradiated on a conveyor belt system during manufacturing. This adaptability and the resulting economic benefits are expected to generate wide industrial applications for this process in the next decade.

23.4.2. Gamma Rays

Many variables affect the kill of microorganisms by gamma radiation. Oxygen tension can enhance the effects of radiation by facilitating peroxide formation and/or ozone generation. Sulfhydryl and reducing compounds act as protective agents because they reduce the oxygen effect. Removal of water may increase resistance by reducing peroxide formation.

In general, vegetative bacteria are the most sensitive to gamma radiation, followed by molds, yeasts, bacterial spores, and viruses. Under most conditions a radiation dose of approximately 2.5 Mrads is sufficient to kill the most resistant organism with an adequate safety factor. Gamma radiation has been successfully used to sterilize items such as hospital supplies, antibiotics, vitamins, hormones, steroids, plastic disposable medical devices, petri dishes, and sutures.

Because of safety requirements and cost associated with gamma sources and electron beam radiation systems, most radiation studies are conducted in centralized industry or at university facilities. A small gamma source is available as a laboratory irradiator (Gammacell 220, Atomic Energy of Canada, Ltd.). Several excellent references are available that review basic theories of radiation sterilization, radiation equipment, and the structure of radiation facilities (1, 5, 12).

23.4.3. Ultraviolet Rays

Ultraviolet (UV) radiation in the 254-μm range has been used for germicidal purposes, but has limited applications because of its poor penetrating power. Organic matter, dust, or other protective coatings encrusting microorganisms can shield them from UV radiation.

Microorganisms are inactivated by the appropriate radiant energy dose. The total dose received by a microorganism is a function of the energy provided by the UV, the distance from the source, and the duration of microbial exposure. Vegetative microorganisms are more susceptible to UV radiation than bacterial spores, the latter being 3 to 10 times more resistant.

23.4.4. Radiofrequency Rays

The food industry pioneered studies of radiofrequency, but it has not been used as a sterilization process. Future applications may be found, especially when it is used in a synergistic manner with disinfectants. However, radiofrequency has limitations: the killing action is species specific, and different frequencies may be required to kill all types of microorganisms. It must be monitored and well shielded to assure personnel safety, and it can interfere with local communication systems.

23.5. FILTRATION

Filtration is one of the oldest methods used to sterilize solutions and is frequently used to re-move microorganisms and/or particulate matter from solutions and gases. As a terminal process, it is less desirable than the use of moist heat because a greater probability exists that a microorganism will pass through a filter.

Filters function by entrapping microorganisms within the porous structure of the filter matrix. Vacuum or pressure is required to move solutions through the filter. There are two basic types: depth filters and membrane filters. Depth filters consist of fibrous or granular materials that are pressed, wound, fired, or bonded into a maze of flow channels. In these, retention of particulates is a matter of a combination of absorption and mechanical entrapment in the filter matrix. Membrane filters have a continuous structure, and entrapment occurs mainly on the basis of particle size. Some advantages and disadvantages of membrane and depth filters are listed in Table 9. A number of excellent filters are available (Millipore Corp., Gelman Instrument Co., and Pall Corp.), and the commercial brochures describe specific assemblies and filter matrices for specific purposes (21, 25, 26, 30–33).

Filtration is particularly applicable for oil emulsions or solutions that are heat labile. Membrane filtration is widely used to sterilize oils, ointments, ophthalmic solutions, intravenous solutions, diagnostic drugs, radiopharmaceuticals, tissue culture medium, and vitamin and antibiotic solutions.

The integrity of membrane filters may be evaluated by "bubble point" testing (30), forward-flow techniques (32), and/or particle retention studies.

23.6. COMMERCIAL SOURCES

American Biological Control Co., Tenafly, NJ 07670
American Sterilizer Co., 2222 W. Grandview Blvd., Erie, PA 16506
Atomic Energy of Canada, Ltd., 275 Slater St., Ottawa 4, Ontario, Canada
BBL Microbiology Systems, P.O. Box 243, Cockeysville, MD 21030

TABLE 9. *Advantages and disadvantages of membrane and depth filters (19)*

Type	Advantages	Disadvantages
Depth filters	High loading capability Retention of a large percentage of particles smaller than the nominal pore size	Some filter material may migrate into the filtrate Subject to microbial grow-through, especially when the filters are heavily loaded Retains some solution No definite characterization of pore size
Membrane filters	Efficiency independent of flow rate and pressure differential Buildup of large particles over the filter can improve filtration No migration of filter material Little or no product retention	Low loading capacity Some particles smaller than the nominal pore size may be retained Fragile compared to depth filters

Castle Co. Div., Sybron Corp., 1777 E. Henrietta Rd., Rochester, NY 14623

EM Labs, Inc., 500 Executive Blvd., Elmsford, NY 10523

FMC, 200 E. Randolph Dr., Chicago, IL 60601

Gelman Instrument Co., 600 S. Wagner St., Ann Arbor, MI 48103

Gibraltar Biological Laboratories, Inc., Fairfield, NJ 07006

Linde Division, Union Carbide Corp., 51 Cragwood Rd., South Plainfield, NJ 07080

Matheson Gas Products (Div. of Will Ross, Inc.), 932 Paterson Plank Rd., East Rutherford, NJ 07073

Millipore Corp., P.O. Box F, Bedford, MA 01730

North American Science Associates, Inc., 2261 Tracy Rd., Northwood, OH 43619

Pall Corp., 30 Sea Cliff Ave., Glen Cove, NY 11542

Pennsylvania Engineering Co., 32nd St. & A.V.R.R., Pittsburgh, PA 15201

Propper Manufacturing Co., 36-04 Skillman Ave., Long Island City, NY 11101

Rexham Corp., 5501 N. Washington Blvd., Sarasota, FL 33580.

3M Co., 3M Center, St. Paul, MN 55101

Vacudyne-Altair, 375 E. Joe Orr Rd., Chicago Heights, IL 60411

23.7. LITERATURE CITED

23.7.1. General References

1. **Alexander, P., and Z. M. Bacq (ed.).** 1961. Fundamentals of radiology, vol. 5. Pergamon Press. The MacMillan Co., New York.
   A good overview of the effects of radiation upon biological systems.
2. **AVI Publishing Co.** 1968. Laboratory manual for food canners and processors. AVI Publishing Co., Westport, Conn.
   Although primarily written for the canning industry, this volume describes key microbiological and engineering methods and concepts relative to moist-heat sterilization. It is a prerequisite as a primer for those entering the field of sterilization technology.
3. **Block, S. (ed.).** 1977. Disinfection, sterilization and preservation. Lea & Febiger, Philadelphia.
   A collection of papers concerning all modes of sterilization. Contains detailed information on many chemical disinfectants. Valuable as a reference text with many excellent references.
4. **Ernst, R. R., and J. E. Doyle.** 1968. Sterilization with gaseous ethylene oxide—a review of chemical and physical factors. Biotechnol. Bioeng. **10:**1–31.
5. **Gaughran, E. R. L., and A. J. Goudie (ed.).** 1974. Technical developments and prospects of sterilization by ionizing radiation, vol. 1. Multiscience Publication, Ltd., Montreal.
   Review of equipment and facilities used for radiation sterilization. Some information concerning dosimetry systems and effects of radiation upon various materials is included.
6. **Keruluk, K., and R. S. Lloyd.** 1974. Ethylene oxide sterilization. American Sterilizer Co., Erie, Pa.
   An excellent primer. Reviews the key principles and practices. Contains many historically significant references.
7. **National Aeronautics and Space Administration.** 1968. Biological handbook for engineers. NASA-CR-61237. George C. Marshall Space Center, Huntsville, Ala.
   A condensed text for the nonbiologist. Serves as a good introductory text for sterilization technology.
8. **Perkins, J. J.** 1976. Principles and methods of sterilization in health sciences, 2nd ed. Charles C Thomas, Publisher, Springfield, Ill.
   Written primarily for hospital personnel, with an emphasis on the equipment and processes associated with moist-heat sterilization. Some consideration is given to dry-heat and ethylene oxide sterilization and to chemical disinfectants.
9. **Pflug, I. J.** 1977. Microbiology and engineering of sterilization processes. University of Minnesota, Department of Food Science and Nutrition and School of Public Health (sponsored by the Parenteral Drug Association, Philadelphia, Pa.).
   A collection of reports dealing with specific aspects of the design, delivery, and monitoring of sterilization.
10. **Pflug, I. J., and R. G. Holcomb.** 1977. Principles of thermal destruction of microorganisms. p. 933–994. *In* S. S. Block (ed.), Disinfection, sterilization, and preservation. Lea & Febiger, Philadelphia.
    Comprehensive treatment of moist heat and dry heat with thorough examples relative to experimental design.
11. **Pflug, I. J., and T. E. Odlaug.** 1977. Syllabus for a series of lectures and workshop sessions on the microbiology and engineering of sterilization processes. Univerisity of Minnesota, Department of Food Science and Nutrition and School of Public Health. University of Minnesota, Minneapolis.
    A workbook concerning calculations frequently used when using moist-heat sterilization. For the serious student of sterilization technology.
12. **Phillips, G. B., and W. S. Miller (ed.).** 1973. Industrial sterilization. Duke University Press, Durham N.C.
    A collection of papers representing current developments in industrial sterilization.
13. **Stumbo, C. R.** 1973. Thermobacteriology in food processing, 2nd ed. Academic Press, Inc., New York.
    A comprehensive treatise concerning moist-heat sterilization. An excellent reference text.

23.7.2. Specific References

14. **Association for the Advancement of Medical Instrumentation.** 1978. AAMI Standard-BIER/steam vessel (proposed). Association for the Advancement of Medical Instrumentation, Arlington, Va.
15. **Bruch, C. W.** 1973. Developments in industrial microbiology, p. 8. American Institute of Biological Sciences, Washington, D.C.

16. **Ernst, R. R.** 1969. Federal regulations and practical control microbiology for disinfectants, drugs, and cosmetics. Special Publication No. 4, p. 55–60. Society for Industrial Microbiology, Linden, N.J.

17. **Ernst, R. R., and J. E. Doyle.** 1969. Dev. Ind. Microbiol. **9:**293–296.

18. **Feldsine, P. T., A. J. Shechtman, and M. S. Korczynski.** 1977. Dev. Ind. Microbiol. **18:**401–407.

19. **Fifield, C. W.** 1977. In S. S. Block (ed.), Disinfection, sterilization and preservation, 2nd ed., p. 562–591. Lea & Febiger, Philadelphia.

20. **Fraenkel-Conrat, H.** 1954. Biochim. Biophys. Acta **15:**307–309.

21. **Gelman Instrument Co.** 1977. Gelman microfiltration systems. Gelman Instrument Co., Ann Arbor, Mich.

22. **Hall, L. B.** 1966. Spacecraft sterilization technology, p. 25–36. NASA SP-108. George C. Marshall Space Center, Huntsville, Ala.

23. **Halvorson, H. O., and R. Ziegler.** 1932. J. Bacteriol. **25:**101.

24. **Han, W. Y., H. I. Zhang, and J. M. Korchta.** 1976. Can. J. Microbiol. **22:**295–300.

25. **Health Industry Manufacturers Association.** 1978. Medical device sterilization monograph. Report 78-4.1. Validation of sterilization systems. Health Industry Manufacturers Association, Washington, D.C.

26. **Health Industry Manufacturers Association.** 1978. Report 78-4.4. Biological and chemical indicators. Health Industry Manufacturers Association, Washington, D.C.

27. **Keruluk, K., R. A. Gammon, and R. S. Lloyd.** 1970. Microbiological aspects of ethylene oxide sterilization. I. Experimental apparatus and methods. Appl. Microbiol. **19:**146–151.

28. **Kaye, S.** 1949. Am. J. Hyg. **50:**289–295.

29. **Kaye, S., and C. R. Phillips.** 1949. Am. J. Hyg. **50:**296–306.

30. **Millipore Corp.** 1965. Detection and analysis of contamination. ADM-30 and ADM-40. Millipore Corp., Bedford, Mass.

31. **Millipore Corp.** 1976. Preparing sterile particle free fluids in the hospital pharmacy. AM303. Millipore Corp., Bedford, Mass.

32. **Pall Corp.** 1974. Field experience in testing membrane integrity by the forward flow test method. Pall Corp., Cortland, N.Y.

33. **Pall Corp.** 1976. The Pall pharmaceutical filter guide. Pall Corp., Cortland, N.Y.

34. **Pflug, I. J.** 1973. In G. B. Phillips and W. S. Miller (ed.), Industrial sterilization, p. 239–282. Duke University Press, Durham, N.C.

35. **Pflug, I. J., and J. Bearman.** 1974. Treatment of sterilization process microbial survivor data in environmental microbiology as related to planetary quarantine. Progress report 9, NASA grant NGL 24-005-160. University of Minnesota, Department of Food Science and Nutrition and School of Public Health. University of Minnesota, Minneapolis.

36. **Pflug, I. J., and C. F. Schmidt.** 1968. Disinfection, sterilization and preservation, 1st ed., p. 63–105. Lea & Febiger, Philadelphia.

37. **Phillips, C. R.** 1949. Am. J. Hyg. **50:**280–288.

38. **Phillips, G. B., and W. S. Miller.** 1975. Remington's pharmaceutical sciences, 15th ed., p. 1389–1404. Mack Publishing, Easton, Pa.

39. **Tulis, J. J.** 1973. In G. B. Phillips and W. S. Miller (ed.), Industrial sterilization, p. 209–238. Duke University Press, Durham, N.C.

40. **Walker, J. F.** 1964. Formaldehyde, vol. 3, Van Nostrand, Reinhold, N.Y.

Chapter 24

Containment and Disinfection

W. EMMETT BARKLEY

The basic tenet of laboratory safety is good practice: awareness of the possible risks associated with the handling of bacteria, knowledge of mechanisms by which exposures may occur, use of safeguards and techniques that reduce the potential for exposures, and vigilance against compromise and error. A body of published information is available for those interested in information more detailed than that presented here. Recommended references include general articles on laboratory safety (1–7), safety guides and manuals (8–10), surveys of laboratory-acquired infections (11, 12), studies of specific laboratory operations (13–18), and general texts on disinfection (19, 20).

24.1. GENERAL LABORATORY PRACTICES

A summary and analysis of 3,921 cases of laboratory-associated infections were reported by Pike (11), and the cases caused by bacteria are summarized in Table 1. Such accounts indicate that individuals working with bacteria risk infection, but do not indicate the magnitude of risk because the reports of infections are incomplete and the population at risk is indeterminate. The evidence of laboratory-acquired bacterial infections, however, is sufficient to require the use of safety measures when handling all bacteria, whether or not they are "pathogenic."

These reported laboratory-acquired bacterial infections are grouped in Table 2 according to proved or probable source. Recognized accidents accounted for less than 25% of the cases; the causative events leading to infection were unknown in 75% of the cases. In some cases information was available to suggest a probable source of infection, but in 50% of the reported cases either no information was available to indicate a source or the information available only indicated that the infected person had worked with the agent.

Known accidents contributing to illness among laboratory workers who handle bacteria are summarized in Table 3. The causes are diverse and no specific ones predominate. However, most bacteriological practices create aerosols, which can be directly inhaled or settle on surfaces. The release of undetected aerosols may explain the absence of information on the source or cause of many laboratory-acquired infections.

Safety precautions that should be observed when handling pathogenic bacteria of known danger groups are summarized in Table 4. Spe-

TABLE 1. *Summary of laboratory-acquired bacterial infections (11)*

Disease or agent	No. of cases			No. of deaths
	U.S.	For-eign	Total	
Brucellosis	347	76	423	5
Typhoid	61	195	256	20
Tularemia	216	9	225	2
Tuberculosis	163	13	176	4
Streptococcus sp.	69	9	78	4
Leptospirosis	24	43	67	10
Shigellosis	49	9	58	0
Salmonellosis	21	27	48	0
Relapsing fever	19	26	45	2
Anthrax	40	5	45	5
Erysipeloid	32	11	43	0
Diphtheria	24	9	33	0
Staphylococcus sp.	26	3	29	1
Rat-bite fever	16	5	21	0
Glanders	9	11	20	7
Syphilis	5	10	15	0
Cholera	4	8	12	4
Plague	4	6	10	4
Neisseria meningitidis	6	2	8	1
Pseudomonas pseudomallei	2	6	8	0
Clostridium	3	3	6	0
Tetanus toxin	1	4	5	0
Diplococcus pneumoniae	—	—	5	0
Serratia marcescens	—	—	5	0
Haemophilus influenzae	—	—	4	0
Neisseria gonorrhoeae	—	—	3	0
Escherichia coli	—	—	2	0
Listeria monocytogenes	—	—	2	0
Pasteurella multocida	—	—	2	0
Klebsiella pneumoniae	—	—	1	0
Mycobacterium leprae	—	—	1	0
Bartonella bacilliformis	—	—	1	0
Fusobacterium fusiforme	—	—	1	0
Vibrio fetus	—	—	1	0
Vibrio parahaemolyticus	—	—	1	0
Mixed infection	4	1	5	0
Mycoplasma	—	—	4	0
Total	1,145	491	1,669	69

TABLE 2. *Distribution of cases according to proved or probable source of bacterial infection (11)*

Source	Cases	
	No.	Percent
Recognized accident	378	22.7
Aerosol	101	6.1
Animal or ectoparasite	149	8.9
Clinical specimens	90	5.4
Discarded glassware	34	2.0
Human autopsy	56	3.4
Intentional infection	14	0.8
Worked with agent	381	22.8
Other	7	0.4
Unknown or not reported	459	27.5
Total	1,669	100

symbol can be obtained commercially (Interex Corp., Waltham, MA 02154; Nuclear Associates Inc., Westbury, NY 11590; Shamrock Scientific Specialities Systems, Inc., Bellwood, IL 60104) and displayed upon a placard that is large enough for the symbol together with other appropriate information. The symbol is a fluorescent orange or an orange-red color. There is no requirement for the background color as long as there is sufficient contrast to permit the symbol to be clearly defined. The symbol should be as prominent as practical, of a size consistent with the size of the equipment or material to which it is affixed, and easily seen from as many directions as possible. The symbol is displayed to signify the actual or potential presence of a biological hazard. Appropriate wording should be used in association with the symbol to indicate the nature or identity of the hazard, the name of the individual responsible for its control, and emergency information.

The following safety practices should be employed routinely when handling all bacteria:

1. Keep laboratory doors closed.

2. Disinfect work surfaces daily and immediately after spills of bacteria.

3. Autoclave or otherwise disinfect contaminated items such as glassware, animal cages, and laboratory equipment before washing or reuse.

4. Use mechanical pipetting devices; do not use mouth pipetting.

5. Do not eat, drink, or smoke in the laboratory.

6. Wash hands frequently and always before leaving the laboratory.

7. Exercise care in all procedures and manipulations to minimize aerosol formation.

8. Post biohazard signs on appropriate laboratory doors. Freezers, refrigerators, and other units used to store pathogenic bacteria should also display the biohazard sign.

9. Wear laboratory gowns, coats, or uniforms

cial protective equipment including gloves and filter masks are indicated when bacteria identified in Groups 3 and 4 are handled. Filter masks are not necessary when such bacteria are contained in either a Class III cabinet or a Class I cabinet with an installed front closure panel equipped with arm-length rubber gloves or when an effective immunization has been given to potentially exposed laboratory workers. Immunization, when available and effective, is generally recommended. Detailed information on immunizations and recommendations for their use can be found in *Laboratory Safety at the Center for Disease Control* (9).

A biohazard symbol (Fig. 1) is used internationally to indicate the actual or potential presence of a biohazard and to identify equipment, containers, rooms, materials, experimental animals, or combinations that contain, or are contaminated with, viable hazardous agents. The

TABLE 3. *Known accidents resulting in laboratory-acquired bacterial infections (11)*

Type of accident	Cases	
	No.	Percent
Puncture or spray from needle and syringe	83	22.1
Contact with infectious material resulting from spills, sprays, etc.	82	21.7
Injury with broken glass or other sharp object	75	19.8
Aspiration through pipette	67	17.7
Bite or scratch of animal or ectoparasite	41	10.8
Other	3	0.8
Not indicated	27	7.1
Total	378	100

TABLE 4. *Operational recommendations for safe handling of pathogenic bateria*

Bacteria	Aseptic techniques and biological safety practices	Biological safety cabinets	Special facilities	Special protective equipment	Immunization available
Group 1[a]	+	−	−	−	−
Group 2[b]	+	−	−	−	+
Group 3[c]	+	+	+	+	−
Group 4[d]	+	+	+	+	+

[a] Group 1 bacteria: *Actinomyces*—all species; *Arizona arizonae*—all serotypes; *Bordetella*—all species; *Clostridium*—all species except *C. botulinum* and *C. tetani*; *Corynebacterium*—all species except *C. diphtheriae*; *Erysipelothrix insidiosa*; *Haemophilus influenzae*; *Klebsiella*—all species; *Leptospira*—all species; *Listeria*—all species; *Mycobacterium*—all species except *M. avium, M. bovis,* and *M. tuberculosis*; *Neisseria gonorrhoeae* and *N. meningitidis*; *Pasteurella*—all species except *P. multocida*; *Salmonella*—all species, except *S. typhi*; *Shigella*—all species; *Staphylococcus aureus*; *Streptobacillus moniliformis*; *Streptococcus pneumoniae*; *Streptococcus* groups A, B, and D; *Treponema pallidum*; *Yersinia enterocolitica*.

[b] Group 2 bacteria: *Clostridium tetani*; *Corynebacterium diphtheriae*; *Salmonella typhi*; *Vibrio cholerae*.

[c] Group 3 bacteria: *Bartonella*—all species; *Brucella*—all species; *Legionella*; *Mycobacterium avium, M. bovis,* and *M. tuberculosis*; *Pasteurella multocida* (type B); *Pseudomonas pseudomallei* and *P. mallei*.

[d] Group 4 bacteria: *Bacillus anthracis*; *Clostridium botulinum, Francisella tularensis*; *Yersinia pestis*.

as appropriate. Do not wear this protective clothing outside the laboratory.

10. Avoid the use of hypodermic needles and syringes when possible.

24.2. SPECIFIC LABORATORY OPERATIONS

24.2.1. Personal Hygiene

Food, candy, gum, and beverages for human consumption should not be stored or consumed in the laboratory. Smoking should not be permitted. A beard is undesirable because it retains particulate contamination and because a clean-shaven face is essential to the adequate fit of a face mask or respirator. Keep hands away from the mouth, nose, eyes, face, and hair to prevent self-inoculation. Keep personal items such as coats, hats, storm rubbers or overshoes, umbrellas, and purses out of the laboratory.

Wash hands promptly after removing protective gloves. Tests show that it is not unusual for bacteria to be present despite the use of gloves, as a consequence of unrecognized small holes, abrasions, or tears, entry at the wrist, or solvent penetration. Wash hands after removing soiled protective clothing, before leaving the laboratory area, before eating or smoking, and throughout the day at intervals dictated by the nature of the work. A disinfectant wash or dip may be desirable, but its use must not be carried to the point of causing roughening, desiccation, or sensitization of the skin.

Work with bacteria should not be allowed by anyone with a new or old cut, an abrasion, a lesion of the skin, or any open wound, including that resulting from a tooth extraction.

24.2.2. Pipetting

Safe pipetting techniques are required to reduce the potential for infection. Aspiration through a pipette was the causative event in more than 17% of all known accidents that resulted in laboratory-acquired bacterial infections.

The most common mode of exposure associated with pipetting procedures involves the application of mouth suction. Bacteria can also be transferred to the mouth if a contaminated finger is placed on the suction end of the pipette. There is also the danger of inhaling aerosols created in the handling of liquid suspensions when using unplugged pipettes, even if no liquid is drawn into the mouth. Additional hazards of exposure to aerosols are created by dropping liquid from a pipette to a work surface, by mixing cultures with alternate suction and blowing, by forcefully ejecting an inoculum onto a culture dish, or by blowing out the last drop.

Good pipetting practice requires the use of pipetting aids to prevent exposures by ingestion and careful technique to reduce the potential for creating aerosols. A variety of pipetting aids are available, ranging from simple bulb- and piston-actuated devices to sophisticated devices which contain their own vacuum pumps. In selecting a pipetting aid, consider the kind of procedure that is to be performed, the ease of operation, and the accuracy needed. Use several pipetting aids; none is available that is appropriate for all procedures.

BIOHAZARD

ADMITTANCE TO AUTHORIZED PERSONNEL ONLY

Hazard identity: _____

Responsible Investigator: _____

In case of emergency call:

Daytime phone _____   Home phone _____

Authorization for entrance must be obtained from
the Responsible Investigator named above.

FIG. 1. *Biohazard symbol for display in hazardous bacteriological areas.*

Develop pipetting techniques that reduce the potential for creating aerosols. Plug pipettes with cotton. Avoid rapid mixing of liquids by alternate suction and expulsion. No material should be forcibly expelled from a pipette, and air should not be bubbled through liquids with a pipette. Pipettes that do not require expulsion of the last drop are preferable. Take care to avoid dropping cultures from the pipette. When pipetting materials containing pathogenic bacteria, place a disinfectant-soaked towel on the working surface. Discharge liquids from pipettes as close as possible to the fluid or agar level of the receiving vessel, or allow the contents to run down the wall of the receiving vessel. Avoid dropping of the contents from a height into vessels. Place contaminated pipettes into a container with enough suitable disinfectant for complete immersion.

24.2.3. Using Hypodermic Syringes and Needles

Accidents involving the needle and syringe account for the largest number of laboratory-acquired bacterial infections resulting from known accidents. To lessen the chance of accidental injection, aerosol production, or spills, use alternate methods. Use a blunt needle or a cannula on the syringe for oral or intranasal inoculations, and never use a syringe and needle as a substitute for a pipette.

The following practices are recommended for hypodermic needles and syringes when used for parenteral injections:

Use a biological safety cabinet, and avoid quick and unnecessary movements of the hand holding the syringe.

Examine glass syringes for chips and cracks, and examine needles for barbs and plugs. Do this prior to sterilization and before use.

Use needle-locking (Luer-Lok type) syringes only, and be sure that the needle is locked securely.

Wear surgical or other rubber gloves.

Fill syringes carefully to minimize air bubbles and frothing.

Expel excess air, liquid, and bubbles vertically into a cotton pledget moistened with a suitable disinfectant or into a small bottle of sterile cotton.

Do not use a syringe to forcefully expel infectious fluid into an open vial or tube for the purpose of mixing. Mixing with a syringe is condoned only if the tip of the needle is held below the surface of the fluid in the tube.

If syringes are filled from test tubes, take care not to contaminate the hub of the needle, as this may result in transfer of infectious material to the fingers.

When removing a syringe and needle from a rubber-stoppered bottle, wrap the needle and stopper in a cotton pledget moistened with a suitable disinfectant. If there is danger of the disinfectant contaminating sensitive experimental materials, a sterile dry pledget may be used and discarded immediately into a disinfectant solution.

Inoculate animals with the hand positioned behind the needle to avoid punctures.

Be sure the animal is properly restrained prior to inoculation, and be alert for any unexpected movements of the animal.

Before and especially after injecting an animal, swab the injection site with a disinfectant.

After use, submerge contaminated nondisposable glass syringes with attached needles in a container of disinfectant prior to removal for autoclaving. To minimize accidental infection, *the needles should remain on the syringes until after autoclaving.* Place the syringes with attached needles in a pan separate from that holding other discard materials.

Discard disposable needles and syringes in a pan of disinfecting solution and then autoclave. Containers marked "SYRINGES AND/OR NEEDLES" may be disposed of in normal refuse containers after proper packaging and destruction of the contents. Destroy needles and syringes by clipping with a mechanical shear after sterilization, to prevent use by drug addicts.

In instances where the protective needle sheath must be replaced after use of a syringe,

always use forceps to minimize the possibility of accidental finger inoculation. Vacuum tubes are preferable for drawing blood samples.

Use separate pans of disinfectant for disposable and nondisposable syringes and needles, to eliminate sorting problems in the service area.

Do not discard syringes and needles into pans containing pipettes or other glassware that must be sorted out from the syringes and needles.

24.2.4. Opening Containers

Tubes containing bacterial suspensions should be manipulated with care. Simple procedures such as removing a tube cap or transferring an inoculum can create aerosols. Clearly mark tubes and racks of tubes containing biohazardous material. Use safety test-tube trays in place of conventional test-tube racks to minimize spillage from broken tubes. A safety test-tube tray is one that has a solid bottom and sides deep enough to hold the liquid should the test tubes break.

Vigorous shaking of liquid cultures creates a heavy aerosol. A swirling action will generally create a homogeneous suspension with a minimum of aerosol. When resuspending a liquid culture, allow a few minutes to elapse before opening the container, to reduce the aerosol.

The insertion of a sterile, hot loop or needle into a liquid or slant culture can cause spattering and release of aerosol. To minimize aerosols, allow the loop to cool in the air or be cooled by touching it to the inside of the container or to the agar surface where no growth is evident prior to contact with the culture or colony. After use of the inoculating loop or needle, preferably sterilize it in an electric or gas incinerator specifically designed for this purpose rather than heating in an open flame. These small incinerators have a shield to contain any material that may spatter from the loop. Disposable inoculating loops are available commercially; rather than decontaminating them with heat, discard them into a disinfectant.

Streaking an inoculum on rough agar results in aerosol production, created by the vibrating loop or needle. This generally does not occur if the operation is performed on smooth agar. Discard all rough agar poured plates intended for streaking with a wire loop.

Water of syneresis in petri dish cultures may contain viable bacteria and can form a film between the rim and lid of the inverted plate. Aerosols result when this film is broken by opening the plate. Vented plastic petri dishes, whose lid touches the rim at only three points, are less likely to create this hazard. Filter papers fitted into the lids reduce, but do not prevent, aerosols.

If plates are obviously wet, open them in a safety cabinet.

Less obvious is the release of aerosols when screw-capped bottles or plugged tubes are opened. This happens when a film of contaminated liquid between the rim and the liner is broken during opening. Removing cotton plugs or other closures from flasks, bottles, and centrifuge tubes immediately after shaking or centrifugation can release aerosols. Removal of wet closures, which can occur if the flask or centrifuge tube is not held in an upright position, is also hazardous. In addition, when using the centrifuge, there may be a small amount of foaming, and the closures may become slightly moistened. Because of these possibilities, open all liquid cultures of infectious material in a safety cabinet, wearing gloves and a long-sleeved laboratory garment.

When a sealed ampoule containing a lyophilized or liquid culture is opened, an aerosol may be created. Open ampoules in a safety cabinet. When recovering the contents of an ampoule, take care not to cut the gloves or hands or to disperse broken glass into the eyes, face, or environment. In addition, the material itself should not be contaminated with foreign organisms or with disinfectants. To accomplish this, work in a safety cabinet and wear gloves. Nick the ampoule with a file near the neck. Wrap the ampoule in disinfectant-wetted cotton. Snap the ampoule open at the nick, being sure to hold the ampoule upright. Alternatively, at the file mark, apply a hot wire or rod to develop a crack. Then wrap the ampoule in disinfectant-wetted cotton, and snap it open. Discard cotton and ampoule tip into disinfectant. Reconstitute the contents of the ampoule by slowly adding fluid to avoid aerosolizing the dried material. Mix contents without bubbling and withdraw them into a fresh container. Commercially available ampoules prescored for easy opening are available. However, prescoring may weaken the ampoule and cause it to break during handling and storage. Open ampoules of liquid cultures in a similar way.

Harvesting cultures from embryonated eggs is a hazardous procedure and leads to heavy contamination of the egg trays and shells, the environment, and the hands of the operator. Conduct operations of this type in a safety cabinet. A suitable disinfectant should be at hand and used frequently.

24.2.5. Centrifuging

Use safety centrifuge trunnion cups when centrifuging infectious bacterial suspensions. When bench-type centrifuges are used in a Class I safety cabinet, the glove panel and gloves should be in place or the ports should be covered.

Fill and open centrifuge tubes and trunnion cups in a Class I safety cabinet. If centrifugation is to be performed outside the cabinet, use the safety trunnion cup. After it is filled and sealed, decontaminate it on the outside or pass it through a disinfectant dunk bath. Since some disinfectants are corrosive, rinse with clean water after an appropriate contact time has elapsed.

Before centrifuging, eliminate tubes with cracks and chipped rims, inspect the inside of the trunnion cups for rough walls caused by erosion or adhering matter, and carefully remove bits of glass and other debris from the rubber cushion.

Add a disinfectant between the tube and trunnion cup to disinfect the materials in case of accidental breakage. This also provides an excellent cushion against shocks that might break the tube. Take care not to contaminate the culture material with the disinfectant. Also, if the tube breaks, the disinfectant may not completely inactivate the infectious material because of the dilution of the disinfectant and because of the high concentration of packing of the cells.

Avoid pouring supernatant fluid from centrifuge tubes. If this must be done, wipe off the outer rim with a disinfectant afterwards; otherwise, in a subsequent step, biohazardous fluid may be spun off as droplets and form an aerosol. Use of a vacuum system with appropriate in-line safety reservoirs and filters is preferable to pouring from centrifuge tubes or bottles.

Avoid filling the centrifuge tube to the point that the rim, cap, or cotton plug becomes wet with culture. Screw caps or caps that fit over the rim are safer than plug-in closures. Some fluid usually collects between a plug-in closure and the rim of the tube. Even screw-capped bottles are not without risk: if the rim is soiled and sealed imperfectly, some fluid will escape down the outside of the tube.

Do not use aluminum foil to form centrifuge tube caps, because they often become detached or rupture during handling and centrifuging.

The balancing of buckets and trunnion cups is often improperly performed. Take care to ensure that matched sets of trunnions, buckets, and plastic inserts do not become mixed. If the components are not inscribed with their weights by the manufacturer, apply colored stains for identification.

High-speed rotor heads are prone to metal fatigue, and where there is a chance that they may be used on more than one machine, each rotor should be accompanied by its own log book indicating the number of hours run at top or derated speeds. Failure to observe this precaution can result in dangerous and expensive disintegration. Frequent inspection, cleaning, and drying are important to ensure the absence of corrosion or other damage that may lead to the development of cracks. If the rotor is treated with a disinfectant, rinse it with clean water and dry it as soon as the disinfectant has adequately decontaminated the rotor. Examine rubber O rings and tube closures for deterioration, and keep them lubricated as recommended by the manufacturers. Where tubes of different materials are provided (e.g., celluloid, polypropylene, stainless steel), take care that tube closures designed specifically for the type of tube are used. Caps are often similar in appearance, but prone to leak if applied to tubes of the wrong material. Properly designed tubes and rotors, when well maintained and handled, should never leak.

24.2.6. Mixing and Disrupting

When using blenders, mixers, ultrasonic disintegrators, colloid mills, jet mills, grinders, and mortars and pestles with bacteria, do the following:

Use a biological safety cabinet.

Use safety blenders designed to prevent leakage from the rotor bearing at the bottom of the bowl. In the absence of a leakproof rotor, inspect the rotor bearing for leakage prior to operation. Test it in a preliminary run with sterile water, saline or methylene blue solution.

Use a towel moistened with disinfectant over the top of the blender. Sterilize the device and residual contents promptly after use.

Do not use glass blender bowls with infectious materials.

Blender bowls sometimes require cooling to prevent destruction of the bearings and to minimize thermal effects on the contents.

Before opening a blender bowl, wait for at least 1 min to allow settling of an aerosol.

24.2.7. Handling Animals

Animals are capable of shedding pathogens in saliva, urine, or feces. In the absence of information to the contrary, all animals should be regarded as shedders. Careful handling procedures should be employed to minimize the dissemination of dust from animal and cage refuse, and used cages should be sterilized by autoclaving. Refuse, bowls, and watering devices should remain in the cage during sterilization.

Wear heavy gloves when feeding, watering, handling, or removing infected animals. Do not place bare hands in the cage to move any object.

When animals are injected with bacteria, wear protective gloves. Animals should be properly restrained (e.g., use a squeeze cage for primates) or tranquilized to avoid accidents that might disseminate biohazardous material or cause injury to the animal or personnel.

House animals exposed to aerosols of bacteria in ventilated cages, in Class III cabinet systems, or in rooms designed for the use of ventilated suits by personnel.

The oversize canine teeth of large monkeys present a biting hazard and are important in the transmission of naturally occurring, very dangerous, monkey virus infections. Such teeth should be blunted or surgically removed by a veterinarian. Many zoonotic diseases, including infectious hepatitis and tuberculosis, can be transmitted from nonhuman primates to humans. Newly imported animals may be naturally infected with these or other diseases, and persons in close contact with such animals may become infected. Personal protective equipment or cage systems designed to contain infectious material should be used.

Keep doors to animal rooms closed at all times, except for necessary entrance and exit. Do not permit unauthorized persons in animal rooms.

Keep a container of disinfectant, prepared fresh each day, in each animal room for disinfecting gloves and hands and for general decontamination, even though no infectious animals are present. Wash hands, floors, walls, and cage racks with an approved disinfectant at the recommended strength at regular and frequent intervals.

Fill floor drains in animal rooms with water or disinfectant periodically to prevent backup of sewer gases. Do not wash shavings and other refuse on floors down the floor drain, because such refuse clogs sewer lines.

Maintain an insect and rodent control program in all animal rooms and in animal food storage areas. Take special care to prevent live animals, especially mice, from finding their way into disposable trash.

Carry out necropsy of infected animals in a safety cabinet. Wear surgeon's gowns over laboratory clothing, and wear rubber gloves. Wet the fur of animals with a suitable disinfectant.

Upon completion of a necropsy, place all potentially biohazardous material in suitable containers and sterilize immediately. Place contaminated instruments in a horizontal tray containing a suitable disinfectant. Disinfect the inside of the safety cabinet as well as other potentially contaminated surfaces. Clean grossly contaminated rubber gloves in disinfectant before removal from the hands, preparatory to sterilization. Place dead animals in proper leakproof containers, autoclave, and tag properly before removing for incineration.

24.2.8. Using Vacuum Systems

Vacuum filtration of suspensions and aspiration of culture media and supernatant liquids from centrifuged samples into primary collection flasks are common laboratory procedures. Provide protection against pulling bacterial aerosols or overflow fluid into the vacuum system by using an air filter in the line leading to the vacuum source and using an overflow flask for liquids between the collection flask and the air filter.

Two techniques of protecting the vacuum system are shown in Fig. 2. In both, a cartridge-type filter provides an effective barrier to passage of aerosols into the house vacuum system. The filter has a capacity to remove airborne particles 450 nm (0.45 μm) or larger in size. Ultipor (DFA 3001 AXPL5; Pall Corp., Courtland, NY 13045) is an example of such a filter. For assembling either apparatus, flexible tubing

of appropriate inside diameter for the fittings and of sufficient wall thickness for the vacuum should be used. Filter flasks of capacities from 250 to 4,000 ml may be used for the overflow flask, depending on the available space and the amount of fluid that could be aspirated. The overflow flasks should contain a disinfectant solution. An antifoam (such as Dow Corning Antifoam A) should be added to the overflow flask, because bubbling of air through the disinfectant may cause foam to reach the filter and shut off the vacuum. If the filter becomes contaminated or requires changing, it and the flask can be safely removed by clamping the line between the filter and vacuum source. The filter and flask should be autoclaved before the filter is discarded. A new filter can then be installed and the assembly replaced.

24.2.9. Miscellaneous Operations

Water baths and Warburg baths used to inactivate, incubate, or test bacteria should contain a disinfectant. For cold-water baths, 70% propylene glycol is recommended. *CAUTION:*

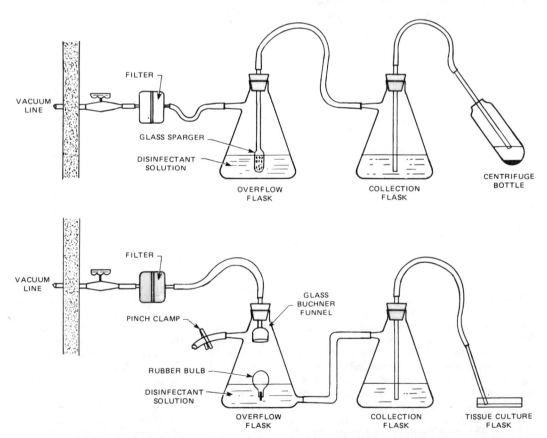

FIG. 2. *Two techniques for protecting vacuum systems from contamination.*

Sodium azide creates an explosive hazard and should not be used as a disinfectant.

Freezers, liquid nitrogen, dry ice chests, and refrigerators should be checked, cleaned, and decontaminated periodically. Use rubber gloves and respiratory protection during cleaning. Label all infectious material stored in refrigerators or freezers.

Evacuating a vacuum steam sterilizer prior to sterilization of contaminated material can create a potential hazard by releasing infectious material to the atmosphere. This hazard can be prevented by installation of a high-efficiency particulate air (HEPA) filter (American Air Filter Co., Inc., Louisville, KY 40201; Cambridge Filter Corp., Syracuse, NY 13221; Flanders Filters Inc., Washington, NC 27889).

Hazardous fluid cultures or viable powdered infectious materials in glass vessels should be transported, incubated, and stored in easily handled, nonbreakable, leakproof containers large enough to contain all the fluid or powder in case of leakage or breakage of the glass vessel.

Inoculated petri plates or other inoculated solid media should be transported and incubated in leakproof pans or leakproof containers.

Exercise care when using membrane filters to obtain sterile filtrates of infectious materials. Because of the fragility of the membrane and other factors, consider the filtrates to be infectious until culture tests have proved their sterility.

Examine shaking machines carefully for potential breakage of flasks or other containers being shaken. Use screw-capped, durable plastic or heavy-walled glass flasks, and securely fasten them to the shaker platform. As an additional precaution, enclose each flask in a plastic bag with or without an absorbent material.

Never work alone on a hazardous operation.

24.3. PHYSICAL CONTAINMENT

Biological safety cabinets are used to limit exposures to bacterial aerosols. They are needed because most laboratory techniques produce inadvertent aerosols, which can be inhaled. Cabinets can also protect the experiment from airborne contamination. Three types of biological safety cabinets are used: Classes I, II, and III, in increasing order of safety.

Class I and II biological safety cabinets should be located in laboratory areas away from doorways, supply air diffusers, and spaces of high activity to reduce the adverse effects of room air currents. Generally, the best location for the cabinet is on a side wall at a position farthest from the door.

Cabinet users should be encouraged to wear long-sleeved gowns with knit cuffs and gloves to reduce the shedding of skin flora into the work area of the cabinet and to protect the hands and arms from contact contamination. The work surface of the cabinet should be decontaminated before equipment and material are placed into the cabinet. Ideally, everything needed for the procedure should be placed in the cabinet before work is started. In Class II cabinets, nothing should be placed over the exhaust grilles at the front and rear of the work surface.

The arrangement of equipment and material on the work surface is an important consideration. Contaminated items should be segregated from clean items and located so that they are not passed over clean items. Discard trays should be located at the rear.

The use of aseptic techniques is important to prevent contact contamination. All infectious items should be contained or disinfected before removal from the cabinet. Trays of discarded pipettes and glassware should be covered before they are removed. After all items have been removed from the cabinet, the work surface should be decontaminated.

24.3.1. Class I Cabinet

The Class I cabinet (Fig. 3) is used in three modes: with a full-width open front, with an installed front panel without gloves, and with the panel equipped with arm-length rubber gloves. Materials may be introduced and removed through the panel opening and, if provided, through a hinged front view panel or a side air lock. Room air entering the cabinet prevents the escape of airborne contaminants. The air flows across the work space, over and under a back wall baffle, and out through a HEPA filter and blower in an overhead duct to the building air exhaust system or outdoors. When operated with the front open, a minimum inward face velocity of 75 ft (22.9 m)/min is needed.

The protection afforded by a Class I cabinet can be compromised by sudden withdrawal of the hands, rapid opening and closing of the room door, or rapid movements past the front of the cabinet. Aerosols created forcefully and in large quantities may escape in spite of the inward flow of air. Also, the cabinet does not protect the hands and arms from contact with hazardous materials. Such protection is dependent on technique and on the use of gloves and other protective clothing.

24.3.2. Class II Cabinets

Class II cabinets (Fig. 4) utilize laminar airflow and have front openings for access to the work space and for introduction and removal of ma-

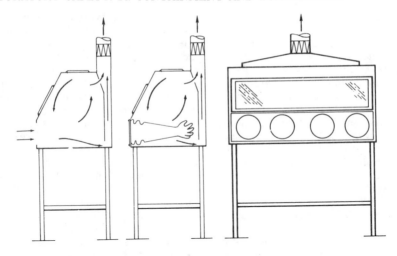

FIG. 3. *Class I biological safety cabinet.*

terials. Airborne contaminants in the cabinet are prevented from escaping by a curtain of air formed by unfiltered air flowing from the room into the cabinet and by HEPA-filtered air supplied from an overhead grille. The air curtain also prevents airborne contaminants in the room air from entering the work space of the cabinet. The air curtain is drawn through a grille at the forward edge of the work surface into a plenum below, where it is HEPA filtered and recirculated through the overhead grille down into the cabinet. A portion of the air is used to maintain the air curtain, and the remainder passes down onto the work surface and is drawn out through grilles at the back edge of the work surface. The

filtered air from the overhead grille flows in a uniform downward movement to minimize turbulence. This air provides and maintains a clear-air work environment. A percentage of air drawn through the front and back grilles of the work surface, which is equal to the flow of room air into the cabinet, is also filtered and exhausted from the cabinet.

There are two types of Class II cabinets, A and B. Type A has a fixed front access opening. The inward face velocity through the opening is at least 75 ft/min. The cabinet operates with a high percentage (approximately 70%) of recirculated air. It can be operated with recirculation of the filtered exhaust air to the room. This

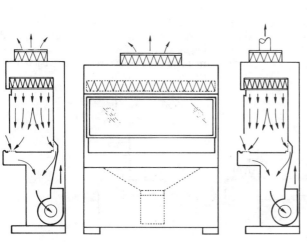
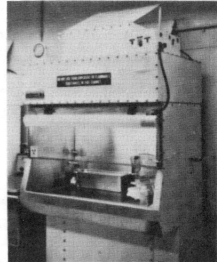

Type A

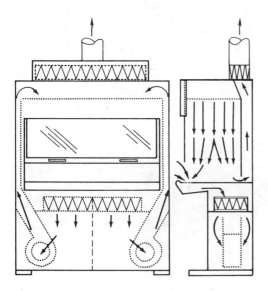

Type B

Fig. 4. *Class II biological safety cabinets.*

minimizes demand on supply and exhaust air systems unless the buildup of heat and odor from the recirculated exhaust air requires otherwise.

Type B cabinets do not recirculate exhaust air and have vertical sliding sashes rather than the fixed opening of the type A cabinets. Inward air velocity of 100 ft (30.5 m)/min is attained at an 8-in. (20-cm) sash opening. The cabinet operates with a low percentage (approximately 30%) of recirculated air.

Type A and B cabinets both provide partial containment. They provide protection to to the user, the environment, and the experiment. Type B cabinets can be used with dilute preparations of chemical carcinogens, low-level radioactive materials, and volatile solvents when the face velocity of 100 ft/min is maintained. With these materials, however, a careful evaluation must be made to determine that concentrations do not reach dangerous levels or cause problems of decontamination. Type A cabinets cannot be used with highly toxic, explosive, flammable, or highly radioactive substances because of the high percentage of recirculated air.

24.3.3. Class III Cabinet

The Class III cabinet (Fig. 5) is a totally enclosed, ventilated unit of gastight construction. Operations are conducted through attached rubber gloves. In use, the cabinet is maintained under a negative air pressure of at least 0.5 in.

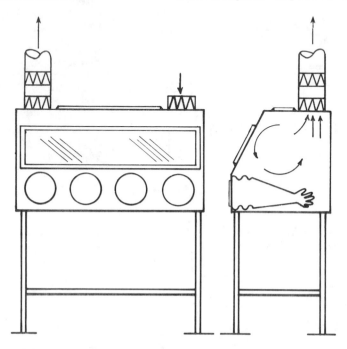

FIG. 5. *Class III biological safety cabinet.*

(1.3 cm) of water. Supply air is drawn into the cabinet through HEPA filters. The exhaust air is passed through two HEPA filters installed in series. The exhaust fan for the Class III cabinet is generally separate from the exhaust fans of the facility ventilation system.

Materials are introduced and removed through attached double-door sterilizers and dunk baths with liquid disinfectants. The usual utility services can be provided, but not gas. Modular designs provide for inclusion of refrigerator, incubator, freezer, centrifuge, and animal-holding and other special cabinet units.

The Class III cabinet provides the highest level of personnel and environmental protection. However, its protection can be compromised by puncture of the gloves or by accidents that create positive pressure. Flammable solvents should not be used unless a careful evaluation has been made to determine that concentrations do not reach dangerous levels. When required and determined safe, these materials should only be introduced into the system in closed, nonbreakable containers and should not be stored in the cabinet. Electric heaters are preferred over portable, canned-gas heaters.

24.4. CHEMICAL DISINFECTION

Disinfection is the removal or destruction of pathogenic microorganisms from inanimate objects or surfaces, usually by use of a chemical agent, and is needed to protect the integrity of bacteriological test results or to prevent the occurrence and spread of disease. Chemical disinfection is necessary because the use of pressurized steam, the most reliable method of sterilization, and other physical methods are not normally feasible for disinfection of large spaces, surfaces, and stationary equipment. Moreover, high temperatures and moisture often damage delicate instruments, particularly those having complex optical and electronic components.

24.4.1. Types of Agents

There are many disinfectants available under a wide variety of trade names. In general, disinfectants are classified as acids or alkalies, halogens, heavy metals, quaternary ammonium compounds, phenolic compounds, aldehydes, ketones, alcohols, amines, and peroxides. Unfortunately, the more active the disinfectant, the more likely it is that it will possess undesirable characteristics, such as toxic or corrosive properties. No disinfectant is equally useful or effective under all conditions. Resistance to the action of chemical disinfectants can be substantially altered by such factors as concentration of active ingredient, duration of contact, pH, temperature, humidity, and presence of organic matter.

Ethyl or isopropyl alcohol in a concentration of 50 to 70% is often used to disinfect surfaces. However, these agents are slow in their germicidal action, several minutes being required, and ineffective against spores.

Formaldehyde for use as a disinfectant is usually marketed as a 37% concentration of the gas in water solution (Formalin) or as a solid polymerized compound (paraformaldehyde). Formaldehyde in a concentration of 5% is an effective liquid disinfectant and is an effective space disinfectant for sterilizing rooms or buildings. However, formaldehyde loses considerable disinfectant activity at refrigeration temperatures, and its pungently irritating odor requires that care be taken when using formaldehyde solutions in the laboratory.

Phenol is not often used by itself as a disinfectant. The odor is unpleasant, and a gummy residue remains on treated surfaces. Phenolic compounds, however, are basic to a number of popular disinfectants which are effective against vegetative bacteria at high dilutions and are essentially odorless (e.g., o-phenylphenol). The phenolics are not effective in ordinary usage against bacterial spores, however.

Quaternary ammonium compounds ("quats") are strongly surface-active, which makes them good surface cleaners. The quats attach to proteins, so in their presence dilute solutions of quats will lose effectiveness. The quats tend to clump bacteria and are neutralized by anionic detergents, such as soap. They are bactericidal at medium concentrations, but they are not tuberculocidal or sporicidal even at high concentrations. The quats have the advantages of being odorless, nonstaining, noncorrosive to metals, stable, inexpensive, and relatively nontoxic.

Chlorine is a universal disinfectant which is active against all bacteria, including bacterial spores, and is effective over a wide range of temperatures. Chlorine combines readily with protein so that an excess of chlorine must be used if proteins are present. Free, available chlorine is the active element. It is a strong oxidizing agent and is corrosive to metals. Chlorine solutions will gradually lose strength, so that fresh solutions must be prepared frequently. Sodium hypochlorite is usually used as a base for chlorine disinfectants. An excellent disinfectant can be prepared from household or laundry bleaches, which usually contain 5.25%, or 52,500 ppm, of available chlorine. If one dilutes them 1:100, the solution will contain 525 ppm of available chlorine, and if 0.7% of a nonionic detergent is added, a very good disinfectant is created.

Iodophors are one of the most popular groups of disinfectants used in the laboratory, and Wescodyne is perhaps the one most widely used. The range of dilution of Wescodyne recommended by the manufacturer (West Chemical Products Inc., New York, NY 11101) is 1 oz in 5 gal of water (giving 25 ppm of available iodine) to 3 oz in 5 gal (giving 75 ppm). At 75 ppm, the concentration of free iodine is 0.0075%. This small amount can be rapidly taken up by extraneous protein present. Clean surfaces or clear water can be effectively treated by 75 ppm of available iodine, but difficulties may be experienced if any appreciable amount of protein is present. For washing the hands or for use as a sporicide, it is recommended that Wescodyne be diluted 1:10 in 50% ethyl alcohol, which will give 1,600 ppm of available iodine, at which concentration relatively rapid inactivation of most bacteria will occur. An alcoholic solution (tincture) of iodine is effective as a disinfectant, but is irritating.

Heavy metals are not recommended because mercury and other heavy metal preparations are toxic and more bacteriostatic than bactericidal.

Peroxides, in the form of weak solutions of hydrogen peroxide, are useful to clean skin surfaces and wounds, but they have negligible antimicrobial activity. A highly concentrated peroxide at a low pH may be useful as a disinfectant.

24.4.2. Selection of Agent

Although the ideal disinfectant does not exist, it is useful to consider what characteristics it should have:

High activity—effectiveness at high dilutions in the presence of organic matter.

Broad spectrum of antimicrobial activity—effectiveness against gram-positive, gram-negative, and acid-fast bacteria and against spores, viruses, and fungi.

Stability—retention of potency after storage for prolonged periods.

Homogeneity—no settling out of active ingredients.

Adequate solubility—solubility in water, fats, and oils for good penetration into microorganisms.

Low surface tension—penetration into cracks and crevices.

Minimum toxicity—lack of acute and chronic toxicity, mutagenicity, carcinogenicity, teratogenicity, allergenicity, irritability, and photosensitization.

Detergent activity—ability to solubilize and remove dirt and debris.

Minimum material effects—low or acceptable effects on metals, wood, plastics, and paint.

Odor control—pleasant odor, odorless, or having deodorizing properties.

Cost—inexpensive in relation to efficiency.

The effectiveness of a disinfection process can be maximized by attention to the following factors:

Select a disinfectant appropriate against the microorganism to be inactivated. If its identity is unknown, select an agent with as broad a spectrum of activity as possible.

Reduce the bioburden and the organic content to a minimum on surfaces or objects to be disinfected.

Take into account the fact that most chemical disinfectants have limited effectiveness against bacterial spores and allow additional exposure time.

Use the disinfectant in the proper concentration; inadequate concentrations may result in lack of disinfection, and excessively concentrated solutions may pose problems of toxicity and effects on materials. The concentration may determine whether the action is static or cidal.

Carefully consider the exposure or contact time necessary for disinfection.

Be sure that the chemical concentration and contact time are compatible with the temperature of disinfection. In general, lower temperatures require longer contact times and higher temperatures increase efficiency by two- to threefold per 10°C rise in temperature.

Make sure that the water used for preparing use dilutions of disinfectants is of the proper quality. With certain chemicals, water hardness (e.g., calcium ions) in excess of 300 to 400 ppm destroys disinfecting ability.

Consider the amount of organic matter present on the object being treated. Organic matter reacts with disinfecting chemicals and in effect removes the active ingredient of the solution.

No single chemical disinfectant or method will be effective or practical for all situations. When selecting a chemical disinfectant, the following questions should be considered:

What is the target organism(s)?

What disinfectants in what form are known to, or can be expected to, inactivate the target organism(s)?

What degree of inactivation is required?

In what menstruum is the organism suspended (i.e., simple or complex, on solid or porous surfaces, or airborne)?

What is the highest concentration of cells anticipated to be encountered?

Can the disinfectant (as either a liquid, a vapor, or a gas) be expected to contact the organisms, and can effective duration of contact be maintained?

What restrictions apply with respect to compatibility of materials?

What is the stability of the disinfectant in use concentrations, and does the anticipated use situation require immediate availability of the disinfectant or will sufficient time be available for preparation of the working concentration shortly before its anticipated use?

Use requirements and important characteristics for several categories of chemical disinfectants most likely to be used in the bacteriology laboratory are summarized in Table 5. Practical concentrations and contact times may differ from the recommendations of manufacturers, because it has been assumed that bacteria will be afforded a high degree of potential protection by organic menstruums. It has not been assumed that a sterile state will result from application of the indicated concentrations and contact times. In actual use situations, the efficacy of the selected disinfectant procedure should be validated by the individual user.

24.4.3. Environmental Surfaces

Routine disinfection of floor surfaces is often a function of the housekeeping department. For laboratories and other areas where infectious material is involved, disinfection should be carried out daily with a suitable phenolic-detergent combination at the specified use dilution, using clean mops and a two-bucket procedure. No more than 500 to 1,000 ft^2 (46.5 to 93 m^2) of floor space should be disinfected with the same solution. The disinfectant used should have a broad spectrum of activity against bacteria, fungi, viruses, and acid-fast organisms and should not be materially affected by hard water, organic matter, or even the cotton mops.

Bench tops, tables, and large equipment may be disinfected by using disinfectant-detergent at the use dilution with a clean cloth, sponge, or disposable towel. Surfaces should be allowed to remain damp, moist, wet, and not rinsed off. If 0.5% hypochlorite solutions are considered, it must be remembered that they are easily neutralized by excess organic matter, may discolor surfaces, and are highly corrosive to metals.

24.4.4. Bacterial Spills

A spill that is confined to the interior of the biological safety cabinet should present little or no hazard to personnel in the area. However, initiate chemical disinfection procedures at once, while the cabinet ventilation system continues to operate, to prevent the escape of contami-

TABLE 5. Summary of practical disinfectants

Type of disinfectant	Practical requirements			Important characteristics							
	Use dilution	Contact time (min)		Effective shelf life of 1 week[a]	Corrosive	Flammable	Residue	Inactivated by organic matter	Skin irritant	Eye irritant	Respiratory irritant
		Vegetative bacteria	Bacterial spores								
Alcohol	70–85%	10	NE[b]	+		+					
Formaldehyde	0.2–8.0%	10	30	+			+		+	+	
Glutaraldehyde	2%	10	30	+			+		+	+	
Phenolic compounds	1.0–5.0%	10	NE	+	+				+	+	
Quaternary ammonium compounds	0.1–2.0%	10	NE					+			
Chlorine compounds	500 ppm[c]	10	30	+	+		+	+	+	+	
Iodophor	25–1,600 ppm[c]	10	30	+	+		+	+	+	+	+

[a] Protected from light and air.
[b] Not effective.
[c] Available halogen.

nants. Spray or wipe walls, work surfaces, and equipment with a disinfectant. A disinfectant with a detergent will help clean the surfaces by removing both dirt and bacteria. A suitable disinfectant is a 3% solution of an iodophor (e.g., Wescodyne) or a 1:100 dilution of a household bleach (e.g., Clorox) with 0.7% of a nonionic detergent. Wear gloves during this procedure. Use sufficient disinfectant solution to ensure that the drain pans and catch basins below the work surface contain the disinfectant. Lift the front exhaust grill and tray, and wipe all surfaces. Wipe the catch basin, and drain the disinfectant into a container. The disinfectant, gloves, wiping cloth, and sponges should be discarded into a pan and autoclaved. This procedure will not disinfect the filters, blower, air ducts, or other interior parts of the cabinet.

If the entire interior of the cabinet is to be disinfected, this can be accomplished with formaldehyde gas. Weigh out 0.3 g of flake paraformaldehyde for each cubic foot (0.028 m^3) of space in the cabinet. Place the paraformaldehyde in an electric frying pan, and place the pan in the cabinet with the power cord run to the outside. Raise the humidity within the cabinet to about 70%; vaporization of water in the frying pan is a convenient technique. Set the thermostat of the frying pan containing the paraformaldehyde at 450°F (292°C). Seal the cabinet opening with sheet plastic and tape. If the cabinet exhaust air is discharged into the room, attach a flexible hose to the cabinet exhaust port and extend the hose to the room exhaust grille. If the building exhaust air recirculates, attach the hose to an open window or door. If the cabinet is exhausted directly into the building system, close the exhaust damper. Turn on the frying pan to depolymerize the paraformaldehyde. After one-half of the paraformaldehyde has been depolymerized, turn on the cabinet fan for about 3 s to allow the formaldehyde gas to reach all areas. After depolymerization is complete, again turn on the cabinet fan for 3 s. Then allow the cabinet to stand for a minimum of 1 h. Afterwards, open the flexible hose on the exhaust damper, slit the plastic covering the opening, and turn on the cabinet fan. Ventilate the cabinet for several hours to remove all traces of formaldehyde.

A spill of potentially hazardous biological material in the laboratory outside a biological safety cabinet can place occupants at risk. The first step in response to such an occurrence is to avoid inhaling any airborne material by holding the breath and leaving the laboratory. Warn others in the area, and go directly to a wash or changeroom area. If clothing is known or suspected to be contaminated, remove it with care, folding the contaminated area inward. Discard clothing into a bag or place it directly into an autoclave. Wash all potentially contaminated areas as well as the arms, face, and hands. Delay reentry into the laboratory for a period of 30 min to allow reduction of the aerosol generated by the spill. When entering the laboratory to clean the spill area, wear protective clothing (rubber gloves, autoclavable footwear, an outer garment, and a respirator). If the spill was on the floor, do not wear a surgical gown that may trail on the floor when bending down. Place a discard container near the spill, and transfer large fragments of material into it; replace the cover. Using a hypochlorite solution containing 1,000 ppm of available chlorine, iodophor solution containing 1,600 ppm of iodine, or other appropriate disinfectant, carefully pour the disinfectant around and into the visible spill. Avoid splashing. Allow a contact time of 15 min. Use paper or cloth towels to wipe up the disinfectant and spill, working toward the center of the spill. Discard towels into a discard container as they are used. Wipe the outside of the discard containers, especially the bottom, with a towel soaked in a disinfectant. Place the discard container and other materials in an autoclave, and sterilize them. Remove shoes, outer clothing, respirator, and gloves, and sterilize by autoclaving. Wash hands, arms, and face; if possible, shower.

24.4.5. Instruments

Contaminated syringes, instruments, pipettes, thermometers, and other glassware should be decontaminated before being reprocessed. Wherever possible, final processing should be carried out with steam for heat-stable items or with ethylene oxide for heat-labile, moisture-sensitive materials. Where decontamination is essential before handling, containers of liquid disinfectant (such as 2% glutaraldehyde) may be used. The items should be completely immersed in the solution, preferably for as long as 20 to 30 min. This allows a safety factor if the item is not extremely clean or if the solution has been used for some period previously. Before reuse, the items should be rinsed carefully with distilled water and, if to be used in or on the body, with sterile distilled water.

Laboratory pipettes may be decontaminated by placing them in a vertical or horizontal container with a disinfectant such as a 2% o-phenylphenol. Glassware should be immersed completely. Before handling and reprocessing glassware, steam sterilization is recommended.

24.4.6. Hands

For routine handwashing in the laboratory when not working with or using infectious agents, a variety of hand cleansers may be used. A good handwashing of 15- to 20-s duration with Wescodyne should reduce the bacterial population by over 50%.

Bar soaps should not be used, not only because of the inherent sloppiness of the soap dish but also because some organisms survive for some time on the soap. Liquid soaps, unless they contain a preservative, may gradually develop large populations of organisms in the reservoir. These should be cleaned out routinely and new soap should be added. Powdered soaps and leaf soaps have the advantage of not being contaminated or allowing organisms to grow in them.

Rapid disinfection of the hands after contamination may be accomplished by the use of one of several procedures, as follows:

Use a Wescodyne-detergent preparation and scrub the hands for 30 to 60 s. Rinse with water.

Use a 4% chlorhexidine-detergent and scrub the hands for 30 to 60 s. Rinse with water.

Use a phenolic disinfectant-detergent for 20 to 30 s and then rinse with water (some phenolics may cause depigmentation of dark skin if used too frequently in too high concentrations).

Use alcohol (50 to 70%) on the hands for 20 to 30 s, followed by a soap scrub of 10 to 15 s and a rinse with water.

ACKNOWLEDGMENT

I acknowledge the contributions of the Special Committee of Safety and Health Experts who prepared the "Laboratory Safety Monograph—A Supplement to the NIH Guidelines for Recombinant DNA Research," July 1978. The information provided in this chapter was based, in large part, on material from this monograph.

24.5. LITERATURE CITED

24.5.1. General Articles

1. **Chatigny, M. A.** 1961. Protection against infection in the microbiological laboratory. Devices and procedures. Adv. Appl. Microbiol. **3**:131–192.
2. **Collins, C. H., E. G. Hartley, and R. Pillsworth.** 1974. The prevention of laboratory acquired infections. Public Health Laboratory Service Monograph Series No. 6. Her Majesty's Stationary Office, London.
3. **Darlow, H. M.** 1969. Safety in the microbiological laboratory, p. 169–204. *In* J. R. Norris and R. W. Ribbons (ed.), Methods in microbiology. Academic Press, Inc., New York.
4. **Phillips, G. B.** 1965. Microbiological hazards in the laboratory. I. Control; II. Prevention. J. Chem. Educ. **42**:A43–A48; **42**:A117–A130.
5. **Phillips, G. B.** 1969. Control of microbiological hazards in the laboratory. Am. Ind. Hyg. Assoc. J. **30**:170–176.

6. **Songer, J. R., and J. F. Sullivan.** 1974. Safety in the biological laboratory. J. Chem. Educ. **51**:A481–A485.
7. **Wedum, A. G.** 1974. Biohazard control, p. 191–210. *In* E. C. Melby, Jr., and N. H. Altman (ed.), Handbook of laboratory animal science, vol. 1. CRC Press, Inc., Cleveland.

24.5.2. Safety Guides and Manuals

8. **U.S. Public Health Service.** 1976. Classification of etiologic agents on the basis of hazard, 4th ed., p. 1–13. Center for Disease Control, Atlanta, Ga.
9. **U.S. Public Health Service.** 1977. Laboratory safety at the Center for Disease Control. Department of Health, Education, and Welfare, Publication No. CDC 77-8118. Center for Disease Control, Atlanta, Ga.
10. **U.S. Public Health Service.** 1974. NIH biohazards safety guide. GPO Stock No. 1740-00383. U.S. Government Printing Office, Washington, D.C.

24.5.3. Laboratory Infections

11. **Pike, R. M.** 1976. Laboratory associated infections: summary and analysis of 3,921 cases. Health Lab. Sci. **13**(2):105–114.
12. **Pike, R. M., S. E. Sulkin, and M. L. Shulze.** 1965. Continuing importance of laboratory-acquired infections. Am. J. Public Health **55**:190–199.

24.5.4. Specific Laboratory Operations

13. **Grieff, D.** 1969. Safe procedure for opening evacuated glass ampoules containing dried pathogens. Appl. Microbiol. **18**:130.
14. **Harvey, R. W. S., T. H. Price, and D. H. M. Joynson.** 1976. Observations on environmental contamination in a microbiological laboratory. J. Hyg. **76**:91–96.
15. **Hellman, A., M. N. Oxman, and R. Pollack (ed.).** 1973. Biohazards in biological research. *In* Proceedings of Conference held at Asilomar Conference Center, Pacific Grove, Calif., 22–24 January. Cold Spring Harbor Laboratory, Cold Spring Harbor, N.Y.
16. **Morris, C. A., and P. H. Everall.** 1972. Safe disposal of air discharged from centrifuges. J. Clin. Pathol. **25**:742.
17. **Phillips, G. B., and J. V. Jemski.** 1963. Biological safety in the animal laboratory. Lab. Anim. Care **13**:13–20.
18. **Reitman, M., and A. G. Wedum.** 1956. Microbiological safety. Public Health Rep. **71**:659–665.

24.5.5. Disinfection

19. **Block, S. S.** 1977. Disinfection, sterilization and preservation, 2nd ed. Lea & Febiger, Philadelphia.
20. **Sykes, G.** 1967. Disinfection and sterilization, 2nd ed. J. B. Lippincott Co., Philadelphia.

Section VII

ADDENDUM

Chapter 25

Diluents and Biomass Measurement

PHILIPP GERHARDT

Several basic methods, applicable to a number of the foregoing chapters, are described in this addendum.

25.1. DILUENTS

Bacterial cells often must be diluted from their original dense concentration to a sparse concentration suitable for observing in a microscope, measuring numbers, analyzing for genetic or metabolic properties, or washing preparatory to study. Whatever the use, the diluted cells must retain their original characteristics. The preservation of viability and metabolic activity often is particularly important. Diluted cells are more likely to be harmed by an unfavorable environment than are cells in dense suspensions, in which the environment contains high levels of materials leached from the cells. Consequently, care must be taken to use a suitable diluting solution.

Distilled or tap water should usually *not* be used, because it is osmotically hypotonic to all bacterial cells (except dormant spores) and unbuffered against pH change. "Physiological saline" (0.85% NaCl) also is usually inadequate, because it is isotonic only to mammalian cells and is unbuffered. Similarly, phosphate buffer by itself is usually inadequate. Viability may be reduced by 50% or more in such diluents.

A common general-purpose diluent is phosphate-buffered saline with gelatin. Presence of the protein stabilizes fragile bacteria and phage, and presence of the phosphate buffers the pH at neutrality. Buffered saline with gelatin contains 8.5 g of NaCl, 0.3 g of anhydrous KH_2PO_4, 0.6 g of anhydrous Na_2HPO_4, and 0.1 g of gelatin per liter of distilled water.

A better general-purpose diluent contains 0.4 M NaCl (or 0.5 M sucrose) for osmotic balance, 50 mM phosphate buffer at the pH of the growth medium (see 6.1.2), 50 mM $MgSO_4$ for preservation of membrane integrity, and 0.01% gelatin as a stabilizing agent (which may be omitted if inappropriate for use of the cells). This diluent is especially desirable for gram-negative bacteria, which are damaged more easily than gram-positive bacteria. Further considerations about diluents for washing bacteria are provided in 10.4.2 and 19.1.1.

For critical or unusual situations, use a diluent adapted to the particular conditions. The growth medium without the carbon source, to arrest further growth, may be the best diluent for cultures. Buffered saline with gelatin may require adjustment of the pH and osmotic tonicity. With many anaerobic bacteria, the diluent must contain a reducing agent and be freed from dissolved oxygen (6.6.3 and 6.6.4). Problems may arise from absorption of bacterial cells onto surfaces of glassware (introduction of Chapter 11). Problems also may arise, in electronic enumeration (11.1.2) and in various other situations, from cells that fail to separate after division or that aggregate afterwards because of surface stickiness or electrostatic charge. Such groups of cells may sometimes be dispersed physically by use of very short bursts of sonic energy or by vigorous blending in a Vortex mixer, with monitoring by viable counts and microscopy to determine maximum dispersion with minimum destruction of the cells. The dispersion of cells can be maintained, or sometimes attained, by addition to the diluent of a chemical agent that absorbs and imparts an electronegative repulsive charge to each cell. A detergent agent (e.g., 0.1% Tween 80) sometimes is suggested (10.3.3), but is effective mainly for mycobacteria and other cells with hydrophobic surfaces and may be harmful to other cells. Polymerized organic salts of sulfonic acids of the alkyl-aryl type (lignin compounds) are effective dispersing agents and apparently are not harmful to representative bacteria (5).

Meynell and Meynell have further discussed the use of diluents (6), and an entire symposium has dwelt on the death and survival of vegetative microbes (3).

25.2. BIOMASS MEASUREMENT

A measure of the mass of bacterial cell constituents frequently is used as the unit basis for measurement of a metabolic activity or of a morphological or chemical constituent. "Biomass" together with cell numbers provide the two basic parameters of bacterial growth, and density provides a basic characteristic of the cell.

The methods for measuring biomass seem obvious and straightforward, but in fact they are complicated if accuracy is sought. Furthermore, the results may be expressed in different ways and the numbers may be more relative than absolute. Consequently, care must be taken in their determination.

25.2.1. Wet Weight

A nominal wet weight of bacterial cells originally in liquid suspension is obtained by weighing a sample in a tared pan after separation and washing of the cells by filtration or centrifugation. In either case, however, diluent is trapped in the interstitial (intercellular) space and contributes to the total weight of the mass. The amount of interstitial diluent may be substantial. A mass of close-packed, rigid spheres contains an interstice of 27% independent of sphere size, and close-packed bacterial cells may contain an interstice of 5 to 30%, depending on their shape and amount of deformation.

One method for obtaining the actual wet weight of the cells themselves is to correct the nominal wet weight by subtracting the experimentally determined weight of diluent in the interstitial space, the method for which is described in 19.1.2. Another method, requiring special equipment, is to partially dry the cells and then to equilibrate them to constant wet weight in a closed weighing chamber in which the atmosphere is equilibrated at 100% relative humidity at constant temperature.

25.2.2. Dry Weight

A nominal dry weight (solids content) of bacterial cells originally in a liquid suspension is obtained by drying a measured wet weight or volume in an oven at 105°C to constant weight. The cells must be washed with water, or a correction must be made for medium or diluent constituents that are dried along with the cells. Separating the cells by filtration poses particular problems (see 10.5). Volatile components of the cells may be lost by oven-drying, and some degradation may occur, evidenced by discoloration (particularly if a higher temperature is used). Some regain of moisture occurs during the transferring and weighing process in room atmosphere, so this should be done quickly within a fixed time for all replicate samples. It is best, of course, to use tared weighing vessels that can be sealed after drying.

Possibly more accurate determination may be made by drying the sample to constant weight in a desiccator vessel with P_2O_5 under oil pump vacuum at 80°C or by lyophilization (see 12.2.1). In my experience, however, results with the three methods varied indistinguishably within 1% (2). An excellent discussion of dry weight procedures and errors is given by Mallette (4).

The dry weight of cells may be expressed on a wet weight basis (grams of solids per gram of wet cells) or on a wet volume basis (grams of solids per cubic centimeter of wet cells or per cubic centimeter of cell suspension).

25.2.3. Water Content

The amount of cell water in fully hydrated cells is equal to the difference between the wet weight (25.2.1) and the dry weight (25.2.2) of the cells themselves.

Water content also can be determined relative to the humidity of the atmosphere or to the water activity of the solution in which the cells occur (6.2). Completely dried cells equilibrate with an atmosphere of controlled, known humidity in successively increasing increments up to saturation (7). The resulting "sorption isotherm" curve typically appears like that shown in Fig. 1. The initial phase of water sorption, at very low humidities, represents tightly bound, monolayer-adsorbed water; the intermediate plateau

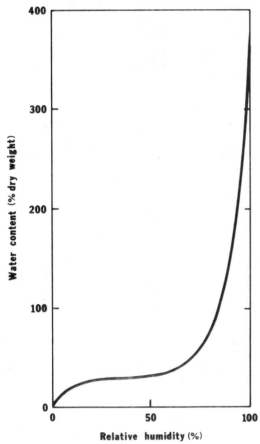

FIG. 1. *Typical water sorption isotherm curve for bacterial cells.*

phase represents loosely bound, multilayer-adsorbed water; and the terminal phase, at high humidities, represents bulk solution, "free" water. The total amount of cell water in fully hydrated cells is obtained at 100% humidity, the intercept for which occurs at a steep rise in the curve and thus is difficult to determine precisely.

Sometimes it is convenient to *estimate* the amount of cell water from interstice-corrected permeability measurements with ^2H_2O, [^{14}C]-urea, or glycerol (19.1.5).

The water content of cells may be expressed on a wet weight basis (grams of water per gram of wet cells), a dry weight basis (grams of water per gram of dry cells), or a wet volume basis (grams of water per cubic centimeter of wet cells). The wet weight basis is perhaps most commonly used by bacteriologists, but the dry weight basis is the most fundamental expression. The two bases are correlated by the equation, $WC_{dry} = WC_{wet}/(1 - WC_{wet})$, where WC_{dry} and

WC_{wet} are the dry and wet water content determinations; e.g., $400\%_{dry} = 80\%_{wet}$.

25.2.4. Volume

The volume of a mass of wet cells (or the average size of a single cell) is best obtained by the procedure described in 19.1.

25.2.5. Wet and Dry Densities

The density (essentially the specific gravity) of a bacterial cell may be obtained as either the wet density, based on the total of the solids and water contents, or the dry ("chemical") density, based on the solids content. Both densities are expressed in units of weight per unit of volume (grams per cubic centimeter).

The wet density may be obtained simply by dividing the cell wet weight (25.2.1) by the cell wet volume (25.2.4).

The dry density may be obtained similarly, by dividing the cell dry weight (25.2.2) by the cell dry volume. Unfortunately, the dry volume is difficult to determine accurately. The cells first are completely dried, e.g., by lyophilization (12.2.1). Occluded gas and residual water vapor are removed by holding a fairly large mass (>2 g) of the dried cells under a high vacuum until the pressure becomes constant (e.g., at <0.01 μmHg [ca. 1.33 mPa]). The volume of an inert and nonadsorbing gas (e.g., helium or nitrogen) displaced by the degassed cells then is measured in a volumetric adsorption apparatus. This procedure was exemplified with bacterial spores by Berlin et al. (1). The equipment may be obtained commercially (Fekrumeter; Gallard-Schlesinger, Inc., Carle Place, NY 11514).

Wet and dry densities may also be *estimated* by use of equilibrium density gradient centrifugation (5.2.3), but the technique is beset with pitfalls in working with cells rather than subcellular components. For determination of the wet density of hydrated cells by this technique (using such solutes as metrizamide or Renografin), the gradient solution varies in osmotic tonicity and water activity and may permeate the water space within the matrix of the cell wall. For determination of the dry density of dried, degassed cells with this technique (using such solvent pairs as perchloroethylene and 4-*tert*-butyltoluene), the gradient solution may extract or dissolve into lipid constituents of the cell.

25.3. LITERATURE CITED

1. **Berlin, E., H. R. Curran, and M. J. Pallansch.** 1963. Physical surface features and chemical

density of dry bacterial spores. J. Bacteriol. **86:** 1030–1036.

2. **Black, S. H., and P. Gerhardt.** 1962. Permeability of bacterial spores. IV. Water content, uptake, and distribution. J. Bacteriol. **83:**960–967.

3. **Gray, T. R. G., and J. R. Postgate (ed.).** 1976. The survival of vegetative microbes. Symp. Soc. Gen. Microbiol. **26:**1–432.

4. **Mallette, M. F.** 1969. Evaluation of growth by physical and chemical means. *In* J. R. Norris and D. W. Ribbons (ed.), Methods in microbiology, vol. 1. Academic Press, Inc., New York.

5. **Marquis, R. E., and P. Gerhardt.** 1959. Polymerized organic salts of sulfonic acids used as dispersing agents in microbiology. Appl. Microbiol. **7:**105–108.

6. **Meynell, G. G., and E. Meynell.** 1970. Theory and practice in experimental bacteriology, p. 25. Cambridge University Press, New York.

7. **Troller, J. A., and J. H. B. Christian.** 1978. Water activity and food. Academic Press, Inc., New York.

AUTHOR INDEX

SUBJECT INDEX

Absorbance, definition of, 194
Absorption photometry
applications and procedures, 286–291
Accidents in the laboratory, 48, 49
Acetate buffer, formula for, 68
Acetobacter aceti, media and cultural requirements, 85
N-Acetyl-L-cysteine-sodium hydroxide reagent, 136
Acetylene reduction method for determination of nitrogenase activity, 428
Acholeplasma laidlawii, media and cultural requirements, 88
Achromatic objectives, 9
Acidaminococcus fermentans, media and cultural requirements, 86
Acid-fast staining, 27–28, 411
Acid hydrolysis, 19
Acidic pH incubation as an enrichment technique, 120
Acidification of carbohydrates, test for, 411
Acids, determination by gas chromatography, 423–424
Acinetobacter calcoaceticus
media and cultural requirements, 86
transformation in, 243, 245–247
Acridine half-mustard mutagenesis, 227
Actinobacillus lignieresii, media and cultural requirements, 86
Actinobifida dichotomica, media and cultural requirements, 87
Actinomyces bovis, media and cultural requirements, 87
Actinomycetes and related organisms, key to media and cultural requirements for, 87–88
Actinomycetes, isolation of, 128
Actinoplanes philippinensis, media and cultural requirements, 87
Adsorption chromatography, 299–300
Adsorption of bacteria on surfaces, 182
Aeration of liquid cultures, 157
Aerobic flask culture, 155–156
Aerococcus viridans, media and cultural requirements, 87
Aeromonas hydrophila, media and cultural requirements, 86
Affinity chromatography, 304–305
Agar for solidifying bacteriological media, 143–144
Agarose-Affi gel, 305
Agar slides for light microscopy, 17–18
Agar substrate for cytophagas, 124
Agitation of liquid cultures, 157–158
Agrobacterium radiobacter, 3-ketolactose from lactose oxidation, 417
Agrobacterium rhizogenes, media and cultural requirements, 85
Agrobacterium tumefaciens, 3-ketolactose from lactose oxidation, 417
Alcian blue staining, 31
Alcohols, determination by gas chromatography, 423–424
Alkali treatment as an enrichment technique, 119–120

Allosteric controls of enzyme activity, 378, 379
Alpha-hemolytic streptococci, routine tests, 412, 420
Alumina grinding as a means of cell breakage, 369
Ames test for mutagenicity of carcinogens, 236–239
Amino acids
quantitative growth-response assay for, 90–91, 93, 96, 100, 102
requirements of bacteria for, 82
p-Aminobenzoic acid, requirement of bacteria for, 80
2-Aminopurine mutagenesis, 227
Ammonia from arginine, test for, 411
Ammonium, determination of, 352–353, 356–358
Ammonium in enrichment for autotrophic nitrifying bacteria, 125
Ammonium ion, electrode measurement of, 293–294
Amoebacter roseus, media and cultural requirements, 85
Amorphosporangium auranticolor, media and cultural requirements, 87
Ampullariella regularis, media and cultural requirements, 87
Anaerobes
API 2A system for, 425
key to media and cultural requirements for, 86
mass culture of, 156, 159
Minitek system for, 425
test for meat digestion by, 418
Anaerobic chambers for stringent anaerobes, 76–77
Anaerobic jars, 74–75
Anaerobiosis
anaerobic chambers for stringent anaerobes, 76
different levels required by bacteria, 72
measurement of degree, 72–73
reducing agents for anaerobic media, 74
techniques for nonstringent anaerobes, 74–75
techniques for stringent anaerobes, 75–76
Analysis of variance, 200
Ancalomicrobium adetum, media and cultural requirements, 85
Animal infection by *Y. pestis*, 127
Animal parasitism by *S. pneumoniae*, 127
Animals, safety in handling, 493
Anthony method for demonstrating capsules and slime layers, 29
Anthrone reaction for total carbohydrates, 333
Antibiotic agar, composition of, 240
Antibiotic inhibition as an enrichment technique, 121–122
Antibiotic resistance mutants, gradient plate technique for, 230
Antibiotics, bioautography of, 148
Antigens, identification by electron microscopy, 46
Antiserum agglutination, enrichment by use of, 119
API systems, 425
Apochromatic objectives, 9
Aquaspirillum
composition of medium for, 130
enrichment and isolation, 122
enrichment for, 117, 124

Nitrogen compounds
  analytical techniques for assay of, 350–354
  procedures for specific compounds, 355–361
Nitrogen fixation, nitrogenase activity and, 428–430
Nitrogen-free media, composition of, 432, 434, 437
Nitrogen requirements of bacteria, 80
Nitrogenase activity, determination of, 428
Nitrosococcus nitrosus, media and cultural requirements, 87
Nitrosoguanidine mutagenesis, 225
Nitrosolobus multiformis, media and cultural requirements, 87
Nitrosomonas europaea, media and cultural requirements, 87
Nitrosospira briensis, media and cultural requirements, 87
Nitrospina gracilis, media and cultural requirements, 87
Nitrous acid mutagenesis, 227
Nocardia
  lipid composition of, 338
  routine tests, 411, 414, 422
Nocardia asteroides, media and cultural requirements, 88
Nonexponential growth, measurement of, 206
Nonionic detergents, use in cell fractionation, 58
Nonsense-type mutations, 223
Nucleic acids (*see also* Deoxyribonucleic acid, Ribonucleic acid)
  cell disruption for isolation of, 451–452
  determination of concentration and purity, 456–457
  electron microscopy of, 44
  estimation and identification in bacterial cells, 348–350
  isolation of DNA, 452–455
  isolation of RNA, 455–456
  molecular weight determination by gel electrophoresis, 324
  separation by ion-exchange chromatography, 297–298
  test for hydrolysis of, 419
Numerical aperture, 9
Numerical taxonomy, 444–449
  computer analysis, 446–447
  data coding, 445–446
  interpretation of results, 447–448
  strain selection, 444–445
  test selection, 445
Nutrient broth and agar, composition of, 240, 435
Nutrient requirements of bacteria, 80–84
Nutritional auxotrophs, detection of, 232–234

O/129 test, 420
O-F medium, composition of, 434–435
Objectives for simple microscopy, 6–7, 9
Oceanospirillum, enrichment and isolation, 122
Oculars for simple microscopy, 6, 9–10
Oil-immersion objectives, 7, 8
ONPG test reagent, preparation of, 439
Open culture systems, 159
Optical density, definition of, 194
Optical rotation, preparation of lactic acid reagents for, 438
Optical rotation determination, 424–425
Optochin test, 420
Orcinol reaction procedure, 349–350
Orcinol test for RNA, 457

Ornithine decarboxylase, test for, 420
Ornithine transcarbamylase
  assay procedure, 375
  repression of, 384
Osmium fixation, 41
Osmium tetroxide fixation, 19, 23–24
Osmotic lysis of bacterial cells, 55, 367–368
Osmotic pressure
  effect on bacterial growth, 69
  measurement and control of, 70
Osmotically sensitive cells, use to study transport, 401–402
Oxidase tests, 334, 420
Oxidation/fermentation, test for, 420
Oxidation-reduction potential
  as a measure of degree of anaerobiosis, 72–73
  measurement of, 73
Oxidoreductase assays, 373–374
Oxi/Ferm system, 426
Oxoid broth, composition of, 239–240
Oxygen
  maintenance in liquid cultures, 71–72
  measuring absorption rate, 72
  measuring dissolved oxygen, 72
Oxygen polarograph, 294–295

Pantothenic acid requirement of bacteria, 81
Paper, preservation of bacteria on, 209–210
Parasitic bacteria, 79
Parasitism
  animal, by *S. pneumoniae,* 127
  bacterial, by bdellovibrios, 127
Pasteurella multocida, media and cultural requirements, 86
PathoTec Rapid I-D system, 425–426
Pediococcus, use in quantitative growth response assays, 90
Pediococcus cerevisiae
  biochemical factors in growth, 81
  media and cultural requirements, 87
Pedomicrobium ferrugineum, media and cultural requirements, 85
Peloscope, use with soil samples, 33
Penassay broth, composition of, 240
Penicillin, use in isolation of plasmid DNA, 268
Penicillin enrichment for mutants, 229
Penicillin inhibition as an enrichment technique, 121
Peptides, requirement by bacteria, 82–83
Peptidoglycan, isolation and characterization, 346–347
Periodate-Schiff staining technique, 30–31
Permanent mounts, 24
Permeability of bacterial cells, 393–397
  assay procedure, 394–396
  definition, 393
  determination of interstitial space, 396
  expression of results, 396–397
Permeabilization of cells for determination of enzymes, 367
Peroxides as disinfectants, 500
Pfennig and Biebl sulfur medium, 136–137
pH in bacterial growth
  continuous control of pH, 69
  definition of pH, 66
  measurement of pH, 66–67
  use of buffers, 67–69
pH indicators, 67, 440

ISBN 0-914826-30-1